DATE LABEL ON PAGE 3

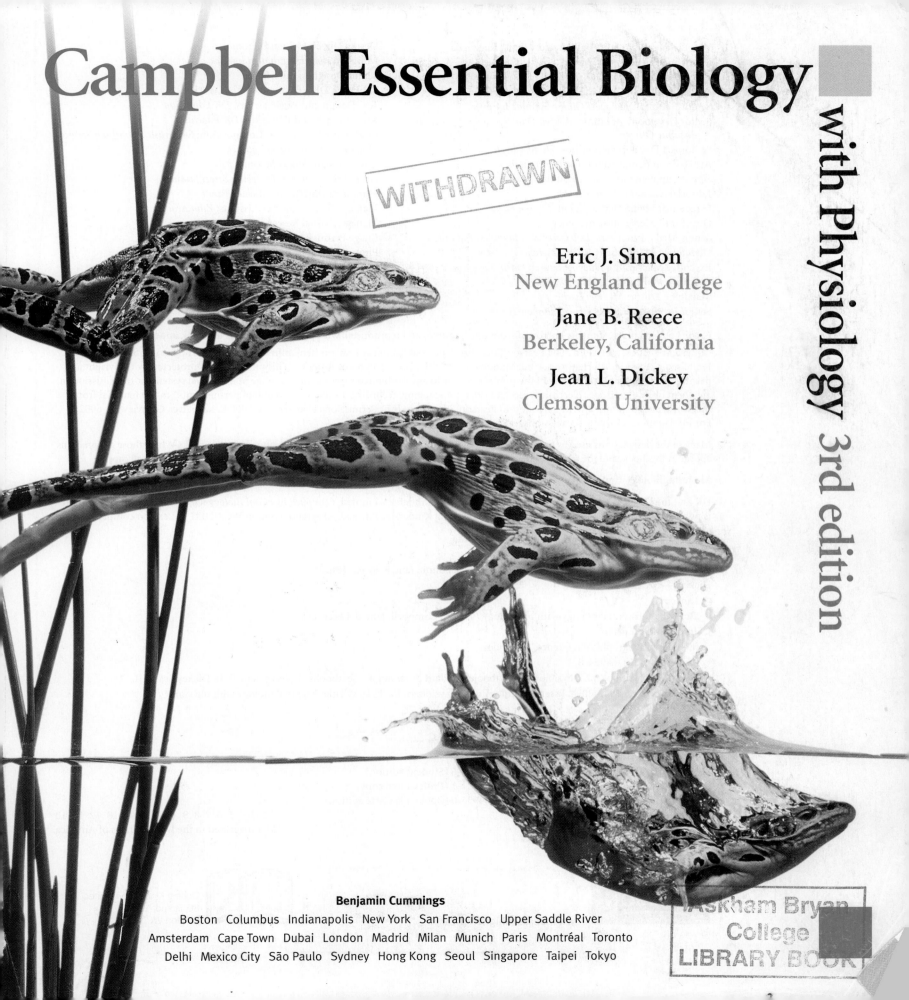

Campbell Essential Biology

with Physiology 3rd edition

Eric J. Simon
New England College

Jane B. Reece
Berkeley, California

Jean L. Dickey
Clemson University

Benjamin Cummings

Boston Columbus Indianapolis New York San Francisco Upper Saddle River
Amsterdam Cape Town Dubai London Madrid Milan Munich Paris Montréal Toronto
Delhi Mexico City São Paulo Sydney Hong Kong Seoul Singapore Taipei Tokyo

Vice President and Editorial Director: *Frank Ruggirello*
Editor-in-Chief: *Beth Wilbur*
Executive Editor: *Chalon Bridges*
Executive Director of Development: *Deborah Gale*
Senior Editorial Manager: *Ginnie Simione Jutson*
Senior Development Editors: *Evelyn Dahlgren, John Burner, and Suzanne Olivier*
Senior Art Development Editor: *Hilair Chism*
Art Developmental Editor: *Carla Simmons*
Senior Supplements Project Editor: *Susan Berge*
Editorial Assistant: *Joshua Taylor*
Director of Media Development: *Lauren Fogel*
Director of Media Strategy: *Stacy Treco*
Senior Media Producers: *Deb Greco and Jonathan Ballard*
Director of Marketing: *Christy Lawrence*
Executive Marketing Manager: *Lauren Harp*
Executive Managing Editor: *Erin Gregg*
Managing Editor: *Michael Early*
Senior Production Supervisor: *Shannon Tozier*

Design Director: *Stuart Jackman*
Design Manager: *Anthony Limerick*
Senior Designer: *Ian Midson*
Publisher, DK Education: *Sophie Mitchell*
Text Design and Page Layouts: *DK Education*
Art Concepts and Direction: *DK Education*
Production Service and Composition: *S4Carlisle Publishing Services*
Illustrations: *Precision Graphics*
Cover Design: *Stuart Jackman*
Cover Production: *Seventeenth Street Studios*
Manufacturing Buyer: *Michael Penne*
Manager, Rights and Permissions: *Zina Arabia*
Manager, Visual Research: *Elaine Soares*
Image Permission Coordinator: *Cynthia Vincenti*
Photo Research: *DK Education and Kristin Piljay*
Senior Photo Editor: *Donna Kalal*
Text and Cover Printer: *Courier Kendallville*
Cover Image: *Frank Greenaway/Dorling Kindersley Media Library*

Credits and acknowledgments borrowed from other sources and reproduced, with permission, in this textbook appear on page A-5.

Library of Congress Cataloging-in-Publication Data
Simon, Eric J. (Eric Jeffrey), 1967-
    Campbell essential biology with physiology / Eric J. Simon, Jane B. Reece, Jean L.
Dickey. — 3rd ed.
        p. cm.
    Rev. ed. of: Essential biology with physiology / Neil A. Campbell, Jane B. Reece, Eric
J. Simon. 2nd ed. c2007.
    Includes bibliographical references and index.
    ISBN 978-0-321-64954-6
    1. Biology—Textbooks. 2. Physiology—Textbooks. 3. Plant physiology—Textbooks. I. Reece, Jane B. II. Dickey, Jean. III.
Campbell, Neil A., 1946-2004. Essential biology with physiology. IV. Title. V. Title: Essential biology with physiology
    QH308.2.C344 2010
    570—dc22

2009027247

ISBN 10: 0-321-64954-0; ISBN 13: 978-0-321-64954-6 (Student edition)
ISBN 10: 0-321-65121-9; ISBN 13: 978-0-321-65121-1 (Professional copy)
ISBN 10: 0-321-64956-7; ISBN 13: 978-0-321-64956-0 (Books a la Carte edition)

2 3 4 5 6 7 8 9 10—CRK—13 12 11 10
Manufactured in the United States of America.

Benjamin Cummings
is an imprint of

www.pearsonhighered.com

Dorling Kindersley Limited,
80 Strand,
London WC2R 0RL.

About the Authors

NEIL A. CAMPBELL

(1946–2004) combined the inquiring nature of a research scientist with the soul of a caring teacher. Over his 30 years of teaching introductory biology to both science majors and nonscience majors, many thousands of students had the opportunity to learn from him and be stimulated by his enthusiasm for the study of life. While he is greatly missed by his many friends throughout the biology community, his coauthors remain inspired by his visionary dedication to education and are committed to searching for ever better ways to engage students in the wonders of biology.

ERIC J. SIMON

is an associate profes___ biology at New En___ College, in Hen___ Hampshire. ___ to science majors and ___ as well as upper-level introducto___ ___s, microbiology, and nonscie___gy. Dr. Simon received a B.A. in cour___mputer science and an M.A. in ___h Wesleyan University and a Ph.D. ___histry from Harvard University. His ___ focuses on innovative ways to use ___logy to improve teaching and learning in ___cience classroom, particularly for ___science majors. Dr. Simon is also a coauthor of *Biology: Concepts & Connections*, 6th Edition.

To the teachers and mentors who sparked and promoted my love of biology, especially Jerome Losty (North Stratfield Elementary), James Donovan (Andrew Warde High School), David Beveridge (Wesleyan University), Stephen Harrison (Harvard University), and Neil Campbell; and to my greatest fan and champion: my mother

JANE B. REECE

has worked in biology publishing since 1978, when she joined the editorial staff of Benjamin Cummings. Her education includes an A.B. in biology from Harvard University (where she was initially a philosophy major), an M.S. in microbiology from Rutgers University, and a Ph.D. in bacteriology from the University of California, Berkeley. At UC Berkeley and later as a postdoctoral fellow in genetics at Stanford University, her research focused on genetic recombination in bacteria. Dr. Reece taught biology at Middlesex County College (New Jersey) and Queensborough Community College (New York). She is the lead author of *Biology*, 8th Edition, and *Biology: Concepts & Connections*, 6th Edition.

To Chalon Bridges, Ginnie Simione Jutson, Evelyn Dahlgren, John Burner, Suzanne Olivier, Hilair Chism, Carla Simmons, and all the other heroes of our editorial team, for their essential contributions to Essential Biology

JEAN L. DICKEY

is a professor of biology at Clemson University, in South Carolina. She had no idea that science was interesting until her senior year in high school, when a scheduling problem landed her in an advanced biology course. Abandoning plans to study English or foreign languages, she enrolled in Kent State University as a biology major. After receiving her B.S. in biology, she went on to earn a Ph.D. in ecology and evolution from Purdue University. Since joining the faculty at Clemson in 1984, Dr. Dickey has specialized in teaching nonscience majors, including a course designed for pre-service elementary teachers and workshops for in-service teachers. She also developed an investigative laboratory curriculum for general biology. Dr. Dickey is the author of *Laboratory Investigations for Biology*, 2nd Edition, and is a coauthor of *Biology: Concepts & Connections*, 6th Edition.

To my mother, who taught me to love learning, and to my daughters Katherine and Jessie, the twin delights of my life

Detailed Contents

16 Plants, Fungi, and the Move onto Land   314
■■■ Chapter Thread: **PLANT-FUNGUS SYMBIOSIS**

17 The Evolution of Animals   336
■■■ Chapter Thread: **HUMAN EVOLUTION**

29 The Working Plant 624
■■■ Chapter Thread: **THE INTERDEPENDENCE OF ORGANISMS**

Appendices

Glossary

Index

Each chapter is built around three features connected by a unifying thread

■■■ CHAPTER THREAD

The chapter thread weaves a single compelling topic throughout the chapter. For example, here the thread is **Life with and without Sex,** and you'll consistently find this thread: at the beginning, middle, and end of each chapter.

The **CHAPTER CONTENTS** gives you an overview of the chapter's features and topics.

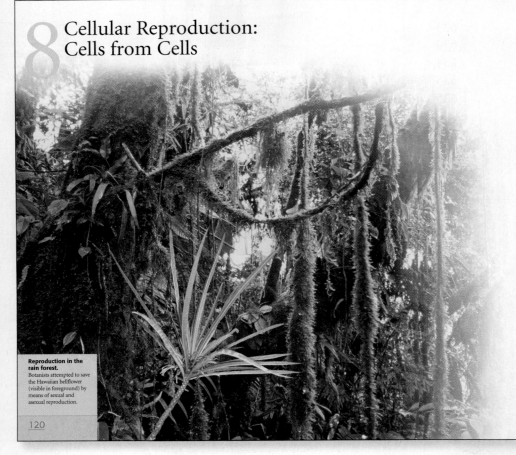

8 Cellular Reproduction: Cells from Cells

Reproduction in the rain forest.
Botanists attempted to save the Hawaiian bellflower (visible in foreground) by means of sexual and asexual reproduction.

120

CHAPTER CONTENTS
■■■ Chapter Thread: LIFE WITH AND WITHOUT SEX

 BIOLOGY AND SOCIETY: Life with and without Sex

Rain Forest Rescue

Deep in a Hawaiian rain forest, the thrumming of a helicopter cut through the dense tropical air. After touching down in a clearing, a pair of rescuers jumped out, one of them carrying a precious bundle. A 20-minute hike through dense foliage brought them to their "patient": the wild *Cyanea kuhihewa*, a plant in the bellflower family. The rescuers were in the middle of a life-or-death battle, armed with the pollen of a species on the very brink of extinction. The stakes couldn't have been higher: The small bellflower before them was the last surviving member of its species in the wild.

The rescuers, scientists from the National Tropical Botanical Garden, were trying to promote sexual reproduction, the reproductive process that involves fertilization, the union of a sperm and an egg. Using a fine brush, the botanists transferred sperm-carrying pollen from a garden-grown plant onto the egg-containing wild bloom. Their hope was to initiate the development of a new seed, and from it, a new plant.

Unfortunately, the bellflower fertilization attempt failed. Even worse, the last remaining wild specimen died in 2003. But hope remains for *C. kuhihewa*. The garden's botanists have also propagated the plant by asexual reproduction, the production of offspring by a single parent. Many plants that normally reproduce sexually can be induced to reproduce asexually in the laboratory. Indeed, work at the botanical garden has produced several new bellflower plants in this way, using stem tissue snipped from wild plants in the 1990s.

Cell division is at the heart of organismal reproduction, whether by sexual or asexual means. The ability of organisms to reproduce their own kind is the one characteristic that best distinguishes living things from nonliving matter. And the perpetuation of life depends on the production of new cells. Plants created from cuttings result from repeated cell divisions, as does the development of a multicellular organism from a fertilized egg. And eggs and sperm themselves result from cell division of a special kind. In this chapter, we'll look at how individual cells reproduce and then see how cell reproduction underlies the process of sexual reproduction. During our discussion, we'll consider examples of asexual and sexual reproduction among both plants and animals.

121

BIOLOGY AND SOCIETY: Rain Forest Rescue

The **BIOLOGY AND SOCIETY** relates biology to your life and interests. For example, in this chapter you'll discover what sex has to do with saving endangered species.

BIOLOGY AND SOCIETY: Life with and without Sex

Rain Forest Rescue

Deep in a Hawaiian rain forest, the thrumming of a helicopter cut through the dense tropical air. After touching down in a clearing, a pair of rescuers jumped out, one of them carrying a precious bundle. A 20-minute hike through dense foliage brought them to their "patient": the wild *Cyanea kuhihewa*, a plant in the bellflower family. The rescuers were in the middle of a life-or-death battle, armed with the pollen of a species on the very brink of extinction. The stakes couldn't have been higher: The small bellflower before them was the last surviving member of its species in the wild.

The rescuers, scientists from the National Tropical Botanical Garden, were trying to promote sexual reproduction, the reproductive process that involves fertilization, the union of a sperm and an egg. Using a fine brush, the botanists transferred sperm-carrying pollen from a garden-grown plant onto the egg-containing wild bloom. Their hope was to initiate the development of a new seed, and from it, a new plant.

Unfortunately, the bellflower fertilization attempt failed. Even worse, the last remaining wild specimen died in 2003. But hope remains for *C. kuhihewa*. The garden's botanists have also propagated the plant by asexual reproduction, the production of offspring by a single parent. Many plants that normally reproduce sexually can be induced to reproduce asexually in the laboratory. Indeed, work at the botanical garden has produced several new bellflower plants in this way, using stem tissue snipped from wild plants in the 1990s.

Cell division is at the heart of organismal reproduction, whether by sexual or asexual means. The ability of organisms to reproduce their own kind is the one characteristic that best distinguishes living things from nonliving matter. And the perpetuation of life depends on the production of new cells. Plants created from cuttings result from repeated cell divisions, as does the development of a multicellular organism from a fertilized egg. And eggs and sperm themselves result from cell division of a special kind. In this chapter, we'll look at how individual cells reproduce and then see how cell reproduction underlies the process of sexual reproduction. During our discussion, we'll consider examples of asexual and sexual reproduction among both plants and animals.

PROCESS OF SCIENCE: Do All Animals Have Sex?

The **PROCESS OF SCIENCE** gives you real-world examples of how the scientific method is applied. For example, in this chapter you'll learn how researchers study an animal that reproduces without sex.

THE PROCESS OF SCIENCE: Life with and without Sex

Do All Animals Have Sex?

As mentioned in the Biology and Society section, many plants can reproduce via both sexual and asexual routes. But while some animal species can also reproduce asexually, very few animals reproduce *only* asexually. In fact, evolutionary biologists have traditionally considered asexual reproduction an evolutionary dead end (for reasons we'll discuss in the Evolution Connection section).

To investigate a case where asexual reproduction seemed to be the norm, researchers from Harvard University studied a group of animals called the bdelloid rotifers (Figure 8.19). This class of nearly microscopic freshwater invertebrates includes over 300 known species. Despite hundreds of years of observations, no one had ever found bdelloid rotifer males or evidence of sexual reproduction. But the possibility remained that bdelloids had sex very infrequently or that the males were impossible to recognize by appearance. Thus, the Harvard research team posed the question, Does this entire class of animals reproduce solely by asexual means?

The researchers formed the hypothesis that bdelloid rotifers have indeed thrived for millions of years despite a lack of sexual reproduction. But how to prove it? In most species, the two versions of a gene in a pair of homologous chromosomes are very similar due to the constant trading of genes during sexual reproduction. If a species has survived without sex for millions of years, the researchers reasoned, then changes in the DNA sequences of homologous genes should accumulate independently, and the two versions of the genes should have significantly diverged from each other over time. This led to the prediction that bdelloid rotifers would display much more variation in their pairs of homologous genes than most organisms.

In a simple but elegant experiment, the researchers compared the sequences of a particular gene in bdelloid and non-bdelloid rotifers. Their results were striking. Among non-bdelloid rotifers that reproduce sexually, the two homologous versions of the gene were nearly identical, differing by only 0.5% on average. In contrast, the two versions of the same gene in bdelloid rotifers differed by 3.5–54%. These data provided strong evidence that bdelloid rotifers have evolved for millions of years without any asexual reproduction.

► Figure 8.19
A bdelloid rotifer.

EVOLUTION CONNECTION: The Advantages of Sex

The **EVOLUTION CONNECTION** concludes the chapter by demonstrating how the theme of evolution runs throughout all of biology. For example, in this chapter you'll learn about the evolutionary advantages of sex.

EVOLUTION CONNECTION: Life with and without Sex

The Advantages of Sex

Throughout this chapter, we've examined cell division within the context of reproduction. Like the Hawaiian bellflower in the Biology and Society section, many plants can reproduce both sexually and asexually (Figure 8.24). An important advantage of asexual reproduction is that there is no need for a partner. Asexual reproduction may thus confer an evolutionary advantage when plants are sparsely distributed and unlikely to be able to exchange pollen. Furthermore, if a plant is superbly suited to a stable environment, asexual reproduction has the advantage of passing on its entire genetic legacy intact.

In contrast to plants, the vast majority of animals reproduce by sexual means. There are exceptions, such as organisms that can regenerate after fragmentation (see Figure 8.1) and the bdelloid rotifers discussed in the Process of Science section. But many animals reproduce only through sex. Therefore, sex must enhance evolutionary fitness. But how? The answer remains elusive. Most hypotheses focus on the unique combinations of genes formed during meiosis and fertilization. By producing offspring of varied genetic makeup, sexual reproduction may enhance survival by speeding adaptation to a changing environment, such as one that contains evolving pathogens. Another idea is that shuffling genes during sexual reproduction might allow a population to rid itself of harmful genes more rapidly. But for now, one of biology's most basic questions—Why have sex?—remains a hotly debated topic that is the focus of much ongoing research and discussion.

◄ Figure 8.24 **Sexual and asexual reproduction.** Many plants, such as this strawberry, have the ability to reproduce both sexually (via flowers that produce fruit) and asexually (via runners).

Quickly orient yourself and find content

FIND THE CHAPTER YOU WANT
using the tabs on the left, and quickly locate units by the color of the tabs.

LOCATE STUDY QUESTIONS
in the margins.

- *CHECKPOINT QUESTIONS help you assess your understanding of the content you just read.*

- *For instant feedback, the answer appears directly under the question.*

- *The CHECKPOINT ICON within the text alerts you to stop and test yourself before you continue reading.*

CHAPTER 8
CELLULAR REPRODUCTION: CELLS FROM CELLS

☑CHECKPOINT

Ordinary cell division produces two daughter cells that are genetically identical. Name three functions of this type of cell division.

Answer: cell replacement, growth of an organism, asexual reproduction of an organism

▶ **Figure 8.1 Three functions of cell division.**

What Cell Reproduction Accomplishes

When you hear the word *reproduction*, you probably think of the birth of new organisms. But reproduction actually occurs much more often at the cellular level. Consider the skin on your arm. Skin cells are constantly reproducing themselves and moving outward toward the surface, replacing dead cells that have rubbed off. This renewal of your skin goes on throughout your life. And when your skin is injured, additional cell reproduction helps heal the wound.

When a cell undergoes reproduction, or **cell division**, the two "daughter" cells that result are genetically identical to each other and to the original "parent" cell. (Biologists traditionally use the word *daughter* in this context; it does not imply gender.) Before the parent cell splits into two, it duplicates its **chromosomes**, the structures that contain most of the organism's DNA. Then, during the division process, one set of chromosomes is distributed to each daughter cell. As a rule, the daughter cells receive identical sets of chromosomes.

As summarized in **Figure 8.1**, cell division plays several important roles in the lives of organisms. For example, within your body, millions of cells must divide every second to replace damaged or lost cells. Another function of cell division is growth. All of the trillions of cells in your body are the result of repeated cell divisions that began in your mother's body with a single fertilized egg cell.

Another vital function of cell division is reproduction of the organism. Single-celled organisms, such as amoebas, reproduce by dividing in half, and the offspring are genetic replicas of the parent. Because it does not involve fertilization of an egg by a sperm, this type of reproduction is called **asexual reproduction**. Offspring produced by asexual reproduction inherit all of their chromosomes from a single parent. Many multicellular organisms can reproduce asexually as well. For example, some sea star species have the ability to grow new individuals from fragmented pieces. And if you've ever grown a houseplant from a clipping, you've observed asexual reproduction in plants. In asexual reproduction, there is one simple principle of inheritance: The lone parent and each of its offspring have identical genes. The type of cell division responsible for asexual reproduction and for the growth and maintenance of multicellular organisms is called mitosis.

Sexual reproduction is different; it requires fertilization of an egg by a sperm. The production of egg and sperm cells involves a special type of cell division called meiosis, which occurs only in reproductive organs (such as testes and ovaries in humans). As we'll discuss later, a sperm or egg cell has only half as many chromosomes as the parent cell that gave rise to it.

Note that two kinds of cell division are involved in the lives of sexually reproducing organisms: meiosis for reproduction and mitosis for growth and maintenance. The remainder of the chapter is divided into two main sections, one for each type of cell division. ☑

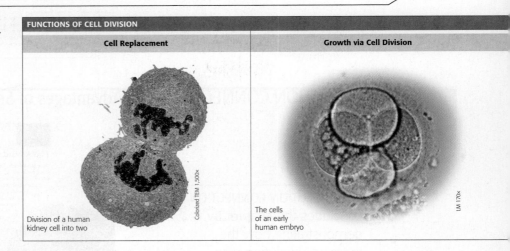

FUNCTIONS OF CELL DIVISION

Cell Replacement	Growth via Cell Division
Division of a human kidney cell into two	The cells of an early human embryo

Colorized TEM 1,500x

LM 170x

MAIN HEADING
at top of a page

The Cell Cycle and Mitosis

Almost all of the genes of a eukaryotic cell—around 21,000 in humans—are located on chromosomes in the cell nucleus. (The main exceptions are genes on small DNA molecules found in mitochondria and chloroplasts.) Because chromosomes are the lead players in cell division, let's focus on them before turning our attention to the cell as a whole.

Eukaryotic Chromosomes

Each eukaryotic chromosome contains one very long DNA molecule, typically bearing thousands of genes. The number of chromosomes in a eukaryotic cell depends on the species (Figure 8.2). Chromosomes are made up of a material called **chromatin**, a combination of DNA and protein molecules. The protein molecules help organize the chromatin and help control the activity of its genes.

Most of the time, the chromosomes exist as a diffuse mass of fibers that are much longer than the nucleus they are stored in. In fact, if stretched out, the DNA in just one of your cells would be taller than you! As a cell prepares to divide, its chromatin

*Bio***Flix**™
Mitosis

MP3 tutor sessions
Mitosis

Species	Number of chromosomes in body cells
Indian muntjac deer	6
Koala	16
Opossum	22
Giraffe	30
Mouse	40
Human	46
Duck-billed platypus	54
Buffalo	60
Dog	78
Red viscacha rat	102

▶ Figure 8.2 **The number of chromosomes in the cells of selected mammals.** Notice that humans have 46 chromosomes and that the number of chromosomes does not correspond to the size or complexity of an organism.

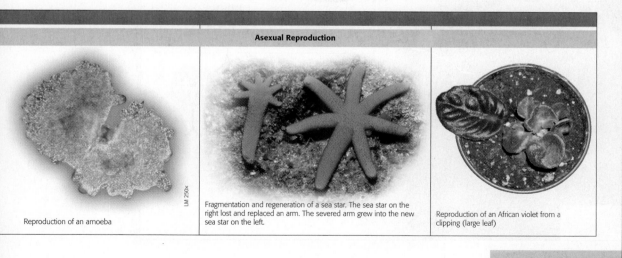

Asexual Reproduction

Reproduction of an amoeba

LM 250x

Fragmentation and regeneration of a sea star. The sea star on the right lost and replaced an arm. The severed arm grew into the new sea star on the left.

Reproduction of an African violet from a clipping (large leaf)

123

REMEMBER THE MAIN TOPIC
you are studying, using the colored tab at the top corner of each right-hand page.

Effortlessly connect images and words

Tools for both visual and verbal learners

FIGURES
and their text descriptions appear on the same page.

GREEN FIGURE NUMBERS
help your eyes move between the text and the figures.

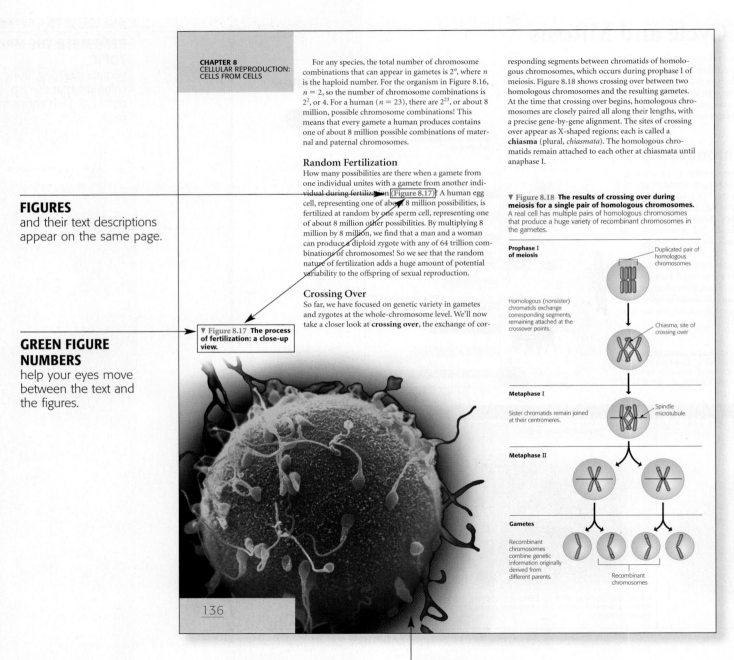

For any species, the total number of chromosome combinations that can appear in gametes is 2^n, where n is the haploid number. For the organism in Figure 8.16, $n = 2$, so the number of chromosome combinations is 2^2, or 4. For a human ($n = 23$), there are 2^{23}, or about 8 million, possible chromosome combinations! This means that every gamete a human produces contains one of about 8 million possible combinations of maternal and paternal chromosomes.

Random Fertilization
How many possibilities are there when a gamete from one individual unites with a gamete from another individual during fertilization (Figure 8.17)? A human egg cell, representing one of about 8 million possibilities, is fertilized at random by one sperm cell, representing one of about 8 million other possibilities. By multiplying 8 million by 8 million, we find that a man and a woman can produce a diploid zygote with any of 64 trillion combinations of chromosomes! So we see that the random nature of fertilization adds a huge amount of potential variability to the offspring of sexual reproduction.

Crossing Over
So far, we have focused on genetic variety in gametes and zygotes at the whole-chromosome level. We'll now take a closer look at **crossing over**, the exchange of cor-

responding segments between chromatids of homologous chromosomes, which occurs during prophase I of meiosis. **Figure 8.18** shows crossing over between two homologous chromosomes and the resulting gametes. At the time that crossing over begins, homologous chromosomes are closely paired all along their lengths, with a precise gene-by-gene alignment. The sites of crossing over appear as X-shaped regions; each is called a **chiasma** (plural, *chiasmata*). The homologous chromatids remain attached to each other at chiasmata until anaphase I.

▼ **Figure 8.17** The process of fertilization: a close-up view.

▼ **Figure 8.18** The results of crossing over during meiosis for a single pair of homologous chromosomes. A real cell has multiple pairs of homologous chromosomes that produce a huge variety of recombinant chromosomes in the gametes.

Prophase I of meiosis — Duplicated pair of homologous chromosomes

Homologous (nonsister) chromatids exchange corresponding segments, remaining attached at the crossover points.

Chiasma, site of crossing over

Metaphase I

Sister chromatids remain joined at their centromeres.

Spindle microtubule

Metaphase II

Gametes

Recombinant chromosomes combine genetic information originally derived from different parents.

Recombinant chromosomes

136

PHOTOS
complement the illustrations.

UNIQUE VISUAL ORGANIZER FIGURES

help you to see important categories at a glance.

IDENTIFY THE SUBJECT OF THE FIGURE
by glancing at the heading.

SPOT THE CATEGORIES
at once by looking at the subheadings.

FUNCTIONS OF CELL DIVISION

Cell Replacement	Growth via Cell Division		Asexual Reproduction	

Division of a human kidney cell into two — Colored TEM 1,500×

The cells of an early human embryo — LM 170×

Reproduction of an amoeba — LM 250×

Fragmentation and regeneration of a sea star. The sea star on the right lost and replaced an arm. The severed arm grew into the new sea star on the left.

Reproduction of an African violet from a clipping (large leaf)

CHAPTER REVIEW
words and images work together to help you review the key concepts of the chapter.

Chapter Review

CHAPTER REVIEW

SUMMARY OF KEY CONCEPTS

(MB) Go to the Study Area at **www.masteringbiology.com** for practice quizzes, myeBook, BioFlix™ 3-D animations, MP3 Tutor Sessions, videos, current events, and more.

What Cell Reproduction Accomplishes

Cell reproduction, also called cell division, produces genetically identical daughter cells:

Duplication of all chromosomes → Distribution via mitosis → Genetically identical daughter cells

Some organisms use mitosis (ordinary cell division) to reproduce. This is called asexual reproduction, and it results in offspring that are genetically identical to the lone parent and to each other. Mitosis also enables multicellular organisms to grow and develop and to replace damaged or lost cells. Organisms that reproduce sexually, by the union of a sperm with an egg cell, carry out meiosis, a type of cell division that yields gametes with only half as many chromosomes as body (somatic) cells.

The Cell Cycle and Mitosis

Eukaryotic Chromosomes

The genes of a eukaryotic genome are grouped into multiple chromosomes in the nucleus. Each chromosome contains one very long DNA molecule, with many genes, that is wrapped around histone proteins. Individual chromosomes are coiled up and therefore visible with a light microscope only when the cell is in the process of dividing; otherwise, they are in the form of thin, loosely packed chromatin fibers. Before a cell starts dividing, the chromosomes duplicate, producing sister chromatids (containing identical DNA) joined together at the centromere.

Chromosome (one long piece of DNA)
Centromere
Sister chromatids
Duplicated chromosome

The Cell Cycle

S phase
DNA synthesis; chromosome duplication

Interphase
Cell growth and chromosome duplication

G₁

Mitotic (M) phase

G₂

Genetically identical "daughter" cells

Cytokinesis (division of cytoplasm)

Mitosis (division of nucleus)

Mitosis and Cytokinesis

Mitosis is divided into four phases: prophase, metaphase, anaphase, and telophase. At the start of mitosis, the chromosomes coil up and the nuclear envelope breaks down (prophase). Then a mitotic spindle made of microtubules moves the chromosomes to the middle of the cell (metaphase). The sister chromatids then separate and are moved to opposite poles of the cell (anaphase), where two new nuclei form (telophase). Cytokinesis overlaps the end of mitosis. In animals, cytokinesis occurs by cleavage, which pinches the cell in two. In plants, a membranous cell plate divides the cell in two. Mitosis and cytokinesis produce genetically identical cells.

Cancer Cells: Growing Out of Control

When the cell cycle control system malfunctions, a cell may divide excessively and form a tumor. Cancer cells may grow to form malignant tumors, invade other tissues (metastasize), and even kill the organism. Surgery can remove tumors, and radiation and chemotherapy are effective as treatments because they interfere with cell division. You can increase the likelihood of surviving some forms of cancer through lifestyle changes and regular screenings.

Meiosis, the Basis of Sexual Reproduction

Homologous Chromosomes

The somatic cells (body cells) of each species contain a specific number of chromosomes; human cells have 46, made up of 23 pairs of homologous chromosomes. The chromosomes of a homologous pair carry genes for the same characteristics at the same places. Mammalian males have X and Y sex chromosomes (only partly homologous), while females have two X chromosomes.

141

MasteringBIOLOGY™

Choose the study tools that fit your learning style

VIDEO TUTOR SESSION TOPICS

MITOSIS AND MEIOSIS
Review the similarities and differences between these two processes.

SEX-LINKED PEDIGREES
Walk through a family pedigree.

DNA STRUCTURE
Review the levels of DNA structure and how they relate to each other.

DNA PROFILING TECHNIQUES
Learn important concepts of DNA technology as you watch a DNA profile be generated.

BIODIVERSITY
Organize information about biodiversity into a coherent framework, and remember different groups by relating them to foods that you eat.

PHYLOGENETIC TREES
Learn how to read these important figures by seeing how one is constructed.

MP3 TUTOR SESSIONS

ERIC SIMON
Author Eric Simon hosts both the **Video Tutor Sessions** and **MP3 Tutor Sessions** that provide on-the-go tutorials focused on key concepts and vocabulary.

Chromosome duplication

Sister chromatids

BioFlix ™

BIOFLIX 3-D ANIMATIONS AND ACTIVITIES

These dynamic 3-D animations and activities help you to visualize and learn the toughest topics in biology. BioFlix topics include Mechanisms of Evolution, DNA Replication, Carbon Cycle, and more.

Water Transport In Plants

Membrane Transport

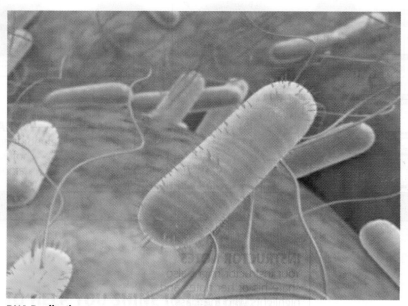
DNA Replication

1 Introduction: Biology Today

Biology is all around us.
Like this hiker in the
Monteverde Cloud Forest
Reserve of Costa Rica, we
are all surrounded by
biology.

CHAPTER CONTENTS

BIOLOGY AND SOCIETY: Biology in Our Everyday Lives

Biology All Around Us

We are living in a golden age of biology. The largest and best-equipped community of scientists in history is beginning to solve biological puzzles that once seemed unsolvable. We are moving ever closer to understanding how a single cell becomes a plant or animal; how plants trap solar energy and store that energy in food; how living creatures form networks in biological communities such as forests and coral reefs; and how the great diversity of life on Earth evolved from the first microbes. Exploring life has never been more exhilarating. Welcome to the big adventure of the 21st century!

Modern biology is as important as it is inspiring, with exciting breakthroughs changing our very culture. Genetics and cell biology are revolutionizing medicine and agriculture. Molecular biology is providing new tools for investigating ancestry and solving crimes. Ecology is helping us evaluate environmental issues, such as the causes and consequences of global climate change. Neuroscience and evolutionary biology are reshaping psychology and sociology. These are just a few examples of how biology is woven into the fabric of society as never before. If you think about it, you can easily find a dozen ways each day that biology affects your life.

We wrote this book to help students who are not biology majors develop an appreciation for the science of life and apply that understanding as they evaluate social issues. We believe that such a biological perspective is essential for any educated person, which is why we named our book *Essential Biology*. So, whatever your reasons for taking this course, even if only to meet your school's science requirement, you'll soon discover that this is the best time ever to study biology.

To help you get started, this first chapter of *Essential Biology* defines biology and then expands on important concepts within this definition. First, we'll survey the properties of life and the scope of life. Next, we'll introduce evolution as the theme that unifies all of biology. Finally, we'll set the study of life in the broader context of science as a process of inquiry. Throughout the chapter, we'll provide examples of how biology intersects everyday life, highlighting the relevance of this topic to society and everyone in it.

The Scope of Life

Biology is the scientific study of life. It's a subject of enormous scope that gets bigger every year. To start our investigation of life, let's look at the properties that are shared by all living things.

The Properties of Life

Our definition of biology as the scientific study of life raises some obvious questions: What is life? What distinguishes living things from nonliving things? The phenomenon of **life** seems to defy a simple, one-sentence definition. Yet almost any child perceives that a dog or a bug or a plant is alive, while a rock is not. We recognize life largely by what living things do.

Figure 1.1 highlights some of the properties and processes associated with life: (a) *Order.* All living things exhibit complex but ordered organization, as seen in the structure of a pinecone. (b) *Regulation.* The environment outside an organism may change drastically, but the organism can adjust its internal environment, keeping it within appropriate limits. When a lizard basks on a rock, its body absorbs solar energy and is warmed to an appropriate internal temperature. (c) *Growth and development.* Information carried by genes controls the pattern of growth and development in all organisms, including the green mamba snake. (d) *Energy utilization.* Organisms take in energy and use it to perform all of life's activities. A puffin obtains energy by eating fish and then uses this energy to power swimming and other work. (e) *Response to the environment.* All organisms respond to environmental stimuli. A Venus flytrap closes its trap rapidly in response to the environmental stimulus of an insect touching sensory hairs. (f) *Reproduction.* Organisms reproduce their own kind. Thus, hippos reproduce only hippos—never snakes or puffins. (g) *Evolution.* Reproduction underlies the capacity of populations to change (evolve) over time. For example,

the appearance of this katydid has evolved in a way that camouflages the animal in its environment. Evolutionary change has been a central, unifying feature of life since life arose nearly 4 billion years ago.

Life at Its Many Levels

In *Essential Biology*, we will probe life all the way down to the microscopic scale of molecules such as DNA, the chemical responsible for inheritance. At the other extreme of biological size and complexity, our exploration will take us up to the global scale of the entire **biosphere**, which consists of all the environments on Earth that support life—including soil; oceans, lakes, and other bodies of water; and the lower atmosphere. **Figure 1.2** takes you on a tour of these levels of biological organization, starting with the biosphere and working down through smaller and smaller levels.

If you follow Figure 1.2 in the opposite direction, you can see life's hierarchy from molecules to the biosphere. It takes many molecules to build a cell, many cells to make a tissue, multiple tissues to make an organ, and so on. At each new level, novel properties emerge, properties that were not part of the components of the preceding level. For example, life emerges at the level of the cell, but a test tube full of molecules is not alive. Such properties illustrate an important theme of biology, called emergent properties. The familiar saying that "the whole is greater than the sum of its parts" captures this idea. The emergent properties of the whole result from the specific arrangement and interactions of the component parts.

From the interactions within the biosphere to the molecular machinery within cells, biologists are investigating life at its many levels. In the next section, we'll take a closer look at two biological levels near opposite ends of the size scale: ecosystems and cells. ☑

☑**CHECKPOINT**

1. Define biology.
2. Which properties of life apply to a car? Which do not?
3. What is the smallest level of biological organization that can display all the characteristics of life?

Answers: **1.** *Biology is the scientific study of life.* **2.** *A car demonstrates order, regulation, energy utilization, and response to the environment. But a car does not grow, reproduce, or evolve.* **3.** *a cell*

▶ Figure 1.1 **Some properties of life.**

(a) Order

(b) Regulation

(c) Growth and development

(d) Energy utilization

▼ Figure 1.2 **Zooming in on life.**

❶ Biosphere
Earth's biosphere includes all the environments on Earth that support life.

❷ Ecosystems
A tide pool in the Galápagos Islands is an example of an ecosystem. An ecosystem consists of all organisms living in a particular area, as well as the nonliving, physical components of the environment that affect the organisms, such as water, air, soil, and sunlight.

❸ Communities
All organisms in the tide pool (iguanas, crabs, seaweed, bacteria, and others) are collectively called a community.

❹ Populations
Within communities are various populations, groups of interacting individuals of one species, such as a group of iguanas.

❺ Organisms
An organism is an individual living thing, like this iguana.

❻ Organ Systems and Organs
An organism's body consists of several organ systems, each of which contains two or more organs. For example, the iguana's circulatory system includes its heart and blood vessels.

❼ Tissues
Each organ is made up of several different tissues, such as the heart muscle tissue shown here. A tissue consists of a group of similar cells performing a specific function.

❽ Cells
The cell is the smallest unit that can display all the characteristics of life.

❾ Organelles
Organelles are functional components of cells, such as the nucleus that houses the DNA.

— Nucleus

❿ Molecules and Atoms
Finally, we reach molecules, the chemical level in the hierarchy. Molecules are clusters of even smaller chemical units called atoms. Each cell consists of an enormous number of chemicals that function together to give the cell the properties we recognize as life. DNA, the molecule of inheritance and the substance of genes, is shown here as a computer graphic. Each sphere in the DNA model represents a single atom.

— Atom

(e) Response to the environment

(f) Reproduction

(g) Evolution

Ecosystems

Life does not exist in a vacuum. Each organism interacts continuously with its environment, which includes other organisms as well as nonliving factors. The roots of a tree, for example, absorb water and minerals from the soil. Leaves take in carbon dioxide gas from the air. Chlorophyll, the green pigment of the leaves, absorbs sunlight, which drives the plant's production of sugar from carbon dioxide and water. This food production is called photosynthesis. The tree releases oxygen to the air, and its roots help form soil by breaking up rocks. Both organism and environment are affected by the interactions between them. The tree also interacts with other living things, including microscopic organisms in the soil that are associated with the plant's roots and animals that eat its leaves and fruit. The interactions between organisms and their environment take place within an **ecosystem**, such as the tide pool shown in Figure 1.2.

The dynamics of any ecosystem depend on two main processes (**Figure 1.3**). The first major process is the cycling of nutrients. For example, minerals that plants take up from the soil can eventually be recycled to the soil by microorganisms that decompose leaf litter and other organic refuse. Producers are photosynthetic organisms, such as plants; consumers are the organisms, such as animals, that feed on plants, either directly (by eating plants) or indirectly (by eating animals that eat plants); decomposers, such as fungi and many bacteria, decompose waste products and the remains of deceased organisms, changing complex dead material into simple nutrients that are recycled. The action of decomposers ensures that nutrients are recycled within an ecosystem.

The second major process in an ecosystem is the flow of energy. In contrast to recycling nutrients, an ecosystem gains and loses energy constantly. Most ecosystems are solar powered. The energy that enters an ecosystem as sunlight is captured by producers when plants and other photosynthesizers absorb the sun's energy and convert it to the chemical energy of sugars and other complex molecules. Chemical energy is then passed through a series of consumers and, eventually, decomposers, powering each organism in turn. In the process of these energy conversions between and within organisms, some energy is converted to heat, which is then lost from the system. Thus, energy flows through an ecosystem, entering as light and exiting as heat.

The biosphere is enriched by a great variety of ecosystems. A tropical rain forest in South America is an ecosystem. Very different from tropical forests are the ecosystems of deserts, such as those of the southwestern United States. A coral reef, such as the Great Barrier Reef off the eastern coast of Australia, is an ecosystem, and so is any small pond that may exist on your campus or in your city. Even a woodland patch in New York's Central Park qualifies as an ecosystem, small and artificial as it is by forest standards and disrupted as it is by human visitors. The fact is, humans are organisms that now have some presence, often disruptive, in all ecosystems. And the collective clout of 6 billion humans and their machines affects the entire biosphere. For example, our fuel burning and forest chopping are changing the atmosphere and the planet's climate in ways that we do not yet fully understand, though we already know that our actions jeopardize the diversity of life on Earth. ☑

Inflow of light energy

ECOSYSTEM

Loss of heat energy

Consumers
(animals)

Chemical energy (food)

Producers
(plants and other photosynthetic organisms)

Cycling of nutrients

Decomposers
(in soil)

◄ **Figure 1.3 Nutrient and energy flow in an ecosystem.** Nutrients are recycled within an ecosystem, whereas energy flows through an ecosystem.

Prokaryotic cell (bacterium)
- Smaller
- Simpler structure
- DNA concentrated in nucleoid region, which is not enclosed by membrane
- Lacks most organelles

Organelles

Nucleoid region

Eukaryotic cell
- Larger
- More complex structure
- Nucleus enclosed by membrane
- Contains many types of organelles

Nucleus

Colorized TEM 40,000×

◀ **Figure 1.4 Two main kinds of cells: prokaryotic and eukaryotic.** The number near the bottom right corner represents the magnification of the image. In this case, the images shown are approximately 40,000 times bigger than the actual cells.

Cells and Their DNA

Let's downsize now from ecosystems to cells. The cell has a special place in the hierarchy of biological organization: It is the level at which the properties of life emerge—the lowest level of structure that can perform all activities required for life.

All organisms are composed of cells. They occur singly as a great variety of unicellular (single-celled) organisms, mostly microscopic. Cells are also the subunits that make up the tissues and organs of plants, animals, and other multicellular organisms. In either case, the cell is the organism's basic unit of structure and function. The ability of cells to divide to form new cells is the basis for all reproduction and for the growth and repair of multicellular organisms, including humans.

We can distinguish two major kinds of cells: prokaryotic and eukaryotic (**Figure 1.4**). The prokaryotic cell is much simpler and usually much smaller than the eukaryotic cell. The cells of the ubiquitous microorganisms called bacteria are prokaryotic. Most other forms of life, including plants and animals, are composed of eukaryotic cells. In contrast to prokaryotic cells, a eukaryotic cell is subdivided by internal membranes into many different functional compartments, or organelles. For example, the nucleus, the largest organelle in most eukaryotic cells, houses DNA, the heritable material that directs the cell's many activities. Prokaryotic cells also have DNA, but it is not packaged within a nucleus.

Though very different in structural complexity, prokaryotic and eukaryotic cells have much in common at the molecular level. Most importantly, all cells use DNA as the chemical material of genes, the discrete units of hereditary information. Of course, bacteria and humans inherit different genes, but that information is encoded in a chemical language common to all organisms. In fact, the language of life has an alphabet of just four letters. The chemical names of DNA's four molecular building blocks are abbreviated as A, G, C, and T (**Figure 1.5**).

An average-sized gene may be hundreds or thousands of chemical "letters" long. A gene's meaning to a cell is encoded in its specific sequence of these letters, just as the message of a sentence is encoded in its arrangement of letters selected from the 26 letters of the English alphabet.

▶ **Figure 1.5 The language of DNA.** Every molecule of DNA is constructed from four kinds of chemical building blocks that are chained together, shown here as simple shapes and letters.

The four chemical building blocks of DNA

A DNA molecule

Evolution: Biology's Unifying Theme

The history of life, as documented by fossils and other evidence, is a saga of a restless Earth billions of years old, inhabited by a changing cast of living forms (**Figure 1.9**). Life evolves. Just as each individual has a family history, each species is one twig of a branching tree of life extending back in time through ancestral species more and more remote. Species that are very similar, such as the brown bear and the polar bear, share a common ancestor that represents a relatively recent branch point on the tree of life (**Figure 1.10**). But through an ancestor that lived much farther back in time, all bears are also related to squirrels, humans, and all other mammals. Hair and milk-producing mammary glands are two uniquely mammalian traits. Such similarities are what we would expect if all mammals descended from a common ancestor, a prototypical mammal. And mammals, reptiles, and all other vertebrates share a common ancestor even more ancient. Evidence of a still broader relationship can be found in similarities that are seen within all eukaryotic cells. Trace life back far enough, and there are only fossils of the primeval prokaryotes

▲ Figure 1.10 **An evolutionary tree of bears.** This tree is based on both the fossil record and a comparison of DNA sequences among modern bears.

that inhabited Earth over 3 billion years ago. All of life is connected. And the basis for this kinship is evolution, the process that has transformed life on Earth from its earliest beginnings to the extensive diversity we see today. Evolution is the theme that unifies all of biology.

The Darwinian View of Life

The evolutionary view of life came into focus in 1859 when British biologist Charles Darwin published *The Origin of Species*. His book developed two main points. First, Darwin marshaled the available evidence in support of the evolutionary view that species living today descended from ancestral species. Darwin called this process "descent with modification." It is an insightful phrase, as it captures the duality of life's unity (descent) and diversity (modification). In the Darwinian view, for example, the diversity of bears is based on different modifications of a common ancestor from which all bears descended. As the second main point in *The Origin of Species*, Darwin proposed a mechanism for descent with modification. He called this process natural selection. ☑

▼ Figure 1.9 **Digging into the past.** A paleontologist carefully excavates a dinosaur bone at Dinosaur National Monument, Utah.

☑CHECKPOINT
What is the modern term for what Darwin called "descent with modification"?

Answer: evolution

Natural Selection

As you'll learn in Chapter 13, Charles Darwin gathered important evidence for his theories during an around-the-world voyage. He was particularly struck by the diversity of animals on the Galápagos Islands, off the coast of Ecuador (**Figure 1.11**). Darwin regarded adaptation to the environment and the origin of new species as closely related processes. If some geographic barrier—an ocean separating islands, for instance—isolated two populations of a single species from each other, the populations could diverge more and more in appearance as each adapted to local environmental conditions. Over many generations, the two populations could become dissimilar enough to be designated separate species. The labeled branch point near the lower right of Figure 1.10 represents the point in time when two populations of one bear species began to diverge from each other, adapting to different climates and resulting in the evolution of the modern brown bear and the polar bear as two distinct species. Darwin realized that explaining such adaptation was the key to understanding descent with modification, or evolution. This focus on adaptation helped Darwin envision his concept of natural selection as the mechanism of evolution.

Darwin's Inescapable Conclusion

Darwin synthesized the theory of natural selection from two observations that by themselves were neither profound nor original. Others had the pieces of the puzzle, but Darwin saw how they fit together. As the evolutionary biologist Stephen Jay Gould put it, Darwin based his mechanism of natural selection on "two undeniable facts and an inescapable conclusion." Let's look at his logic:

Observation 1: Overproduction and competition.
Any population of a species has the potential to produce far more offspring than the environment can possibly support with available resources such as food and shelter. This overproduction leads to competition among the varying individuals of a population for these limited resources.

Observation 2: Individual variation.
Individuals in a population of any species vary in many inherited traits. No two individuals in a population are exactly alike. You know this variation to be true of human populations; careful observers find variation in populations of all species.

Conclusion: Unequal reproductive success.
In the struggle for existence, those individuals with traits best suited to the local environment will, on average, have the greatest reproductive success: They will leave the greatest number of surviving, fertile offspring. Therefore, the very traits that enhance survival and reproductive success will be disproportionately represented in succeeding generations of a population.

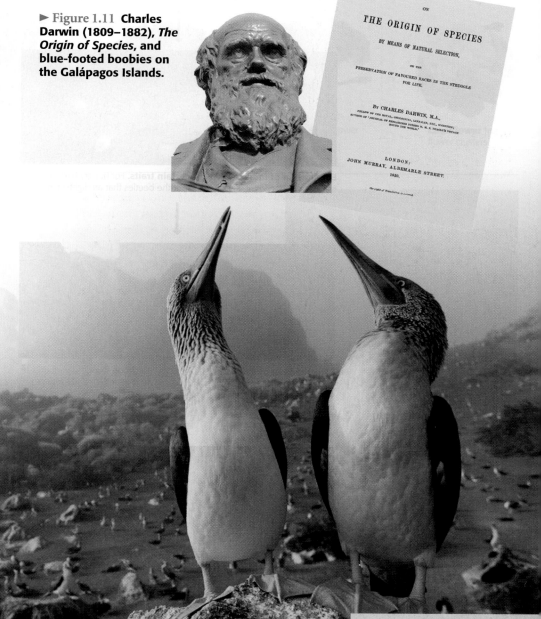

▶ **Figure 1.11 Charles Darwin (1809–1882),** *The Origin of Species,* **and blue-footed boobies on the Galápagos Islands.**

to foods (in particular, margarine, shortening, and foods cooked in partially hydrogenated oils) in increasing amounts during the 20th century.

Dietary fat was one component examined in the Nurses' Health Study, a landmark study of over 120,000 female U.S. nurses begun in 1976. Every two years, the study participants completed questionnaires about their health status and dietary habits. In 1994, researchers examining the trove of collected data discovered that study participants who ate high levels of trans fat had nearly double the risk of heart disease compared to participants who ate little trans fat. This is a good example of discovery science: examination of a set of data gathered without a preconceived notion of what it might reveal.

While the data from the Nurses' Health Study were suggestive, some researchers remained skeptical about the link between trans fat and heart health. In a hypothesis-driven study published in 2004, an Australian research team started with the **observation** that human body fat (adipose tissue) retains traces of consumed dietary fats. This observation raised a **question**: Would the adipose tissue of heart attack patients be measurably different from adipose tissue in a similar group of healthy patients? The researchers' **hypothesis** was that it would be, leading to the **prediction** that healthy patients' body fat would contain less trans fat than the body fat in heart attack victims.

The researchers then set up an **experiment** to determine the amounts of different kinds of fat in the adipose tissue of 79 patients who had experienced a heart attack. These amounts were compared with data for 167 patients who had not experienced a heart attack but were similar to the first group in most other respects. Since trans fat is not produced by the body, the researchers assumed that any trans fat found in the adipose tissue had been consumed in the diet. Their **results** showed significantly higher levels of trans fat in the bodies of the heart attack patients than in the healthier group (**Figure 1.16**). These results added to the growing body of evidence that trans fats are unhealthy and should not be added to foods. Indeed, several countries (such as Australia and Denmark) have banned trans fats altogether, and some American municipalities (such as New York City) have followed suit. You would do well to read nutrition labels and avoid trans fat as much as possible in your own diet.

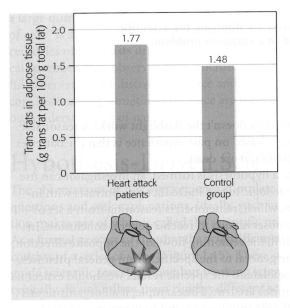

▲ Figure 1.16 **Levels of trans fat.** Results of an experiment that measured the levels of trans fat in the adipose tissue of 79 patients who had experienced a heart attack and 167 patients who had not (control group).

The trans fat study described in the Process of Science section is an example of a **controlled experiment**. An experiment of this type is designed to compare an experimental group (the patients who had a heart attack, in this case) with a control group (patients who did not have a heart attack). Ideally, a control group and an experimental group differ only in one variable—in our example, the occurrence of a heart attack. Other factors—such as age, weight, and sex of the patients—were matched between the two sets of patients. Thus, the control group canceled out the effects of all variables other than the one being tested. The use of a controlled experiment enabled researchers to draw conclusions about the effect of just one variable—trans fat—on heart attack risk. ☑

Theories in Science

Many people associate facts with science, but accumulating facts is not the primary goal of science. A telephone book is an impressive catalog of factual information, but it has little to do with science. It is true that facts, in the form of verifiable observations and repeatable experimental results, are the prerequisites of science. What really advances science, however, are new theories that tie together a number of observations that previously seemed unrelated. The cornerstones of science are the explanations that apply to the greatest variety of phenomena. People like Newton, Darwin, and Einstein stand out in the history of science not because they discovered a great many facts, but because their theories had such broad explanatory power.

What is a scientific theory, and how is it different from a hypothesis? A **theory** is much broader in scope than a hypothesis. This is a hypothesis: "White fur is an evolutionary adaptation that helps polar bears survive in an arctic habitat." But this is a theory: "Adaptations evolve by natural selection."

Because theories are so comprehensive, they only become widely accepted in science if they are supported by an accumulation of extensive and varied evidence. The use of the term *theory* in science for a comprehensive explanation supported by abundant evidence contrasts with our everyday usage, which equates theories more with speculations or hypotheses. Natural selection qualifies as a scientific theory because of its broad application and because it has been validated by a huge number of observations and experiments.

Scientific theories are not the only way of "knowing nature," of course. A comparative religion course would be a good place to learn about the diverse legends that tell of a supernatural creation of Earth and its life. Science and religion are two very different ways of trying to make sense of nature. Art is still another way. A broad education should include exposure to these different ways of viewing the world. Each of us synthesizes our worldview by integrating our life experiences and multi-disciplinary education. As a science textbook and part of that broad education, *Essential Biology* showcases life in the scientific context of evolution, the one theme that continues to hold all of biology together, no matter how big and complex the subject becomes. ✓

The Culture of Science

It is not unusual for several scientists to ask the same questions. Such convergence contributes to the progressive and self-correcting qualities of science. Scientists build on what has been learned from earlier research, and they pay close attention to contemporary scientists working on the same problem. Scientists share information through publications, seminars, meetings, and personal communication. The Internet has added a new medium for this exchange of ideas and data.

Both cooperation and competition characterize the scientific culture (**Figure 1.17**). Scientists working in the same research field subject one another's work to careful scrutiny. It is common for scientists to check the conclusions of others by attempting to repeat experiments. This obsession with evidence and confirmation helps characterize the scientific style of inquiry. Scientists are generally skeptics.

We have seen that science has two key features that distinguish it from other styles of inquiry: (1) a dependence on observations and measurements that others can

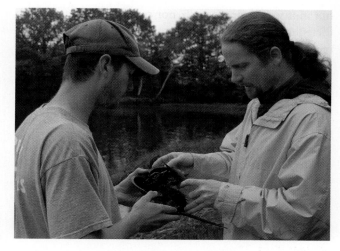

▲ **Figure 1.17 Science as a social process.** New England College neurobiologist Jim Newcomb (right) mentors a student in the methods of marine biology.

verify and (2) the requirement that ideas (hypotheses) are testable by experiments that others can repeat.

Science, Technology, and Society

Science and technology are interdependent. New technologies, such as more powerful microscopes and computers, advance science. And scientific discoveries can lead to new technologies. In most cases, technology applies scientific discoveries to the development of new goods and services. For example, just over 50 years ago, two scientists, James Watson and Francis Crick, discovered the structure of DNA through the process of science. Their discovery eventually led to a variety of DNA technologies, including the genetic engineering of microorganisms to mass-produce human insulin (see Figure 1.6) and the use of DNA profiling for investigating crimes (**Figure 1.18**). Perhaps Watson and Crick envisioned that their discovery would someday inform new technologies, but that probably did

▶ **Figure 1.18 DNA technology and the law.** The stained bands visible in this photograph represent fragments of DNA extracted from body tissue collected at a crime scene. The pattern of bands varies from person to person.

In what way does modern genetic engineering depend on the common ancestry shared by all life on Earth?

Answer: Genetic engineering is made possible by the fact that all life uses chemically identical DNA, a result of our evolution from a common ancestor.

not motivate their research, nor could they have predicted exactly what the applications would be. The direction technology takes depends less on the curiosity that drives basic science than it does on the current needs of humans and the changing cultural climate.

Technology has improved our standard of living in many ways, but it is a double-edged sword. Technology that keeps people healthier has enabled the population to grow more than tenfold in the past three centuries and to more than double to 6.6 billion in just the past 40 years. The environmental consequences are sometimes devastating. Acid rain, deforestation, global climate change, nuclear accidents, toxic wastes, and extinction of species are just a few of the repercussions of more and more people wielding more and more technology. Science can help us identify such problems and provide insight about what course of action may prevent further damage. But solutions to these problems have as much to do with politics, economics, culture, and the values of societies as with science and technology. Now that science and technology have become such powerful functions of society, each of us has the responsibility to become a "citizen scientist" by developing a reasonable amount of scientific and technological literacy. ✔

EVOLUTION CONNECTION: Biology in Our Everyday Lives

Evolution in Our Everyday Lives

Evolution is the core theme of biology. To emphasize the centrality of evolution to every area of biology, we end each chapter of *Essential Biology* with an Evolution Connection section. But is evolution connected to our everyday lives? And if so, in what ways? Evolution teaches us that the environment is a powerful selective force. Our world is rich with examples of such natural selection in action. One example that may directly affect your life is the development of resistance to antibiotics in disease-causing bacteria.

Antibiotics are drugs that help cure certain infections by impairing the bacteria that cause them. Antibiotics have saved millions of human lives, but there's a dark side to the widespread use of antibiotics: It has driven the evolution of antibiotic-resistant populations of the very bacteria the drugs are meant to kill.

When an antibiotic is taken by a patient, it will usually kill most, but not all, of the infecting bacteria. The environment, which now contains the antibiotic, selects among the varying bacteria of a population for those individuals that can survive the drug. In some cases, the mechanism of resistance in the bacteria is an ability to destroy the antibiotic—or even use the drug as food! Even though the drug kills most of the bacteria, those few bacteria that are resistant may soon multiply and become the norm in the population rather than the exception.

The evolution of antibiotic-resistant bacteria is a huge problem in public health. For example, there are now some strains of tuberculosis-causing bacteria that are resistant to all three of the antibiotics currently used to treat the disease (Figure 1.19). Unfortunate people infected with one of these resistant strains have no better chance of surviving than did tuberculosis patients a century ago.

The problem of antibiotic-resistant bacteria is influencing the way many physicians prescribe antibiotics. Doctors who understand that the abuse of these drugs is speeding the evolution of resistant bacteria are less likely to prescribe antibiotics needlessly—for instance, for a patient complaining of a common cold or flu, diseases caused by viruses (not bacteria), against which antibacterial drugs are powerless. Additionally, many farmers are reducing the use of antibiotics in animal feed, and many consumers are choosing to buy meats from animals raised without antibiotics.

It is important to note that adaptation of bacteria to an environment containing an antibiotic does not mean that the drug *created* the antibiotic resistance trait. Instead, the environment screened the heritable variations that already existed among individuals of a population and favored the individuals best suited to the present conditions. Throughout *Essential Biology*, you will learn more about how natural selection works and see other examples of how natural selection affects your life.

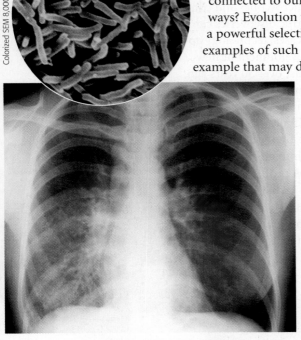

Colorized SEM 8,000x

▲ Figure 1.19 **Natural selection in action.** Antibiotic-resistant strains of tuberculosis-causing bacteria (top) have made the disease a threat again in the United States. The colorized X-ray above shows the lungs of a tuberculosis patient. The infection is shown in red.

Chapter Review

SUMMARY OF KEY CONCEPTS

 Go to the Study Area at **www.masteringbiology.com** for practice quizzes, eText, BioFlix™ 3-D animations, MP3 Tutor Sessions, videos, current events, and more.

The Scope of Life

Biology is the scientific study of life. The study of biology encompasses a wide scale of size and a huge variety of life, both past and present.

The Properties of Life

All life displays a common set of characteristics:

Order

Regulation

Growth and development

Energy utilization

Response to the environment

Reproduction

Evolution

Life at Its Many Levels

Within ecosystems, nutrients are recycled, but energy flows through. Cells are either prokaryotic (simple, small, and lacking membrane-bounded organelles, such as the nucleus) or eukaryotic (more complex, larger, and containing a nucleus and other membrane-bounded organelles).

Life in Its Diverse Forms

Biologists organize living organisms into three domains. The domain Eukarya is further divided into three kingdoms (distinguished partly by their means of obtaining food) and one catch-all group:

Life			
Prokaryotes		**Eukaryotes**	
		Plantae Fungi Animalia | Protists	
		Three kingdoms	
Domain Bacteria	**Domain Archaea**	**Domain Eukarya**	

Evolution: Biology's Unifying Theme

The Darwinian View of Life

Charles Darwin established the ideas of evolution ("descent with modification") via natural selection (unequal reproductive success) in his 1859 publication *The Origin of Species*.

Natural Selection

Darwin formulated the theory of natural selection through a logical inference: Natural selection leads to adaptations to the environment, which—when passed from generation to generation—is the mechanism of evolution.

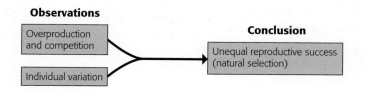

Observations

Overproduction and competition

Individual variation

Conclusion

Unequal reproductive success (natural selection)

The Process of Science

Discovery Science

Describing the natural world with verifiable data is the hallmark of discovery science.

Hypothesis-Driven Science

A scientist formulates a hypothesis (tentative explanation) to explain the natural world. The hypothesis may then be tested via the steps of the scientific method:

Revise and repeat

Observation → Question → Hypothesis → Prediction → Experiment

Theories in Science

A theory is a broad and comprehensive statement about the world that is supported by the accumulation of a great deal of verifiable evidence.

The Culture of Science

Scientists build on each other's work by always remaining skeptical and by seeking reproducible evidence to confirm ideas.

Science, Technology, and Society

Scientific advances promote new technologies, which in turn enable further scientific advances and useful applications to our lives.

SELF-QUIZ

1. Which of the following is *not* a characteristic of all living organisms?
   a. capable of self-reproduction
   b. composed of multiple cells
   c. complex yet organized
   d. energy utilization

2. Place the following levels of biological organization in order from smallest to largest: atom, biosphere, cell, ecosystem, molecule, organ, organism, population, tissue. Which is the smallest level capable of demonstrating all of the characteristics of life?

3. Plants use the process of photosynthesis to convert the energy in sunlight to chemical energy in the form of sugar. While doing so, they consume carbon dioxide and water and release oxygen. Explain how this process functions in both the cycling of chemical nutrients and the flow of energy through an ecosystem.

4. For each of the following organisms, match its description to its most likely domain and/or kingdom:

a. A foot-tall organism capable of producing its own food from sunlight

1. Bacteria

b. A microscopic, simple, nucleus-free organism found growing in a riverbed

2. Eukarya/Animalia

c. An inch-tall organism growing on the forest floor that consumes material from dead leaves

3. Eukarya/Fungi

d. A thimble-sized organism that feeds on algae growing in a pond

4. Eukarya/Plantae

5. How does natural selection cause a population to become adapted to its environment over time?

6. Why is it difficult to draw a conclusion from an experiment that does not include a control group?

7. Which of the following best describes the logic of the scientific method?

a. If I generate a testable hypothesis, tests and observations will support it.

b. If my prediction is correct, it will lead to a testable hypothesis.

c. If my observations are accurate, they will support my hypothesis.

d. If my hypothesis is correct, I can expect certain test results.

8. Which of the following statements best distinguishes hypotheses from theories in science?

a. Theories are hypotheses that have been proved.

b. Hypotheses are tentative guesses; theories are correct answers to questions about nature.

c. Hypotheses usually are narrow in scope; theories have broad explanatory power.

d. Hypotheses and theories mean essentially the same thing in science.

9. _____ is the core idea that unifies all areas of biology.

10. Match each of the following terms to the phrase that best describes it.

a. Natural selection
b. Evolution
c. Hypothesis
d. Biosphere

1. A testable idea
2. Descent with modification
3. Unequal reproductive success
4. All life-supporting environments on Earth

Answers to the Self-Quiz questions can be found in Appendix D.

THE PROCESS OF SCIENCE

11. The graph below shows the results of an experiment in which mice learned to run through a maze.

a. State a hypothesis and prediction that you think this experiment may have been designed to test.

b. Which was the control group and which the experimental group? Why was a control group needed?

c. List some variables that must have been controlled so as not to affect the results.

d. Do the data support the hypothesis you chose? Explain.

12. The fruits of wild species of tomato are tiny compared to the giant beefsteak tomatoes available today. This difference in fruit size is almost entirely due to the larger number of cells in the domesticated fruits. Plant biologists have recently discovered genes that are responsible for controlling cell division in tomatoes. Why would such a discovery be important to producers of other kinds of fruits and vegetables? To the study of human development and disease? To our basic understanding of biology?

BIOLOGY AND SOCIETY

13. The news media and popular magazines frequently report stories that are connected to biology. In the next 24 hours, record all such stories you hear or read about from three different sources and briefly describe the biological connections you perceive in each story.

14. If you pay attention, you will find yourself conducting many hypothesis-driven experiments each day. Over the next day, try to think of a good example where you performed a simple experiment to test a hypothesis about some observation. Write the experience out in a standard narrative paragraph, and then rewrite it using the steps of the scientific method (observation, question, hypothesis, prediction, and so on).

Unit 1
Cells

Chapter 2: **Essential Chemistry for Biology**

Chapter Thread: **Water as the Chemical of Life**

Chapter 3: **The Molecules of Life**

Chapter Thread: **Lactose Intolerance**

Chapter 4: **A Tour of the Cell**

Chapter Thread: **Bacterial Diseases and Cellular Structures**

Chapter 5: **The Working Cell**

Chapter Thread: **Molecular Technologies**

Chapter 6: **Cellular Respiration: Obtaining Energy from Food**

Chapter Thread: **Life with and without Oxygen**

Chapter 7: **Photosynthesis: Using Light to Make Food**

Chapter Thread: **Harnessing Solar Energy**

2 Essential Chemistry for Biology

Ice, water.
Water, shown here in its liquid and solid forms, is essential to all life.

CHAPTER CONTENTS

■■■ Chapter Thread: **WATER AS THE CHEMICAL OF LIFE**

BIOLOGY AND SOCIETY: Water as the Chemical of Life

Fluoride in the Water

One of the first things that people notice about you is your smile. But you might not realize that healthy teeth are much more common today than they were several decades ago. One reason for the decrease in tooth decay is the addition of fluoride-containing chemicals to drinking water and dental products.

Although the surface enamel of your teeth is lifeless, it is part of a dynamic, living system that includes the bacteria in your mouth. These oral bacteria reside in plaque, a sticky combination of food, saliva, and dead cells that coats your teeth. As these bacteria grow, they release acids that can dissolve the dental surface, causing pain and, eventually, tooth loss. Fluoride helps to prevent cavities in two ways: It affects the metabolism of oral bacteria, and it promotes the replacement of lost minerals on the surface of the teeth.

Frequent exposure to small amounts of fluoride is the best way to prevent tooth decay. How do we get fluoride in our diet? Fluoride is a common ingredient in Earth's crust. As subsurface water trickles through the crust, it dissolves fluoride (as well as other minerals). In some areas, this natural process delivers enough fluoride in drinking water to help maintain healthy teeth. But in many areas of the United States and other industrialized nations, the water supply is fluoridated—meaning that it is supplemented with small amounts of fluoride during municipal water treatment. In addition, fluoride is frequently added to dental products, such as toothpaste and mouthwash.

The ability to dissolve fluoride and many other substances is just one of several properties that make water vital for life on Earth. In fact, molecules of water participate in many chemical reactions necessary to sustain life. Living organisms are, at their most basic level, chemical systems. Many questions about life reduce to questions about chemicals and their interactions, making knowledge of chemistry essential to understanding biology. In this chapter, you will learn some essential chemistry that you can apply throughout your study of life. In the last section, we'll return to the crucial role that water plays in sustaining life on Earth.

CHAPTER 2
ESSENTIAL CHEMISTRY
FOR BIOLOGY

Some Basic Chemistry

Why would a biology textbook include a chapter on chemistry? Well, take any biological system apart, and you eventually end up at the chemical level. In fact, you could think of your body as one big container of chemicals undergoing a continuous series of chemical reactions. Beginning at this basic biological level, let's explore the chemistry of life.

Matter: Elements and Compounds

All organisms and everything around them are made of matter, the physical "stuff" of the universe. Defined more formally, **matter** is anything that occupies space and has mass. Matter is found on Earth in three physical states: solid, liquid, and gas.

Matter is composed of chemical **elements**, substances that cannot be broken down into other substances. There are 92 naturally occurring elements on Earth; examples are carbon, oxygen, and gold. Each element has a symbol, the first letter or two of its English, Latin, or German name. For instance, the symbol for gold, Au, is from the Latin word *aurum*, while the symbol O stands for the English word *oxygen*. All the elements—the 92 that occur naturally and several dozen that are human-made—are listed in the periodic table of the elements, a familiar fixture in any chemistry or biology lab (**Figure 2.1**; see Appendix B for a full version).

Of the 92 naturally occurring elements, 25 are essential to life. Four of these elements—oxygen (O), carbon

Mercury (Hg)

Copper (Cu)

Lead (Pb)

▶ Figure 2.1 **Abbreviated periodic table of the elements.** In the full periodic table (see Appendix B), each entry contains the element symbol in the center, with the atomic number above and the mass number below (both are discussed on the facing page). The element highlighted here is carbon (C).

▲ Figure 2.2 **Chemical composition of the human body by weight.**

(C), hydrogen (H), and nitrogen (N)—make up about 96% of the weight of the human body, as well as most other living matter (**Figure 2.2**). Much of the remaining 4% is accounted for by 7 elements, most of which are probably familiar to you, such as calcium (Ca) and phosphorus (P). Calcium, important for building strong bones and teeth, is found abundantly in milk and dairy products as well as sardines and green, leafy vegetables (collards, kale, and broccoli, for example). Phosphorus, a component of DNA and other important biological molecules, can be obtained by eating eggs, beans, and nuts.

Less than 0.01% of your weight is made up of 14 trace elements (see Figure 2.2). **Trace elements** are required in only very small amounts, but you cannot live without them. The average human, for example, needs only a tiny speck of iodine, about 0.15 milligram (mg) each day. Iodine is an essential ingredient of a hormone produced by the thyroid gland, located in the neck. An iodine deficiency in the diet causes the thyroid gland to enlarge, a condition called goiter. Foods naturally rich in iodine include kelp and dairy products. The addition of iodine to table salt (iodized salt) has nearly eliminated goiter in many nations. Unfortunately, goiter still affects many thousands of people in developing countries (**Figure 2.3**). Fluorine, whose "fluoride" form was discussed in the Biology and Society section, is another example of a trace element; it is a needed component of healthy bones and teeth.

Elements can combine to form **compounds**, substances that contain two or more elements in a fixed ratio. In everyday life, compounds are much more common than pure elements. Familiar examples are relatively simple compounds such as table salt and water. Table salt is sodium chloride, NaCl, consisting of equal parts of the elements sodium (Na) and chlorine (Cl). A molecule of water, H_2O, has two atoms of hydrogen and one atom of oxygen. Most of the compounds in living organisms contain several different elements. DNA, for example, contains carbon, nitrogen, oxygen, hydrogen, and phosphorus. ☑

Atoms

Each element consists of one kind of atom, which is different from the atoms of other elements. An **atom**, named from a Greek word meaning "indivisible," is the smallest unit of matter that still retains the properties of an element. In other words, the smallest amount of the element carbon is one carbon atom. Just how small is this "piece" of carbon? It would take about a million carbon atoms to stretch across the period at the end of this sentence.

The Structure of Atoms

Atoms are composed of subatomic particles, of which the three most important are protons, electrons, and neutrons. A **proton** is a subatomic particle with a single unit of positive electrical charge (+). An **electron** is a subatomic particle with a single unit of negative electrical charge (−). A **neutron** is electrically neutral (has no electrical charge).

Let's look at the structure of an atom of the element helium (He), the "lighter-than-air" gas used in party balloons (**Figure 2.4**). Each atom of helium has 2 neutrons (⦾) and 2 protons (⊕) tightly packed into the **nucleus**, the atom's central core. Two electrons (⊖) move around

▼ Figure 2.4 **Two simplified models of a helium atom.**

2 ⊕ Protons
2 ⦾ Neutrons — Nucleus
2 ⊖ Electrons

Nucleus

Cloud of negative charge (2 electrons)

This model shows the subatomic particles in an atom of helium.

This model, slightly more realistic, shows the electrons as a spherical cloud of negative charge surrounding the nucleus.

▼ Figure 2.3 **Why is salt "iodized"?**

Goiter, an enlargement of the thyroid gland, shown here in a Malaysian woman, can occur when a person's diet does not include enough iodine, a trace element.

Eating iodine-rich foods can prevent goiter.

the nucleus in a spherical cloud at nearly the speed of light. The electron cloud is much bigger than the nucleus; in fact, if the electron cloud were the size of a football stadium, the nucleus would be the size of a fly on the field. The attraction between the negatively charged electrons and the positively charged protons keeps the electrons moving. When an atom has an equal number of protons and electrons (as helium does), its net electrical charge is zero and so the atom is neutral.

Elements differ in the number of subatomic particles in their atoms. The number of protons in an atom, called the **atomic number**, determines which element it is. For example, helium always has 2 protons and thus an atomic number of 2. **Mass** is a measure of the amount of material in an object. A proton and a neutron have nearly identical mass. An electron has so little mass—only about 1/2,000 the mass of a proton—that it is usually approximated as zero. Therefore, an atom's **mass number** is just the sum of the number of protons and neutrons in its nucleus. For helium, the mass number is 4 (2 protons + 2 neutrons). Both the atomic number and the mass number can be read from the periodic table (see Figure 2.1 and Appendix B).

Isotopes

Some elements can exist in different forms called isotopes. The different **isotopes** of an element have the same numbers of protons and electrons but different numbers of neutrons; in other words, isotopes are

Table 2.1	Isotopes of Carbon		
	Carbon-12	**Carbon-13**	**Carbon-14**
Protons	6 ⎱ mass number	6 ⎱ mass number	6 ⎱ mass number
Neutrons	6 ⎰ 12	7 ⎰ 13	8 ⎰ 14
Electrons	6	6	6

forms of an element that differ in mass. **Table 2.1** shows the numbers of subatomic particles in the three isotopes of carbon. Carbon-12 (named for its mass number 12), with 6 neutrons and 6 protons, makes up about 99% of all naturally occurring carbon. Most of the other 1% consists of carbon-13, with 7 neutrons and 6 protons. A third isotope, carbon-14, with 8 neutrons and 6 protons, occurs in minute quantities. Notice that all three isotopes have 6 protons—otherwise, they would not be carbon. Both carbon-12 and carbon-13 are stable isotopes, meaning that their nuclei remain intact more or less forever. The isotope carbon-14, on the other hand, is unstable, or radioactive. A **radioactive isotope** is one in which the nucleus decays, giving off particles and energy.

Radioactive isotopes have many uses in biological research and medicine. Cells use radioactive isotopes the same way they use nonradioactive isotopes of the same element. Once the cell takes up the radioactive isotopes, the location and concentration of the isotopes can be detected because of the radiation they emit. This makes radioactive isotopes useful as tracers—biological spies, in effect—for monitoring the fate of atoms in living organisms. For example, a medical diagnostic tool called a PET scan works by detecting small amounts of radioactive materials introduced into the body. PET scans can diagnose heart disorders and some cancers.

Although radioactive isotopes have many beneficial uses, uncontrolled exposure to them can harm living organisms by damaging cellular molecules, especially DNA. In 1986, the explosion of a nuclear reactor at Chernobyl, Ukraine, released large amounts of radioactive isotopes, killing 30 people within a few weeks and exposing thousands to an increased risk of developing cancer. In fact, the incidence of thyroid cancer among Ukrainian children increased tenfold in the decade following the accident.

Natural sources of radiation can also pose a threat. Radon, a radioactive gas, may cause lung cancer. Radon can contaminate buildings in regions where underlying rocks naturally contain the radioactive element uranium. Homeowners can buy a radon detector or hire a company to test their home to ensure that radon levels are safe.

Electron Arrangement and the Chemical Properties of Atoms

Of the three subatomic particles we've discussed—protons, neutrons, and electrons—electrons are the ones that primarily determine how an atom behaves when it encounters other atoms. Electrons vary in the amount of energy they possess. The farther an electron is from the nucleus, the greater its energy. Electrons do not move around an atom at just any energy level, but only at specific levels called electron shells. Depending on the number of electrons, atoms may have one, two, or more electron shells, with electrons in the outermost shell having the highest energy. Each shell can accommodate up to a specific number of electrons. The innermost shell is full with only 2 electrons, while the second and third shells can each hold up to 8 electrons.

The number of electrons in the outermost shell determines the chemical properties of an atom. Atoms whose outer shells are not full tend to interact with other atoms—that is, to participate in chemical reactions. **Figure 2.5** shows the electron shells of four biologically important elements. Because the outer shells of all four atoms are not filled, these atoms react readily with other atoms. The hydrogen atom is highly reactive because it has only 1 electron in its single electron shell, which can accommodate 2 electrons. Atoms of carbon, nitrogen, and oxygen are also highly reactive because their outer shells, which can hold 8 electrons, are not filled. In contrast, the helium atom in Figure 2.4 has a single, first-level shell that is full with 2 electrons. As a result, helium is chemically unreactive. ☑

Hydrogen (H)
Atomic number = 1

Carbon (C)
Atomic number = 6

Nitrogen (N)
Atomic number = 7

Oxygen (O)
Atomic number = 8

◄ **Figure 2.5 Atoms of the four elements most abundant in living matter.** All four atoms are chemically reactive because their outermost electron shells are not filled. The small empty circles (o) in these diagrams represent unfilled "spaces" in the outer electron shells.

Chemical Bonding and Molecules

Chemical reactions enable atoms to give up or acquire electrons, thereby completing their outer shells. Atoms do this by either transferring or sharing outer electrons. These interactions usually result in atoms staying close together, held by attractions called **chemical bonds**. In this section, we will discuss three types of chemical bonds: ionic, covalent, and hydrogen bonds.

Ionic Bonds

Table salt is an example of how the transfer of electrons can bond atoms together. The two ingredients of table salt are the elements sodium (Na) and chlorine (Cl). When a chlorine atom strips an electron from a sodium atom, the electron transfer results in both atoms having full outer shells of electrons (**Figure 2.6**). Before the electron transfer, each of these atoms is electrically neutral. Because electrons are negatively charged particles, the electron transfer moves one unit of negative charge from sodium to chlorine. Both atoms are now **ions**, the term for atoms that are electrically charged as a result of gaining or losing electrons. The loss of an electron gives the sodium ion a charge of $+1$, while chlorine's gain of an electron gives it a charge of -1. The sodium ion (Na^+) and chloride ion (Cl^-) are held together by an **ionic bond**, the attraction between oppositely charged ions. Compounds, such as table salt, that are held together by ionic bonds are called ionic compounds. Fluorine in Earth's crust is often found in

the form of ionic compounds such as calcium fluoride (CaF_2), the result of bonds between calcium ions (Ca^{2+}) and fluoride ions (F^-). (Note that negatively charged ions often have names ending in "-ide," like "chloride" or "fluoride.") ☑

Covalent Bonds

In contrast to the complete *transfer* of electrons that leads to ionic bonds, a **covalent bond** forms when two atoms *share* one or more pairs of outer-shell electrons. Atoms that are held together by covalent bonds form a **molecule**. For example, a covalent bond connects each hydrogen atom to the carbon in the molecule CH_4, a common gas called methane. In **Figure 2.7**, you can see that each of the four hydrogen atoms in a molecule of

▼ **Figure 2.6 Electron transfer and ionic bonding.** When a sodium atom and a chlorine atom meet, the electron transfer between the two atoms results in two ions with opposite charges.

Outer shell has 1 electron

The outer electron is stripped from sodium and completes the chlorine atom's outer shell.

Outer shell has 7 electrons

Complete outer shells

Na
Sodium atom

Cl
Chlorine atom

Na⁺
Sodium ion

Cl⁻
Chloride ion

The attraction between the ions—an ionic bond—holds them together.

Sodium chloride (NaCl)

Name (molecular formula)	Electron configuration Shows how each atom completes its outer shell by sharing electrons	Structural formula Represents each covalent bond (a pair of shared electrons) with a line	Space-filling model Shows the shape of a molecule by symbolizing atoms with color-coded balls	Ball-and-stick model Represents atoms with "balls" and bonds with "sticks"
Hydrogen gas (H_2)	(H)(H)	H—H Single bond (a pair of shared electrons)		
Oxygen gas (O_2)	(O)(O)	O=O Double bond (two pairs of shared electrons)		
Methane (CH_4)	(H)(H)(C)(H)(H)	H—C—H (with H above and below)		

◄ **Figure 2.7 Alternative ways to represent molecules.** A molecular formula, such as CH_4 or H_2O, tells you the number of each kind of atom in a molecule but not how they are attached together. This figure shows four common ways of representing the arrangement of atoms in molecules.

methane shares one pair of electrons with the single carbon atom.

The number of covalent bonds an atom can form is equal to the number of additional electrons needed to fill its outer shell. Note in Figure 2.7 that hydrogen (H) can form one covalent bond, oxygen (O) can form two, and carbon (C) can form four. The single covalent bond in H_2 completes the outer shells of both hydrogen atoms. In contrast, an oxygen atom needs 2 electrons to complete its outer shell. In an O_2 molecule, the two oxygen atoms share two pairs of electrons, forming a double covalent bond.

Hydrogen Bonds

Water (H_2O) is also a compound. Its structure consists of two hydrogen atoms joined to one oxygen atom by single covalent bonds (the "sticks" represent bonds between the atoms, which are shown as "balls"):

However, the electrons of the covalent bonds are not shared equally between the oxygen and hydrogen. The two yellow arrows in the following diagram indicate the stronger pull on the shared electrons that oxygen has compared with its hydrogen partners:

(slightly +)   (slightly +)

(slightly −)

The unequal sharing of negatively charged electrons, combined with its V shape, makes a water molecule polar. A **polar molecule** has opposite charges on opposite ends. In the case of water, the oxygen end of the molecule has a slight negative charge, while the region around the two hydrogen atoms is slightly positive.

The polarity of water results in weak electrical attractions between neighboring water molecules. The molecules tend to orient such that the hydrogen atom of one molecule is near the oxygen atom of an adjacent water molecule. These weak attractions

are called **hydrogen bonds** (Figure 2.8). As you will see later in this chapter, the ability of water to form hydrogen bonds has many implications for life.

Chemical Reactions

The chemistry of life is dynamic. Your cells are constantly rearranging molecules by breaking existing chemical bonds and forming new ones. Such changes in the chemical composition of matter are called **chemical reactions**. A simple example is the reaction between oxygen gas and hydrogen gas that forms water (this is an explosive reaction, which, fortunately, does not occur in your cells):

2 H_2 + O_2 ⟶   2 H_2O

Hydrogen   Oxygen   Water
gas   gas

Reactants   **Products**

Let's translate the chemical shorthand: Two molecules of hydrogen gas (2 H_2) react with one molecule of oxygen gas (O_2) to form two molecules of water (2 H_2O). The arrows indicate the conversion of the starting materials, the **reactants** (2 H_2 and O_2), to the **products** (2 H_2O).

Notice that the same numbers of hydrogen and oxygen atoms are present in reactants and products, although they are grouped differently. Chemical reactions cannot create or destroy matter, but only rearrange it. These rearrangements usually involve the breaking of chemical bonds in reactants and the forming of new bonds in products.

The water molecules we have built here are a good conclusion to this section on basic chemistry. Water is a substance so important in biology that we'll take a closer look at its life-supporting properties in the next section. ✓

Hydrogen bond

(−)
(+)

H
|
O—H

(+)---(−)   (+)---(−)
(−)
(+)

◄ **Figure 2.8 Hydrogen bonding in water.** The charged regions of the polar water molecules are attracted to oppositely charged areas of neighboring molecules. Each molecule can hydrogen-bond to a maximum of four partners.

✓**CHECKPOINT**

Predict the formula for the compound that results when a molecule of sulfur trioxide (SO_3) combines with a molecule of water to produce a single molecule of product. (*Hint*: In chemical reactions, no atoms are gained or lost.)

Answer: H_2SO_4 (sulfuric acid)

Water and Life

The Properties of Water

Life on Earth began in water and evolved there for 3 billion years before spreading onto land. Modern life, even land-dwelling life, is still tied to water. You've had personal experience with this dependence on water every time you seek liquids to quench your thirst and replenish your body's water. Inside your body, your cells are surrounded by a fluid that's composed mostly of water, and your cells themselves range from 70% to 95% in water content.

The abundance of water is a major reason that Earth is habitable. Water is so common in our environment that it's easy to overlook its extraordinary behavior (Figure 2.9). We can trace water's unique life-supporting properties to the structure and interactions of its molecules.

Water's Life-Supporting Properties

The polarity of water molecules and the hydrogen bonding that results (see Figure 2.8) explain most of water's life-supporting properties. We'll explore four of those properties here: the cohesive nature of water, the ability of water to moderate temperature, the biological significance of ice floating, and the versatility of water as a solvent.

The Cohesion of Water

Water molecules stick together as a result of hydrogen bonding. Hydrogen bonds between molecules of liquid water last for only a few trillionths of a second, yet at any instant, many of the molecules are hydrogen-bonded to others. This tendency of molecules of the same kind to stick together, called **cohesion**, is much stronger for water than for most other liquids. The cohesion of water is important in the living world. Trees, for example, depend on cohesion to help transport water from their roots to their leaves (Figure 2.10).

Related to cohesion is surface tension, a measure of how difficult it is to stretch or break the surface of a liquid. Hydrogen bonds give water unusually high surface tension, making it behave as though it were coated with an invisible film (Figure 2.11).

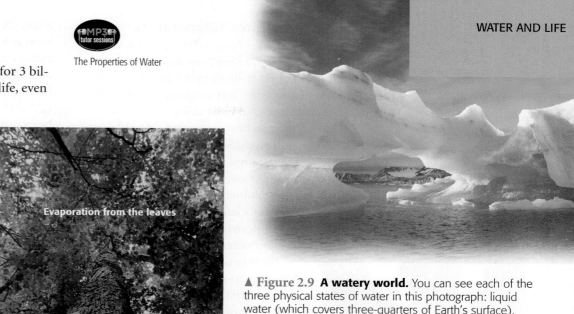

▲ Figure 2.9 **A watery world.** You can see each of the three physical states of water in this photograph: liquid water (which covers three-quarters of Earth's surface), ice, and water vapor.

Evaporation from the leaves

Flow of water

▲ Figure 2.10 **Cohesion and water transport in plants.** The evaporation of water from leaves pulls water upward from the roots through microscopic tubes in the trunk of the tree. Because of cohesion, the pulling force is relayed through the tubes all the way down to the roots. As a result, water rises against the force of gravity.

Microscopic tubes

Cohesion due to hydrogen bonds between water molecules

Colorized SEM 150×

▼ Figure 2.11 **A raft spider walking on water.** The cumulative strength of hydrogen bonds between water molecules allows this spider to walk on pond water without breaking the surface.

2

How Water Moderates Temperature

If you've ever burned your finger on a metal pot while waiting for the water in it to boil, you know that water heats up much more slowly than metal. In fact, because of hydrogen bonding, water has a stronger resistance to temperature change than most other substances.

Temperature and heat are related, but different. A swimmer crossing San Francisco Bay has a higher temperature than the water, but the bay contains far more heat because of its immense volume. **Heat** is the amount of energy associated with the movement of the atoms and molecules in a body of matter. **Temperature** measures the intensity of heat—that is, the average speed of molecules rather than the total amount of heat energy in a body of matter.

When water is heated, the heat energy first disrupts hydrogen bonds and then makes water molecules jostle around faster. The temperature of the water doesn't go up until the water molecules start to speed up. Because heat is first used to break hydrogen bonds rather than raise the temperature, water absorbs and stores a large amount of heat while warming up only a few degrees. Conversely, when water cools, hydrogen bonds form, a process that releases heat. Thus, water can release a relatively large amount of heat to the surroundings while the water temperature drops only slightly.

Earth's giant water supply—the oceans, seas, lakes, and rivers—enables temperatures to stay within limits that permit life by storing a huge amount of heat from the sun during warm periods and giving off heat to warm the air during cold conditions. That's why coastal areas generally have milder climates than inland regions. Water's resistance to temperature change also stabilizes ocean temperatures, creating a favorable environment for marine life.

Another way that water moderates temperature is by **evaporative cooling**. When a substance evaporates (changes physical state from a liquid to a gas), the surface of the liquid remaining behind cools down. This occurs because the molecules with the greatest energy (the "hottest" ones) tend to vaporize first. It's as if the five fastest runners on your track team quit school, lowering the average speed of the remaining team. On a global scale, surface evaporation cools tropical oceans. On the scale of individual organisms, evaporative cooling prevents some land-dwelling creatures from overheating. It's why sweating helps you maintain a constant body temperature, even when exercising on a hot day (**Figure 2.12**). And the old expression "It's not the heat, it's the humidity" has its basis in the difficulty of sweating water into air that is already saturated with water vapor.

▲ Figure 2.12 **Sweating as a mechanism of evaporative cooling.**

The Biological Significance of Ice Floating

When most liquids get cold, their molecules move closer together. If the temperature is cold enough, the liquid freezes and becomes a solid. Water, however, behaves differently. When water molecules get cold enough, they move apart, forming ice. A chunk of ice has fewer molecules than an equal volume of liquid water; it floats because it is less dense than the liquid water around it. Like water's other life-supporting properties, floating ice is a consequence of hydrogen bonding. In contrast to the short-lived hydrogen bonds in liquid water, those in solid ice last longer, with each molecule bonded to four neighbors. As a result, ice is a spacious crystal (**Figure 2.13**).

Liquid water
Hydrogen bonds constantly break and re-form.

Ice
Stable hydrogen bonds hold molecules apart, making ice less dense than water.

Hydrogen bond

◀ Figure 2.13 **Why ice floats.** Compare the tightly packed molecules in liquid water with the spaciously arranged molecules in the ice crystal. The less dense ice floats atop the denser water.

How does the fact that ice floats help support life on Earth? Imagine what would happen if ice sank. All ponds, lakes, and even the oceans would eventually freeze solid. During summer, only the upper few inches of the oceans would thaw. Instead, when a deep body of water cools, the floating ice insulates the liquid water below, allowing life to persist under the frozen surface.

Water as the Solvent of Life

If you've ever enjoyed a glass of sweetened ice tea or added salt to soup, you know that you can dissolve sugar or salt in water. This results in a mixture known as a **solution**, a liquid consisting of a homogeneous mixture of two or more substances. The dissolving agent is called the **solvent**, and a substance that is dissolved is called a

solute. When water is the solvent, the resulting solution is called an **aqueous solution**.

The fluids of organisms are aqueous solutions. Water can dissolve an enormous variety of solutes necessary for life. Water is the solvent inside all cells, in blood, and in plant sap. As a solvent, it is a medium for chemical reactions. Water can dissolve salt ions, as shown in **Figure 2.14**. Each ion becomes surrounded by oppositely charged regions of water molecules. Solutes that are polar molecules, such as sugars, dissolve by orienting locally charged regions of their molecules toward water molecules in a similar way.

We have discussed four special properties of water. Next, we'll consider one way that water's unique structure can reveal important clues about the human brain. ☑

☑**CHECKPOINT**

1. Explain why, if you pour very carefully, you can actually "stack" water slightly above the rim of a cup.
2. Explain why ice floats.

Answers: 1. Surface tension due to water's cohesion will keep the water from spilling over. 2. Ice is less dense than liquid water because the more stable hydrogen bonds lock the molecules into a spacious crystal.

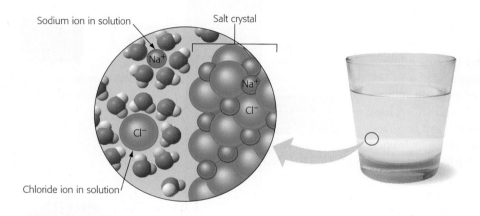

Sodium ion in solution

Salt crystal

Na⁺

Na⁺

Cl⁻

Cl⁻

Chloride ion in solution

◄ Figure 2.14 **A crystal of table salt (NaCl) dissolving in water.** As a result of electrical charge attractions, H_2O molecules surround the sodium and chloride ions, dissolving the crystal in the process.

THE PROCESS OF SCIENCE: Water as the Chemical of Life

Can Exercise Boost Your Brain Power?

Biologists use a number of imaging techniques to reveal the body's inner structure. One such technique, magnetic resonance imaging (MRI), depends on the behavior of the hydrogen atoms in water molecules. MRI first aligns the hydrogen nuclei with powerful magnets, then knocks the nuclei out of alignment with a brief pulse of radio waves. In response, the hydrogen atoms give out faint radio signals of their own, which are picked up by the MRI scanner and translated by computer into an image. Since water is a major component of all of our soft tissues, MRI allows doctors to see regions of the body that are invisible to X-rays.

MRI has been particularly useful in studying the brain. One recent study began with the **observation** that the human brain shrinks as we age: About 20% of

our brain matter is lost between ages 30 and 90, and the shrinkage is most prominent in brain areas responsible for memory and learning. Because exercise had been previously associated with improved memory, researchers from the University of Illinois asked the **question**, Can aerobic exercise slow or even reverse brain loss? Their **hypothesis** was that MRI scans would reveal differences between the brains of people who regularly exercised and those who did not. They made the **prediction** that the brains of active people would shrink less than the brains of less active people.

In their **experiment**, 29 subjects in their 60s and 70s participated for six months in an exercise program of three 1-hour aerobic training sessions per week. A control group of 29 subjects (matched for age) engaged in a program of non-aerobic whole-body stretching lasting the same duration. At the start and end of the program, MRI was used to scan the brains of all participants. The **results** of the experiment showed significant increases in brain volume for the aerobic group compared to the non-aerobic controls (**Figure 2.15**). The researchers concluded that there may be a link between cardiovascular fitness and the retention of memory in aging adults. In other words, regular aerobic exercise may benefit both your body and your mind.

▶ **Figure 2.15 The effect of aerobic exercise on brain size.** The colors highlight brain regions that grew in aerobic exercise participants compared to a control group of non-aerobic exercisers.

Acids, Bases, and pH

In the aqueous solutions within organisms, most of the water molecules are intact. However, some of the water molecules actually break apart into hydrogen ions (H^+) and hydroxide ions (OH^-). A balance of these two ions is critical for the proper functioning of chemical processes within organisms.

A chemical compound that releases H^+ to a solution is called an **acid**. One example of a strong acid is hydrochloric acid (HCl), the acid in your stomach. In solution, HCl breaks apart into the ions H^+ and Cl^-. A **base** (or alkali) is a compound that accepts H^+ and removes it from solution. Some bases, such as sodium hydroxide (NaOH), do this by releasing OH^-, which combines with H^+ to form H_2O.

To describe the acidity of a solution, chemists use the **pH scale**, a measure of the hydrogen ion (H^+) concentration in a solution. The scale ranges from 0 (most acidic) to 14 (most basic). Each pH unit represents a tenfold change in the concentration of H^+ (**Figure 2.16**). For example, lemon juice at pH 2 has 100 times more H^+ than an equal amount of tomato juice at pH 4. Pure water and aqueous solutions that are neither acidic nor basic are said to be neutral; they have a pH of 7. They do contain H^+ and OH^-, but the concentrations of the two ions are equal. The pH of the solution inside most living cells is close to 7.

Even a slight change in pH can be harmful to an organism because the molecules in cells are extremely sensitive to H^+ and OH^- concentrations. Biological fluids contain **buffers**, substances that prevent harmful changes in pH by accepting H^+ when that ion is in excess and donating H^+ when it is depleted. This buffering process, however, is not

▶ **Figure 2.16 The pH scale.** A solution having a pH of 7 is neutral, meaning that its H^+ and OH^- concentrations are equal. The lower the pH below 7, the more acidic the solution, or the greater its excess of H^+ compared with OH^-. The higher the pH above 7, the more basic the solution, or the greater the deficiency of H^+ relative to OH^-.

foolproof, and changes in environmental pH can profoundly affect ecosystems. For example, much of the human-made CO_2 released into the atmosphere is absorbed by the oceans. When it combines with seawater, a chemical reaction produces an acid. The resulting ocean acidification can greatly change marine environments. This principle was illustrated by a 2008 study of the Mediterranean Sea near Naples, Italy. There, undersea volcanoes release high levels of CO_2 (**Figure 2.17**). Researchers found that the resulting acidification caused coral reef collapse and widespread death of marine organisms. This natural system suggests that high levels of dissolved human-made CO_2 could cause similar ecological disaster.

The effects of ocean acidification are daunting reminders that the chemistry of life is linked to the chemistry of the environment. It reminds us, too, that chemistry happens on a global scale, since industrial processes in one region of the world often cause acid precipitation to fall in another part of the world. ☑

◄ **Figure 2.17 The release of volcanic CO_2, causing undersea bubbling off the coast of Italy.**

☑**CHECKPOINT**
Compared with a solution of pH 8, the same volume of a solution at pH 5 has _____ times more hydrogen ions (H$^+$). This second solution is considered a(n)_____.

Answer: 1,000; acid

EVOLUTION CONNECTION: Water as the Chemical of Life

The Search for Extraterrestrial Life

Throughout this chapter, we have highlighted the importance of water to living systems. In fact, water is the substance that makes life as we know it possible.

One of the great questions facing humankind is, Has life evolved elsewhere in the universe? If it has, and if this life is at all similar to ours, then it, too, would depend on water. This explains why the search for water on other planets (particularly Mars) is such a priority for the U.S. space program.

Researchers at the National Aeronautics and Space Administration (NASA) have accumulated data that suggest that water was once abundant on Mars. (Although Mars has clearly visible polar ice caps, no liquid water is known to flow there at present.) In January 2004, NASA landed two golf-cart-sized rovers on Mars. These robotic geologists determined the composition of rocks and sent back images of rock formations. One of the rovers detected a mineral that is formed only in the presence of water. Other chemical evidence indicated that water once permeated Martian rocks. And pictures revealed physical evidence of the effects of liquid water flowing over the Martian surface in the past.

Images taken by NASA's *Mars Global Surveyor* spacecraft indicate that not only did the red planet have a water-filled ancient past, but liquid water may have flowed recently on its surface. In 2008, an analysis of Martian soil by the *Phoenix Mars Lander* (**Figure 2.18**) provided evidence that components of the soil had at one time been dissolved in water. As this chapter has shown, where there is water, there may indeed be life. The evidence for water on Mars raises the tantalizing possibility that microbial life may exist below the Martian surface.

▲ **Figure 2.18 The *Phoenix Mars Lander* probing for evidence of water on the Martian surface.**

Chapter Review

SUMMARY OF KEY CONCEPTS

 Go to the Study Area at **www.masteringbiology.com** for practice quizzes, eText, BioFlix™ 3-D animations, MP3 Tutor Sessions, videos, current events, and more.

Some Basic Chemistry

Matter: Elements and Compounds

Matter consists of elements and compounds, which are combinations of two or more elements. Of the 25 elements essential for life, oxygen, carbon, hydrogen, and nitrogen are the most abundant in living matter.

Atoms

Proton
• Positive charge
• Determines element

Electron
• Negative charge
• Participates in chemical reactions
• Outer-shell electrons determine chemical behavior

Neutron
• No charge
• Determines isotope

Atom

Nucleus
• Consists of neutrons and protons

Chemical Bonding and Molecules

Transfer of one or more electrons produces attractions between oppositely charged ions:

Attraction between oppositely charged ions holds the ions in an ionic bond.

Lithium ion   Fluoride ion

Lithium fluoride (ionic compound)

A molecule consists of two or more atoms connected by covalent bonds, which are formed by electron sharing:

Electron sharing

Atoms joined into a molecule via covalent bonds

Water is a polar molecule; the slightly positively charged H atoms in one water molecule may be attracted to the partial negative charge of O atoms in neighboring water molecules, forming weak but important hydrogen bonds:

Hydrogen bond

Chemical Reactions

By breaking bonds in reactants and forming new bonds in products, chemical reactions rearrange matter.

Water and Life

Water's Life-Supporting Properties

The ability of leaves to pull water up microscopic tubes within trunks and stems is an example of how water's cohesion supports life. Water moderates temperature by absorbing heat in warm environments and releasing heat in cold environments. Evaporative cooling also helps stabilize the temperatures of oceans and organisms. The fact that ice floats because it is less dense than liquid water prevents the oceans from freezing solid. Blood and other biological fluids are aqueous solutions with a diversity of solutes dissolved in water, a versatile solvent.

Acids, Bases, and pH

Basic — 14 Lower H^+ concentration

Neutral $[H^+] = [OH^-]$ — 7 H^+ concentration equal to OH^- concentration

Acidic — 0 Greater H^+ concentration

pH scale

SELF-QUIZ

1. An atom can be changed into an ion by adding or removing _____. An atom can be changed into a different isotope by adding or removing _____. But if you change the number of _____, the atom becomes a different element.

2. A nitrogen atom has 7 protons, and the most common isotope of nitrogen has 7 neutrons. A radioactive isotope of nitrogen has 9 neutrons. What are the atomic numbers and mass numbers of the stable and radioactive forms of nitrogen?

3. Why are radioactive isotopes useful as tracers in research on the chemistry of life?

4. A sulfur atom has 6 electrons in its third (outermost) shell, which can hold 8 electrons. As a result, it forms _____ covalent bonds with other atoms. (Provide a number.)

5. What is chemically nonsensical about this structure?

    H—C≡C—H

6. Why is it unlikely that two neighboring water molecules would be arranged like this?

7. Which of the following is not a chemical reaction?
    a. Sugar ($C_6H_{12}O_6$) and oxygen gas (O_2) combine to form carbon dioxide (CO_2) and water (H_2O).
    b. Sodium metal and chlorine gas unite to form sodium chloride.
    c. Hydrogen gas combines with oxygen gas to form water.
    d. Ice melts to form liquid water.

8. Some people in your study group say they don't understand what a polar molecule is. You explain that a polar molecule
    a. is slightly negative at one end and slightly positive at the other end.
    b. has an extra electron, giving it a positive charge.
    c. has an extra electron, giving it a negative charge.
    d. has covalent bonds.

9. Explain how the unique properties of water result from the fact that water is a polar molecule.

10. A can of cola consists mostly of sugar dissolved in water, with some carbon dioxide gas that makes it fizzy and makes the pH less than 7. Describe the cola using the following terms: solute, solvent, acidic, aqueous solution.

Answers to the Self-Quiz questions can be found in Appendix D.

THE PROCESS OF SCIENCE

11. Animals obtain energy through a series of chemical reactions in which sugar ($C_6H_{12}O_6$) and oxygen gas (O_2) are reactants. This process produces water (H_2O) and carbon dioxide (CO_2) as waste products. How might you use a radioactive isotope to find out whether the oxygen in CO_2 comes from sugar or oxygen gas?

12. The following diagram shows the arrangement of electrons around the nucleus of a fluorine atom (left) and a potassium atom (right). Predict what would happen if a fluorine atom and a potassium atom came into contact. What kind of bond do you think they would form?

Fluorine atom   Potassium atom

BIOLOGY AND SOCIETY

13. Critically evaluate this statement: "It's paranoid and ignorant to worry about industry or agriculture contaminating the environment with chemical wastes; this stuff is just made of the same atoms that were already present in our environment."

14. A major source of the CO_2 that causes ocean acidification is emissions from coal-burning power plants. One way to reduce these emissions is to use nuclear power to produce electricity. The proponents of nuclear power contend that it is the only way that the United States can increase its energy production while reducing air pollution, because nuclear power plants emit little or no acid-precipitation-causing pollutants. What are some of the benefits of nuclear power? What are the possible costs and dangers? Do you think we ought to increase our use of nuclear power to generate electricity? Why or why not? If a new power plant were to be built near your home, would you prefer it to be a coal-burning plant or a nuclear plant? Why?

3 The Molecules of Life

Ice cream and lactose.
Lactose digestion results
from interactions between
several classes of the body's
molecules.

BIOLOGY AND SOCIETY: Lactose Intolerance

Got Lactose?

A milk mustache is meant to represent good health. Indeed, milk is a very healthy food because it's rich in protein, minerals, and vitamins. But for the majority of adults in the world, a glass of milk or a serving of milk-containing foods (such as ice cream or cheese) can cause bloating, gas, and other discomforts. Such people are exhibiting symptoms of lactose intolerance, the inability to properly digest lactose, the main sugar found in milk.

For people with lactose intolerance, the problem starts when lactose enters the small intestine. To absorb this sugar, digestive cells there must produce a molecule called lactase. Lactase is an enzyme, a protein that helps drive chemical reactions—in this case, the breakdown of lactose into smaller sugars. People with lactose intolerance produce insufficient amounts of the enzyme. Lactose that is not broken down in the small intestine passes into the large intestine, where bacteria feed on it and belch out gaseous by-products, producing uncomfortable symptoms. Sufficient lactase can thus mean the difference between delight and discomfort when someone feasts on an ice cream sundae.

What options are there for people with lactose intolerance? There is no treatment for the underlying cause: the underproduction of lactase. But symptoms can be controlled through diet. The first option is avoiding lactose-containing foods. Many substitutes are available, such as milk made from soy or milk that has been pretreated with lactase. Also, lactase in pill form can be taken along with food to ease digestion by artificially providing the enzyme that the body naturally lacks.

Lactose intolerance illustrates one way the interplay of biological molecules can affect human health. Such molecular interactions, repeated in countless variations, drive all biological processes. In this chapter, we'll explore the structure and function of large molecules that are essential to life. We'll start with an overview and then examine four classes of molecules: carbohydrates, lipids, proteins, and nucleic acids. Along the way, we'll emphasize where these molecules occur in your diet and the important roles they play in your body.

the polymer each contribute atoms to form a water molecule. The monomers replace the lost covalent bonds with a bond to each other.

by adding a water molecule, which breaks the bond between two monomers.

Organic Compounds

A cell is mostly water, but the rest of it consists mainly of ~~carbon-based molecules. Carbon is unparalleled in its~~ only carbon and hydrogen atoms. And the simplest ~~hydrocarbon is methane, a single carbon atom bonded~~

Large Biological Molecules

The remainder of the chapter will introduce you to the four categories of large biological molecules: carbohydrates, lipids, proteins, and nucleic acids. For each category, you'll explore the structure and function of these large molecules by first learning about the smaller molecules used to build them.

Carbohydrates

Carbohydrates, commonly known as "carbs," are sugars or sugar polymers. Some examples are the small sugar molecules dissolved in soft drinks and the long starch molecules in pasta and potatoes. In animals, carbohydrates are a primary source of dietary energy; in plants, they serve as a building material for much of the plant body.

Monosaccharides

Simple sugars, or **monosaccharides** (from the Greek *mono*, single, and *sacchar*, sugar), cannot be broken down by hydrolysis into smaller sugars. Common examples are glucose, found in sports drinks, and fructose, found in fruit. Both of these simple sugars are also in honey (**Figure 3.5**). The molecular formula for glucose is $C_6H_{12}O_6$. Fructose has the same formula, but its atoms are arranged differently. Glucose and fructose are examples of **isomers**, molecules that have the same

molecular formula but different structures. (Isomers are like anagrams—words that contain the same letters in a different order, such as *heart* and *earth*.) Because shape is so important, seemingly minor differences in the arrangement of atoms give isomers different properties. In this case, the rearrangement of functional groups makes fructose taste considerably sweeter than glucose.

It is convenient to draw sugars as if their carbon skeletons were linear. In water, however, many monosaccharides form rings when one part of the molecule forms a bond with another part of the molecule, as shown for glucose in **Figure 3.6**.

Monosaccharides, particularly glucose, are the main fuel molecules for cellular work. Analogous to an automobile engine consuming gasoline, your cells break down glucose molecules and extract their stored energy, giving off carbon dioxide as "exhaust." The rapid conversion of glucose to cellular energy is why an aqueous solution of glucose (often called dextrose) is injected into the bloodstream of sick or injured patients; the glucose provides an immediate energy source to tissues in need of repair. In addition to their use as an energy source, monosaccharides provide cells with carbon skeletons that can be used as a raw material for manufacturing other kinds of organic compounds. ✓

► **Figure 3.5 Monosaccharides (simple sugars).** These molecules have the two trademarks of sugars: several hydroxyl groups (—OH) and a carbonyl group $\ce{>C=O}$ (C=O). Glucose and fructose, which make honey sweet, are isomers.

Glucose
$C_6H_{12}O_6$

Fructose
$C_6H_{12}O_6$

Isomers
(same formula, different structures)

▼ **Figure 3.6 The ring structure of glucose.**

(a) Linear and ring structures. The carbon atoms are numbered so you can relate the linear and ring versions of the molecule. As the double arrows indicate, ring formation is a reversible process, but at any instant in an aqueous solution, most glucose molecules are rings.

(b) Abbreviated ring structure. From now on, we'll use the abbreviated ring symbol for glucose. Each unmarked corner represents a carbon and its attached atoms.

Disaccharides

A **disaccharide**, or double sugar, is constructed from two monosaccharides through a dehydration reaction. In the Biology and Society section, you learned about the disaccharide lactose, sometimes called "milk sugar." Lactose is made from the monosaccharides glucose and galactose (**Figure 3.7**). Another common disaccharide is maltose, naturally found in germinating seeds. It is used in making beer, malt whiskey and liquor, malted milk shakes, and malted milk ball candy. A molecule of maltose consists of two glucose monomers joined together.

The most common disaccharide is sucrose (table sugar), which consists of a glucose linked to a fructose. Sucrose is the main carbohydrate in plant sap, and it nourishes all the parts of the plant. Sucrose is extracted from the stems of sugarcane or the roots of sugar beets. However, sucrose is rarely used as a sweetener in processed foods in the U.S. Much more common is high-fructose corn syrup, made through a commercial process that converts natural glucose in corn syrup to the much sweeter fructose. If you read the label on a soft drink can or bottle, you're likely to find that high-fructose corn syrup is one of the first ingredients listed (**Figure 3.8**).

The United States is one of the world's leading markets for sweeteners, with the average American consuming about 45 kilograms (kg)—a whopping 100 pounds—per year, mainly as sucrose and high-fructose corn syrup. This national "sweet tooth" persists in spite of our growing awareness about how sugar can negatively affect our health. Sugar is a major cause of tooth decay. Moreover, high sugar consumption tends to replace eating more varied and nutritious foods. The description of sugars as "empty calories" is accurate in the sense that most sweeteners contain only negligible amounts of nutrients other than carbohydrates. For good health, we also require proteins, fats, vitamins, and minerals. And we need to include substantial amounts of complex carbohydrates—that is, polysaccharides—in our diet. Let's examine these macromolecules next.

◄ **Figure 3.8 High-fructose corn syrup.** Many processed foods include high-fructose corn syrup, an artificial sweetener made by chemically treating sugars extracted from corn.

processed to extract

Starch

broken down into

Glucose

converted to sweeter

Fructose

added to foods as high-fructose corn syrup

Ingredients: carbonated water, high-fructose corn syrup, caramel color, phosphoric acid, natural flavors.

▼ **Figure 3.7 Disaccharide (double sugar) formation.** To form a disaccharide, two simple sugars are joined by a dehydration reaction, in this case forming a bond between monomers of glucose and galactose to make the double sugar lactose. (The slight structural difference between glucose and galactose is not shown here.)

Glucose   Galactose

H_2O

Lactose

Polysaccharides

Complex carbohydrates, or **polysaccharides**, are long chains of sugar units—polymers of monosaccharides. One familiar example of a polysaccharide is starch, found in plants. **Starch** consists of many glucose monomers strung together (**Figure 3.9a**). Plant cells store starch in granules, where it is available as a sugar stockpile that can be broken down as needed to provide energy and raw material for building other molecules. Potatoes and grains, such as wheat, corn, and rice, are the major sources of starch in the human diet. Humans and most other animals can use starch as food because their digestive systems break the bonds between glucose monomers through hydrolysis.

Animals store excess sugar in the form of a polysaccharide called **glycogen**. Glycogen and starch are similar in structure: They are both polymers of glucose monomers. But glycogen is more extensively branched (**Figure 3.9b**). Most of our glycogen is stored as granules in our liver and muscle cells, which break down the glycogen to release glucose when it is needed for energy. This is the basis for "carbo loading," the consumption of large amounts of starchy foods the night before an athletic event. The starch is converted to glycogen, which is then available for rapid use during physical activity the next day.

In addition to playing an important role in nutrition, certain polysaccharides serve as structural components. **Cellulose**, the most abundant organic compound on Earth, forms cable-like fibrils in the tough walls that enclose plant cells and is a major component of wood (**Figure 3.9c**). We take advantage of that structural strength when we use lumber as a building material.

Cellulose resembles starch and glycogen in being a polymer of glucose, but its glucose monomers are linked together in a different orientation. Unlike the glucose linkages in starch and glycogen, those in cellulose cannot be broken by most animals. The cellulose in plant foods, which passes unchanged through our digestive tract, is commonly known as dietary "fiber" or "roughage." Because it remains undigested, fiber does not serve as a nutrient, but it does appear to help keep our digestive system healthy. Most Americans do not get the recommended levels of fiber in their diet. Foods rich in fiber include fruits and vegetables, whole grains, bran, and beans. Grazing animals and wood-eating insects such as termites, which do derive nutrition from cellulose, have prokaryotes inhabiting their digestive tracts that break down the cellulose.

Monosaccharides (such as glucose or fructose) and disaccharides (such as sucrose or lactose) dissolve readily in water, forming sugary solutions, as in soft drinks. In contrast, cellulose does not dissolve in water. In spite of this difference, almost all carbohydrates are **hydrophilic** ("water-loving") molecules that adhere water to their surface. It is the hydrophilic quality of cellulose that makes a fluffy bath towel so water absorbent. ✔

► Figure 3.9
Polysaccharides.

Starch granules in potato tuber cells
(a) Starch

Glycogen granules in muscle tissue
(b) Glycogen

Cellulose fibril in a plant cell wall

Cellulose molecules
(c) Cellulose

Glucose monomer

Lipids

In contrast to carbohydrates and most other biological molecules, **lipids** are organic compounds that are **hydrophobic** ("water-fearing"); they do not mix with water. You have probably observed this chemical behavior in an unshaken bottle of salad dressing: The oil, which is a type of lipid, separates from the vinegar, which is mostly water (**Figure 3.10**). If you shake the bottle, you can force a temporary mixture long enough to douse your salad with dressing, but what remains in the bottle will quickly separate again when you stop shaking it.

Unlike carbohydrates, proteins, and nucleic acids, lipids are neither macromolecules nor polymers. Whereas all carbohydrates, for example, are made of sugar monomers, lipids vary in structure. That is, they aren't all made of similar "building blocks." We'll look at two main types of lipids: fats and steroids.

Fats

A typical **fat** consists of a glycerol molecule joined with three fatty acid molecules via dehydration reactions (**Figure 3.11**). The resulting fat is called a **triglyceride**, a term you may see in the results of medical tests for fat in the blood. The major portion of a fatty acid is a long hydrocarbon that stores a lot of energy, like the hydrocarbons of gasoline. In fact, a pound of fat packs more than twice as much energy as a pound of carbohydrate. The downside to this energy efficiency is that it is very difficult for a person trying to lose weight to "burn off" excess fat. It is important to understand that a reasonable amount of body fat is both normal and healthy as a fuel reserve. We stock these long-term food stores in specialized reservoirs called adipose cells, which swell and shrink when we deposit and withdraw fat from them. This adipose tissue, or "body fat," not only stores energy but also cushions vital organs and insulates us, helping maintain a warm body temperature even when the outside air is cold.

Notice in Figure 3.11b that the bottom fatty acid bends where there is a double bond in the carbon skeleton. That fatty acid is said to be **unsaturated** because it has fewer than the maximum number of hydrogens at the location of the double bond. The other two fatty acids in the fat molecule lack double bonds in their hydrocarbon portions. Those fatty acids are **saturated**, meaning that they contain the maximum number of hydrogen atoms. A saturated fat is one with all three of its fatty acid tails saturated. If one or more of the fatty acids is unsaturated, then it's an unsaturated fat, like the one in Figure 3.11b. A polyunsaturated fat has several double bonds within its fatty acids.

▼ **Figure 3.10 The separation of hydrophobic and hydrophilic components in salad dressing.**

Oil (hydrophobic)

Vinegar (hydrophilic)

▼ Figure 3.11 **The synthesis and structure of a fat, or triglyceride.**

Fatty acid

Glycerol

(a) A dehydration reaction linking a fatty acid to glycerol

(b) A fat molecule with a glycerol "head" and three energy-rich hydrocarbon fatty acid "tails"

Most animal fats, such as lard and butter, have a relatively high proportion of saturated fatty acids. The linear shape of saturated fatty acids allows these molecules to stack easily, so saturated fats tend to be solid at room temperature, as displayed by the foods on the left side of **Figure 3.12**. Diets rich in saturated fats may contribute to cardiovascular disease by promoting **atherosclerosis**. In this condition, lipid-containing deposits called plaque build up along the inside walls of blood vessels, reducing blood flow and increasing risk of heart attacks and strokes.

In contrast to red meat and dairy products, plant and fish fats are relatively high in unsaturated fatty acids. The bent shape of unsaturated fatty acids makes them less likely to form solids, so most unsaturated fats are liquid at room temperature, as displayed by the plant oils in Figure 3.12. Oils are liquid fats. Examples of oils that are primarily unsaturated include vegetable oils (such as corn and canola oil) and fish oils (such as cod liver oil).

Although plant oils tend to be low in saturated fat, tropical plant fats are an exception. Cocoa butter, a main ingredient in chocolate, contains a mix of saturated and unsaturated fat that gives it a melting point near body temperature. Thus, chocolate stays solid at room temperature but melts in the mouth. This pleasing "mouth feel" is one of the reasons chocolate is so appealing.

Sometimes a food manufacturer wants to use a vegetable oil but needs the food product to be solid, as in the case of margarine or some peanut butters. To achieve this, the manufacturer can convert unsaturated fats to saturated fats by adding hydrogen, a process called **hydrogenation**. Unfortunately, hydrogenation also creates **trans fat**, a type of unsaturated fat that is even less healthy than saturated fat. Since 2006, the FDA has required trans fats to be listed in nutrition labels. But even if a label on a hydrogenated product says 0 grams (g) of trans fat per serving, the product might still have up to 0.5 g per serving. Any product with partially hydrogenated vegetable oil or vegetable shortening contains trans fat, as displayed by the processed foods in the center of Figure 3.12. (See the Process of Science section in Chapter 1—"Is Trans Fat Bad for You?"—for an investigation of the health effects of trans fats.) Due to their unhealthy nature, trans fats are becoming less common as food manufacturers substitute other forms of fat.

Although saturated and trans fats should generally be avoided, it is not true that *all* fats are unhealthy. In fact, some fats perform important functions within the body and are beneficial and even essential to a healthy diet. For example, fats containing omega-3 fatty acids have been shown to reduce the risk of coronary heart disease and relieve the symptoms of arthritis and inflammatory bowel disease. Some sources are nuts and oily fish such as salmon, as displayed on the right side of Figure 3.12. ☑

☑CHECKPOINT

What are "unsaturated fats"? What two kinds of fat are least healthy? What fats are especially healthy?

Answer: fats without the maximum number of hydrogens because of double bonds between some carbons; trans fats and saturated fats; fats built from omega-3 fatty acids

▼ **Figure 3.12 Types of fats.**

TYPES OF FATS	
Saturated Fats (unhealthy fats found primarily in meat and full-fat dairy products; solid at room temperature)	**Unsaturated Fats** (fats found primarily in fish and plants; usually liquid at room temperature)

Margarine

INGREDIENTS: SOYBEAN OIL, FULLY HYDROGENATED COTTONSEED OIL, PARTIALLY HYDROGENATED COTTONSEED AND SOYBEAN OILS, MONO AND DIGLYCERIDES, TBHO AND CITRIC ACID (ANTIOXIDANTS).

Plant oils (unhydrogenated; usually liquid at room temperature)

Trans fats (unhealthy fats in hydrogenated processed foods; solid at room temperature)

Omega-3 fats (beneficial fats in some fish and plant oils; liquid at room temperature)

Steroids

Classified as lipids because they are hydrophobic, **steroids** are very different from fats in structure and function. All steroids have a carbon skeleton that is bent to form four fused rings. Different steroids vary in the functional groups attached to this core set of rings, and these variations affect their function. One common steroid is cholesterol, which has a bad reputation because of its association with cardiovascular disease. However, cholesterol is an essential molecule in your body. It is a key component of the membranes that surround your cells. It is also the "base steroid" from which your body produces other steroids, such as the sex hormones estrogen and testosterone (**Figure 3.13**).

The controversial drugs called anabolic steroids are synthetic variants of testosterone, the male sex hormone. Testosterone causes a general buildup in muscle and bone mass during puberty in males and maintains masculine traits throughout life. Because anabolic steroids structurally resemble testosterone, they also mimic some of its effects. Some athletes use anabolic steroids to build up their muscles quickly and enhance their performance.

In 2003, the discovery that some athletes were using a new anabolic steroid called THG rocked the sports world. THG is a chemically modified ("designer") steroid intended to avoid detection by drug tests (hence its nickname, "the clear"). New tests revealed widespread use of THG and other steroids among athletes in many sports (**Figure 3.14**).

Using anabolic steroids is indeed a fast way to increase body size beyond what hard work alone can produce. But at what cost? Steroid abuse can cause serious physical and mental problems, including violent mood swings ("roid rage"), depression, liver damage, high cholesterol, shrunken testicles, reduced sex drive, and infertility. These last symptoms occur because anabolic steroids often cause the body to reduce its normal output of sex hormones. Most athletic organizations now ban the use of anabolic steroids because of their many potential health hazards coupled with the unfairness of an artificial advantage. ✔

▼ Figure 3.13 **Examples of steroids.** The molecular structures of the steroids shown here are abbreviated by omitting all the atoms that make up the rings. The subtle difference between testosterone and estrogen influences the development of the anatomical and physiological differences between male and female mammals, including lions and humans.

Cholesterol

can be converted by the body to

Testosterone

A type of estrogen

▼ Figure 3.14 **Steroids and the modern athlete.** Baseball player Jose Canseco (left), an admitted user of performance-enhancing steroids, testifies before Congress during an inquiry into the abuse of drugs by professional athletes. Track-and-field star Marion Jones (right) admitted in 2007 that she had taken performance-enhancing drugs. As a result, she was stripped of her five Olympic medals. Note the similarity in chemical structure of testosterone (shown above) and THG (shown below).

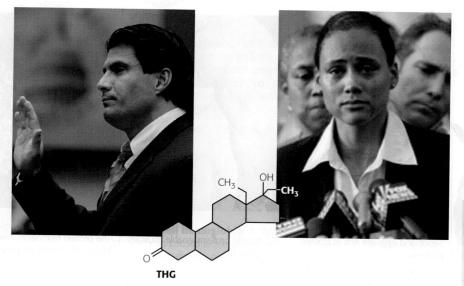

THG

✔CHECKPOINT

How are dietary fats similar to human sex hormones?

Answer: Both are lipids, which are hydrophobic.

Cell Organelles

The Nucleus and Ribosomes: Genetic Control of the Cell

If we think of the cell as a factory, then the nucleus is its executive boardroom. The top managers are the genes, the inherited DNA molecules that direct almost all the business of the cell. As you read in Chapter 3, each gene is a stretch of DNA that stores the information necessary to produce a particular protein. Proteins then do most of the actual work of the cell.

Structure and Function of the Nucleus

The nucleus is bordered by a double membrane called the **nuclear envelope** (**Figure 4.8**). Each membrane of the nuclear envelope is similar in structure to the plasma membrane. Pores in the envelope allow certain materials to pass between the nucleus and the cytoplasm. Within the nucleus, long DNA molecules and associated proteins form fibers called **chromatin**. Each long chromatin fiber constitutes one **chromosome** (**Figure 4.9**). The number of chromosomes in a cell depends on the species; for example, each human body cell has 46 chromosomes, whereas rice cells have 24 and dog cells have 78 (see Figure 8.2 for more examples). The **nucleolus**, a prominent structure within the nucleus, is the site where the components of ribosomes are made. We'll examine ribosomes next. ☑

▼ Figure 4.8 **The nucleus.**

Ribosomes   Chromatin   Nuclear envelope   Nucleolus   Pore

Surface of nuclear envelope

Nuclear pores

▼ Figure 4.9 **The relationship between DNA, chromatin, and a chromosome.**

DNA molecule

Proteins

Chromatin fiber

Chromosome

Ribosomes

The small dots in the cells in Figure 4.5 and outside the nucleus in Figure 4.8 are the ribosomes. As you'll see in the next section, ribosomes are responsible for protein synthesis (**Figure 4.10**). In eukaryotic cells, the components of ribosomes are made in the nucleus, then transported through the pores of the nucleus into the cytoplasm. It is in the cytoplasm that the ribosomes begin their work. Some are suspended in the fluid of the cytoplasm, making proteins that remain in the fluid (**Figure 4.11**). Other ribosomes are attached to the outside of an organelle called the endoplasmic reticulum, making proteins that are incorporated into membranes or secreted by the cell. Recall from the Biology and Society section that there are some important structural differences between bacterial and eukaryotic ribosomes, making them a good target for antibiotic drugs.

▼ **Figure 4.10** **Computer model of a ribosome synthesizing a protein.**

Ribosome

mRNA

Protein

▼ **Figure 4.11** **Locations of ribosomes.**

TEM 50,000×

Ribosomes in cytoplasm

Ribosomes attached to endoplasmic reticulum

How DNA Directs Protein Production

How do the DNA "executives" in the nucleus direct the "workers" in the cytoplasm? **Figure 4.12** shows the sequence of events in a eukaryotic cell (with the DNA and other structures being shown disproportionately large in relation to the nucleus). **❶** DNA programs protein production in the cytoplasm by transferring its coded information to a molecule called messenger RNA (mRNA). Like a middle manager, the RNA molecule then carries the order to "build this type of protein" from the nucleus to the cytoplasm. **❷** The mRNA exits through pores in the nuclear envelope and travels to the cytoplasm, where it then binds to ribosomes. **❸** As a ribosome moves along the mRNA, the genetic message is translated into a protein with a specific amino acid sequence. You'll learn how the message is translated in Chapter 10. ☑

▼ **Figure 4.12** **DNA → RNA → Protein.** Inherited genes in the nucleus control protein production and hence the activities of the cell.

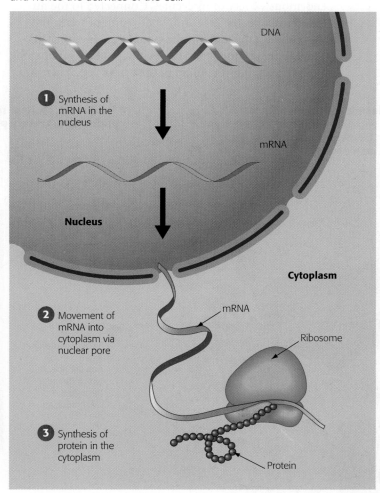

DNA

1 Synthesis of mRNA in the nucleus

mRNA

Nucleus

Cytoplasm

2 Movement of mRNA into cytoplasm via nuclear pore

mRNA

Ribosome

3 Synthesis of protein in the cytoplasm

Protein

☑**CHECKPOINT**

1. What is the function of ribosomes?
2. What is the role of mRNA in making a protein?

*Answers: **1.** protein synthesis **2.** A molecule of mRNA carries the genetic message from a gene (DNA) to ribosomes that translate it into protein.*

The Endomembrane System: Manufacturing and Distributing Cellular Products

The cytoplasm of a eukaryotic cell is partitioned by organelle membranes (see Figure 4.5). Some of the organelles are connected to each other, either directly by their membranes or by transfer of membrane segments between them. Together, these organelles form the **endomembrane system**. This system includes the nuclear envelope, the endoplasmic reticulum, the Golgi apparatus, lysosomes, and vacuoles.

The Endoplasmic Reticulum

The **endoplasmic reticulum (ER)** is one of the main manufacturing facilities within a cell. It produces an enormous variety of molecules. Connected to the nuclear envelope, the ER forms a labyrinth of tubes and sacs running throughout the cytoplasm (**Figure 4.13**). The ER membrane separates its internal compartment from the surrounding fluid in the cytoplasm. There are two components that make up the ER: rough ER and smooth ER. These two types of ER are physically connected but differ in structure and function.

Rough ER

The "rough" in **rough ER** refers to the appearance of this organelle in electron micrographs (see the bottom left of Figure 4.13). The roughness is due to ribosomes that stud the outside of the ER membrane. These ribosomes produce membrane proteins and secretory proteins. Some newly manufactured membrane proteins are embedded right in the rough ER membrane. Thus, one function of rough ER is to produce new membrane. Secretory proteins are those that are exported (secreted) to the fluid outside the cell. Cells that secrete a lot of protein—such as the cells of your salivary glands, which secrete enzymes into your mouth—are especially rich in rough ER. Some products manufactured by rough ER are dispatched to other locations in the cell by means of **transport vesicles**, membranous spheres that bud from the rough ER (**Figure 4.14**).

▼ **Figure 4.13 Endoplasmic reticulum (ER).** In this drawing (top) and micrograph (bottom), the flattened sacs of rough ER and the tubes of smooth ER are connected. Notice that the ER is also connected to the nuclear envelope.

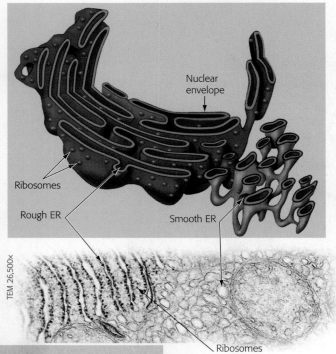

Nuclear envelope

Ribosomes

Rough ER

Smooth ER

TEM 26,500×

Ribosomes

▼ **Figure 4.14 How rough ER manufactures and packages secretory proteins.**

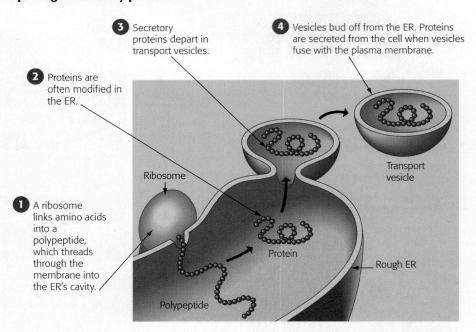

3 Secretory proteins depart in transport vesicles.

4 Vesicles bud off from the ER. Proteins are secreted from the cell when vesicles fuse with the plasma membrane.

2 Proteins are often modified in the ER.

1 A ribosome links amino acids into a polypeptide, which threads through the membrane into the ER's cavity.

Ribosome

Protein

Transport vesicle

Rough ER

Polypeptide

Smooth ER

The "smooth" in **smooth ER** refers to the fact that this organelle lacks the ribosomes that populate the surface of rough ER (see Figure 4.13). A diversity of enzymes built into the smooth ER membrane enables this organelle to perform many functions. One is the synthesis of lipids, including steroids (see Figure 3.13). For example, the cells in ovaries or testes that produce the steroid sex hormones are enriched with smooth ER. In liver cells, enzymes of the smooth ER detoxify circulating sedatives such as barbiturates, stimulants such as amphetamines, and some antibiotics (which is why they don't persist in the bloodstream after combating an infection). As liver cells are exposed to a drug, the amounts of smooth ER and its detoxifying enzymes increase. This can strengthen the body's tolerance of the drug, meaning that higher doses will be required in the future to achieve the desired effect. The growth of smooth ER in response to one drug can also increase tolerance of other drugs. Barbiturate use, for example, may make certain antibiotics less effective by accelerating their breakdown in the liver. Furthermore, increased tolerance of drugs is one of the hallmarks of addiction—a potentially serious consequence of the continued use of certain drugs.

The Golgi Apparatus

The **Golgi apparatus**, an organelle named for its discoverer (Italian scientist Camillo Golgi), is a refinery, warehouse, and shipping center. Working in close partnership with the ER, the Golgi apparatus receives, refines, stores, and distributes chemical products of the cell **(Figure 4.15)**. Products made in the ER reach the Golgi in transport vesicles. One side of a Golgi stack serves as a receiving dock for these vesicles. The shipping side of a Golgi stack is a depot from which finished products can be carried in transport vesicles to other organelles or to the plasma membrane. Vesicles that bind with the plasma membrane transfer proteins to it or secrete finished products to the outside of the cell. Products of the ER are usually modified by enzymes during their transit from the receiving to the shipping side of the Golgi. For example, molecular identification tags, such as phosphate groups, may be added that serve to mark and sort protein molecules into different batches for different destinations. ☑

☑**CHECKPOINT**

1. What makes rough ER rough?
2. What is the relationship between the Golgi apparatus and the ER in a protein-secreting cell?

Answers: 1. ribosomes attached to the membrane 2. The Golgi receives proteins from the ER via vesicles, finishes processing the proteins, and then dispatches them in vesicles.

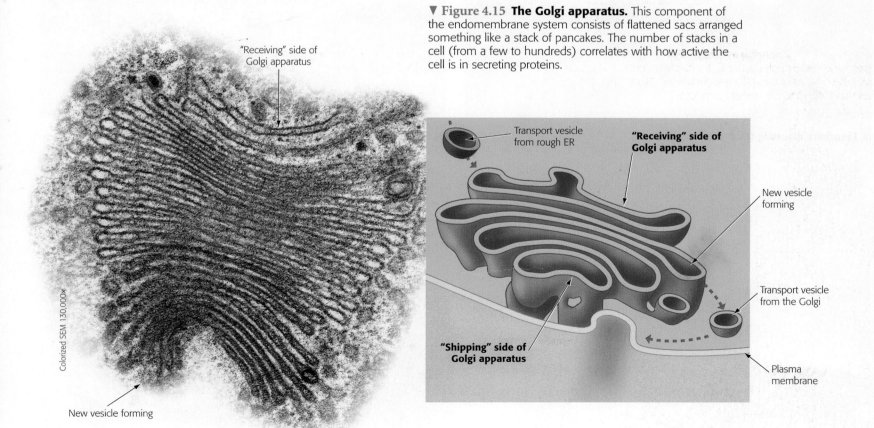

▼ Figure 4.15 **The Golgi apparatus.** This component of the endomembrane system consists of flattened sacs arranged something like a stack of pancakes. The number of stacks in a cell (from a few to hundreds) correlates with how active the cell is in secreting proteins.

"Receiving" side of Golgi apparatus

New vesicle forming

Colorized SEM 130,000×

Transport vesicle from rough ER

"Receiving" side of Golgi apparatus

New vesicle forming

Transport vesicle from the Golgi

"Shipping" side of Golgi apparatus

Plasma membrane

Lysosomes

A **lysosome** is a sac of digestive enzymes found in animal cells. (Lysosomes are absent from most plant cells.) Lysosomes develop from vesicles that bud off from the Golgi. Enzymes within a lysosome can break down large molecules such as proteins, polysaccharides, fats, and nucleic acids. The lysosome provides a compartment where the cell can digest these molecules safely, without unleashing these digestive enzymes on the cell itself.

Lysosomes have several types of digestive functions. Many cells engulf nutrients into tiny cytoplasmic sacs called **food vacuoles**. Lysosomes fuse with the food vacuoles, exposing the food to enzymes that digest it (**Figure 4.16a**). Small molecules that result from this digestion, such as amino acids, leave the lysosome and nourish the cell. Lysosomes also help destroy harmful bacteria. Our white blood cells ingest bacteria into vacuoles, and lysosomal enzymes that are emptied into these vacuoles rupture the bacterial cell walls. Additionally, lysosomes break down the large molecules of damaged organelles. Without harming the cell, a lysosome can engulf and digest parts of another organelle, making its molecules available for the construction of new organelles (**Figure 4.16b**). Lysosomes also have sculpturing functions in embryonic development. In an early human embryo, lysosomes release enzymes that digest webbing between fingers of the developing hand.

The importance of lysosomes to cell function and human health is made strikingly clear by hereditary disorders called lysosomal storage diseases. A person with such a disease is missing one or more of the digestive enzymes normally found within lysosomes. The abnormal lysosomes become engorged with indigestible substances, and this eventually interferes with other cellular functions. Most of these diseases are fatal in early childhood. In Tay-Sachs disease, lysosomes lack a lipid-digesting enzyme. As a result, nerve cells die as they accumulate excess lipids, ravaging the nervous system. Fortunately, storage diseases are rare. ☑

▼ Figure 4.16 **Two functions of lysosomes.**

(a) Lysosome digesting food

(b) Lysosome breaking down the molecules of damaged organelles

Vacuoles

Vacuoles are sacs that bud from the ER, Golgi, or plasma membrane. They come in different sizes and have a variety of functions. For example, Figure 4.16a shows a food vacuole budding from the plasma membrane. Certain freshwater protists have contractile vacuoles that pump out excess water that flows into the cell from the outside environment (**Figure 4.17a**).

Another type of vacuole is a **central vacuole**, which can account for more than half the volume of a mature plant cell (**Figure 4.17b**). The central vacuole of a plant cell is a versatile compartment. It stores organic nutrients, such as proteins stockpiled in the vacuoles of seed cells. It also contributes to plant growth by absorbing water and causing cells to expand. In the cells of flower petals, central vacuoles may contain pigments that attract pollinating insects. Central vacuoles may also contain poisons that protect against plant-eating animals.

▼ **Figure 4.17** **Two types of vacuoles.**

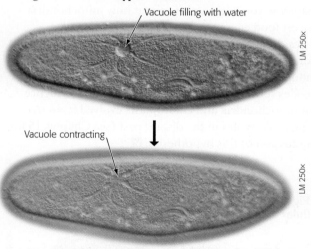

Vacuole filling with water

Vacuole contracting

LM 250x

LM 250x

(a) Contractile vacuole in *Paramecium*. A contractile vacuole fills with water and then contracts to pump the water out of the cell.

Central vacuole

Colorized TEM 1,700x

(b) Central vacuole in a plant cell. The central vacuole (colorized blue in this micrograph) is often the largest organelle in a mature plant cell.

Figure 4.18 will help you review how organelles of the endomembrane system are related. Note that a product made in one part of the endomembrane system may exit the cell or become part of another organelle without crossing a membrane. Also note that membrane made by the ER can become part of the plasma membrane through the fusion of a transport vesicle. In this way, even the plasma membrane is related to the endomembrane system. ☑

▼ **Figure 4.18** **Review of the endomembrane system.** The arrows show some of the pathways of cell product distribution and membrane migration via transport vesicles.

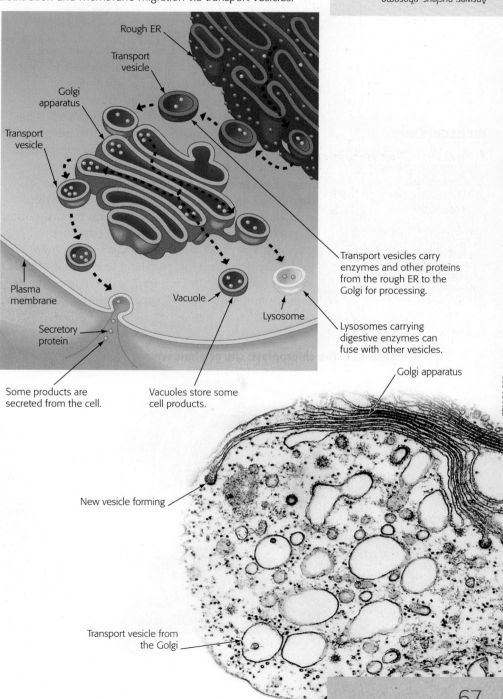

Rough ER

Transport vesicle

Golgi apparatus

Transport vesicle

Plasma membrane

Secretory protein

Vacuole

Lysosome

Transport vesicles carry enzymes and other proteins from the rough ER to the Golgi for processing.

Lysosomes carrying digestive enzymes can fuse with other vesicles.

Some products are secreted from the cell.

Vacuoles store some cell products.

Golgi apparatus

New vesicle forming

Transport vesicle from the Golgi

TEM 31,000x

☑**CHECKPOINT**

Place the following cellular structures in the order they would be used in the production and secretion of a protein: Golgi apparatus, nucleus, plasma membrane, ribosome, transport vesicle.

Answer: nucleus, ribosome, transport vesicle, Golgi apparatus, plasma membrane

Chloroplasts and Mitochondria: Energy Conversion

A cell requires a continuous energy supply to perform the work of life. Two organelles act as cellular power stations: chloroplasts and mitochondria.

Chloroplasts

Most of the living world runs on the energy provided by photosynthesis, the conversion of light energy from the sun to the chemical energy of sugar and other organic molecules. **Chloroplasts**, which are unique to the photosynthetic cells of plants and algae, are the organelles that perform photosynthesis.

A chloroplast is partitioned into three major compartments by internal membranes (**Figure 4.19**). One compartment is the space between the two membranes that envelop the chloroplast. The **stroma**, a thick fluid within the chloroplast, is the second compartment. Suspended in that fluid, the interior of a network of membrane-enclosed disks and tubes forms the third compartment. Notice in Figure 4.19 that the disks occur in interconnected stacks called **grana** (singular, *granum*). The grana are a chloroplast's solar power packs, the structures that trap light energy and convert it to chemical energy. In Chapter 7, you will learn the details of photosynthesis as it occurs within a chloroplast.

Mitochondria

Mitochondria (singular, *mitochondrion*) are the sites of cellular respiration, a process that harvests energy from sugars and other food molecules and converts it to another form of chemical energy called ATP. Cells use molecules of ATP as the direct energy source for most of their work. In contrast to chloroplasts, mitochondria are found in almost all eukaryotic cells, including your own. An envelope of two membranes encloses the mitochondrion, which contains a thick fluid called the **matrix (Figure 4.20)**. The inner membrane of the envelope has numerous infoldings called **cristae**. Many of the enzymes and other molecules that function in cellular respiration are built into the inner membrane. By increasing the surface area of this membrane, the cristae maximize ATP output. In Chapter 6, you'll learn more about how mitochondria convert food energy to ATP energy.

Besides their ability to provide cellular energy, mitochondria and chloroplasts share another feature unique among eukaryotic organelles: They contain DNA that encodes some of their proteins. This DNA is evidence that mitochondria and chloroplasts evolved from free-living prokaryotes in the distant past (see Chapter 15 for a discussion of this hypothesis). ☑

▼ Figure 4.19 **The chloroplast: site of photosynthesis.**

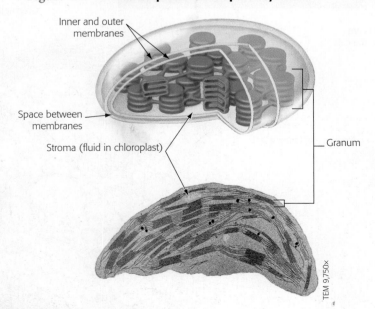

Inner and outer membranes

Space between membranes

Stroma (fluid in chloroplast)

Granum

TEM 9,750x

▼ Figure 4.20 **The mitochondrion: site of cellular respiration.**

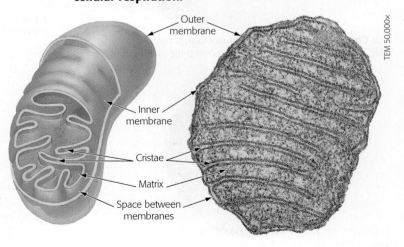

Outer membrane

Inner membrane

Cristae

Matrix

Space between membranes

TEM 50,000x

The Cytoskeleton: Cell Shape and Movement

If someone asked you to describe a house, you would most likely mention the various rooms and their locations. You probably would not think to mention the foundation and beams that support the house. Yet these structures perform an extremely important function. Similarly, cells have an infrastructure called the **cytoskeleton**, a network of fibers extending throughout the cytoplasm. The cytoskeleton serves as both skeleton and "muscles" for the cell, functioning in support and movement.

Maintaining Cell Shape

One function of the cytoskeleton is to give mechanical support to the cell and maintain its shape. This is especially important for animal cells, which lack rigid cell walls. The cytoskeleton contains several types of fibers made from different types of protein. One important type of fiber is **microtubules** (**Figure 4.21a**). Microtubules are straight, hollow tubes composed of proteins. The other kinds of cytoskeletal fibers, called intermediate filaments and microfilaments, are thinner and solid.

Just as the bony skeleton of your body helps fix the positions of your organs, the cytoskeleton provides anchorage and reinforcement for many organelles in a cell. For instance, the nucleus is often held in place by a cytoskeletal cage of filaments. Other organelles move along tracks made from microtubules. For example, a lysosome might reach a food vacuole by moving along a microtubule. Microtubules also guide the movement of chromosomes when cells divide.

A cell's cytoskeleton is dynamic: It can quickly dismantle in one part of the cell by removing protein subunits and re-form in a new location by reattaching the subunits. Such rearrangement can provide rigidity in a new location, change the shape of the cell, or even cause the whole cell or some of its parts to move. This process contributes to the amoeboid (crawling) movements of the protist *Amoeba* (**Figure 4.21b**) and some of our white blood cells. ☑

☑ CHECKPOINT

How does the cytoskeleton function in both support and movement within the cell?

Answer: The cytoskeleton serves as an anchor onto which organelles can attach and also provides tracks along which organelles can move.

▼ Figure 4.21 **The cytoskeleton.**

(a) **Microtubules in the cytoskeleton.** In this micrograph of animal cells, the cytoskeleton microtubules are labeled with a fluorescent yellow dye.

LM 650×

(b) **Microtubules and movement.** The crawling movement of an *Amoeba* is due to the rapid degradation and rebuilding of microtubules.

LM 220×

Cilia and Flagella

In some eukaryotic cells, a specialized arrangement of microtubules functions in the beating of flagella and cilia. Cilia and flagella are motile appendages—extensions from a cell that aid in movement. Eukaryotic **flagella** (singular, *flagellum*) propel the cell by an undulating whiplike motion. They often occur singly, such as in the sperm cells of humans and other animals (**Figure 4.22a**). **Cilia** (singular, *cilium*) are generally shorter and more numerous than flagella and promote movement by a coordinated back-and-forth motion, like the rhythmic oars of an eight-person crew team. Both cilia and flagella propel various protists through water (**Figure 4.22b**). Though different in length, number per cell, and beating pattern, cilia and flagella have the same basic architecture, with a core of microtubules wrapped in an extension of the plasma membrane. (Although eukaryotic flagella look much like the prokaryotic flagella shown in Figure 4.4, they have a distinct internal architecture.)

Some cilia extend from nonmoving cells that are part of a tissue layer. There they move fluid over the tissue's surface. For example, the ciliated lining of your windpipe helps cleanse your respiratory system by sweeping mucus with trapped debris out of your lungs (**Figure 4.22c**). Tobacco smoke can inhibit or destroy these cilia, interfering with the normal cleansing mechanisms and allowing more toxin-laden smoke particles to reach the lungs. Frequent coughing—common in heavy smokers—then becomes the body's attempt to cleanse the respiratory system.

Because human sperm rely on flagella for movement, it's easy to understand why problems with flagella can lead to male infertility. Sperm with malfunctioning flagella cannot travel up the female reproductive tract to fertilize an egg. Interestingly, some men with a type of hereditary sterility also suffer from respiratory problems. The explanation lies in the similarities between flagella (found in sperm) and cilia (found lining the respiratory tract). Because of a defect in the structure of their flagella and cilia, their sperm do not swim (causing sterility) and their cilia do not sweep mucus out of their lungs (causing recurrent respiratory infections). ✓

✓CHECKPOINT

Compare and contrast cilia and flagella.

Answer: Cilia and flagella have the same basic structure and help move cells or move fluid over cells. Cilia are short and numerous and move back and forth. Flagella are longer, often occurring singly, and they undulate.

Colorized SEM 1,300x

▼ Figure 4.22 **Flagella and cilia.**

(a) Flagellum of a human sperm cell. A eukaryotic flagellum undulates in a snakelike motion, driving a cell such as this sperm cell through its fluid environment.

Colorized SEM 500x

(b) Cilia on a protist. Cilia are shorter and more numerous than flagella and move with a back-and-forth motion. As shown here, a dense nap of beating cilia covers *Paramecium*, a freshwater protist that can dart rapidly through its watery home.

Colorized SEM 3,000x

(c) Cilia lining the respiratory tract. The cilia lining your respiratory tract sweep mucus with trapped debris out of your lungs.

The Evolution of Antibiotic Resistance

As you read in the Biology and Society section, many antibiotics disrupt the cellular structures of invading microorganisms. This mechanism is quite effective. In fact, when first introduced in the 1940s, penicillin appeared to be such a "wonder drug" that some doctors predicted the end of infections in humans (**Figure 4.23**).

Why hasn't that optimistic prediction come true? The answer is both simple and profound: The prediction does not take into account the force of evolution. Within a population of bacteria, random genetic changes result in slightly different proteins being produced in different bacteria. Some bacteria may produce a protein that reduces the effectiveness of an antibiotic. For example, a random change in a nucleotide sequence might lead to the production of an enzyme that deactivates or breaks down penicillin.

When an environment changes, natural selection favors individuals with beneficial genetic changes. In this case, the introduction of an antibiotic creates a pressure that favors resistant bacteria. Those individuals will survive and multiply, producing more of the same type of bacteria. Over time, drug-resistant bacteria, such as the MRSA strain discussed in the Process of Science section, become more common.

In what ways do we contribute to the problem of antibiotic resistance? Livestock producers add antibiotics to animal feed to promote growth and prevent illness. These practices may favor bacteria that resist standard antibiotics. Some doctors overprescribe antibiotics—for example, to patients with viral infections, which do not respond to such treatment. And patients misuse prescribed antibiotics by prematurely stopping their medication. This misuse allows mutant bacteria that may be killed more slowly by the drug to survive and multiply.

The evolution of antibiotic-resistant bacteria is a serious public health concern. Nearly 100,000 people die each year in the United States from infections they contract in the hospital, often from antibiotic-resistant bacteria. Penicillin, effective against many bacterial infections in the 1940s, is virtually useless today in its original form. New drugs are developed but also continue to be rendered ineffective as resistant bacteria evolve. The medical community is engaged in a race against the powerful force of bacterial evolution.

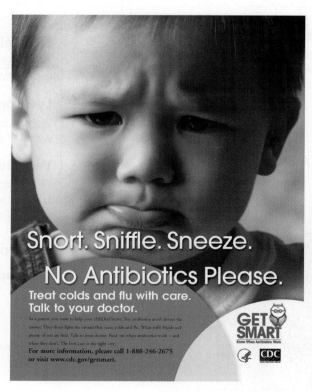

◄ Figure 4.23 **The changing role of antibiotics.** These two posters illustrate how the early promise of antibiotics (depicted in a poster from 1944) has given way to caution about their overuse (depicted in a poster from 2008).

Chapter Review

The Microscopic World of Cells

Microscopes as Windows on the World of Cells

Using early microscopes, biologists discovered that all organisms are made of cells. Resolving power limits the useful magnification of microscopes. A light microscope (LM) has useful magnifications up to about 1,000×. Electron microscopes, both scanning (SEM) and transmission (TEM), are much more powerful.

The Two Major Categories of Cells

CATEGORIES OF CELLS	
Prokaryotic Cells	**Eukaryotic Cells**
• Smaller • Simpler • Most do not have organelles • Found in bacteria and archaea	• Larger • More complex • Have organelles • Found in protists, plants, fungi, animals

An Overview of Eukaryotic Cells

Many cellular functions are partitioned by membranes in the complex organization of eukaryotic cells. The largest organelle is usually the nucleus. Other organelles are located in the cytoplasm, the region between the nucleus and the plasma membrane.

Membrane Structure

The Plasma Membrane: A Fluid Mosaic of Lipids and Proteins

Cell Surfaces

Most cells secrete an extracellular coat that helps protect and support the cell. The walls that encase plant cells support plants against the pull of grav-

ity and also prevent cells from absorbing too much water. Animal cells are coated by a sticky extracellular matrix.

The Nucleus and Ribosomes: Genetic Control of the Cell

Structure and Function of the Nucleus

An envelope consisting of two membranes encloses the nucleus. Within the nucleus, DNA and proteins make up chromatin fibers; each very long fiber is a single chromosome. The nucleus also contains the nucleolus, which produces components of ribosomes.

Ribosomes

Ribosomes produce proteins in the cytoplasm.

How DNA Directs Protein Production

Genetic messages are transmitted to the ribosomes via messenger RNA, which travels from the nucleus to the cytoplasm.

The Endomembrane System: Manufacturing and Distributing Cellular Products

The Endoplasmic Reticulum

The ER consists of membrane-enclosed tubes and sacs within the cytoplasm. Rough ER, named for the ribosomes attached to its surface, makes membrane and secretory proteins. The functions of smooth ER include lipid synthesis and detoxification.

The Golgi Apparatus

The Golgi refines certain ER products and packages them in transport vesicles targeted for other organelles or export from the cell.

Lysosomes

Lysosomes, sacs containing digestive enzymes, aid digestion and recycling within the cell.

Vacuoles

These organelles include the contractile vacuoles that expel water from certain freshwater protists and the large, multifunctional central vacuoles of plant cells.

Chloroplasts and Mitochondria: Energy Conversion

Chloroplasts and Mitochondria

The Cytoskeleton: Cell Shape and Movement

Maintaining Cell Shape

Straight, hollow microtubules are an important component of the cytoskeleton, an organelle that gives support to and maintains the shape of cells.

Cilia and Flagella

Cilia and eukaryotic flagella are both motile appendages made primarily of microtubules. Cilia are short and numerous and move by coordinated beating. Flagella are long, often occur singly, and propel a cell through whiplike movements.

SELF-QUIZ

1. If you wanted to film the movement of chromosomes during cell division, the best choice for a microscope would be a
   a. light microscope, because of its magnifying power.
   b. transmission electron microscope, because of its resolving power.
   c. scanning electron microscope, because the chromosomes are on the cell surface.
   d. light microscope, because the specimen must be kept alive.

2. Using a light microscope to examine a thin section of a large spherical cell, you find that the cell is 0.3 mm in diameter. The nucleus is about one-fourth as wide. What would be the diameter of the nucleus in micrometers? (*Hint*: See Figure 4.3.)

3. You look into a light microscope and view an unknown cell. What might you see that would tell you whether the cell is prokaryotic or eukaryotic?
   a. a rigid cell wall
   b. a nucleus
   c. a plasma membrane
   d. ribosomes

4. Explain how each word in the term *fluid mosaic* describes the structure of a membrane.

5. Identify which one of the following structures includes all the others in the list: rough ER, smooth ER, endomembrane system, the Golgi apparatus.

6. The ER has two distinct regions that differ in structure and function. Lipids are synthesized within the _____, and proteins are synthesized within the _____.

7. A type of cell called a lymphocyte makes proteins that are exported from the cell. You can track the path of these proteins within the cell from production through export by labeling them with radioactive isotopes. Identify which of the following structures would be radioactively labeled in your experiment, listing them in the order in which they would be labeled: chloroplasts, Golgi apparatus, plasma membrane, smooth ER, rough ER, nucleus, mitochondria.

8. Name two similarities in the structure or function of chloroplasts and mitochondria. Name two differences.

9. Match the following organelles with their functions:
   a. ribosomes
   b. microtubules
   c. mitochondria
   d. chloroplasts
   e. lysosomes

   1. movement
   2. photosynthesis
   3. protein synthesis
   4. digestion
   5. cellular respiration

10. DNA controls the cell by transmitting genetic messages that result in protein production. Place the following organelles in the order that represents the flow of genetic information from the DNA through the cell: nuclear pores, ribosomes, nucleus, rough ER, Golgi apparatus.

Answers to the Self-Quiz questions can be found in Appendix D.

THE PROCESS OF SCIENCE

11. The cells of plant seeds store oils in the form of droplets enclosed by membranes. Unlike the membranes you learned about in this chapter, the oil droplet membrane consists of a single layer of phospholipids rather than a bilayer. Draw a model for a membrane around an oil droplet. Explain why this arrangement is more stable than a bilayer.

12. Imagine that you are a pediatrician and one of your patients is a newborn who may have a lysosomal storage disease. You remove some cells from the patient and examine them under the microscope. What would you expect to see? Design a series of tests that could reveal whether the patient is indeed suffering from a lysosomal storage disease.

BIOLOGY AND SOCIETY

13. Doctors at a university medical center removed John Moore's spleen, which is a standard treatment for his type of leukemia. The disease did not recur. Researchers kept the spleen cells alive in a nutrient medium. They found that some cells produced a blood protein that showed promise as a treatment for cancer and AIDS. The researchers patented the cells. Moore sued, claiming a share in profits from any products derived from his cells. The U.S. Supreme Court ruled against Moore, stating that his lawsuit "threatens to destroy the economic incentive to conduct important medical research." Moore argued that the ruling left patients "vulnerable to exploitation at the hands of the state." Do you think Moore was treated fairly? What else would you like to know about this case that might help you decide?

5 The Working Cell

Cellular structures.
These human pancreatic
cells have been stained to
highlight various cellular
structures.

CHAPTER CONTENTS

BIOLOGY AND SOCIETY: Molecular Technologies

Natural Nanotechnology

One of the most fascinating applications of modern science is nanotechnology, the manipulation of materials at the molecular scale. When designing devices of such small size, researchers often turn to living cells for inspiration. After all, you can think of a cell as a machine that continuously and efficiently performs a variety of functions. Let's consider one example of cell-based nanotechnology and see how it relates to three of the main concerns in the lives of working cells: energy, enzymes, and the plasma membrane.

All living processes depend on energy. Researchers at Cornell University are attempting to harvest the energy-producing capability of a human sperm cell. Like other cells, a sperm cell generates energy by breaking down sugars and other molecules that pass through its plasma membrane. Enzymes within the cell carry out a process called glycolysis. During glycolysis (discussed in Chapter 6), the energy released from the breakdown of glucose is used to produce molecules of ATP. Within a living sperm, the ATP produced during glycolysis and other processes provides the energy that propels the sperm through the female reproductive tract. In an attempt to harness this energy-producing system, the Cornell researchers attached three glycolysis enzymes to a computer chip. The enzymes continued to function in this artificial system. The hope is that a larger set of enzymes can eventually be used to power microscopic robots. Such nanorobots could use glucose from the bloodstream to power the delivery of drugs to body tissues.

This example of cell-based nanotechnology highlights the three main topics of this chapter. We'll explore how a cell uses energy, enzymes, and the plasma membrane to carry out the work of controlling its internal chemical environment. Along the way, we'll look at some other technologies that mimic the natural activities of living cells.

Basic Energy Concepts

Some Basic Energy Concepts

Energy makes the world go round—both the cellular world and the larger world outside. But what exactly is energy? Our first step in understanding the working cell is to learn a few basic concepts about energy.

Conservation of Energy

Energy is defined as the capacity to perform work. Work is performed whenever an object is moved against an opposing force. In other words, work moves things in ways they would not move if left alone.

For example, imagine a diver climbing to the top of a platform and diving off **(Figure 5.1)**. To get the diver to the top of the platform, work must be performed to overcome the opposing force of gravity. In the diver climbing up the steps to the diving platform, chemical energy from food is being converted to **kinetic energy**, the energy of motion. In this case, the kinetic energy takes the form of muscle movement.

What happens to the kinetic energy when the diver reaches the top of the platform? Has it disappeared?

The answer is no. You may be familiar with the principle of conservation of matter, which states that matter cannot be created or destroyed but can only be converted from one form to another. A similar principle, known as **conservation of energy**, states that it is not possible to destroy or create energy. Like matter, energy can only be converted from one form to another. A power plant, for example, does not make energy; it merely converts it from one form (such as energy stored in coal) to a more convenient form (such as electricity). That's what happens in the diver's climb up the steps. The kinetic energy of muscle movement is now stored in a form called potential energy. **Potential energy** is energy that an object has because of its location or structure, such as the energy contained by water behind a dam or by a compressed spring. In our example, the diver at the top of the platform has potential energy because of his elevated location. The act of diving off the platform into the water converts the potential energy back to kinetic energy. Life depends on the conversion of energy from one form to another. ✔

▼ Figure 5.1 **Energy conversions during a dive.**

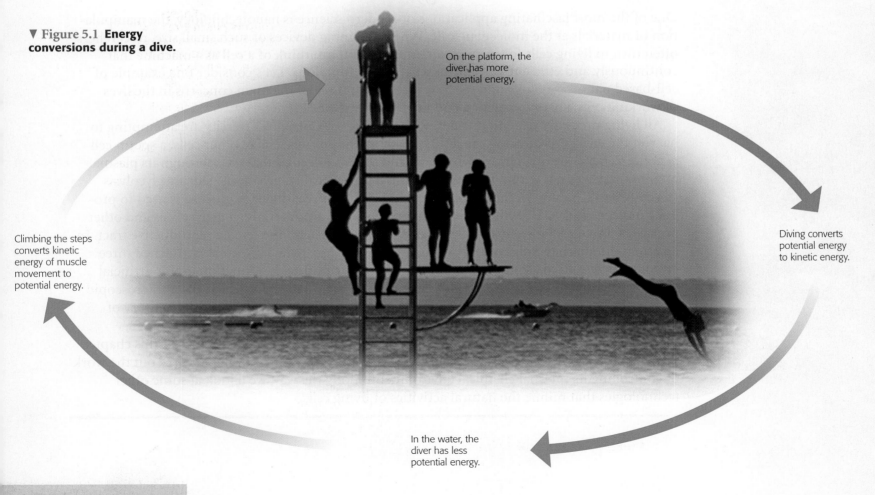

On the platform, the diver has more potential energy.

Climbing the steps converts kinetic energy of muscle movement to potential energy.

Diving converts potential energy to kinetic energy.

In the water, the diver has less potential energy.

Entropy

If energy cannot be destroyed, where has it gone when the diver hits the water? It has been converted to **heat**, a type of kinetic energy contained in the random motion of atoms and molecules. The friction between the body and its surroundings generated heat in the air and then in the water.

All energy conversions generate some heat. Although heat production does not destroy energy, it does make it less useful. Heat, of all energy forms, is the most difficult to "tame"—the most difficult to harness for useful work. Heat is energy in its most chaotic form, the energy of aimless molecular movement.

Entropy is a measure of the amount of disorder, or randomness, in a system. Every time energy is converted from one form to another, entropy increases. The energy conversions during the climb up the ladder and the dive from the platform increased entropy because all of the diver's potential energy was lost to the surroundings as heat. To climb up the steps again for another dive, the diver must use additional stored food energy. ☑

Chemical Energy

How can molecules derived from the food we eat provide energy for our working cells? The molecules of food, gasoline, and other fuels have a special form of potential energy called **chemical energy**, which arises from the arrangement of atoms. Carbohydrates, fats, and gasoline have structures that make them especially rich in chemical energy.

Living cells and automobile engines use the same basic process to make the chemical energy stored in their fuels available for work (**Figure 5.2**). In both cases, this process breaks the organic fuel into smaller waste molecules that have much less chemical energy than the fuel molecules did, thereby releasing energy that can be used to perform work.

For example, the engine of an automobile mixes oxygen with gasoline in an explosive chemical reaction that breaks down the fuel molecules and pushes the pistons that eventually move the wheels. The waste products emitted from the car's exhaust pipe are mostly carbon dioxide and water. About 25% of the energy that an automobile engine extracts from its fuel is converted to

◄ Figure 5.2 **Energy conversions in a car and a cell.** In both a car and a cell, the chemical energy of organic fuel molecules is harvested using oxygen. This chemical breakdown releases energy stored in the fuel molecules and produces carbon dioxide and water. The released energy can be used to perform work.

the kinetic energy of the car's movement. Most of the rest is converted to heat—so much that the engine would melt if the car's radiator and fan did not disperse excess heat into the atmosphere.

Your cells also use oxygen to help harvest chemical energy. As in a car engine, the "exhaust" is mostly carbon dioxide and water. The "combustion" of fuel in cells is called cellular respiration, which is a more gradual and efficient "burning" of fuel compared with the explosive combustion in an automobile engine. Cellular respiration is the energy-releasing chemical breakdown of fuel molecules and the storage of that energy in a form the cell can use to perform work. We will discuss the details of cellular respiration in the next chapter. You convert about 40% of your food energy to useful work, such as the contraction of your muscles. About 60% of the energy released by the breakdown of fuel molecules generates body heat. Humans and many other animals can use this heat to keep the body at an almost constant temperature (37°C, or 98.6°F, in the case of humans), even when the surrounding air is much colder. The liberation of heat energy also explains why you feel hot after exercise. Sweating and other cooling mechanisms enable your body to lose the excess heat, much as a car's radiator keeps the engine from overheating.

☑CHECKPOINT

According to Figure 5.3, how long would you have to ride your bicycle to burn off the energy in three slices of pepperoni pizza?

Answer: a little over 1 hour (1 slice = 181 Calories, so 3 slices = 543 Calories; 1 hour of cycling consumes 490 Calories)

Food Calories

Read any packaged food label and you'll find the number of calories in each serving of that food. Calories are units of energy. A **calorie** (cal) is the amount of energy that can raise the temperature of 1 gram (g) of water by 1°C. You could actually measure the caloric content of a peanut by burning it under a container of water to convert all of the stored chemical energy to heat and then measuring the temperature increase of the water.

Calories are tiny units of energy, so using them to describe the fuel content of foods is not practical. Instead, it's conventional to use kilocalories (kcal), units of 1,000 calories. In fact, the Calories (capital C) on a food package are actually kilocalories. That's a lot of energy. For example, one peanut has about 5 kcal. That's enough energy to increase the temperature of 1 kg (a little more than a quart) of water by 5°C in our peanut-burning experiment. And just a handful of peanuts contains enough Calories, if converted to heat, to boil 1 kg of water. In living organisms, of course, food isn't used to boil water but instead to fuel the activities of life. **Figure 5.3** shows the number of Calories in several foods and how many Calories are burned off by some typical activities. ☑

▼ Figure 5.3 **Some caloric accounting.**

Food	Food Calories
Cheeseburger	295
Spaghetti with sauce (1 cup)	241
Baked potato (plain, with skin)	220
Fried chicken (drumstick)	193
Bean burrito	189
Pizza with pepperoni (1 slice)	181
Peanuts (1 ounce)	166
Apple	81
Garden salad (2 cups)	56
Popcorn (plain, 1 cup)	31
Broccoli (1 cup)	25

(a) Food Calories (kilocalories) in various foods

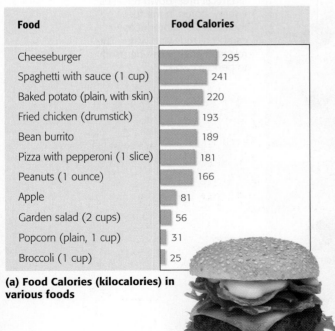

Activity	Food Calories consumed per hour by a 150-pound person*
Running (7 min/mi)	979
Dancing (fast)	510
Bicycling (10 mph)	490
Swimming (2 mph)	408
Walking (3 mph)	245
Dancing (slow)	204
Playing the piano	73
Driving a car	61
Sitting (writing)	28

*Not including energy necessary for basic functions, such as breathing and heartbeat

(b) Food Calories (kilocalories) we burn in various activities

ATP and Cellular Work

The carbohydrates, fats, and other fuel molecules we obtain from food do not drive the work in our cells directly. Instead, the chemical energy released by the breakdown of organic molecules during cellular respiration is used to generate molecules of ATP. These molecules of ATP then power cellular work. ATP acts like an energy shuttle, storing energy obtained from food and then releasing it as needed at a later time.

The Structure of ATP

The abbreviation ATP stands for adenosine triphosphate. **ATP** consists of an organic molecule called adenosine plus a tail of three phosphate groups (P) (**Figure 5.4**). The triphosphate tail is the "business" end of ATP, the part that provides energy for cellular work. Each phosphate group is negatively charged. Negative charges repel each other. The crowding of negative charges in the triphosphate tail contributes to the potential energy of ATP. It's analogous to storing energy by compressing a spring; if you release the spring, it will "relax," and you can use that springiness to do some useful work. For ATP power, it is release of the phosphate at the tip of the triphosphate tail that makes energy available to working cells. What remains is **ADP,** adenosine diphosphate (two phosphate groups instead of three; see Figure 5.4).

Phosphate Transfer

When ATP drives work in cells, phosphate groups don't just fly off into space. ATP energizes other molecules in cells by transferring phosphate groups to those molecules. This transfer of phosphate groups helps cells perform three main kinds of work: mechanical work, transport work, and chemical work.

Imagine a bicyclist pedaling up a hill. In the muscle cells of the rider's legs, ATP transfers phosphate groups to motor proteins. The proteins then change shape, causing the muscle cells to contract and perform mechanical work (**Figure 5.5a**). ATP also enables the transport of ions and other solutes across the membranes of the rider's brain cells (**Figure 5.5b**). The transport of ions prepares the brain cells to transmit signals to muscles and other tissues. And ATP drives the chemical work of making some of a cell's large molecules (**Figure 5.5c**). Notice again in Figure 5.5 that all these types of work occur when target molecules accept a phosphate group from ATP.

▼ **Figure 5.4 ATP power.** Each (P) in the triphosphate tail of ATP represents a phosphate group, a phosphorus atom bonded to oxygen atoms. The transfer of a phosphate from the triphosphate tail to other molecules provides energy for cellular work.

▼ **Figure 5.5 How ATP drives cellular work.** Each type of work shown here is powered when enzymes transfer phosphate from ATP to a recipient molecule.

(a) Motor protein performing mechanical work

(b) Transport protein performing transport work

(c) Chemical reactants performing chemical work

☑CHECKPOINT

1. Explain how ATP powers cellular work.
2. What is the source of energy for regenerating ATP from ADP?

Answers: 1. ATP transfers a phosphate group to another molecule, increasing that molecule's energy content. 2. chemical energy harvested from sugars and other organic fuels via cellular respiration

The ATP Cycle

Your cells spend ATP continuously. Fortunately, it is a renewable resource. ATP can be restored by adding a phosphate group back to ADP. That takes energy, like recompressing a spring. And that's where food reenters the story. The chemical energy that cellular respiration harvests from sugars and other organic fuels is put to work regenerating a cell's supply of ATP. Cellular work spends ATP, which is recycled when ADP and phosphate are combined using energy released by cellular respiration (**Figure 5.6**). Thus, energy from processes that yield energy, such as the breakdown of organic fuels, is transferred to processes that consume energy, such as muscle contraction and other cellular work. The third phosphate group acts as an energy shuttle within the ATP cycle.

▼ Figure 5.6 **The ATP cycle.**

The ATP cycle runs at an astonishing pace. A working muscle cell recycles all of its ATP about once every minute. That's about 10 million ATP molecules spent and regenerated per second per cell. ☑

Enzymes

As you've seen, a living organism contains a vast collection of chemicals, with countless chemical reactions constantly changing the organism's molecular makeup. In a sense, a living organism is a complex "chemical square dance," with the molecular "dancers" continually changing partners via chemical reactions. The total of all the chemical reactions in an organism is called **metabolism**. Interestingly, almost no metabolic reactions occur without help. Most require the assistance of **enzymes**, proteins that speed up chemical reactions. All living cells contain thousands of different enzymes, each promoting a different chemical reaction.

Activation Energy

For a chemical reaction to begin, chemical bonds in the reactant molecules must be broken. (The first step in swapping partners during a square dance is to let go of your current partner's hand.) This process requires that the molecules absorb energy from their surroundings. This energy is called **activation energy** because it activates the reactants and triggers the chemical reaction.

Enzymes enable metabolism to occur by reducing the amount of activation energy required to break the bonds of reactant molecules. If you think of the requirement for activation energy as a barrier to a chemical reaction, an enzyme's function is to lower that barrier (**Figure 5.7**). It does so by binding to reactant molecules and putting them under physical or chemical stress, making it easier to break their bonds and start a reaction. ☑

☑CHECKPOINT

How does an enzyme affect the activation energy of a chemical reaction?

Answer: An enzyme lowers the activation energy.

▼ Figure 5.7 **Enzymes and activation energy.**

(a) Without enzyme. A reactant molecule must overcome the activation energy barrier before a chemical reaction can break the molecule into products.

(b) With enzyme. An enzyme speeds the process by lowering the activation energy barrier.

Can Enzymes Be Engineered?

Like all other proteins, enzymes are encoded by genes. **Observations** of genetic sequences suggest that many of our genes were formed through a type of molecular evolution: One ancestral gene randomly duplicated, and the two copies diverged over time via genetic mutation, eventually becoming two distinct genes coding for two enzymes with different functions.

The natural evolution of enzymes raises a **question**: Can laboratory methods mimic this process through artificial selection? In 1997, a group of researchers at two California biotechnology companies formed the **hypothesis** that an artificial process could be used to modify the gene that codes for the enzyme lactase into a new gene coding for a new enzyme with a new function. Recall from Chapter 3 that lactase breaks down the sugar lactose. Their **experiment** used a procedure called directed evolution. In this process, many copies of the gene for the starting lactase enzyme were mutated at random (**Figure 5.8**). The researchers tested the enzymes resulting from these mutated genes to determine which enzymes best displayed a new activity (in this case, breaking down a different but closely related sugar). The genes for the enzymes that did show the new activity were then subjected to several more rounds of duplication, mutation, and screening.

After seven rounds, the **results** were clear: Directed evolution had produced a new enzyme with a novel function. Researchers have used similar methods to produce many artificial enzymes with desired properties. Directed evolution is another example of how scientists can mimic the natural processes of cells for useful purposes.

▼ Figure 5.8 **Directed evolution of an enzyme.** During seven rounds of directed evolution, the lactase enzyme gradually gained a new function.

Gene for lactase

Gene **duplicated** and **mutated** at random

Mutated genes
(mutations shown in orange)

Mutated genes **screened** by testing new enzymes

Genes coding for enzymes that show new activity

Genes coding for enzymes that do *not* show new activity

Genes **duplicated** and **mutated** at random

Mutated genes **screened** by testing new enzymes

After seven rounds, some genes code for enzymes that can efficiently perform new activity.

Ribbon model showing the polypeptide chains of the enzyme lactase

Induced Fit

An enzyme is very selective in the reaction it catalyzes. This specificity is based on the enzyme's ability to recognize a certain reactant molecule, which is called the enzyme's **substrate**. And the ability of the enzyme to recognize and bind to its specific substrate depends on the enzyme's shape. A region of the enzyme called the **active site** has a shape and chemistry that fit the substrate molecule. When a substrate slips into this docking station, the active site changes shape slightly to embrace the substrate and catalyze the reaction. This interaction is called **induced fit** because the entry of the substrate induces the enzyme to change shape slightly, making the fit between substrate and active site snugger. Think of a handshake: As your hand makes contact with another hand, it changes shape slightly to make a better fit.

After the products are released from the active site, the enzyme can accept another molecule of its substrate. In fact, the ability to function repeatedly is a key characteristic of enzymes. **Figure 5.9** follows the action of the enzyme sucrase, which hydrolyzes the disaccharide sucrose (the substrate). Like sucrase, many enzymes are named for their substrates, but with an *-ase* ending.

Enzyme Inhibitors

Certain molecules can inhibit a metabolic reaction by binding to an enzyme and disrupting its function (**Figure 5.10**). Some of these **enzyme inhibitors** are actually substrate imposters that plug up the active site (see Figure 5.10b). (You can't shake a person's hand if someone else puts a banana in it first!) Other inhibitors bind to the enzyme at a site remote from the active site, but the binding changes the enzyme's shape so that

▼ **Figure 5.9 How an enzyme works.** Our example is the enzyme sucrase, named for its substrate, sucrose.

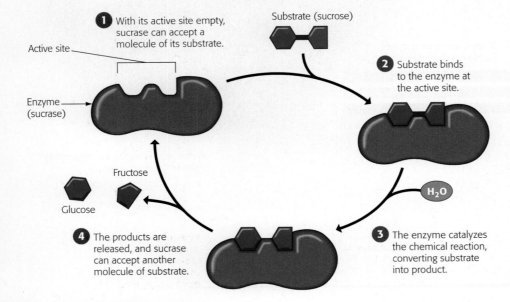

1 With its active site empty, sucrase can accept a molecule of its substrate.

Active site

Enzyme (sucrase)

Substrate (sucrose)

2 Substrate binds to the enzyme at the active site.

H_2O

3 The enzyme catalyzes the chemical reaction, converting substrate into product.

4 The products are released, and sucrase can accept another molecule of substrate.

Glucose

Fructose

▼ **Figure 5.10 Enzyme inhibitors.**

Substrate
Active site

Enzyme

(a) Enzyme and substrate binding normally

Inhibitor
Active site

Substrate

Enzyme

(b) Enzyme inhibition by a substrate imposter

Active site

Substrate

Inhibitor

Enzyme

(c) Enzyme inhibition by a molecule that causes the active site to change shape

the active site no longer accepts the substrate (see Figure 5.10c). (Imagine being unable to shake hands because someone else is tickling your ribs, causing you to clench your hand.) In some cases, the binding is reversible, enabling certain inhibitors to regulate metabolism. For example, if metabolism is producing more of a certain product than a cell needs, that product may reversibly inhibit an enzyme required for its production. This **feedback regulation** keeps the cell from wasting resources that could be put to better use.

Many antibiotics work by inhibiting enzymes of disease-causing bacteria. Penicillin, for example, inhibits an enzyme that bacteria use in making their cell walls (see Chapter 4). Many cancer drugs inhibit enzymes that promote cell division. In addition to drugs, many toxins and poisons are irreversible inhibitors. Nerve gases bind to the active site of an enzyme vital to transmitting nerve impulses. The inhibition of this enzyme leads to rapid paralysis of vital functions and death. Many pesticides are toxic to insects because they irreversibly inhibit this same enzyme. ☑

Membrane Function

So far, we have discussed how cells control the flow of energy and the pace of chemical reactions. They must also regulate the flow of materials to and from the environment. In Chapter 4, you learned about the structure of the plasma membrane (see Figure 4.6). **Figure 5.11** describes the major functions of the proteins embedded within the phospholipid bilayer. In this section, we'll focus mainly on how the plasma membrane regulates the passage of materials into and out of the cell. Transport proteins, such as the one shown in Figure 5.11, are critical to this task. **Transport proteins** are membrane proteins that help move substances across a cell membrane. In this section, you'll learn about the most important mechanisms of transport across membranes.

Passive Transport: Diffusion across Membranes

Membrane Transport

Molecules are restless. The heat energy they contain makes them vibrate and wander randomly. One result of this motion is **diffusion**, the movement of molecules of any substance so that they spread out into the available space. Each molecule moves randomly, and yet the overall diffusion of a population of molecules is usually directional, from a region where they are more concentrated to where they are less concentrated. For example, imagine many molecules of perfume inside a bottle. If you remove the bottle top, every molecule of perfume

▼ Figure 5.11 **Primary functions of membrane proteins.** In this diagram of two adjacent membranes, all six types of membrane proteins are shown for convenience. An actual cell may have just a few of these types of proteins.

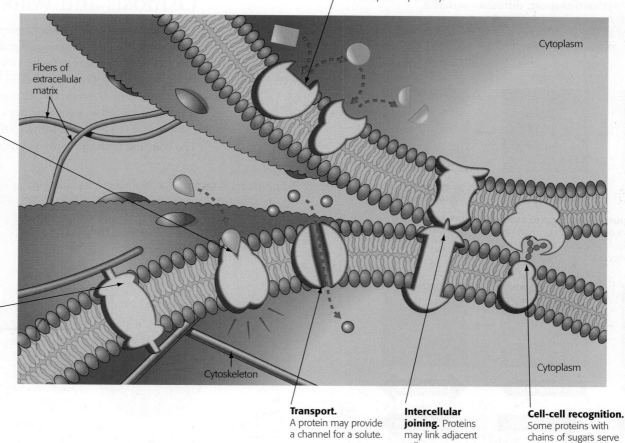

Enzymatic activity. This protein and the one next to it are enzymes, having an active site that fits a substrate. Enzymes may form a team that carries out steps of a pathway.

Cytoplasm

Fibers of extracellular matrix

Cell signaling. A binding site fits the shape of a chemical messenger. The messenger may cause a change in the protein that relays the message to the inside of the cell.

Attachment to the cytoskeleton and extracellular matrix. Such proteins help maintain cell shape and coordinate changes.

Cytoskeleton

Cytoplasm

Transport. A protein may provide a channel for a solute.

Intercellular joining. Proteins may link adjacent cells.

Cell-cell recognition. Some proteins with chains of sugars serve as identification tags recognized by other cells.

will move randomly about, but the net direction of the molecules will be out of the bottle to fill the room. You could, with great effort, return the perfume to its bottle, but the molecues would never return spontaneously.

For an example closer to a living cell, imagine a membrane separating pure water from a solution of a dye dissolved in water (**Figure 5.12**). Assume that this membrane is permeable to the dye molecules—meaning that they can pass through the membrane. Although each dye molecule moves randomly, there will be a net migration across the membrane to the side that began as pure water. The spreading of the dye across the membrane will continue until both solutions have equal concentrations of the dye. When that point is reached, there will be a dynamic equilibrium, with as many dye molecules moving per second across the membrane in one direction as the other.

Diffusion across a membrane is an example of **passive transport**—passive because the cell does not expend any energy for it to happen. However, the cell membrane does play a regulatory role by being selectively permeable. For example, small molecules such as carbon dioxide (CO_2) and oxygen (O_2) generally pass

through more readily than larger molecules such as amino acids. But the membrane is relatively impermeable to even some very small substances, such as hydrogen ions (H^+) and other inorganic ions, which are too hydrophilic to pass through the phospholipid bilayer. Passive transport is extremely important to all cells. In passive transport, a substance diffuses down its **concentration gradient**, a region in which the substance's density changes. That is, the substance moves from where it is more concentrated to where it is less concentrated. In our lungs, for example, passive transport along concentration gradients is the sole means by which O_2, essential for metabolism, enters the blood and CO_2, a metabolic waste, passes out of it.

Substances that do not cross membranes spontaneously can be transported via **facilitated diffusion** by specific transport proteins that act as selective corridors (see Figure 5.11). Without the protein, the substance does not cross the membrane or diffuses across it too slowly to be useful to the cell. Facilitated diffusion is a type of passive transport because it does not require energy. As in all passive transport, the driving force is the concentration gradient. ✓

▼ **Figure 5.12 Passive transport: diffusion across a membrane.** A substance will diffuse from where it is more concentrated to where it is less concentrated. Put another way, a substance tends to diffuse down its concentration gradient.

Molecules of dye   Membrane

Net diffusion   Net diffusion   Equilibrium

(a) Passive transport of one type of molecule. The membrane is permeable to these dye molecules, which diffuse down their concentration gradient. At equilibrium, the molecules are still restless, but the rate of transport is equal in both directions.

Net diffusion   Net diffusion   Equilibrium

Net diffusion   Net diffusion   Equilibrium

(b) Passive transport of two types of molecules. If solutions have two or more solutes, each will diffuse down its own concentration gradient.

Osmosis and Water Balance

The diffusion of water across a selectively permeable membrane is called **osmosis** (**Figure 5.13**). Consider the case of a membrane separating two solutions with different concentrations of a solute—say, the sugar sucrose. The membrane is permeable to water but not to the solute. The solution with a higher concentration of solute is said to be **hypertonic** to the other solution. The solution with the lower solute concentration is said

▼ **Figure 5.13 Osmosis.** A membrane separates two solutions with different sugar concentrations. Water molecules can pass through the membrane, but sugar molecules cannot.

Hypotonic solution (lower concentration of solute)   **Hypertonic solution** (higher concentration of solute)   **Isotonic solutions** (equal concentration of solute)

Sugar molecule (solute)

Selectively permeable membrane

Osmosis (net movement of water)

Osmosis reduces the difference in sugar concentrations and changes the volumes of the two solutions.

to be **hypotonic** to the other. Note that the hypotonic solution, by having the lower solute concentration, has the higher water concentration. Therefore, water will diffuse across the membrane along its concentration gradient from an area of higher water concentration (hypotonic solution) to one of lower water concentration (hypertonic solution). This reduces the difference in solute concentrations and changes the volumes of the two solutions.

People can take advantage of osmosis to preserve foods. Salt is often applied to meats to cure them; the salt causes water to move out of the food toward the region of greater solute (salt) concentration. Food can also be preserved in honey because a high sugar concentration draws water out of food.

When the solute concentrations are the same on both sides of a membrane, water molecules will move at the same rate in both directions, so there will be no net change in solute concentration. Solutions of equal solute concentration are said to be **isotonic** to each other. You may have seen this term on bottles of contact lens saline solution, which is formulated to have the same solute concentration as the fluid at the surface of the human eye, making it nonirritating.

Water Balance in Animal Cells

The survival of a cell depends on its ability to balance water uptake and loss. When an animal cell, such as a red blood cell, is immersed in an isotonic solution, the cell's volume remains constant because the cell gains water at the same rate that it loses water (**Figure 5.14a**, top). In this case, the cell is isotonic to its surroundings because the two solutions have the same total concentration of solutes. Many marine animals, such as sea stars and crabs, are isotonic to seawater. But what happens if an animal cell is in contact with a hypotonic solution, which has a lower solute concentration than the cell? Due to osmosis, the cell would gain water, swell, and possibly burst (lyse) like an overfilled water balloon (**Figure 5.14b**, top). A hypertonic environment is also harsh on an animal cell; the cell shrivels and can die from water loss (**Figure 5.14c**, top).

For an animal to survive a hypotonic or hypertonic environment, the animal must have a way to balance an excessive uptake or excessive loss of water. The control of water balance is called **osmoregulation**. For example, a freshwater fish, whose environment is hypotonic to its body, has kidneys and gills that work constantly to prevent an excessive buildup of water in the body. And if you look back to Figure 4.17a, you'll see *Paramecium*'s contractile vacuole, which bails out the excess water that continuously enters the cell from the hypotonic pond water.

▼ Figure 5.14 **The behavior of animal and plant cells in different osmotic environments.**

Animal cell

H_2O → H_2O — Normal

H_2O — Lysing

H_2O — Shriveled

Plant cell

H_2O → H_2O — Flaccid (wilts)

H_2O → H_2O — Turgid

Plasma membrane — H_2O — Shriveled

(a) **Isotonic solution**   (b) **Hypotonic solution**   (c) **Hypertonic solution**

Water Balance in Plant Cells

Problems of water balance are somewhat different for plant cells because of their rigid cell walls. A plant cell immersed in an isotonic solution is flaccid (floppy), and a plant wilts in this situation (Figure 5.14a, bottom). In contrast, a plant cell is turgid (firm) and healthiest in a hypotonic environment, with a net inflow of water (Figure 5.14b, bottom). Although the elastic cell wall expands a bit, the back pressure it exerts prevents the cell from taking in too much water and bursting, as an animal cell would in this environment. Turgor is necessary for plants to retain their upright posture and the extended state of their leaves (**Figure 5.15**). However, in a hypertonic environment, a plant cell is no better off than an animal cell. As a plant cell loses water, it shrivels, and its plasma membrane pulls away from the cell wall (Figure 5.14c, bottom). This process, called **plasmolysis**, usually kills the cell. ☑

▼ Figure 5.15 **Plant turgor.**
A wilted plant regains its turgor when watered.

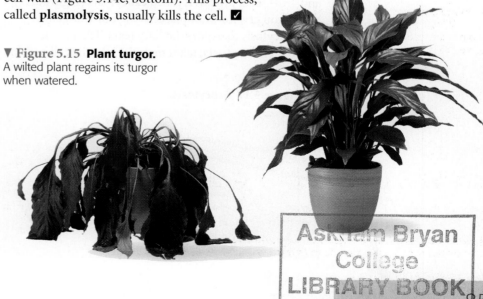

☑**CHECKPOINT**

1. An animal cell shrivels when it is _____ compared with its environment.
2. The cells of a wilted plant are _____ compared with their environment.

Answers: 1. hypotonic 2. isotonic

85

Active Transport: The Pumping of Molecules across Membranes

In contrast to passive transport, **active transport** requires that a cell expend energy to move molecules across a membrane. In active transport, cellular energy is used to drive a transport protein that actively pumps a solute across a membrane *against* the solute's concentration gradient—that is, away from the side where it is less concentrated and toward the side where it is more concentrated (**Figure 5.16**). Membrane proteins usually use ATP as their energy source for active transport.

Active transport enables cells to maintain internal concentrations of small solutes that differ from environmental concentrations. For example, compared with its surroundings, an animal nerve cell has a much higher concentration of potassium ions and a much lower concentration of sodium ions. The plasma membrane helps maintain these differences by pumping sodium out of the cell and potassium into the cell. This particular case of active transport (called the sodium-potassium pump) is vital in the propagation of nerve signals. ☑

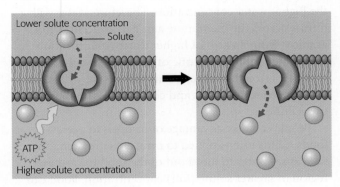

▲ Figure 5.16 **Active transport.** Like enzymes, transport proteins are specific in their recognition of atoms or molecules. This transport protein (purple) has a binding site that accepts only a certain solute. Using energy from ATP, the protein pumps the solute against its concentration gradient.

Exocytosis and Endocytosis: Traffic of Large Molecules

So far, we've focused on how water and small solutes enter and leave cells by moving through the plasma membrane. The story is different for macromolecules such as proteins, which are much too big to fit through the membrane itself. Their traffic into and out of the cell depends on the ability of the membrane to form sacs, thereby packaging larger molecules into vesicles. You have already seen an example in the packaging and secretion of proteins. During protein production by the cell, secretory proteins exit the cell from transport vesicles that fuse with the plasma membrane, spilling the contents outside the cell (see Figures 4.14 and 4.18). That process is called **exocytosis** (**Figure 5.17**). When you cry, for example, cells in your tear glands use exocytosis to export the salty tears. The reverse process, **endocytosis**, takes material into the cell within vesicles that bud inward from the plasma membrane (**Figure 5.18**).

There are three types of endocytosis. In **phagocytosis** ("cellular eating"), a cell engulfs a particle and packages it within a food vacuole. In **pinocytosis** ("cellular drinking"), the cell "gulps" droplets of fluid by forming vesicles. Both phagocytosis and pinocytosis are nonspecific in the substances transported. In contrast, **receptor-mediated endocytosis** is very specific: It is triggered by the binding of certain external molecules to specific receptor proteins built into the plasma membrane. This binding causes the local region of the membrane to form a vesicle that transports the specific substance into the cell. Human liver cells use receptor-mediated endocytosis to take up cholesterol particles from the blood.

▼ Figure 5.17 **Exocytosis.**

▼ Figure 5.18 **Endocytosis.**

The Role of Membranes in Cell Signaling

In addition to transport, the plasma membrane plays a key role in conveying signals from the external environment into the cell and also between cells (see Figure 5.11). Communication begins when a receptor protein in the plasma membrane receives a stimulus from the outside environment or from a signaling molecule, such as a hormone. This stimulus triggers a chain reaction in one or more molecules that function in transduction (passing the signal along). The proteins and other molecules of this **signal transduction pathway** relay the signal and convert it to chemical forms that can function within the cell. This signal may lead to various responses (see also Figure 11.8).

Figure 5.19 shows an example of cell signaling: signal reception, transduction, and response. When a person gets "psyched up" for an athletic contest, certain cells in the adrenal glands secrete a hormone called epinephrine (also called adrenaline) into the bloodstream. When that hormone reaches muscle cells, it is recognized by receptor proteins in the plasma membrane. This recognition triggers responses (such as the breakdown of glycogen into glucose) in the muscle cells without the hormone even entering. This chain of events is part of the "fight-or-flight" response that enables you to attack or run when in danger—or keep alert during an intense competition. ☑

☑ **CHECKPOINT**
How can a molecule change a cell without entering it?

Answer: It can bind to a membrane protein that triggers a signal transduction pathway.

▼ Figure 5.19 **An example of cell signaling.**

Outside of cell   Cytoplasm

Reception   **Transduction**   **Response**

Receptor protein

Proteins of signal transduction pathway

Hydrolysis of glycogen releases glucose for energy

Epinephrine (adrenaline) from adrenal glands

Plasma membrane

EVOLUTION CONNECTION: Molecular Technologies

The Origin of Membranes

Because all cells have a plasma membrane, it is logical to infer that membranes first formed early in the evolution of life on Earth. Phospholipids, their key ingredients, were probably among the first organic compounds that formed before life emerged. Once formed, they could self-assemble into simple membranes. When a mixture of phospholipids and water is shaken, for example, the phospholipids organize into bilayers, forming water-filled bubbles of membrane (**Figure 5.20**). This assembly requires neither genes nor other information beyond the properties of the phospholipids.

The tendency of lipids in water to spontaneously form membranes has led biomedical engineers to produce artificial vesicles called liposomes that can encase particular chemicals. In the future, these liposomes may deliver medications to specific sites within the body. Thus, membranes—like the other cellular components discussed in the Biology and Society and the Process of Science sections—have inspired novel technologies.

▶ Figure 5.20
The spontaneous formation of membranes: a key step in the origin of life.

Chapter Review

SUMMARY OF KEY CONCEPTS

 Go to the Study Area at **www.masteringbiology.com** for practice quizzes, eText, BioFlix™ 3-D animations, MP3 Tutor Sessions, videos, current events, and more.

Some Basic Energy Concepts

Conservation of Energy

Machines and organisms can transform kinetic energy (energy of motion) to potential energy (stored energy) and vice versa. In all such energy transformations, total energy is conserved. Energy cannot be created or destroyed.

Entropy

Entropy is a measure of disorder, or randomness. Every energy conversion releases some randomized energy in the form of heat.

Chemical Energy

Molecules store varying amounts of potential energy in the arrangement of their atoms. Organic compounds are relatively rich in such chemical energy.

Food Calories

Food Calories, actually kilocalories, are units used to measure the amount of energy in our foods and the amount of energy we expend in various activities.

ATP and Cellular Work

Enzymes

Activation Energy

Enzymes are biological catalysts that speed up metabolic reactions by lowering the activation energy required to break the bonds of reactant molecules.

Induced Fit

The entry of a substrate into the active site of an enzyme causes the enzyme to change shape slightly, allowing for a better fit and thereby promoting the interaction of enzyme with substrate.

Enzyme Inhibitors

Enzyme inhibitors are molecules that can disrupt metabolic reactions by binding to enzymes, either at the active site or elsewhere.

Membrane Function

Proteins embedded in the plasma membrane perform a wide variety of functions, including regulating transport.

Passive Transport, Osmosis, and Active Transport

Most animal cells require an isotonic environment. Plant cells need a hypotonic environment, which keeps walled cells turgid. Osmoregulation is the control of water balance within a cell or organism.

Exocytosis and Endocytosis: Traffic of Large Molecules

Exocytosis is the secretion of large molecules within vesicles. The three kinds of endocytosis are phagocytosis ("cellular eating"), pinocytosis ("cellular drinking"), and receptor-mediated endocytosis, which enables the cell to take in specific large molecules.

The Role of Membranes in Cell Signaling

Receptors on the cell surface trigger signal transduction pathways that control processes within the cell.

SELF-QUIZ

1. Describe the energy transformations that occur when you climb to the top of a stairway.

2. _____ is the capacity to perform work, while _____ is a measure of randomness.

3. The label on a candy bar says that it contains 150 Calories. If you could convert all of that energy to heat, you could raise the temperature of how much water by 15°C?

4. Why does removing a phosphate group from the triphosphate tail in a molecule of ATP release energy?

5. Your digestive system uses a variety of enzymes to break down large food molecules into smaller ones that your cells can assimilate. A generic name for a digestive enzyme is hydrolase. What is the chemical basis for that name? (*Hint*: Review Figure 3.4.)

6. Explain how an inhibitor can disrupt an enzyme's action without binding to the active site.

7. If someone at the other end of a restaurant smokes a cigarette, you may breathe in some smoke. The movement of smoke is similar to what type of transport?
   a. osmosis
   b. diffusion
   c. facilitated diffusion
   d. active transport

8. The total solute concentration in a red blood cell is about 2%. Sucrose cannot pass through a red blood cell's plasma membrane, but water and urea can. Osmosis will cause such a cell to shrink the most when the cell is immersed in which of the following?
   a. a hypertonic sucrose solution
   b. a hypotonic sucrose solution
   c. a hypertonic urea solution
   d. a hypotonic urea solution

9. Explain why it is not enough just to say that a solution is hypertonic.

10. What is the primary difference between passive and active transport in terms of concentration gradients?

11. Which of these types of cellular transport require(s) energy?
   a. facilitated diffusion
   b. active transport
   c. osmosis
   d. a and b

12. A _____ is a process that links the reception of a cell signal to a response within the cell.

Answers to the Self-Quiz questions can be found in Appendix D.

THE PROCESS OF SCIENCE

13. HIV, the virus that causes AIDS, depends on an enzyme called reverse transcriptase in order to multiply. Reverse transcriptase reads a molecule of RNA and creates a molecule of DNA from it. A molecule of AZT, the first drug approved to treat AIDS, has a shape very similar to that of the DNA base thymine. Propose a model for how AZT inhibits HIV.

14. Gaining and losing weight are matters of caloric accounting: Calories in the food you eat minus Calories that you spend in activity. One pound of human body fat contains approximately 3,500 Calories. Using Figure 5.3, compare ways you could burn off those Calories. How far would you have to run, swim, or walk to burn the equivalent of 1 pound of fat, and how long would it take? Which method of burning Calories appeals the most to you? The least?

BIOLOGY AND SOCIETY

15. Obesity is a serious health problem for many Americans. Several popular diet plans advocate low-carbohydrate diets. Most low-carb dieters compensate by eating more protein and fat. What are the advantages and disadvantages of such a diet? Should the government regulate the claims of diet books? How should the claims be tested? Should diet proponents be required to obtain and publish data before making claims?

16. Lead acts as an enzyme inhibitor, and it can interfere with the development of the nervous system. One manufacturer of lead-acid batteries instituted a "fetal protection policy" that banned female employees of childbearing age from working in areas where they might be exposed to high levels of lead. These women were transferred to lower-paying jobs in lower-risk areas. Some employees challenged the policy in court, claiming that it deprived women of job opportunities available to men. The U.S. Supreme Court ruled the policy illegal. But many people are uncomfortable about the "right" to work in an unsafe environment. What rights and responsibilities of employers, employees, and government agencies are in conflict? What criteria should be used to decide who can work in a particular environment?

6 Cellular Respiration: Obtaining Energy from Food

Muscles in action.
Sprinters, like all athletes, depend on cellular respiration to power their muscles.

BIOLOGY AND SOCIETY: Life with and without Oxygen

Marathoners versus Sprinters

Track-and-field athletes usually have a favorite event in which they excel. For some runners, this event is a sprint, a short race of only 100 or 200 meters. For others, it may be a longer race of 1,500, 5,000, or even 10,000 m. It is unusual to find a runner who competes equally well in both 100-m and 10,000-m races; most runners just seem to feel more comfortable running races of particular lengths.

It turns out that there is a biological basis for such preferences. The muscles that move our legs contain two main types of muscle fibers: slow-twitch and fast-twitch. Slow-twitch muscle fibers can contract many times over a longer period but don't generate a lot of quick power for the body. They perform better in endurance exercises requiring slow, steady muscle activity (marathons). Fast-twitch muscle fibers can contract more quickly and powerfully than slow-twitch fibers but fatigue much more quickly. They function best in short bursts of intense activity (sprints).

All human muscles contain both slow-twitch and fast-twitch fibers, but the percentage of each fiber type in a particular muscle varies from person to person. For example, the thigh muscles of most marathon runners contain more slow-twitch fibers, whereas sprinters have more fast-twitch fibers. These differences, which are genetically determined, undoubtedly help account for varying athletic capabilities. Thus, to a certain degree, champion marathoners and sprinters are born, not trained!

What makes these two types of muscle fibers perform so differently? An important part of the answer is that they use different processes for making ATP, the molecule that supplies the energy for muscle contraction. While the cells in both types of muscle fiber break down glucose to make chemical energy for ATP production, slow-twitch fibers do it using oxygen (O_2), while fast-twitch fibers can work without oxygen.

The series of chemical reactions that provides energy to muscles is also used by other cells. In fact, all your body cells need a continuous supply of energy for you to walk, talk, and think—in short, to stay alive. In this chapter, you'll learn how cells harvest food energy and put it to work.

Energy Flow and Chemical Cycling in the Biosphere

Humans and other animals depend on plants to convert the energy of sunlight to the chemical energy of sugars and other organic molecules we consume as food. This is accomplished via photosynthesis. **Photosynthesis** uses light energy from the sun to power a chemical process that builds organic molecules (as we'll discuss in Chapter 7). But we depend on plants for more than our food. You're probably wearing clothing made of another product of photosynthesis—cotton. Most of our homes are framed with lumber, which is wood produced by photosynthetic trees. Even the text you are now reading is printed on a material (paper) that can be traced to photosynthesis in plants. But from an animal's point of view, photosynthesis is primarily about providing food.

Producers and Consumers

Plants and other **autotrophs** ("self-feeders") are organisms that make all their own organic matter—including carbohydrates, lipids, proteins, and nucleic acids—from nutrients that are entirely inorganic: carbon dioxide from the air and water and minerals from the soil. In contrast, humans and other animals are **heterotrophs** ("other-feeders"), organisms that cannot make organic molecules from inorganic ones. Therefore, we must eat organic material to get our nutrients.

Most ecosystems depend entirely on photosynthesis for food. For this reason, biologists refer to plants and other autotrophs as **producers**. Heterotrophs, in contrast, are **consumers**, because they obtain their food by eating plants or by eating animals that have eaten plants (**Figure 6.1**). We animals and other heterotrophs depend on autotrophs for organic fuel and for the raw organic materials we need to build our cells and tissues. ✔

Chemical Cycling between Photosynthesis and Cellular Respiration

The chemical ingredients for photosynthesis are carbon dioxide (CO_2), a gas that passes from the air into a plant via tiny pores, and water (H_2O), which is absorbed from the soil by the plant's roots (**Figure 6.2**). Chloroplasts in the cells of the leaves use light energy to rearrange the atoms of these ingredients to produce sugars—most importantly glucose ($C_6H_{12}O_6$)—and

▼ Figure 6.1 **Producer and consumer.** A parrot (consumer) eats a fruit produced by a photosynthetic plant (producer).

other organic molecules. A by-product of photosynthesis is oxygen gas (O_2).

Both animals and plants use the organic products of photosynthesis as sources of energy. A chemical process called cellular respiration harvests energy that is stored in sugars and other organic molecules. Cellular respiration uses O_2 to help convert energy extracted from organic fuel to another source of chemical energy, called ATP. Cells expend ATP for almost all their work. In both plants and animals, the production of ATP during cellular respiration occurs mainly in the organelles called mitochondria (see Figure 4.20).

Notice in Figure 6.2 that the waste products of cellular respiration are

CO_2 and H_2O—the very same ingredients used for photosynthesis. Plants store chemical energy via photosynthesis and then harvest this energy via cellular respiration. However, plants usually make more organic molecules than they need for fuel. This photosynthetic surplus provides the organic material for the plant to grow. It is also the source of food for humans and other consumers. (Note that plants perform *both* photosynthesis to produce fuel molecules *and* cellular respiration to burn them, while animals perform *only* cellular respiration.) Analyze nearly any food chain, and you can trace the energy and raw materials for growth back to solar-powered photosynthesis. ☑

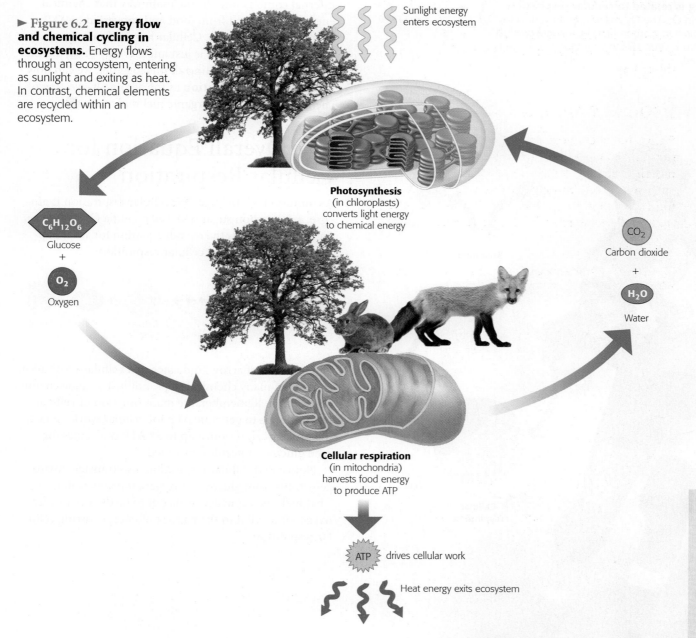

► Figure 6.2 **Energy flow and chemical cycling in ecosystems.** Energy flows through an ecosystem, entering as sunlight and exiting as heat. In contrast, chemical elements are recycled within an ecosystem.

Sunlight energy enters ecosystem

Photosynthesis (in chloroplasts) converts light energy to chemical energy

$C_6H_{12}O_6$
Glucose
+
O_2
Oxygen

CO_2
Carbon dioxide
+
H_2O
Water

Cellular respiration (in mitochondria) harvests food energy to produce ATP

ATP drives cellular work

Heat energy exits ecosystem

BioFlix™
Cellular Respiration

MP3 tutor sessions

Cellular Respiration Part 1: Glycolysis

Cellular Respiration Part 2: Citric Acid
Cycle and Electron Transport Chain

Cellular Respiration:
Aerobic Harvest of Food Energy

We sometimes use the word *respiration* to mean breath-ing. While respiration on the organismal level should not be confused with cellular respiration, the two processes are closely related (**Figure 6.3**). Cellular res-piration requires a cell to exchange two gases with its surroundings. The cell takes in oxygen in the form of the gas O_2. It gets rid of waste in the form of the gas carbon dioxide, or CO_2. Breathing results in the exchange of these same gases between your blood and the outside air. Oxygen present in the air you inhale diffuses across the lining of your lungs and into your bloodstream. And the CO_2 in your bloodstream diffuses into your lungs and exits when you exhale.

Internal combustion engines, like the ones found in cars, use O_2 (via the air intakes) to break down gasoline. A cell also requires O_2 to break down its fuel (see Figure 5.2). Cellular respiration—a living version of internal combustion—is the main way that chemical energy is harvested from food and converted to ATP energy (see Figure 5.6). Cellular respiration is an **aerobic** process, which is just another way of saying that it requires oxygen. Putting all this together, we can now define **cellular respiration** as the aerobic harvesting of chemical energy from organic fuel molecules. ✓

The Overall Equation for Cellular Respiration

A common fuel molecule for cellular respiration is glu-cose, a six-carbon sugar with the formula $C_6H_{12}O_6$ (see Figure 3.6). Here is the overall equation for what hap-pens to glucose during cellular respiration:

$$C_6H_{12}O_6 + 6\ O_2 \rightarrow\rightarrow\rightarrow 6\ CO_2 + 6\ H_2O + ATP$$

Glucose Oxygen Carbon dioxide Water Energy

The series of arrows indicates that cellular respiration consists of many chemical steps, not just a single chemi-cal reaction. Remember, the main function of cellular respiration is to generate ATP for cellular work. In fact, the process can produce up to 38 ATP molecules for each glucose molecule consumed.

Notice that cellular respiration also transfers hydro-gen atoms from glucose to oxygen, forming water. That hydrogen transfer turns out to be the key to why oxygen is so vital to the harvest of energy during cellu-lar respiration.

▼ Figure 6.3 **How breathing is related to cellular respiration.**
When you inhale, you breathe in O_2. The O_2 is delivered to your cells, where it is used in cellular respiration. Carbon dioxide, a waste product of cellular respiration, diffuses from your cells to your blood and travels to your lungs, where it is exhaled.

O₂   CO₂

Breathing

Lungs

O₂   CO₂

Muscle cells

Cellular respiration

✓CHECKPOINT

At both the organismal and cellular levels, respiration involves taking in the gas _____ and expelling the gas _____.

Answer: O_2; CO_2

The Role of Oxygen in Cellular Respiration

In tracking the transfer of hydrogen from sugar to oxygen, we are also following the transfer of electrons. The atoms of sugar and other molecules are bonded together by shared electrons (see Figure 2.7). During cellular respiration, hydrogen and its bonding electrons change partners from sugar to oxygen, forming water as a product.

Redox Reactions

Chemical reactions that transfer electrons from one substance to another substance are called oxidation-reduction reactions, or **redox reactions** for short. The loss of electrons during a redox reaction is called **oxidation**. Glucose is oxidized during cellular respiration, losing electrons to oxygen. The acceptance of electrons during a redox reaction is called **reduction**. (Note that *adding* electrons is called *reduction*; negatively charged electrons added to an atom *reduce* the amount of positive charge of that atom.) Oxygen is reduced during cellular respiration, accepting electrons (and hydrogen) lost from glucose:

Oxidation
Glucose loses electrons (and hydrogens)

$C_6H_{12}O_6$ + 6 O_2 → → → 6 CO_2 + 6 H_2O

Glucose — Oxygen — Carbon dioxide — Water

Reduction
Oxygen gains electrons (and hydrogens)

Water isn't the only thing produced when hydrogen and its bonding electrons change partners from glucose to oxygen; energy is released, too.

Why does electron transfer to oxygen release energy? In redox reactions, oxygen is an "electron grabber." An oxygen atom attracts electrons more strongly than almost any other type of atom. When electrons move (along with hydrogen) from glucose to oxygen, it is as though they are falling. They are not really falling in the sense of an apple dropping from a tree. However, in both cases, potential energy is unlocked. Instead of gravity, it is the attraction of electrons to oxygen that causes the "fall" and energy release during cellular respiration.

Figure 6.4 shows a simple example of a redox reaction: the reaction between hydrogen gas (H_2) and oxygen gas (O_2), producing water (H_2O). The reaction releases a large amount of energy as the electrons of the hydrogen "fall" into their new bonds with oxygen. This reaction is actually quite explosive, and it would be difficult for a cell to capture such a burst of energy and put it to useful work. As you'll see shortly, cellular respiration is a more controlled "fall" of electrons—more like a stepwise cascade of electrons down an energy staircase. Instead of liberating food energy in a burst, cellular respiration unlocks chemical energy in smaller amounts that cells can put to productive use. ☑

NADH and Electron Transport Chains

Let's take a closer look at the path that electrons take on their way from glucose to oxygen (**Figure 6.5**). The first stop is a positively charged electron acceptor called NAD^+ (<u>n</u>icotinamide <u>a</u>denine <u>d</u>inucleotide). The transfer of electrons from organic fuel (food) to NAD^+ reduces the NAD^+ to **NADH** (the H represents the transfer of hydrogen along with the electrons). In our staircase analogy, the electrons have now taken one baby step down in their trip from glucose to oxygen. The rest of the staircase consists of an **electron transport chain**.

▼ **Figure 6.4 A simple redox reaction.** An all-at-once redox reaction, such as the reaction of hydrogen and oxygen to form water, releases a burst of energy.

H_2 $\frac{1}{2}$ O_2

Release of heat energy

H_2O

▼ **Figure 6.5 The role of oxygen in harvesting food energy.** In cellular respiration, electrons (e^-) "fall" in small steps from food to oxygen, producing water. NADH transfers electrons from food to an electron transport chain. Oxygen "pulls" the electrons down the chain.

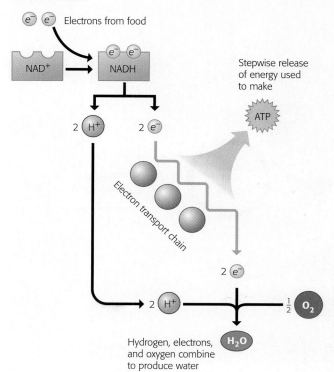

e^- e^- Electrons from food

NAD^+ → e^- e^- NADH

Stepwise release of energy used to make

ATP

2 H^+ 2 e^-

Electron transport chain

2 e^-

2 H^+ $\frac{1}{2}$ O_2

H_2O

Hydrogen, electrons, and oxygen combine to produce water

Each link in an electron transport chain is actually a molecule, usually a protein. In a series of redox reactions, each member of the chain first accepts and then donates electrons. With each transfer, the electrons give up a small amount of energy that can then be used indirectly to generate ATP. The first molecule of the chain accepts electrons from NADH. Thus, NADH carries electrons from glucose and other fuel molecules and deposits them at the top of an electron transport chain. The electrons cascade down the chain, from molecule to molecule, like an electron bucket brigade. The molecule at the bottom of the chain finally "drops" the electrons to oxygen. The oxygen also picks up hydrogen, forming water.

The overall effect of all this transfer of electrons during cellular respiration is a "downward" trip for electrons from glucose to NADH to an electron transport chain to oxygen. During the stepwise release of chemical energy during electron transport, our cells make most of their ATP. It is actually oxygen, the "electron grabber," that makes it all possible. By pulling electrons down the

transport chain from fuel molecules, oxygen functions somewhat like gravity pulling objects downhill. This is how the oxygen we breathe functions in our cells and why we cannot survive more than a few minutes without it. Viewed this way, drowning is deadly because it deprives cells of the final "electron grabbers" (oxygen) needed to drive cellular respiration.

An Overview of Cellular Respiration

Cellular respiration is an example of a metabolic pathway. That means that it is not a single chemical reaction, but a series of reactions. A specific enzyme catalyzes each reaction in a metabolic pathway. More than two dozen reactions are involved in cellular respiration. We can group them into three main metabolic stages: glycolysis, the citric acid cycle, and electron transport (which you've already encountered).

Figure 6.6 is a map that will help you follow the three stages of respiration and see where each stage occurs in your cells. During **glycolysis**, a molecule of glucose is split into two molecules of a compound called pyruvic acid. The enzymes for glycolysis are located in the cytoplasm. The **citric acid cycle** (also called the Krebs cycle) completes the breakdown of glucose all the way to CO_2, one of the waste products of cellular respiration. The enzymes for the citric acid cycle are dissolved in the fluid within mitochondria. Glycolysis and the citric acid cycle generate a small amount of ATP directly. They generate much more ATP indirectly, via redox reactions that transfer electrons from fuel molecules to NAD^+, forming NADH. The third stage of cellular respiration is **electron transport**. Electrons captured from food by the NADH formed in the first two stages "fall" down electron transport chains to oxygen. The proteins and other molecules that make up electron transport chains are embedded within the inner membrane of the mitochondria. Electron transport from NADH to oxygen releases the energy your cells use to make most of their ATP. ☑

☑CHECKPOINT

Which stages of cellular respiration take place in the mitochondria?

Answer: the citric acid cycle and electron transport

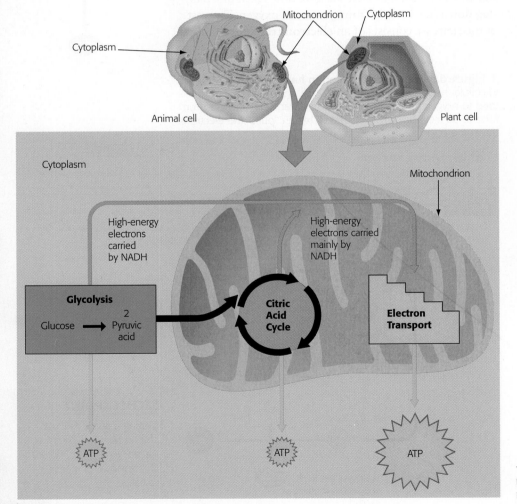

▲ Figure 6.6 **A road map for cellular respiration.**

The Three Stages of Cellular Respiration

Now that you have a big-picture view of cellular respiration, let's examine the process in more detail. A small version of Figure 6.6 will help you keep the overall process of cellular respiration in plain view as we take a closer look at its three stages.

Stage 1: Glycolysis

The word *glycolysis* means "splitting of sugar." That is exactly what happens (**Figure 6.7**). **1** During glycolysis, a six-carbon glucose molecule is broken in half, forming two three-carbon molecules. Notice in Figure 6.7 that the initial split requires an energy investment of two ATP molecules per glucose. **2** The three-carbon molecules then donate high-energy electrons to NAD$^+$, the electron carrier, forming NADH. **3** In addition to NADH, glycolysis also makes four ATP molecules directly when enzymes transfer phosphate

groups from fuel molecules to ADP (**Figure 6.8**). Glycolysis thus produces a net of two molecules of ATP per molecule of glucose. (This fact will become important during our discussion of fermentation.) What remains of the fractured glucose at the end of glycolysis are two molecules of pyruvic acid. The pyruvic acid still holds most of the energy of glucose, and that energy is harvested in the second stage of cellular respiration, the citric acid cycle.

▼ **Figure 6.8 ATP synthesis by direct phosphate transfer.** Glycolysis generates ATP when enzymes transfer phosphate groups directly from fuel molecules to ADP.

▼ **Figure 6.7 Glycolysis.** In glycolysis, a team of enzymes splits glucose, eventually forming two molecules of pyruvic acid. After investing 2 ATP at the start, glycolysis generates 4 ATP directly. More energy will be harvested later from high-energy electrons used to form NADH.

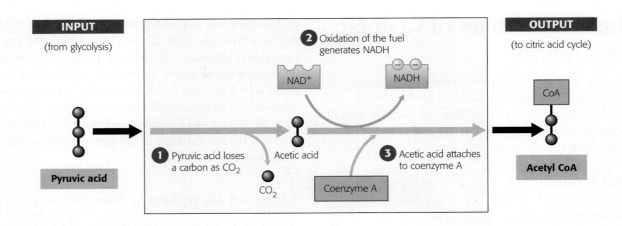

▶ **Figure 6.9 The link between glycolysis and the citric acid cycle: the conversion of pyruvic acid to acetyl CoA.** Remember that one molecule of glucose is split into two molecules of pyruvic acid. Therefore, the process shown here occurs twice.

Stage 2: The Citric Acid Cycle

Pyruvic acid, the fuel that remains after glycolysis, is not quite ready for the citric acid cycle. (Remember that each molecule of glucose produces two molecules of pyruvic acid.) The pyruvic acid must be "prepped"—converted to a form the citric acid cycle can use (**Figure 6.9**). **1** First, pyruvic acid loses a carbon as CO_2. This is the first of this waste product we've seen so far in the breakdown of glucose. The remaining fuel molecules, each with only two carbons left, are called acetic acid (the acid that's in vinegar). **2** Oxidation of the fuel generates NADH. **3** Finally, the acetic acid is attached to a molecule called coenzyme A (CoA) to form acetyl CoA. The CoA escorts the acetic acid into the first reaction of the citric acid cycle. The CoA is then stripped and recycled.

The citric acid cycle finishes extracting the energy of sugar by breaking the acetic acid molecules all the way down to CO_2 (**Figure 6.10**). **1** Acetic acid joins a four-carbon acceptor molecule to form a six-carbon product called citric acid (for which the cycle is named). For every acetic acid molecule that enters the cycle as fuel, **2** two CO_2 molecules eventually exit as a waste product. Along the way, the citric acid cycle harvests energy from the fuel. **3** Some of the energy is used to produce ATP directly. However, the cycle captures much more energy in the form of **4** NADH and **5** a second, closely related electron carrier, $FADH_2$. **6** All the carbon atoms that entered the cycle as fuel are accounted for as CO_2 exhaust, and the four-carbon acceptor molecule is recycled. We have tracked only one acetic acid molecule through the citric acid cycle here. But since glycolysis splits glucose in two, the citric acid cycle actually turns twice for each glucose molecule that fuels a cell. ✔

▶ **Figure 6.10 The citric acid cycle.**

Stage 3: Electron Transport

The molecules of electron transport chains are built into the inner membranes of mitochondria (see Figure 4.18). An electron transport chain functions as a chemical machine that uses the energy released by the "fall" of electrons to pump hydrogen ions (H^+) across the inner mitochondrial membrane. This pumping causes ions to become more concentrated on one side of the membrane than on the other. Such a concentration gradient stores potential energy.

The energy stored by electron transport behaves something like the elevated reservoir of water behind a dam. There is a tendency for hydrogen ions to gush back to where they are less concentrated, just as there is a tendency for water to flow downhill. The inner membrane, analogous to the dam, temporarily restrains hydrogen ions.

The energy of dammed water can be harnessed to perform work. Gates in a dam allow the water to rush downhill, turning giant turbines as it goes. The spinning turbines perform work that can be used to generate electricity. Your mitochondria have structures that act like turbines. Each of these miniature machines, called an **ATP synthase**, is constructed from protein built into

the inner mitochondrial membrane, adjacent to the proteins of the electron transport chains. **Figure 6.11** shows a simplified view of how the energy previously stored in NADH and $FADH_2$ can now be used to generate ATP. ① NADH and ② $FADH_2$ transfer electrons to an electron transport chain. ③ The electron transport chain uses this energy supply to pump H^+ across the inner mitochondrial membrane. ④ Oxygen pulls electrons down the transport chain. ⑤ The H^+ concentrated on one side of the membrane rushes back "downhill" through an ATP synthase. This action spins a component of the ATP synthase, just as water turns the turbines in a dam. ⑥ The rotation activates parts of the synthase molecule that attach phosphate groups to ADP molecules to generate ATP.

The poison cyanide produces its deadly effect by binding to one of the protein complexes in the electron transport chain (marked by ☠ in Figure 6.11). When bound there, cyanide blocks the passage of electrons to oxygen. This blockage is like turning off a faucet; electrons cease to flow through the "pipe." As a result, no H^+ gradient is generated, and no ATP is made. Cells stop working, and the organism dies. Cyanide was the lethal agent in an infamous case of product tampering: the Tylenol murders of 1982. Seven people in the Chicago area died after ingesting Tylenol capsules that had been laced with cyanide. The perpetrator of that crime was never caught. ✔

✔CHECKPOINT

What is the potential energy source that drives ATP production by ATP synthase?

Answer: a concentration gradient of H^+ across the inner membrane of a mitochondrion

▼ Figure 6.11 **How electron transport drives ATP synthase machines.**

The Versatility of Cellular Respiration

We have seen that glucose can provide the energy to make the ATP our cells use for all their work. The importance of glucose is underscored by the severity of diseases in which glucose balance is disturbed. Diabetes, which affects over 20 million Americans, is caused by an inability to properly regulate glucose levels in the blood due to problems with the hormone insulin. If left untreated, a glucose imbalance can lead to cardiovascular disease, coma, and even death.

But even though we have concentrated on glucose as the fuel that is broken down during cellular respiration, respiration is a versatile metabolic furnace that can "burn" many other kinds of food molecules. **Figure 6.12** diagrams some metabolic routes for the use of carbohydrates, fats, and proteins as fuel for cellular respiration.

Adding Up the ATP from Cellular Respiration

Taking cellular respiration apart to see how all the molecular nuts and bolts of its metabolic machinery work, it's easy to lose sight of its overall function: generating up to 38 molecules of ATP per molecule of glucose (the actual number can vary by a few). **Figure 6.13** will help you add up the ATP molecules. Glycolysis and the citric acid cycle each contribute 2 ATP by direct synthesis. All of the rest of the ATP molecules are produced by the ATP synthase machines, powered by the "fall" of elec-

trons from food to oxygen. The electrons are carried from the organic fuel to electron transport chains by NADH and $FADH_2$. Each electron pair "dropped" down a transport chain from NADH can power the synthesis of up to 3 ATP. Each electron pair transferred to an electron transport chain from $FADH_2$ is worth up to 2 ATP. Next, we'll see what happens when cells harvest food energy without the help of oxygen. ☑

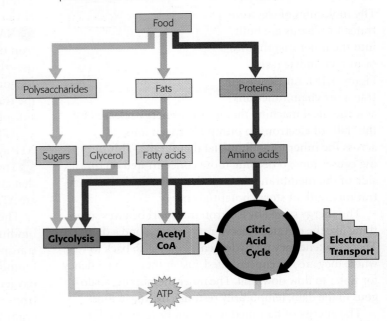

▼ **Figure 6.12 Energy from food.** The monomers from carbohydrates (polysaccharides and sugars), fats, and proteins can all serve as fuel for cellular respiration.

▼ **Figure 6.13 A summary of ATP yield during cellular respiration.**

☑**CHECKPOINT**

Which stage of cellular respiration produces the majority of ATP?

Answer: electron transport

Fermentation: Anaerobic Harvest of Food Energy

Although you must breathe to stay alive, some of your cells can work for short periods without oxygen. This **anaerobic** ("without oxygen") harvest of food energy is called **fermentation**.

Fermentation in Human Muscle Cells

When you walk between classes, your leg muscles require a constant supply of ATP, which is generated by cellular respiration. To keep this process going, blood provides your muscle cells with enough O_2 to keep electrons "falling" down transport chains in your mitochondria. But if you start to run because you're late for class, your muscles are forced to work under anaerobic conditions. That's because they are spending ATP at a rate that outpaces your bloodstream's delivery of O_2 from your lungs to your muscles.

After functioning anaerobically for about 15 seconds, muscle cells will begin to generate ATP by the process of fermentation. Fermentation relies on glycolysis, the same metabolic pathway that functions as the first stage of cellular respiration. Glycolysis does not require O_2 but does produce 2 ATP molecules for each glucose molecule broken down to pyruvic acid. That isn't very efficient compared with the 38 or so ATP molecules each glucose generates during cellular respiration, but it can energize your leg muscles long enough for you to make it to class. However, your cells will have to consume more glucose fuel per second, since so much less ATP per glucose molecule is generated under anaerobic conditions.

To harvest food energy during glycolysis, NAD^+ must be present as an electron acceptor (see Figure 6.7). This is no problem under aerobic conditions, because the cell regenerates NAD^+ when NADH drops its electron cargo down electron transport chains to O_2 (see Figure 6.5). However, this recycling of NAD^+ cannot occur under anaerobic conditions because there is no O_2 to accept the electrons. Instead, NADH disposes of electrons by adding them to the pyruvic acid produced by glycolysis (**Figure 6.14**). This restores NAD^+ and keeps glycolysis working.

The addition of electrons to pyruvic acid produces a waste product called lactic acid. The lactic acid by-product is eventually transported to the liver, where liver cells convert it back to pyruvic acid. Exercise physiologists (biologists who study the effects of exercise on the body) have long speculated about the role that lactic acid plays in muscle fatigue, as you'll see next. ✔

✔ CHECKPOINT

How many molecules of ATP can be produced from one molecule of glucose during fermentation?

Answer: 2

▼ Figure 6.14 **Fermentation: producing lactic acid.** Glycolysis produces ATP even in the absence of O_2. This process requires a continuous supply of NAD^+ to accept electrons from glucose. The NAD^+ is regenerated when NADH transfers the electrons it removed from food to pyruvic acid, thereby producing lactic acid (or other waste products, depending on the species of organism).

INPUT

2 ADP + 2 P   2 ATP

Glycolysis

2 NAD⁺   2 NADH   2 NADH   2 NAD⁺

Glucose   2 Pyruvic acid + 2 H⁺

OUTPUT

2 Lactic acid

Does Lactic Acid Buildup Cause Muscle Burn?

You may have heard that the burn you feel after hard exercise ("Feel the burn! It's a *good* burn!") is due to the buildup of lactic acid in your muscles. This idea originated with the work of a British biologist named A.V. Hill. Considered one of the founders of the field of exercise physiology, Hill won a 1922 Nobel Prize for his work on understanding muscle contraction.

In 1929, Hill performed a classic experiment that began with the **observation** that muscles produce lactic acid under anaerobic conditions. Hill asked the **question**, Does the buildup of lactic acid cause muscle fatigue? To find out, Hill developed a technique for electrically stimulating dissected frog muscles in a laboratory solution. He formed the **hypothesis** that a buildup of lactic acid would cause muscle activity to stop.

Hill's **experiment** tested frog muscles under two different sets of conditions (**Figure 6.15**). First, he showed that muscle performance declined when lactic acid could not diffuse away from the muscle tissue. Next, he showed that when lactic acid was allowed to diffuse away, performance improved significantly. These **results** led Hill to the conclusion that "lactic acid accumulation is the primary cause of failure" in muscle tissue.

Given his scientific stature (he was considered the world's leading authority on muscle activity), Hill's conclusion went unchallenged for many decades. Gradually, however, evidence began to accumulate that contradicted Hill's results. For example, the effect that Hill demonstrated did not appear to occur at human body temperature. And certain individuals who are unable to accumulate lactic acid have muscles that fatigue *more* rapidly, which is the opposite of what you would expect. Recent experiments have directly refuted Hill's conclusions. Today, the role of lactic acid in muscle fatigue remains a hotly debated topic.

The changing view of lactic acid's role in muscle fatigue illustrates an important point about the process of science: It is dynamic and subject to constant adjustment as new evidence is uncovered. This would not have surprised Hill, who himself wrote that the "built-in obsolescence" of scientific hypotheses is a necessary feature for the advancement of science.

▼ **Figure 6.15 A. V. Hill's apparatus for measuring muscle fatigue.**

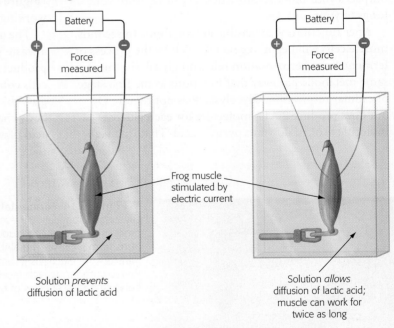

Battery

Force measured

Frog muscle stimulated by electric current

Solution *prevents* diffusion of lactic acid

Battery

Force measured

Solution *allows* diffusion of lactic acid; muscle can work for twice as long

Fermentation in Microorganisms

Our muscles cannot function by lactic acid fermentation for very long. However, the two ATP molecules produced per glucose molecule during fermentation is enough to sustain many microorganisms. We have domesticated such microbes to transform milk into cheese, sour cream, and yogurt. These foods owe their sharp or sour flavor mainly to lactic acid. The food industry also uses fermentation to produce soy sauce from soybeans, to pickle cucumbers, olives, and cabbage, and to produce meat products like sausage, pepperoni, and salami.

Yeast, a microscopic fungus, is capable of both cellular respiration and fermentation. If you keep yeast cells in an anaerobic environment, they are forced to ferment sugars and other foods to stay alive. When yeast ferment,

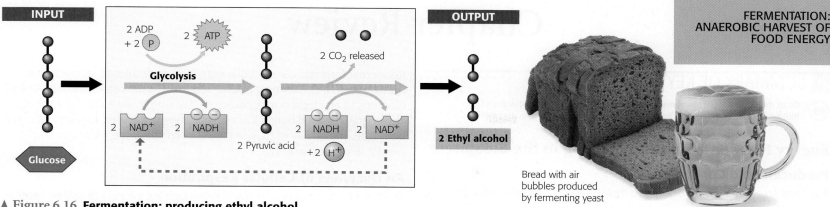

▲ Figure 6.16 **Fermentation: producing ethyl alcohol.**

Bread with air
bubbles produced
by fermenting yeast

Beer fermentation

they produce ethyl alcohol as a waste product instead of lactic acid (**Figure 6.16**). This alcoholic fermentation also releases CO_2. For thousands of years, humans have put yeast to work producing alcoholic beverages such as beer and wine. And as every baker knows, the CO_2 bubbles from fermenting yeast also cause bread dough to rise. (The alcohol produced in fermenting bread is released during baking.) ☑

☑CHECKPOINT

What kind of acid builds up in human muscle during strenuous activity?

Answer: lactic acid

EVOLUTION CONNECTION: Life with and without Oxygen

Life before and after Oxygen

In the Biology and Society and Process of Science sections, we have seen that living cells can generate ATP either aerobically (with oxygen via cellular respiration) or anaerobically (without oxygen via fermentation). Both of these processes start with glycolysis, the splitting of glucose to form pyruvic acid. Glycolysis is thus the universal energy-harvesting process of life.

The role of glycolysis in both respiration and fermentation has an evolutionary basis. Ancient prokaryotes probably used glycolysis to make ATP long before oxygen was present in Earth's atmosphere. The oldest known fossils of bacteria date back more than 3.5 billion years, but significant levels of O_2 did not accumulate in the atmosphere until about 2.7 billion years ago (**Figure 6.17**). For almost a billion years, prokaryotes must have generated ATP exclusively from glycolysis.

The fact that glycolysis occurs in almost all organisms suggests that it evolved very early in ancestors common to all the domains of life. The location of glycolysis within the cell also implies great antiquity; the pathway does not require any of the membrane-bounded organelles of the eukaryotic cell, which evolved more than a billion years after the prokaryotic cell. Glycolysis is a metabolic heirloom from early cells that continues to function in fermentation and as the first stage in the breakdown of organic molecules by cellular respiration.

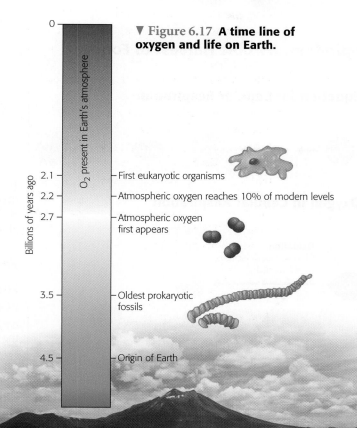

▼ Figure 6.17 **A time line of oxygen and life on Earth.**

O_2 present in Earth's atmosphere

Billions of years ago

2.1 — First eukaryotic organisms

2.2 — Atmospheric oxygen reaches 10% of modern levels

2.7 — Atmospheric oxygen first appears

3.5 — Oldest prokaryotic fossils

4.5 — Origin of Earth

Chapter Review

SUMMARY OF KEY CONCEPTS

Go to the Study Area at **www.masteringbiology.com** for practice quizzes, eText, BioFlix™ 3-D animations, MP3 Tutor Sessions, videos, current events, and more.

Energy Flow and Chemical Cycling in the Biosphere

Producers and Consumers

Autotrophs (producers) make organic molecules from inorganic nutrients via photosynthesis. Heterotrophs (consumers) must consume organic material and obtain energy via cellular respiration.

Chemical Cycling between Photosynthesis and Cellular Respiration

Cellular Respiration: Aerobic Harvest of Food Energy

The Overall Equation for Cellular Respiration

The Role of Oxygen in Cellular Respiration

Redox reactions transfer electrons from food molecules to an electron acceptor called NAD$^+$, forming NADH. The NADH then passes the high-energy electrons to an electron transport chain that eventually "drops" them to O$_2$. The energy released during this electron transport is used to regenerate ATP from ADP. The affinity of oxygen for electrons keeps the redox reactions of cellular respiration working.

An Overview of Cellular Respiration

You can follow the flow of molecules through the process of cellular respiration in the following diagram:

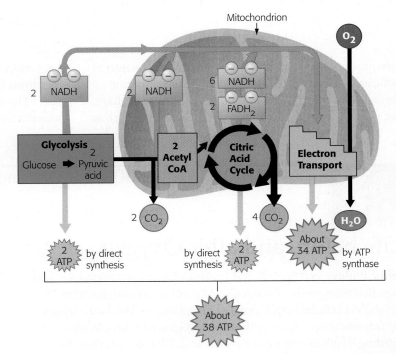

The electron transport chains pump H$^+$ across the inner mitochondrial membrane as electrons flow stepwise from NADH to oxygen. Backflow of H$^+$ across the membrane powers the ATP synthases, which attach phosphate to ADP to make ATP.

Fermentation: Anaerobic Harvest of Food Energy

Fermentation in Human Muscle Cells

When muscle cells consume ATP faster than O$_2$ can be supplied for cellular respiration, they regenerate ATP by fermentation. The waste product under these anaerobic conditions is lactic acid. The ATP yield per glucose is much lower during fermentation (2 ATP) than during cellular respiration (up to 38 ATP).

Fermentation in Microorganisms

Yeast and some other organisms can survive with or without O$_2$. Wastes from fermentation can be ethyl alcohol, lactic acid, or other compounds, depending on the species.

SELF-QUIZ

1. Which of the following statements is a correct distinction between autotrophs and heterotrophs?
   a. Only heterotrophs require chemical compounds from the environment.
   b. Cellular respiration is unique to heterotrophs.
   c. Only heterotrophs have mitochondria.
   d. Only autotrophs can live on nutrients that are entirely inorganic.

2. Why are plants called producers? Why are animals called consumers?

3. How is your breathing related to your cellular respiration?

4. Of the three stages of cellular respiration, which produces the most ATP molecules per glucose?

5. In glycolysis, _____ is oxidized and _____ is reduced.

6. The final electron acceptor of electron transport chains in mitochondria is _____.

7. The poison cyanide acts by blocking a key step in the electron transport chain. Knowing this, explain why cyanide kills so quickly.

8. Cells can harvest the most chemical energy from which of the following?
   a. an NADH molecule
   b. a glucose molecule
   c. six carbon dioxide molecules
   d. two pyruvic acid molecules

9. _____ is a metabolic pathway common to both fermentation and cellular respiration.

10. Sports physiologists at an Olympic training center wanted to monitor athletes to determine at what point their muscles were functioning anaerobically. They could do this by checking for a buildup of
    a. ADP.
    b. lactic acid.
    c. carbon dioxide.
    d. oxygen.

11. A glucose-fed yeast cell is moved from an aerobic environment to an anaerobic one. For the cell to continue to generate ATP at the same rate, approximately how much glucose must it consume in the anaerobic environment compared with the aerobic environment?

Answers to the Self-Quiz questions can be found in Appendix D.

THE PROCESS OF SCIENCE

12. Your body makes NAD^+ from two B vitamins, niacin and riboflavin. You need only tiny amounts of these vitamins. The U.S. Food and Drug Administration's recommended dietary allowances are 20 mg daily for niacin and 1.7 mg daily for riboflavin. These amounts are thousands of times less than the amount of glucose your body needs each day to fuel its energy requirements. How many NAD^+ molecules are needed for the breakdown of each glucose molecule? Why do you think your daily requirement for these substances is so small?

BIOLOGY AND SOCIETY

13. Nearly all human societies use fermentation to produce alcoholic drinks such as beer and wine. The technology dates back to the earliest civilizations. Suggest a hypothesis for how humans first discovered fermentation. In preindustrial cultures, why do you think wine was a more practical beverage than the grape juice from which it was made?

14. The consumption of alcohol by a pregnant woman can cause a series of birth defects called fetal alcohol syndrome (FAS). Symptoms of FAS include head and facial irregularities, heart defects, mental retardation, and behavioral problems. The U.S. Surgeon General's Office recommends that pregnant women abstain from drinking alcohol, and the government has mandated that a warning label be placed on liquor bottles. Imagine you are a server in a restaurant. An obviously pregnant woman orders a strawberry daiquiri. How would you respond? Is it the woman's right to make those decisions about her unborn child's health? Do you bear any responsibility in the matter? Is a restaurant responsible for monitoring the dietary habits of its customers?

7 Photosynthesis: Using Light to Make Food

Capturing solar energy.
Sunlight drives the
process of photosynthesis,
which plants use to
produce sugars.

CHAPTER CONTENTS

■■■ Chapter Thread: HARNESSING SOLAR ENERGY

BIOLOGY AND SOCIETY: Harnessing Solar Energy

Green Energy

What's old is sometimes new again. Throughout human history, people have burned wood to produce heat and light. Then, as societies became industrialized, wood was largely displaced as an energy source by fossil fuels such as coal, gas, and oil. But now, as fossil fuel supplies dwindle and pollution accumulates, scientists are researching better ways to use wood and other forms of *biomass* (living material) as efficient and renewable energy sources. Some researchers focus on burning plant matter directly, while others focus on using recently harvested wood or even plankton to produce *biofuels* (such as ethanol) that can be burned. When we derive energy from biomass or biofuels, we are actually tapping into the energy of the sun, which drives the process of photosynthesis in plants.

Certain types of fast-growing trees have shown promise as a source of wood fuel. Willows, for example, grow very quickly and can be cut every three years. The harvested wood is sent to power plants, where it is burned to generate electricity. The willows then resprout. Willows are a good source of biomass because each crop can produce about ten times as much wood per acre as natural forests. Other fast-growing tree species are also being tested as fuel sources.

Fuel from fast-growing trees is not only a renewable energy source; it can also be environmentally friendlier than fossil fuels. Wood has very little of the sulfur compounds (present in fossil fuels) that contribute to acid rain. Fast-growing trees also provide habitat for wildlife, reduce erosion, and help farmers diversify. Perhaps most significantly, vigorously growing young plants remove a lot of carbon dioxide from the air, potentially reducing the levels of atmospheric gases that cause global climate change. Today, biomass energy accounts for only about 4% of all energy consumed in the United States. But this renewable, relatively low-emissions resource may help satisfy our future energy needs.

Biomass fuels are just one way that people take advantage of plants' ability to capture solar energy. The topic of this chapter is photosynthesis, the process whereby plants use light to make sugars from carbon dioxide—sugars that are food for the plant and the starting point for most of our own food. Because photosynthesis can seem complex, we'll examine some basic concepts before looking at the specific mechanisms involved.

BioFlix™

Photosynthesis

MP3 tutor sessions

Photosynthesis

The Basics of Photosynthesis

Photosynthesis is a process whereby plants, algae (which are protists), and certain bacteria transform light energy into chemical energy, using carbon dioxide and water as starting materials. The chemical energy produced via photosynthesis is stored in the bonds of sugar molecules. As you read in Chapter 6, organisms that generate their own organic matter from inorganic ingredients are called autotrophs. Plants and other organisms that do this by photosynthesis—photosynthetic autotrophs—are the producers for most ecosystems (**Figure 7.1**). This section presents an overview of photosynthesis, focusing on plants. Later, we'll take a closer look at some details of this process.

Chloroplasts: Sites of Photosynthesis

In Chapter 4, you learned that photosynthesis in plants and algae occurs within light-absorbing organelles called **chloroplasts**. All green parts of a plant have chloroplasts and can carry out photosynthesis. In most plants, however, the leaves have the most chloroplasts (about 500,000 per square millimeter of leaf surface) and are therefore the major locations of photosynthesis. Their green color is from **chlorophyll**, a light-absorbing pigment in the chloroplasts that plays a central role in converting solar energy to chemical energy.

Chloroplasts are concentrated in the interior cells of leaves (**Figure 7.2**). Carbon dioxide (CO_2) enters, and oxygen (O_2) exits, by way of tiny pores called **stomata** (singular, *stoma*, meaning "mouth"). In addition to carbon dioxide, photosynthesis requires water, which is absorbed by the plant's roots and transported to the leaves, where veins carry it to the photosynthetic cells.

Membranes within the chloroplast form the framework where many of the reactions of photosynthesis occur. Like a mitochondrion, a chloroplast has a double-membrane envelope (see right side of Figure 7.2). The chloroplast's inner membrane encloses a compartment filled with **stroma**, a thick fluid. Suspended in the stroma are interconnected membranous sacs called **thylakoids**. The thylakoids are concentrated in stacks called **grana** (singular, *granum*). The chlorophyll molecules that capture light energy are built into the thylakoid membranes. The structure of a chloroplast—with its stacks of disks— aids its function by providing a large surface area for the reactions of photosynthesis. ☑

▼ Figure 7.1 **Photosynthetic autotrophs: producers for most ecosystems.**

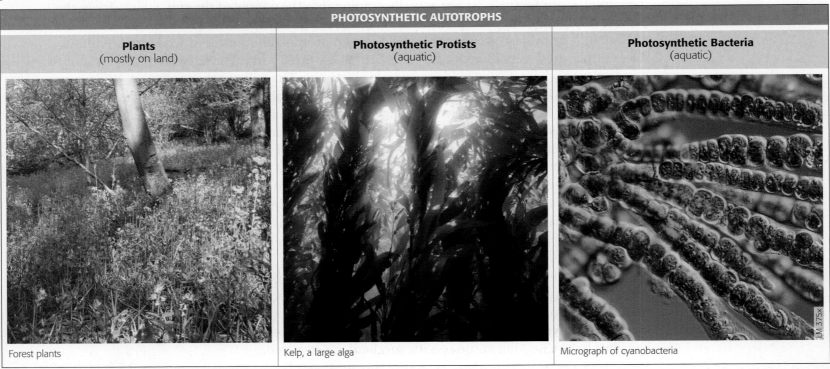

PHOTOSYNTHETIC AUTOTROPHS		
Plants (mostly on land)	**Photosynthetic Protists** (aquatic)	**Photosynthetic Bacteria** (aquatic)
Forest plants	Kelp, a large alga	Micrograph of cyanobacteria

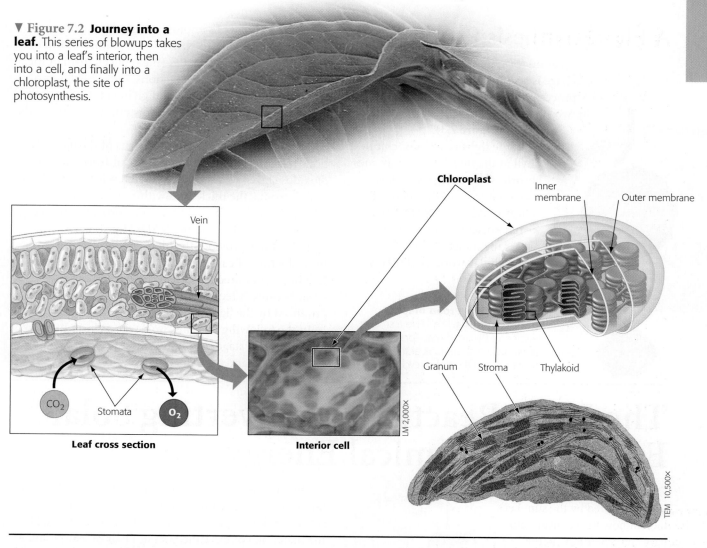

▼ Figure 7.2 **Journey into a
leaf.** This series of blowups takes
you into a leaf's interior, then
into a cell, and finally into a
chloroplast, the site of
photosynthesis.

The Overall Equation for Photosynthesis

The following chemical equation, simplified to highlight
the relationship between photosynthesis and cellular
respiration, provides a summary of the reactants and
products of photosynthesis:

Notice that the reactants of photosynthesis, carbon
dioxide (CO_2) and water (H_2O), are also the waste
products of cellular respiration (see Figure 6.2). And
photosynthesis produces what respiration uses, namely
glucose ($C_6H_{12}O_6$) and oxygen (O_2). In other words,

photosynthesis takes the "exhaust" of cellular respiration
and rearranges its atoms to produce food and oxygen. It's
a chemical transformation that requires much energy, and
sunlight absorbed by chlorophyll provides that energy.

You learned in Chapter 6 that cellular respiration is a
process of electron transfer, or reduction and oxidation
(redox). A "fall" of electrons from food molecules to
oxygen to form water releases the energy that mitochondria can use to make ATP (see Figure 6.5). The
opposite occurs in photosynthesis: Electrons are boosted
"uphill" and added to carbon dioxide to produce sugar.
Hydrogen is moved along with the electrons, so the
redox process takes the form of hydrogen transfer from
water to carbon dioxide. This transfer of hydrogen requires the chloroplast to split water molecules into hydrogen and oxygen. The hydrogen is transferred along
with electrons to carbon dioxide to form sugar. The oxygen escapes through stomata into the atmosphere as O_2,
a waste product of photosynthesis. ☑

A Photosynthesis Road Map

The equation for photosynthesis on the previous page is a simple summary of a complex process. Actually, photosynthesis occurs in two stages: the light reactions and the Calvin cycle (**Figure 7.3**).

In the **light reactions**, chlorophyll in the thylakoid membranes absorbs solar energy, which is then converted to the chemical energy of ATP and NADPH. As we have discussed previously, ATP is the molecule that drives most cellular work. **NADPH** is an electron carrier. In photosynthesis, light drives electrons from water to NADP$^+$ (the oxidized form of the carrier) to form NADPH (the reduced form of the carrier). During the light reactions, water is split, providing a source of electrons and giving off O_2 gas as a by-product. Notice that no sugar is produced during the light reactions.

The **Calvin cycle** uses the products of the light reactions to power the production of sugar from carbon dioxide. The enzymes for the Calvin cycle are dissolved in the stroma, the thick fluid within the chloroplast. ATP generated by the light reactions provides the energy for sugar synthesis. And the NADPH produced by the light reactions provides the high-energy electrons for the reduction of carbon dioxide to glucose. Thus, the Calvin cycle indirectly depends on light to produce sugar because it requires the supply of ATP and NADPH produced by the light reactions.

In the following sections, we'll take a closer look at how these two stages of photosynthesis work: the light reactions first and then the Calvin cycle. ✔

◀ **Figure 7.3 A road map for photosynthesis.** We'll use a smaller version of this road map for orientation as we take a closer look at the light reactions and the Calvin cycle.

The Light Reactions: Converting Solar Energy to Chemical Energy

Chloroplasts are chemical factories powered by the sun. Let's look at how they convert sunlight into chemical energy.

The Nature of Sunlight

Sunlight is a type of energy called radiation, or electromagnetic energy. Electromagnetic energy travels through space as rhythmic waves analogous to the ripples that are made by a pebble dropped into a pond. The distance between the crests of two adjacent waves is called a **wavelength**. The full range of radiation, from the very short wavelengths of gamma rays to the very long wavelengths of radio signals, is called the **electromagnetic spectrum** (**Figure 7.4**). Visible light is only a small fraction of the spectrum; it consists of the wavelengths that our eyes see as different colors.

When sunlight shines on a pigmented material, certain wavelengths (colors) of the visible light are absorbed and disappear from the light that is reflected by the material. For example, we see a pair of jeans as blue because pigments in the fabric absorb the other colors, leaving only light in the blue part of the spectrum to be reflected from the fabric to our eyes.

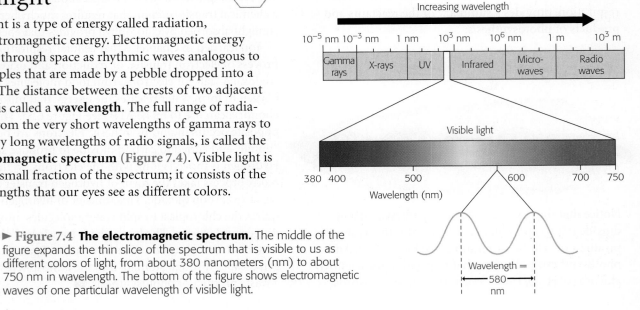

▶ **Figure 7.4 The electromagnetic spectrum.** The middle of the figure expands the thin slice of the spectrum that is visible to us as different colors of light, from about 380 nanometers (nm) to about 750 nm in wavelength. The bottom of the figure shows electromagnetic waves of one particular wavelength of visible light.

✔CHECKPOINT

Name the two stages of photosynthesis in their proper order.

Answer: light reactions, Calvin cycle

110

THE PROCESS OF SCIENCE: Harnessing Solar Energy

What Colors of Light Drive Photosynthesis?

In 1883, German biologist Theodor Engelmann made the **observation** that certain bacteria living in water tend to cluster in areas with higher oxygen concentrations. He already knew that light passed through a prism would separate into the different wavelengths (colors). Engelmann soon began to **question** whether he could use this information to determine which wavelengths of light work best for photosynthesis.

Engelmann's **hypothesis** was that oxygen-seeking bacteria would congregate near regions of algae performing the most photosynthesis (and hence producing the most oxygen). Engelmann began his **experiment** by laying a string of freshwater algal cells within a drop of water on a microscope slide. He then added oxygen-sensitive bacteria to the drop. Next, using a prism, he created a spectrum of light and shined it on the slide. His **results**, summarized in **Figure 7.5**, showed that most bacteria congregated around algae exposed to red-orange and blue-violet light, with very few bacteria moving to the area of green light. Other experiments have since verified that chloroplasts absorb light mainly in the blue-violet and red-orange part of the spectrum and that those wavelengths of light are the ones mainly responsible for photosynthesis.

▲ **Figure 7.5** **Investigating how light wavelength affects photosynthesis.** When algal cells are suspended in water on a microscope slide, oxygen-seeking bacteria migrate toward algae exposed to certain colors of light. These results suggest that blue-violet and orange-red wavelengths best drive photosynthesis, while green wavelengths hardly do so.

The selective absorption of light by leaves explains why they appear green to us; light of that color is poorly absorbed by chloroplasts and is thus reflected or transmitted toward the observer (**Figure 7.6**). Since energy cannot be destroyed, the absorbed energy must be converted to other forms. Chloroplasts contain pigments that drive the conversion of some of the solar energy they absorb to chemical energy.

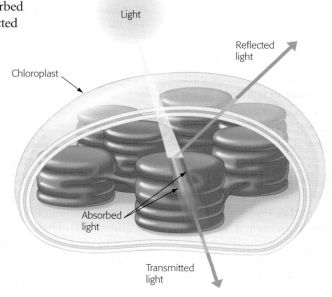

► **Figure 7.6** **Why are leaves green?** Chlorophyll and other pigments in chloroplasts reflect or transmit green light while absorbing other colors.

Chloroplast Pigments

Different pigments absorb light of different wavelengths, and chloroplasts contain several kinds of pigments. One, **chlorophyll *a***, absorbs mainly blue-violet and red light. Chlorophyll *a* is the pigment that participates directly in the light reactions. A very similar molecule, chlorophyll *b*, absorbs mainly blue and orange light. Chlorophyll *b* does not participate directly in the light reactions, but it broadens the range of light that a plant can use by conveying absorbed energy to chlorophyll *a*, which then puts the energy to work in the light reactions.

Chloroplasts also contain a family of yellow-orange pigments called carotenoids, which absorb mainly blue-green light. Some pass energy to chlorophyll *a*. Other carotenoids have a protective function: They absorb and dissipate excessive light energy that would otherwise damage chlorophyll. (Similar carotenoids, which we obtain from carrots and certain other plants, may help protect our eyes from bright light.) The spectacular colors of fall foliage in some parts of the world are due partly to the yellow-orange light reflected from carotenoids (Figure 7.7). The falling autumn temperatures cause a decrease in the levels of chlorophyll, allowing the colors of the longer-lasting carotenoids to show through.

All of these chloroplast pigments are built into the thylakoid membranes (see Figure 7.2). There the pigments are organized into light-harvesting complexes called photosystems. ☑

How Photosystems Harvest Light Energy

The theory of light as waves explains most of light's properties. However, light also behaves as discrete packets of energy called photons. A **photon** is a fixed quantity of light energy. The shorter the wavelength of light, the greater the energy of a photon. A photon of violet light, for example, packs nearly twice as much energy as a photon of red light.

When a pigment molecule absorbs a photon, one of the pigment's electrons gains energy, and we say that the electron has become "excited"; that is, the electron has been raised from a ground state to an excited state. The excited state is highly unstable, so an excited electron usually loses its excess energy and falls back to its

▼ Figure 7.7
Photosynthetic pigments.
Falling autumn temperatures cause a decrease in the levels of green chlorophyll within the leaves of deciduous trees. This decrease allows the colors of the carotenoids to be seen.

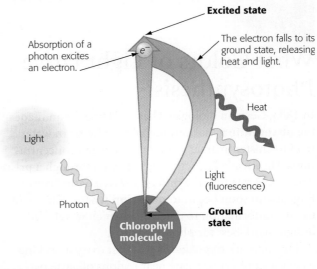

(a) Absorption of a photon

(b) Fluorescence of a glow stick. Breaking a glass vial within a glow stick starts a chemical reaction that excites electrons within a fluorescent dye. As the electrons fall from their excited state to the ground state, the excess energy is emitted as light.

ground state almost immediately (Figure 7.8a). Most pigments merely release heat energy as their light-excited electrons fall back to their ground state. (That's why a dark surface, such as a black automobile hood, gets so hot on a sunny day.) But some pigments emit light as well as heat after absorbing photons. The fluorescent light emitted by a glow stick is caused by a chemical reaction that excites electrons of a fluorescent dye (Figure 7.8b). The excited electrons quickly fall back down to their ground state, releasing energy in the form of fluorescent light.

☑CHECKPOINT

What is the specific name of the pigment that absorbs energy during the light reactions?

Answer: *chlorophyll a*

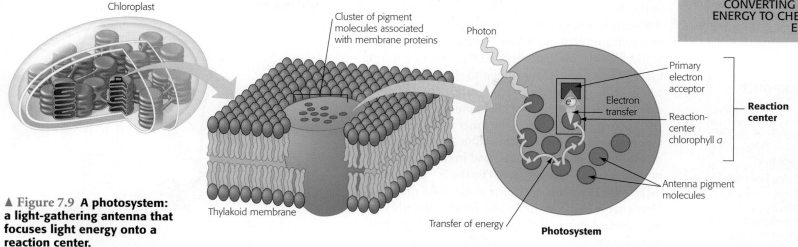

▲ Figure 7.9 **A photosystem: a light-gathering antenna that focuses light energy onto a reaction center.**

In the thylakoid membrane, chlorophyll is organized with other molecules into photosystems. Each **photosystem** has a cluster of a few hundred pigment molecules, including chlorophylls *a* and *b* and some carotenoids (**Figure 7.9**). This cluster of pigment molecules functions as a light-gathering antenna. When a photon strikes one of the pigment molecules, the energy jumps from molecule to molecule until it arrives at the reaction center of the photosystem. The reaction center consists of a chlorophyll *a* molecule that sits next to another molecule called a **primary electron acceptor**. This primary electron acceptor traps the light-excited electron (e⁻) from the chlorophyll *a* in the reaction center. Another team of molecules built into the thylakoid membrane then uses that trapped energy to make ATP and NADPH. ☑

☑**CHECKPOINT**

What is the role of a reaction center during photosynthesis?

Answer: A reaction center transfers a light-excited photon from pigment molecules to molecules that can use this trapped energy to drive chemical reactions.

How the Light Reactions Generate ATP and NADPH

Two types of photosystems cooperate in the light reactions (**Figure 7.10**). ❶ Photons excite electrons in the chlorophyll of the water-splitting photosystem. These photons are then trapped by the primary electron acceptor. The water-splitting photosystem replaces its light-excited electrons by extracting electrons from water. This is the step that releases O_2 during photosynthesis. ❷ Energized electrons from the water-splitting photosystem pass down an electron transport chain to the NADPH-producing photosystem. The chloroplast uses the energy released by this electron "fall" to make ATP. ❸ The NADPH-producing photosystem transfers its light-excited electrons to $NADP^+$, reducing it to NADPH.

▶ Figure 7.10 **The light reactions of photosynthesis.** The orange arrows trace a light-driven flow of electrons from H_2O to NADPH. These electrons also produce ATP.

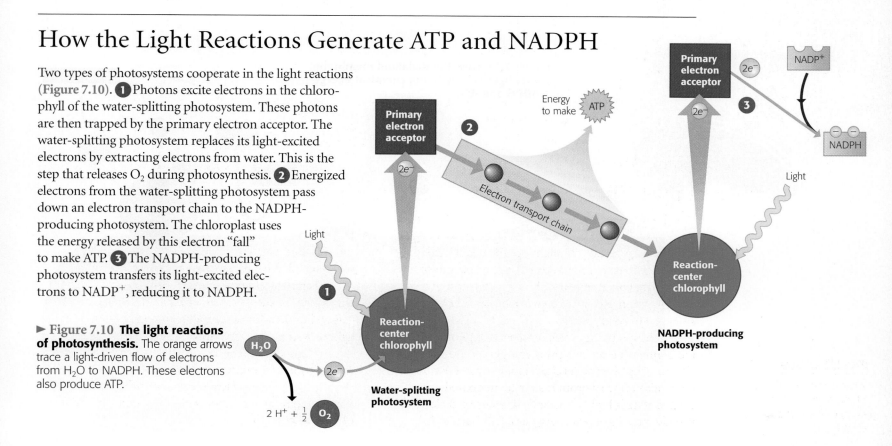

☑**CHECKPOINT**

1. Why is water required as a reactant in photosynthesis? (*Hint*: Review Figures 7.10 and 7.11.)

2. In addition to conveying electrons from the water-splitting photosystem to the NADPH-producing photosystem, the electron transport chains of chloroplasts provide the energy for the synthesis of _____.

Answers: **1.** The splitting of water provides electrons for converting NAD⁺ to NADPH (which will be used in converting CO₂ to sugar in the Calvin cycle). **2.** ATP

Figure 7.11 shows the location of the light reactions in the thylakoid membrane. The two photosystems and the electron transport chain that connects them transfer electrons from H_2O to $NADP^+$, reducing it to NADPH. Notice that the mechanism of ATP production during the light reactions is very similar to the mechanism we saw in cellular respiration (see Figure 6.11). In both cases, an electron transport chain pumps hydrogen ions (H^+) across a membrane—the inner mitochondrial membrane in the case of respiration and the thylakoid membrane in photosynthesis. And in both cases, ATP synthases use the energy stored by the H^+ gradient to make ATP. The main difference is that food provides the high-energy electrons in cellular respiration, whereas light-excited electrons flow down the transport chain during photosynthesis. The traffic of electrons shown in Figures 7.10 and 7.11 is analogous to the cartoon in **Figure 7.12**.

We have seen how the light reactions absorb solar energy and convert it to the chemical energy of ATP and NADPH. Notice again, however, that the light reactions produce no sugar. That's the job of the Calvin cycle, as we'll see next. ☑

▼ Figure 7.12 **A hard-hat analogy for the light reactions.**

Water-splitting photosystem

NADPH-producing photosystem

▼ Figure 7.11 **How the thylakoid membrane converts light energy to the chemical energy of NADPH and ATP.**

The Calvin Cycle: Making Sugar from Carbon Dioxide

The Calvin cycle functions like a sugar factory within the stroma of a chloroplast. It is called a cycle because, like the citric acid cycle in cellular respiration, the starting material is regenerated with each turn of the cycle. And with each turn, there are chemical inputs and outputs. The inputs are CO_2 from the air as well as ATP and NADPH produced by the light reactions. Using carbon from CO_2, energy from ATP, and high-energy electrons from NADPH, the Calvin cycle constructs an energy-rich sugar molecule called glyceraldehyde 3-phosphate (G3P). The plant cell can then use G3P as the raw material to make the glucose and other organic compounds (such as cellulose and starch) that it needs. **Figure 7.13** presents the basics of the Calvin cycle, emphasizing inputs and outputs. Each ● represents a carbon atom and each ⓟ represents a phosphate group. ✓

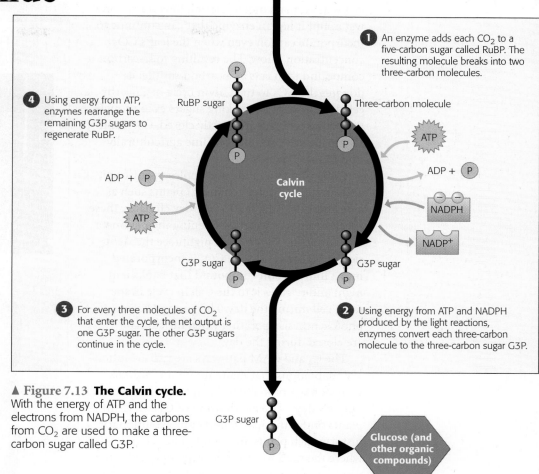

1 An enzyme adds each CO_2 to a five-carbon sugar called RuBP. The resulting molecule breaks into two three-carbon molecules.

4 Using energy from ATP, enzymes rearrange the remaining G3P sugars to regenerate RuBP.

3 For every three molecules of CO_2 that enter the cycle, the net output is one G3P sugar. The other G3P sugars continue in the cycle.

2 Using energy from ATP and NADPH produced by the light reactions, enzymes convert each three-carbon molecule to the three-carbon sugar G3P.

▲ **Figure 7.13 The Calvin cycle.** With the energy of ATP and the electrons from NADPH, the carbons from CO_2 are used to make a three-carbon sugar called G3P.

EVOLUTION CONNECTION: Harnessing Solar Energy

Solar-Driven Evolution

Throughout this chapter, you've studied how plants convert solar energy to chemical energy via photosynthesis. In the Process of Science section, you learned that plants can best use certain wavelengths of light to drive these reactions. Sunlight strikes Earth differently in different geographic regions. Since natural selection promotes the evolution of adaptations that best suit a local environment, it's not surprising that different modes of photosynthesis have evolved in different climates.

So far, we've focused on plants that use CO_2 directly from the air to drive the Calvin cycle. Such plants are called **C_3 plants** because the first organic compound produced in the Calvin cycle is a three-carbon molecule (see Figure 7.13). C_3 plants are widely distributed and include such common crops as soybeans, oats, wheat, and rice. One problem that farmers face in growing C_3 plants, however, is that plants close their stomata on hot, dry days. This adaptation reduces water loss but also prevents CO_2 from entering the leaves. As a result, CO_2 levels get very low in the leaves, and sugar production ceases. Thus, when the weather turns hot, the productivity of C_3 crops slows significantly because of the need to save water.

In hot, dry climates, alternative modes of incorporating carbon from CO_2 have evolved in some plants, allowing them to save water without shutting down photosynthesis. These species are categorized as either C_4 plants or CAM plants.

☑ **CHECKPOINT**

What is the function of NADPH in the Calvin cycle?

Answer: It provides the high-energy electrons that are added to CO_2 to form G3P sugar.

C$_4$ plants are so named because they incorporate carbon from CO$_2$ into a four-carbon compound before proceeding to the Calvin cycle (**Figure 7.14**, left side). When the weather is hot and dry, a C$_4$ plant keeps its stomata mostly closed, thus conserving water. But it has an enzyme that can continue to incorporate carbon even when the leaf's CO$_2$ concentration is low. The resulting four-carbon compound then acts as a carbon shuttle; it donates the CO$_2$ to the Calvin cycle in a nearby cell, which keeps on making sugars even though the plant's stomata are mostly closed. Corn and sugarcane are examples of some agriculturally important C$_4$ plants.

CAM plants include pineapples, many cacti, and succulent (water-retaining) plants such as aloe and jade. Adapted to very dry climates, these species conserve water by opening their stomata and admitting CO$_2$ only at night (see the right side of Figure 7.14). The CO$_2$ is incorporated into a four-carbon compound that banks it at night and releases it to the Calvin cycle in the same cell during the day. This process keeps photosynthesis operating when the stomata are closed during the day.

The C$_4$ and CAM pathways are two evolutionary adaptations that maintain photosynthesis with stomata partially or completely closed on hot, dry days. The distribution of C$_4$ and CAM plants on Earth today reflects this evolutionary history; such plants are found in hot climates where the need to conserve water outweighs the need for continuous sugar production. During the long evolution of plants in diverse environments, natural selection has refined photosynthetic adaptations that enable these plants to continue producing food even in arid conditions.

▼ **Figure 7.14 C$_4$ and CAM photosynthesis.** The C$_4$ and CAM pathways are two evolutionary adaptations that maintain photosynthesis with stomata partially or completely closed on hot, dry days.

ALTERNATIVE PHOTOSYNTHETIC PATHWAYS

C$_4$ Pathway
(example: sugarcane)

CAM Pathway
(example: pineapple)

Cell type 1 — CO$_2$ — **Four-carbon compound** — Cell type 2 — CO$_2$ — **Calvin cycle** — Sugar

C$_4$ plant

CO$_2$ — Night — **Four-carbon compound** — CO$_2$ — **Calvin cycle** — Sugar — Day

CAM plant

Carbon incorporation and the Calvin cycle occur in different types of cells.

Carbon incorporation and the Calvin cycle occur in the same cells at different times.

Chapter Review

SUMMARY OF KEY CONCEPTS

Go to the Study Area at **www.masteringbiology.com** for practice quizzes, eText, BioFlix™ 3-D animations, MP3 Tutor Sessions, videos, current events, and more.

The Basics of Photosynthesis

Photosynthesis is a process whereby light energy is transformed into chemical energy, which is stored as bonds in sugars made from carbon dioxide and water.

Chloroplasts: Sites of Photosynthesis

Chloroplasts contain a thick fluid called stroma surrounding a network of membranes called thylakoids.

The Overall Equation for Photosynthesis

A Photosynthesis Road Map

The Light Reactions: Converting Solar Energy to Chemical Energy

The Nature of Sunlight

Visible light is part of the spectrum of electromagnetic energy. It travels through space as waves.

Chloroplast Pigments

Pigment molecules absorb light energy of certain wavelengths and reflect other wavelengths. We see the reflected wavelengths as the color of the pigment. Several chloroplast pigments absorb light of various wavelengths, but it is the green pigment chlorophyll *a* that participates directly in the light reactions.

How Photosystems Harvest Light Energy; How the Light Reactions Generate ATP and NADPH

The Calvin Cycle: Making Sugar from Carbon Dioxide

Within the stroma (fluid) of the chloroplast, carbon dioxide from the air and ATP and NADPH produced during the light reactions are used to produce G3P, an energy-rich sugar molecule that can be used to make glucose and other organic molecules.

Evolution Connection: Solar-Driven Evolution

The photosynthetic adaptations of C_4 and CAM plants enable sugar production to continue even when stomata are closed, thereby reducing water loss in arid environments.

SELF-QUIZ

1. The light reactions take place in the structures of the chloroplast called the _____, while the Calvin cycle takes place in the _____.

2. In terms of the spatial organization of photosynthesis within the chloroplast, what is the advantage of the light reactions producing NADPH and ATP on the stroma side of the thylakoid membrane?

3. Which of the following are inputs to photosynthesis? Which are outputs?
   a. CO_2
   b. O_2
   c. sugar
   d. H_2O
   e. light

4. What color of light is the least effective in driving photosynthesis? Why?

5. When light strikes chlorophyll molecules, they lose electrons, which are ultimately replaced by splitting molecules of _____.

6. Which of the following are produced by reactions that take place in the thylakoids and are consumed by reactions in the stroma?
   a. CO_2 and H_2O
   b. $NADP^+$ and ADP
   c. ATP and NADPH
   d. glucose and O_2

7. The reactions of the Calvin cycle are not directly dependent on light, and yet they usually do not occur at night. Why?

8. Why is it difficult for most plants to carry out photosynthesis in very hot, dry environments, such as deserts?

9. What is the primary advantage offered by the C_4 and CAM pathways?

10. Of the following metabolic processes, which one is common to photosynthesis and cellular respiration?
    a. reactions that convert light energy to chemical energy
    b. reactions that split H_2O molecules and release O_2
    c. reactions that store energy by pumping H^+ across membranes
    d. reactions that convert CO_2 to sugar

Answers to the Self-Quiz questions can be found in Appendix D.

THE PROCESS OF SCIENCE

11. Tropical rain forests cover only about 3% of Earth's surface, but they are estimated to be responsible for more than 20% of global photosynthesis. For this reason, rain forests are often referred to as the "lungs" of the planet, providing O_2 for life all over Earth. However, most experts believe that rain forests make little or no net contribution to global O_2 production. From your knowledge of photosynthesis and cellular respiration, can you explain why they might think this? (*Hint*: What happens to the energy stored as sugars in the body of a plant when that plant dies or parts of it are eaten by animals?)

12. Suppose you wanted to discover whether the oxygen atoms in the glucose produced by photosynthesis come from H_2O or CO_2. Explain how you could use a radioactive isotope to find out.

BIOLOGY AND SOCIETY

13. There is strong evidence that Earth is getting warmer because of an intensified greenhouse effect resulting from increased CO_2 emissions from industry, vehicles, and the burning of forests. Global warming could influence agriculture, melt polar ice, and flood coastal regions. In response to these threats, 178 countries have accepted the Kyoto agreement, which calls for mandatory reductions of greenhouse gas emissions in 30 industrialized nations by 2012. As of the start of 2009, the United States had not signed the agreement, instead proposing a more modest set of voluntary goals allowing businesses to decide whether they wish to participate and providing tax incentives to encourage them to do so. The reasons given for rejecting the agreement are that it might hurt the American economy and that some less industrialized countries (such as India) are exempted from it, even though they produce a lot of pollution. Do you agree with this decision? In what ways might efforts to reduce greenhouse gases hurt the economy? How can those costs be weighed against the costs of global warming? Should poorer nations carry an equal burden to reduce their emissions?

14. As discussed in the Biology and Society section, burning biomass to produce electricity avoids many of the problems associated with gathering, refining, transporting, and burning fossil fuels. Yet the use of biomass as fuel is not without its own set of problems. What challenges might arise from a large-scale conversion to biomass energy? How do these challenges compare with those encountered with fossil fuels? Which set of challenges do you think is more likely to be overcome? Does one energy source have more benefits and fewer costs than the others? Explain.

Unit 2
Genetics

Chapter 8: **Cellular Reproduction: Cells from Cells**

Chapter Thread: **Life with and without Sex**

Chapter 9: **Patterns of Inheritance**

Chapter Thread: **Dog Breeding**

Chapter 10: **The Structure and Function of DNA**

Chapter Thread: **The Flu**

Chapter 11: **How Genes Are Controlled**

Chapter Thread: **Cancer**

Chapter 12: **DNA Technology**

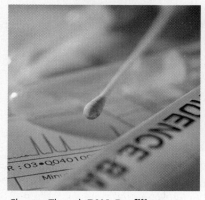

Chapter Thread: **DNA Profiling**

8 Cellular Reproduction: Cells from Cells

Reproduction in the rain forest.
Botanists attempted to save the Hawaiian bellflower (visible in foreground) by means of sexual and asexual reproduction.

CHAPTER CONTENTS

■■■ Chapter Thread: **LIFE WITH AND WITHOUT SEX**

BIOLOGY AND SOCIETY: Life with and without Sex

Rain Forest Rescue

Deep in a Hawaiian rain forest, the thrumming of a helicopter cut through the dense tropical air. After touching down in a clearing, a pair of rescuers jumped out, one of them carrying a precious bundle. A 20-minute hike through dense foliage brought them to their "patient": the wild *Cyanea kuhihewa*, a plant in the bellflower family. The rescuers were in the middle of a life-or-death battle, armed with the pollen of a species on the very brink of extinction. The stakes couldn't have been higher: The small bellflower before them was the last surviving member of its species in the wild.

The rescuers, scientists from the National Tropical Botanical Garden, were trying to promote sexual reproduction, the reproductive process that involves fertilization, the union of a sperm and an egg. Using a fine brush, the botanists transferred sperm-carrying pollen from a garden-grown plant onto the egg-containing wild bloom. Their hope was to initiate the development of a new seed, and from it, a new plant.

Unfortunately, the bellflower fertilization attempt failed. Even worse, the last remaining wild specimen died in 2003. But hope remains for *C. kuhihewa*. The garden's botanists have also propagated the plant by asexual reproduction, the production of offspring by a single parent. Many plants that normally reproduce sexually can be induced to reproduce asexually in the laboratory. Indeed, work at the botanical garden has produced several new bellflower plants in this way, using stem tissue snipped from wild plants in the 1990s.

Cell division is at the heart of organismal reproduction, whether by sexual or asexual means. The ability of organisms to reproduce their own kind is the one characteristic that best distinguishes living things from nonliving matter. And the perpetuation of life depends on the production of new cells. Plants created from cuttings result from repeated cell divisions, as does the development of a multicellular organism from a fertilized egg. And eggs and sperm themselves result from cell division of a special kind. In this chapter, we'll look at how individual cells reproduce and then see how cell reproduction underlies the process of sexual reproduction. During our discussion, we'll consider examples of asexual and sexual reproduction among both plants and animals.

What Cell Reproduction Accomplishes

When you hear the word *reproduction*, you probably think of the birth of new organisms. But reproduction actually occurs much more often at the cellular level. Consider the skin on your arm. Skin cells are constantly reproducing themselves and moving outward toward the surface, replacing dead cells that have rubbed off. This renewal of your skin goes on throughout your life. And when your skin is injured, additional cell reproduction helps heal the wound.

When a cell undergoes reproduction, or **cell division**, the two "daughter" cells that result are genetically identical to each other and to the original "parent" cell. (Biologists traditionally use the word *daughter* in this context; it does not imply gender.) Before the parent cell splits into two, it duplicates its **chromosomes**, the structures that contain most of the organism's DNA. Then, during the division process, one set of chromosomes is distributed to each daughter cell. As a rule, the daughter cells receive identical sets of chromosomes.

As summarized in **Figure 8.1**, cell division plays several important roles in the lives of organisms. For example, within your body, millions of cells must divide every second to replace damaged or lost cells. Another function of cell division is growth. All of the trillions of cells in your body are the result of repeated cell divisions that began in your mother's body with a single fertilized egg cell.

Another vital function of cell division is reproduction of the organism. Single-celled organisms, such as amoebas, reproduce by dividing in half, and the offspring are genetic replicas of the parent. Because it does not involve fertilization of an egg by a sperm, this type of reproduction is called **asexual reproduction**. Offspring produced by asexual reproduction inherit all of their chromosomes from a single parent. Many multicellular organisms can reproduce asexually as well. For example, some sea star species have the ability to grow new individuals from fragmented pieces. And if you've ever grown a houseplant from a clipping, you've observed asexual reproduction in plants. In asexual reproduction, there is one simple principle of inheritance: The lone parent and each of its offspring have identical genes. The type of cell division responsible for asexual reproduction and for the growth and maintenance of multicellular organisms is called mitosis.

Sexual reproduction is different; it requires fertilization of an egg by a sperm. The production of egg and sperm cells involves a special type of cell division called meiosis, which occurs only in reproductive organs (such as testes and ovaries in humans). As we'll discuss later, a sperm or egg cell has only half as many chromosomes as the parent cell that gave rise to it.

Note that two kinds of cell division are involved in the lives of sexually reproducing organisms: meiosis for reproduction and mitosis for growth and maintenance. The remainder of the chapter is divided into two main sections, one for each type of cell division. ✔

☑CHECKPOINT

Ordinary cell division produces two daughter cells that are genetically identical. Name three functions of this type of cell division.

Answer: cell replacement, growth of an organism, asexual reproduction of an organism

▶ Figure 8.1 **Three functions of cell division.**

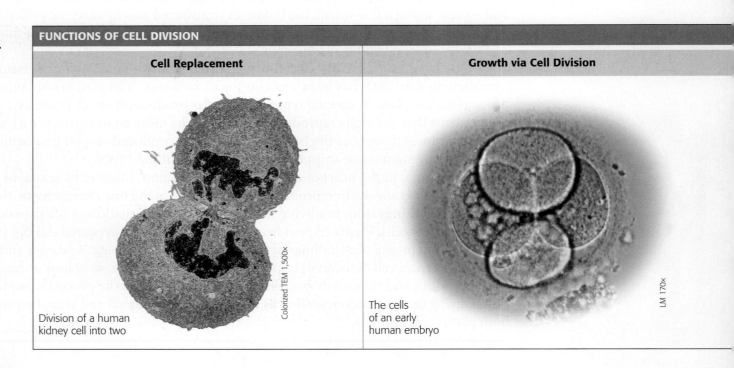

FUNCTIONS OF CELL DIVISION

Cell Replacement	Growth via Cell Division
Division of a human kidney cell into two	The cells of an early human embryo

Colorized TEM 1,500x

LM 170x

The Cell Cycle and Mitosis

Almost all of the genes of a eukaryotic cell—around 21,000 in humans—are located on chromosomes in the cell nucleus. (The main exceptions are genes on small DNA molecules found in mitochondria and chloroplasts.) Because chromosomes are the lead players in cell division, let's focus on them before turning our attention to the cell as a whole.

Eukaryotic Chromosomes

Each eukaryotic chromosome contains one very long DNA molecule, typically bearing thousands of genes. The number of chromosomes in a eukaryotic cell depends on the species (**Figure 8.2**). Chromosomes are made up of a material called **chromatin**, a combination of DNA and protein molecules. The protein molecules help organize the chromatin and help control the activity of its genes.

Most of the time, the chromosomes exist as a diffuse mass of fibers that are much longer than the nucleus they are stored in. In fact, if stretched out, the DNA in just one of your cells would be taller than you! As a cell prepares to divide, its chromatin

BioFlix ™

Mitosis

MP3
tutor sessions

Mitosis

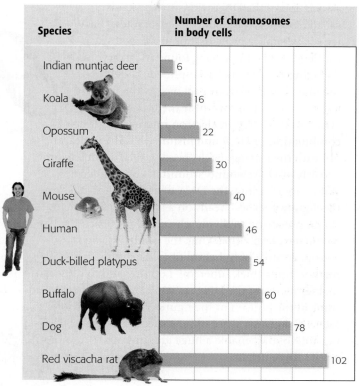

Species	Number of chromosomes in body cells
Indian muntjac deer	6
Koala	16
Opossum	22
Giraffe	30
Mouse	40
Human	46
Duck-billed platypus	54
Buffalo	60
Dog	78
Red viscacha rat	102

► Figure 8.2 **The number of chromosomes in the cells of selected mammals.** Notice that humans have 46 chromosomes and that the number of chromosomes does not correspond to the size or complexity of an organism.

Asexual Reproduction

LM 250x

Reproduction of an amoeba

Fragmentation and regeneration of a sea star. The sea star on the right lost and replaced an arm. The severed arm grew into the new sea star on the left.

Reproduction of an African violet from a clipping (large leaf)

fibers coil up, forming compact chromosomes. When they are in this state, chromosomes are clearly visible under the light microscope, as shown in the plant cell in **Figure 8.3**. When a cell is not dividing, the chromosomes are too thin to be seen in a light micrograph.

Such long molecules of DNA can fit into the tiny nucleus because within each chromosome the DNA is packed into an elaborate, multilevel system of coiling and folding. A crucial aspect of DNA packing is the association of the DNA with small proteins called **histones**, found only in eukaryotes. (Bacteria have similar proteins, but prokaryotes lack the degree of DNA packing found in eukaryotes.)

Figure 8.4 presents a simplified model for the main levels of DNA packing. At the first level of packing, shown near the top, histones attach to the DNA. In electron micrographs, the combination of DNA and histones has the appearance of beads on a string. Each "bead," called a **nucleosome**, consists of DNA wound around histone molecules. When not dividing, the DNA of active genes takes on this lightly packed, "beads on a string" arrangement. When preparing to divide, chromosomes pack further. At the next level of packing, the beaded string is wrapped into a tight helical fiber. Then this fiber coils further into a thick supercoil. Looping and folding can further compact the DNA, as you can see in the chromosome at the bottom of the figure. Viewed as a whole, Figure 8.4 gives a sense of how successive levels of coiling and folding enable a huge amount of DNA to fit into a cell's tiny nucleus.

Before a cell begins the division process, it duplicates all of its chromosomes. The DNA molecule of each chromosome is copied through the process of DNA replication (see Chapter 10), and new protein molecules

▼ Figure 8.3 **A plant cell just before division (colored by stains).**

Chromosomes

▼ Figure 8.4 **DNA packing in a eukaryotic chromosome.** Successive levels of coiling of DNA and associated proteins ultimately results in highly compacted chromosomes.

DNA double helix

Histones

"Beads on a string"

Nucleosome

Tight helical fiber

Looped domains

Duplicated chromosomes (sister chromatids)

Centromere

attach as needed. The result is that each chromosome now consists of two copies called **sister chromatids**, which contain identical genes. At the bottom of Figure 8.4, you can see an electron micrograph of a human chromosome that has duplicated. The fuzzy appearance comes from the intricate twists and folds of the chromatin fibers. The two sister chromatids are joined together tightly at a narrow "waist" called the **centromere**.

When the cell divides, the sister chromatids of a duplicated chromosome separate from each other, as shown in the simple diagram in **Figure 8.5**. Once separated from its sister, each chromatid is considered a full-fledged chromosome, and it is identical to the original chromosome. One of the new chromosomes goes to one daughter cell, and the other goes to the other daughter cell. In this way, each daughter cell receives a complete and identical set of chromosomes. A dividing human skin cell, for example, has 46 duplicated chromosomes, and each of the two daughter cells that result from it has 46 single chromosomes.

The Cell Cycle

How do chromosome duplication and cell division fit into the life of a cell? The rate at which a cell divides depends on its role within the organism's body. Some cells divide once a day, others less often, and highly specialized cells, such as mature muscle cells, not at all.

The ordered sequence of events that extends from the time a cell is first formed from a dividing parent cell until its own division into two cells is called the **cell cycle**. As **Figure 8.6** shows, most of the cell cycle is spent in **interphase**. This is a time when a cell performs its normal functions within the organism. For example, a cell in your stomach lining might make and release enzyme molecules that aid in digestion. During interphase, a cell roughly doubles everything in its cytoplasm. It increases its supply of proteins, increases the number of many of its organelles (such as mitochondria and ribosomes),

▲ **Figure 8.5 Duplication and distribution of a single chromosome.** During cell reproduction, the cell duplicates each chromosome and distributes the two copies to the daughter cells.

and grows in size. Typically, interphase lasts for at least 90% of the cell cycle.

From the standpoint of cell reproduction, the most important event of interphase is chromosome duplication, when the DNA in the nucleus is precisely doubled. This occurs approximately in the middle of interphase, and the period when it is occurring is called the S phase (for DNA *synthesis*). The interphase periods before and after the S phase are called the G_1 and G_2 phases, respectively (G stands for *gap*). During G_2, each chromosome in the cell consists of two identical sister chromatids, and the cell is preparing to divide.

The part of the cell cycle when the cell is actually dividing is called the **mitotic (M) phase**. It includes two overlapping processes, mitosis and cytokinesis. In **mitosis**, the nucleus and its contents, notably the duplicated chromosomes, divide and are evenly distributed, forming two daughter nuclei. In **cytokinesis**, the cytoplasm is divided in two. Cytokinesis usually begins before mitosis is completed. The combination of mitosis and cytokinesis produces two genetically identical daughter cells, each with a single nucleus, the surrounding cytoplasm with organelles, and a plasma membrane. ☑

☑**CHECKPOINT**

1. A duplicated chromosome consists of two sister _____ joined together at the _____.

2. The cell cycle is divided into _____, during which the cell grows and performs its normal function, and the _____ phase, during which the cell divides. The latter stage includes _____, when sister chromatids separate and two nuclei form, and _____, when the cytoplasm divides.

Answers: **1.** *chromatids; centromere* **2.** *interphase; mitotic; mitosis; cytokinesis*

▼ **Figure 8.6 The eukaryotic cell cycle.** The cell cycle extends from the "birth" of a cell (just after the dark blue arrow at the bottom of the cycle), resulting from cell reproduction, to the time the cell itself divides in two. (During interphase, the chromosomes are diffuse masses of thin fibers; they do not actually appear in the rodlike form you see here.)

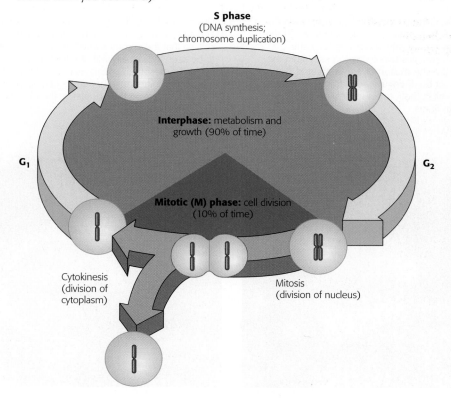

S phase
(DNA synthesis; chromosome duplication)

Interphase: metabolism and growth (90% of time)

G_1

G_2

Mitotic (M) phase: cell division (10% of time)

Cytokinesis (division of cytoplasm)

Mitosis (division of nucleus)

Mitosis and Cytokinesis

Figure 8.7 illustrates the cell cycle for an animal cell using drawings, texts, and photos. The micrographs running along the bottom of the page show dividing cells from a salamander, with chromosomes appearing in blue. The drawings in the top row (which are simplified to include just four chromosomes) include details that are not visible in the micrographs. The text within the figure describes the events occurring at each stage. As you study the figure, you'll see striking changes in the nucleus and other cellular structures.

▼ **Figure 8.7 Cell reproduction: A dance of the chromosomes.** After the chromatin doubles during interphase, the elaborately choreographed stages of mitosis—prophase, metaphase, anaphase, and telophase—distribute the duplicate sets of chromosomes to two separate nuclei. Cytokinesis then divides the cytoplasm, yielding two genetically identical daughter cells.

INTERPHASE	PROPHASE

Interphase is the period of cell growth, when the cell makes new molecules and organelles. At the point shown here, late interphase (G_2), the cytoplasm contains two centrosomes. Within the nucleus, the chromosomes are duplicated, but they cannot be distinguished individually because they are still in the form of loosely packed chromatin fibers.

During prophase, changes occur in both nucleus and cytoplasm. In the nucleus, the chromatin fibers coil, so that the chromosomes become thick enough to be seen individually with the light microscope. Each chromosome appears as two identical sister chromatids joined together at the narrow "waist" of the centromere. In the cytoplasm, the mitotic spindle

begins to form as microtubules grow out from the centrosomes, which are moving away from each other. Late in prophase, the nuclear envelope breaks up. The spindle microtubules attach to the centromeres of the chromosomes and move the chromosomes toward the center of the cell.

LM 375x

Mitosis is a continuum, but biologists distinguish four main stages: **prophase**, **metaphase**, **anaphase**, and **telophase**. The chromosomes are the stars of the mitotic drama, and their movements depend on the **mitotic spindle**, a football-shaped structure of microtubules that guides the separation of the two sets of daughter chromosomes (see Figure 4.21a to review microtubules; in the micrographs of Figure 8.7, they appear green). The spindle microtubules grow from two **centrosomes**, clouds of cytoplasmic material that in animal cells contain centrioles. (Centrioles are can-shaped structures made of microtubules; see Figure 4.5.)

METAPHASE	ANAPHASE	TELOPHASE AND CYTOKINESIS

Spindle

Daughter chromosomes

Nuclear envelope forming

Cleavage furrow

The mitotic spindle is now fully formed. The centromeres of all the chromosomes line up on an imaginary plate between the two poles of the spindle. For each chromosome, the spindle microtubules attached to the two sister chromatids pull toward opposite poles. This tug of war keeps the chromosomes in the middle of the cell.

Anaphase begins suddenly when the sister chromatids of each chromosome separate. Each is now considered a full-fledged (daughter) chromosome. The chromosomes move toward opposite poles of the cell as the spindle microtubules shorten. Simultaneously, the microtubules not attached to chromosomes lengthen, pushing the poles farther apart and elongating the cell.

Telophase begins when the two groups of chromosomes have reached opposite ends of the cell. Telophase is the reverse of prophase: Nuclear envelopes form, the chromosomes uncoil, and the spindle disappears. Mitosis, the division of one nucleus into two genetically identical daughter nuclei, is now finished. Cytokinesis, the division of the cytoplasm, usually occurs with telophase. In animals, a cleavage furrow pinches the cell in two, producing two daughter cells.

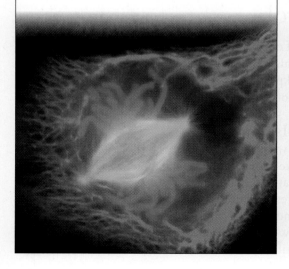

▼ Figure 8.8 **Cytokinesis in animal and plant cells.**

SEM 100x

LM 850x

Wall of parent cell   Cell plate forming   Daughter nucleus

Cleavage furrow

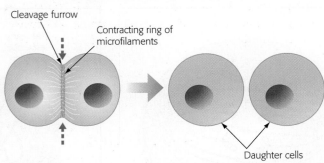

Cleavage furrow

Contracting ring of microfilaments

Daughter cells

(a) Animal cell cytokinesis

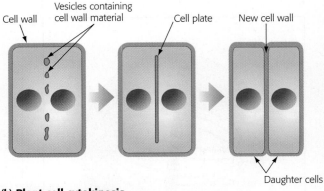

Cell wall   Vesicles containing cell wall material   Cell plate   New cell wall

Daughter cells

(b) Plant cell cytokinesis

Cytokinesis, the division of the cytoplasm into two genetically identical daughter cells, typically occurs during telophase. In animal cells, the cytokinesis process is known as cleavage. The first sign of cleavage is the appearance of a **cleavage furrow**, an indentation at the equator of the cell. A ring of microfilaments in the cytoplasm just under the plasma membrane contracts, like the pulling of a drawstring, deepening the furrow and pinching the parent cell in two (**Figure 8.8a**).

Cytokinesis in a plant cell occurs differently. Membranous vesicles containing cell wall material collect at the middle of the cell. The vesicles gradually fuse, forming a membranous disk called the **cell plate**. The cell plate grows outward, accumulating more cell wall material as more vesicles join it. Eventually, the membrane of the cell plate fuses with the plasma membrane, and the cell plate's contents join the parental cell wall. The result is two daughter cells (**Figure 8.8b**). ✓

Cancer Cells: Growing Out of Control

For a plant or animal to grow and maintain its tissues normally, it must be able to control the timing of cell division. The sequential events of the cell cycle are directed by a **cell cycle control system** that consists of specialized proteins within the cell. These proteins integrate information from the environment and from other body cells and send "stop" and "go-ahead" signals at certain key points during the cell cycle using signal transduction pathways (see Figure 5.19). For example, the cell cycle normally halts within the G_1 phase of interphase unless the cell receives a go-ahead signal via certain cell cycle control proteins. If that signal never arrives, the cell will switch into a permanently nondividing state.

Some of our nerve and muscle cells, for example, are arrested this way. If the go-ahead signal is received and the G_1 checkpoint is passed, the cell will usually complete the rest of the cycle.

What Is Cancer?

Cancer, which currently claims the lives of one out of every five people in the United States and other industrialized nations, is a disease of the cell cycle. Cancer cells do not respond normally to the cell cycle control system; they divide excessively and can invade other tissues of the body. If unchecked, cancer cells may continue to divide until they kill the host.

The abnormal behavior of cancer cells begins when a single cell undergoes transformation, a process whereby a normal cell converts to a cancer cell. Transformation occurs after a genetic change (mutation) in one or more genes that encode for proteins in the cell cycle control system. Because a transformed cell grows abnormally, the immune system normally recognizes and destroys it. However, if the cell evades destruction, it may proliferate to form a **tumor**, an abnormally growing mass of body cells. If the abnormal cells remain at the original site, the lump is called a **benign tumor**. Benign tumors can cause problems if they grow large and disrupt certain organs, such as the brain, but often they can be completely removed by surgery.

In contrast, a **malignant tumor** can spread into neighboring tissues and other parts of the body, displacing normal tissue and interrupting organ function (**Figure 8.9**). An individual with a malignant tumor is said to have **cancer**. Cancer cells may separate from the original tumor or secrete signal molecules that cause blood vessels to grow toward the tumor. A few tumor cells may then enter the circulatory or lymphatic system (blood or lymph vessels) and move to other parts of the body, where they may proliferate and form new tumors. The spread of cancer cells beyond their original site is called **metastasis**. Cancers are named according to where they originate. Liver cancer, for example, always begins in liver tissue and may spread from there.

Cancer Treatment

Once a tumor starts growing in the body, how can it be treated? The three main types of cancer treatment are sometimes referred to as "slash, burn, and poison." Surgery to remove a tumor ("slash") is usually the first step. "Burn" and "poison" refer to treatments that attempt to stop cancer cells from dividing. In **radiation therapy** ("burn"), parts of the body that have cancerous tumors are exposed to concentrated beams of high-energy radiation, which can often destroy cancer cells without seriously injuring the normal cells of the body. However, there is sometimes enough damage to normal body cells to produce unwanted side effects, such as nausea and hair loss.

Chemotherapy ("poison") uses the same basic strategy as radiation; in this case, drugs are administered that disrupt cell division. These drugs work in a variety of ways. Some prevent cell division by interfering with the mitotic spindle. For example, paclitaxel (trade name Taxol) freezes the spindle after it forms, keeping it from functioning. Paclitaxel is made from a chemical found in the bark of the Pacific yew, a tree found mainly in the northwestern United States. It has fewer side effects than many other anticancer drugs and seems to be effective against some hard-to-treat cancers of the ovary and breast. Another drug, vinblastine, prevents the mitotic spindle from forming in the first place. Vinblastine was first obtained from the periwinkle plant, which is native to the tropical rain forests of Madagascar.

Cancer Prevention and Survival

Although cancer can strike anyone, there are certain lifestyle changes you can make to reduce your chances of developing cancer or increase your chances of surviving it. Not smoking, exercising adequately, avoiding overexposure to the sun, and eating a high-fiber, low-fat diet can all help reduce the likelihood of getting cancer. Seven types of cancer can be easily detected: skin and oral (via physical exam), breast (via self-exams and mammograms for higher-risk women), prostate (via rectal exam), cervical (via Pap smear), testicular (via self-exam), and colon (via colonoscopy). Regular visits to the doctor can help identify tumors early, which is the best way to increase the chance of successful treatment. ☑

▼ Figure 8.9 **Growth and metastasis of a malignant tumor of the breast.**

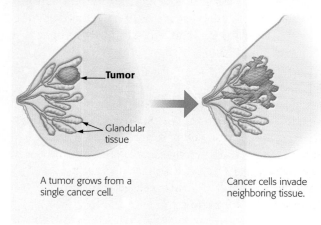

A tumor grows from a single cancer cell.

Cancer cells invade neighboring tissue.

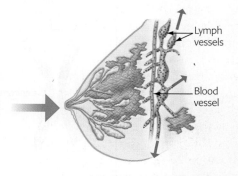

Lymph vessels

Blood vessel

Metastasis: Cancer cells spread through lymph and blood vessels to other parts of the body.

☑ CHECKPOINT

What differentiates a benign tumor from a malignant tumor?

Answer: A benign tumor remains at its point of origin, whereas a malignant tumor spreads.

Meiosis, the Basis of Sexual Reproduction

Only maple trees produce more maple trees, only goldfish make more goldfish, and only people make more people. These simple facts of life have been recognized for thousands of years and are reflected in the age-old saying "Like begets like." But in a strict sense, "Like begets like" applies only to asexual reproduction, such as the reproduction of the African violet in Figure 8.1. Because the offspring of asexual reproduction inherit all their DNA from a single parent, they are exact genetic replicas of that one parent and of each other, and their appearances are very similar.

The family photo in **Figure 8.10** makes the point that in a sexually reproducing species, like does not exactly beget like. You probably resemble your parents more closely than you resemble a stranger, but you do not look exactly like your parents or your siblings. Each offspring of sexual reproduction inherits a unique combination of genes from its two parents, and this combined set of genes programs a unique combination of traits. As a result, sexual reproduction can produce tremendous variety among offspring.

Sexual reproduction depends on the cellular processes of meiosis and fertilization. But before discussing these processes, we need to return to chromosomes and their role in the life cycles of sexually reproducing organisms.

▼ **Figure 8.10 The varied products of sexual reproduction.** Every child inherits a unique combination of genes from his or her parents and displays a unique combination of traits.

Homologous Chromosomes

If we examine cells from different individuals of a single species—sticking to one gender, for now—we find that they have the same number and types of chromosomes. Viewed with a microscope, your chromosomes would look just like those of Angelina Jolie (if you're a woman) or Brad Pitt (if you're a man).

In humans, a typical body cell, called a **somatic cell**, has 46 chromosomes. If we break open a human cell in metaphase of mitosis, stain the chromosomes with dyes, take a picture with a microscope, and arrange them in matching pairs, we produce a display called a **karyotype** (**Figure 8.11**). Notice that each chromosome is duplicated, consisting of two sister chromatids joined at the centromere. Notice also that every (or almost every) chromosome has a twin that resembles it in length and centromere position. The two chromosomes of such a matching pair, called **homologous chromosomes**, carry genes controlling the same inherited characteristics. If a gene influencing eye color is located at a particular place on one chromosome—for example, within the yellow band in the drawing in Figure 8.11—then the homologous chromosome has that same gene in the same location. However, the two homologous chromosomes may have different versions of the same gene. (Let's restate this concept, since it can be confusing: A pair of homologous chromosomes has two nearly identical chromo-

▼ **Figure 8.11 Pairs of homologous chromosomes in a male karyotype.** This karyotype shows 22 completely homologous pairs (autosomes) and a 23rd pair that consists of an X chromosome and a Y chromosome (sex chromosomes). With the exception of X and Y, the homologous chromosomes of each pair match in size, centromere position, and staining pattern.

Pair of homologous chromosomes

Centromere

One duplicated chromosome

Sister chromatids

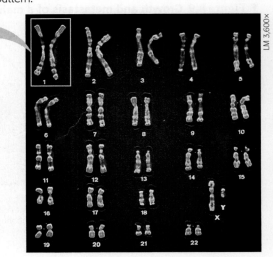

LM 3,600x

somes, each of which consists of two identical sister chromatids after chromosome duplication; see the drawing in Figure 8.11.) Altogether, we humans have 23 pairs of homologous chromosomes. Other species have different numbers of chromosomes (see Figure 8.2), but those, too, usually match in pairs.

For a human female, the 46 chromosomes fall neatly into 23 homologous pairs, with the members of each pair essentially identical in appearance. For a male, however, the chromosomes in one pair do not look alike. This nonmatching pair, only partly homologous, is the male's sex chromosomes. **Sex chromosomes** determine a person's sex (male versus female). As in all mammals, human males have one X chromosome and one Y chromosome. Females have two X chromosomes. (Other organisms have different systems; in this chapter, we focus on humans.) The remaining chromosomes, found in both males and females, are called **autosomes**. For both autosomes and sex chromosomes, we inherit one chromosome of each pair from our mother and the other from our father.

Gametes and the Life Cycle of a Sexual Organism

The **life cycle** of a multicellular organism is the sequence of stages leading from the adults of one generation to the adults of the next. Having two sets of chromosomes, one inherited from each parent, is a key factor in the life cycle of humans and all other species that reproduce sexually. **Figure 8.12** shows the human life cycle, emphasizing the number of chromosomes.

Humans (as well as most other animals and many plants) are said to be **diploid** organisms because all body cells contain pairs of homologous chromosomes. The total number of chromosomes, 46 in humans, is the diploid number, represented as $2n$. The exceptions are the egg and sperm cells, known as **gametes**. Made by meiosis in an ovary or testis, each gamete has a single set of chromosomes: 22 autosomes plus a sex chromosome, X or Y. A cell with a single chromosome set is called a **haploid** cell; it has only one member of each homologous pair. For humans, the haploid number, n, is 23.

In the human life cycle, a haploid sperm cell from the father fuses with a haploid egg cell from the mother in a process called **fertilization**. The resulting fertilized egg, called a **zygote**, is diploid. It has two sets of homologous chromosomes, one set from each parent. The life cycle is completed as a sexually mature adult develops from the zygote. Mitotic cell division ensures that all somatic

cells of the human body receive a copy of all of the zygote's 46 chromosomes. Thus, every one of the trillions of cells in your body can trace its ancestry back through mitotic divisions to the single zygote produced when your father's sperm and mother's egg fused about nine months before you were born.

All sexual life cycles involve an alternation of diploid and haploid stages. Producing haploid gametes by meiosis keeps the chromosome number from doubling in every generation. To illustrate, **Figure 8.13** tracks one pair of homologous chromosomes. ❶ Each of the chromosomes is duplicated during interphase (before mitosis). ❷ The first division, meiosis I, segregates the two chromosomes of the homologous pair, packaging them in separate (haploid) daughter cells. But each chromosome is still doubled. ❸ Meiosis II separates the sister chromatids. Each of the four daughter cells is haploid and contains only a single chromosome from the homologous pair.

▶ Figure 8.12 **The human life cycle.** In each generation, the doubling of chromosome number that results from fertilization is offset by the halving of chromosome number during meiosis.

▼ Figure 8.13 **How meiosis halves chromosome number.**

❶ Chromosomes duplicate.

❷ Homologous chromosomes separate.

❸ Sister chromatids separate.

Pair of homologous chromosomes in diploid parent cell

Duplicated pair of homologous chromosomes

Sister chromatids

INTERPHASE BEFORE MEIOSIS

MEIOSIS I

MEIOSIS II

The Process of Meiosis

Meiosis, the process that produces haploid daughter cells in diploid organisms, resembles mitosis, but with two special features. The first is that the number of chromosomes is reduced to half. In meiosis, a cell that has duplicated its chromosomes undergoes two consecutive divisions, called meiosis I and meiosis II. Because one duplication of the chromosomes is followed by two divisions, each of the four daughter cells resulting from meiosis has a haploid set of chromosomes—only half as many chromosomes as the starting cell.

The second special feature of meiosis is an exchange of genetic material—pieces of chromosomes—between homologous chromosomes. This exchange, called crossing over, occurs during the first prophase of meiosis. We'll look more closely at crossing over later. For now, study **Figure 8.14**, including the text below it, which describes the stages of meiosis in detail for an animal cell containing four chromosomes.

As you go through Figure 8.14, keep in mind the difference between homologous chromosomes and sister chromatids: The two chromosomes of a homologous pair are individual chromosomes that were inherited

▼ Figure 8.14 **The stages of meiosis.**

MEIOSIS I: HOMOLOGOUS CHROMOSOMES SEPARATE

INTERPHASE	PROPHASE I	METAPHASE I	ANAPHASE I
			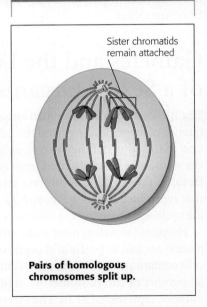
Centrosomes (with centriole pairs); Nuclear envelope; Chromatin	Sites of crossing over; Spindle; Sister chromatids; Pair of homologous chromosomes	Microtubules attached to chromosome; Centromere	Sister chromatids remain attached
Chromosomes duplicate.	**Homologous chromosomes pair up and exchange segments.**	**Pairs of homologous chromosomes line up.**	**Pairs of homologous chromosomes split up.**

Like mitosis, meiosis is preceded by an interphase during which the chromosomes duplicate. Each chromosome then consists of two identical sister chromatids.

Prophase I As the chromosomes coil up, special proteins cause the homologous chromosomes to stick together in pairs. The resulting structure has four chromatids. Within each set, chromatids of the homologous chromosomes exchange corresponding segments—they "cross over." Crossing over rearranges genetic information.

As prophase I continues, the chromosomes coil up further, a spindle forms, and the homologous pairs are moved toward the center of the cell.

Metaphase I At metaphase I, the homologous pairs are aligned in the middle of the cell. The sister chromatids of each chromosome are still attached at their centromeres, where they are anchored to spindle microtubules. Notice that for each chromosome pair, the spindle microtubules attached to one homologous chromosome come from one pole of the cell, and the microtubules attached to the other chromosome come from the opposite pole. With this arrangement, the homologous chromosomes are poised to move toward opposite poles of the cell.

Anaphase I The attachment between the homologous chromosomes of each pair breaks, and the chromosomes now migrate toward the poles of the cell. *In contrast to mitosis, the sister chromatids migrate as a pair instead of splitting up.* They are separated not from each other, but from their homologous partners.

from different parents, one from the mother and one from the father. The members of a homologous pair of chromosomes in Figure 8.14 (and later figures) are colored red and blue to remind you that they differ in this way. In the interphase just before meiosis, each chromosome duplicates to form sister chromatids that remain together until anaphase of meiosis II. Before crossing over occurs, sister chromatids are identical and carry the same versions of all their genes. ✔

LM 800x

Metaphase II in
a lily cell

☑**CHECKPOINT**

If a single diploid cell with 18 chromosomes undergoes meiosis and produces sperm, the result will be _____ sperm, each with _____ chromosomes. (Provide two numbers.)

Answer: *four; nine*

TELOPHASE I AND CYTOKINESIS

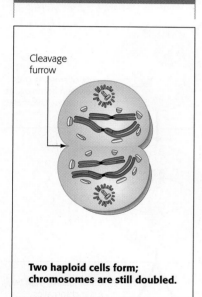

Cleavage
furrow

**Two haploid cells form;
chromosomes are still doubled.**

Telophase I and Cytokinesis In telophase I, the chromosomes arrive at the poles of the cell. When they finish their journey, each pole has a haploid chromosome set, although each chromosome is still in duplicate form. Usually, cytokinesis occurs along with telophase I, and two haploid daughter cells are formed.

MEIOSIS II: SISTER CHROMATIDS SEPARATE

PROPHASE II	METAPHASE II	ANAPHASE II	TELOPHASE II AND CYTOKINESIS

Sister chromatids
separate

Haploid daughter cells
forming

**During another round of cell division, the sister chromatids finally separate;
four haploid daughter cells result, containing single chromosomes.**

The Process of Meiosis II Meiosis II is essentially the same as mitosis. The important difference is that meiosis II starts with a haploid cell that has *not* undergone a chromosome duplication during the preceding interphase.

During prophase II, a spindle forms and moves the chromosomes toward the middle of the cell. During metaphase II, the chromosomes are aligned as they are in mitosis, with the microtubules attached to the sister chromatids of each chromosome coming from opposite poles. In anaphase II, the centromeres of sister chromatids separate, and the sister chromatids of each pair move toward opposite poles of the cell. In telophase II, nuclei form at the cell poles, and cytokinesis occurs at the same time. There are now four daughter cells, each with the haploid number of single chromosomes.

Review: Comparing Mitosis and Meiosis

We have now described the two ways that cells of eukaryotic organisms divide. Mitosis, which provides for growth, tissue repair, and asexual reproduction, produces daughter cells that are genetically identical to the parent cell. Meiosis, needed for sexual reproduction, yields genetically unique haploid daughter cells—cells with only one member of each homologous chromosome pair.

For both mitosis and meiosis, the chromosomes duplicate only once, in the preceding interphase. Mitosis involves one division of the nucleus and cytoplasm, producing two diploid cells. Meiosis entails two nuclear and cytoplasmic divisions, yielding four haploid cells.

Figure 8.15 compares mitosis and meiosis, tracing these two processes for a diploid parent cell with four

▶ Figure 8.15 **Comparing mitosis and meiosis.** The events unique to meiosis occur during meiosis I: In prophase I, duplicated homologous chromosomes pair along their lengths, and crossing over occurs between homologous (nonsister) chromatids. In metaphase I, pairs of homologous chromosomes (rather than individual chromosomes) are aligned at the center of the cell. During anaphase I, sister chromatids of each chromosome stay together and go to the same pole of the cell as homologous chromosomes separate. At the end of meiosis I, there are two haploid cells, but each chromosome still has two sister chromatids.

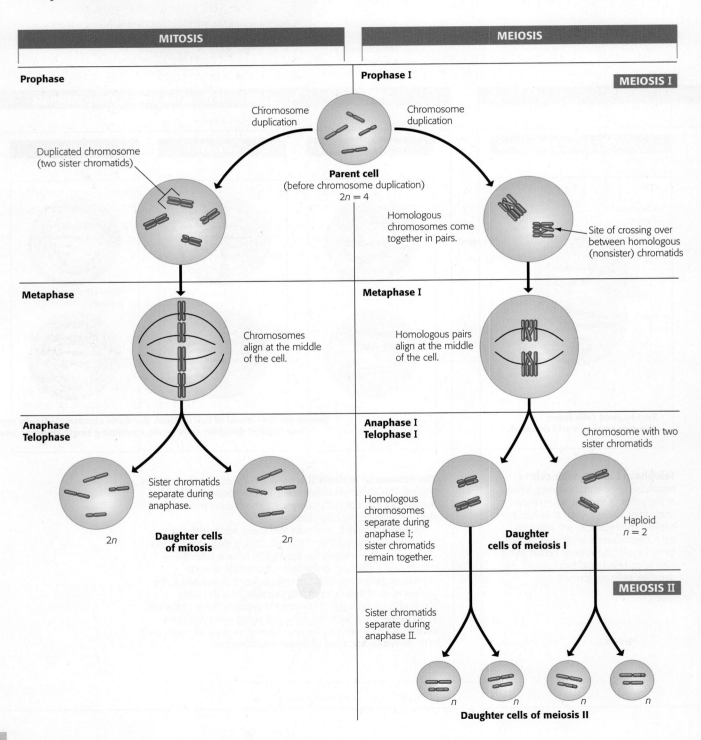

chromosomes. As before, homologous chromosomes are those matching in size. (Imagine that the red chromosomes were inherited from the mother and the blue chromosomes from the father.) Notice that all the events unique to meiosis occur during meiosis I. Meiosis II is virtually identical to mitosis in that it separates sister chromatids. But unlike mitosis, meiosis II yields daughter cells with a haploid set of chromosomes. ☑

The Origins of Genetic Variation

As we discussed earlier, offspring that result from sexual reproduction are genetically different from their parents and from one another. When we discuss evolution in Unit Three, we'll see that this genetic variety in offspring is the raw material for natural selection. For now, let's take another look at meiosis and fertilization to see how genetic variety arises.

Independent Assortment of Chromosomes

Figure 8.16 illustrates one way in which meiosis contributes to genetic variety. The figure shows how the arrangement of homologous chromosome pairs at metaphase of meiosis I affects the resulting gametes.

Once again, our example is from a diploid organism with four chromosomes (two homologous pairs), with colors used to differentiate homologous chromosomes (red for chromosomes inherited from the mother and blue for those from the father).

When aligned during metaphase I, the side-by-side orientation of each homologous pair of chromosomes (consisting of two sister chromatids) is a matter of chance—either the red or blue chromosome may be on the left or right. Thus, in this example, there are two possible ways that the chromosome pairs can align during metaphase I. In possibility 1, the chromosome pairs are oriented with both red chromosomes on the same side (blue/red and blue/red). In this case, each of the gametes produced at the end of meiosis II has only red or only blue chromosomes (combinations a and b). In possibility 2, the chromosome pairs are oriented differently (blue/red and red/blue). This arrangement produces gametes with one red and one blue chromosome (combinations c and d). Thus, with the two possible arrangements shown in this example, the organism will produce gametes with four different combinations of chromosomes. For a species with more than two pairs of chromosomes, such as the human, every chromosome pair orients independently of all the others at metaphase I. (Chromosomes X and Y behave as a homologous pair in meiosis.)

☑ **CHECKPOINT**

True or false: Both mitosis and meiosis are preceded by chromosome duplication.

Answer: true

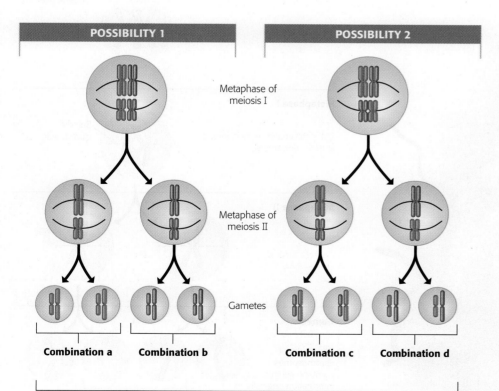

◀ **Figure 8.16 Results of alternative arrangements of chromosomes at metaphase of meiosis I.** The arrangement of chromosomes at metaphase I determines which chromosomes will be packaged together in the haploid gametes.

POSSIBILITY 1

POSSIBILITY 2

Metaphase of meiosis I

Metaphase of meiosis II

Gametes

Combination a   **Combination b**   **Combination c**   **Combination d**

Because possibilities 1 and 2 are equally likely, the four possible types of gametes will be made in approximately equal numbers.

For any species, the total number of chromosome combinations that can appear in gametes is 2^n, where n is the haploid number. For the organism in Figure 8.16, $n = 2$, so the number of chromosome combinations is 2^2, or 4. For a human ($n = 23$), there are 2^{23}, or about 8 million, possible chromosome combinations! This means that every gamete a human produces contains one of about 8 million possible combinations of maternal and paternal chromosomes.

Random Fertilization

How many possibilities are there when a gamete from one individual unites with a gamete from another individual during fertilization (**Figure 8.17**)? A human egg cell, representing one of about 8 million possibilities, is fertilized at random by one sperm cell, representing one of about 8 million other possibilities. By multiplying 8 million by 8 million, we find that a man and a woman can produce a diploid zygote with any of 64 trillion combinations of chromosomes! So we see that the random nature of fertilization adds a huge amount of potential variability to the offspring of sexual reproduction.

Crossing Over

So far, we have focused on genetic variety in gametes and zygotes at the whole-chromosome level. We'll now take a closer look at **crossing over**, the exchange of cor-

responding segments between chromatids of homologous chromosomes, which occurs during prophase I of meiosis. **Figure 8.18** shows crossing over between two homologous chromosomes and the resulting gametes. At the time that crossing over begins, homologous chromosomes are closely paired all along their lengths, with a precise gene-by-gene alignment. The sites of crossing over appear as X-shaped regions; each is called a **chiasma** (plural, *chiasmata*). The homologous chromatids remain attached to each other at chiasmata until anaphase I.

▼ Figure 8.17 **The process of fertilization: a close-up view.**

▼ Figure 8.18 **The results of crossing over during meiosis for a single pair of homologous chromosomes.** A real cell has multiple pairs of homologous chromosomes that produce a huge variety of recombinant chromosomes in the gametes.

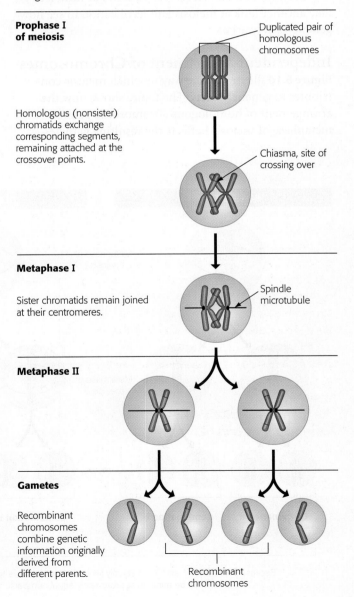

Prophase I of meiosis

Duplicated pair of homologous chromosomes

Homologous (nonsister) chromatids exchange corresponding segments, remaining attached at the crossover points.

Chiasma, site of crossing over

Metaphase I

Sister chromatids remain joined at their centromeres.

Spindle microtubule

Metaphase II

Gametes

Recombinant chromosomes combine genetic information originally derived from different parents.

Recombinant chromosomes

The exchange of segments between nonsister chromatids—one maternal chromatid and one paternal chromatid of a homologous pair—adds to the genetic variety resulting from sexual reproduction. In Figure 8.18, if there were no crossing over, meiosis could produce only two types of gametes. These would be the ones ending up with the "parental" types of chromosomes, either all blue or all red (as in Figure 8.16). With crossing over, gametes arise that have chromosomes that are part red and part blue. These chromosomes are called "recombinant" because they result from **genetic recombination**, the production of gene combinations different from those carried by the parental chromosomes.

Because most chromosomes contain thousands of genes, a single crossover event can affect many genes. When we also consider that multiple crossovers can occur in each pair of homologous chromosomes, it's not surprising that gametes and the offspring that result from them can be so varied. ☑

☑ **CHECKPOINT**

Name two events during meiosis that contribute to genetic variety among gametes. During what stages of meiosis does each occur?

Answer: crossing over between homologous chromosomes during prophase 1 and independent orientation/assortment of the pairs of homologous chromosomes at metaphase 1

THE PROCESS OF SCIENCE: Life with and without Sex

Do All Animals Have Sex?

As mentioned in the Biology and Society section, many plants can reproduce via both sexual and asexual routes. But while some animal species can also reproduce asexually, very few animals reproduce *only* asexually. In fact, evolutionary biologists have traditionally considered asexual reproduction an evolutionary dead end (for reasons we'll discuss in the Evolution Connection section).

To investigate a case where asexual reproduction seemed to be the norm, researchers from Harvard University studied a group of animals called the bdelloid rotifers **(Figure 8.19)**. This class of nearly microscopic freshwater invertebrates includes over 300 known species. Despite hundreds of years of **observations**, no one had ever found bdelloid rotifer males or evidence of sexual reproduction. But the possibility remained that bdelloids had sex very infrequently or that the males were impossible to recognize by appearance. Thus, the Harvard research team posed the **question**, Does this entire class of animals reproduce solely by asexual means?

The researchers formed the **hypothesis** that bdelloid rotifers have indeed thrived for millions of years despite a lack of sexual reproduction. But how to prove it? In most species, the two versions of a gene in a pair of homologous chromosomes are very similar due to the constant trading of genes during sexual reproduction. If a species has survived without sex for millions of years, the researchers reasoned, then changes in the DNA sequences of homologous genes should accumulate independently, and the two versions of the genes should have significantly diverged from each other over time. This led to the **prediction** that bdelloid rotifers would display much more variation in their pairs of homologous genes than most organisms.

In a simple but elegant **experiment**, the researchers compared the sequences of a particular gene in bdelloid and non-bdelloid rotifers. Their **results** were striking. Among non-bdelloid rotifers that reproduce sexually, the two homologous versions of the gene were nearly identical, differing by only 0.5% on average. In contrast, the two versions of the same gene in bdelloid rotifers differed by 3.5–54%. These data provided strong evidence that bdelloid rotifers have evolved for millions of years without any sexual reproduction.

▶ Figure 8.19
A bdelloid rotifer.

LM 300×

When Meiosis Goes Awry

So far, our discussion of meiosis has focused on the process as it normally and correctly occurs. But what happens when an error occurs in the process? Such a mistake can result in genetic abnormalities that range from mild to severe to fatal.

How Accidents during Meiosis Can Alter Chromosome Number

Within the human body, meiosis occurs repeatedly as the testes or ovaries produce gametes. Almost always, the meiotic spindle distributes chromosomes to daughter cells without error. But occasionally there is a mishap, called a **nondisjunction**, in which the members of a chromosome pair fail to separate at anaphase. Nondisjunction can occur during meiosis I or II (**Figure 8.20**). In either case, gametes with abnormal numbers of chromosomes are the result.

Figure 8.21 shows what can happen when an abnormal gamete produced by nondisjunction unites with a normal gamete during fertilization. When a normal sperm fertilizes an egg cell with an extra chromosome,

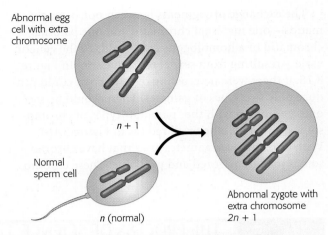

Abnormal egg cell with extra chromosome

$n + 1$

Normal sperm cell

n (normal)

Abnormal zygote with extra chromosome $2n + 1$

▲ Figure 8.21 **Fertilization after nondisjunction in the mother.**

the result is a zygote with a total of $2n + 1$ chromosomes. Mitosis then transmits the abnormality to all embryonic cells. If the organism survives, it will have an abnormal karyotype and probably a syndrome of disorders caused by the abnormal number of genes. ☑

▼ Figure 8.20 **Two types of nondisjunction.** In both examples in the figure, the cell at the top is diploid ($2n$), with two pairs of homologous chromosomes.

☑CHECKPOINT

Explain how nondisjunction in meiosis could result in a diploid gamete.

Answer: A diploid gamete would result if there were nondisjunction of all the chromosomes during meiosis I or II.

Down Syndrome: An Extra Chromosome 21

Figure 8.11 showed a normal human complement of 23 pairs of chromosomes. Compare it with **Figure 8.22**; besides having two X chromosomes (because it's from a female), the karyotype in Figure 8.22 has three number 21 chromosomes, making 47 chromosomes in total. This condition is called **trisomy 21**.

In most cases, a human embryo with an atypical number of chromosomes develops so abnormally that it is spontaneously aborted (miscarried) long before birth.

▼ **Figure 8.22 Trisomy 21 and Down syndrome.** This child displays the characteristic facial features of Down syndrome. The karyotype (bottom) shows trisomy 21; notice the three copies of chromosome 21.

Chromosome 21

However, some aberrations in chromosome number seem to upset the genetic balance less drastically, and individuals with such abnormalities can survive. These people usually have a characteristic set of symptoms, called a syndrome. A person with trisomy 21, for instance, is said to have **Down syndrome** (named after John Langdon Down, who described it in 1866).

Affecting about 1 out of every 700 children, trisomy 21 is the most common chromosome number abnormality and the most common serious birth defect in the United States. Down syndrome includes characteristic facial features—frequently a fold of skin at the inner corner of the eye, a round face, and a flattened nose—as well as short stature, heart defects, and susceptibility to leukemia and Alzheimer's disease. People with Down syndrome usually have a life span shorter than normal. They also exhibit varying degrees of mental retardation. However, individuals with the syndrome may live to middle age or beyond, and many are socially adept and can function well within society.

As indicated in **Figure 8.23**, the incidence of Down syndrome in the offspring of normal parents increases markedly with the age of the mother. Down syndrome affects less than 0.05% of children (fewer than 1 in 2,000) born to women under age 30. The risk climbs to 1% (10 in 1,000) for mothers at age 40 and is even higher for older mothers. Because of this relatively high risk, pregnant women over 35 are candidates for fetal testing for trisomy 21 and other chromosomal abnormalities (see Chapter 9). ☑

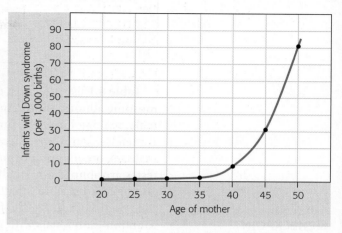

▲ **Figure 8.23 Maternal age and Down syndrome.** The chance of having a baby with Down syndrome rises with the age of the mother.

Abnormal Numbers of Sex Chromosomes

Nondisjunction in meiosis does not affect just autosomes, such as chromosome 21. It can also lead to abnormal numbers of sex chromosomes (X and Y). Unusual numbers of sex chromosomes seem to upset the genetic balance less than unusual numbers of autosomes. This may be because the Y chromosome is very small and carries relatively few genes. Furthermore, mammalian cells normally operate with only one functioning X chromosome because other copies of the chromosome become inactivated in each cell (see Chapter 11).

Table 8.1	Abnormalities of Sex Chromosome Number in Humans		
Sex Chromosomes	**Syndrome**	**Origins of Nondisjunction**	**Frequency in Population**
XXY	Klinefelter syndrome (male)	Meiosis in egg or sperm formation	$\frac{1}{2,000}$
XYY	None (normal male)	Meiosis in sperm formation	$\frac{1}{2,000}$
XXX	None (normal female)	Meiosis in egg or sperm formation	$\frac{1}{1,000}$
XO	Turner syndrome (female)	Meiosis in egg or sperm formation	$\frac{1}{5,000}$

Table 8.1 lists the most common human sex chromosome abnormalities. An extra X chromosome in a male, making him XXY, produces a condition called Klinefelter syndrome. Men with this disorder have male sex organs and normal intelligence, but the testes are abnormally small, the individual is sterile, and he often has breast enlargement and other feminine body contours. Klinefelter syndrome is also found in individuals with more than three sex chromosomes, such as XXYY, XXXY, or XXXXY. These abnormal numbers of sex chromosomes result from multiple nondisjunctions.

Human males with a single extra Y chromosome (XYY) do not have any well-defined syndrome, although they tend to be taller than average. Females with an extra X chromosome (XXX) cannot be distinguished from XX females except by karyotype.

Females who are lacking an X chromosome are designated XO; the O indicates the absence of a second sex chromosome. These women have Turner syndrome. They have a characteristic appearance, including short stature and often a web of skin extending between the neck and shoulders. Women with Turner syndrome are sterile because their sex organs do not fully mature at adolescence, and they have poor development of breasts and other secondary sex characteristics. However, they are usually of normal intelligence. The XO condition is the sole known case where having only 45 chromosomes is not fatal in humans.

The sex chromosome abnormalities described here illustrate the crucial role of the Y chromosome in determining a person's sex. In general, a single Y chromosome is enough to produce "maleness," regardless of the number of X chromosomes. The absence of a Y chromosome results in "femaleness." ☑

EVOLUTION CONNECTION: Life with and without Sex

The Advantages of Sex

Throughout this chapter, we've examined cell division within the context of reproduction. Like the Hawaiian bellflower in the Biology and Society section, many plants can reproduce both sexually and asexually (Figure 8.24). An important advantage of asexual reproduction is that there is no need for a partner. Asexual reproduction may thus confer an evolutionary advantage when plants are sparsely distributed and unlikely to be able to exchange pollen. Furthermore, if a plant is superbly suited to a stable environment, asexual reproduction has the advantage of passing on its entire genetic legacy intact.

In contrast to plants, the vast majority of animals reproduce by sexual means. There are exceptions, such as organisms that can regenerate after fragmentation (see Figure 8.1) and the bdelloid rotifers discussed in the Process of Science section. But many animals reproduce only through sex. Therefore, sex must enhance evolutionary fitness. But how? The answer remains elusive. Most hypotheses focus on the unique combinations of genes formed during meiosis and fertilization. By producing offspring of varied genetic makeup, sexual reproduction may enhance survival by speeding adaptation to a changing environment, such as one that contains evolving pathogens. Another idea is that shuffling genes during sexual reproduction might allow a population to rid itself of harmful genes more rapidly. But for now, one of biology's most basic questions—Why have sex?—remains a hotly debated topic that is the focus of much ongoing research and discussion.

◀ **Figure 8.24 Sexual and asexual reproduction.** Many plants, such as this strawberry, have the ability to reproduce both sexually (via flowers that produce fruit) and asexually (via runners).

Chapter Review

SUMMARY OF KEY CONCEPTS

 Go to the Study Area at **www.masteringbiology.com** for practice quizzes, eText, BioFlix™ 3-D animations, MP3 Tutor Sessions, videos, current events, and more.

What Cell Reproduction Accomplishes

Cell reproduction, also called cell division, produces genetically identical daughter cells:

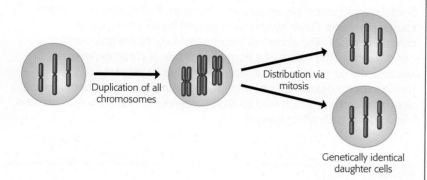

Duplication of all chromosomes

Distribution via mitosis

Genetically identical daughter cells

Some organisms use mitosis (ordinary cell division) to reproduce. This is called asexual reproduction, and it results in offspring that are genetically identical to the lone parent and to each other. Mitosis also enables multicellular organisms to grow and develop and to replace damaged or lost cells. Organisms that reproduce sexually, by the union of a sperm with an egg cell, carry out meiosis, a type of cell division that yields gametes with only half as many chromosomes as body (somatic) cells.

The Cell Cycle and Mitosis

Eukaryotic Chromosomes

The genes of a eukaryotic genome are grouped into multiple chromosomes in the nucleus. Each chromosome contains one very long DNA molecule, with many genes, that is wrapped around histone proteins. Individual chromosomes are coiled up and therefore visible with a light microscope only when the cell is in the process of dividing; otherwise, they are in the form of thin, loosely packed chromatin fibers. Before a cell starts dividing, the chromosomes duplicate, producing sister chromatids (containing identical DNA) joined together at the centromere.

Chromosome (one long piece of DNA)

Centromere

Sister chromatids

Duplicated chromosome

The Cell Cycle

S phase
DNA synthesis; chromosome duplication

Interphase
Cell growth and chromosome duplication

G₁

G₂

Mitotic (M) phase

Genetically identical "daughter" cells

Cytokinesis (division of cytoplasm)

Mitosis (division of nucleus)

Mitosis and Cytokinesis

Mitosis is divided into four phases: prophase, metaphase, anaphase, and telophase. At the start of mitosis, the chromosomes coil up and the nuclear envelope breaks down (prophase). Then a mitotic spindle made of microtubules moves the chromosomes to the middle of the cell (metaphase). The sister chromatids then separate and are moved to opposite poles of the cell (anaphase), where two new nuclei form (telophase). Cytokinesis overlaps the end of mitosis. In animals, cytokinesis occurs by cleavage, which pinches the cell in two. In plants, a membranous cell plate divides the cell in two. Mitosis and cytokinesis produce genetically identical cells.

Cancer Cells: Growing Out of Control

When the cell cycle control system malfunctions, a cell may divide excessively and form a tumor. Cancer cells may grow to form malignant tumors, invade other tissues (metastasize), and even kill the organism. Surgery can remove tumors, and radiation and chemotherapy are effective as treatments because they interfere with cell division. You can increase the likelihood of surviving some forms of cancer through lifestyle changes and regular screenings.

Meiosis, the Basis of Sexual Reproduction

Homologous Chromosomes

The somatic cells (body cells) of each species contain a specific number of chromosomes; human cells have 46, made up of 23 pairs of homologous chromosomes. The chromosomes of a homologous pair carry genes for the same characteristics at the same places. Mammalian males have X and Y sex chromosomes (only partly homologous), while females have two X chromosomes.

Gametes and the Life Cycle of a Sexual Organism

Human Life Cycle

Key
■ Haploid (n)
■ Diploid (2n)

Haploid gametes (n = 23)

n
Egg cell

n
Sperm cell

MEIOSIS

FERTILIZATION

Male and female diploid adults
(2n = 46)

Diploid zygote
(2n = 46)
2n

MITOSIS
and development

The Process of Meiosis

Meiosis, like mitosis, is preceded by chromosome duplication. But in meiosis, the cell divides twice to form four daughter cells. The first division, meiosis I, starts with the pairing of homologous chromosomes. In crossing over, homologous chromosomes exchange corresponding segments. Meiosis I separates the members of the homologous pairs and produces two daughter cells, each with one set of (duplicated) chromosomes. Meiosis II is essentially the same as mitosis; in each of the cells, the sister chromatids of each chromosome separate.

Review: Comparing Mitosis and Meiosis

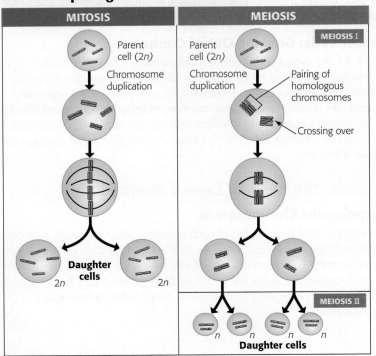

MITOSIS	MEIOSIS

Parent cell (2n)

Chromosome duplication

Daughter cells
2n   2n

MEIOSIS I

Parent cell (2n)

Chromosome duplication

Pairing of homologous chromosomes

Crossing over

MEIOSIS II

n   n   n   n
Daughter cells

The Origins of Genetic Variation

Because the chromosomes of a homologous pair come from different parents, they carry different versions of many of their genes. The large number of possible arrangements of chromosome pairs at metaphase of meiosis I leads to many different combinations of chromosomes in eggs and sperm. Random fertilization of eggs by sperm greatly increases the variation. Crossing over during prophase of meiosis I increases variation still further.

When Meiosis Goes Awry

Sometimes a person has an abnormal number of chromosomes, which causes problems. Down syndrome is caused by an extra copy of chromosome 21. The abnormal chromosome count is a product of nondisjunction, the failure of a homologous pair of chromosomes to separate during meiosis I or of sister chromatids to separate during meiosis II. Nondisjunction can also produce gametes with extra or missing sex chromosomes, which lead to varying degrees of malfunction in humans but do not usually affect survival.

SELF-QUIZ

1. Which of the following is not a function of mitosis in humans?
   a. repair of wounds
   b. growth
   c. production of gametes from diploid cells
   d. replacement of lost or damaged cells

2. In what sense are the two daughter cells produced by mitosis identical?

3. Why is it difficult to observe individual chromosomes during interphase?

4. A biochemist measures the amount of DNA in cells growing in the laboratory. The quantity of DNA in a cell would be found to double
   a. between prophase and anaphase of mitosis.
   b. between the G_1 and G_2 phases of the cell cycle.
   c. during the M phase of the cell cycle.
   d. between prophase I and prophase II of meiosis.

5. Which two phases of mitosis are essentially opposites in terms of changes in the nucleus?

6. Complete the following table to compare mitosis and meiosis:

	Mitosis	Meiosis
a. Number of chromosomal duplications		
b. Number of cell divisions		
c. Number of daughter cells produced		
d. Number of chromosomes in daughter cells		
e. How chromosomes line up during metaphase		
f. Genetic relationship of daughter cells to parent cells		
g. Functions performed in the human body		

7. A chemical that disrupts microfilament formation would interfere with
   a. DNA replication.
   b. formation of the mitotic spindle.
   c. cleavage.
   d. crossing over.

8. If an intestinal cell in a dog contains 78 chromosomes, a dog sperm cell would contain _____ chromosomes.

9. A micrograph of a dividing cell from a mouse shows 19 chromosomes, each consisting of two sister chromatids. During which stage of meiosis could this picture have been taken? (Explain your answer.)

10. Tumors that remain at their site of origin are called _____, while tumors from which cells migrate to other body tissues are called _____.

11. A fruit fly somatic cell contains eight chromosomes. This means that _____ different combinations of chromosomes are possible in its gametes.

12. Although nondisjunction is a random event, there are many more individuals with an extra chromosome 21, which causes Down syndrome, than individuals with an extra chromosome 3 or chromosome 16. Propose an explanation for this.

Answers to the Self-Quiz questions can be found in Appendix D.

THE PROCESS OF SCIENCE

13. A mule is the offspring of a horse and a donkey. A donkey sperm contains 31 chromosomes and a horse egg 32 chromosomes, so the zygote contains a total of 63 chromosomes. The zygote develops normally. The combined set of chromosomes is not a problem in mitosis, and the mule combines some of the best characteristics of horses and donkeys. However, a mule is sterile; meiosis cannot occur normally in its testes or ovaries. Explain why mitosis is normal in cells containing both horse and donkey chromosomes but the mixed set of chromosomes interferes with meiosis.

14. You prepare a slide with a thin slice of an onion root tip. You see the following view in a light microscope. Identify the stage of mitosis for each of the outlined cells, a–d.

LM 600×

(a)   (b)   (c)   (d)

BIOLOGY AND SOCIETY

15. Every year, about a million Americans are diagnosed with cancer. This means that about 75 million Americans now living will eventually have cancer, and one in five will die of the disease. There are many kinds of cancers and many causes of the disease. For example, smoking causes most lung cancers. Overexposure to ultraviolet rays in sunlight causes most skin cancers. There is evidence that a high-fat, low-fiber diet is a factor in breast, colon, and prostate cancers. And agents in the workplace, such as asbestos and vinyl chloride, are also implicated as causes of cancer. Hundreds of millions of dollars are spent each year in the search for effective treatments for cancer, yet far less money is spent on preventing cancer. Why might this be true? What kinds of lifestyle changes could we make to help prevent cancer? What kinds of prevention programs could be initiated or strengthened to encourage these changes? What factors might impede such changes and programs? Should we devote more of our resources to treating cancer or preventing it? Defend your position.

16. The practice of buying and selling gametes, particularly eggs from fertile women, is becoming increasingly common in the United States and other industrialized countries. Do you have any objections to this type of transaction? Would you be willing to sell your gametes? At any price? Whether you are willing to do so or not, do you think that other people should be restricted from doing so?

9 Patterns of Inheritance

Canine genetics.
Dogs, such as these
Siberian huskies, are one
of humankind's longest-
running genetic
experiments.

CHAPTER CONTENTS

 BIOLOGY AND SOCIETY: Dog Breeding

A Matter of Breeding

The hard-working dogs shown on the facing page are Siberian huskies. Each member of the team looks like the others because they share a similar pedigree (genetic history). When these dogs are bred, they are expected to produce others that inherit the desirable traits that distinguish the breed, such as a thick double coat of fur and triangular well-insulated ears. This is a reasonable expectation because a purebred dog's well-documented pedigree includes parents, grandparents, and great-grandparents with similar genetic makeup and appearance.

But whether dogs are purebreds or mutts, we can ultimately explain their inborn traits through genetics, the scientific study of heredity. Because their patterns of inheritance are relatively simple, purebred dogs are important in genetic research. However, after generations of inbreeding, purebreds often suffer from serious genetic defects. For instance, a type of hereditary blindness called progressive retinal atrophy (PRA) is common among Siberian huskies and several other breeds. Studies of such defects may be useful for veterinary and human medicine because many genetic disorders are similar in dogs and humans.

Geneticists are also studying dogs to help shed light on the relationship between genetic makeup and behavior. Certain breeds have behavioral tendencies—perhaps you've witnessed the way border collies and sheepdogs will herd any animal!—though genetics clearly isn't everything. A dog's behavior is influenced not only by its genes but also by its environment and care.

In this chapter, you will learn the basic rules of how genetic traits are passed from generation to generation and how the behavior of chromosomes (the topic of Chapter 8) accounts for these rules. We will look at several inheritance patterns and see how to predict the ratios of offspring with particular traits. At several points in the chapter, we will return to the subject of dog breeding to help illustrate genetic principles.

Heritable Variation and Patterns of Inheritance

Heredity is the transmission of traits from one generation to the next. **Genetics**, the scientific study of heredity, began in the 1860s, when an Augustinian monk named Gregor Mendel deduced its fundamental principles by breeding garden peas (**Figure 9.1**). Mendel lived and worked in an abbey in Brunn, Austria (now Brno, in the Czech Republic). Strongly influenced by his study of physics, mathematics, and chemistry at the University of Vienna, his research was both experimentally and mathematically rigorous, and these qualities were largely responsible for his success.

▲ **Figure 9.1**
Gregor Mendel.

In a paper published in 1866, Mendel correctly argued that parents pass on to their offspring discrete "heritable factors" that are responsible for inherited traits, such as purple flowers or round seeds in pea plants. (It is interesting to note that Mendel's publication came just seven years after Darwin's 1859 publication of *The Origin of Species*, making the 1860s a banner decade in the development of modern biology.) In his paper, Mendel stressed that the heritable factors (today called genes) retain their individual identities generation after generation, no matter how they are mixed up or temporarily masked.

reach the carpel. When he wanted cross-fertilization (fertilization of one plant by pollen from a different plant), he pollinated the plants by hand, as shown in **Figure 9.3**. Thus, whether Mendel let a pea plant self-fertilize or cross-fertilized it with a known source of pollen, he could always be sure of the parentage of his new plants.

Each of the characters Mendel chose to study, such as flower color, occurred in two distinct forms. Mendel worked with his plants until he was sure he had **true-breeding** (purebred) varieties—that is, varieties for which self-fertilization produced offspring all identical to the parent. For instance, he identified a purple-flowered variety

In an Abbey Garden

Mendel probably chose to study garden peas because they were easy to grow and they came in many readily distinguishable varieties. For example, one variety has purple flowers, and another variety has white flowers. A heritable feature that varies among individuals, such as flower color, is called a **character**. A variant of a character, such as purple or white flowers, is called a **trait**.

Perhaps the most important advantage of pea plants as an experimental model was that Mendel could strictly control their reproduction. The petals of the pea flower almost completely enclose the egg- and sperm-producing parts—the carpel and stamens, respectively (**Figure 9.2**). Consequently, in nature, pea plants usually self-fertilize because sperm-carrying pollen grains released from the stamens land on the tip of the egg-containing carpel of the same flower. Mendel could ensure self-fertilization by covering a flower with a small bag so that no pollen from another plant could

▼ **Figure 9.2 The structure of a pea flower.** To reveal the reproductive organs—the stamens and carpel—one of the petals has been removed in this drawing.

Petal

Stamen (makes sperm-producing pollen)

Carpel (produces eggs)

▼ **Figure 9.3 Mendel's technique for cross-fertilizing pea plants.**

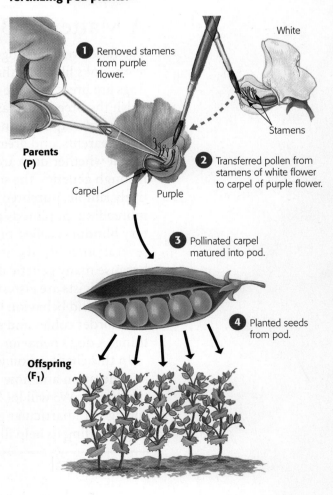

White

1 Removed stamens from purple flower.

Stamens

Parents (P)

2 Transferred pollen from stamens of white flower to carpel of purple flower.

Carpel   Purple

3 Pollinated carpel matured into pod.

4 Planted seeds from pod.

Offspring (F₁)

that, when self-fertilized, always produced offspring plants that had all purple flowers.

Now Mendel was ready to ask what would happen when he crossed different true-breeding varieties with each other. For example, what offspring would result if plants with purple flowers and plants with white flowers were cross-fertilized as shown in Figure 9.3? In the language of breeders and geneticists, the offspring of two different true-breeding varieties are called **hybrids**, and the cross-fertilization itself is referred to as a genetic **cross**. The parental plants are called the **P generation**, and their hybrid offspring are the **F₁ generation** (F for *filial*, from the Latin for "son" or "daughter"). When F₁ plants self-fertilize or fertilize each other, their offspring are the **F₂ generation**. ✓

Mendel's Law of Segregation

Mendel performed many experiments in which he tracked the inheritance of characters, such as flower color, that occur as two alternative traits (**Figure 9.4**). The results led him to formulate several hypotheses about inheritance. Let's look at some of his experiments and follow the reasoning that led to his hypotheses.

▼ Figure 9.4 **The seven characters of pea plants studied by Mendel.** Each character comes in the two alternative traits shown here.

	Dominant	Recessive
Flower color	Purple	White
Flower position	Axial	Terminal
Seed color	Yellow	Green
Seed shape	Round	Wrinkled
Pod shape	Inflated	Constricted
Pod color	Green	Yellow
Stem length	Tall	Dwarf

HERITABLE VARIATION AND PATTERNS OF INHERITANCE

✓**CHECKPOINT**

Why was the development of true-breeding pea plant varieties critical to Mendel's work?

Answer: True-breeding varieties allowed Mendel to predict the outcome of specific crosses and therefore to run controlled experiments.

Monohybrid Crosses

Figure 9.5 shows a cross between a true-breeding pea plant with purple flowers and a true-breeding pea plant with white flowers. This is called a **monohybrid cross** because the parent plants differ in only one character. Mendel saw that the F₁ plants all had purple flowers. Was the heritable factor for white flowers now lost as a result of the cross? By mating the F₁ plants with each other, Mendel found the answer to be no. Of the 929 F₂ plants, about three-fourths (705) had purple flowers and one-fourth (224) had white flowers; that is, there were about three purple F₂ plants for every white plant, or a 3:1 ratio of purple to white. Mendel concluded that the heritable factor for white flowers did not disappear in the F₁ plants, but was somehow hidden or masked when the purple-flower factor was present. He also deduced that the F₁ plants must have carried two factors for the flower-color character, one for purple and one for white. From these results and others, Mendel developed four hypotheses. Using modern terminology (including "gene" instead of "heritable factor"), here are his hypotheses:

1. *There are alternative versions of genes, the units that determine heritable traits.* For example, the gene for flower color in pea plants exists in one form for purple and another for white. The alternative versions of a gene are called **alleles**.

2. *For each inherited character, an organism inherits two alleles, one from each parent.* These alleles may be the same or different. An organism that has two identical alleles for a gene is said to be **homozygous** for that gene (and is called a homozygote). An organism that has two different alleles for a gene is said to be **heterozygous** for that gene (and is called a heterozygote).

3. *If the two alleles of an inherited pair differ, then one determines the organism's appearance and is called the* **dominant allele**; *the other has no noticeable effect on the organism's appearance and is called the* **recessive allele**. We use uppercase italic letters to represent dominant alleles and lowercase italic letters to represent recessive alleles.

4. *A sperm or egg carries only one allele for each inherited character because the two members of an allele pair segregate (separate) from each other during the production of gametes.* This statement is now known as the **law of segregation**. When sperm and egg unite at fertilization, each contributes its alleles, restoring the paired condition in the offspring.

Do Mendel's hypotheses account for the 3:1 ratio he observed in the F₂ generation? **Figure 9.6** illustrates Mendel's law of segregation, which explains the inheritance pattern shown in Figure 9.5. His hypotheses predict that when alleles segregate during gamete formation in the F₁ plants, half the gametes will receive a purple-flower allele (P) and the other half a white-flower allele (p). During pollination among the F₁ plants, the gametes unite randomly. An egg with a purple-flower allele has an equal chance of being fertilized by a sperm with a purple-flower allele or one with a white-flower allele (that is, a P egg may fuse with a P sperm or a p sperm). Because the same is true for an egg with a white-flower allele (a p egg with a P sperm or p sperm), there are a total of four equally likely combinations of sperm and egg.

The diagram at the bottom of Figure 9.6, called a **Punnett square**, repeats the cross shown in Figure 9.5 in a way that highlights the four possible combinations of gametes and the offspring that result from each. Each square represents an equally probable product of fertilization. For example, the box in the upper right corner of the Punnett square shows the genetic combination resulting from a p sperm fertilizing a P egg.

According to the Punnett square, what will be the physical appearance of these F₂ offspring? One-fourth of the plants have two alleles specifying purple flowers (PP); clearly, these plants will have purple flowers. One-half (two-fourths) of the F₂ offspring have

Mendel attending his peas

▼ **Figure 9.5 Mendel's cross tracking one character (flower color).** Note the 3:1 ratio of purple flowers to white flowers in the F₂ generation.

P Generation
(true-breeding parents)

Purple flowers × White flowers

F₁ Generation

All plants have purple flowers

Fertilization among F₁ plants (F₁ × F₁)

F₂ Generation

$\frac{3}{4}$ of plants have purple flowers

$\frac{1}{4}$ of plants have white flowers

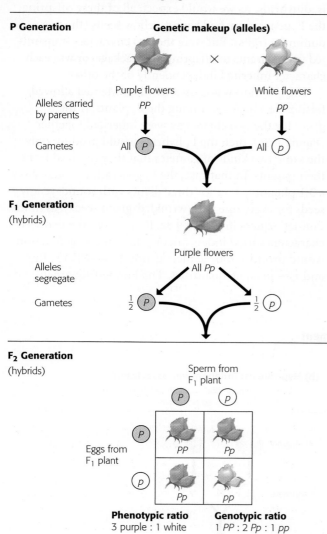

▼ Figure 9.6 **The law of segregation.**

P Generation

Genetic makeup (alleles)

Purple flowers
PP

×

White flowers
pp

Alleles carried
by parents

Gametes   All (P)   All (p)

F₁ Generation
(hybrids)

Purple flowers
All *Pp*

Alleles
segregate

Gametes   ½ (P)   ½ (p)

F₂ Generation
(hybrids)

Sperm from
F₁ plant

(P)   (p)

Eggs from
F₁ plant

(P)   *PP*   *Pp*

(p)   *Pp*   *pp*

Phenotypic ratio
3 purple : 1 white

Genotypic ratio
1 *PP* : 2 *Pp* : 1 *pp*

Mendel found that each of the seven characters he studied had the same inheritance pattern: One parental trait disappeared in the F₁ generation, only to reappear in one-fourth of the F₂ offspring. The underlying mechanism is stated by Mendel's law of segregation: Pairs of alleles segregate (separate) during gamete formation; the fusion of gametes at fertilization creates allele pairs again. Research since Mendel's day has established that the law of segregation applies to all sexually reproducing organisms, including dogs and humans.

Genetic Alleles and Homologous Chromosomes

Before continuing with Mendel's experiments, let's see how some of the concepts we discussed in Chapter 8 fit with what we've said about genetics so far. The diagram in **Figure 9.7** shows a pair of homologous chromosomes—chromosomes that carry alleles of the same genes. Recall from Chapter 8 that every diploid individual, whether pea plant or human, has chromosomes in homologous pairs. One member of each pair comes from the organism's female parent; the other member of each pair comes from the male parent. The labeled bands on the chromosomes in the figure represent three gene **loci** (singular, *locus*), specific locations of genes along the chromosome. You can see the connection between Mendel's law of segregation and homologous chromosomes: Alleles (alternative versions) of a gene reside at the same locus on homologous chromosomes. However, the two chromosomes may bear either identical alleles or different ones at any one locus. In other words, the organisms may be homozygous or heterozygous for the gene at that locus. We will return to the chromosomal basis of Mendel's law later in the chapter. ✔

inherited one allele for purple flowers and one allele for white flowers (*Pp*); like the F₁ plants, these plants will also have purple flowers, the dominant trait. (Note that *Pp* and *pP* are equivalent and usually written as the former.) Finally, one-fourth of the F₂ plants have inherited two alleles specifying white flowers (*pp*) and will express this recessive trait. Thus, Mendel's model accounts for the 3:1 ratio that he observed in the F₂ generation.

Geneticists distinguish between an organism's physical traits, called its **phenotype** (such as purple or white flowers), and its genetic makeup, called its **genotype** (in our example, *PP*, *Pp*, or *pp*). Now we can see that Figure 9.5 shows only the phenotypes and Figure 9.6 both the genotypes and phenotypes in our sample cross. For the F₂ plants, the ratio of plants with purple flowers to those with white flowers (3:1) is called the phenotypic ratio. The genotypic ratio is 1(*PP*):2(*Pp*):1(*pp*).

▼ Figure 9.7 **The relationship between alleles and homologous chromosomes.** The matching colors of corresponding loci highlight the fact that homologous chromosomes carry alleles for the same genes at the same positions along their lengths.

Gene loci

Dominant allele

Homologous chromosomes

P   *a*   *B*

P   *a*   *b*

Recessive allele

Genotype: *PP*   *aa*   *Bb*

Homozygous
for the
dominant allele

Homozygous
for the
recessive allele

Heterozygous

Mendel's Law of Independent Assortment

Two other pea plant characters Mendel studied were seed shape and seed color. Mendel's seeds were either round or wrinkled in shape and either yellow or green in color. From tracking these characters one at a time in monohybrid crosses, Mendel knew that the allele for round shape (designated R) was dominant to the allele for wrinkled shape (r) and that the allele for yellow seed color (Y) was dominant to the allele for green seed color (y). What would result from a **dihybrid cross**, the crossing of parental varieties differing in two characters? Mendel crossed homozygous plants having round-yellow seeds (genotype $RRYY$) with plants having wrinkled-green seeds ($rryy$). As shown in **Figure 9.8**, the union of RY and ry gametes from the P generation yielded

hybrids heterozygous for both characters ($RrYy$)—that is, dihybrids. As we would expect, all of these offspring, the F_1 generation, had round-yellow seeds (the two dominant traits). But were the two characters transmitted from parents to offspring as a package, or was each character inherited independently of the other?

The question was answered when Mendel allowed fertilization to occur among the F_1 plants. If the genes for the two characters were inherited together (Figure 9.8a), then the F_1 hybrids would produce only the same two kinds of gametes that they received from their parents. In that case, the F_2 generation would show a 3:1 phenotypic ratio (three plants with round-yellow seeds for every one with wrinkled-green seeds), as in the Punnett square in Figure 9.8a. If, however, the two seed characters sorted independently, then the F_1 generation would produce four gamete genotypes—RY, rY, Ry, and ry—in equal quantities. The Punnett square in

▼ **Figure 9.8 Testing alternative hypotheses for gene assortment in a dihybrid cross.**

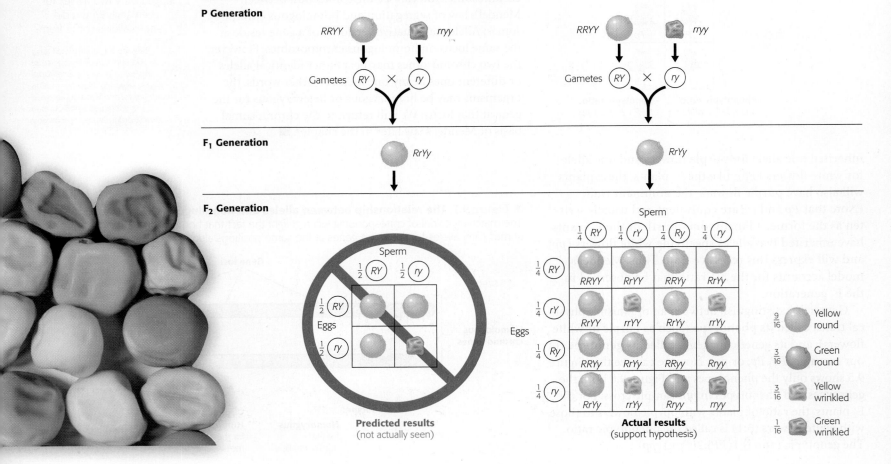

(a) Hypothesis: Dependent assortment
Leads to prediction that F_2 plants will have seeds that match the parents, either round-yellow or wrinkled-green

(b) Hypothesis: Independent assortment
Leads to prediction that F_2 plants will have four different seed phenotypes

P Generation

F₁ Generation

F₂ Generation

Predicted results
(not actually seen)

Actual results
(support hypothesis)

$\frac{9}{16}$ Yellow round

$\frac{3}{16}$ Green round

$\frac{3}{16}$ Yellow wrinkled

$\frac{1}{16}$ Green wrinkled

Figure 9.8b shows all possible combinations of alleles that can result in the F_2 generation from the union of four kinds of sperm with four kinds of eggs. If you study the Punnett square, you'll see that it predicts nine different genotypes in the F_2 generation. These nine genotypes will produce four different phenotypes in a ratio of 9:3:3:1.

The Punnett square in Figure 9.8b also reveals that a dihybrid cross is equivalent to two monohybrid crosses occurring simultaneously. From the 9:3:3:1 ratio, we can see that the ratio of plants with round seeds to those with wrinkled seeds is 12:4, as is the ratio of yellow-seeded plants to green-seeded ones. These 12:4 ratios each reduce to 3:1, which is the F_2 ratio for a monohybrid cross. Mendel tried his seven pea characters in various dihybrid combinations and always observed a 9:3:3:1 ratio (or two simultaneous 3:1 ratios) of phenotypes in the F_2 generation. These results supported the hypothesis that *each pair of alleles assorts independently of the other pairs of alleles during gamete formation*. In other words, the inheritance of one character has no effect on the inheritance of another. This is called the **law of independent assortment**.

For another application of the law of independent assortment, examine **Figure 9.9**. The inheritance of two hereditary characters in Labrador retrievers is controlled by separate genes: black versus chocolate coat color, and normal vision versus the eye disorder progressive retinal atrophy (PRA). Black Labs have at least one copy of an allele called *B*, which gives their hairs densely packed granules of a dark pigment. The *B* allele is dominant to *b*, which leads to a less tightly packed distribution of pigment granules. As a result, the coats of dogs with genotype *bb* are chocolate in color. The allele that causes PRA, called *n*, is recessive to allele *N*, which is necessary for normal vision. Thus, only dogs of genotype *nn* become blind from PRA. If you mate two doubly heterozygous (*BbNn*) Labs (bottom of Figure 9.9), the phenotypic ratio of the offspring (F_2) is 9:3:3:1. These results resemble the F_2 results in Figure 9.8, demonstrating that the coat color and PRA genes are inherited independently. ✓

✓CHECKPOINT
Looking at the Punnett square in Figure 9.8b, what is the ratio of yellow seeds to green seeds? Of round seeds to wrinkled seeds?

Answer: 3:1; 3:1

▼ **Figure 9.9 Independent assortment of genes in Labrador retrievers.**
Blanks in the genotypes indicate alleles that can be either dominant or recessive.

Phenotypes	Black coat, normal vision	Black coat, blind (PRA)	Chocolate coat, normal vision	Chocolate coat, blind (PRA)
Genotypes	B_N_	B_nn	bbN_	bbnn

(a) Possible phenotypes of Labrador retrievers

Mating of double heterozygotes
(black coat, normal vision)
BbNn ⎯ × ⎯ BbNn

| Phenotypic ratio of offspring | 9 black coat, normal vision | 3 black coat, blind (PRA) | 3 chocolate coat, normal vision | 1 chocolate coat, blind (PRA) |

(b) A Labrador dihybrid cross

Using a Testcross to Determine an Unknown Genotype

Suppose you have a Labrador retriever with a chocolate coat. Consulting Figure 9.9, you can tell that its genotype must be *bb*, the only combination of alleles that produces the chocolate-coat phenotype. But what if you have a black Lab? It could have one of two possible genotypes—*BB* or *Bb*—and there is no way to tell which is correct by looking at the dog. To determine your dog's genotype, you could perform a **testcross**, a mating between an individual of dominant phenotype but unknown genotype (your black Lab) and a homozygous recessive individual—in this case, a *bb* chocolate Lab.

Figure 9.10 shows the offspring that could result from such a mating. If, as shown on the left, the black parent's genotype is *BB*, we would expect all the offspring to be black, because a cross between genotypes *BB* and *bb* can produce only *Bb* offspring. On the other hand, if the black parent is *Bb*, we would expect both black (*Bb*) and chocolate (*bb*) offspring in a 1:1 phenotypic ratio. Thus, the appearance of the offspring may reveal the original black dog's genotype. ☑

▼ **Figure 9.10 A Labrador retriever testcross.** To determine the genotype of a black Lab, it can be crossed with a chocolate Lab (homozygous recessive, *bb*). If all the offspring are black, the black parent most likely had genotype *BB*. If half the offspring are chocolate, the black parent must be heterozygous (*Bb*).

Testcross		
Genotypes	*B_*	*bb*

Two possible genotypes for the black dog:

BB or *Bb*

Gametes	*B*		*B*	*b*
Offspring	**Bb**		**Bb**	**bb**
	All black		1 black : 1 chocolate	

The Rules of Probability

Mendel's strong background in mathematics served him well in his studies of inheritance. For instance, he understood that genetic crosses obey the rules of probability—the same rules that apply to the tossing of coins, the rolling of dice, and the drawing of cards. An important lesson we can learn from coin tossing is that for each and every toss of the coin, the probability of heads is $\frac{1}{2}$. Even if heads has landed five times in a row, the probability of the next toss coming up heads is still $\frac{1}{2}$. In other words, the outcome of any particular toss is unaffected by what has happened on previous attempts. Each toss is an independent event.

If two coins are tossed simultaneously, the outcome for each coin is an independent event, unaffected by the other coin. What is the chance that both coins will land heads-up? The probability of such a dual event is the product of the separate probabilities of the independent events—for the coins, $\frac{1}{2} \times \frac{1}{2} = \frac{1}{4}$. This is called the **rule of multiplication**, and it holds true for independent events that occur in genetics as well as coin tosses, as shown in **Figure 9.11**. In our dihybrid cross of Labradors (see

▼ **Figure 9.11 Segregation of alleles and fertilization as chance events.** When a heterozygote (*Bb*) forms gametes, segregation of alleles during sperm and egg formation is like two separately tossed coins (that is, two independent events).

F₁ Genotypes

Bb female × *Bb* male

Formation of eggs   Formation of sperm

F₂ Genotypes

Male gametes

Female gametes

$\frac{1}{4}$ $\left(\frac{1}{2} \times \frac{1}{2}\right)$ $\frac{1}{4}$

$\frac{1}{4}$ $\frac{1}{4}$

Figure 9.9), the genotype of the F₁ dogs for coat color was Bb. What is the probability that a particular F₂ dog will have the bb genotype? To produce a bb offspring, both egg and sperm must carry the b allele. The probability that an egg from a Bb dog will have the b allele is $\frac{1}{2}$, and the probability that a sperm will have the b allele is also $\frac{1}{2}$. By the rule of multiplication, the probability that two b alleles will come together at fertilization is $\frac{1}{2} \times \frac{1}{2} = \frac{1}{4}$. This is exactly the answer given by the Punnett square in Figure 9.11. If we know the genotypes of the parents, we can predict the probability for any genotype among the offspring. By applying the rules of probability to segregation and independent assortment, we can solve some rather complex genetics problems. ☑

Family Pedigrees

Mendel's laws apply to the inheritance of many human traits. **Figure 9.12** illustrates alternative forms of three human characters that are each thought to be determined by simple dominant-recessive inheritance at one gene locus. (The genetic basis of many other human characters—such as eye and hair color—are much more complex and poorly understood.) If we call the dominant allele of any such gene A, the dominant

phenotype results from either the homozygous genotype AA or the heterozygous genotype Aa. Recessive phenotypes always result from the homozygous genotype aa. In genetics, the word *dominant* does not imply that a phenotype is either normal or more common than a recessive phenotype; **wild-type traits** (those seen most often in nature) are not necessarily specified by dominant alleles. In genetics, dominance means that a heterozygote (Aa), carrying only a single copy of a dominant allele, displays the dominant phenotype. By contrast, the phenotype of the corresponding recessive allele is seen only in a homozygote (aa). Recessive traits can be more common in the population than dominant ones. For example, the absence of freckles is more common than their presence.

How do we know how particular human traits are inherited? Researchers working with pea plants or Labrador retrievers can perform testcrosses. But geneticists who study humans obviously cannot control the mating of their subjects. Instead, they must analyze the results of matings that have already occurred. First, the geneticist collects as much information as possible about a family's history for the trait. Then the researcher assembles this information into a family tree—the family **pedigree**. (You may associate pedigrees with purebred animals such as racehorses and champion dogs, but they

☑ **CHECKPOINT**

Using a standard 52-card deck, what is the probability of being dealt an ace? What about being dealt an ace or a king? What about being dealt an ace and then another ace?

Answer: $\frac{1}{13}$ (4 aces / 52 cards); $\frac{2}{13}$ ($\frac{4}{52} + \frac{4}{52}$); $\frac{4}{52} \times \frac{3}{51}$ (since there are 3 aces left in a deck with 51 cards remaining) = 0.0045, or $\frac{1}{221}$

DOMINANT TRAITS

Freckles

Widow's peak

Free earlobe

RECESSIVE TRAITS

No freckles

Straight hairline

Attached earlobe

◄ **Figure 9.12 Examples of inherited traits in humans thought to be controlled by a single gene.**

can represent human matings just as well.) To analyze a pedigree, the geneticist uses Mendel's concept of dominant and recessive alleles and his law of segregation.

Let's apply this approach to the example in **Figure 9.13**, which shows a pedigree tracing the incidence of free versus attached earlobes. The letter *F* stands for the dominant allele for free earlobes, and *f* symbolizes the recessive allele for attached earlobes. In the pedigree, □ represent males, ○ represent females, and colored symbols (■ and ●) indicate that the person has the trait under investigation (in this case, attached earlobes). The earliest generation studied is at the top of the pedigree. By applying Mendel's laws, we can deduce that the attached allele is recessive because that is the only way that one of the third-generation children (at the bottom of the pedigree) could have attached earlobes when both parents did not. We can therefore label all the individuals with attached earlobes in the pedigree (that is, all those with colored circles or squares) as homozygous recessive (*ff*). Mendel's laws also enable us to deduce the genotypes for most of the people in the pedigree. For example, both of the second-generation parents must have carried the *f* allele (which they passed on to the affected child) along with the *F* allele that gave them free earlobes. The same must be true of the set of grandparents on the left because they both had free earlobes but two of their sons had attached earlobes. Notice that we cannot deduce the genotype of every member of the pedigree. For example, the sister with the free earlobes must

have at least one *F* allele, but it is possible that she could be *FF* or *Ff*. We cannot distinguish between these two possibilities using the available data. ☑

Human Disorders Controlled by a Single Gene

The human genetic disorders listed in **Table 9.1** are known to be inherited as dominant or recessive traits controlled by a single gene. These disorders therefore show simple inheritance patterns like the ones Mendel studied in pea plants. The genes involved are all located on autosomes, chromosomes other than the sex chromosomes X and Y.

Recessive Disorders

Most human genetic disorders are recessive. They range in severity from relatively harmless to life-threatening. Most people who have recessive disorders are born to normal parents who are both heterozygotes—that is, who are **carriers** of the recessive allele for the disorder but appear normal themselves.

Using Mendel's laws, we can predict the fraction of affected offspring that is likely to result from a marriage between two carriers. Consider a form of inherited deafness caused by a recessive allele. Suppose two heterozygous carriers (*Dd*) have a child. What is the probability that the child will be deaf? As the Punnett square in

► Figure 9.13 **A family pedigree showing inheritance of free versus attached earlobes.**

First generation
(grandparents)

Ff *Ff* *ff* *Ff*

Second generation
(parents, aunts, and uncles)

FF or Ff *ff* *ff* *Ff* *Ff* *ff*

Third generation
(brother and sister)

ff *FF or Ff*

Female   Male

● ■ Attached

○ □ Free

Table 9.1	Some Autosomal Disorders in Humans	
Disorder	**Major Symptoms**	**Incidence**
Recessive Disorders		
Albinism	Lack of pigment in skin, hair, and eyes	$\frac{1}{22,000}$
Cystic fibrosis	Excess mucus in lungs, digestive tract, liver; increased susceptibility to infections; death in early childhood unless treated	$\frac{1}{1,800}$ European-Americans
Phenylketonuria (PKU)	Accumulation of phenylalanine in blood; lack of normal skin pigment; mental retardation unless treated	$\frac{1}{10,000}$ in U.S. and Europe
Sickle-cell disease	Sickled red blood cells; damage to many tissues	$\frac{1}{500}$ African-Americans
Tay Sachs disease	Lipid accumulation in brain cells; mental deficiency; blindness; death in childhood	$\frac{1}{3,500}$ European Jews
Dominant Disorders		
Achondroplasia	Dwarfism	$\frac{1}{25,000}$
Alzheimer's disease (one type)	Mental deterioration; usually strikes late in life	Not known
Huntington's disease	Mental deterioration and uncontrollable movements; strikes in middle age	$\frac{1}{25,000}$
Hypercholesterolemia	Excess cholesterol in blood; heart disease	$\frac{1}{500}$

Figure 9.14 shows, each child of two carriers has a $\frac{1}{4}$ chance of inheriting two recessive alleles. Thus, we can say that about one-fourth of the children of this marriage are likely to be deaf. We can also say that a hearing child from such a marriage has a $\frac{2}{3}$ chance of being a carrier (that is, on average, two out of three of the offspring with the hearing phenotype will be *Dd*). We can apply this same method of pedigree analysis and prediction to any genetic trait controlled by a single gene locus.

The most common lethal genetic disease in the United States is **cystic fibrosis**. Affecting about 30,000 people in the United States and about 70,000 people worldwide, the cystic fibrosis allele is recessive and carried by about 1 in 25 people of European ancestry. A person with two copies of this allele has cystic fibrosis, which is characterized by an excessive secretion of very thick mucus from the lungs, pancreas, and other organs. This mucus can interfere with breathing, digestion, and liver function and makes the person vulnerable to recurrent bacterial infections. Although there is no cure for this fatal disease, a special diet, antibiotics to prevent infection, frequent pounding of the chest and back to clear the lungs, and other treatments can greatly extend life. Once invariably fatal in childhood, advances in cystic fibrosis treatment have raised the median survival age of Americans with cystic fibrosis to 37.

Like cystic fibrosis, most genetic disorders are not evenly distributed across all ethnic groups. Such uneven distribution is the result of prolonged geographic isolation of certain populations. For example, the isolated lives of the early inhabitants of Martha's Vineyard (an island off the coast of Massachusetts) led to frequent marriages between close relatives. Consequently, the frequency of an allele that caused deafness was high, and the deafness allele was rarely transmitted to outsiders.

With the increased mobility in most societies today, it is relatively unlikely that two carriers of a rare, harmful allele will meet and mate. However, the probability increases greatly if close blood relatives marry and have children. People with recent common ancestors are more likely to carry the same recessive alleles than are unrelated people. Therefore, a mating between close blood relatives, called **inbreeding**, is more likely to produce offspring homozygous for a harmful recessive trait.

▼ **Figure 9.14 Predicted offspring when both parents are carriers for a recessive disorder.**

Parents

Hearing
Dd × Hearing
Dd

Offspring

	Sperm	
	D	*d*
D	*DD* Hearing	*Dd* Hearing (carrier)
d	*Dd* Hearing (carrier)	*dd* Deaf

Eggs

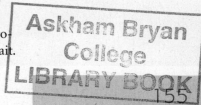

Geneticists have observed increased incidence of harmful recessive traits among many types of inbred animals. Most lines of purebred dogs have known genetic defects. The Siberian huskies discussed in the Biology and Society section, for example, are susceptible to the same form of blindness (PRA) as the retrievers shown in Figure 9.9. The detrimental effects of inbreeding are also seen in some endangered species, such as cheetahs (see Chapter 13).

Dominant Disorders

A number of human disorders are caused by dominant alleles. Some are nonlethal conditions, such as extra fingers and toes or fingers and toes that are webbed. A serious but nonlethal disorder caused by a dominant allele is **achondroplasia**, a form of dwarfism in which the head and torso develop normally, but the arms and legs are short (**Figure 9.15**). The homozygous dominant genotype causes death of the embryo, and therefore only heterozygotes, individuals with a single copy of the defective allele, have this disorder. This also means that a person with achondroplasia has a 50% chance of passing the condition on to any children (**Figure 9.16**). Therefore, all those who do not have achondroplasia, more than 99.99% of the population, are homozygous for the recessive allele. This example makes it clear that a dominant allele is not necessarily more plentiful in a population than the corresponding recessive allele.

Dominant alleles that are lethal are much less common than lethal recessive alleles. One example is the allele that causes **Huntington's disease**, a degeneration of the nervous system that usually does not begin until middle age. Once the deterioration of the nervous system begins, it is irreversible and inevitably fatal. Because the allele for Huntington's disease is dominant, any child born to a parent with the allele has a 50% chance of inheriting the allele and the disorder. This example makes it clear that a dominant allele is not necessarily "better" than the corresponding recessive allele.

The allele that causes Huntington's disease is dominant but lethal when inherited in two copies. Let's return to the chapter thread—dog breeding—to investigate a striking dog trait with a similar inheritance pattern.

▼ **Figure 9.15**
Achondroplasia, a dominant trait. Dr. Michael C. Ain, a pediatric orthopedic surgeon at the Johns Hopkins Children's Center, specializes in the repair of bone defects caused by achondroplasia and related disorders.

▼ **Figure 9.16 A family with and without achondroplasia.** Amy Roloff (who has achondroplasia) and her husband Matt (who has an unrelated form of dwarfism) have four children. Zachary received Amy's achondroplasia gene, while twin brother Jeremy, sister Molly Jo, and brother Jake did not. The hypothetical Punnett square shows the offspring expected from a mating of a wild-type male with a dwarf female.

Parents

Normal
(no achondroplasia)   ×   Drawf
(achondroplasia)
dd   *Dd*

Sperm (*d*)   (*d*)

Eggs

	d	*d*
D	*Dd* Dwarf	*Dd* Dwarf
d	*dd* Normal	*dd* Normal

Molly Jo   Matt   Amy   Zachary   Jake   Jeremy

THE PROCESS OF SCIENCE: Dog Breeding

What Is the Genetic Basis of Hairless Dogs?

You've probably made the **observation** that compared to most mammals, dogs come in a wide variety of physical types. For instance, some dogs come in both hairless and thick-coated varieties (**Figure 9.17**). Breeders know from experience that the hairless phenotype is inherited as a dominant trait and that, as in achondroplasia, the homozygous dominant condition is lethal.

In 2005, the complete genome of a dog—the sequence of all the DNA of a female boxer named Tasha—was published. Since that time, canine geneticists have added a wealth of data from other breeds to that first genome. In 2008, an international group of researchers set out to investigate the **question** of the genetic basis for the hairless phenotype. The research team proposed the **hypothesis** that a comparison of genes of coated and hairless dogs would identify the gene or genes responsible. They made the **prediction** that a mutation in a single gene could account for the hairless appearance. Their **experiment** compared DNA sequences in 140 hairless dogs from 3 breeds with similar sequences in 87 coated

◄ Figure 9.17 **Hairless versus coated Chinese crested dogs.** The Chinese crested dog comes in both naturally hairless (left) and thick-coated (right) varieties.

dogs from 22 breeds. The **results** were clear: Every one of the hairless dogs, but none of the coated dogs, had a change in a single gene on chromosome 17. Due to the insertion of seven extra DNA bases, the chromosome 17 mutation knocks out a single protein that is involved in embryonic development. This experiment shows how the extreme range of phenotypes in dogs combined with the availability of genome sequences can be used to provide insight into interesting genetic questions.

Genetic Testing

Until relatively recently, the onset of symptoms was the only way to know if a person had inherited an allele that might lead to disease. Today, there are many tests that can detect the presence of disease-causing alleles in an individual's genome.

The most common forms of genetic testing are performed during pregnancy. Some prospective parents are aware that they have an increased risk of having a baby with a genetic disease. Genetic testing before birth usually requires the collection of fetal cells. In amniocentesis, a physician uses a needle to extract about 2 teaspoonfuls of the fluid that bathes the developing fetus. In chorionic villus sampling, a physician inserts a narrow, flexible tube through the mother's vagina and into her uterus, removing some placental tissue. Once cells are obtained, they can be screened for genetic diseases.

Because amniocentesis and chorionic villus sampling have risks of complications, these techniques are usually reserved for situations in which the possibility of a genetic disease is significantly higher than average. Newer

genetic screening procedures involve isolating tiny amounts of fetal cells or DNA released into the mother's bloodstream. Although few reliable tests are yet available using this method, this promising and complication-free technology may soon replace more invasive procedures.

As genetic testing becomes more routine, geneticists are working to make sure that the tests do not cause more problems than they solve. Geneticists stress that patients seeking genetic testing should receive counseling both before and after to explain the test and to help them cope with the results. But will sufficient numbers of genetic counselors be available to help individuals understand their test results? If fetal tests reveal a serious disorder, the parents must choose between terminating the pregnancy and preparing themselves for a baby with severe problems. Identifying a genetic disease early can give families time to prepare—emotionally, medically, and financially. Advances in biotechnology offer possibilities for reducing human suffering, but not before key ethical issues are resolved. The dilemmas posed by human genetics reinforce one of this book's themes: the immense social implications of biology. ☑

Variations on Mendel's Laws

Mendel's two laws explain inheritance in terms of discrete factors—genes—that are passed along from generation to generation according to simple rules of probability. These laws are valid for all sexually reproducing organisms, including garden peas, Siberian huskies, and human beings. But just as the basic rules of musical harmony cannot account for all the rich sounds of a symphony, Mendel's laws stop short of explaining some patterns of genetic inheritance. In particular, they do not explain the many cases of inherited characters that exist in more than two clear-cut variants—such as flower color that can be red, white, or pink or human skin color in all its range of shades. In fact, for most sexually reproducing organisms, cases where Mendel's rules can strictly account for the patterns of inheritance are relatively rare. More often, the observed inheritance patterns are more complex. We will now add several extensions to Mendel's laws that help account for this complexity.

Incomplete Dominance in Plants and People

The F_1 offspring of Mendel's pea crosses always looked like one of the two parent plants. In such situations, the dominant allele has the same effect on the phenotype whether present in one or two copies. But for some characters, the F_1 hybrids have an appearance in between the phenotypes of the two parents, an effect called **incomplete dominance**. For instance, when red snapdragons are crossed with white snapdragons, all the F_1 hybrids have pink flowers (**Figure 9.18**). And in the F_2 generation, the genotypic ratio and the phenotypic ratio are the same: 1:2:1.

We also see examples of incomplete dominance in humans. One case involves a recessive allele (h) that causes **hypercholesterolemia**, a condition characterized by dangerously high levels of cholesterol in the blood. Normal individuals are homozygous dominant, HH. Heterozygotes (Hh) have blood cholesterol levels about twice normal. They are unusually prone to cholesterol buildup in artery walls and may have heart attacks from blocked heart arteries by their mid-30s. Hypercholesterolemia is even more serious in homozygous individuals (hh). These homozygotes have about five times the normal amount of blood cholesterol and may have heart attacks as early as age 2. If we look at the molecular basis for hypercholesterolemia, we can understand the intermediate phenotype of heterozygotes (**Figure 9.19**). The H allele specifies a cell-surface receptor protein that

▼ **Figure 9.18 Incomplete dominance in snapdragons.**
Compare this diagram with Figure 9.6, where one of the alleles displays complete dominance.

P Generation

Red
RR × White
rr

Gametes (*R*) (*r*)

F₁ Generation

Pink
Rr

Gametes ½ (*R*) ½ (*r*)

F₂ Generation

Sperm

½ (*R*) ½ (*r*)

Eggs

½ (*R*) | *RR* | *Rr* |
½ (*r*) | *Rr* | *rr* |

▼ **Figure 9.19 Incomplete dominance in human hypercholesterolemia.**
LDL receptors promote the breakdown of cholesterol carried in the bloodstream by LDL (low-density lipoproteins). This process helps prevent the accumulation of cholesterol in the arteries. Having too few receptors allows dangerous levels of LDL to build up in the blood.

GENOTYPE

| ***HH*** | ***Hh*** | ***hh*** |
| Homozygous for ability to make LDL receptors | Heterozygous | Homozygous for inability to make LDL receptors |

PHENOTYPE

LDL

LDL receptor

Cell

Normal | Mild disease | Severe disease

certain cells use to mop up excess low-density lipo-protein (LDL, or "bad cholesterol") from the blood. With only half as many receptors as *HH* individuals, heterozygotes can remove much less excess cholesterol.

ABO Blood Groups: An Example of Multiple Alleles and Codominance

So far, we have discussed inheritance patterns involving only two alleles per gene (*H* versus *h*, for example). But most genes occur in more than two forms, known as multiple alleles. Although each individual carries, at most, two different alleles for a particular gene, in cases of multiple alleles, more than two possible alleles exist in the population.

The **ABO blood groups** in humans are an example of multiple alleles. There are three common alleles for the character of ABO blood type, which in various combinations produce four phenotypes: A person's blood type may be A, B, AB, or O. These letters refer to two carbohydrates, designated A and B, that may be found on the surface of red blood cells (**Figure 9.20**). A person's red blood cells may be coated with carbohydrate A (giving them type A blood), carbohydrate B (type B), both (type AB), or neither (type O). Matching compatible blood groups is critical for safe blood transfusions. Our immune system produces blood proteins called antibodies that can bind specifically to the blood cell carbohydrates we lack. If a donor's blood cells have a carbohydrate that is foreign to the recipient, then the recipient's antibodies will cause the donated blood cells to clump together. This clumping can kill the recipient. The clumping reaction is also the basis of a blood-typing lab test.

The four blood groups result from various combinations of the three different alleles: I^A (for the ability to make substance A), I^B (for B), and i (for neither A nor B). Each person inherits one of these alleles from each parent. Because there are three alleles, there are six possible genotypes, as listed in Figure 9.20. Both the I^A and I^B alleles are dominant to the i allele. Thus, I^AI^A and I^Ai people have type A blood, and I^BI^B and I^Bi people have type B. Recessive homozygotes (ii) have type O blood; they make neither the A nor the B carbohydrate. Finally, people of genotype I^AI^B make *both* carbohydrates. In other words, the I^A and I^B alleles exhibit **codominance**, meaning that both alleles are expressed in heterozygous individuals (I^AI^B), who have type AB blood. Be careful to distinguish codominance (the expression of both alleles) from incomplete dominance (the expression of one intermediate trait). ✔

▼ Figure 9.20 **Multiple alleles for the ABO blood groups.** The three versions of the gene responsible for blood type may produce carbohydrate A (allele I^A), carbohydrate B (allele I^B), or neither carbohydrate (allele i). Because each person carries two alleles, six genotypes are possible that result in four different phenotypes. The clumping reaction that occurs between antibodies and foreign blood cells is the basis of blood-typing (shown in the photograph at right) and of the adverse reaction that occurs when someone receives a transfusion of incompatible blood.

Blood Group (Phenotype)	Genotypes	Red Blood Cells	Antibodies Present in Blood	Reactions When Blood from Groups Below Is Mixed with Antibodies from Groups at Left			
				O	A	B	AB
A	I^AI^A or I^Ai	Carbohydrate A	Anti-B				
B	I^BI^B or I^Bi	Carbohydrate B	Anti-A				
AB	I^AI^B		—				
O	ii		Anti-A Anti-B				

✔CHECKPOINT

1. Why is a testcross unnecessary to determine whether a snapdragon with red flowers is homozygous or heterozygous?
2. Maria has type O blood, and her sister has type AB blood. What are the genotypes of the girls' parents?

Answer: 1. Only plants homozygous for the dominant allele have red flowers; heterozygotes have pink flowers. 2. One parent is I^Ai, and the other parent is I^Bi.

Pleiotropy and Sickle-Cell Disease

Our genetic examples to this point have been cases in which each gene specifies only one hereditary character. But in many cases, one gene influences several characters. The impact of a single gene on more than one character is called **pleiotropy**.

An example of pleiotropy in humans is **sickle-cell disease** (also sometimes called sick-cell anemia), a disorder characterized by a diverse set of symptoms. The direct effect of the sickle-cell allele is to make red blood cells produce abnormal hemoglobin proteins (see Figure 3.19). These abnormal molecules tend to link together and crystallize, especially when the oxygen content of the blood is lower than usual because of high altitude, overexertion, or respiratory ailments. As the hemoglobin crystallizes, the normally disk-shaped red blood cells deform to a sickle shape with jagged edges **(Figure 9.21)**. Sickled cells are destroyed rapidly by the body, and their destruction may cause anemia and general weakening of the body. Also, because of their angular shape, sickled cells do not flow smoothly in the blood and tend to accumulate and clog tiny blood vessels. Blood flow to body parts is reduced, resulting in

multiple symptoms, such as periodic fever, pain, and damage to various organs, including the heart, brain, and kidneys. Blood transfusions and drugs may relieve some of the symptoms, but there is no cure, and sickle-cell disease kills about 100,000 people in the world annually.

In most cases, only people who are homozygous for the sickle-cell allele have sickle-cell disease. Heterozygotes, who have one sickle-cell allele and one normal allele, are usually healthy; hence, the disease is considered recessive. However, at the molecular level, the two alleles are actually codominant: Both alleles are expressed in heterozygous individuals, and their red blood cells contain both normal and abnormal hemoglobin. A simple blood test can distinguish homozygotes from heterozygotes. (See the Evolution Connection section in Chapter 13 for further discussion of sickle-cell disease.) ☑

Polygenic Inheritance

Mendel studied genetic characters that could be classified on an either-or basis, such as purple or white flower color. However, many characters, such as human skin color and height, vary along a continuum in a population. Many such features result from **polygenic inheritance**, the additive effects of two or more genes on a single phenotypic character. (This is the converse of pleiotropy, in which one gene affects several characters.)

Let's consider a hypothetical model of how polygenic inheritance might work. Assume that skin pigmentation in humans is controlled by three genes that are inherited separately, like Mendel's pea genes. (Actually, genetic evidence indicates that *at least* three genes control this character.) The dark-skin allele for each gene (*A*, *B*, and *C*) contributes one "unit" of darkness to the phenotype and is incompletely dominant to the other alleles (*a*, *b*, and *c*). A person who is *AABBCC* would be very dark, while an *aabbcc* individual would be very light. An

▼ Figure 9.21 **Sickle-cell disease: multiple effects of a single human gene.**

Individual homozygous for sickle-cell allele

Sickle-cell (abnormal) hemoglobin

Abnormal hemoglobin crystallizes into long flexible chains, causing red blood cells to become sickle-shaped.

Colorized SEM 4,000×

Sickled cells can lead to a cascade of symptoms, such as weakness, pain, organ damage, and paralysis.

AaBbCc person would have skin of an intermediate shade. Because the alleles have an additive effect, the genotype *AaBbCc* would produce the same skin color as any other genotype with just three dark-skin alleles, such as *AABbcc*. The Punnett square in **Figure 9.22**

▼ **Figure 9.22 A model for polygenic inheritance of skin color.** The seven bars in the graph at the bottom of the figure depict the relative numbers of each of the phenotypes in the F₂ generation. The bell-shaped curve indicates the distribution of an even greater variety of skin shades in the population that might result from the combination of heredity and environmental effects, such as sun-tanning.

P Generation

aabbcc
(very light)

×

AABBCC
(very dark)

F₁ Generation

AaBbCc

×

AaBbCc

F₂ Generation

Sperm

Eggs

Fraction of population

Skin pigmentation

shows all possible genotypes of F₂ offspring. The row of squares below the Punnett square shows the seven skin pigmentation phenotypes that would theoretically result. This hypothetical example shows how inheritance of three genes could lead to seven levels of pigmentation at the frequencies indicated by the bars in the graph. ✔

The Role of Environment

If we examine a real human population for the skin-color phenotype, we would see more shades than just seven. The true range might be similar to the entire spectrum of color under the bell-shaped curve in Figure 9.22. In fact, no matter how carefully we characterize the genes for skin color, a purely genetic description will always be incomplete. This is because some intermediate shades of skin color result from the effects of environmental factors, such as exposure to the sun.

Many phenotypic characters result from a combination of heredity and environmental factors. For example, although a single tree is locked into its inherited genotype, its leaves vary in size, shape, and color, depending on exposure to wind and sun and the tree's nutritional state. For humans, exercise alters build; experience improves performance on intelligence tests; and social and cultural factors can greatly affect appearance. As geneticists learn more and more about our genes, it's becoming clear that many human characters—such as a person's susceptibility to heart disease, cancer, alcoholism, and schizophrenia—are influenced by both genes and environment.

Whether human characters are more influenced by genes or by the environment—nature or nurture—is a very old and hotly contested issue. For some characters, such as the ABO blood group, a given genotype mandates a very specific phenotype. In contrast, a person's blood count of red and white cells varies quite a bit, depending on such factors as the altitude, the customary level of physical activity, and the presence of infectious agents.

Simply spending time with identical twins will convince anyone that environment, and not just genes, affects a person's traits (**Figure 9.23**). However, there is an important difference between these two sources of variation: Only genetic influences are inherited. Any effects of the environment are not passed on to the next generation.

☑**CHECKPOINT**

Based on the skin-tone model in Figure 9.22, put the following individuals in order by skin tone, from lightest to darkest: *AAbbCC*, *aaBBcc*, *AabBCc*, *Aabbcc*, *AaBBCC*.

Answer: Aabbcc, aaBBcc, AabBCc, AABBCC, AAbbCC

▼ **Figure 9.23 As a result of environmental influences, even identical twins can look different.**

Chromosomal Basis
of Inheritance

The Chromosomal Basis of Inheritance

Mendel published his results in 1866, but not until long after he died did biologists understand the significance of his work. Cell biologists worked out the processes of mitosis and meiosis (see Chapter 8) in the late 1800s. Then, around 1900, researchers began to notice parallels between the behavior of chromosomes and the behavior of Mendel's "heritable factors" (what we now call genes). One of biology's most important concepts—the chromosome theory of inheritance—began to emerge.

The **chromosome theory of inheritance** states that genes are located at specific positions on chromosomes and that the behavior of chromosomes during meiosis and fertilization accounts for inheritance patterns. Indeed, it is chromosomes that undergo segregation and independent assortment during meiosis and thus account for Mendel's laws. **Figure 9.24** correlates the results of the dihybrid cross in Figure 9.8b with the movement of chromosomes through meiosis. Starting with two true-breeding parental plants, the diagram follows two genes on different chromosomes—one for seed shape (alleles R and r) and one for seed color (alleles Y and y)—through the F_1 and F_2 generations. ☑

► **Figure 9.24 The chromosomal basis of Mendel's laws.**

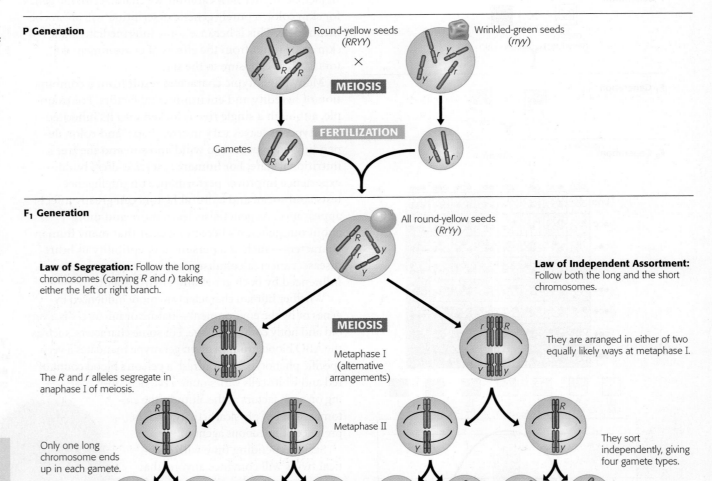

Linked Genes

Realizing that genes are *on chromosomes* when they segregate enables us to extend Mendel's work in several important ways. For example, alleles that start out together on the same chromosome would be expected to travel together during meiosis and fertilization. In general, genes that are located close together on a chromosome, called **linked genes**, tend to be inherited as a set and therefore do not follow Mendel's law of independent assortment.

The most important early studies into linked genes took place in the New York City laboratory of American embryologist Thomas Hunt Morgan. In the early 1900s, Morgan and his colleagues studied the fruit fly *Drosophila melanogaster* (**Figure 9.25**). Often seen flying around overripe fruit, *Drosophila* is a good research animal for studies of inheritance because it is easily and inexpensively grown and can produce several generations in a matter of months.

Normal, or wild-type, *Drosophila* flies have gray bodies (genotype *GG*) and long wings (*LL*). Morgan cultivated true-breeding mutant (non-wild-type) fruit flies that had black bodies (*gg*) and short, underdeveloped wings (*ll*). He then performed a dihybrid testcross, crossing doubly heterozygous flies (*GgLl*) with double mutants (*ggll*). Morgan expected that the offspring would show equal numbers of all four possible phenotypes ($\frac{1}{4}$ gray-long, $\frac{1}{4}$ gray-short, $\frac{1}{4}$ black-long, $\frac{1}{4}$ black-short).

In fact, Morgan's testcross produced equal numbers of two of the four expected phenotypes but far smaller numbers of the other two phenotypes. He found that 83% of the flies displayed the phenotypes of the parents (gray-long and black-short), while only 17% showed nonparental phenotypes (gray-short and black-long). Morgan reasoned that the genes for body color and wing shape were linked—in this case, *G* with *L* and *g* with *l*—causing them to be inherited together. Such linkage would mean that meiosis in the heterozygous fruit flies would yield gametes with only two genotypes (*GL* and *gl*, but not *Gl* or *gL*). The large numbers of flies with gray-long and black-short traits in the experiment resulted from fertilization among the *GL* and *gl* gametes. ☑

Genetic Recombination: Crossing Over

But what of the smaller numbers of Morgan's gray-short and black-long flies? How could they be produced if the two genes involved are linked? In Chapter 8, we saw that during meiosis, crossing over between homologous chromosomes results in shuffled chromosome segments in the haploid daughter cells (see Figure 8.18), thereby producing new combinations of alleles. **Figure 9.26** reviews crossing over, showing how two linked genes can give rise to four different gamete genotypes. Two of the gamete genotypes reflect the presence of parental-type chromosomes, which have not been altered by crossing

Dihybrid testcross

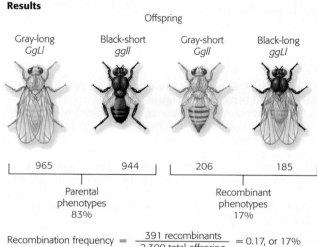

Results

Offspring

| Gray-long *GgLl* | Black-short *ggll* | Gray-short *Ggll* | Black-long *ggLl* |

| 965 | 944 | 206 | 185 |

Parental phenotypes 83% | Recombinant phenotypes 17%

$$\text{Recombination frequency} = \frac{391 \text{ recombinants}}{2{,}300 \text{ total offspring}} = 0.17, \text{ or } 17\%$$

◄ **Figure 9.25 Thomas Morgan's experiment and results.** Among the 2,300 offspring of this dihybrid testcross, Morgan observed more than the expected parental phenotypes (gray-long and black-short) and fewer than expected nonparental phenotypes (gray-short and black-long).

▼ **Figure 9.26 Review: Crossing over can produce recombinant gametes.**

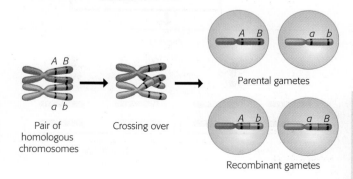

Pair of homologous chromosomes

Crossing over

Parental gametes

Recombinant gametes

over. In contrast, the other two gamete genotypes are recombinant (nonparental). The chromosomes of these gametes carry new combinations of alleles that result from the exchange of chromosome segments in crossing over.

Morgan hypothesized that crossing over accounted for the observed nonparental offspring, those with the recombinant phenotypes (gray-short and black-long). Furthermore, the 17% of offspring with recombinant phenotypes must have resulted from fertilization involving recombinant gametes (**Figure 9.27**). The percentage of recombinant offspring among the total is called the **recombination frequency** (see bottom of Figure 9.25).

▼ Figure 9.27 **Explaining the unexpected results from the dihybrid testcross in Figure 9.25.** The disproportionately large numbers of the parental phenotypes among the offspring is explained by gene linkage, the fact that the gene loci for the two characters are located nearby on the same chromosome and tend to remain together during meiosis and fertilization (*G* with *L* and *g* with *l*). As a result of crossing over, some of the gametes end up with recombinant chromosomes, carrying new combinations of alleles, either *Gl* or *gL*. When the recombinant gametes participate in fertilization, recombinant offspring can result.

Linkage Maps

While working with *Drosophila*, Alfred H. Sturtevant, one of Morgan's students, developed a way to use crossover data to map gene loci. This technique is based on the assumption that the chance of crossing over is approximately equal at all points along a chromosome. He hypothesized that the farther apart two genes are on a chromosome, the higher the probability that a crossover will occur between them. His reasoning was elegantly simple: The greater the distance between two genes, the more points there are between them where crossing over can occur. (This assumption is not entirely accurate, but it is good enough to provide useful data.)

By applying Sturtevant's reasoning, geneticists can use recombination data to assign genes to relative positions on chromosomes—that is, to map genes. **Figure 9.28** shows some of the data used to map three linked genes (*g*, *c*, and *l*) that reside on one of the *Drosophila* chromosomes. Under the chromosome are the actual recombination frequencies (crossover frequencies) between these genes, taken two at a time: 17% between *g* and *l*, 9% between *g* and *c*, and 9.5% between *l* and *c*. Geneticists reasoned that these values represent the relative distances between the genes. Because the recombination frequencies between *g* and *c* and between *l* and *c* are approximately half the frequency between *g* and *l*, gene *c* must lie roughly midway between *g* and *l*. Thus, the sequence of these genes on their chromosome must be *g-c-l* (or the equivalent *l-c-g*). Such a diagram of relative gene locations is called a **linkage map**.

The linkage-mapping method has proved extremely valuable in establishing the relative positions of many genes in many organisms. The real beauty of the technique is that a wealth of information about genes can be learned simply by breeding and observing the organisms; no fancy equipment is required. ☑

▼ Figure 9.28 **Using crossover data to map genes.** The three recessive alleles specify black body (*g*), cinnabar eyes (*c*), and short wings (*l*). The corresponding dominant alleles specify the wild-type traits of gray body, red eyes, and long wings.

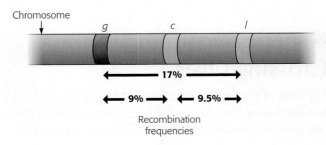

Sex Chromosomes and Sex-Linked Genes

The patterns of inheritance we've discussed so far have involved genes located only on autosomes, not on the sex chromosomes. We're now ready to look at the role of sex chromosomes in humans and the inheritance patterns exhibited by the characters they control.

Sex Determination in Humans

A number of organisms, including nearly all mammals, have a pair of sex chromosomes—designated X and Y—that determine an individual's sex (**Figure 9.29**). Individuals with one X chromosome and one Y chromosome are males; XX individuals are females. Human males and females both have 44 autosomes (chromosomes other than sex chromosomes). As a result of chromosome segregation during meiosis, each gamete contains one sex chromosome and a haploid set of autosomes (22 in humans). All eggs contain a single X chromosome. Of the sperm cells, half contain an X chromosome and half contain a Y chromosome. An offspring's sex depends on whether the sperm cell that fertilizes the egg bears an X or a Y. ☑

Sex-Linked Genes

Besides bearing genes that determine sex, the so-called sex chromosomes also contain genes for characters unrelated to maleness or femaleness. Any gene located on a sex chromosome is called a **sex-linked gene**. The X chromosome contains many more genes than the Y; therefore, most sex-linked genes are found on the X chromosome.

A number of human conditions, including red-green colorblindness, hemophilia, and a type of muscular dystrophy, result from sex-linked recessive alleles. **Red-green colorblindness** is a common sex-linked disorder characterized by a malfunction of light-sensitive cells in the eyes. (It is actually a class of disorders involving several sex-linked genes.) A person with normal color vision can see more than 150 colors. In contrast, someone with red-green colorblindness can see fewer than 25. For some affected people, red hues appear gray; others see gray instead of green; still others are green-weak or red-weak, tending to confuse shades of these colors. **Figure 9.30** shows a simple test for red-green colorblindness.

▲ **Figure 9.30 A test for red-green colorblindness.** Can you see a green numeral 7 against the reddish background? If not, you probably have some form of red-green colorblindness, a sex-linked trait.

▼ Figure 9.29 **The chromosomal basis of sex determination in humans.** The micrograph at right shows human X and Y chromosomes in a duplicated state.

Male
Somatic cells
44 + XY

Female
44 + XX

Sperm
22 + X
22 + Y

Egg
22 + X

Offspring
44 + XX — Female
44 + XY — Male

Colorized SEM 41,500×

Y

X

☑CHECKPOINT

At the moment of fertilization in humans, what determines the sex of the offspring?

Answer: whether the egg is fertilized by a sperm bearing an X chromosome (producing a female offspring) or a Y chromosome (producing a male)

▼ Figure 9.31 **Inheritance of colorblindness, a sex-linked recessive trait.** We use the uppercase letter *N* for the dominant, normal color vision allele and *n* for the recessive, colorblind allele. To indicate that these alleles are on the X chromosome, we show them as superscripts to the letter X. The Y chromosome does not have a gene locus for vision; therefore, the male's phenotype results entirely from the sex-linked gene on his single X chromosome.

Key

◻ Unaffected individual

◻ Carrier

◼ Colorblind individual

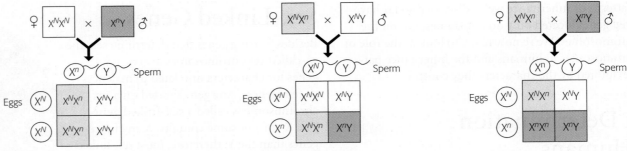

(a) **Normal female × colorblind male** (b) **Carrier female × normal male** (c) **Carrier female × colorblind male**

Because they are located on the sex chromosomes, sex-linked genes exhibit unusual inheritance patterns. **Figure 9.31a** illustrates what happens when a colorblind male has offspring with a homozygous female with normal color vision. All the children have normal color vision, suggesting that the allele for wild type (normal color vision) is dominant. If a female carrier mates with a male who has normal color vision, the classic 3:1 phenotypic ratio of normal color vision to colorblindness appears among the children (**Figure 9.31b**). However, there is a surprising twist: The colorblind trait shows up only in males. All the females have normal vision, while half the males are colorblind and half are normal. This is because the gene involved in this inheritance pattern is located exclusively on the X chromosome; there is no corresponding locus on the Y. Thus, females (XX) carry two copies of the gene for this character, while males (XY) carry only one. Because the colorblindness allele is recessive, a female will be colorblind only if she receives that allele on both X chromosomes (**Figure 9.31c**). For a male, however, a single copy of the recessive allele confers colorblindness. For this reason, recessive sex-linked traits are expressed much more frequently in men than in women. For example, colorblindness is about 20-fold more common among males than among females.

Hemophilia is a sex-linked recessive trait with a long, well-documented history. Hemophiliacs bleed excessively when injured because they have inherited an abnormal allele for a factor involved in blood clotting. The most seriously affected individuals may bleed to death after relatively minor bruises or cuts. A high incidence of hemophilia has plagued the royal families of Europe. The first royal hemophiliac seems to have been a son of Queen Victoria (1819–1901) of England. It is likely that the hemophilia allele arose through a mutation in one of the gametes of Victoria's mother or father, making Victoria a carrier of the deadly allele. Victoria later passed the allele on to her daughter and granddaughter, both carriers. Hemophilia was eventually introduced into the royal families of Prussia, Russia, and Spain through the marriages of two of Victoria's daughters who were carriers. In this way, the age-old practice of strengthening international alliances by marriage effectively spread hemophilia through the royal families of several nations (**Figure 9.32**). ✔

▼ Figure 9.32 **Hemophilia in the royal family of Russia.** The photograph shows Queen Victoria's granddaughter Alexandra, her husband Nicholas, who was the last czar of Russia, their son Alexis, and their daughters. In the pedigree, half-colored symbols represent heterozygous carriers of the hemophilia allele and fully colored symbols represent a person with hemophilia.

✔CHECKPOINT

1. What is meant by a sex-linked gene?
2. White eye color is a recessive sex-linked trait in fruit flies. If a white-eyed *Drosophila* female is mated with a red-eyed (wild-type) male, what do you predict for the numerous offspring?

Answers: 1. a gene that is located on a sex chromosome, usually the X chromosome 2. All female offspring will be heterozygous ($X^R X^r$), with red eyes; all male offspring will be white-eyed ($X^r Y$).

Barking Up
the Evolutionary Tree

As we've seen throughout this chapter, dogs are more than man's best friend—they are also one of our longest-running genetic experiments. About 15,000 years ago, in East Asia, humans began to cohabit with ancestral canines that were predecessors of both modern wolves and dogs. As people settled into permanent, geographically isolated settlements, populations of canines were separated from one another and eventually became inbred.

Different groups of people chose dogs with different traits, depending on their needs. Herders selected dogs that were good at controlling flocks of animals. Hunters chose dogs that were good at retrieving prey. Continued over millennia, such genetic tinkering has resulted in a diverse array of dog body types and behaviors, from tiny, feisty Chihuahuas to huge, docile St. Bernards. In each breed, a distinct genetic makeup results in a distinct set of physical and behavioral traits.

As discussed in the Process of Science section, our understanding of canine evolution took a big leap forward when researchers sequenced the complete genome of a dog. Using the genome sequence and a wealth of other data, canine geneticists produced an evolutionary tree of dog breeds (Figure 9.33). Published in 2006, the tree is based on genetic analysis of 414 dogs from 85 recognized breeds. The analysis shows that the canine family tree includes a series of well-defined branch points. Each fork represents artificial selection that produced a genetically distinct subpopulation with specific desired traits.

The genetic tree shows that the most ancient breeds, those most closely related to the wolf, are Asian species such as the shar-pei and Akita. Subsequent genetic splits created distinct breeds in Africa (basenji), the Arctic (Alaskan malamute and Siberian husky), and the Middle East (Afghan hound and saluki). The remaining breeds, primarily of European ancestry, can be grouped by genetic makeup into those bred for guarding (for example, the rottweiler), herding (such as various sheepdogs), and hunting (including the golden retriever and beagle).

The formulation of an evolutionary tree for the domestic dog shows that new technologies can provide answers to important genetic and evolutionary questions without ever performing controlled matings (as Mendel had to do with his peas). It also shows that the study of genetics can provide important insights into the evolutionary history of life on Earth.

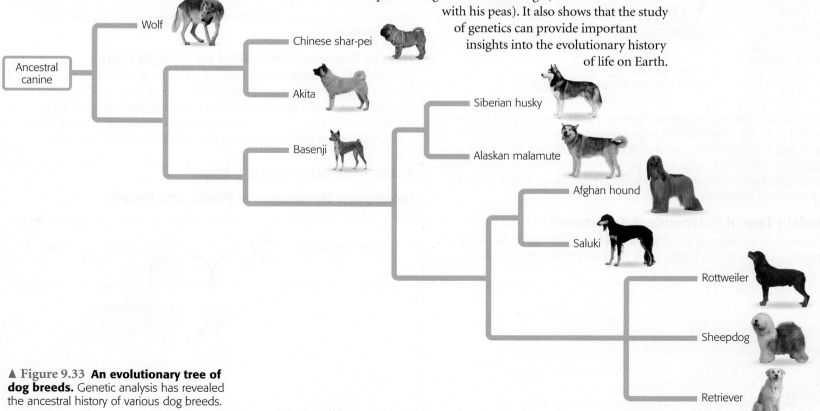

▲ Figure 9.33 **An evolutionary tree of dog breeds.** Genetic analysis has revealed the ancestral history of various dog breeds.

Chapter Review

SUMMARY OF KEY CONCEPTS

(MB) Go to the Study Area at **www.masteringbiology.com** for practice quizzes, eText, BioFlix™ 3-D animations, MP3 Tutor Sessions, videos, current events, and more.

Heritable Variation and Patterns of Inheritance

Gregor Mendel was the first to study genetics, the science of heredity, by analyzing patterns of inheritance. He emphasized that heritable factors (genes) retain permanent identities.

In an Abbey Garden

Mendel started with true-breeding varieties of pea plants representing two alternative variants of a hereditary character, such as flower color. He then crossed the different varieties and traced the inheritance of traits from generation to generation.

Mendel's Law of Segregation

Pairs of alleles separate during gamete formation; fertilization restores the pairs.

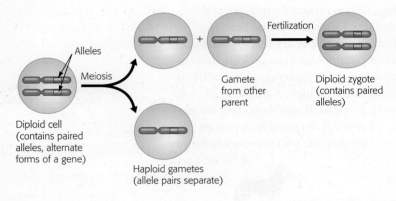

If an individual's genotype (genetic makeup) has two different alleles for a gene and only one influences the organism's phenotype (appearance), that allele is said to be dominant and the other allele recessive. Alleles of a gene reside at the same locus, or position, on homologous chromosomes. When the allele pair match, the organism is homozygous; when they're different, the organism is heterozygous.

Mendel's Law of Independent Assortment

By following two characters at once, Mendel found that the alleles of a pair segregate independently of other allele pairs during gamete formation.

Using a Testcross to Determine an Unknown Genotype

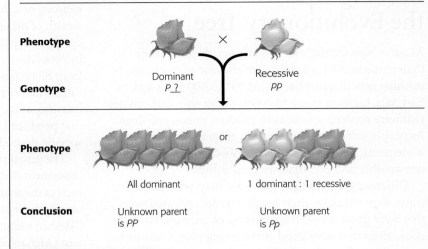

The Rules of Probability

Inheritance follows the rules of probability. The chance of inheriting a recessive allele from a heterozygous parent is $\frac{1}{2}$. The chance of inheriting it from both of two heterozygous parents is $\frac{1}{2} \times \frac{1}{2} = \frac{1}{4}$, illustrating the rule of multiplication for calculating the probability of two independent events.

Family Pedigrees

The inheritance of many human traits, from freckles to genetic diseases, follows Mendel's laws and the rules of probability. Geneticists can use family pedigrees to determine patterns of inheritance and individual genotypes among humans.

Human Disorders Controlled by a Single Gene

For traits that vary within a population, the one most commonly found in nature is called the wild type. Many inherited disorders in humans are controlled by a single gene (represented by two alleles). Most of these disorders, such as cystic fibrosis, are caused by autosomal recessive alleles. A few, such as Huntington's disease, are caused by dominant alleles.

Variations on Mendel's Laws

Incomplete Dominance in Plants and People

ABO Blood Groups: An Example of Multiple Alleles and Codominance

Within a population, there are often multiple kinds of alleles for a character, such as the three alleles for the ABO blood groups. The alleles determining the A and B blood factors are codominant; that is, both are expressed in a heterozygote.

Pleiotropy and Sickle-Cell Disease

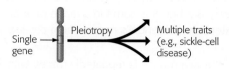

The presence of two copies of the sickle-cell allele at a single gene locus brings about the many symptoms of sickle-cell disease.

Polygenic Inheritance

The Role of Environment

Many phenotypic characters result from a combination of genetic and environmental effects, but only genetic influences are biologically heritable.

The Chromosomal Basis of Inheritance

Genes are located on chromosomes. The behavior of chromosomes during meiosis and fertilization accounts for inheritance patterns.

Linked Genes

Certain genes are linked: They tend to be inherited together because they lie close together on the same chromosome.

Genetic Recombination: Crossing Over

Crossing over can separate linked alleles, producing gametes with recombinant chromosomes and offspring with recombinant phenotypes.

Linkage Maps

The fact that crossing over between linked genes is more likely to occur between genes that are farther apart enables geneticists to map the relative positions of genes on chromosomes.

Sex Chromosomes and Sex-Linked Genes

Sex Determination in Humans

In humans, sex is determined by whether a Y chromosome is present. A person who inherits two X chromosomes develops as a female. A person who inherits one X and one Y chromosome develops as a male.

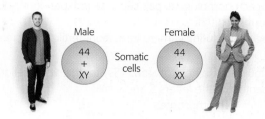

Sex-Linked Genes

Genes on the sex chromosomes (most often the X) are said to be sex-linked. Their inheritance pattern reflects the fact that females have two homologous X chromosomes, but males have only one. Most sex-linked human disorders, such as red-green colorblindness and hemophilia, are due to recessive alleles and are seen mostly in males. A male receiving a single sex-linked recessive allele from his mother will have the disorder; a female has to receive the allele from both parents to be affected.

Sex-Linked Traits				
Female: Two alleles	Genotype	$X^N X^N$	$X^N X^n$	$X^n X^n$
	Phenotype	Normal female	Carrier female	Affected female (rare)
Male: One allele	Genotype	$X^N Y$		$X^n Y$
	Phenotype	Normal male		Affected male

SELF-QUIZ

1. The genetic makeup of an organism is called its _____, while the physical traits of an organism are called its _____.

2. Which of Mendel's laws is represented by each statement?
  a. Alleles of each homologous pair separate independently during gamete formation.
  b. Alleles segregate during gamete formation; fertilization creates pairs of alleles once again.

3. Edward was found to be heterozygous (*Ss*) for the sickle-cell trait. The alleles represented by the letters *S* and *s* are
  a. on the X and Y chromosomes.
  b. linked.
  c. on homologous chromosomes.
  d. both present in each of Edward's sperm cells.

4. Whether an allele is dominant or recessive depends on
  a. how common the allele is, relative to other alleles.
  b. whether it is inherited from the mother or the father.
  c. whether it or another allele determines the phenotype when both are present.
  d. whether or not it is linked to other genes.

5. Two fruit flies with eyes of the usual red color are crossed, and their offspring are as follows: 77 red-eyed males, 71 ruby-eyed males, 152 red-eyed females. The gene that controls whether eyes are red or ruby is _____, and the allele for ruby eyes is _____.
  a. autosomal (carried on an autosome); dominant
  b. autosomal; recessive
  c. sex-linked; dominant
  d. sex-linked; recessive

6. All the offspring of a white hen and a black rooster are gray. The simplest explanation for this pattern of inheritance is
  a. pleiotropy.
  b. sex linkage.
  c. codominance.
  d. incomplete dominance.

7. A man who has type B blood and a woman who has type A blood could have children of which of the following phenotypes? (*Hint*: Review Figure 9.20.)
  a. A, B, or O
  b. AB only
  c. AB or O
  d. A, B, AB, or O

8. Duchenne muscular dystrophy is a sex-linked recessive disorder characterized by a progressive loss of muscle tissue. Neither Rudy nor Carla has Duchenne muscular dystrophy, but their first son does have it. If the couple has a second child, what is the probability that he or she will also have the disease?

9. In fruit flies, the genes for wing shape and body stripes are linked. In a fly whose genotype is *WwSs*, *W* is linked to *S*, and *w* is linked to *s*. Show how this fly can produce gametes containing four different combinations of alleles. Which are parental-type gametes? Which are recombinant gametes? What process produces recombinant gametes?

10. Adult height in humans is at least partially hereditary; tall parents tend to have tall children. But humans come in a range of sizes, not just tall or short. What extension of Mendel's model could produce this variation in height?

11. A true-breeding brown mouse is repeatedly mated with a true-breeding white mouse, and all their offspring are brown. If two of these brown offspring are mated, what fraction of the F_2 mice will be brown?

12. How could you determine the genotype of one of the brown F_2 mice in problem 11? How would you know whether a brown mouse is homozygous? Heterozygous?

13. Tim and Jan both have freckles (a dominant trait), but their son Michael does not. Show with a Punnett square how this is possible. If Tim and Jan have two more children, what is the probability that *both* of them will have freckles?

14. Incomplete dominance is seen in the inheritance of hypercholesterolemia. Mack and Toni are both heterozygous for this character, and both have elevated levels of cholesterol. Their daughter Katerina has a cholesterol level six times normal; she is apparently homozygous, *hh*. What fraction of Mack and Toni's children are likely to have elevated but not extreme levels of cholesterol, like their parents? If Mack and Toni have one more child, what is the probability that the child will suffer from the more serious form of hypercholesterolemia seen in Katerina?

15. A female fruit fly with forked bristles on her body is mated with a male fly with normal bristles. Their offspring are 121 females with normal bristles and 138 males with forked bristles. Explain the inheritance pattern for this trait.

16. Both parents of a boy are phenotypically normal, but their son suffers from hemophilia, a sex-linked recessive disorder. Draw a pedigree that shows the genotypes of the three individuals. What fraction of the couple's children are likely to suffer from hemophilia? What fraction are likely to be carriers?

17. Heather was surprised to discover that she suffered from red-green colorblindness. She told her biology professor, who said, "Your father is colorblind too, right?" How did her professor know this? Why did her professor not say the same thing to the colorblind males in the class?

Answers to Self-Quiz questions can be found in Appendix D.

THE PROCESS OF SCIENCE

18. In 1981, a stray cat with unusual curled-back ears was adopted by a family in Lakewood, California. Hundreds of descendants of this cat have since been born, and cat fanciers hope to develop the "curl" cat into a show breed. The curl allele is apparently dominant and carried on an autosome. Suppose you owned the first curl cat and wanted to develop a true-breeding variety. Describe tests that would determine whether the curl gene is dominant or recessive and whether it is autosomal or sex-linked.

19. Imagine that you have a large collection of fruit flies divided into ten different strains. Each strain is true-breeding (homozygous) and differs from wild-type flies in just one character. The only special equipment available is a magnifying glass that lets you determine the sex and readily observable traits of a fly, a large number of bottles to perform controlled matings, and an anesthetic liquid that enables you to examine and sort live flies. Using only this equipment, how much could you learn about the genetic makeup of the flies? Describe a series of experiments that would give you knowledge about fruit fly genetics.

BIOLOGY AND SOCIETY

20. Gregor Mendel never saw a gene, yet he concluded that "heritable factors" were responsible for the patterns of inheritance he observed in peas. Similarly, maps of *Drosophila* chromosomes (and the very idea that genes are carried on chromosomes) were conceived by observing the patterns of inheritance of linked genes, not by observing the genes directly. Is it legitimate for biologists to claim the existence of objects and processes they cannot actually see? How do scientists know whether an explanation is correct?

21. Many infertile couples turn to in vitro fertilization to try to have a baby. In this technique, sperm and ova are collected and used to create eight-cell embryos for implantation into a woman's uterus. At the eight-cell stage, one of the fetal cells can be removed without causing harm to the developing fetus. Once removed, the cell can be genetically tested. Some couples may know that a particular genetic disease runs in their family. They might wish to avoid implanting any embryos with the disease-causing genes. Do you think this is an acceptable use of genetic testing? What if a couple wanted to use genetic testing to select embryos for traits unrelated to disease, such as freckles? Do you think that couples undergoing in vitro fertilization should be allowed to perform whatever genetic tests they wish? Or do you think that there should be limits on what tests can be performed? How do you draw the line between genetic tests that are acceptable and those that are not?

10 The Structure and Function of DNA

The influenza virus.
Although it can be deadly, influenza is caused by a relatively simple virus consisting of little more than protein and nucleic acid.

CHAPTER CONTENTS
■■■ Chapter Thread: **THE FLU**

BIOLOGY AND SOCIETY: The Flu

Tracking a Killer

On February 18, 2009, a group of elite investigators met in a conference room outside Washington, DC. Their mission was to track a serial killer and, by anticipating its next moves, prevent future deaths. The decisions they would make would affect hundreds of millions of people and save, or lose, tens of thousands of lives.

On that day, the Vaccines and Related Biological Products Advisory Committee of the U.S. Food and Drug Administration selected the influenza virus strains to be used in the upcoming 2009–2010 U.S. flu vaccine program. The committee evaluated the patterns of recent infections and predicted the three most dangerous strains of flu for the coming year. Based on their recommendations, pharmaceutical manufacturers produced hundreds of millions of doses of flu vaccine to be distributed during the fall and winter of 2009–2010. The precise formulation of the vaccine was nothing less than a matter of life and death.

While you may think of the flu as merely a seasonal inconvenience, the influenza virus is actually one of the deadliest pathogens known to science. Each year in the United States, over 20,000 people die from influenza infection, mostly the elderly or people with chronic diseases. And that is in a good year. Once every few decades, a new strain of flu explodes on the scene that causes global epidemics and widespread death. The worst of these was the flu of 1918–1919. In just 18 months, about 40 million died worldwide—that's more people than have died of AIDS since it was discovered over 25 years ago. Health officials are always on the lookout for new flu strains that have the potential to cause similar devastation. They are concerned that we may be overdue—the last major flu outbreak was the Hong Kong flu of 1968–1969—and that the new avian flu (discussed in the Evolution Connection section) could be particularly deadly.

The flu virus, like all viruses, consists of a relatively simple structure of nucleic acid (RNA in this case) and protein. The search for an effective flu vaccine depends on a detailed understanding of life at the molecular level. In this chapter, you will explore the structure of DNA, how it replicates and mutates, and how it controls the cell by directing the synthesis of RNA and protein.

Plant Viruses

Viruses that infect plant cells can stunt plant growth and diminish crop yields. Most known plant viruses have RNA rather than DNA as their genetic material. Many of them, like the tobacco mosaic virus shown in **Figure 10.27**, are rod-shaped with a spiral arrangement of proteins surrounding the nucleic acid.

▼ Figure 10.27 **Tobacco mosaic virus.** The photo shows the mottling of leaves in tobacco mosaic disease. The rod-shaped virus causing the disease has RNA as its genetic material.

Tobacco mosaic virus

RNA

Protein

To infect a plant, a virus must first get past the plant's outer protective layer of cells (the epidermis). For this reason, a plant damaged by wind, chilling, injury, or insects is more susceptible to infection than a healthy plant. Some insects carry and transmit plant viruses, and farmers and gardeners may spread plant viruses through the use of pruning shears and other tools. Also, infected plants may pass viruses to their offspring.

There is no cure for most viral plant diseases, and agricultural scientists focus on preventing infection and on breeding or genetically engineering varieties of crop plants that resist viral infection. In Hawaii, for example, the spread of papaya ringspot potyvirus (PRSV) by aphids wiped out the papaya (Hawaii's second largest crop) in certain island regions. But since 1998, farmers have been able to plant a genetically engineered PRSV-resistant strain of papaya, and papayas have been reintroduced into their old habitats. ✔

☑**CHECKPOINT**
What are three ways that
viruses can get into a plant?

*Answer: through lesions caused by
injuries, transfer by insects that feed
on the plant, and contaminated
farming or gardening tools*

Animal Viruses

Viruses that infect animal cells are common causes of disease. As discussed in the Biology and Society section, no virus is a greater human health threat than the influenza (flu) virus (**Figure 10.28**). Like many animal viruses, this one has an outer envelope made of phospholipid membrane, with projecting spikes of protein. The envelope enables the virus to enter and leave a cell. As mentioned earlier, flu viruses have RNA as their genetic material. Other RNA viruses include those that cause the common cold, measles, mumps, AIDS, and polio. Diseases caused by DNA viruses include hepatitis, chicken pox, and herpes infections.

Protein spike

Membranous envelope

RNA

Protein coat

► Figure 10.28 **An influenza virus.** The genetic material of this virus consists of eight separate molecules of RNA, each wrapped in a protein coat.

Figure 10.29 shows the reproductive cycle of an enveloped RNA virus: the mumps virus, which causes a once-common childhood disease that has been almost eliminated by vaccines. When the virus contacts a susceptible cell, protein spikes on its outer surface attach to receptor proteins on the cell's plasma membrane. ❶ The viral envelope fuses with the cell's membrane, allowing the protein-coated RNA to enter the cytoplasm. ❷ Enzymes then remove the protein coat. ❸ An enzyme that entered the cell as part of the virus uses the virus's RNA genome as a template for making complementary strands of RNA. The new strands have two functions: ❹ They serve as mRNA for the synthesis of new viral proteins, and ❺ they serve as templates for synthesizing new viral genome RNA. ❻ The new coat proteins assemble around the new viral RNA. ❼ Finally, the viruses leave the cell by cloaking themselves in plasma membrane. In other words, the virus obtains its envelope from the cell, leaving the cell without necessarily lysing it.

Not all animal viruses reproduce in the cytoplasm. For example, the viruses called herpesviruses—which cause chicken pox, shingles, cold sores, and genital herpes—are enveloped DNA viruses that reproduce in a cell's nucleus, and they get their envelopes from the cell's nuclear membrane. Copies of the herpesvirus DNA usually remain behind as mini-chromosomes in the nuclei of certain nerve cells. There they remain latent until some sort of physical stress, such as a cold or sunburn, or emotional stress triggers virus production, resulting in unpleasant symptoms. Once acquired, herpes infections may flare up repeatedly throughout a person's life. Over 75% of American adults are thought to carry herpes simplex 1 (which causes cold sores), and over 20% carry herpes simplex 2 (which causes genital herpes).

The amount of damage a virus causes the body depends partly on how quickly the immune system responds to fight the infection and partly on the ability of the infected tissue to repair itself. We usually recover completely from colds because our respiratory tract tissue can efficiently replace damaged cells by mitosis. In contrast, the poliovirus attacks nerve cells, which are not usually replaceable. The damage to such cells by polio, unfortunately, is permanent. In such cases, the only medical option is to prevent the disease with vaccines.

How effective are vaccines? Next, we'll look at the example of the flu vaccine. ☑

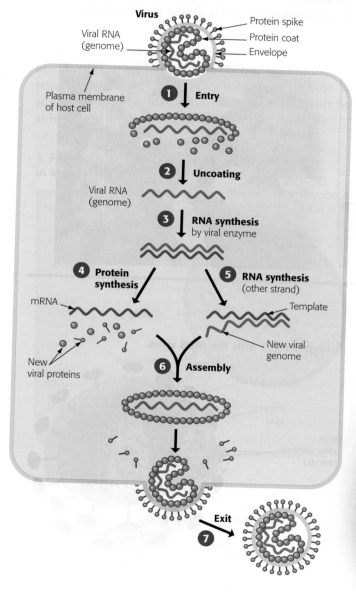

▼ **Figure 10.29 The reproductive cycle of an enveloped virus.** This virus is the one that causes mumps. Like the flu virus, it has a membranous envelope with protein spikes, but its genome is a single molecule of RNA.

CHECKPOINT

Why is infection by herpesvirus permanent?

Answer: because herpesvirus leaves DNA in the nuclei of nerve cells

Mumps virus

Protein spike

Envelope

Colorized TEM 294,000×

HIV (red dots) infecting a white blood cell

SEM 5,500×

OH H

Part of a T nucleotide

N₃ H

AZT

Do Flu Vaccines Protect

and deaths during the winter after vaccination. Their
experiment followed tens of thousands of people over

11 How Genes Are Controlled

Smoking and cancer.
Cancer, like the orange-
tinted lung cancer shown
in this X-ray, is caused by
a breakdown of gene
control.

BIOLOGY AND SOCIETY: Cancer

Tobacco's Smoking Gun

When European explorers returned from their first voyages to the Americas, they brought back tobacco, a common trade item among Native Americans. It quickly became popular in Europe, and the southern United States developed into a major tobacco producer. By the 1950s, about half of all Americans smoked over a pack of cigarettes each day. Little credence was paid to health risks; in fact, cigarette advertising often touted the "health benefits" of tobacco. Smoking was pervasive in American society.

But doctors began to notice a disturbing trend: As tobacco use increased, so did the rate of lung cancer. This disease was rare in 1930, but by 1955 it had become the deadliest form of cancer among American men. In fact, by 1990, lung cancer was killing over twice as many men each year as any other type of cancer. But a few skeptics doubted the link between smoking and cancer. They pointed out that the evidence was purely statistical or based on animal studies; no direct proof had been found that tobacco smoke caused cancer in humans.

The "smoking gun" of proof was found in 1996 when researchers added one component of tobacco smoke, called BPDE, to human lung cells growing in the lab. In these cells, the researchers showed, BPDE binds to DNA within a gene called *p53*, which codes for a protein that normally helps suppress the formation of tumors. The researchers proved that BDPE causes mutations in the *p53* gene that deactivate the protein and lead to tumors. This work directly linked a chemical in tobacco smoke with the formation of human lung tumors.

How can a mutation in a gene lead to cancer? It turns out that many cancer-associated genes encode proteins that turn other genes on or off within a cell. When these proteins malfunction, the cell may become cancerous. In fact, the ability to properly control which genes are active at any given time is crucial to normal cell function. How genes are controlled and how the regulation of genes affects cells and organisms are the subjects of this chapter. As you will see, these issues touch on some of the most interesting topics in all of biology.

Control of Gene Expression

How and Why Genes Are Regulated

In a multicellular organism, every somatic cell (every cell except the gametes) is produced by repeated rounds of mitosis, starting from a zygote. Since a cell's genes are duplicated with each round of the cell cycle, each somatic cell has the same DNA as the zygote. Simply put, every somatic cell contains every gene.

If every somatic cell contains identical genetic instructions, how do cells become different from one another? Cells with the same genetic information can develop into cells with different structures and functions only if gene activity is regulated. Control mechanisms must turn *on* certain genes, while other genes remain turned *off* in a particular cell. In other words, individual cells must undergo **cellular differentiation**—that is, they must become specialized in structure and function. It is **gene regulation**, the turning on and off of genes, that leads to this specialization.

Patterns of Gene Expression in Differentiated Cells

What does it mean to say that genes are active or inactive, turned on or off? As discussed in Chapter 10, genes determine the nucleotide sequence of specific mRNA molecules, and mRNA in turn determines the sequence of amino acids in proteins (DNA→RNA→ protein). A gene that is turned on is being transcribed into mRNA, and that message is being translated into specific proteins. The overall process by which genetic

information flows from genes to proteins—that is, from genotype to phenotype—is called **gene expression**.

Because all the differentiated cells in an individual organism contain the same genes, the differences among the cells must result from the selective expression of genes—that is, from the pattern of genes turned on in a given cell at a given time. Such regulation of gene expression plays a central role in the development of a unicellular zygote into a multicellular organism. During embryonic growth, groups of cells follow diverging developmental pathways, and each group becomes a particular kind of tissue. In the mature organism, each cell type—nerve or pancreas, for instance—has a different pattern of turned-on genes.

Figure 11.1 shows the patterns of gene expression for four genes in three different specialized cells of an adult human. Note that the genes for "housekeeping" enzymes, such as those that provide energy via glycolysis, are "on" in all the cells. In contrast, the genes for some proteins, such as insulin and hemoglobin, are expressed only by particular kinds of cells. One protein, hemoglobin, is not expressed in any of the cell types shown in the figure. ☑

Gene Regulation in Bacteria

To understand how a cell can regulate gene expression, consider the relatively simple case of bacteria. In the course of their lives, bacteria must regulate their genes in response to environmental changes. For example,

► Figure 11.1 **Patterns of gene expression in three types of human cells.** Different types of cells express different combinations of genes. The specialized proteins whose genes are represented here are a glycolysis enzyme; an antibody, which aids in fighting infection; insulin, a hormone made in the pancreas; and the oxygen transport protein hemoglobin, which is expressed only in red blood cells.

	Pancreas cell	White blood cell	Nerve cell
Gene for a glycolysis enzyme	✓	✓	✓
Antibody gene		✓	
Insulin gene	✓		
Hemoglobin gene			

Key

✓
Active gene

when a nutrient is plentiful, bacteria do not squander resources to make it from scratch. Bacterial cells that can conserve resources and energy have a survival advantage over cells that are unable to do so. Thus, natural selection has favored bacteria that express only the genes whose products are needed by the cell.

Imagine an *Escherichia coli* bacterium living in your intestines. It will be bathed in various nutrients, depending on what you eat. If you drink a milk shake, for example, there will be a sudden rush of the sugar lactose. In response, *E. coli* will express three genes for enzymes that enable the bacterium to absorb and digest this sugar. After the lactose is gone, *E. coli* turns the genes off; it does not waste its energy continuing to produce these enzymes. Thus, a bacterium can adjust its gene expression to changes in the environment.

How does the presence or absence of lactose influence the activity of the genes that code for lactose enzymes? The key is the way the three genes are organized: They are adjacent in the DNA and regulated (turned on and off) as a single unit. This regulation is achieved through control sequences, short stretches of DNA that help turn all three genes on and off, coordinating their expression. Such a cluster of genes with related functions, along with the control sequences, is called an **operon** (Figure 11.2). The operon considered here, the *lac* (short for lactose) operon, was first described in the 1960s by French biologists François Jacob and Jacques Monod. The *lac* operon illustrates principles of gene regulation that apply to a wide variety of prokaryotic genes.

How do DNA control sequences turn genes on or off? One control sequence, called a **promoter** (green in the figure), is the site where the enzyme RNA polymerase attaches and initiates transcription—in our example, transcription of the three genes for lactose enzymes. Between the promoter and the enzyme genes, a DNA segment called an **operator** (yellow) acts as a switch that is turned on or off, depending on whether a specific protein is bound there. The operator and protein together determine whether RNA polymerase can attach to the promoter and start transcribing the genes (light blue). In the *lac* operon, when the operator switch is turned on, all the enzymes needed to metabolize lactose are made at once.

The top half of Figure 11.2 shows the *lac* operon in "off" mode, its status when there is no lactose around. Transcription is turned off because a protein called a **repressor** (🔲) ❶ binds to the operator (🔲) and ❷ physically blocks the attachment of RNA polymerase (🔲) to the promoter (🔲).

The bottom half of Figure 11.2 shows the operon in "on" mode, when lactose is present. The lactose (◯)

interferes with attachment of the *lac* repressor to the operator by ❶ binding to the repressor and ❷ changing the repressor's shape. In its new shape (🔲), the repressor cannot bind to the operator, and the operator switch remains on. ❸ RNA polymerase is no longer blocked, so it can now bind to the promoter and from there ❹ transcribe the genes for the lactose enzymes into mRNA. ❺ Translation produces all three lactose enzymes (purple).

Many operons have been identified in bacteria. Some are quite similar to the *lac* operon, while others have somewhat different mechanisms of control. For example, operons that control amino acid synthesis cause bacteria to stop making these molecules when they are already present in the environment, saving materials and energy for the cells. In these cases, the amino acid *activates* the repressor. Armed with a variety of operons, *E. coli* and other prokaryotes can thrive in frequently changing environments. ✓

▼ Figure 11.2 **The *lac* operon of *E. coli*.**

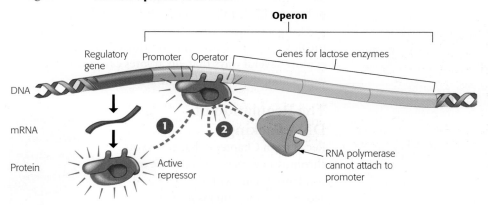

Operon turned off (lactose absent)

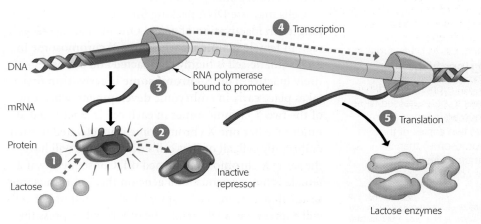

Operon turned on (lactose inactivates repressor)

Gene Regulation in Eukaryotic Cells

Eukaryotes, especially multicellular ones, have more sophisticated mechanisms than bacteria for regulating the expression of their genes. This is not surprising because a prokaryote, being a single cell, does not require the elaborate regulation of gene expression that leads to cell specialization in multicellular eukaryotic organisms.

The pathway from gene to functioning protein in eukaryotic cells is a long one, providing a number of points where the process can be regulated—turned on or off, speeded up or slowed down. Picture the series of pipes that carry water from your local water supply to a faucet in your home. At various points, valves control the flow of water. We use this analogy in **Figure 11.3** to illustrate the flow of genetic information from a eukaryotic chromosome—a reservoir of genetic information—to an active protein that has been made in the cell's cytoplasm. The multiple mechanisms that control gene expression are analogous to the control valves in your water pipes. In the figure, each control knob indicates a gene expression "valve." All these knobs represent possible control points, although only one or a few control points are likely to be important for a typical protein.

Using a reduced version of Figure 11.3 as a guide to the flow of genetic information through the cell, we will explore several ways that eukaryotes can control gene expression, starting within the cell nucleus.

The Regulation of DNA Packing

Recall from Chapter 8 that eukaryotic chromosomes may be in a more or less condensed state, with the DNA and accompanying proteins more or less tightly wrapped together. DNA packing tends to prevent gene expression by preventing RNA polymerase and other transcription proteins from binding to the DNA.

Cells may use DNA packing for the long-term inactivation of genes. One intriguing case is seen in female mammals, where one X chromosome in each somatic cell is highly compacted and almost entirely inactive. This **X chromosome inactivation** first takes place early in embryonic development, when one of the two X chromosomes in each cell is inactivated at random. After one X chromosome is inactivated in each embryonic cell, all of that cell's descendants will have the same X chromosome turned off. Consequently, if a female is heterozygous for a gene on the X chromosome (a sex-linked gene; see Chapter 9), about half of her cells will express one allele, while the others will express the alternate allele (**Figure 11.4**). ☑

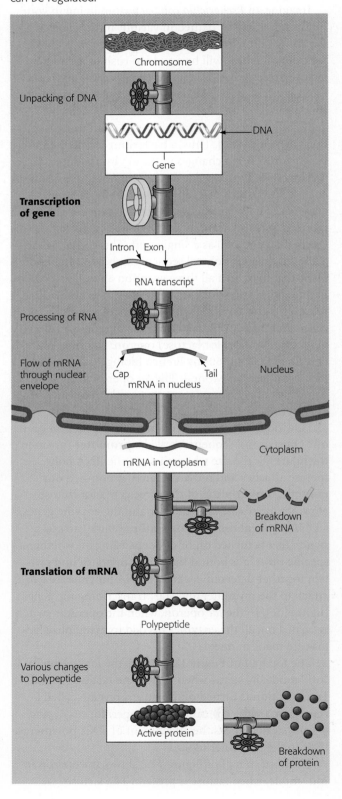

▼ **Figure 11.3 The gene expression "pipeline" in a eukaryotic cell.** Each valve in the pipeline represents a stage at which the pathway from chromosome to functioning protein can be regulated.

Chromosome

Unpacking of DNA

DNA

Gene

Transcription of gene

Intron   Exon

RNA transcript

Processing of RNA

Flow of mRNA through nuclear envelope

Cap   Tail
mRNA in nucleus

Nucleus

Cytoplasm

mRNA in cytoplasm

Breakdown of mRNA

Translation of mRNA

Polypeptide

Various changes to polypeptide

Active protein

Breakdown of protein

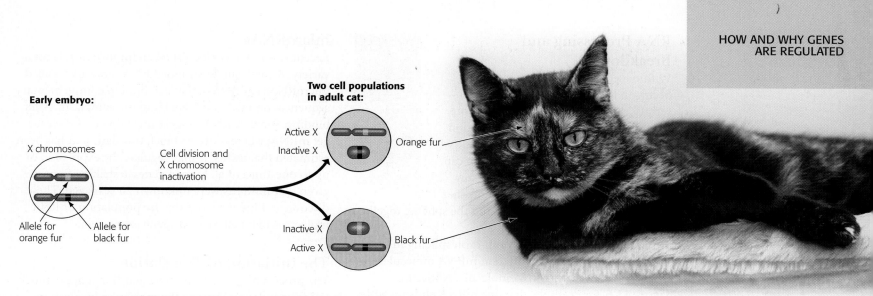

▲ **Figure 11.4 X chromosome inactivation: the tortoiseshell pattern on a cat.** The tortoiseshell gene is on the X chromosome, and the tortoiseshell phenotype requires the presence of two different alleles, one for orange fur and one for non-orange (black) fur. If a female is heterozygous for the tortoiseshell gene, orange patches are formed by populations of cells in which the X chromosome with the orange allele is active; black patches have cells in which the X chromosome with the non-orange allele is active.

The Initiation of Transcription

The initiation of transcription (whether transcription starts or not) is the most important stage for regulating gene expression. In both prokaryotes and eukaryotes, regulatory proteins bind to DNA and turn the transcription of genes on and off. Unlike prokaryotic genes, however, most eukaryotic genes have individual promoters and other control sequences. That is, eukaryotic genes are generally not organized into groups as operons.

As seen in **Figure 11.5**, transcriptional regulation in eukaryotes is complex, typically involving many proteins (collectively called **transcription factors**, purple in the figure) acting in concert to bind to DNA sequences called **enhancers** (yellow) and to the promoter (green). The DNA-protein assembly promotes the binding of RNA polymerase (orange) to the promoter. Genes coding for related enzymes, such as those in a metabolic pathway, may share a specific kind of enhancer (or collection of enhancers), allowing these genes to be activated at the same time. Not shown in the figure are repressor proteins, which may bind to DNA sequences called **silencers**, inhibiting the start of transcription.

In fact, repressor proteins that turn genes off are less common in eukaryotes than **activators**, proteins that turn genes on by binding to DNA. Activators act by making it easier for RNA polymerase to bind to the promoter. The use of activators is efficient because a typical animal or plant cell needs to turn on (transcribe) only a small percentage of its genes, those required for the cell's specialized structure and function. The "default" state for most genes in multicellular eukaryotes seems to be "off," with the exception of "housekeeping" genes for routine activities such as glucose metabolism. ✔

☑**CHECKPOINT**

Of all the control points of DNA expression shown in Figure 11.3, which is under the tightest regulation?

Answer: the initiation of transcription

▼ **Figure 11.5 A model for turning on a eukaryotic gene.** A large assembly of proteins and several control sequences in the DNA are involved in initiating the transcription of a eukaryotic gene.

RNA Processing and Breakdown

Within a eukaryotic cell, transcription occurs in the nucleus, where RNA transcripts are processed into mRNA before moving to the cytoplasm for translation by the ribosomes (see Figure 10.20). RNA processing includes the addition of a cap and a tail, as well as the removal of any introns—noncoding DNA segments that interrupt the genetic message—and the splicing together of the remaining exons.

Sometimes a cell can carry out exon splicing in more than one way, generating different mRNA molecules from the same starting RNA molecule. Notice in **Figure 11.6**, for example, that one mRNA ends up with the green exon and the other with the brown exon. With this sort of **alternative RNA splicing**, an organism can get more than one type of polypeptide from a single gene. Alternative RNA splicing is very common in humans. In one case, the RNA transcript from a gene can be spliced to encode seven different versions of a cellular protein.

After an mRNA is produced in its final form, its "lifetime" can be highly variable, from hours to weeks to months. Controlling the timing of mRNA breakdown provides another opportunity for control. But all mRNAs are eventually broken down and their parts recycled.

☑ CHECKPOINT

After a gene is transcribed in the nucleus, how is the transcript converted to mRNA? After the mRNA reaches the cytoplasm, what are four control mechanisms that can regulate the amount of active protein in the cell?

Answer: by RNA processing, including the addition of cap and tail, and RNA splicing; control by microRNAs, initiation of translation, activation of the protein, and breakdown of the protein

microRNAs

Recent research has established an important role for a variety of small single-stranded RNA molecules, called microRNAs (miRNAs), that can bind to complementary sequences on mRNA molecules in the cytoplasm. After binding, some miRNAs trigger breakdown of their target mRNA, whereas others block translation. It has been estimated that miRNAs may regulate the expression of up to one-third of all human genes, a striking figure given that miRNAs were unknown 20 years ago. These discoveries hint at a large, diverse population of RNA molecules that help regulate gene expression.

The Initiation of Translation

The process of translation offers additional opportunities for regulation. Among the molecules involved in translation are many regulatory proteins. Red blood cells, for instance, have a protein that prevents the translation of hemoglobin mRNA unless the cell has a supply of heme, an iron-containing chemical group essential for hemoglobin function.

Protein Activation and Breakdown

The final opportunities for regulating gene expression occur after translation. Post-translational control mechanisms in eukaryotes often involve cutting polypeptides into smaller, active final products. The hormone insulin, for example, is synthesized as one long, inactive polypeptide (**Figure 11.7**). After translation, an enzyme removes an interior section, leaving two shorter chains that constitute the active insulin molecule.

Another control mechanism operating after translation is the selective breakdown of proteins. Some proteins that trigger metabolic changes in cells are broken down within a few minutes or hours. This regulation allows a cell to adjust the kinds and amounts of its proteins in response to changes in its environment. ☑

▼ **Figure 11.6 Alternative RNA splicing: producing two different mRNAs from the same gene.** Two different cells can use a given DNA gene to synthesize somewhat different mRNAs and proteins. The cells might be from different tissues or the same type of cell at different times in the organism's life. In this example, one mRNA has ended up with exon 3 (brown) and the other with exon 4 (green). These two mRNAs can then be translated into different but related proteins.

▼ **Figure 11.7 The formation of an active insulin molecule.** Only in its final form does insulin act as a hormone.

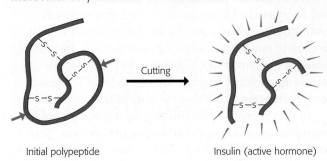

Initial polypeptide   Cutting   Insulin (active hormone)

Cell Signaling

So far, we have considered gene regulation only within a single cell. In a multicellular organism, the process can cross cell boundaries. A cell can produce and secrete chemicals, such as hormones, that affect gene regulation in another cell.

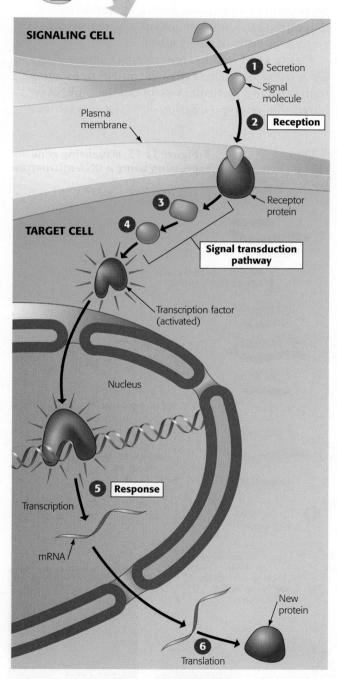

▲ Figure 11.8 **A cell-signaling pathway that turns on a gene.** The coordination of cellular activities in a multicellular organism depends on cell-to-cell signaling that helps regulate genes.

As you saw in Figure 5.19, a signal molecule can act by binding to a receptor protein and initiating a signal transduction pathway, a series of molecular changes that converts a signal to a specific response inside the target cell. **Figure 11.8** shows an example of cell-to-cell signaling in which the target cell's response is the transcription (turning on) of a gene. **1** First, the signaling cell secretes the signal molecule (⬤). **2** This molecule binds to a specific receptor protein (⬤) embedded in the target cell's plasma membrane. **3** The binding activates a signal transduction pathway consisting of a series of relay proteins (green) within the target cell. Each relay molecule activates another. **4** The last relay molecule in the pathway activates a transcription factor (⬤) that **5** triggers the transcription of a specific gene. **6** Translation of the mRNA produces a protein. ☑

Homeotic Genes

Cell-to-cell signaling and the control of gene expression are most important during early embryonic development, when a zygote develops into a multicellular organism. Master control genes called **homeotic genes** regulate groups of other genes that determine what body parts will develop in which locations. For example, one set of homeotic genes in fruit flies instructs cells in the embryonic head and thorax (midbody) to form antennae and legs, respectively. Elsewhere, these homeotic genes remain turned off, while others are turned on. Mutations in homeotic genes can produce bizarre effects. For example, fruit flies with mutations in homeotic genes may have extra sets of wings or have legs growing from their head (**Figure 11.9**).

► Figure 11.9 **The effect of homeotic genes.** The strange mutant fruit flies shown at the bottom result from mutations in homeotic (master control) genes.

Normal fruit fly

Normal head

Mutant fly with extra wings

Mutant fly with extra legs growing from head

One of the exciting biological discoveries in recent years is that similar homeotic genes help direct embryonic development in nearly every eukaryotic organism examined so far, including yeasts, plants, earthworms, frogs, chickens, mice, and humans. **Figure 11.10** highlights some striking similarities in the chromosomal locations and developmental roles of homeotic genes in two quite different animals: a fruit fly and a mouse. The colored segments represent homeotic genes that are very similar in both animals. Notice that the fruit fly and mouse chromosomes have the same order of genes, corresponding to analogous body regions in both animals. These similarities suggest that the original version of these homeotic genes arose very early in the history of life and that the genes have remained remarkably unchanged over eons of animal evolution. ☑

DNA Microarrays: Visualizing Gene Expression

Scientists who study gene regulation can use a DNA microarray to visualize patterns of gene expression. A **DNA microarray** is a glass slide with thousands of single-stranded DNA fragments attached to wells in a tightly spaced array (grid). Each DNA fragment is obtained from a particular gene; a single microarray thus carries DNA from thousands of genes, perhaps even all the genes of an organism.

Figure 11.11 outlines how microarrays are used.
❶ A researcher collects all of the mRNA transcribed from genes in a particular type of cell. This mRNA is mixed with reverse transcriptase, a viral enzyme that

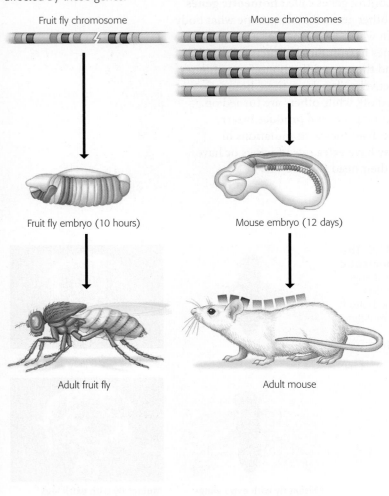

▼ **Figure 11.10 Homeotic genes in two different animals.** At the top of the figure are portions of chromosomes that carry homeotic genes in the fruit fly and the mouse. The colored boxes represent homeotic genes that have especially close DNA sequences in fruit flies and mice. The same color coding identifies the parts of the animals affected by these genes.

Fruit fly chromosome

Mouse chromosomes

Fruit fly embryo (10 hours)

Mouse embryo (12 days)

Adult fruit fly

Adult mouse

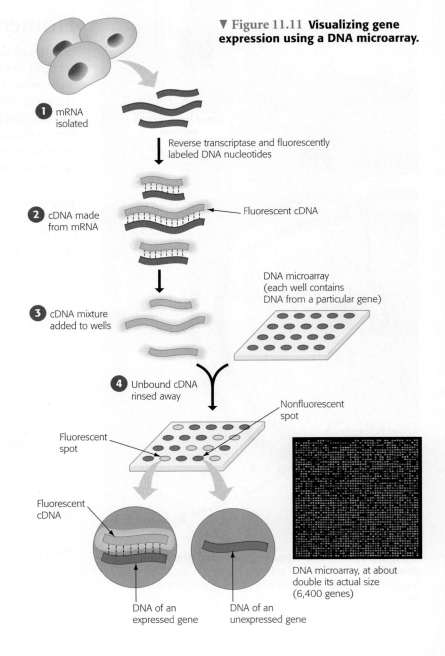

▼ **Figure 11.11 Visualizing gene expression using a DNA microarray.**

❶ mRNA isolated

Reverse transcriptase and fluorescently labeled DNA nucleotides

❷ cDNA made from mRNA

Fluorescent cDNA

❸ cDNA mixture added to wells

DNA microarray (each well contains DNA from a particular gene)

❹ Unbound cDNA rinsed away

Nonfluorescent spot

Fluorescent spot

Fluorescent cDNA

DNA of an expressed gene

DNA of an unexpressed gene

DNA microarray, at about double its actual size (6,400 genes)

2 catalyzes synthesis of DNA that is complementary to each mRNA sequence. This **complementary DNA (cDNA)** is synthesized using nucleotides that have been modified to fluoresce (glow). The fluorescent cDNA collection thus represents all of the genes being actively transcribed in the cell. 3 A small amount of the fluorescently labeled cDNA mixture is added to the DNA fragment in each well of the microarray. If a molecule in the cDNA mixture is complementary to a DNA fragment at a particular location on the microarray, the cDNA molecule binds to it, becoming fixed there. 4 After unbound cDNA is rinsed away, the remaining cDNA glows in the microarray. The pattern of glowing spots enables the researcher to determine which genes were being transcribed in the starting cells. Researchers can thus learn which genes are active in different tissues or in tissues from individuals in different states of health. ✓

☑CHECKPOINT

What is cDNA? Do you have any in your cells?

Answer: cDNA is DNA synthesized from mRNA. cDNA is artificially produced and does not exist naturally in our cells.

Cloning Plants and Animals

Now that we have examined how gene expression is regulated, we will devote the rest of this chapter to how gene regulation affects two important processes: cloning and cancer.

The Genetic Potential of Cells

One of the most important "take-home lessons" from this chapter is that differentiated cells express only a small percentage of their genes. So how do we know that all the genes are still present? And if all the genes are still there, do differentiated cells retain the potential to express them?

One way to approach these questions is to see if a differentiated cell can be dedifferentiated and stimulated to generate a whole new organism. This sort of genetic potential can be readily demonstrated in plants. For example, if you have ever grown a plant from a small cutting, you've seen evidence that a differentiated plant cell can undergo cell division and give rise to all the tissues of an entire plant. On a larger scale, the technique described in **Figure 11.12** can be used to produce hundreds or thousands of genetically identical organisms—clones—from the cells of a single plant.

Plant cloning is now used extensively in agriculture. For some plants, such as orchids, cloning is the only commercially practical means of reproducing plants. In other cases, cloning has been used to reproduce a plant with specific desirable traits, such as high fruit yield or resistance to disease. The success of plant cloning shows that cell differentiation in plants does not result in irreversible changes in the DNA.

A similar, naturally occurring process in animals is **regeneration**, the regrowth of lost body parts. When a salamander loses a leg, for example, certain cells in the leg stump reverse their differentiated state, divide, and then differentiate again to give rise to a new leg. Many other animals, especially among the invertebrates, can regenerate lost parts, and isolated pieces of a few relatively simple animals can dedifferentiate and then develop into an entirely new organism (see Figure 8.1).

▶ Figure 11.12 **Test-tube cloning of a carrot plant.** A single cell removed from the root of a carrot plant and placed in growth medium may begin dividing and eventually grow into an adult plant. The new plant is a genetic duplicate of the parent plant. This process proves that mature plant cells can reverse their differentiation and develop into all the specialized cells of an adult plant.

Root of carrot plant

Root cells in growth medium

Single cell

Cell division in culture

Young plant

Adult plant

Reproductive Cloning of Animals

Animal cloning is achieved through a procedure called **nuclear transplantation** (Figure 11.13). First performed in the 1950s on frog embryos, nuclear transplantation involves replacing the nucleus of an egg cell or a zygote with a nucleus removed from an adult body cell. The recipient cell may then begin to divide. Repeated cell divisions form a blastocyst, a hollow ball of about 100 cells that is an early stage in normal animal development. At this point, the blastocyst may be used for different purposes, as indicated by the two branches in Figure 11.13.

If the animal to be cloned is a mammal, further development requires implanting the blastocyst into the uterus of a surrogate mother (Figure 11.13, upper branch). The resulting animal will be genetically identical to the donor of the nucleus—a "clone" of the donor. This type of cloning is called **reproductive cloning** because it results in the birth of a new animal.

Scottish researcher Ian Wilmut and his colleagues used reproductive cloning to produce the celebrated sheep Dolly in 1997. The researchers used an electric shock to fuse specially treated adult sheep udder cells with 277 eggs from which they had removed the nuclei. After several days of growth, 29 of the resulting embryos were implanted in the uteruses of surrogate mothers. One of the embryos developed into the world-famous Dolly. As expected, Dolly resembled her genetic parent, the nucleus donor, not the egg donor or the surrogate mother.

Practical Applications of Reproductive Cloning

In the years since Dolly's landmark birth, researchers have cloned over a dozen other species of mammals, including mice, horses, dogs, mules, cows, pigs, rabbits, ferrets, and cats (Figure 11.14a). Why would anyone want to do this? In agriculture, farm animals with specific sets of desirable traits might be cloned to produce herds with these traits. In research, genetically identical animals may provide perfect "control animals" for experiments. The pharmaceutical industry is experimenting with cloning animals for potential medical use. For example, the pigs in Figure 11.14b are clones that lack a gene for a protein that can cause immune system rejection in humans. Organs from such pigs may one day be used in human patients who need transplants.

Perhaps the most intriguing use of reproductive cloning is to restock populations of endangered animals. Among the rare animals that have been cloned are a wild mouflon (a small European sheep), a banteng (a Javanese cow), a gaur (an Asian ox), and gray wolves (Figure 11.14c). The 2003 cloning of a banteng, whose numbers have dwindled to just a few in the wild, was a remarkable case. Using frozen cells from a zoo-raised

▼ Figure 11.13 **Cloning by nuclear transplantation.** In nuclear transplantation, an adult somatic nucleus is injected into a nucleus-free egg cell. The resulting embryo may then be used to produce a new organism (reproductive cloning, shown in the upper branch) or to provide stem cells (therapeutic cloning, lower branch). The lamb in the photograph is the famous Dolly, shown with her surrogate mother.

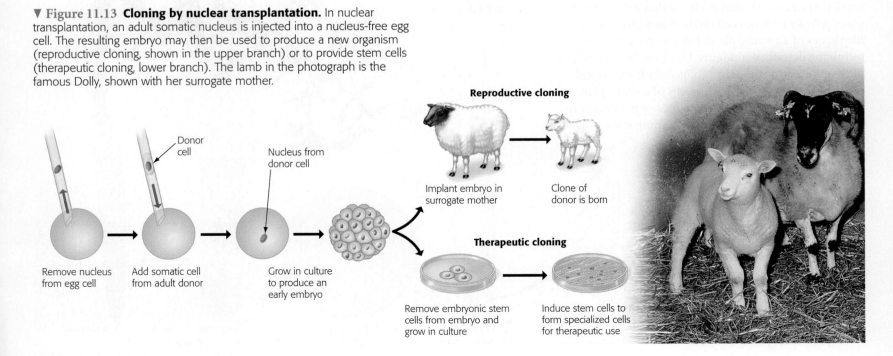

banteng that had died 23 years prior, scientists transplanted nuclei from the frozen cells into nucleus-free eggs from dairy cows. The resulting embryos were implanted into surrogate cows, leading to the birth of a healthy baby banteng. This success shows that it is possible to produce a baby even when a female of the donor species is unavailable. Scientists may someday be able to use similar cross-species methods to clone an animal from a recently extinct species.

The use of cloning to repopulate endangered species holds tremendous promise. However, cloning may also create new problems. Conservationists object that cloning may detract from efforts to preserve natural habitats. They correctly point out that cloning does not increase genetic diversity and is therefore not as beneficial to endangered species as natural reproduction. Cloned animals are also less healthy than those arising from a fertilized egg. In 2003, Dolly was euthanized

after suffering complications from a lung disease that is normally seen only in much older sheep. Other cloned animals have exhibited defects such as susceptibility to obesity, pneumonia, liver failure, and premature death.

Human Cloning

The cloning of various mammals has heightened speculation that humans could be cloned. Critics point out practical and ethical objections to human cloning. Practically, cloning of mammals is extremely difficult and inefficient. Only a small percentage of cloned embryos develop normally, and they appear less healthy than naturally born kin. Ethically, creating human embryos for cloning raises troubling questions about the status of the human blastocyst. Consensus on this topic is unlikely anytime soon. Meanwhile, the research and the debate continue. ☑

☑ **CHECKPOINT**

Suppose a nucleus from an adult body cell of a black mouse is injected into an egg removed from a white mouse, and then the embryo is implanted into a brown mouse. What would be the color of the resulting cloned mice?

Answer: black, the color of the nucleus donor

▼ Figure 11.14 **Reproductive cloning of mammals.**

(a) **The first cloned cat.** CC ("Copy Cat") and her lone parent (left).

(b) **Cloning for medical use.**
These piglets are clones of a pig that was genetically modified to lack a protein that causes transplant rejection in humans.

(c) **Clones of endangered animals**

Mouflon calf with mother   Banteng   Gaur   Gray wolf

Therapeutic Cloning and Stem Cells

The lower branch of Figure 11.13 (p. 208) depicts an alternative cloning outcome called therapeutic cloning. The purpose of **therapeutic cloning** is not to produce a viable organism but to produce embryonic stem cells.

Embryonic Stem Cells

In mammals, **embryonic stem cells (ES cells)** are derived from blastocysts. During development, embryonic stem cells in the blastocyst differentiate, giving rise to all the specialized cells in the body. When removed from an early embryo and grown in laboratory culture, embryonic stem cells can divide indefinitely. The right conditions—such as the presence of certain growth-stimulating proteins—can induce changes in gene expression that cause the cells to develop into a particular cell type (**Figure 11.15**). If scientists can discover the right conditions, they may be able to grow cells for the repair of injured or diseased organs. The use of embryonic stem cells in therapeutic cloning is controversial, however, because the removal of ES cells destroys the embryo.

Adult Stem Cells

Embryonic stem cells are not the only stem cells available to researchers. **Adult stem cells** are cells in adult tissues that generate replacements for nondividing differentiated cells. Unlike ES cells, adult stem cells are partway along the road to differentiation; in the body, they usually give rise to only a few related types of specialized cells. For example, stem cells in bone marrow can generate the different kinds of blood cells (see Figure 11.15). Adult stem cells are much more difficult to grow in culture than ES cells, and they are not as versatile, but researchers have had some success. Because no embryonic tissue is involved in their harvest, adult stem cells may provide an ethically less problematic route for human tissue and organ replacement than ES cells. However, some researchers think that ES cells are the only cells likely to lead to groundbreaking advances in human health.

Umbilical Cord Blood Banking

Another source of stem cells is blood collected from the umbilical cord and placenta at birth. Such stem cells appear to be partially differentiated, less so than adult stem cells, more so than embryonic stem cells. To obtain these cells, a physician inserts a needle into the umbilical cord and extracts 1/4 to 1/2 cup of blood (**Figure 11.16**). The cells are then frozen and kept in a blood bank until needed for later medical treatment. In 2005, doctors reported that an infusion of umbilical cord blood stem cells from a compatible (but unrelated) donor appeared to cure some babies of Krabbe's disease, a usually fatal inherited disorder of the nervous system. Other people have received cord blood as a treatment for leukemia. To date, however, most attempts at umbilical cord blood therapy have not been successful. At present, the American Academy of Pediatrics recommends cord blood banking only for babies born into families with a known genetic risk. So far, the promise of cord blood banking vastly exceeds the accomplishments. ✔

✔CHECKPOINT

How do the results of reproductive cloning and therapeutic cloning differ?

Answer: Reproductive cloning results in the production of a live individual; therapeutic cloning produces stem cells.

▼ **Figure 11.15 Differentiation of embryonic stem cells in culture.** Scientists hope to discover growth conditions that will stimulate cultured stem cells to differentiate into specialized cells.

Adult stem cells in bone marrow

Cultured **embryonic stem cells**

Different culture conditions

Blood cells

Nerve cells

Heart muscle cells

Different types of **differentiated cells**

▼ **Figure 11.16 Umbilical cord blood banking.** Just after birth, a doctor may collect blood from a newborn's umbilical cord. The umbilical cord blood (inset), rich in stem cells, is stored for possible future use.

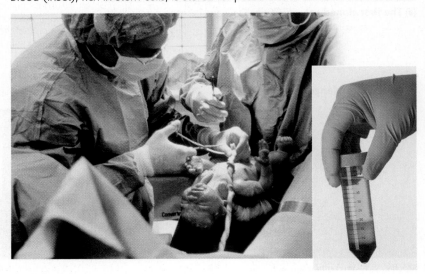

The Genetic Basis of Cancer

Chapter 8 introduced cancer as a variety of diseases in which cells escape from the control mechanisms that normally limit their growth and division. This escape involves changes in gene expression.

Genes That Cause Cancer

One of the earliest clues to the role of genes in cancer was the discovery in 1911 of a virus that causes cancer in chickens. Recall that viruses are simply "genes in a box," molecules of DNA or RNA surrounded by protein and in some cases a membrane. Viruses that cause cancer can become permanent residents in host cells by inserting their nucleic acid into the DNA of host chromosomes. Researchers have identified a number of viruses that harbor cancer-causing genes. When inserted into a host cell, these genes can make the cell cancerous. A gene that causes cancer is called an **oncogene** ("tumor gene").

Oncogenes and Tumor-Suppressor Genes

In 1976, American molecular biologists J. Michael Bishop, Harold Varmus, and their colleagues made a startling discovery. They found that the virus that causes cancer in chickens contains an oncogene that is an altered version of a normal chicken gene. Subsequent research has shown that the chromosomes of many animals, including humans, contain genes that can be converted to oncogenes. A normal gene with the potential to become an oncogene is called a **proto-oncogene**. (These terms can be confusing, so they bear repeating: a *proto-oncogene* is a normal gene that, if changed, can become a cancer-causing *oncogene*.) A cell can acquire an oncogene from a virus or from the conversion of one of its own proto-oncogenes.

How can a change in a gene cause cancer? Searching for the normal roles of proto-oncogenes in the cell, researchers found that many of these genes code for **growth factors**—proteins that stimulate cell division—or for other proteins that affect the cell cycle. When all these proteins are functioning normally, in the right amounts at the right times, they help keep the rate of cell division at an appropriate level. When they malfunction—if a growth factor becomes hyperactive, for example—cancer (uncontrolled cell growth) may result.

For a proto-oncogene to become an oncogene, a mutation must occur in the cell's DNA. **Figure 11.17** illustrates three kinds of changes in DNA that can produce active oncogenes. In all three cases, abnormal gene expression stimulates the cell to divide excessively.

Changes in genes whose products inhibit cell division are also involved in cancer. These genes are called **tumor-suppressor genes** because the proteins they encode normally help prevent uncontrolled cell growth (**Figure 11.18**). Any mutation that keeps a normal growth-inhibiting protein from being made or from functioning may contribute to development of cancer. Researchers have identified many mutations in both tumor-suppressor and growth factor genes that are associated with cancer, as we'll discuss next.

▼ Figure 11.17 **How a proto-oncogene can become an oncogene.**

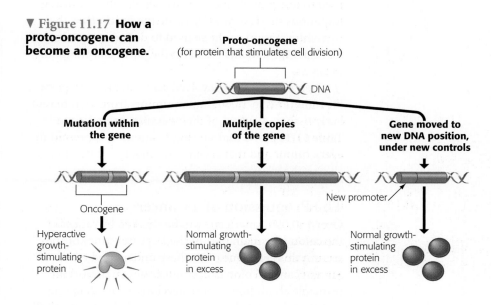

▼ Figure 11.18 **Tumor-suppressor genes.**

(a) **Normal cell growth.** A tumor-suppressor gene normally codes for a protein that inhibits cell growth and division. In this way, the gene helps prevent cancerous tumors from arising.

(b) **Uncontrolled cell growth (cancer).** When a mutation in a tumor-suppressor gene makes its protein defective, cells that are usually under the control of the normal protein may divide excessively, forming a tumor.

Can Cancer Therapy Be Personalized?

Medical researchers have made many **observations** of specific mutations that can lead to cancer. But it is reasonable to **question** whether this information can be used to help patients. A research team from the Dana Farber Cancer Institute in Boston formed the **hypothesis** that DNA-sequencing technology can be used to test tumors and identify which cancer-causing mutations they carry. They made the **prediction** that such information could be used to design customized therapies for patients based on the particular mutations involved.

Their **experiment** involved screening for 238 possible mutations in 1,000 human tumors from 18 different body tissues. A subset of their **results**, presented in **Table 11.1**, shows that no single mutation is present in every tumor and that each tumor involves different mutations. (In this table, a ✓ indicates that a given mutation was found in at least one tumor of that type.) These results reveal that it is possible to cheaply and accurately determine which mutations are present in a given cancer patient. Indeed, there are now a number of cancer types (including leukemia, breast cancer, and lung cancer) for which personalized therapies—based on genetic screening—are available. Personalized cancer treatment remains an active area of research.

Table 11.1	Cancer-Causing Mutations			
Tumor location	**A**	**B**	**C**	**D**
Pancreas		✓		
Ovary	✓	✓		✓
Lung	✓		✓	✓
Colon	✓	✓	✓	✓
Breast	✓			✓

The Progression of a Cancer

Over 150,000 Americans will be stricken by cancer of the colon (the main part of the large intestine) or the rectum this year. One of the best-understood types of human cancer, colon cancer illustrates an important principle about how cancer develops: More than one mutation is needed to produce a full-fledged cancer cell. As in many cancers, the development of colon cancer is a gradual process.

As shown in **Figure 11.19**, **1** colon cancer begins when an oncogene arises or is activated through mutation, causing unusually frequent division of normal-looking cells in the colon lining. **2** Later, additional DNA mutations (such as the inactivation of a tumor-suppressor gene) cause the growth of a small benign tumor (polyp) in the colon wall. **3** Further mutations eventually lead to formation of a malignant tumor—a tumor that has the potential to metastasize (spread).

▼ Figure 11.19 **Stepwise development of a typical colon cancer.**

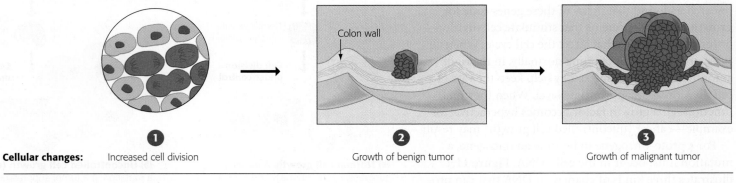

	1	**2**	**3**
Cellular changes:	Increased cell division	Growth of benign tumor	Growth of malignant tumor
DNA changes:	Oncogene activated	Tumor-suppressor gene inactivated	Second tumor-suppressor gene inactivated

The development of a malignant tumor is accompanied by a gradual accumulation of mutations that convert proto-oncogenes to oncogenes and knock out tumor-suppressor genes (Figure 11.20). The requirement for several DNA mutations—usually four or more—explains why cancers can take a long time to develop. This requirement may also help explain why the incidence of cancer increases greatly with age; the longer we live, the more likely we are to accumulate mutations that cause cancer.

"Inherited" Cancer

Most mutations that lead to cancer arise in the organ where the cancer starts—the colon, for example. Because these mutations do not affect the cells that give rise to eggs or sperm, they are not passed from parent to child. Sometimes, however, a cancer-causing mutation occurs in a cell that gives rise to gametes and is therefore passed on from generation to generation. Such mutations predispose the people who inherit them to developing cancer. Such cancer is called familial or inherited. But even familial cancers don't appear unless the person acquires additional mutations in the susceptible tissue.

One well-studied inherited cancer gene is *BRCA1* (pronounced "braca-1"), whose normal allele encodes a tumor-suppressor protein. Certain alleles of *BRCA1* are associated with breast cancer, a disease that strikes one out of every ten American women (Figure 11.21). Some *BRCA1* mutations put a woman at high risk for both breast cancer and ovarian cancer—a more than 80% risk

▲ Figure 11.21 **Breast cancer.** In this mammogram, normal breast tissue appears as gray and white, while the tumor is highlighted in yellow and red.

of developing cancer during her lifetime. In recent years, tests for *BRCA1* mutations have become available. Unfortunately, these tests are of limited use because surgical removal of the breasts and/or ovaries is the only preventive option currently available to women who carry the mutant genes. ☑

▼ Figure 11.20 **Accumulation of mutations in the development of a cancer cell.** Mutations leading to cancer accumulate in a lineage of cells. In this figure, colors distinguish the normal cells from cells with one or more mutations, leading to increased cell division and cancer. Once a cancer-promoting mutation occurs (orange band on chromosome), it is passed to all the descendants of the cell carrying it.

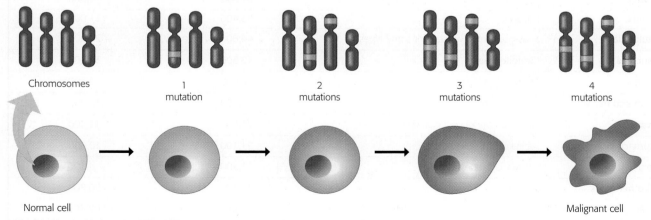

Chromosomes

1 mutation

2 mutations

3 mutations

4 mutations

Normal cell

Malignant cell

Cancer Risk and Prevention

Cancer is the second-leading cause of death in most industrialized countries (after heart disease). Death rates due to certain forms of cancer have decreased in recent years, but the overall cancer death rate is still on the rise, currently increasing at about 1% per decade.

Most cancers arise from mutations that are caused by **carcinogens**, cancer-causing agents found in the environment. Mutations often result from decades of exposure to carcinogens. One of the most potent carcinogens is ultraviolet (UV) radiation. Excessive exposure to UV radiation from the sun can cause skin cancer, including a deadly type called melanoma.

The one substance known to cause more cases and types of cancer than any other is tobacco. More people die from lung cancer (over 160,000 Americans in 2008) than from any other form of cancer. Most tobacco-related cancers are due to smoking, but the passive inhalation of secondhand smoke also poses a risk. As **Table 11.2** indicates, tobacco use, sometimes in combination with alcohol consumption, causes some types of cancer. In nearly all cases, cigarettes are the main culprit, but smokeless tobacco products (snuff and chewing tobacco) are linked to cancer of the mouth and throat.

Exposure to some of the most lethal carcinogens is often a matter of individual choice: Tobacco use, the consumption of alcohol, and excessive time spent in the sun are all avoidable behaviors that affect cancer risk.

There is increasing evidence that some food choices significantly reduce a person's cancer risk. For instance, eating 20–30 g of plant fiber daily (about twice the amount the average American consumes) while eating less animal fat may help prevent colon cancer. Some studies suggest that certain substances in fruits and vegetables, including vitamins C and E and certain compounds related to vitamin A, may help protect against a variety of cancers. Cabbage and its relatives, such as broccoli and cauliflower, are thought to be especially rich in substances that help prevent cancer, although some of the specific substances have not yet been identified. Determining how diet influences cancer has become an important focus of nutrition research.

The battle against cancer is being waged on many fronts, and there is reason for optimism in the progress being made. It is especially encouraging that we can help reduce our risk of acquiring some of the most common forms of cancer by the choices we make in our daily lives. ☑

Table 11.2	Cancer in the United States (Ranked by Number of Cases)		
Cancer	**Known or Likely Carcinogens or Factors**	**Estimated Cases (2009)**	**Estimated Deaths (2009)**
Lung	Cigarette smoke	219,000	159,000
Breast	Estrogen; possibly dietary fat	194,000	40,600
Prostate	Testosterone; possibly dietary fat	192,000	27,000
Colon and rectum	High dietary fat; low dietary fiber	147,000	49,900
Skin	Ultraviolet light	74,600	11,600
Lymphomas	Viruses (for some types)	74,500	20,800
Bladder	Cigarette smoke	71,000	14,300
Kidney	Cigarette smoke	57,800	13,000
Uterus	Estrogen	53,400	11,800
Leukemias	X-rays; benzene; viruses (for some types)	44,800	21,900
Pancreas	Cigarette smoke	42,500	35,200
Liver	Alcohol; hepatitis viruses	22,600	18,200
Brain and nerve	Trauma; X-rays	22,100	12,900
Ovary	Large number of ovulation cycles	21,600	14,600
Stomach	Table salt; cigarette smoke	21,100	10,600
Cervix	Viruses; cigarette smoke	11,300	4,100
All other types		210,100	96,800
Total		1,479,400	562,300

Source: *Cancer Facts and Figures 2009* (American Cancer Society Inc.)

☑**CHECKPOINT**

Of all known behavioral factors, which one causes the most cancer cases and deaths?

Answer: *tobacco use*

EVOLUTION CONNECTION: Cancer

The Evolution of Cancer in the Body

The theory of evolution describes natural selection acting on populations of organisms. Recently, medical researchers have been using an evolutionary perspective to gain insight into the development of tumors, such as the bone tumor shown in **Figure 11.22**. Evolution drives the growth of a tumor—which can be thought of as a population of cancer cells—and also affects how those cells respond to cancer treatments.

Recall from Chapter 1 that there are several assumptions behind Darwin's theory of natural selection. Let's consider how each one can be applied to cancer. First, all evolving populations have the potential to produce more offspring than can be supported by the environment. Cancer cells, with their uncontrolled growth, clearly demonstrate such overproduction. Second, there must be variation among individuals of the population.

Studies of tumor cell DNA, like the one described in the Process of Science section, show genetic variability within tumors. Finally, variations in the population must affect survival and reproductive success. Indeed, the accumulation of mutations in cancer cells renders them less susceptible to normal mechanisms of reproductive control. Mutations that enhance survival of malignant cancer cells are passed on to that cell's descendants. In short, a tumor evolves, much like a population of organisms.

Viewing the progression of cancer through the lens of evolution helps explain why there is no easy "cure" for cancer but may also pave the way for novel therapies. For example, some researchers are attempting to "prime" tumors for treatment by increasing the reproductive success of only those cells that will be susceptible to a chemotherapy drug. Our understanding of cancer, like all other aspects of biology, benefits from an evolutionary perspective.

▼ **Figure 11.22 X-ray of shoulder and upper arm, revealing a large bone tumor.**

Chapter Review

SUMMARY OF KEY CONCEPTS

 Go to the Study Area at **www.masteringbiology.com** for practice quizzes, eText, BioFlix™ 3-D animations, MP3 Tutor Sessions, videos, current events, and more.

How and Why Genes Are Regulated

Patterns of Gene Expression in Differentiated Cells

The various cell types in a multicellular organism are a consequence of different combinations of genes being turned on and off via gene regulation.

Gene Regulation in Bacteria

An operon is a cluster of genes with related functions together with their promoter and other DNA sequences that control their transcription. The *lac* operon allows *E. coli* to produce enzymes for lactose use only when the sugar is present.

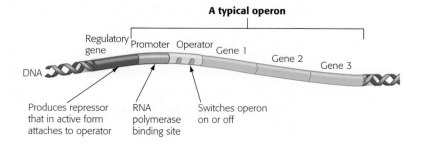

A typical operon

Regulatory gene — Produces repressor that in active form attaches to operator

Promoter — RNA polymerase binding site

Operator — Switches operon on or off

Gene 1

Gene 2

Gene 3

DNA

Gene Regulation in Eukaryotic Cells

In the nucleus of eukaryotic cells, there are several possible control points in the pathway of gene expression.

- DNA packing tends to block gene expression, presumably by preventing access of transcription proteins to the DNA. An extreme example is X chromosome inactivation in the cells of female mammals.

- The most important control point in both eukaryotes and prokaryotes is at gene transcription. Various regulatory proteins interact with DNA and with each other to turn the transcription of eukaryotic genes on or off.

- There are also opportunities for the control of eukaryotic gene expression after transcription, when introns are cut out of the RNA and a cap and tail are added to process RNA transcripts into mRNA.

- In the cytoplasm, presence of microRNAs may block the translation of an mRNA, and various proteins may regulate the start of translation.

- Finally, the cell may activate the finished protein in various ways (for instance, by cutting out portions). Eventually, the protein may be selectively broken down.

DNA unpacking

Transcription

RNA processing

RNA transport

mRNA breakdown

Translation

Protein activation

Protein breakdown

Cell Signaling

Cell-to-cell signaling is key to the development and functioning of multicellular organisms. Signal transduction pathways convert molecular messages to cell responses, often the transcription of particular genes.

Homeotic Genes

Evidence for the evolutionary importance of gene regulation is apparent in homeotic genes, master genes that regulate other genes that in turn control embryonic development.

DNA Microarrays: Visualizing Gene Expression

DNA microarrays can be used to determine which genes are turned on in a particular cell type.

Cloning Plants and Animals

The Genetic Potential of Cells

Most differentiated cells retain a complete set of genes, so a carrot plant, for example, can be made to grow from a single carrot cell. Under special conditions, animals can also be cloned.

Reproductive Cloning of Animals

Nuclear transplantation is a procedure whereby a donor cell nucleus is inserted into a nucleus-free egg. First demonstrated in frogs in the 1950s, reproductive cloning was used in 1997 to clone a sheep from an adult mammary cell and has since been used to create many other cloned animals.

Nucleus from donor cell → Early embryo resulting from nuclear transplantation → Embryo implanted in surrogate mother → Clone of nucleus donor

Therapeutic Cloning and Stem Cells

The purpose of therapeutic cloning is to produce embryonic stem cells for medical uses. Both embryonic and adult stem cells show promise for future therapeutic uses.

Nucleus from donor cell → Early embryo resulting from nuclear transplantation → Embryonic stem cells in culture → Specialized cells

The Genetic Basis of Cancer

Genes That Cause Cancer

Cancer cells, which divide uncontrollably, can result from mutations in genes whose protein products regulate the cell cycle.

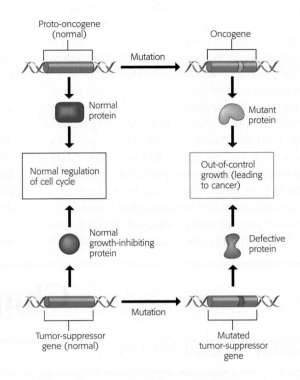

Proto-oncogene (normal) — Mutation → Oncogene

Normal protein — Mutant protein

Normal regulation of cell cycle — Out-of-control growth (leading to cancer)

Normal growth-inhibiting protein — Defective protein

Tumor-suppressor gene (normal) — Mutation → Mutated tumor-suppressor gene

Many proto-oncogenes and tumor-suppressor genes code for proteins active in signal transduction pathways regulating cell division. Mutations of these genes cause malfunction of the pathways. Cancers result from a series of genetic changes in a cell lineage. Researchers have gained insight into the genetic basis of breast cancer by studying families in which a disease-predisposing mutation is inherited.

Cancer Risk and Prevention

Reducing exposure to carcinogens (which induce cancer-causing mutations) and making other healthful lifestyle choices can help reduce cancer risk.

SELF-QUIZ

1. Your bone cells, muscle cells, and skin cells look different because
   a. different kinds of genes are present in each kind of cell.
   b. they are present in different organs.
   c. different genes are active in each kind of cell.
   d. different mutations have occurred in each kind of cell.

2. A group of prokaryotic genes with related functions that are regulated as a single unit, along with the control sequences that perform this regulation, is called a(n) _____.

3. The regulation of gene expression must be more complex in multicellular eukaryotes than in prokaryotes because
   a. eukaryotic cells are much larger.
   b. in a multicellular eukaryote, different cells are specialized for different functions.
   c. prokaryotes are restricted to stable environments.
   d. eukaryotes have fewer genes, so each gene must do several jobs.

4. A eukaryotic gene was inserted into the DNA of a bacterium. The bacterium then transcribed this gene into mRNA and translated the mRNA into protein. The protein produced was useless and contained many more amino acids than the protein made by the eukaryotic cell. Why?
   a. The mRNA was not spliced as it is in eukaryotes.
   b. Eukaryotes and prokaryotes use different genetic codes.
   c. Repressor proteins interfered with transcription and translation.
   d. Ribosomes were not able to bind to tRNA.

5. How does dense packing of DNA in chromosomes prevent gene expression?

6. What evidence demonstrates that differentiated cells in a plant or animal retain their full genetic potential?

7. The most common procedure for cloning an animal is _____.

8. What is learned from a DNA microarray?

9. Which of the following is a valid difference between embryonic stem cells and the stem cells found in adult tissues?
   a. In laboratory culture, only adult stem cells are immortal.
   b. In nature, only embryonic stem cells give rise to all the different types of cells in the organism.
   c. Only adult stem cells can be made to differentiate in the laboratory.
   d. Only embryonic stem cells are in every tissue of the adult body.

10. Name three potential sources of stem cells.

11. What is the difference between oncogenes and proto-oncogenes? How can one turn into the other? What function do proto-oncogenes serve?

12. A mutation in a single gene may cause a major change in the body of a fruit fly, such as an extra pair of legs or wings. Yet it takes many genes to produce a wing or leg. How can a change in just one gene cause such a big change in the body? What are such genes called?

Answers to the Self-Quiz questions can be found in Appendix D.

THE PROCESS OF SCIENCE

13. Study the depiction of the *lac* operon in Figure 11.2. Normally, the genes are turned off when lactose is not present. Lactose activates the genes, which code for enzymes that enable the cell to use lactose. Mutations can alter the function of this operon; in fact, the effects of various mutations enabled Jacob and Monod to figure out how the operon works. Predict how the following mutations would affect the function of the operon in the presence and absence of lactose:
    a. mutation of regulatory gene; repressor will not bind to lactose
    b. mutation of operator; repressor will not bind to operator
    c. mutation of regulatory gene; repressor will not bind to operator
    d. mutation of promoter; RNA polymerase will not attach to promoter

14. The human body has a far greater variety of proteins than genes, a fact that seems to highlight the importance of alternative RNA splicing, which allows several different mRNAs to be made from a single gene. Suppose you have samples of two types of adult cells from one person. Design an experiment using microarrays to determine whether or not the different gene expression is due to alternative RNA splicing.

15. Because a cat must have both orange and non-orange alleles to be tortoiseshell (see Figure 11.4), we would expect only female cats, which have two X chromosomes, to be tortoiseshell. Normal male cats (XY) can carry only one of the two alleles. Male tortoiseshell cats are rare and usually sterile. What might you guess their genotype to be?

16. Design a DNA microarray experiment that measures the difference in gene expression between normal colon cells and cells from a colon tumor.

BIOLOGY AND SOCIETY

17. A chemical called dioxin is produced as a by-product of certain chemical manufacturing processes. Trace amounts of this substance were present in Agent Orange, a defoliant sprayed on vegetation during the Vietnam War. There has been a continuing controversy over its effects on soldiers exposed to Agent Orange during the war. Animal tests have suggested that dioxin can cause cancer, liver and thymus damage, immune system suppression, and birth defects; at high dosage it can be lethal. But such animal tests are inconclusive; a hamster is not affected by a dose that can kill a much larger guinea pig, for example. Researchers have discovered that dioxin enters a cell and binds to a protein that in turn attaches to the cell's DNA. How might this mechanism help explain the variety of dioxin's effects on different body systems and in different animals? How might you determine whether a particular individual became ill as a result of exposure to dioxin? Do you think this information is relevant in the lawsuits of soldiers suing over exposure to Agent Orange? Why or why not?

18. There are genetic tests available for several types of "inherited cancer." The results from these tests cannot usually predict that someone will get cancer within a particular amount of time. Rather, they indicate only that a person has an increased risk of developing cancer. For many of these cancers, lifestyle changes cannot decrease a person's risk. Therefore, some people consider the tests useless. If your close family had a history of cancer and a test were available, would you want to get screened? Why or why not? What would you do with this information? If a sibling decided to get screened, explain whether you would want to know the results.

12 DNA Technology

A DNA profile.
Even minuscule bits of
evidence can provide a
DNA profile.

CHAPTER CONTENTS
■■■ Chapter Thread: **DNA PROFILING**

BIOLOGY AND SOCIETY: DNA Profiling

DNA, Guilt, and Innocence

Early in the morning of May 3, 1992, Christine Jackson, a 3-year-old girl sleeping in her Mississippi home, was abducted, raped, murdered, and thrown into a creek. A tiny semen sample was recovered at the crime scene. Police arrested the mother's boyfriend, who had been babysitting the girl that night. He was convicted of the horrifying crime and sentenced to death.

The cells of every person (except identical twins) contain unique DNA. DNA profiling is the analysis of DNA samples to determine whether they come from the same individual. Although the crime scene semen sample was insufficient for DNA testing in 1992, new methods allowed a DNA profile to be determined in 2001. The results were conclusive: The boyfriend did not leave his semen at the crime scene.

DNA profiling can provide evidence of guilt as well as innocence. Several years later, the DNA profile obtained from the crime scene was matched to a man who confessed to the crime and to the nearly identical 1990 murder of 3-year-old Courtney Smith, for which a third man was serving jail time. After years of legal maneuvers, the two wrongly convicted men were freed in 2008, exonerated based on the DNA evidence. They were assisted by a nonprofit legal organization called the Innocence Project. To date, this organization has used DNA profiling data to exonerate 227 prisoners, including 17 death row inmates.

Because of its unbiased nature, DNA profiling has rapidly transformed crime investigation. Beyond the courtroom, DNA technology has led to some of the most remarkable scientific advances in recent years: Corn has been genetically modified to produce its own insecticide; human genes are being compared with those of other animals; and significant advances have been made toward curing fatal genetic diseases. This chapter will describe these and other uses of DNA technology and explain how various techniques are performed. We'll also examine some social, legal, and ethical issues that these new technologies raise.

Recombinant DNA Technology

DNA Technology

You may think of **biotechnology**, the manipulation of organisms or their components to make useful products, as a modern field, but it actually dates back to the dawn of civilization. Consider such ancient practices as the use of yeast to make bread and the selective breeding of livestock. But when people use the term *biotechnology* today, they are usually referring to **DNA technology**, methods for studying and manipulating genetic material. Using these techniques, scientists can modify specific genes and move them between organisms as different as bacteria, plants, and animals.

In the 1970s, the field of biotechnology exploded with the invention of methods for making recombinant DNA in the laboratory. **Recombinant DNA** results when scientists combine nucleotide sequences (pieces of DNA) from two different sources—often from different species—to form a single DNA molecule. Recombinant DNA technology is widely used in **genetic engineering**, the direct manipulation of genes for practical purposes. Scientists have genetically engineered bacteria to mass-produce a variety of useful chemicals, from cancer drugs to pesticides. Scientists have also transferred genes from bacteria to plants and from one animal species to another (**Figure 12.1**). ☑

☑CHECKPOINT

What is biotechnology? What is recombinant DNA?

Answer: the manipulation of organisms or their parts to produce a useful product; a molecule containing DNA from two different sources, often different species

▼ **Figure 12.1 Glowing fish.** Genetic engineers produced glowing fish by transferring a gene for a fluorescent protein originally obtained from jellies ("jellyfish").

Applications: From Humulin to Foods to "Pharm" Animals

By transferring the gene for a desired protein into a bacterium, yeast, or other kind of cell that is easy to grow in culture, scientists can produce large quantities of proteins that are present naturally in only small amounts. In this section, you'll learn about some applications of recombinant DNA technology.

Making Humulin

Humulin is human insulin produced by genetically modified bacteria (**Figure 12.2**). In humans, insulin is a protein normally made by the pancreas. Functioning as a hormone, insulin helps regulate the level of glucose in the blood. If the body fails to produce enough insulin, the result is type 1 diabetes. There is no cure, so people with this disease must inject themselves with daily doses of insulin for the rest of their lives.

Since human insulin is not readily available, diabetes was historically treated using cow and pig insulins. This treatment was not without problems, however. Pig and cow insulins can cause allergic reactions because their chemical structures differ slightly from that of human insulin. In addition, by the 1970s, the supply of beef and pork pancreas available for insulin extraction could not keep up with the demand. A new source was needed.

In 1978, scientists working at the biotechnology company Genentech chemically synthesized two genes, one for each of the polypeptides of the active form of human insulin (see Figure 11.7). Since the amino acid sequences of the two insulin polypeptides were already known, it was easy to use the genetic code (see Figure 10.11) to determine nucleotide sequences that would encode for them. Researchers synthesized DNA fragments and linked them

► Figure 12.2
Humulin, human insulin produced by genetically modified bacteria.

to form the insulin genes. In 1979, they succeeded in inserting these artificial genes into *E. coli* host cells. Under proper growing conditions, these bacteria cranked out large quantities of the human protein.

In 1982, Humulin hit the market as the world's first genetically engineered pharmaceutical product. Today, it is produced in gigantic fermentation vats filled with a liquid culture of bacteria; the vats operate around the clock. Each day, more than 4 million people with diabetes use the insulin collected, purified, and packaged at such facilities (**Figure 12.3**).

Insulin is just one of many human proteins produced by genetically modified bacteria. Another example is human growth hormone (HGH). Abnormally low levels of this hormone during childhood and adolescence can cause a form of dwarfism. Because growth hormones from other animals are not effective in humans, HGH was an early target of genetic engineers. Before genetically engineered HGH became available in 1985, children with an HGH deficiency could only be treated with scarce and expensive supplies of HGH obtained from human cadavers.

Besides bacteria, yeast and mammalian cells can also be used to produce medically valuable human proteins. For example, genetically modified mammalian cells growing in laboratory cultures are currently used to produce a hormone called EPO that stimulates production of red blood cells. EPO is used to treat anemia; unfortunately, some athletes abuse the drug to seek the advantage of artificially high levels of oxygen-carrying red blood cells (called "blood doping").

DNA technology is also helping medical researchers develop vaccines. A **vaccine** is a harmless variant or derivative of a disease-causing microbe—such as a bacterium or virus—that is used to prevent an infectious disease. When a person is inoculated, the vaccine stimulates the immune system to develop lasting defenses against the microbe. For many viral diseases, the only way to prevent serious harm from the illness is to use vaccination to prevent the illness in the first place. One approach to vaccine production is to use genetically engineered yeast cells to make large amounts of a protein found on the microbe's outer surface. The vaccine against hepatitis B, a disabling and sometimes fatal liver disease, is made in this way.

Genetically Modified (GM) Foods

Since ancient times, humans have selectively bred agricultural crops to make them more useful (see Figure 1.13a). Today, DNA technology is quickly replacing traditional breeding programs as scientists work to improve the productivity of agriculturally important plants and animals. Scientists have produced many types of **genetically modified (GM) organisms**, organisms that have acquired one or more genes by artificial means. If the newly acquired gene is from another organism, typically of another species, the recombinant organism is called a **transgenic organism**.

In the United States today, roughly half the corn crop and over three-quarters of the soybean and cotton crops are genetically modified. **Figure 12.4** shows corn that has been genetically engineered to resist attack by an insect called the European corn borer. Growing insect-resistant plants reduces the need for chemical insecticides. In another example, modified strawberry plants produce bacterial proteins that act as a natural antifreeze, protecting the plants from cold weather, which can harm the delicate crop. Potatoes and rice have been experimentally modified to produce harmless proteins derived from the cholera bacterium; researchers hope that these modified foods will one day serve as an edible vaccine against cholera, a disease that kills thousands of children in less-industrialized nations every year. In India, the insertion of a natural but rare saltwater-resistance gene has enabled new varieties of rice to grow in water three times as salty as seawater.

▼ Figure 12.3 **A factory that produces genetically engineered insulin.**

▼ Figure 12.4 **Genetically modified corn.** The corn plants in this field carry a bacterial gene that helps prevent infestation by the European corn borer (inset).

▲ **Figure 12.5 Genetically modified rice.** "Golden rice," shown here alongside ordinary rice, has been genetically modified to produce high levels of beta-carotene, a molecule that the body converts to vitamin A.

Scientists are also using genetic engineering to improve the nutritional value of crop plants. One example is "golden rice," a transgenic variety of rice that carries two daffodil genes. This rice produces yellow grains containing beta-carotene, which our body uses to make vitamin A (**Figure 12.5**). This rice could help prevent vitamin A deficiency and resulting blindness. However, controversy surrounds the use of GM foods, as we'll discuss at the end of the chapter. ☑

"Pharm" Animals

Figure 12.6 shows a transgenic pig that carries a gene for human hemoglobin. The pig-produced hemoglobin can be isolated and used in human blood transfusions. Because transgenic animals are difficult to produce, researchers may create a single transgenic animal and then breed or clone it. The resulting herd of transgenic animals, all carrying a recombinant human gene, could then serve as a grazing pharmaceutical factory—"pharm" animals.

DNA technology may eventually replace traditional animal breeding. Scientists might, for example, identify a gene that causes the development of larger muscles (which make up most of the meat we eat) in one variety of cattle and transfer it to other cattle or even to chickens. In 2006, University of Pittsburgh researchers genetically modified pigs to carry a roundworm gene whose protein converts less healthy fatty acids to omega-3 fatty acids. Meat from the modified pigs contains four to five times as much healthy omega-3 fat as regular pork. Unlike transgenic plants, however, transgenic animals are currently used only to produce potentially useful proteins; at present, no transgenic animals are sold as food.

Recombinant DNA technology serves many roles today and will certainly play an even larger part in our future. In the next section, you'll learn how scientists create and manipulate recombinant DNA.

Recombinant DNA Techniques

Although recombinant DNA techniques can use a variety of cell types, bacteria are the workhorses of modern biotechnology. To manipulate genes in the laboratory, biologists often use bacterial **plasmids**, which are small, circular DNA molecules that replicate (duplicate) separately from the much larger bacterial chromosome (**Figure 12.7**). Because plasmids can carry virtually any gene and are passed on from one generation of bacteria to the next, they are key tools for **gene cloning**, the production of multiple identical copies of a gene-carrying piece of DNA. Gene-cloning methods are central to the production of useful products from genetically engineered organisms. Consider a typical genetic engineering challenge: A molecular biologist identifies a gene of interest that codes for a valuable protein. The biologist wants to manufacture the protein on a large scale. **Figure 12.8** illustrates a way to accomplish this by using recombinant DNA techniques.

To start, the biologist isolates two kinds of DNA: ➊ bacterial plasmids that will serve as **vectors** (gene carriers) and ➋ DNA from another organism that includes the gene of interest, along with other genes. This other DNA may be from any type of organism, even a human. ➌ The researcher uses an enzyme to cut the two kinds of DNA. Each plasmid is cut in only one place; the other DNA is cut into many fragments, one of which carries the gene of interest. The figure shows the processing of just three of these DNA fragments and three plasmids, but actually millions of plasmids and

◄ **Figure 12.6 A genetically modified swine.** This pig has been genetically modified to carry a gene for human hemoglobin.

▼ **Figure 12.7 Bacterial plasmids.** The micrograph shows a bacterial cell that has been ruptured, revealing one long chromosome and several smaller plasmids. The inset is an enlarged view of a single plasmid.

Plasmids

Bacterial chromosome

Remnant of bacterium

Colorized TEM 2,700x

▼ Figure 12.8 **Using recombinant DNA technology to produce useful products.**

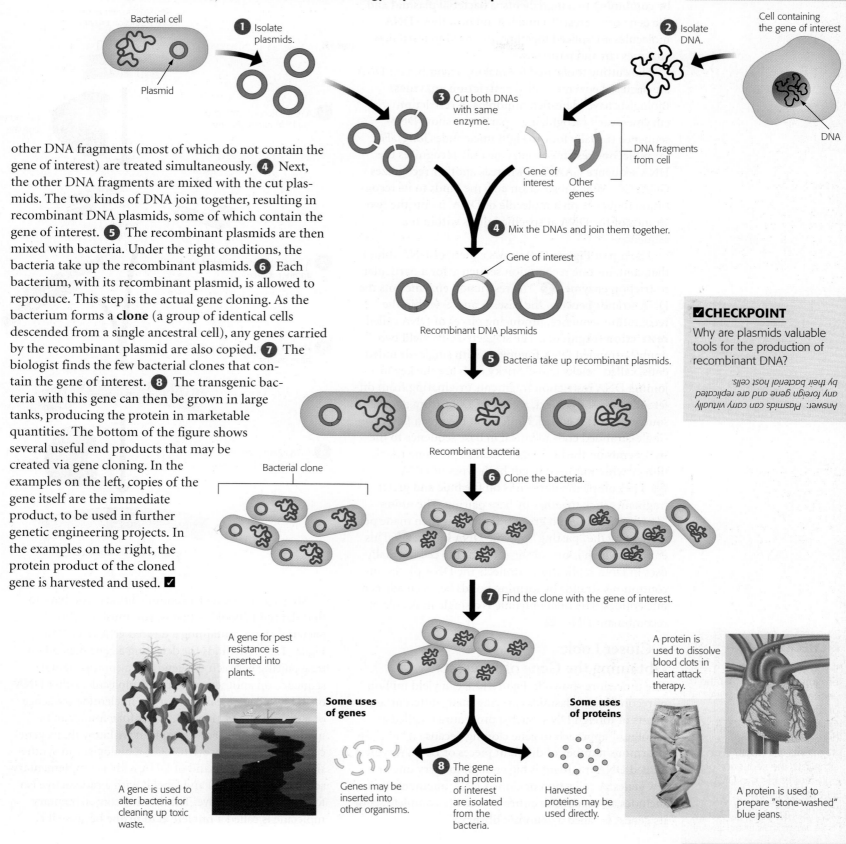

other DNA fragments (most of which do not contain the gene of interest) are treated simultaneously. **4** Next, the other DNA fragments are mixed with the cut plasmids. The two kinds of DNA join together, resulting in recombinant DNA plasmids, some of which contain the gene of interest. **5** The recombinant plasmids are then mixed with bacteria. Under the right conditions, the bacteria take up the recombinant plasmids. **6** Each bacterium, with its recombinant plasmid, is allowed to reproduce. This step is the actual gene cloning. As the bacterium forms a **clone** (a group of identical cells descended from a single ancestral cell), any genes carried by the recombinant plasmid are also copied. **7** The biologist finds the few bacterial clones that contain the gene of interest. **8** The transgenic bacteria with this gene can then be grown in large tanks, producing the protein in marketable quantities. The bottom of the figure shows several useful end products that may be created via gene cloning. In the examples on the left, copies of the gene itself are the immediate product, to be used in further genetic engineering projects. In the examples on the right, the protein product of the cloned gene is harvested and used. ✓

☑CHECKPOINT

Why are plasmids valuable tools for the production of recombinant DNA?

Answer: Plasmids can carry virtually any foreign gene and are replicated by their bacterial host cells.

Image labels:

- Bacterial cell
- Plasmid
- **1** Isolate plasmids.
- **2** Isolate DNA.
- Cell containing the gene of interest
- DNA
- **3** Cut both DNAs with same enzyme.
- Gene of interest
- Other genes
- DNA fragments from cell
- **4** Mix the DNAs and join them together.
- Gene of interest
- Recombinant DNA plasmids
- **5** Bacteria take up recombinant plasmids.
- Recombinant bacteria
- Bacterial clone
- **6** Clone the bacteria.
- **7** Find the clone with the gene of interest.
- A gene for pest resistance is inserted into plants.
- A gene is used to alter bacteria for cleaning up toxic waste.
- **Some uses of genes**
- Genes may be inserted into other organisms.
- **8** The gene and protein of interest are isolated from the bacteria.
- **Some uses of proteins**
- Harvested proteins may be used directly.
- A protein is used to dissolve blood clots in heart attack therapy.
- A protein is used to prepare "stone-washed" blue jeans.

A Closer Look: Cutting and Pasting DNA with Restriction Enzymes

As you saw in Figure 12.8, recombinant DNA is created by combining two ingredients: a bacterial plasmid and the gene of interest. To understand how these DNA molecules are spliced together, you need to learn how enzymes cut and paste DNA.

The cutting tools used for making recombinant DNA are bacterial enzymes called **restriction enzymes**. Biologists have identified hundreds of restriction enzymes, each recognizing a particular short DNA sequence (usually four to eight nucleotides long). For example, one restriction enzyme only recognizes the DNA sequence GAATTC, whereas another recognizes GGATCC. After a restriction enzyme binds to its recognition sequence on a molecule of DNA, it cuts the two strands of the DNA at specific points within the sequence.

The top of **Figure 12.9** shows a piece of DNA (blue) that contains one recognition sequence for a particular restriction enzyme. ❶ The restriction enzyme cuts the DNA strands between the bases A and G within the recognition sequence, producing pieces of DNA called **restriction fragments**. The staggered cuts yield two double-stranded DNA fragments with single-stranded ends, called "sticky ends." Sticky ends are the key to joining DNA restriction fragments originating from different sources. ❷ Next, a piece of DNA from another source (green) is added. Notice that the green DNA has single-stranded ends identical in base sequence to the sticky ends on the blue DNA because the same restriction enzyme was used to cut both types of DNA. ❸ The complementary ends on the blue and green fragments stick together by base pairing. The union between the blue and green fragments is then made permanent by the "pasting" enzyme **DNA ligase**. ❹ This enzyme, which is one of the proteins the cell normally uses in DNA replication, connects the DNA pieces into continuous strands by forming bonds between adjacent nucleotides. The final outcome is a single molecule of recombinant DNA. ✓

A Closer Look: Obtaining the Gene of Interest

The procedure shown in Figure 12.8 can yield millions of recombinant plasmids carrying many different segments of foreign DNA. Such a procedure is called a "shotgun" approach to gene cloning because it "hits" an enormous number of different pieces of DNA. A typical cloned DNA fragment is big enough to carry one or a few genes. A collection of cloned DNA fragments that includes an organism's entire genome (a complete set of its genes) is called a **genomic library**.

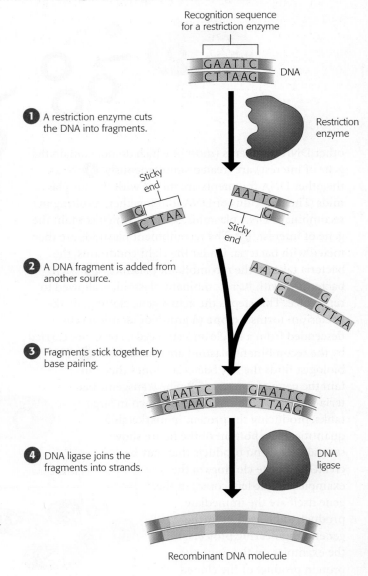

▼ Figure 12.9 **Cutting and pasting DNA.** The production of recombinant DNA requires two enzymes: a restriction enzyme, which cuts the original DNA molecules into pieces, and DNA ligase, which pastes the pieces together.

Recognition sequence for a restriction enzyme

GAATTC
CTTAAG — DNA

❶ A restriction enzyme cuts the DNA into fragments.

Restriction enzyme

Sticky end

G   AATTC
CTTAA   G

Sticky end

❷ A DNA fragment is added from another source.

AATTC
G   G
CTTAA

❸ Fragments stick together by base pairing.

GAATTC   GAATTC
CTTAAG   CTTAAG

❹ DNA ligase joins the fragments into strands.

DNA ligase

Recombinant DNA molecule

After you've created a genomic library, you have to find the right "book"—that is, you must identify the bacterial clone containing a desired gene (step 7 in Figure 12.8). Methods for detecting a gene depend on base pairing between the gene and a complementary sequence on another nucleic acid molecule, either DNA or RNA. When at least part of the nucleotide sequence of a gene is already known, this information can be used to advantage. For example, if we know that a gene contains the sequence TAGGCT, a biologist can synthesize a short single strand of DNA with a complementary sequence (ATCCGA) and label it with a radioactive isotope or fluorescent dye. This labeled complementary molecule is called a **nucleic acid probe** because it is used to find a specific gene or other nucleotide sequence within a mass of DNA. (In actual practice, probe molecules are considerably longer than six nucleotides.) When a radioactive DNA probe is added to the DNA of various clones, it tags the correct molecule—finds the right book in the library—by base-pairing to the complementary sequence in the gene of interest (**Figure 12.10**). After a probe detects the desired clone within a library, more of the tagged cells can be grown, resulting in the production of large quantities of the gene of interest.

Another approach to obtaining a gene of interest is to synthesize it. One method uses reverse transcriptase, a viral enzyme that can synthesize DNA by using an mRNA template (see Chapter 10). **Figure 12.11** shows the steps involved. A eukaryotic cell ❶ transcribes the gene of interest and ❷ processes the transcript, removing introns and splicing exons together to produce mRNA. A researcher then ❸ isolates the mRNA in a test tube and ❹ makes single-stranded DNA from it using reverse transcriptase. ❺ DNA polymerase is then used to synthesize a second DNA strand.

When the researcher starts with an mRNA mixture from a particular cell type, the DNA that results from this procedure, called complementary DNA (cDNA), represents only those genes that were actually transcribed in the starting cells. Because cDNA molecules lack introns, they are shorter than the full version of the genes and therefore easier to work with.

A final approach is to synthesize a gene of interest from scratch. Automated DNA-synthesizing machines can accurately and rapidly produce customized DNA molecules of any sequence up to lengths of a few hundred nucleotides (**Figure 12.12**). ✓

▼ Figure 12.11 **Making a gene from eukaryotic mRNA.** Using the enzyme reverse transcriptase (green), a researcher can produce an artificial DNA gene (cDNA, in blue) from a molecule of messenger RNA (mRNA, in red).

Cell nucleus

Exon Intron Exon Intron Exon

DNA of eukaryotic gene

❶ Transcription

RNA transcript

❷ Introns removed and exons spliced together

mRNA

Test tube

❸ Isolation of mRNA from cell and addition of reverse transcriptase

Reverse transcriptase

❹ Synthesis of cDNA strand

cDNA strand being synthesized

❺ Synthesis of second DNA strand by DNA polymerase

cDNA of gene without introns

☑**CHECKPOINT**

Name three ways a gene of interest can be obtained.

Answer: The gene can be isolated from a genomic library, produced from mRNA using reverse transcriptase, or synthesized from scratch.

▼ Figure 12.10 **How a DNA probe tags a gene.** The probe is a short, radioactive, single-stranded molecule of DNA or RNA. When it is mixed with single-stranded DNA from a gene with a complementary sequence, it attaches by hydrogen bonds (shown in red), "labeling" the gene.

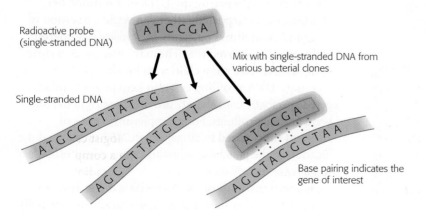

Radioactive probe (single-stranded DNA)

ATCCGA

Mix with single-stranded DNA from various bacterial clones

Single-stranded DNA

ATGCGCTTATCG

AGCCTTATGCAT

ATCCGA
AGGTAGGCTAA

Base pairing indicates the gene of interest

► Figure 12.12 **A DNA synthesizer.** This synthesizer constructs molecules of DNA from a programmed sequence, using solutions of the four DNA nucleotides (stored in the bottles beneath the machine).

DNA Profiling and Forensic Science

As discussed in the Biology and Society section, the analysis of DNA samples to determine whether they come from the same individual, known as **DNA profiling**, can help in criminal investigations. Indeed, DNA profiling has rapidly transformed the field of **forensics**, the scientific analysis of evidence for crime scene investigations and other legal proceedings. To produce a DNA profile, scientists compare **genetic markers**, sequences in the genome that vary from person to person. Like a gene (which is a type of genetic marker), a noncoding genetic marker is more likely to be a match between relatives than between unrelated individuals.

Figure 12.13 presents an overview of a typical investigation using DNA profiling. **1** First, DNA samples from the crime scene, suspects, victims, or other evidence are isolated. **2** Next, the selected markers from each DNA sample are amplified (copied many times) to produce a large sample of DNA fragments. **3** Finally, the amplified DNA markers are compared, proving which samples are from the same individual and which samples are unique. ✓

Investigating Murder, Paternity, and Ancient DNA

Since its introduction in 1986, DNA profiling has become a standard tool of forensics and has provided crucial evidence in many famous cases. The first well-known use of DNA profiling was in the O. J. Simpson murder trial. In this case, DNA analysis proved that blood in Simpson's car belonged to the victims and that blood at the crime scene belonged to Simpson. (The jury did not find the DNA evidence alone to be sufficient to convict the suspect, and Simpson was found not guilty.) During the investigation that led up to his impeachment, President Bill Clinton repeatedly denied that he had sexual relations with Monica Lewinsky—until DNA profiling proved that his semen was on her dress. Of course, DNA evidence can prove innocence as well as guilt. As discussed in the Biology and Society section, DNA profiling has helped lawyers at the Innocence Project exonerate over 240 convicted criminals, including some on death row. In more than a third of these cases, DNA profiling has also identified the true perpetrators.

DNA profiling can also be used to identify crime victims. The largest such effort in history took place after the World Trade Center attack on September 11, 2001. Forensic scientists in New York City worked for years to identify over 20,000 samples of victims' remains. DNA profiles of tissue samples from the disaster site were matched to DNA profiles from tissue known to be from the victims. If no sample of a victim's DNA was available, blood samples from close relatives were used to confirm identity through near matches. Over half of the victims identified at the World Trade Center site were recognized solely by DNA evidence, providing closure to many grieving families.

The use of DNA profiling extends beyond crimes. For instance, comparing the DNA of a mother, her child, and the purported father can settle a question of paternity. Sometimes paternity is of historical interest: DNA profiling proved that Thomas Jefferson or a close male relative fathered a child with his slave Sally Hemings. DNA profiling can also help protect endangered species by conclusively proving the origin of contraband animal products. For example, an analysis of elephant tusks seized in Singapore in 2007 pinpointed the location of the poaching, allowing enforcement officials to increase surveillance.

Modern methods of DNA profiling are so specific and powerful that the DNA samples can be in a partially

▶ Figure 12.13 **Overview of DNA profiling.** In this example, DNA from suspect 1 does not match DNA found at the crime scene, but DNA from suspect 2 does match.

Crime scene   Suspect 1   Suspect 2

1 DNA isolated

2 DNA amplified

3 DNA compared

degraded state. This allows DNA analysis to be applied in a great number of ways. In evolution research, the technique has been used to study DNA pieces recovered from an ancient mummified human and from a 30-million-year-old plant fossil. A 2005 study determined that DNA extracted from a 27,000-year-old Siberian mammoth was 98.6% identical to DNA from modern African elephants. One of the strangest cases of DNA profiling is that of Cheddar Man, a 9,000-year-old skeleton found in a cave near Cheddar, England (Figure 12.14). DNA was extracted from his tooth and used to construct a DNA profile. The results suggested that Cheddar Man was a direct ancestor (through approximately 300 generations) of a present-day schoolteacher who lived only half a mile from the cave!

► Figure 12.14 **Cheddar Man.** Analysis of DNA extracted from "Cheddar Man"—a 9,000-year-old skeleton found in an English cave—suggested that he was a direct ancestor of a local schoolteacher (shown above).

DNA Profiling Techniques

In this section, you'll learn about techniques for making a DNA profile (steps 2 and 3 in Figure 12.13).

The Polymerase Chain Reaction (PCR)

The **polymerase chain reaction (PCR)** is a technique by which a specific segment of DNA can be targeted and copied quickly and precisely. Through PCR, a scientist can obtain enough DNA from even minute amounts of blood or other tissue to allow a DNA profile to be constructed.

In principle, PCR is simple. A DNA sample is mixed with nucleotides, the DNA replication enzyme DNA polymerase, and a few other ingredients. The solution is then exposed to cycles of heating (to separate the DNA strands) and cooling (to allow double-stranded DNA to re-form). During these cycles, specific regions of each molecule of DNA are replicated, doubling the amount of that DNA (Figure 12.15). The key to automated PCR is an unusually heat-stable DNA polymerase, first isolated from prokaryotes living in hot springs. Unlike most proteins, this enzyme can withstand the heat at the start of each cycle. Beginning with a single DNA molecule, automated PCR can generate hundreds of billions of copies in a few hours. ☑

Initial
DNA
segment

| 1 | 2 | 4 | 8 |

Number of DNA molecules

◄ Figure 12.15 **DNA amplification by PCR.** The polymerase chain reaction (PCR) is a method for making many copies of a specific segment of DNA. Each round of PCR, performed on a tabletop thermal cycler (shown at top), doubles the total quantity of DNA.

☑CHECKPOINT

Why is only the slightest trace of DNA at a crime scene often sufficient for forensic analysis?

Answer: because PCR can be used to produce enough molecules for analysis

Short Tandem Repeat (STR) Analysis

How do you prove that two samples of DNA come from the same person? You could compare the entire genomes found in the two samples. But such an approach would be extremely impractical, requiring a lot of time and money. Instead, forensic scientists typically compare about a dozen short segments of noncoding repetitive DNA. **Repetitive DNA**, which makes up much of the DNA that lies between genes in humans, consists of nucleotide sequences that are present in multiple copies in the genome. Some of this DNA consists of short sequences repeated many times tandemly (one after another); such a series of repeats is called a **short tandem repeat (STR)**. For example, one person might have the sequence AGAT repeated 12 times in a row at one place in the genome, the sequence GATA repeated 35 times at a second place, and so on; another person is likely to have the same sequences at the same places but with a different number of repeats. These stretches of repetitive DNA, like any genetic marker, are more likely to be an exact match between relatives than between unrelated individuals.

STR analysis is a method of DNA profiling that compares the lengths of STR sequences at certain sites in the genome. Most commonly, STR analysis compares the number of repeats of specific four-nucleotide DNA sequences at 13 sites scattered throughout the genome. Each repeat site, which typically contains from 3 to 50 four-nucleotide repeats in a row, varies widely from person to person. In fact, some STRs used in the standard procedure have up to 80 variations in the number of repeats. In the United States, the number of repeats at each site is entered into a database called CODIS (Combined DNA Index System) administered by the Federal Bureau of Investigation.

Consider the two samples of DNA shown in **Figure 12.16**. Imagine that the top DNA segment was obtained at a crime scene and the bottom from a suspect's blood. The two segments have the same number of repeats at the first site: 7 repeats of the four-nucleotide DNA sequence AGAT (orange). Notice, however, that they differ in the number of repeats at the second site: 8 repeats of GATA (purple) in the crime scene DNA, compared with 13 repeats in the suspect's DNA. To create a DNA profile, a scientist uses PCR to specifically amplify the regions of DNA that include these STR sites. The resulting fragments are then compared. Next we'll look at how this comparison is made. ☑

▼ **Figure 12.16 Short tandem repeat (STR) sites.** Scattered throughout the genome, STR sites contain tandem repeats of four-nucleotide sequences. The number of repetitions at each site can vary from individual to individual. In this figure, both DNA samples have the same number of repeats (7) at the first STR site, but different numbers (8 versus 13) at the second.

Gel Electrophoresis

The lengths of DNA fragments are compared using **gel electrophoresis**, a method for sorting macromolecules—usually proteins or nucleic acids—primarily by their electrical charge and size. **Figure 12.17** shows how gel electrophoresis separates DNA fragments obtained from

▼ **Figure 12.17 Gel electrophoresis of DNA molecules.**

☑**CHECKPOINT**

What are STRs, and why are they useful for DNA profiling?

Answer: STRs (short tandem repeats) are nucleotide sequences repeated many times in a row within the human genome. STRs are valuable for DNA profile identification because different people have different numbers of repeats at the various STR sites.

different sources. A DNA sample from each source is placed in a separate well (hole) at one end of a flat, rectangular gel, a thin slab of jellylike material that acts as a molecular sieve. A negatively charged electrode is then attached to the DNA-containing end of the gel and a positive electrode to the other end. Because the phosphate (PO_4^-) groups of nucleotides give DNA fragments a negative charge, the fragments move through the gel toward the positive pole. However, longer DNA fragments move more slowly through the thicket of polymer fibers in the gel than do shorter DNA fragments. Over time, shorter molecules move farther through the gel than longer molecules. Gel electrophoresis thus separates DNA fragments by length. When the current is turned off, a series of bands is left in each "lane" of the gel. Each band is a collection of DNA fragments of the same length. The bands can be made visible by staining, by exposure onto photographic film (if the DNA is radioactively labeled), or by measuring fluorescence (if the DNA is labeled with a fluorescent dye).

Figure 12.18 shows the gel that would result from using gel electrophoresis to separate the DNA fragments from the example in Figure 12.16. The differences in the locations of the bands reflect the different lengths of the DNA fragments. This gel would provide evidence that the crime scene DNA did not come from the suspect. Because gel electrophoresis reveals similarities and differences between DNA samples, DNA profiling can provide evidence of either guilt or innocence. ☑

RFLP Analysis

Gel electrophoresis has many uses besides STR analysis. One application is RFLP analysis. RFLP (pronounced "rif-lip") stands for restriction fragment length polymorphism. In this method, the DNA molecules to be compared are exposed to a restriction enzyme (**Figure 12.19**). The resulting restriction fragments are separated and made visible on a gel. In this case, both the number and location of bands indicate whether the original DNA samples had identical nucleotide sequences at the sites shown.

☑CHECKPOINT

You use a restriction enzyme to cut a DNA molecule that has three copies of the enzyme's recognition sequence clustered near one end. When you separate the restriction fragments by gel electrophoresis, how do you expect the bands to appear?

Answer: three bands near the positive pole at the bottom of the gel (small fragments) and one band near the negative pole at the top of the gel (large fragment)

▼ **Figure 12.19 RFLP analysis.** The DNA segments shown have nucleotide sequences that differ at one base pair (orange boxes). A particular restriction enzyme may therefore cut the segments at different places. In this case, the difference in DNA sequence results in three restriction fragments from the first DNA sample and two from the second. This difference, revealed by gel electrophoresis, indicates that the two DNA samples come from different individuals.

▼ **Figure 12.18 Visualizing STR fragment patterns.** This figure shows the bands that would result from gel electrophoresis of the STR sites illustrated in Figure 12.16. Notice that one of the bands from the crime scene DNA doesn't match one of the bands from the suspect's DNA. (A gel from an actual DNA profile would typically contain more than just two bands in each lane.)

Genomics and Proteomics

In 1995, a team of scientists announced that they had determined the nucleotide sequence of the entire genome of *Haemophilus influenzae*, a bacterium that can cause several human diseases, including pneumonia and meningitis. **Genomics**, the study of complete sets of genes (genomes), was born.

The first targets of genomics research were bacteria, which have relatively little DNA. The *H. influenzae* genome, for example, contains 1.8 million nucleotides and 1,709 genes. But soon the attention of genomics researchers turned toward more complex organisms with much larger genomes (**Table 12.1**). As of 2009, the genomes of nearly 1,000 species have been published, and thousands more are in progress. The majority of organisms sequenced to date are prokaryotes, including *E. coli* and several hundred other bacteria (some of medical importance) and a few dozen archaea. Over 100 eukaryotic genomes have been completed.

Baker's yeast (*Saccharomyces cerevisiae*) was the first eukaryote to have its full sequence determined, and the roundworm *Caenorhabditis elegans* was the first multicellular organism. Other sequenced animals include the fruit fly (*Drosophila melanogaster*) and lab rat (*Ratus norvegicus*), model organisms for genetics. Among the sequenced plants are *Arabidopsis thaliana*, a type of mustard plant used as a model organism, and rice (*Oryza sativa*) and sorghum (*Sorghum bicolor*), two of the world's most economically important grain crops. Next we'll discuss a particularly notable example of a sequenced animal genome—our own.

The Human Genome Project

The **Human Genome Project** was a massive scientific endeavor to determine the nucleotide sequence of all the DNA in the human genome and to identify the location and sequence of every gene. The project began in 1990 as an effort by government-funded researchers from six countries. Several years into the project, private companies, chiefly Celera Genomics, joined the effort. At the completion of the project, over 99% of the genome had been determined to 99.999% accuracy. (There remain a few hundred gaps of unknown sequence that will require special methods to figure out.) This ambitious project has provided a wealth of data that may illuminate the genetic basis of what it means to be human.

The chromosomes in the human genome (22 autosomes plus the X and Y sex chromosomes) contain approximately 3.2 billion nucleotide pairs of DNA. To try to get a sense of this quantity of DNA, imagine that its nucleotide sequence is printed in letters (A, T, C, and G) in books the size of this textbook. The sequence would fill a stack of books 18 stories high! However, the biggest surprise from the Human Genome Project is the small number of human genes—only about 20,000—very close to the number found in a roundworm! This number is well below estimates made before the project began. In fact, a friendly betting pool set up in the year 2000 by the world's leading genome researchers included a wide range of guesses—from 26,000 to 150,000. Every single genome expert overestimated!

Our genome was a major challenge to sequence because, like the genomes of most complex eukaryotes, only a small amount of our total DNA consists of genes that code for proteins, tRNAs, or rRNAs. Most complex eukaryotes have a huge amount of noncoding DNA—about 98% of human DNA is of this type. Some of this

Table 12.1	Some Important Sequenced Genomes		
Organism	**Year Completed**	**Size of Genome (in base pairs)**	**Approximate Number of Genes**
Haemophilus influenzae (bacterium)	1995	1.8 million	1,700
Saccharomyces cerevisiae (yeast)	1996	12 million	5,800
Escherichia coli (bacterium)	1997	4.6 million	4,400
Caenorhabditis elegans (roundworm)	1998	97 million	19,100
Drosophila melanogaster (fruit fly)	2000	180 million	13,700
Arabidopsis thaliana (mustard plant)	2000	120 million	25,500
Oryza sativa (rice)	2002	430 million	40,000
Homo sapiens (human)	2003	3.2 billion	20,000
Rattus norvegicus (lab rat)	2004	2.8 billion	20,000
Pan troglodytes (chimpanzee)	2005	3.1 billion	20,000
Macaca mulatta (macaque)	2007	2.9 billion	22,000
Ornithorhynchus anatinus (duck-billed platypus)	2008	1.8 billion	18,500
Sorghum bicolor (sorghum)	2009	730 million	34,500

noncoding DNA is made up of gene control sequences such as promoters, enhancers, and microRNAs (see Chapter 11). Other noncoding regions include introns (whose total length in a gene may be ten times greater than the total length of the exons) and repetitive DNA (some of which is used in DNA profiling). Some noncoding DNA is important to our health, with certain regions known to carry several disease-causing mutations. But the function (if any) of most noncoding DNA remains unknown.

The potential benefits of having a complete map of the human genome are enormous. For instance, hundreds of disease-associated genes have already been identified. One example is the gene that is mutated in an inherited type of Parkinson's disease, a debilitating brain disorder that causes tremors of increasing severity. Half a million Americans suffer from this disorder (**Figure 12.20**). Until recently, Parkinson's disease was thought to have only an environmental basis; there was no evidence of a hereditary component. But data from the Human Genome Project mapped some cases of Parkinson's disease to a specific gene. Interestingly, an altered version of the protein encoded by this gene has also been tied to Alzheimer's disease, suggesting a link between these two brain disorders. Moreover, the same gene is also found in rats, where it plays a role in the sense of smell, and in zebra finches, where it is thought to be involved in song learning. Cross-species comparisons such as these may uncover clues about the role played by the normal version of the protein in the human brain. And such knowledge could eventually lead to treatment for Parkinson's disease. ☑

Tracking the Anthrax Killer

In October 2001, a 63-year-old Florida man died from inhalation anthrax, a disease caused by breathing spores of the bacterium *Bacillus anthracis*. As the first victim of this disease in the United States since 1976, his death was immediately suspicious. By the end of the year, four more people had died from anthrax. Law enforcement officials realized that someone was sending anthrax spores through the mail (**Figure 12.21**). The United States was facing an unprecedented bioterrorist attack.

In the investigation that followed, one of the most helpful clues turned out to be the anthrax spores themselves. Investigators compared the genomes of the mailed anthrax spores with several laboratory strains. They quickly established that all of the mailed spores were genetically identical, suggesting that a single perpetrator was behind all the attacks. Furthermore, they were able to match the deadly spores with a laboratory subtype isolated at the U.S. Army Medical Research Institute of Infectious Diseases in Fort Detrick, Maryland. A second, more comprehensive whole-genome analysis of the spores used in the attack was completed in 2008. This analysis found four unique mutations in the mailed anthrax and traced the mutations to a single flask at the army facility. Based in part on this evidence, the FBI named an army research scientist as a suspect in the case.

☑ **CHECKPOINT**

1. Approximately how many nucleotides and genes are contained in the human genome?
2. Name three types of DNA that do not code for another molecule.

Answers: **1.** *about 3.2 billion nucleotides and 20,000 genes* **2.** *introns, repetitive DNA, and gene control sequences*

▼ **Figure 12.20 The fight against Parkinson's disease.** Actor Michael J. Fox and boxer Muhammad Ali—both of whom have Parkinson's disease—testify before the Senate on the status of federal funding for Parkinson's research.

▼ **Figure 12.21 The 2001 anthrax attacks.** In 2001, envelopes containing anthrax spores caused five deaths.

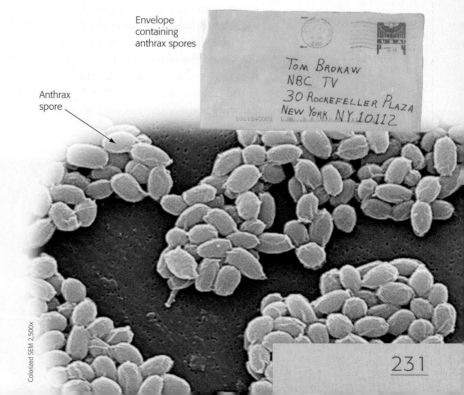

Envelope containing anthrax spores

Anthrax spore

Tom Brokaw
NBC TV
30 Rockefeller Plaza
New York, NY 10112

Colorized SEM 2,500x

The anthrax investigation is just one example of the new field of comparative genomics, the comparison of whole genomes. In 1991, sequence data provided strong evidence that a Florida dentist transmitted HIV to several patients. In 1993, after a cult released anthrax spores in downtown Tokyo, genomic analysis showed why their attack didn't kill anyone: They had used a harmless veterinary vaccine strain. And investigation of the West Nile virus outbreak in 1999 proved that a single natural strain of virus was infecting both birds and humans. Comparative genomics has even allowed geneticists to produce a family tree of dog breeds (see the Evolution Connection section that ends Chapter 9).

Comparative genomics can also reveal similarities and differences in organisms more closely related to humans. In 2005, researchers completed the genome sequence for our closest living relative on the evolutionary tree of life, the chimpanzee (*Pan troglodytes*). Comparisons with human DNA revealed that we share 96% of our genome with our closest animal relative. Genomic scientists are currently finding and studying the important differences, shedding scientific light on the age-old question of what makes us human.

Genome-Mapping Techniques

Genomes are most often sequenced using the *whole-genome shotgun method*. The first step in this method is to chop the entire genome into fragments using restriction enzymes. Next, the fragments are cloned and sequenced. Finally, computers running specialized mapping software reassemble the millions of overlapping short sequences into a single continuous sequence for every chromosome—an entire genome (Figure 12.22).

The DNA sequences determined by the Human Genome Project have been deposited in a database that is available via the Internet. (You can browse it yourself at the website for the National Center for Biotechnology Information.) Scientists use software to scan and analyze the sequences for genes, control elements, and other features. The result is a genetic map containing all the genes and their locations on chromosomes. Now comes the most exciting challenge: figuring out the functions of the genes and other sequences and how they work together to direct the structure and function of a living organism. This challenge and the applications of the new knowledge should keep scientists busy well into the twenty-first century.

One interesting question to ask about the Human Genome Project is whose genome was sequenced. The human genome sequenced by government-funded scientists was actually a reference genome compiled from a group of individuals. The genome sequenced by Celera consisted primarily of DNA from scientist

▼ Figure 12.22 **Genome sequencing.** In the photo at the bottom, a technician performs a step in the whole-genome shotgun method (depicted in the diagram).

Chromosome

Chop up with restriction enzyme

DNA fragments

Sequence fragments

AATC TTAATGTA TCGGAC GACGATTA

Align fragments

AATC GACGATTA
TCGGAC TTAATGTA

Reassemble full sequence

AATCGGACGATTAATGTA

Craig Venter, the company's president. These representative sequences will serve as standards so that comparisons of individual differences and similarities can be made. Starting in 2007, the genomes of a few other individuals—the first was James Watson, codiscoverer of the structure of DNA—have also been sequenced. These sequences are part of a larger effort called the Human Variome Project (where "variome" refers to variations in the genome). Begun in 2006, this project seeks to collect information on all of the genetic variations that affect human health. As the amount of sequence data grows, the small differences that account for individual variation within our species will come to light. We may soon enter an age of "personal genomics," where individual genetic differences among people will be put to routine medical use. Some of these seemingly minuscule differences can actually be a matter of life or death, as we'll see next.

THE PROCESS OF SCIENCE: DNA Profiling

Can Genomics Cure Cancer?

Lung cancer, which kills more Americans every year than any other type of cancer, has long been the target of searches for effective chemotherapy drugs. One drug used to treat lung cancer, called gefitinib, targets the protein encoded by the *EGFR* gene. This protein is found on the surface of cells that line the lungs and is also found in lung cancer tumors.

Unfortunately, gefitinib is ineffective for many patients. While studying the effectiveness of gefitinib, researchers at the Dana-Farber Cancer Institute in Boston made the **observation** that a few patients actually responded quite positively to the drug. This posed a **question**: Are genetic differences among lung cancer patients responsible for the differences in gefitinib's effectiveness? The researchers' **hypothesis** was that mutations in the *EGFR* gene were causing the different responses to gefitinib. The team made the **prediction** that DNA profiles focusing on the *EGFR* gene would reveal different DNA sequences in the tumors of responsive patients compared with the tumors of unresponsive patients. The researchers' **experiment** involved sequencing the *EGFR* gene in cells extracted from the tumors of five patients who responded to the drug and four who did not.

The **results**, published in 2004, were quite striking: All five tumors from gefitinib-responsive patients had mutations in *EGFR*, whereas none of the other four tumors did (**Figure 12.23**). These results suggest that doctors can use DNA profiling techniques to screen lung cancer patients for those who are most likely to benefit from treatment with this drug. In broader terms, this work suggests that genomics may bring about a revolution in the treatment of disease by allowing therapies to be custom-tailored to the genetic makeup of each patient.

▲ Figure 12.23 **The *EGFR* protein: Fighting cancer with genomics.** Mutations (located at sites indicated by black arrows) in the *EFGR* protein can affect the ability of a cancer-fighting drug to destroy lung tumors. Here, the amino acid backbone of the protein is shown in green, with some important regions highlighted in orange, blue, and red.

Proteomics

The successes in the field of genomics have encouraged scientists to attempt similar systematic studies of the full protein sets (proteomes) that genomes encode, an approach called **proteomics**. The number of different proteins in humans far exceeds the number of different genes (about 100,000 proteins versus about 20,000 genes). And since proteins, not genes, actually carry out the activities of the cell, scientists must study when and where proteins are produced and how they interact in order to understand the functioning of cells and organisms.

Genomics and proteomics are enabling biologists to approach the study of life from an increasingly holistic perspective. Biologists are now in a position to compile catalogs of genes and proteins—that is, listings of all the "parts" that contribute to the operation of cells, tissues, and organisms. With such catalogs in hand, researchers are shifting their attention from the individual parts to how these parts function as a whole in biological systems. ☑

Human Gene Therapy

Human gene therapy is a recombinant DNA procedure intended to treat disease by altering an afflicted person's genes. In some cases, a mutant version of a gene may be replaced or supplemented with the normal allele. This could potentially correct a genetic disorder, perhaps permanently. In other cases, genes are inserted and expressed only long enough to treat a medical problem.

Figure 12.24 summarizes one approach to human gene therapy. ❶ A gene from a normal individual is isolated and cloned by recombinant DNA techniques. ❷ The gene is inserted into a vector, such as a harmless virus. ❸ The virus is then injected into the patient. The virus inserts a copy of its genome, including the human gene, into the DNA of the patient's cells. The normal gene is then transcribed and translated within the patient's body, producing the desired protein. Ideally, the nonmutant version of the gene would be inserted into cells that multiply throughout a person's life. Bone marrow cells, which include the stem cells that give rise to all the types of blood cells, are prime candidates. If the procedure succeeds, the cells will multiply through-out the patient's life and express the normal gene. The engineered cells will supply the missing protein, and the patient will be cured.

Gene therapy has been used to treat severe combined immunodeficiency (SCID), a fatal inherited disease caused by a defective gene. Absence of the enzyme encoded by this gene prevents the development of the immune system, requiring patients to remain isolated within protective "bubbles." Unless treated with a bone marrow transplant (effective just 60% of the time), SCID patients quickly die from infections by microbes that most of us easily fend off.

Since the year 2000, gene therapy has cured 22 chil-dren with inborn SCID, providing the first strong evi-dence of the effectiveness of gene therapy. As part of this treatment, researchers periodically removed immune system cells from the patients' blood, infected them with a virus engineered to carry the normal allele of the defective gene, then reinjected the blood into the patient. The celebrations of this medical breakthrough were short-lived, however; four of the patients devel-oped leukemia, and one died. Apparently, the retrovirus used as a vector activated an oncogene (see Chapter 11), creating cancerous blood cells. Thus, although gene therapy remains promising, there is very little evidence of safe and effective application. Active research contin-ues, with new, tougher safety guidelines. ☑

▼ Figure 12.24 **One approach to human gene therapy.**

Healthy person

❶ Normal human gene isolated and cloned

Harmless virus (vector)

❷ Normal human gene inserted into virus

Virus containing normal human gene

Bone marrow

Bone of person with disease

❸ Virus injected into patient with abnormal gene

☑ CHECKPOINT

Why are bone marrow stem cells ideally suited as targets for gene therapy?

Answer: because bone marrow stem cells multiply throughout a person's life

Safety and Ethical Issues

As soon as scientists realized the power of DNA technology, they began to worry about potential dangers. Early concerns focused on the possibility of creating hazardous new disease-causing organisms. What might happen, for instance, if cancer-causing genes were transferred into infectious bacteria or viruses? To address such concerns, scientists developed a set of guidelines that have become formal government regulations in the United States and some other countries.

One safety measure is a set of strict laboratory procedures to protect researchers from infection by engineered microbes and to prevent microbes from accidentally leaving the laboratory (Figure 12.25). In addition, strains of microbes to be used in recombinant DNA experiments are genetically crippled to ensure that they cannot survive outside the laboratory. Finally, certain obviously dangerous experiments have been banned. Today, most public concern about possible hazards centers not on recombinant microbes but on genetically modified (GM) foods.

The Controversy over Genetically Modified Foods

GM strains account for a significant percentage of several crops. Controversy about the safety of these foods is an important political issue (Figure 12.26). For example, the European Union has suspended the introduction of new GM crops and considered banning the import of all GM foodstuffs. In the United States and other countries where the GM revolution has proceeded more quietly than in Europe, the labeling of GM foods is now being debated but has not yet become law.

Advocates of a cautious approach fear that crops carrying genes from other species might harm the environment or be hazardous to human health (by, for example, introducing new allergens, molecules that can cause allergic reactions, into foods). A major concern is that transgenic plants might pass their new genes to close relatives in nearby wild areas. We know that lawn and

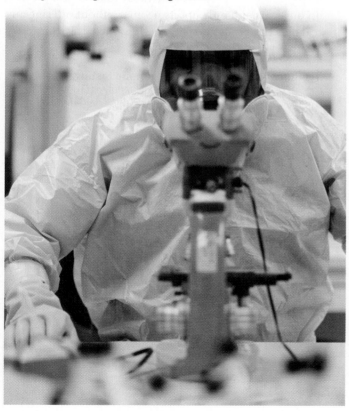

▼ Figure 12.25 **Maximum-security laboratory.** A scientist in a high-containment laboratory wears a biohazard suit, used for working with dangerous microorganisms.

► Figure 12.26 **Opposition to genetically modified organisms (GMOs).** This photo shows tractors destroying a Belgian field of genetically modified rape, a forage crop used to feed sheep, birds, and hogs and also to make cooking oil. The Belgian Minister for Consumer Protection ordered the destruction as part of a widespread opposition to GM crops. The sign reads "GMO, no thanks! Yes to biodiversity."

crop grasses, for example, commonly exchange genes with wild relatives via pollen transfer. If domestic plants carrying genes for resistance to herbicides, diseases, or insect pests pollinated wild plants, the offspring might become "superweeds" that would be very difficult to control. However, researchers may be able to prevent the escape of such plant genes in various ways—for example, by engineering plants so they cannot hybridize. Concern has also been raised that the widespread use of GM seeds may reduce natural genetic diversity, leaving crops susceptible to catastrophic die-offs in the event of a sudden change to the environment or introduction of a new pest (see, for example, the Biology and Society section in Chapter 16).

The U.S. National Academy of Sciences released a study finding no scientific evidence that transgenic crops pose any special health or environmental risks. But the authors of the study also recommended more stringent long-term monitoring to watch for unanticipated environmental impacts.

In 2000, negotiators from 130 countries (including the United States) agreed on a Biosafety Protocol that requires exporters to identify GM organisms present in bulk food shipments and allows importing countries to decide whether the shipments pose environmental or health risks. This agreement has been hailed as a breakthrough by environmentalists.

Today, governments and regulatory agencies throughout the world are grappling with how to facilitate the use of biotechnology in agriculture, industry, and medicine while ensuring that new products and procedures are safe. In the United States, all projects are evaluated for potential risks by a number of regulatory agencies, including the Food and Drug Administration, the Environmental Protection Agency, the National Institutes of Health, and the Department of Agriculture. ✔

Ethical Questions Raised by DNA Technology

DNA technology raises legal and ethical questions—few of which have clear answers. Consider, for example, the treatment of dwarfism with injections of human growth hormone (HGH) produced by genetically engineered cells. Should parents of short but hormonally normal children be able to seek HGH treatment to make their kids taller? If not, who decides which children are "tall enough" to be excluded from treatment?

Genetic engineering of gametes (sperm or ova) and zygotes has been accomplished in lab animals. It has not been attempted in humans because such a procedure would raise very difficult ethical questions. Should we try to eliminate genetic defects in our children and their descendants? Should we interfere with evolution in this way? From a long-term perspective, the elimination of unwanted versions of genes from the gene pool could backfire. Genetic variety is a necessary ingredient for the adaptation of a species as environmental conditions change with time. Genes that are damaging under some conditions may be advantageous under others (one example is the sickle-cell allele—see the Evolution Connection section at the end of Chapter 13). Are we willing to risk making genetic changes that could be detrimental to our species in the future? We may have to face such questions soon.

Advances in genetic profiling raise privacy issues. If we were to create a DNA profile of every person at birth, then we could theoretically match nearly every violent crime to a perpetrator because it is virtually impossible for someone to commit a violent crime without leaving behind DNA evidence. But are we, as a society, prepared to sacrifice our genetic privacy, even for such a worthwhile goal?

As more information becomes available about our personal genetic makeup, ethicists question whether greater access to this information is always beneficial. For example, mail-in kits have become available that can tell healthy people their relative risk of developing various diseases (such as Parkinson's and Crohn's) later in life (**Figure 12.27**). Some argue that such information helps families to prepare. Others say that the test preys on our fears and only upsets people without offering any real benefit because certain diseases, such as Parkinson's,

▼ **Figure 12.27 Personalized genetic testing.** This kit can be used to send saliva for genetic analysis. The results can indicate a person's risk of developing certain diseases.

are not currently preventable or treatable. Other tests (such as ones for breast cancer risk) may help a person make changes that can prevent disease. How can we identify truly useful tests?

There is a danger that information about disease-associated genes could be abused. One issue is the possibility of discrimination and stigmatization. Would you, as an employer, want to know if a potential employee had an increased risk of schizophrenia? Should you be able to find out? Should insurance companies have the right to screen applicants for disease genes? People might be coerced into taking a DNA test in order to be considered for a job or an insurance policy. How do we prevent genetic information from being used in a discriminatory manner?

A much broader ethical question is how do we really feel about wielding one of nature's powers—the evolution of new organisms? Some might ask if we have any right to alter an organism's genes—or to create new organisms.

The benefits to people and the environment must also be considered. For example, bacteria are being engineered to clean up mining wastes and other pollutants that threaten the soil, water, and air. Since these organisms may be the only feasible solutions to some of our most pressing environmental problems, many people think this type of genetic engineering should be encouraged.

DNA technologies raise many complex issues that have no easy answers. It is up to you, as a "citizen scientist," to make informed choices. ☑

☑**CHECKPOINT**

Why does genetically modifying a human gamete raise different ethical questions than genetically modifying a human somatic (body) cell?

Answer: A genetically modified somatic cell will affect only the patient. Modifying a gamete will affect an unborn individual as well as all of his or her descendants.

EVOLUTION CONNECTION: DNA Profiling

Profiling the Y Chromosome

Barring mutations, the human Y chromosome passes essentially intact from father to son. By comparing Y DNA, researchers can learn about the ancestry of human males. DNA profiling can thus provide data about recent human evolution.

In 2003, geneticists discovered that about 8% of males currently living in central Asia have Y chromosomes of striking genetic similarity. Further analysis traced their common genetic heritage to a single man living about 1,000 years ago. In combination with historical records, the data led to the speculation that the Mongolian ruler Genghis Khan (**Figure 12.28**) may be responsible for the spread of the chromosome to nearly 16 million men living today. A similar study of Irish men in 2006 suggested that nearly 10% of them were descendants of Niall of the Nine Hostages, a warlord who lived during the 1400s. Another study of Y DNA seemed to confirm the claim by the Lemba people of southern Africa that they are descended from ancient Jews. Sequences of Y DNA distinctive of the Jewish priestly caste called Kohanim are found at high frequencies among the Lemba.

Comparison of Y chromosome DNA profiles is part of a larger effort to learn more about the human genome. Other research efforts are extending genomic studies to many more species. These studies will advance our understanding of all aspects of biology, including health and ecology as well as evolution. In fact, comparisons of the completed genome sequences of bacteria, archaea, and eukaryotes first supported the theory that these are the three fundamental domains of life—a topic we discuss further in the next unit, "Evolution and Diversity."

▲ Figure 12.28 **Genghis Khan.**

Chapter Review

SUMMARY OF KEY CONCEPTS

 Go to the Study Area at **www.masteringbiology.com** for practice quizzes, eText, BioFlix™ 3-D animations, MP3 Tutor Sessions, videos, current events, and more.

Recombinant DNA Technology

DNA technology, the manipulation of genetic material, is a new branch of biotechnology, the use of organisms to make helpful products. DNA technology often involves the use of recombinant DNA, the combination of nucleotide sequences from two different sources.

Applications: From Humulin to Foods to "Pharm" Animals

Recombinant DNA techniques have been used to create genetically modified organisms, ones that carry artificially introduced genes. Nonhuman cells have been engineered to produce human proteins, genetically modified food crops, and transgenic farm animals. A transgenic organism is one that carries artificially introduced genes, typically from a different species.

Recombinant DNA Techniques

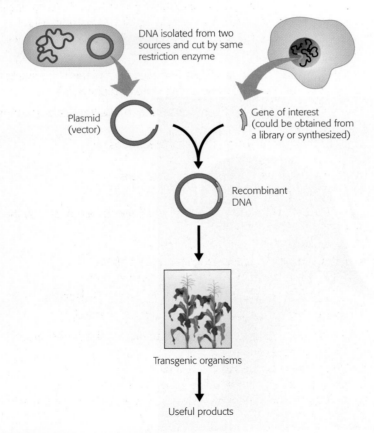

DNA isolated from two sources and cut by same restriction enzyme

Plasmid (vector)

Gene of interest (could be obtained from a library or synthesized)

Recombinant DNA

Transgenic organisms

Useful products

DNA Profiling and Forensic Science

Forensics, the scientific analysis of legal evidence, has been revolutionized by DNA technology. DNA profiling is used to determine whether two DNA samples come from the same individual.

Investigating Murder, Paternity, and Ancient DNA

DNA profiling can be used to establish innocence or guilt of a criminal suspect, identify victims, determine paternity, and contribute to basic research.

DNA Profiling Techniques

Short tandem repeat (STR) analysis compares DNA fragments using the polymerase chain reaction (PCR) and gel electrophoresis.

Crime scene   Suspect 1   Suspect 2

DNA

Polymerase chain reaction (PCR) amplifies STR sites

Longer DNA fragments

Gel

Shorter DNA fragments

DNA fragments compared by gel electrophoresis
(Bands of shorter fragments move faster toward the positive pole.)

Genomics and Proteomics

The Human Genome Project

The nucleotide sequence of the human genome is providing a wealth of useful data. The 24 different chromosomes of the human genome contain about 3.2 billion nucleotide pairs and 20,000 genes. The majority of the genome consists of noncoding DNA.

Tracking the Anthrax Killer

Comparing genomes can aid criminal investigations and basic research.

Genome-Mapping Techniques

The whole-genome shotgun method involves sequencing DNA fragments from an entire genome and then assembling the sequences.

Proteomics

Success in genomics has given rise to proteomics, the systematic study of the full set of proteins found in organisms.

Human Gene Therapy

Normal human gene

Virus

Bone marrow

Normal human gene is transcribed and translated in patient, potentially curing genetic disease permanently

Gene therapy trials have focused on SCID, an inherited immune disease, with some success and some setbacks.

Safety and Ethical Issues

The Controversy over Genetically Modified Foods

The debate about genetically modified crops centers on whether they might harm humans or damage the environment by transferring genes through cross-pollination with other species.

Ethical Questions Raised by DNA Technology

We as a society and as individuals must become educated about DNA technologies to address the ethical questions raised by their use.

SELF-QUIZ

1. Suppose you wish to create a large batch of the protein lactase using recombinant DNA. Place the following steps in the order you would have to perform them.
   a. Find the clone with the gene for lactase.
   b. Insert the plasmids into bacteria and grow the bacteria into clones.
   c. Isolate the gene for lactase.
   d. Create recombinant plasmids, including one that carries the gene for lactase.

2. Why is an artificial gene that is made using reverse transcriptase often shorter than the natural form of the gene?

3. A carrier that moves DNA from one cell to another, such as a plasmid, is called a _____.

4. In making recombinant DNA, what is the benefit of using a restriction enzyme that cuts DNA in a staggered fashion?

5. A paleontologist has recovered a bit of organic material from the 400-year-old preserved skin of an extinct dodo. She would like to compare DNA from the sample with DNA from living birds. The most useful method for initially increasing the amount of dodo DNA available for testing is _____.

6. Why do DNA fragments containing STR sites from different people tend to migrate to different locations during gel electrophoresis?

7. What feature of a DNA fragment causes it to move through a gel during electrophoresis?
   a. the electrical charges of its phosphate groups
   b. its nucleotide sequence
   c. the hydrogen bonds between its base pairs
   d. its double helix shape

8. After a gel electrophoresis procedure is run, the pattern of bars in the gel shows
   a. the order of bases in a particular gene.
   b. the presence of various-sized fragments of DNA.
   c. the order of genes along particular chromosomes.
   d. the exact location of a specific gene in a genomic library.

9. Name the steps of the whole-genome shotgun method.

10. Put the following steps of human gene therapy in the correct order.
    a. Virus is injected into patient.
    b. Human gene is inserted into a virus.
    c. Normal human gene is isolated and cloned.
    d. Normal human gene is transcribed and translated in the patient.

Answers to the Self-Quiz questions can be found in Appendix D.

THE PROCESS OF SCIENCE

11. A biochemist hopes to find a gene in human liver cells that codes for a blood-clotting protein. The nucleotide sequence of a small part of the gene is CTGGACTGACA. Briefly explain how to obtain the desired gene.

12. Some scientists once joked that when the DNA sequence of the human genome was complete, "we can all go home" because there would be nothing left for genetic researchers to discover. Why haven't they all "gone home"?

BIOLOGY AND SOCIETY

13. In the not-too-distant future, gene therapy may be used to treat many inherited disorders. What do you think are the most serious ethical issues to face before human gene therapy is used on a large scale? Explain.

14. Today, it is fairly easy to make transgenic plants and animals. What are some safety and ethical issues raised by this use of recombinant DNA technology? What are some dangers of introducing genetically engineered organisms into the environment? What are some reasons for and against leaving such decisions to scientists? Who should make these decisions?

15. In October 2002, the government of the African nation of Zambia announced that it was refusing to distribute 15,000 tons of corn donated by the United States, enough corn to feed 2.5 million Zambians for three weeks. The government rejected the corn because it was likely to contain genetically modified kernels. The government made the decision after its scientific advisers concluded that the studies of the health risks posed by GM crops "are inconclusive." Do you agree with this assessment? Do you think that it is a good justification for refusing the donated corn? At the time of the government's decision, Zambia was facing food shortages, and 35,000 Zambians were expected to die from starvation over the next six months. In light of this, do you think it was morally acceptable for the government to refuse the food? How do the relative risks posed by GM crops compare with the relative risks posed by starvation?

16. In 1983, a 10-year-old girl was kidnapped from her home, raped, and murdered. A jury convicted a local teenager of the crimes and sentenced him to death for the brutal killing. In 1995, DNA analysis proved that semen found near the scene could not have come from the man accused. After 12 years on death row, he was exonerated and released from prison. His case, which took place in Illinois, was far from unique. From 1977 to 2000, 12 convicts were executed and 13 exonerated. In 2000, the governor of Illinois declared a moratorium on all executions in his state because the death penalty system was "fraught with errors." Do you support the Illinois governor's decision? What rights should death penalty inmates have with regard to DNA testing of old evidence? Who should pay for this additional testing?

Unit 3
Evolution and Diversity

Chapter 13: **How Populations Evolve**

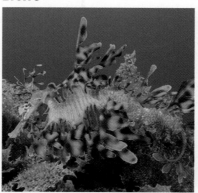

Chapter Thread: **Evolution in Action**

Chapter 14: **How Biological Diversity Evolves**

Chapter Thread: **Mass Extinctions**

Chapter 15: **The Evolution of Microbial Life**

Chapter Thread: **The Origin of Life**

Chapter 16: **Plants, Fungi, and the Move onto Land**

Chapter Thread: **Plant-Fungus Symbiosis**

Chapter 17: **The Evolution of Animals**

Chapter Thread: **Human Evolution**

13 How Populations Evolve

Evolutionary adaptations. Through natural selection, species such as the leafy sea dragon can become highly adapted to their environments.

CHAPTER CONTENTS

■■■ Chapter Thread: **Evolution in Action**

 BIOLOGY AND SOCIETY: Evolution in Action

Persistent Pests

In the 1960s, the World Health Organization (WHO) began a campaign to eradicate the mosquitoes that transmit the disease malaria. It was a noble goal because malaria kills an estimated 3 million people each year in tropical regions. The effort focused on spraying the mosquitoes' habitat with a pesticide called DDT. Early results were promising. Progress soon stalled, however, and the plan was dropped. How could a tiny mosquito thwart the best efforts of well-funded scientists?

Such failures, it turns out, are common. Pesticides often have encouraging early results: A relatively small amount dusted onto a crop may kill 99% of the insects. However, the few surviving insects have alleles (alternate forms of a gene) that enable them to resist the chemical attack. For example, the alleles may code for enzymes that destroy the pesticide. When the survivors reproduce, their offspring may inherit these alleles. In each generation, the proportion of resistant insects in the population increases, making subsequent sprayings less effective.

Pesticide-resistant insects are just one example of evolution in action. The sea dragon on the facing page demonstrates how well natural selection can drive the evolutionary adaptations of species to their surroundings.

An understanding of evolution informs all of biology, from exploring life's molecules to analyzing ecosystems. And applications of evolutionary biology are transforming medicine, agriculture, biotechnology, and conservation biology. This unit of chapters features mechanisms of evolution and traces the history of life on Earth. This chapter starts with the story of how Charles Darwin formulated his ideas. Next, it presents some evidence in support of evolution, followed by a closer look at Darwin's theory of natural selection. It then focuses on the genetic basis of evolution and the way evolution proceeds. Throughout the discussion, you'll read about evolution in action: verifiable, measurable examples of evolution that affect our world.

Charles Darwin and *The Origin of Species*

Biology came of age on November 24, 1859, the day Charles Darwin published *On the Origin of Species by Means of Natural Selection*. Darwin's book presented two main concepts. First, he argued convincingly from several lines of evidence that contemporary species arose from a succession of ancestors through a process of "descent with modification," his phrase for evolution. Darwin's second concept in *The Origin of Species* was a mechanism for how life evolves: natural selection.

Natural selection is a process in which organisms with certain inherited traits are more likely to survive and reproduce than are individuals with other traits (see Figure 1.12). As a result of natural selection, a **population**—a group of individuals of the same species living in the same place at the same time—can change over generations. Natural selection leads to **evolutionary adaptation**, a population's increase in the frequency of traits suited to the environment (**Figure 13.1**). (The term *adaptation* can also refer to the trait itself; for example, an insect's camouflage is an adaptation that helps it avoid predators.) In modern terms, we would say that the genetic composition of the population has changed over time, and that is one way of defining **evolution**. But we can also use the term *evolution* on a much grander scale to mean all of biological history, from the earliest microbes to the enormous diversity of organisms that live on the Earth today.

Darwin's book drew a cohesive picture of life by connecting the dots between a bewildering array of seemingly unrelated facts. *The Origin of Species* focused biologists' attention on the great diversity of organisms—their origins and relationships, their similarities and differences, their geographic distribution, and their adaptations to surrounding environments. ☑

Darwin's Cultural and Scientific Context

Before we examine how natural selection works and how Darwin derived the idea, let's place the Darwinian revolution in its historical context. As you read, you can refer to the time line in **Figure 13.2**.

The view of life developed in *The Origin of Species* contrasts sharply with the prevailing view during Darwin's lifetime. Many scientists of his day thought Earth was relatively young and populated by a huge number of unrelated species. *The Origin of Species* challenged that widely held notion. It was truly radical for its time—not only challenging the current scientific views but also shaking the deepest roots of Western culture.

▶ Figure 13.1 **Camouflage as an example of evolutionary adaptation.** Related species of insects called mantids have diverse shapes and colors that evolved in different environments.

A Trinidad tree mantid that mimics dead leaves

A flower mantid in Malaysia

A leaf mantid in Costa Rica

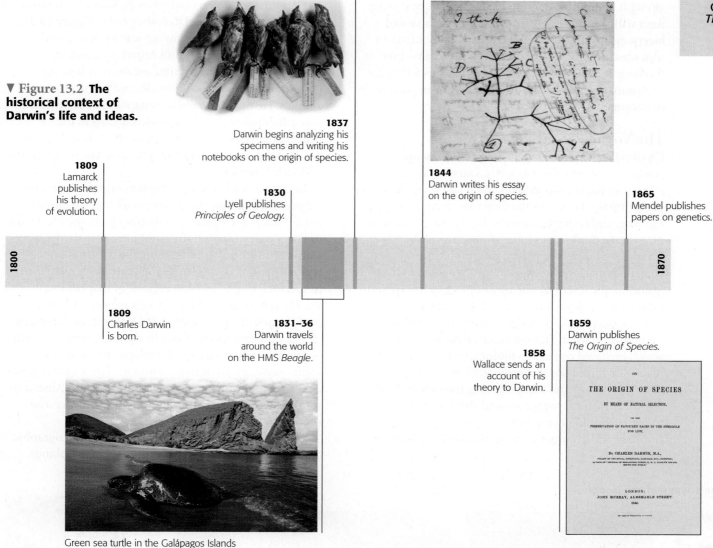

▼ Figure 13.2 **The historical context of Darwin's life and ideas.**

1837
Darwin begins analyzing his specimens and writing his notebooks on the origin of species.

1809
Lamarck publishes his theory of evolution.

1830
Lyell publishes *Principles of Geology.*

1844
Darwin writes his essay on the origin of species.

1865
Mendel publishes papers on genetics.

1800

1870

1809
Charles Darwin is born.

1831–36
Darwin travels around the world on the HMS *Beagle.*

1858
Wallace sends an account of his theory to Darwin.

1859
Darwin publishes *The Origin of Species.*

Green sea turtle in the Galápagos Islands

The Idea of Fixed Species

The Greek philosopher Aristotle, whose views had an enormous impact on Western culture, generally held that species are fixed, or permanent, and do not evolve. Judeo-Christian culture fortified this idea with a literal interpretation of the biblical book of Genesis, which tells the story of each form of life being individually created in its present-day form. The idea that all living species are unchanging in form and inhabit an Earth that is only about 6,000 years old dominated the intellectual climate of the Western world for centuries.

Lamarck and Evolutionary Adaptations

In the mid-1700s, the study of **fossils**—imprints or remains of organisms that lived in the past—led French naturalist Georges Buffon to suggest that Earth might be much older than 6,000 years. He also observed similarities between particular fossils and living animals. In 1766, Buffon proposed that certain fossil forms might

be ancient versions of similar living species. Then, in the early 1800s, French naturalist Jean-Baptiste de Lamarck suggested that the best explanation for this relationship of fossils to current organisms is that life evolves. Lamarck explained evolution as the refinement of traits that equip organisms to perform successfully in their environments. For example, some birds have powerful beaks that enable them to crack tough seeds.

We remember Lamarck mainly for his erroneous view of *how* species evolve. He proposed that by using or not using its body parts, an individual may develop certain traits that it passes on to its offspring. In other words, Lamarck proposed that acquired traits are inherited. He suggested, for example, that the strong beaks of seed-cracking birds are the cumulative result of ancestors exercising their beaks during feeding and passing that acquired beak power on to offspring. However, simple observations provide evidence against the inheritance of acquired traits: A carpenter who builds up strength and stamina through a lifetime of pounding nails with a heavy hammer will not pass enhanced biceps on to children. However, Lamarck's mistaken idea obscures the important fact that he helped set the stage for Darwin by proposing that species evolve as a result of interactions between organisms and their environments.

The Voyage of the *Beagle*

Charles Darwin was born in 1809, on the same day as Abraham Lincoln. Even as a boy, Darwin's consuming interest in nature was evident. When he was not reading nature books, he was in the fields and forests fishing, hunting, and collecting insects. His father, an eminent physician, could see no future for a naturalist and sent Charles to the University of Edinburgh to study medicine. But Charles, who was only 16 years old at the time, found medical school boring and distasteful. He left Edinburgh without a degree and then enrolled at Christ College at Cambridge University, intending to become a minister. Darwin received his B.A. degree, which included courses in biology, in 1831. Soon after, his botany professor recommended him to Captain Robert FitzRoy, who was preparing the survey ship HMS *Beagle* for a voyage around the world. It was a tour that would have a profound effect on Darwin's thinking and eventually on the thinking of the entire world.

Darwin was 22 years old when he sailed from Great Britain on the *Beagle* in December 1831 (**Figure 13.3**). The main mission of the voyage was to chart poorly known stretches of the South American coastline. Darwin spent most of his time on shore collecting thousands of specimens of fossils and living plants and animals. He noted the unique adaptations of organisms that inhabited such diverse environments as the Brazilian jungles, the grasslands of the Argentine pampas, and the desolate and frigid lands at the southern tip of South America.

In spite of their unique adaptations, the plants and animals throughout the continent all had a definite South American stamp, very distinct from the life-forms of Europe. That in itself may not have surprised Darwin. But the plants and animals living in temperate regions of South America seemed more closely related to species living in tropical regions of that continent than to species living in temperate regions of Europe. And the South American fossils Darwin found, though clearly different species from living ones, were distinctly South American in their resemblance to the living plants and animals of that continent. These observations led Darwin to wonder if contemporary South American species owed their features to descent from ancestral species on that continent.

Darwin was particularly intrigued by the geographic distribution of organisms on the Galápagos Islands.

▼ Figure 13.3 **The voyage of the *Beagle*.**

Darwin in 1840

HMS *Beagle*

The Galápagos are relatively young volcanic islands about 900 kilometers (about 540 miles) off the Pacific coast of South America. Most of the animals that inhabit these remote islands are found nowhere else in the world, but they resemble species living on the South American mainland (**Figure 13.4**).

While on his voyage, Darwin was strongly influenced by the newly published *Principles of Geology*, by Scottish geologist Charles Lyell. The book presented the case for an ancient Earth sculpted by gradual geologic processes that continue today. Having witnessed an earthquake that raised part of the coastline of Chile almost a meter, Darwin realized that natural forces gradually changed Earth's surface and that these forces still operate. Thus, the growth of mountains as a result of earthquakes could account for the presence of marine snail fossils he collected on mountaintops in the Andes. Darwin would eventually apply this principle of *gradualism* to the evolution of Earth's life.

By the time Darwin returned to Great Britain five years after the *Beagle* set sail (a full three years longer than was originally planned), his experiences and reading had led him to seriously doubt that Earth and all its living organisms had been specially created only a few thousand years earlier. Darwin had come to realize that Earth was very old and constantly changing. He began to analyze his collections and to discuss them with colleagues. He continued to read, correspond with other scientists, and maintain extensive journals of his observations, studies, and thoughts.

Descent with Modification

By the early 1840s, Darwin had composed a long essay describing the major features of his theory of evolution. He realized that his ideas would cause a social furor, however, and therefore delayed publishing his essay. Then, in the mid-1850s, Alfred Wallace, a British naturalist doing fieldwork in Indonesia, developed a theory almost identical to Darwin's. When Wallace sent Darwin a manuscript describing his own ideas on natural selection, Darwin wrote, "All my originality … will be smashed." However, in 1858, two of Darwin's colleagues presented Wallace's paper and excerpts from Darwin's earlier essay together to the scientific community. With the publication in 1859 of *The Origin of Species*, Darwin presented the world with an avalanche of evidence and a strong, logical argument for evolution.

As noted earlier, Darwin made two main points in *The Origin of Species*. First, he presented evidence that each living species descended from a succession of ancestral species. In the first edition of his book, he did not use the word *evolution*, referring instead to "descent with modification." Darwin hypothesized that as the descendants of the earliest organisms spread into various habitats over millions of years, they accumulated modifications, or adaptations, to diverse ways of life. Darwin's second main point was that natural selection is the mechanism for descent with modification. Next, we'll examine the evidence for evolution and then look more closely at Darwin's theory of natural selection. ☑

▼ **Figure 13.4 A marine iguana (right), an example of the unique species inhabiting the Galápagos.** Darwin noticed that Galápagos marine iguanas—with webbed feet and flattened tail that aid in swimming—are similar to, but distinct from, land-dwelling iguanas on the islands and on the South American mainland (left).

Evidence of Evolution

Evolution leaves observable signs. Such clues to the past are essential to any historical science. Historians of human civilization can study written records from earlier times. But they can also piece together the evolution of societies by recognizing vestiges of the past in modern cultures. Even if we did not know from written documents that Spaniards colonized the Americas, we would deduce this from the Hispanic stamp on Latin American culture. Similarly, biological evolution has left evidence in today's organisms, as well as in fossils.

In this section, we will examine five of the many lines of evidence in support of evolution. One of them—fossils—is a historical record. The other four—biogeography, comparative anatomy, comparative embryology, and molecular biology—encompass historical vestiges of evolution evident in modern life.

The Fossil Record

Over millions of years, sand and silt that eroded from the land were carried by rivers and deposited in the oceans, piling up and compressing older deposits below into rock. Some dead organisms that settled along with the sediments left imprints in the rocks. Thus, each rock layer, or stratum (plural, *strata*), bears a unique set of fossils representing a local sampling of the organisms that lived and died when that sediment was deposited. Younger strata are on top of older ones, so the positions of fossils in the strata reveal their relative age. (The ages of fossils can be confirmed using radiometric dating—see Figure 14.15.) The **fossil record** is this ordered sequence of fossils as they appear in the rock layers, marking the passage of geologic time (**Figure 13.5**).

The fossil record reveals the appearance of organisms in a historical sequence. The oldest known fossils, dating from about 3.5 billion years ago, are prokaryotes. This fossil evidence fits with the molecular and cellular evidence that prokaryotes are the ancestors of all life. Fossils in younger layers of rock reveal the evolution of various groups of eukaryotic organisms.

Paleontologists (scientists who study fossils) have discovered many transitional forms that link past and present. For example, a series of transitional fossils provides evidence that birds descended from one branch of dinosaurs. Another example is a series of transitional whale fossils connecting these aquatic mammals to four-legged land mammals (**Figure 13.6**). Whales living today have forelegs in the form of flippers and small bones that may be remnants of ancestral hind legs and feet. In the past few decades, a series of remarkable fossils of extinct mammals have been discovered in Pakistan, Egypt, and North America that document the transition from life on land to life in the sea. Some whale ancestors had a type of anklebone that is otherwise unique to the group of land mammals that includes pigs, hippos, cows, camels, and deer. The anklebone similarity strongly suggests that whales (as well as dolphins and porpoises) are most closely related to this group of land mammals. ☑

☑CHECKPOINT

Why are older fossils generally in deeper rock layers than younger fossils?

Answer: Sedimentation places younger rock layers on top of older ones.

▼ Figure 13.5 **Strata of sedimentary rock at the Grand Canyon.** The Colorado River has cut through over 2,000 m of rock, exposing sedimentary strata that are like huge pages from the book of life. Each stratum entombs fossils that represent some of the organisms from that period of Earth's history.

▼ Figure 13.6 **A transitional fossil linking past and present.** The hypothesis that whales evolved from terrestrial (land-dwelling) ancestors predicts a four-limbed beginning for whales. Paleontologists digging in Egypt and Pakistan have identified extinct whales that had hind limbs. Shown here are fossilized leg bones of *Basilosaurus*, one of those ancient whales. These whales were already aquatic animals that no longer used their legs to support their weight.

Biogeography

It was the geographic distribution of species, called **biogeography**, that first suggested to Darwin that today's organisms evolved from ancestral forms. Consider, for example, Darwin's visit to the Galápagos Islands. Darwin noted that the Galápagos animals resembled species of the South American mainland more than they resembled animals on similar but distant islands. The logical explanation was that the Galápagos species evolved from animals that had migrated from South America, each species adapting to its new environment.

Many other examples from biogeography seem baffling without an evolutionary perspective, such as the diversity of marsupials in Australia (**Figure 13.7**). Why is Australia home to so many kinds of marsupials (mammals that complete embryonic development outside the uterus, typically in a mother's pouch) but relatively few placental mammals (mammals that complete embryonic development in the uterus)? It is not because Australia is inhospitable to placental mammals. Humans have introduced rabbits, foxes, and many other placental mammals to Australia, where these introduced species have thrived to the point of becoming ecological and economic nuisances. The prevailing hypothesis is that the unique Australian wildlife evolved on that island continent in isolation from regions where early placental mammals diversified.

The geographic distribution of species makes little sense if we imagine that species were individually placed in suitable environments. In the Darwinian view, we find species where they are because they evolved from ancestors that inhabited those regions.

▼ Figure 13.7 **Biogeography.** The continent of Australia is home to many unique plants and animals, such as these marsupials, mammals that evolved in relative isolation from other continents where placental mammals diversified.

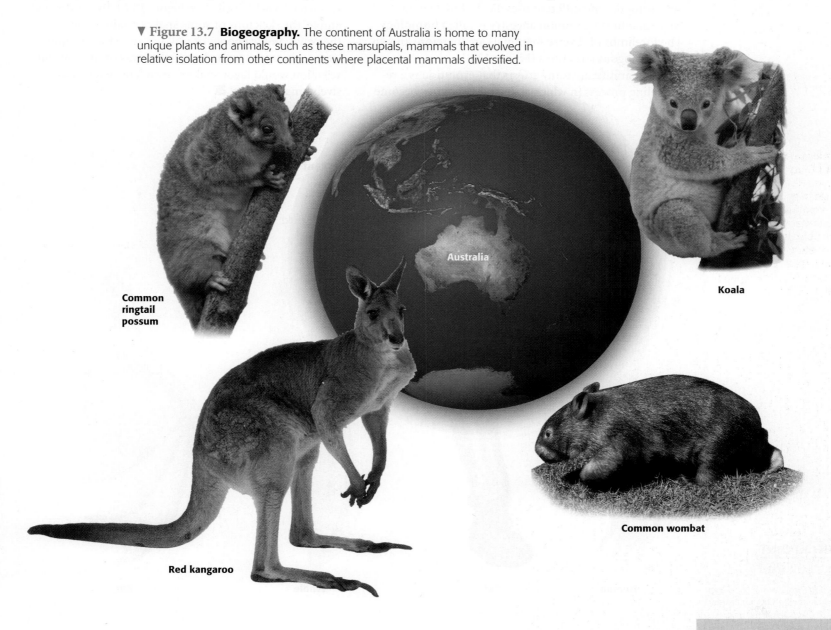

Common ringtail possum

Koala

Australia

Red kangaroo

Common wombat

Comparative Anatomy

The comparison of body structures in different species is called **comparative anatomy**. Certain anatomical similarities among species are signs of evolutionary history. For example, the same skeletal elements make up the forelimbs of humans, cats, whales, and bats, even though the functions of these forelimbs differ greatly; clearly, a whale's flipper does not do the same job as a bat's wing. If these limbs had been uniquely engineered in their current forms, we would expect a variety of basic designs that reflect their unique tasks. It is more logical that the arms, forelegs, flippers, and wings of these different mammals are variations on anatomical structures of an ancestral organism, structures that over millions of years have become adapted to different functions. These adaptations account for the observed similarities in structures despite different uses. Such similarity in structure due to common ancestry is called **homology**. The forelimbs of diverse mammals are therefore known as homologous structures (**Figure 13.8**).

Comparative anatomy attests that evolution is a remodeling process in which ancestral structures become modified as they take on new functions—the kind of process that Darwin referred to as descent with modification. The historical constraints of this modification are evident in anatomical imperfections. For example, the human spine and knee joint were derived from ancestral structures that supported four-legged mammals. Consequently, almost none of us will reach old age without experiencing knee or back problems. If these structures had first taken form specifically to support our bipedal posture, we would expect them to be less subject to sprains, spasms, and other common injuries.

Some of the most interesting homologous structures are "leftover" structures of marginal, if any, importance to the organism. These **vestigial structures** are remnants of features that served important functions in the organism's ancestors, such as the rear leg bones evident in ancient whale fossils (see Figure 13.6). In another example, the skeletons of some snakes retain vestiges of the pelvis and leg bones of walking ancestors. If limbs were a hindrance to ancient snakes' way of life, natural selection would favor snake descendants with successively smaller limbs. ✔

▼ **Figure 13.8 Homologous structures: anatomical signs of descent with modification.** The forelimbs of all mammals are constructed from the same skeletal elements. (Homologous bones in each of these four mammals are colored the same.) The hypothesis that all mammals descended from a common ancestor predicts that their forelimbs, though diversely adapted, would be variations on a common anatomical theme.

Human   **Cat**   **Whale**   **Bat**

✔**CHECKPOINT**

What is homology?

Answer: similarity between species that is due to shared ancestry

Comparative Embryology

Comparing early stages of development in different animal species reveals additional homologies not visible in adult organisms. For example, all vertebrate embryos have a developmental stage in which structures called pharyngeal pouches appear on the sides of the throat. At this stage, the embryos of fishes, frogs, snakes, birds, and apes—indeed, all vertebrates—look more alike than different (Figure 13.9). The different classes of vertebrates take on more distinctive features as development progresses. For example, pharyngeal pouches develop into gills in fishes but into parts of the ear and throat in humans.

Molecular Biology

As you saw in Chapter 10, the hereditary background of an organism is documented in its DNA and in the proteins encoded by the DNA. If two species have genes with nucleotide sequences that match closely (and thus proteins with amino acid sequences that match closely), biologists conclude that these sequences are homologous and must have been inherited from a relatively recent common ancestor (Figure 13.10). In contrast, the greater the number of sequence differences between species, the less likely they share a close common ancestor. Molecular comparisons between diverse organisms have allowed biologists to develop hypotheses about the evolutionary divergence of branches on the tree of life. For example, genetic analyses first suggested that the domain Archaea is more closely related to the eukaryotes than it is to the domain Bacteria.

Darwin's boldest hypothesis was that *all* forms of life are related to some extent through branching evolution from the earliest organisms. About 100 years after Darwin made his claim, molecular biology began providing strong evidence for evolution: All forms of life use the same genetic language of DNA and RNA, and the genetic code (how RNA triplets are translated into amino acids) is nearly universal. This genetic language has been passed along through all the branches of evolution since its beginnings in an early form of life. ☑

▼ Figure 13.9 **Evolutionary signs from comparative embryology.** At the early stage of development shown here, the kinship of vertebrates is unmistakable. Notice, for example, the pharyngeal pouches and tails in both the chicken embryo and the human embryo.

Pharyngeal pouches

Post-anal tail

Chicken embryo   **Human embryo**

▼ Figure 13.10 **Genetic relationships among some primates.** The bars in this diagram show the percent of selected DNA sequences that match between a chimpanzee and other primates, the animal group that includes monkeys, apes, and humans. For example, note that the selected DNA sequences of chimps and humans are better than a 98% match. In contrast, the DNA sequences of chimps and Old World monkeys (such as macaques, mandrills, baboons, and rhesus monkeys) have less than a 93% match.

Primate	Percent of selected DNA sequences that match a chimpanzee's DNA
	92% 96% 100%
Chimpanzee	
Human	
Gorilla	
Orangutan	
Gibbon	
Old World monkey	

Natural Selection

Natural Selection

Darwin perceived adaptation to the environment and the origin of new species as closely related processes. Imagine, for example, that an animal species from a mainland colonizes a chain of distant, relatively isolated islands. In the Darwinian view, populations on the different islands may diverge more and more in appearance as each population adapts to its local environment. Over many generations, the populations on different islands could become dissimilar enough to be designated as separate species. Evolution of the finches on the Galápagos Islands is an example. It is a reasonable hypothesis that the islands were colonized by finches that strayed from elsewhere and then diversified on the different islands. Among the differences between the Galápagos finches are their beaks, which are adapted to the specific foods available on each species' home island (**Figure 13.11**). Darwin anticipated that explaining how such adaptations arise is the key to understanding evolution. And his theory of natural selection remains our best explanation for the formation of new adaptations. ☑

Darwin's Theory of Natural Selection

Darwin based his theory of natural selection on two key observations, both of which had already been noted by scientists. First, he recognized that all species tend to produce excessive numbers of offspring (**Figure 13.12**). Darwin deduced that because natural resources are limited, the production of more individuals than the environment can support leads to a struggle for existence among the individuals of a population. In most cases, only a small percentage of offspring will survive in each generation. Many eggs are laid, young born, and seeds spread, but only a tiny fraction complete their development and leave offspring of their own. The rest are starved, eaten, frozen, diseased, unmated, or unable to reproduce for other reasons.

The second key observation that led Darwin to natural selection was his awareness of variation among individuals of a population. Just as no two people in a

▼ Figure 13.11 **Galápagos finches with beaks adapted for specific diets.**

(a) The large ground finch. This species of Galápagos finch has a large beak specialized for cracking seeds that fall from plants to the ground.

(b) The small tree finch. The smaller beak of the small tree finch is used to grasp insects.

(c) The woodpecker finch. The long, narrow beak of the woodpecker finch allows it to hold tools such as cactus spines to probe for wood-boring insects.

▼ Figure 13.12 **Overproduction of offspring.** A cloud of millions of spores is exploding from these puffballs, a type of fungus. (Each puffball in this photo is about 2 cm across.) Only a tiny fraction of the spores will actually give rise to offspring that survive and reproduce.

Spore cloud

▲ **Figure 13.13 Color variations within a single species of Asian lady beetles.**

human population are alike, individual variation abounds in all species (**Figure 13.13**). Much of this variation is heritable. Siblings share more traits with each other and with their parents than they do with less closely related members of the population.

From these two observations, Darwin arrived at the conclusion that defines natural selection: Individuals whose inherited traits are best suited to the local environment are more likely than less fit individuals to survive and reproduce. In other words, the individuals that function best should leave the most surviving offspring. Darwin's genius was in connecting two observations that anyone could make and drawing a conclusion that could explain how adaptations evolve.

Observation 1: Overproduction and competition.
Populations of all species have the potential to produce many more offspring than the environment can possibly support with food, space, and other resources. This overproduction makes a struggle for existence among individuals inevitable.

Observation 2: Individual variation.
Individuals in a population vary in many heritable traits.

Conclusion: Unequal reproductive success.
Those individuals with heritable traits best suited to the local environment generally leave a larger share of surviving, fertile offspring.

Darwin's insight was both simple and profound. The environment screens a population's inherent variability. Unequal success in reproduction (natural selection) leads to an accumulation of the favored traits in the population over generations (evolution). In other words, natural selection promotes evolutionary adaptations.

It is important to emphasize three key but subtle points about evolution by natural selection. The first point is that individuals do not evolve, even though natural selection occurs through interactions between individual organisms and their environment. Evolution refers to generation-to-generation changes in populations (as you'll see in the next section).

The second key point is that natural selection can amplify or diminish only heritable traits. An organism may become modified through interactions with the environment during its lifetime, and those acquired traits may help it survive. But unless these traits are coded for in the genes of the organism's gametes (sperm or egg), they cannot be passed on to offspring and therefore will not affect the reproductive success of the next generation.

The third key point is that evolution is not goal directed; it does not lead to perfectly adapted organisms. Natural selection is the result of environmental factors that vary from place to place and from time to time. A trait that is favorable in one situation may be useless—or even detrimental—in different circumstances. For example, some genetic mutations that happen to endow mosquitoes with resistance to the pesticide DDT also reduce a mosquito's growth rate. Before DDT was introduced, the gene for resistance was a handicap. But after DDT became part of the environment, the mutant alleles were advantageous, and natural selection increased their frequency in mosquito populations. This example shows that significant evolutionary change can occur in a short time—something we'll stress again in the next section. ☑

Natural Selection in Action

Natural selection and evolution are observable. You learned about one unsettling example in the Biology and Society section: the evolution of pesticide resistance within insect species. Pesticides control insects and prevent them from eating crops, transmitting diseases such as malaria, or just annoying us. But widespread use has led to the unintended evolution of pesticide-resistant insect populations (**Figure 13.14**). Chapter 1 described another example of natural selection in action: the evolution of antibiotic-resistant bacteria. More recently, doctors have documented an increase in drug-resistant strains of HIV, the virus that causes AIDS. The unifying thread sewn throughout this chapter is that evolution and natural selection did not operate only in the distant past, but are observable today. Let's look at another example of evolution in action.

☑CHECKPOINT
Explain why the following statement is incorrect: "Pesticides cause pesticide resistance in insects."

Answer: An environmental factor does not create new traits such as pesticide resistance, but favors traits that are already represented in the population.

▼ **Figure 13.14 Evolution of pesticide resistance in insect populations.** By spraying crops with poisons to kill insect pests, people have unwittingly favored the reproductive success of insects with inherent resistance to the poisons.

Insecticide application

Chromosome with gene conferring resistance to pesticide

Survivors

Reproduction

Additional applications of the same pesticide will be less effective, and the frequency of resistant insects in the population will grow.

The Modern Synthesis: Darwinism Meets Genetics

Natural selection requires hereditary processes that Darwin could not explain. How do the variations that are the raw material for natural selection arise in a population? And how are these variations passed along from parents to offspring? Darwin and Gregor Mendel lived and worked at the same time. In fact, by breeding peas in his abbey garden, Mendel illuminated the very hereditary processes required for natural selection to work. However, Mendel's discoveries went unappreciated by the scientific community during his lifetime. Mendelism and Darwinism finally came together in the mid-1900s, decades after both scientists had died. This fusion of genetics with evolutionary biology came to be known as the **modern synthesis** (here, the term *synthesis* means "combination"). One of its key elements is an emphasis on the biology of populations.

Populations as the Units of Evolution

As noted earlier, individual organisms do not evolve during their lifetimes. Although natural selection affects an organism's survival and reproductive success, the evolutionary impact is only apparent in the changes in a population over time.

One population may be isolated from other populations of the same species, with little interbreeding and thus little exchange of genes between them. Such isolation is common for populations confined to widely separated islands, unconnected lakes, or mountain ranges separated by lowlands. However, populations are not usually so isolated, and they rarely have sharp boundaries (**Figure 13.17a**). One population center may blur into another in a region of overlap, where members of both populations are present but less numerous. Nevertheless, individuals are more concentrated in the population centers and are more likely to breed with other locals (**Figure 13.17b**). Therefore, organisms of a population are generally more closely related to one another than to members of other populations.

A population is the smallest biological unit that can evolve. In the Biology and Society section and Figure 13.14, you saw that natural selection can affect pesticide resistance in populations of insects. That impact is measured by the change in the relative numbers of resistant insects over a span of generations, not by the survivability of any individual insect.

▼ Figure 13.17 **Populations.**

(a) Two dense populations of trees separated by a lake. Interbreeding occurs when wind blows pollen between the populations. Nevertheless, trees are more likely to breed with members of the same population than with trees on the other side of the lake.

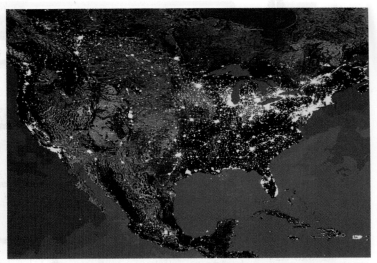

(b) A nighttime satellite view of North America. Notice the lights of human population centers, or cities. People move around, of course, and there are suburban and rural communities between cities, but people are more likely to choose mates locally.

In studying evolution at the population level, biologists focus on what is called the **gene pool**, the total collection of alleles in a population at any one time (that is, the total of all the alleles in all the individuals making up the population). For many genes, there are two or more alleles (versions of the gene) in the gene pool. For example, in an insect population, there may be one allele that codes for an enzyme that breaks down a certain pesticide and one for a version of the enzyme that does not. In fields sprayed with pesticide, the first allele will increase in frequency and the other allele will decrease in frequency. When the relative frequencies of alleles in a population change like this over a number of generations, evolution is occurring on its smallest scale. Later we'll look at the causes of evolution within a population. But first let's explore the sources of genetic variation. ☑

phenotypes; there are no in-between types. Next, we'll focus on the sources of different alleles.

Sources of Genetic Variation

Mutations and sexual reproduction, both involving random processes, produce genetic variation. Let's examine each of these processes in detail.

Mutations, random changes in the nucleotide sequence of DNA (see Figure 10.22), can result in new alleles. For example, a mutation in a gene may substitute one nucleotide for another. If such a change affects the protein's function, the mutation will probably be harmful. A random mutation is like a shot in the dark; it is not likely to improve a genome any more than shooting a bullet through the hood of a car is likely to improve engine performance.

☑**CHECKPOINT**

What is the smallest biological unit that can evolve?

Answer: a population

Genetic Variation in Populations

You have no trouble recognizing your friends in a crowd. People vary in appearance, reflecting the individual differences in their genomes. Individual variation abounds in all species (**Figure 13.18**). In addition to visible differences, most populations vary greatly at the molecular level. For example, you cannot tell someone's blood group (A, B, AB, or O) just by looking at that person, but biochemical analyses can reveal such variations.

Not all variation in a population is heritable. Phenotype results from a combination of the genotype, which is inherited, and many environmental influences. For instance, a strength-training program can build up your muscle mass beyond what would naturally occur from your genetic makeup. However, you would not pass this environmentally induced physique on to your offspring. Only the genetic component of variation is relevant to natural selection.

Many of the variable traits in a population result from the combined effect of several genes. This polygenic ("many genes") inheritance produces traits that vary more or less continuously—in human height, for instance, from very short individuals to very tall ones. By contrast, other features, such as human blood group, are determined by a single gene locus, with different alleles producing one of only a few distinct

▼ Figure 13.18 **Variation in a garter snake population.** These four garter snakes, which belong to the same species, were all captured in one Oregon field. The behavior of each physical type is correlated with its coloration. When approached, spotted snakes, which blend in with their background, generally freeze. In contrast, snakes with stripes, which make it difficult to judge the speed of motion, usually flee rapidly when approached.

On rare occasions, however, a mutant allele may actually enhance reproductive success. This kind of effect is more likely when the environment is changing in such a way that alleles that were once disadvantageous are favorable under the new conditions. We already considered one example, the changing fortunes of the DDT resistance allele in insect populations.

A new mutation that is transmitted in gametes can immediately change the gene pool of a population by replacing one allele with another. For any given gene locus, mutation alone has little effect on a large population in a single generation. This is because a mutation at any given gene locus is a very rare event. However, the cumulative impact of mutations across the genome can be significant because an individual has thousands of genes, and many populations have thousands or millions of individuals.

Organisms with very short generation spans, such as bacteria, can evolve rapidly with mutation as the only source of genetic variation. Bacteria multiply so quickly that natural selection can increase a population's frequency of a beneficial mutation in just hours or days. For most animals and plants, however, their long generation times prevent new mutations from significantly affecting overall genetic variation in the short term. Consequently, in sexually reproducing organisms, most of the genetic variation in a population results from the unique combination of alleles that each individual inherits. (Of course, the origin of that allele variation is past mutation.)

As you saw in Chapter 8, fresh assortments of existing alleles arise every generation from three random components of sexual reproduction: independent orientation of homologous chromosomes at metaphase I of meiosis (see Figure 8.16), random fertilization, and crossing over (see Figure 8.18). During prophase I of meiosis, pairs of homologous chromosomes, one set inherited from each parent, trade some of their genes by crossing over, and then each pair separate into gametes independently of other chromosome pairs. Gametes from one individual vary extensively in their genetic makeup, and each zygote made by a mating pair has a unique assortment of alleles resulting from the random union of sperm and egg.

While the processes that generate genetic variation—mutation and events during sexual reproduction—are random, natural selection (and hence evolution) is not. The environment selectively promotes the propagation of those genetic combinations that enhance survival and reproductive success. ☑

Analyzing Gene Pools

As described earlier, a gene pool consists of all the alleles in a population at any one time. The gene pool is the reservoir from which the next generation of organisms draws its genes.

Imagine a wildflower population with two varieties contrasting in flower color (**Figure 13.19**). An allele for red flowers, which we will symbolize by R, is dominant to an allele for white flowers, symbolized by r. These are the only two alleles for flower color in the gene pool of this hypothetical plant population. Now, let's say that 80%, or 0.8, of all flower-color loci in the gene pool have the R allele. We'll use the letter p to represent the relative frequency of the R allele in the population. Thus, $p = 0.8$. Because there are only two alleles in this example, the r allele must be present at the other 20% (0.2) of the gene pool's flower-color loci. Let's use the letter q for the frequency of the r allele in the population. For the wildflower population, $q = 0.2$. And since there are only two alleles for flower color, we know that we can express their frequencies as follows:

$$p \quad + \quad q \quad = \quad 1$$

Frequency of   Frequency of
one allele   alternate allele

Notice that if we know the frequency of either allele in the gene pool, we can subtract it from 1 to calculate the frequency of the other allele.

From the frequencies of alleles, we can also calculate the frequencies of different genotypes in the population if the gene pool is completely stable (not evolving). In the wildflower population, what is the probability of producing an RR individual by "drawing" two R alleles from the pool of gametes? Here we apply the rule of multiplication that you learned in Chapter 9 (review Figure 9.11). The probability of drawing an R sperm

▼ Figure 13.19 **A population of wildflowers with two varieties of color.**

multiplied by the probability of drawing an *R* egg is $p \times p = p^2$, or $0.8 \times 0.8 = 0.64$. In other words, 64% of the plants in the population will have the *RR* genotype. Applying the same math, we also know the frequency of *rr* individuals in the population: $q^2 = 0.2 \times 0.2 = 0.04$. Thus, 4% of the plants are *rr*, giving them white flowers. Calculating the frequency of heterozygous individuals, *Rr*, is trickier. That's because the heterozygous genotype can form in two ways, depending on whether the sperm or egg supplies the dominant allele. So the frequency of the *Rr* genotype is $2pq$, which is $2 \times 0.8 \times 0.2 = 0.32$. In our imaginary wildflower population, 32% of the plants are *Rr*, with red flowers. **Figure 13.20** reviews these calculations graphically.

Now we can write a general formula for calculating the frequencies of genotypes in a gene pool from the frequencies of alleles, and vice versa:

$$p^2 \quad + \quad 2pq \quad + \quad q^2 \quad = \quad 1$$

Frequency of homozygotes for one allele | Frequency of heterozygotes | Frequency of homozygotes for alternate allele

Notice that the frequencies of all genotypes in the gene pool must add up to 1. This formula is called the Hardy-Weinberg formula, named for the two scientists who derived it in 1908.

▼ **Figure 13.20 A mathematical swim in the gene pool.** Each of the four boxes in the grid corresponds to an equally probable "draw" of alleles from the gene pool.

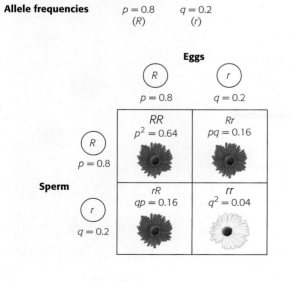

Allele frequencies $p = 0.8$ $q = 0.2$
           (*R*)   (*r*)

Eggs

R   *r*
$p = 0.8$ $q = 0.2$

Sperm

R $p = 0.8$
    RR $p^2 = 0.64$ | *Rr* $pq = 0.16$

r $q = 0.2$
    rR $qp = 0.16$ | *rr* $q^2 = 0.04$

Genotype frequencies $p^2 = 0.64$ $2pq = 0.32$ $q^2 = 0.04$
                     (*RR*)   (*Rr*)   (*rr*)

Population Genetics and Health Science

Public health scientists use the Hardy-Weinberg formula to calculate the percentage of a human population that carries the allele for certain inherited diseases. Consider phenylketonuria (PKU), which is an inherited inability to break down the amino acid phenylalanine. If untreated, the disorder causes severe mental retardation. PKU occurs in about 1 out of 10,000 babies born in the United States. Newborn babies are now routinely tested for PKU, and symptoms can be prevented if individuals living with the disease follow a strict diet (**Figure 13.21**).

PKU is caused by a recessive allele (that is, one that must be present in two copies to produce the phenotype). Thus, we can represent the frequency of individuals in the U.S. population born with PKU with the q^2 term in the Hardy-Weinberg formula. For one PKU occurrence per 10,000 births, $q^2 = 0.0001$. Therefore, *q*, the frequency of the recessive allele in the population, equals the square root of 0.0001, or 0.01. And *p*, the frequency of the dominant allele, equals $1 - q$, or 0.99.

Now let's calculate the frequency of carriers, who are heterozygous individuals who carry the PKU allele in a single copy and may pass it on to offspring. Carriers are represented in the formula by $2pq$: $2 \times 0.99 \times 0.01$, or 0.0198. Thus, the Hardy-Weinberg formula tells us that about 2% of the U.S. population are carriers for the PKU allele. Estimating the frequency of a harmful allele is essential for any public health program dealing with genetic diseases.

Microevolution as Change in a Gene Pool

As stated earlier, evolution can be measured as changes in the genetic composition of a population over time. It helps, as a basis of comparison, to know what to expect if a population is *not* evolving. A nonevolving population is in genetic equilibrium, which is also known as **Hardy-Weinberg equilibrium**. The population's gene pool remains constant. From generation to generation, the frequencies of alleles (p and q) and genotypes (p^2, $2pq$, and q^2) are unchanged. Sexual shuffling of genes cannot by itself change a large gene pool.

One of the products of the modern synthesis was a definition of evolution that is based on the genetics of populations: Evolution is a generation-to-generation change in a population's frequencies of alleles. Because this is evolution viewed on the smallest scale, it is sometimes referred to as **microevolution**. ✓

INGREDIENTS: SORBITOL, MAGNESIUM STEARATE, ARTIFICIAL FLAVOR, **ASPARTAME† (SWEETENER)**, ARTIFICIAL COLOR (YELLOW 5 LAKE, BLUE 1 LAKE), ZINC GLUCONATE.
†PHENYLKETONURICS: CONTAINS PHENYLALANINE

▲ **Figure 13.21 A warning to individuals with PKU.** People with PKU (phenylketonuria) must strictly regulate their dietary intake of the amino acid phenylalanine. In addition to natural sources, phenylalanine is found in aspartame, a common artificial sweetener. The frequency of the PKU allele is high enough to warrant warnings on foods that contain phenylalanine.

☑**CHECKPOINT**

1. Which term in the Hardy-Weinberg formula ($p^2 + 2pq + q^2 = 1$) corresponds to the frequency of individuals with *no* alleles for the recessive disease PKU?
2. Define microevolution.

Answers: 1. p^2 2. Microevolution is a change in a population's frequencies of alleles.

BioFlix™

Mechanisms of Evolution

Mechanisms of Evolution

Now that we understand that evolution can be evident in changes in a population's genetic makeup from generation to generation (microevolution), we come to an obvious question: What mechanisms can change a gene pool? The three main causes of evolutionary change are genetic drift, gene flow, and natural selection.

Genetic Drift

Flip a coin 1,000 times, and a result of 700 heads and 300 tails would make you very suspicious about that coin. But flip a coin 10 times, and an outcome of 7 heads and 3 tails would seem within reason. With a smaller sample, there is a greater chance of deviation from an idealized result—in this case, an equal number of heads and tails.

Let's apply coin toss logic to a population's gene pool. If a new generation draws its alleles at random from the previous generation, then the larger the population (the sample size), the better the new generation will represent the gene pool of the previous generation. Thus, one requirement for a gene pool to maintain the status quo is a large population size. The gene pool of a small population may not be accurately represented in the next

generation because of sampling error. The changed gene pool is analogous to the erratic outcome from a small sample of coin tosses.

Figure 13.22 applies this concept of sampling error to a small population of wildflowers. Chance causes the frequencies of the alleles for red (R) and white (r) flowers to change over the generations. And that fits our definition of microevolution. This evolutionary mechanism, a change in the gene pool of a population due to chance, is called **genetic drift**. But what would cause a population to shrink down to a size where there is genetic drift? Two ways this can occur are the bottleneck effect and the founder effect.

The Bottleneck Effect

Disasters such as earthquakes, floods, and fires may kill large numbers of individuals, producing a small surviving population that is unlikely to have the same genetic makeup as the original population. By chance, certain alleles may be overrepresented among the survivors. Other alleles may be underrepresented. And some alleles may be eliminated. Chance may continue to change the gene pool for many generations until the population is again large enough for sampling errors to be insignificant.

▼ **Figure 13.22 Genetic drift.** This hypothetical wildflower population consists of only ten plants. Due to random change over the generations, genetic drift can eliminate some alleles, as is the case for the *r* allele in generation 3 of this imaginary population.

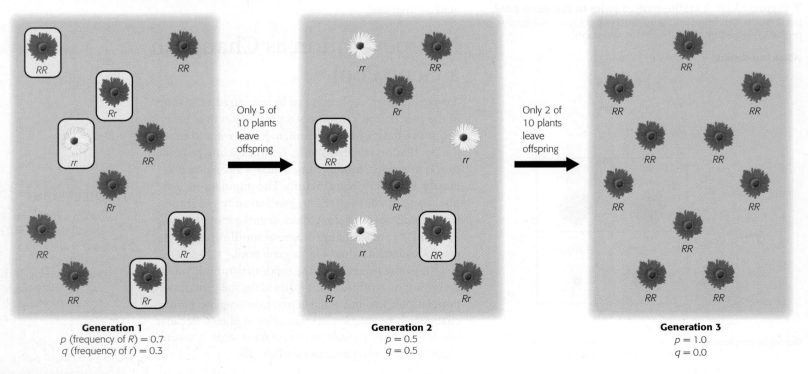

Generation 1
p (frequency of R) = 0.7
q (frequency of r) = 0.3

Only 5 of 10 plants leave offspring

Generation 2
p = 0.5
q = 0.5

Only 2 of 10 plants leave offspring

Generation 3
p = 1.0
q = 0.0

The analogy in **Figure 13.23** illustrates why genetic drift due to a drastic reduction in population size is called the **bottleneck effect**.

Bottlenecking usually reduces the overall genetic variability in a population because at least some alleles are likely to be lost from the gene pool. An important application of this concept is the potential loss of individual variation, and hence adaptability, in bottlenecked populations of endangered species, such as the cheetah (**Figure 13.24**). The fastest of all running animals, cheetahs are magnificent cats that were once widespread in Africa and Asia. Like many African mammals, the number of cheetahs fell drastically during the last ice age (around 10,000 years ago). At that time, the species suffered a severe bottleneck, possibly as a result of disease, human hunting, and periodic droughts. Evidence suggests that the South African cheetah population suffered a second bottleneck during the 1800s, when farmers hunted the animals to near extinction. Today, only a few small populations of cheetahs exist in the wild. Genetic variability in these populations is very low compared with populations of other mammals. This lack of variability, coupled with an increasing loss of habitat, makes the cheetah's future precarious. The cheetahs remaining in Africa are being crowded into nature preserves and parks as human demands on the land increase. Along with crowding comes an increased potential for the spread of disease. With so little variability, the cheetah has a reduced capacity to adapt to such environmental challenges. Captive breeding programs are already under way and may be required for the cheetah's long-term survival. ✔

▼ Figure 13.23 **The bottleneck effect.** The colored marbles in this analogy represent three alleles in an imaginary population. Shaking just a few of the marbles through the bottleneck is like drastically reducing the size of a population struck by some environmental disaster. By chance, purple marbles are overrepresented in the new population, green marbles are underrepresented, and orange marbles are absent. Similarly, bottlenecking a population of organisms tends to reduce variability.

Original population   Bottlenecking event   Surviving population

✔CHECKPOINT

Would you expect modern cheetahs to have more genetic variation or less genetic variation than cheetahs did 1,000 years ago?

Answer: less, because the bottleneck effect reduces genetic variability

▼ Figure 13.24 **Implications of the bottleneck effect in conservation biology.** Some endangered species, such as the cheetah, have low genetic variability. As a result, they are less adaptable to environmental changes, such as new diseases, than are species with a greater resource of genetic variation.

The Founder Effect

Genetic drift is also likely when a few individuals colonize an isolated island, lake, or other new habitat. The smaller the colony, the less its genetic makeup will match the gene pool of the larger population from which the colonists emigrated. If the colony succeeds, random drift will continue to affect the frequency of alleles until the population is large enough for genetic drift to be minimal.

The type of genetic drift resulting from the establishment of a small, new population whose gene pool differs from that of the parent population is called the **founder effect**. The founder effect undoubtedly contributed to the evolutionary divergence of the finches and other organisms that arrived as strays on the remote Galápagos Islands that Darwin visited.

The founder effect also explains the relatively high frequency of certain inherited disorders in some small human populations. In 1814, a group of 15 people founded a British colony on Tristan da Cunha, a cluster of small islands in the middle of the Atlantic Ocean (**Figure 13.25**). Apparently, one of the colonists carried a recessive allele for retinitis pigmentosa, a progressive form of blindness. Of the 240 descendants who still lived on the islands in the 1960s, 4 had retinitis pigmentosa, and 9 others were heterozygous carriers with one copy of the recessive allele. The frequency of the retinitis pigmentosa allele was ten times higher on Tristan da Cunha than in the British population from which the founders came.

▼ **Figure 13.25 Residents of Tristan da Cunha in the early 1900s.** The island of Tristan da Cunha, located in the middle of the Atlantic Ocean, is listed in the *Guinness World Records* as the world's most remote inhabited island. The genetic isolation of the island's residents resulted in a disproportionately high rate of hereditary blindness.

Gene Flow

Another source of evolutionary change, separate from genetic drift, is **gene flow**, where a population may gain or lose alleles when fertile individuals move into or out of the population or when gametes (such as plant pollen) are transferred between populations. Gene flow tends to reduce differences between populations. For example, because humans today move more freely about the world than in the past, gene flow has become an important agent of evolutionary change in previously isolated human populations (**Figure 13.26**). ☑

▲ **Figure 13.26 Human gene flow.** Human migration is transferring alleles between populations that were once isolated. This magazine cover celebrates our changing gene pools and culture with a computer-generated image blending facial features from several races.

Natural Selection: A Closer Look

Genetic drift, gene flow, and even mutation can cause evolution, but they do not necessarily lead to adaptation. In fact, random drift, migrant alleles, and shot-in-the-dark mutations are unlikely to improve a population's fit to its environment. Of all causes of evolution, only natural selection promotes adaptation. And such evolutionary adaptation, remember, is a blend of chance and sorting—chance in the random generation of genetic variability, and sorting in the unequal reproductive success among the varying individuals. Darwin explained the basics of natural selection. But it took the modern synthesis to fill in the details.

Hardy-Weinberg equilibrium, which defines a nonevolving population, demands that all individuals in a population be equal in their ability to survive and reproduce. This condition is probably never actually met. On average, individuals that function best in the environment leave the most offspring and therefore have a disproportionate impact on the gene pool. When farmers began spraying their fields with pesticides, resistant pests started outreproducing other members of the insect populations. This spraying increased the frequency of alleles for pesticide resistance in gene pools. As a result, evolution by natural selection occurred.

Darwinian Fitness

The phrases "struggle for existence" and "survival of the fittest" are misleading if we take them to mean direct competitive contests between individuals. There *are* animal species in which individuals lock horns or otherwise fight one another to determine mating privilege. But

Africa

South
America

Tristan da Cunha

▲ Figure 13.27 **Darwinian fitness of some flowering plants depends in part on competition in attracting pollinators.**

reproductive success is generally more subtle and passive. Plants in a wildflower population, for example, may differ in reproductive success because some attract more pollinators—perhaps the result of slight differences in flower color, shape, or fragrance (**Figure 13.27**). A frog may produce more eggs than her neighbors because she is better at catching insects for food. These examples point to a biological definition of **fitness**: the contribution an individual makes to the gene pool of the next generation relative to the contributions of other individuals. Thus, the fittest individuals in the context of evolution are those that produce the largest numbers of viable, fertile offspring and thus pass on the most genes to the next generation. ☑

Three General Outcomes of Natural Selection

Imagine a population of mice with individuals ranging in fur color from very light to very dark gray. If we graph the number of mice in each color category, we get a bell-shaped curve, like the curve some of your instructors draw after grading an exam. If natural selection favors certain fur-color phenotypes over others, the population of mice will change over the generations. Three general outcomes are possible, depending on which phenotypes are favored. These three modes of natural selection are called directional selection, disruptive selection, and stabilizing selection.

Directional selection shifts the overall makeup of a population by selecting in favor of one extreme phenotype—the darkest mice, for example (**Figure 13.28a**). Directional selection

is most common when the local environment changes or when organisms migrate to a new environment. An actual example is the shift of insect populations toward a greater frequency of pesticide-resistant individuals.

Disruptive selection can lead to a balance between two or more contrasting phenotypes in a population (**Figure 13.28b**). A patchy environment, which favors different phenotypes in different patches, is one situation associated with disruptive selection. The variations seen in the snake population in Figure 13.18 result from disruptive selection.

Stabilizing selection favors intermediate phenotypes (**Figure 13.28c**). It typically occurs in relatively stable environments, where conditions tend to reduce physical variation. This evolutionary conservatism works by selecting against the more extreme phenotypes. For example, stabilizing selection keeps the majority of human birth weights between 3 and 4 kg (approximately 6.5 to 9 pounds). For babies much lighter or heavier than this, infant mortality is greater.

◀ Figure 13.28 **Three possible outcomes for selection working on fur color in imaginary populations of mice.** The large downward arrows symbolize the pressure of natural selection working against certain phenotypes.

(a) **Directional selection** shifts the overall makeup of the population by favoring variants at one extreme. In this case, the trend is toward darker color, perhaps because the landscape has been shaded by the growth of trees, making darker mice less noticeable to predators.

(b) **Disruptive selection** favors variants at opposite extremes over intermediate individuals. Here, the relative frequencies of very light and very dark mice have increased. Perhaps the mice have colonized a patchy habitat where a background of light soil is studded with dark rocks.

(c) **Stabilizing selection** culls extreme variants from the population, in this case eliminating individuals that are unusually light or dark. The trend is toward reduced phenotypic variation and maintenance of the status quo.

Of the three selection modes, stabilizing selection prevails most of the time, resisting change in well-adapted populations. Evolutionary spurts occur when a population is stressed by a change in the environment or by migration to a new place. When challenged with a new set of environmental problems, a population either adapts through natural selection or dies off in that locale. The fossil record tells us that extinction is the most common result. Those populations that do survive crises may change enough to be designated new species, as we will see in Chapter 14. ☑

Sexual Selection

Darwin was the first to explore the implications of **sexual selection**, a form of natural selection in which individuals with certain traits are more likely than other individuals to obtain mates. The males and females of an animal species obviously have different reproductive organs. But they may also have secondary sexual traits, noticeable differences not directly associated with reproduction or survival. This distinction in appearance, called **sexual dimorphism**, is often manifested in a size difference. Among male vertebrates, sexual dimorphism may also be evident in adornment, such as manes on lions, antlers on deer, or colorful plumage on peacocks and other birds (**Figure 13.29a**).

In some species, secondary sex structures may be used to compete with members of the same sex (usually males) for mates. Contests may involve physical combat, but are more often ritualized displays (**Figure 13.29b**). Such selection is common in species where the winning individual acquires a harem of mates (an obvious boost to that male's evolutionary fitness).

In a more common type of sexual selection, individuals of one sex (usually females) are choosy in selecting their mates. Males with the largest or most colorful adornments are often the most attractive to females. The extraordinary feathers of a peacock's tail are an example of this sort of "choose me!" statement. Every time a female chooses a mate based on a certain appearance or behavior, she perpetuates the alleles that caused her to make that choice and allows a male with that particular phenotype to perpetuate his alleles.

What is the advantage to females of being choosy? One hypothesis is that females prefer male traits that are correlated with "good" genes. In several bird species, research has shown that traits preferred by females, such as bright beaks or long tails, are related to overall male health. This link—between genes and health—brings us to this chapter's final example of evolution in action.

▼ Figure 13.29 **Sexual dimorphism.**

(a) Sexual dimorphism in a finch species. Among vertebrates, including this pair of green-winged Pytilia (native to Africa), males (right) are usually the showier sex.

(b) Competing for mates. Male Spanish ibex engage in nonlethal combat for the right to mate with females.

The Genetics of the Sickle-Cell Allele

Throughout this book, we emphasize evolution as biology's unifying theme. Within this chapter, we have highlighted examples of evolution in action, observable instances of evolution by natural selection in our world today. Let's continue that thread by considering one illness in an evolutionary context: sickle-cell disease.

About 1 out of every 400 African-Americans has sickle-cell disease, a genetic disorder in which oxygen delivery by the blood is impaired due to abnormally shaped red blood cells. Red blood cell sickling causes periodic painful episodes as well as some potentially life-threatening complications, such as stroke. Sickle-cell disease is caused by a recessive allele. Only homozygous individuals, who inherit the recessive allele from both parents, have the disorder. About one in ten African-Americans has a single copy of the sickle-cell allele (as do a very small number from other ethnic groups). These heterozygous individuals do not have sickle-cell disease but can pass the allele for the disorder on to their children.

Why is the sickle-cell allele so much more common in African-Americans than in the general U.S. population? And how can we explain such a high frequency among African-Americans for an allele with the potential to shorten life (and hence reduce reproductive success)? Evolutionary biology holds the answers.

In the African tropics, the sickle-cell allele is both boon and bane. It is true that when two copies are inherited, the allele causes a life-threatening disease. But heterozygous individuals, who have just one copy of the sickle-cell allele, are relatively resistant to the symptoms of malaria. This is an important advantage in tropical regions where malaria, caused by a parasitic microorganism, is a major cause of death. The frequency of the sickle-cell allele in Africa is generally highest in areas where the malaria parasite is most common (**Figure 13.30**). The representation of the allele among black Americans is thus a vestige of African roots.

This example of the intersection of evolutionary biology and health science is a reminder that biology is the foundation of all medicine. And evolution is the foundation of all biology.

▼ Figure 13.30 **Mapping malaria and the sickle-cell allele.** The inset shows sickled red blood cells.

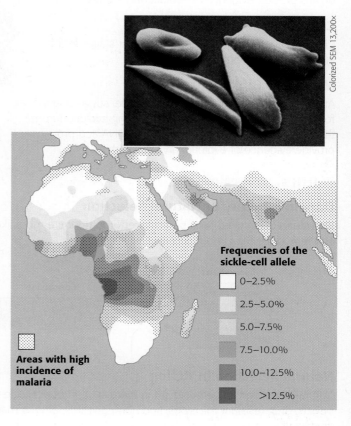

Colorized SEM 13,200×

Frequencies of the sickle-cell allele
- 0–2.5%
- 2.5–5.0%
- 5.0–7.5%
- 7.5–10.0%
- 10.0–12.5%
- >12.5%

Areas with high incidence of malaria

Chapter Review

SUMMARY OF KEY CONCEPTS

(MB) Go to the Study Area at **www.masteringbiology.com** for practice quizzes, eText, BioFlix™ 3-D animations, MP3 Tutor Sessions, videos, current events, and more.

Charles Darwin and *The Origin of Species*

Charles Darwin established the ideas of evolution and natural selection in his 1859 publication *On the Origin of Species by Means of Natural Selection*.

Darwin's Cultural and Scientific Context

During his around-the-world voyage on the *Beagle*, Darwin observed adaptations of organisms that inhabited diverse environments. In particular, Darwin was struck by the geographic distribution of organisms on the Galápagos Islands, off the South American coast. When Darwin considered his observations in light of new evidence for a very old Earth that changed slowly, he arrived at ideas that were at odds with the long-held notion of a young Earth populated by unrelated and unchanging species.

Descent with Modification

Darwin made two proposals in *The Origin of Species*:
(1) Existing species descended from ancestral species, and
(2) natural selection is the mechanism of evolution.

Evidence of Evolution

The Fossil Record

The fossil record shows that organisms have appeared in a historical sequence, and many fossils link ancestral species with those living today.

Biogeography

Biogeography, the study of the geographic distribution of species, suggests that species evolved from ancestors that inhabited the same region.

Comparative Anatomy

Homologous structures among species and vestigial organs provide evidence of evolutionary history.

Comparative Embryology

Closely related species often have similar stages in their embryonic development.

Molecular Biology

All species share a common genetic code, suggesting that all forms of life are related through branching evolution from the earliest organisms. Comparisons of DNA and proteins provide evidence of evolutionary relationships.

Natural Selection

Darwin's Theory of Natural Selection

Individuals best suited for a particular environment are more likely to survive and reproduce than less fit individuals.

Natural Selection in Action

Natural selection can be observed in the evolution of pesticide-resistant insects, drug-resistant microbes, and horned lizards, among many other organisms.

Evolutionary Trees

An evolutionary tree represents a succession of related species, with the most recent at the tips of the branches. Each branch point represents a common ancestor of all species that radiate from it.

The Modern Synthesis: Darwinism Meets Genetics

The modern synthesis fused genetics (Mendelism) and evolutionary biology (Darwinism) in the mid-1900s.

Populations as the Units of Evolution

A population, members of the same species living in the same time and place, is the smallest biological unit that can evolve. Population genetics emphasizes the extensive genetic variation within populations and tracks the genetic makeup of populations over time.

Genetic Variation in Populations

Polygenic ("many genes") inheritance produces traits that vary continuously, whereas traits determined by one genetic locus may be present in two or more distinct forms. Mutation and sexual reproduction produce genetic variation. Individual mutations have little short-term effect on a large gene pool, but in the long term, mutation is the source of genetic variation.

Analyzing Gene Pools

A gene pool consists of all the alleles in all the individuals making up a population. The Hardy-Weinberg formula can be used to calculate the frequencies of genotypes in a gene pool from the frequencies of alleles, and vice versa:

Population Genetics and Health Science

The Hardy-Weinberg formula can be used to estimate the frequency of a harmful allele, which is useful information for public health programs dealing with genetic diseases.

Microevolution as Change in a Gene Pool

Microevolution is generation-to-generation change in a population's frequencies of alleles.

Mechanisms of Evolution

Genetic Drift

Genetic drift is a change in the gene pool of a small population due to chance. Bottlenecking (a drastic reduction in population size) and the founder effect (occurring in a new population started by a few individuals) are two situations leading to genetic drift.

Gene Flow

A population may gain or lose alleles by gene flow, which is genetic exchange with another population.

Natural Selection: A Closer Look

Of all causes of evolution, only natural selection promotes evolutionary adaptations. Darwinian fitness is the contribution an individual makes to the gene pool of the next generation relative to the contributions of other individuals. The outcome of natural selection may be directional, disruptive, or stabilizing.

Sexual Selection

Secondary sexual traits (such as sex-specific plumage or behaviors) can promote sexual selection, a type of natural selection where inherited traits determine mating preferences.

SELF-QUIZ

1. Which of the following is *not* an observation or inference on which Darwin's theory of natural selection is based?
   a. There is heritable variation among individuals.
   b. Poorly adapted individuals never produce offspring.
   c. Because excessive numbers of offspring are produced, there is a competition for limited resources.
   d. Individuals whose inherited traits best fit them to the environment will generally produce more offspring.

2. Which of the following is a true statement about Charles Darwin?
   a. He was the first to discover that living things can change, or evolve.
   b. He based his theory on the inheritance of acquired traits.
   c. He proposed natural selection as the mechanism of evolution.
   d. He was the first to realize that Earth is more than 6,000 years old.

3. What is gradualism? How did Darwin apply that idea to the evolution of life?

4. In a population with two alleles for a particular genetic locus, B and b, the allele frequency of B is 0.7. If this population is in Hardy-Weinberg equilibrium, the frequency of heterozygotes is _____, the frequency of homozygous dominants is _____, and the frequency of homozygous recessives is _____.

5. Define fitness from an evolutionary perspective.

6. The processes of _____ and _____ generate variation, and _____ produces adaptation to the environment.
   a. sexual reproduction . . . natural selection . . . mutation
   b. mutation . . . sexual reproduction . . . genetic drift
   c. genetic drift . . . mutation . . . sexual reproduction
   d. mutation . . . natural selection . . . sexual reproduction
   e. mutation . . . sexual reproduction . . . natural selection

7. As a mechanism of evolution, natural selection can be most closely equated with
   a. random mating.
   b. genetic drift.
   c. unequal reproductive success.
   d. gene flow.

8. Compare and contrast how bottlenecking and a founder event can lead to genetic drift.

9. In a particular bird species, individuals with average-sized wings survive severe storms more successfully than other birds in the same population with longer or shorter wings. Of the three general outcomes of natural selection (directional, disruptive, or stabilizing), this example illustrates _____.

10. Which of the following statements is (are) true about a population in Hardy-Weinberg equilibrium? (More than one may be true.)
    a. The population is quite small.
    b. The population is not evolving.
    c. Gene flow between the population and surrounding populations does not occur.
    d. Natural selection is not occurring.

11. What environmental factor accounts for the relatively high frequency of the sickle-cell allele in tropical Africa?

Answers to the Self-Quiz questions can be found in Appendix D.

THE PROCESS OF SCIENCE

12. A population of snails has recently become established in a new region. The snails are preyed on by birds that break the snails open on rocks, eat the soft bodies, and leave the shells. The snails occur in both striped and unstriped forms. In one area, researchers counted both live snails and broken shells. Their data are summarized here:

	Striped shells	Unstriped shells
Number of live snails	264	296
Number of broken snail shells	486	377
Total	750	673

Based on these data, which snail form is more subject to predation by birds? Predict how the frequencies of striped and unstriped individuals might change over the generations.

13. Imagine that the presence or absence of stripes on the snails from the previous question is determined by a single gene locus, with the dominant allele (S) producing striped snails and the recessive allele (s) producing unstriped snails. Combining the data from both the living snails and broken shells, calculate the following: the frequency of the dominant allele, the frequency of the recessive allele, and the number of heterozygotes in the observed groups.

BIOLOGY AND SOCIETY

14. To what extent are humans in a technological society exempt from natural selection? Explain your answer.

14

How Biological Diversity Evolves

An endangered species.
These mountain gorillas, photographed in Rwanda, are just one of many species threatened with extinction.

CHAPTER CONTENTS

BIOLOGY AND SOCIETY: Mass Extinctions

The Sixth Mass Extinction

The fossil record reveals that the evolutionary history of life on Earth has been episodic, with long, relatively stable periods punctuated by brief, cataclysmic ones. During these upheavals, new species formed and others died out in great numbers.

Extinctions are inevitable in a changing world, but the fossil record reveals a few instances of great change, times when the majority of life on Earth—between 50% and 90% of living species—suddenly died out, disappearing forever. Scientists have documented five such mass extinctions during the last 500 million years. We'll discuss the causes and effects of these ancient mass extinctions later in this chapter.

Today, many biologists believe that we are witnessing a sixth mass extinction. Human activities are modifying the global environment to such an extent that many species are currently threatened with extinction. In the past 400 years—a very short time on a geologic scale—more than a thousand species are known to have become extinct. Scientists estimate that this is 100 to 1,000 times the extinction rate seen in most of the fossil record.

Do these events represent the beginning of a mass extinction? This question is difficult to answer, partly because it is hard to document both the total number of species on Earth (estimates range from 10 million to 200 million) and the number of extinctions. Although losses have not reached the levels of the other "big five" extinctions, the populations of many species—such as the mountain gorillas pictured on the facing page—are declining at an alarming rate, suggesting that a sixth (human-caused) mass extinction could occur within the next few centuries. And, as with prior mass extinctions, life on Earth may take millions of years to recover.

But the fossil record also shows a creative side to the destruction. Mass extinctions can pave the way for the evolution of many diverse species from a common ancestor. Accordingly, we'll begin this chapter by discussing the birth of new species and then examine how biologists trace the evolution of biological diversity and classify living organisms.

Macroevolution and the Diversity of Life

When Darwin traveled to the Galápagos Islands, he realized that he was visiting a place of origins. Though the volcanic islands were geologically young, they were already home to many plants and animals known nowhere else in the world. Among these unique inhabitants were giant tortoises, for which the Galápagos are named (*galápago* means "tortoise" in Spanish). After visiting the islands, Darwin wrote in his diary: "Both in space and time, we seem to be brought somewhat near to that great fact—that mystery of mysteries—the first appearance of new beings on this Earth."

To understand the "appearance of new beings," it is not enough to explain microevolution, the generation-to-generation change of allele frequencies within a population. If that were all that ever happened, then Earth would be populated only by a highly adapted version of one species. The evolutionary mechanisms you read about in Chapter 13 must also explain **macroevolution**, the major changes in the history of life, which are usually evident in the fossil record. Macroevolution includes the origin of new species, which generates biological diversity; the origin of evolutionary novelty, such as the wings and feathers of birds and the large brains of humans; the explosive diversification following some evolutionary breakthroughs, such as the origin of thousands of plant species after the flower evolved; and mass extinctions, which clear the way for new adaptive explosions, such as the diversification of mammals following the disappearance of most of the dinosaurs.

The discussion of macroevolution in this chapter focuses on the origin of species. Sometimes a whole population may change significantly through adaptation to a changing environment. Such linear, nonbranching evolution does not create a new species (**Figure 14.1**, left). The formation of new species, called **speciation**, occurs when one species evolves into two or more species (see Figure 14.1, right).

Each time speciation occurs, the diversity of life increases. Beginning 3.6 billion years ago, an ancestral species gave rise to two or more species, which then gave rise to new lineages, which in turn branched again and again, populating Earth with the millions of species that have ever existed and that exist today. A new species often closely resembles the species that gave rise to it. Occasionally, however, a new species has properties novel enough to define a major branch in the tree of life, such as the flowers of flowering plants. But how is a species recognized and defined? We'll explore that question next. ✔

► Figure 14.1
Two patterns of evolution.

PATTERNS OF EVOLUTION	
Nonbranching Evolution (no new species)	**Branching Evolution** (results in speciation)

Nonbranching evolution can transform a population significantly but does not create a new species.

Branching evolution splits a lineage into two or more species, thereby increasing the total number of species.

The Origin of Species

Species is a Latin word meaning "kind" or "appearance." Indeed, we learn to distinguish between the kinds of plants and animals—between dogs and cats, for example, or between roses and dandelions—from differences in their appearance. Although the basic idea of species as distinct life-forms seems intuitive, devising a more formal definition is not so easy.

What Is a Species?

One way of defining a species (and the main definition used in this book) is the **biological species concept**. It defines a **species** as a group of populations whose members have the potential to interbreed with one another in nature to produce fertile offspring (offspring that can reproduce) (**Figure 14.2**). Geography and culture may conspire to keep a Manhattan businesswoman and a Mongolian dairyman apart. But if the two did meet and mate, they could have viable babies that develop into fertile adults because all humans belong to the same species. In contrast, humans and chimpanzees, despite having a shared evolutionary history, remain distinct species because they do not interbreed.

We cannot apply the biological species concept to all situations. For example, basing the definition of species on reproductive compatibility excludes organisms that only reproduce asexually (producing offspring from a single parent). Fossils aren't doing any sexual reproduction either, so they cannot be evaluated by the biological species concept.

In response to such challenges, biologists have developed several other ways to define species. For example, most of the 1.8 million species named to date have been classified based on measurable physical traits. Another approach identifies species in terms of ecological niches, focusing on unique adaptations to particular roles in a biological community. Yet another classification scheme defines a species as the smallest group of individuals sharing a common ancestor and forming one branch on the tree of life.

Each species concept is useful, depending on the situation and the questions being asked. The biological species concept, however, is particularly useful when focusing on how discrete groups of organisms may arise and be maintained by reproductive isolation. Next, we'll look at reproductive isolation more closely because it is such an essential factor in the evolution of so many species. ☑

Speciation

☑ **CHECKPOINT**

By defining a species by its reproductive _____ from other populations, the biological species concept can only be applied to organisms that reproduce _____.

Answer: isolation; sexually

▼ Figure 14.2 **The biological species concept is based on reproductive compatibility rather than physical similarity.**

Similarity between different species. The eastern meadowlark (left) and the western meadowlark (right) are very similar in appearance, but they are separate species and do not interbreed.

Diversity within one species. Humans, as diverse in appearance as we are, belong to a single species (*Homo sapiens*) and can interbreed.

Reproductive Barriers between Species

Clearly, a fly will not mate with a frog or a fern. But what prevents closely related species from interbreeding? What, for example, maintains the species boundary between the eastern meadowlark and the western meadowlark (shown in Figure 14.2)? Their geographic ranges overlap in the Great Plains region, and they are so similar that only expert birders can tell them apart. And yet, these two bird species do not interbreed.

A **reproductive barrier** is anything that prevents individuals of closely related species from interbreeding. Let's examine the different kinds of reproductive barriers that isolate the gene pools of species (**Figure 14.3**). We can classify reproductive barriers as either prezygotic or postzygotic, depending on whether they block interbreeding before or after the formation of zygotes (fertilized eggs).

Prezygotic barriers prevent mating or fertilization between species (**Figure 14.4**). The barrier may be time based (temporal isolation). For example, western spotted skunks breed in the fall, but the eastern species breeds in late winter. Temporal isolation keeps the species from mating even where they coexist on the Great Plains. In other cases, species live in the same region but not in the same habitats (habitat isolation). For example, one species of North American garter snake lives mainly in water, while a closely related species lives on land. Traits that enable individuals to recognize potential mates, such as a particular odor, coloration, or courtship ritual, can also function as reproductive barriers (behavioral isolation). In many bird species, for example, courtship

Figure 14.3 Reproductive barriers between closely related species.

INDIVIDUALS OF DIFFERENT SPECIES

Prezygotic Barriers

Temporal isolation: Mating or fertilization occurs at different seasons or times of day.

Habitat isolation: Populations live in different habitats and do not meet.

Behavioral isolation: Little or no sexual attraction exists between populations.

MATING ATTEMPT

Mechanical isolation: Structural differences prevent fertilization.

Gametic isolation: Female and male gametes fail to unite in fertilization.

FERTILIZATION (ZYGOTE FORMS)

Postzygotic Barriers

Reduced hybrid viability: Hybrid zygotes fail to develop or fail to reach sexual maturity.

Reduced hybrid fertility: Hybrids fail to produce functional gametes.

Hybrid breakdown: Hybrids are feeble or sterile.

VIABLE, FERTILE OFFSPRING

► Figure 14.4 **Prezygotic barriers.** Prezygotic barriers prevent mating or fertilization.

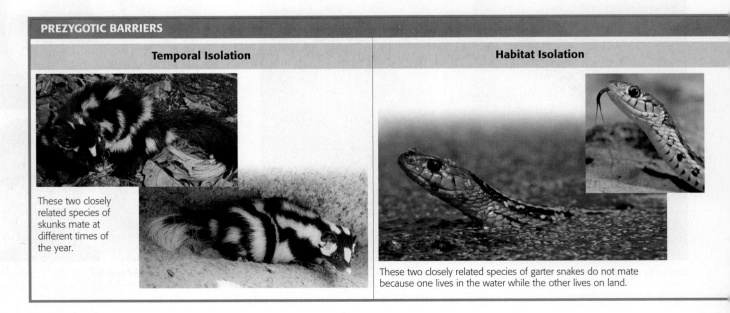

PREZYGOTIC BARRIERS

Temporal Isolation

These two closely related species of skunks mate at different times of the year.

Habitat Isolation

These two closely related species of garter snakes do not mate because one lives in the water while the other lives on land.

behavior is so elaborate that individuals are unlikely to mistake a bird of a different species as one of their kind. In still other cases, the egg-producing and sperm-producing structures of different species are anatomically incompatible (mechanical isolation). For example, the two closely related species of snails in Figure 14.4 cannot join their male and female sex organs because their shells spiral in opposite directions. In still other cases, gametes (eggs and sperm) of different species are incompatible, preventing fertilization (gametic isolation). Gametic isolation is very important when fertilization is external. Male and female sea urchins of many species release eggs and sperm into the sea, but fertilization occurs only if species-specific molecules on the surface of egg and sperm attach to each other.

Postzygotic barriers operate if interspecies mating actually occurs and results in hybrid zygotes (Figure 14.5). (In this context, "hybrid" means that the egg comes from one species and the sperm from another species.) In some cases, hybrid offspring die before reaching reproductive maturity (reduced hybrid viability). For example, although certain closely related salamander species will hybridize, the offspring fail to develop normally because of genetic incompatibilities between the two species. In other cases of hybridization, offspring may become vigorous adults, but are infertile (reduced hybrid fertility). A mule, for example, is the hybrid offspring of a female horse and a male donkey. Because mules are sterile, there is no avenue for gene transfer between the two parental species, horse and donkey. In other cases, the first-generation hybrids are viable and fertile, but when these hybrids mate with one another or with either parent species, the offspring are feeble or

sterile (hybrid breakdown). For example, different species of cotton plants can produce fertile hybrids, but the offspring of the hybrids do not survive.

In summary, reproductive barriers form the boundaries around closely related species. In most cases, it is not a single reproductive barrier but some combination of two or more that keeps species isolated. Next, we examine situations that make reproductive isolation and speciation possible. ☑

▼ Figure 14.5 **Postzygotic barriers.** Postzygotic barriers prevent development of fertile adults.

POSTZYGOTIC BARRIERS

Reduced Hybrid Viability

Some species of salamander can hybridize, but their offspring do not develop fully or, like this one, are frail.

Reduced Hybrid Fertility

Horse

Donkey

Mule

The hybrid offspring of a donkey and a horse is a mule, which is sterile.

Hybrid Breakdown

The rice hybrids at the left and right are fertile, but plants of the next generation (middle) are small and sterile.

Behavioral Isolation

Galápagos blue-footed boobies mate only after a specific ritual of high stepping that advertises a mate's bright blue feet.

Mechanical Isolation

Because these snails' shells spiral in opposite directions, their genital openings (indicated by arrows) cannot be aligned and mating cannot occur.

Gametic Isolation

Gametes of these red and purple urchins are unable to fuse because proteins on the surface of the eggs and sperm cannot bind to one another.

273

Mechanisms of Speciation

A key event in the origin of many species occurs when a population is somehow severed from other populations of the parent species. With its gene pool isolated, the splinter population can follow its own evolutionary course. Changes in its allele frequencies caused by genetic drift and natural selection will not be diluted by alleles entering from other populations (gene flow). Such reproductive isolation can result from two general scenarios: allopatric ("different country") speciation and sympatric ("same country") speciation (**Figure 14.6**). In **allopatric speciation**, the initial block to gene flow is a geographic barrier that physically isolates the splinter population. In contrast, **sympatric speciation** is the origin of a new species without geographic isolation. The splinter population becomes reproductively isolated even though it is right in the midst of the parent population.

Allopatric Speciation

Several geologic processes can fragment a population into two or more isolated populations. A mountain range may emerge and gradually split a population of organisms that can inhabit only lowlands. A land bridge, such as the Isthmus of Panama, may form and separate the marine life on either side. A large lake may subside and form several smaller lakes, with their populations now isolated. Allopatric speciation can also occur if individuals colonize a new, geographically remote area and become isolated from the parent population. An example is the speciation that occurred on the Galápagos Islands following initial colonization by immigrant organisms.

How formidable must a geographic barrier be to keep allopatric populations apart? The answer depends partly on the ability of the organisms to move about. Birds, mountain lions, and coyotes can cross mountain ranges, rivers, and canyons. Nor do such barriers hinder the windblown pollen of pine trees or the spread of seeds carried by animals. In contrast, small rodents may find a deep canyon or a wide river an impassable barrier (**Figure 14.7**).

Allopatric speciation is more common for a small, isolated population because it is more likely than a large population to have its gene pool changed substantially by both genetic drift and natural selection. For example, in less than 2 million years, the few foreign animals and plants that first colonized the Galápagos Islands gave rise to all the species now found there. But for each

▼ Figure 14.6 **Two modes of speciation.**

Allopatric speciation occurs when a population forms a new species after being geographically isolated from its parent population.

Sympatric speciation occurs when part of a population becomes a new species while in the midst of its parent population.

Ammospermophilus harrisii

Ammospermophilus leucurus

▲ **Figure 14.7 Allopatric speciation of antelope squirrels on opposite rims of the Grand Canyon.** Harris's antelope squirrel (*Ammospermophilus harrisii*) is found on the south rim of the Grand Canyon. Just a few miles away on the north rim is the closely related white-tailed antelope squirrel (*Ammospermophilus leucurus*). Birds and other organisms that can disperse easily across the canyon have not diverged into different species on opposite rims.

small, isolated population that becomes a new species, many more simply perish in their new environment. Life on the frontier is harsh, and most pioneer populations become extinct.

Even if a small, isolated population survives, it does not necessarily evolve into a new species. The population may adapt to its local environment and begin to look very different from the ancestral population, but

that doesn't necessarily make it a new species. Speciation occurs with the evolution of reproductive barriers between the isolated population and its parent population. In other words, if speciation occurs during geographic separation, the new species cannot breed with its ancestral population, even if the two populations should come back into contact at some later time (**Figure 14.8**). ☑

► Figure 14.8 **Has speciation occurred during geographic isolation?** In this diagram, the arrows track populations over time. The mountain symbolizes a period of geographic isolation. The two parts of the figure show the two possibilities when populations come back together after a long period of separation.

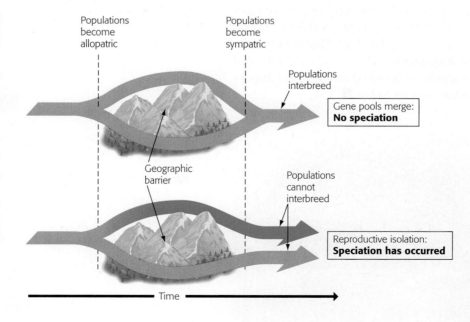

Populations become allopatric

Populations become sympatric

Populations interbreed

Gene pools merge: **No speciation**

Geographic barrier

Populations cannot interbreed

Reproductive isolation: **Speciation has occurred**

Time

Sympatric Speciation

In sympatric speciation, a new species emerges from an old one, even though they live in the same time and place. How can a subpopulation become reproductively isolated while in the midst of its parent population? In some cases, subgroups of a population evolve adaptations for exploiting food sources in different habitats (such as the shallow versus deep habitats of a lake). In other cases, a type of sexual selection in which females choose mates based on color can also contribute to rapid reproductive isolation. But the most frequently observed mechanism of sympatric speciation involves large-scale genetic changes that occur in a single generation.

Biologists have documented many examples of plant species that originated from accidents during cell division. Such accidents can produce organisms with extra sets of chromosomes. A new species that evolves this way has polyploid cells, meaning that each cell has more than two sets of chromosomes. The new species cannot produce fertile hybrids with its parent species. This is a good example of sympatric speciation because reproductive isolation has occurred in a single generation, without geographic isolation.

Polyploids do not always come from a single parent species. In fact, most polyploid species arise from the hybridization of two parent species. This mechanism of sympatric speciation accounts for many of the plant species we grow for food, including oats, potatoes, bananas, peanuts, barley, plums, apples, sugarcane, coffee, and wheat. Wheat makes a good case study. The most widely cultivated plant in the world, what we call wheat is actually represented by 20 species. Humans began domesticating wheat from wild grasses at least 11,000 years ago in the Middle East. **Figure 14.9** traces the evolutionary path from that first cultivated wheat species to bread wheat, the most important wheat species today. ☑

☑CHECKPOINT

Each speciation episode in the evolution of bread wheat (see Figure 14.9) is an example of _____ speciation, which is the origin of a new species without geographic isolation from the parent species.

Answer: sympatric

▼ Figure 14.9 **The evolution of wheat.** The uppercase letters in this diagram represent not alleles but sets of chromosomes that can be traced in the evolution of wheat, or *Triticum*. Interestingly, the AABB hybrid, emmer wheat (*Triticum turgidum*), is still grown widely in Eurasia and western North America and is used mainly for making pasta.

Wild *Triticum*
(14 chromosomes)

Domesticated
Triticum monococcum
(14 chromosomes)

AA × BB

1 Hybridization occurred between a domesticated wheat species and a wild relative that probably grew as a weed at the edges of cultivated fields.

AB

Sterile hybrid
(14 chromosomes)

2 Chromosome sets A and B would not have been able to pair at meiosis, making the AB hybrid sterile. However, an error in cell division, followed by self-fertilization, produced a new species.

AA BB × DD

T. turgidum
Emmer wheat
(28 chromosomes)

Wild
T. tauschii
(14 chromosomes)

3 Over 8,000 years ago, cultivated emmer wheat hybridized with another closely related wild species.

ABD

Sterile hybrid
(21 chromosomes)

4 This sterile hybrid gave rise to bread wheat, with two each of the three ancestral sets of chromosomes.

AA BB DD

T. aestivum
Bread wheat
(42 chromosomes)

What Is the Tempo of Speciation?

Biologists continue to make field observations and devise experiments to study evolution in progress. However, much of the evidence of evolution comes from the fossil record, the chronicle of extinct organisms engraved in layers of rock over millions of years of geologic time. What does this record say about the process of speciation?

Many fossil species appear suddenly in a layer of rock and persist essentially unchanged through several layers (strata) until disappearing as suddenly as they appeared. The term **punctuated equilibria** describes these long periods of little change, or equilibrium, punctuated by abrupt episodes of speciation.

Figure 14.10 shows two models for the tempo of speciation: the punctuated model and the graduated model. In the punctuated model, there are no transitional stages in the lineages. The new species change little, if at all, once they appear. In the graduated model, differences evolve little by little in populations as the species become adapted to their local environments, and new species (represented by the two butterflies at the far right) evolve gradually from the ancestral population. According to the graduated model, big changes (speciations) occur by the steady accumulation of many small changes.

What do punctuated and graduated patterns tell us about how long it takes new species to form? Suppose that a species survived for 5 million years, but most of the changes in its body features occurred during the first 50,000 years of its existence, just 1% of its overall history. Time periods this short often cannot be distinguished in fossil strata. Thus, based on its fossils, the species would seem to have appeared suddenly and continued with little change before becoming extinct. Even though such a species may have originated more slowly than its fossils suggest, a punctuated pattern indicates that speciation occurred relatively rapidly. For species whose fossils show much more gradual change, we cannot tell exactly when a new biological species formed. It is likely that speciation in such groups occurred relatively slowly, perhaps taking millions of years.

How often do species evolve abruptly and then remain essentially unchanged for most of their existence? Rapid speciation is certainly known to occur. As we saw earlier, abrupt speciation can occur by polyploidy in plants and even occasionally in a few animals. And genetic drift and natural selection can significantly alter the gene pool of a small, isolated population in just a few hundred generations.

Regardless of the tempo of speciation, as species diverge, differences accumulate and become more pronounced, eventually leading to the formation of new groups of organisms that differ greatly from their ancestors. Furthermore, as one group produces many new species, another group may lose species to extinction. The cumulative effects of multiple speciation and extinction events have helped to shape the dramatic changes documented in the fossil record. This type of macroevolution is the subject of the next section. ☑

▼ Figure 14.10 **Two models for the tempo of evolution.**

Punctuated model. A new species changes most as it first branches from a parent species. After this speciation episode, there is little change for the rest of the species' existence.

Time →

Graduated model. Species that are descended from a common ancestor diverge gradually in form as they acquire unique adaptations.

The Evolution of Biological Novelty

The two squirrels in Figure 14.7 are different species, but they are very similar animals that live very much the same way. When most people think of evolution, they envision much more dramatic transformation. How can we account for such evolutionary products as flight in birds and braininess in humans?

Adaptation of Old Structures for New Functions

Birds are derived from a lineage of earthbound reptiles (**Figure 14.11**). How could flying animals evolve from flightless ancestors? More generally, how do major novelties of biological structure and function evolve? One way evolutionary novelty arises is when structures that originally played one role gradually acquire a different role. Structures that evolve in one context but become adapted for another function are called exaptations. This term suggests that a structure can become adapted to alternative functions; it does not mean that a structure somehow evolves in anticipation of future use. New features can arise gradually via intermediate stages, each functioning in the organism's current context.

Exaptation can account for the evolution of novel structures. Consider the evolution of birds from a dinosaur ancestor. Birds have lightweight skeletons with honeycombed bones, a feature found in their dinosaur ancestors. Since the fossil record clearly indicates that light bones predated flight, they must have had some function on the ground. The ancestors of birds were small, agile, bipedal dinosaurs that would have benefited from a light frame. Moreover, a winglike form with feathers would have increased the surface area of the forelimbs. These enlarged forelimbs were adapted for flight after functioning in some other capacity, such as thermal regulation, courtship displays, or camouflage (roles that wings still play today). The first flights may have been only glides or extended hops as the animal pursued prey or fled from a predator. As flight evolved, animals with better-adapted feathers and wings would have had a selective advantage.

According to Harvard zoologist Karel Liem, "Evolution is like modifying a machine while it's running." The concept that biological novelties can evolve by the remodeling of old structures is in the Darwinian tradition of large changes being an accumulation of many small changes crafted by natural selection. ☑

☑ **CHECKPOINT**

Explain why the concept of exaptation does not imply that a structure evolves in anticipation of some future environmental change.

Answer: Although an exaptation has new or additional functions in a new environment, it existed earlier because it worked as an adaptation to the old environment.

► Figure 14.11 **An extinct bird.** Called *Archaeopteryx* ("ancient wing"), this animal lived near tropical lagoons in central Europe about 150 million years ago. Like birds today, it had flight feathers, but otherwise it was more like some small bipedal dinosaurs of its era. *Archaeopteryx* probably relied mainly on gliding from trees. Despite its feathers, *Archaeopteryx* is not considered an ancestor of today's birds. Instead, it probably represents an extinct side branch of the bird lineage.

Fossil

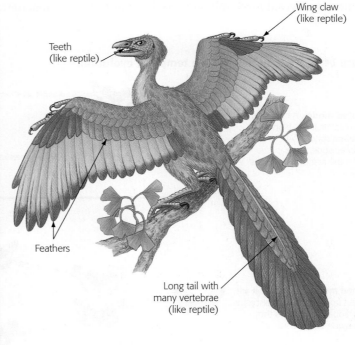

Teeth
(like reptile)

Wing claw
(like reptile)

Feathers

Long tail with
many vertebrae
(like reptile)

Artist's reconstruction

Evo-Devo: Development and Evolutionary Novelty

Gradual evolutionary remodeling, such as the accumulation of flight adaptations in birds, probably involves a large number of genetic changes in populations. In other cases of macroevolution, just a few genetic changes can become magnified into major structural differences between organisms. Such cases are of particular interest to scientists working in the field of **evo-devo**, or evolutionary developmental biology, which studies the evolution of developmental processes in multicellular organisms.

Genes that program development control the rate, timing, and spatial pattern of changes in an organism's form as it develops from a zygote into an adult. (Chapter 11 provides a closer look at how genes control development.) A subtle change in a species' developmental program can have profound effects. The same or very similar genes are involved in the development of form across multiple lineages. Changes in the number, nucleotide sequence, and regulation of these genes have led to the huge diversity in body forms. Many striking evolutionary transformations are the result of a change in the rate or timing of developmental events.

For example, the animal in **Figure 14.12**, a salamander called an axolotl, illustrates a phenomenon called **paedomorphosis**, which is the retention into adulthood of features that were solely juvenile in ancestral species. The axolotl grows to full size and reproduces without losing its external gills, a juvenile feature in most species of salamanders.

Paedomorphosis has also been important in human evolution. Humans and chimpanzees are even more alike in body form as fetuses than they are as adults. In the fetuses of both species, the skulls are rounded and the jaws are small, making the face rather flat (**Figure 14.13**). As development proceeds, uneven bone growth makes the chimpanzee skull sharply angular, with heavy browridges and massive jaws. The adult chimpanzee has much greater jaw strength than we have, and its teeth are proportionately larger. In contrast, the adult human has a skull with decidedly rounded, more fetus-like contours. Put another way, our skull is paedomorphic; it retains fetal features even after we are mature. Our large skull is one of our most distinctive features. Our large, complex brain, which fills that bulbous skull, is another. The human brain is proportionately larger than the chimpanzee brain because growth of the organ is switched off much later in human development. Compared with the brain of a chimpanzee, our brain continues to grow for several more years, which can be interpreted as the prolonging of a juvenile process.

Homeotic genes, the master control genes described in Chapter 11, determine such basic features as where a pair of wings or legs will develop on a fruit fly. Changes in such genes or in where such genes are expressed can profoundly affect body form. Consider, for example, the evolution of tetrapods (animals with backbones and four limbs) from fishes. The location within a developing limb where certain homeotic genes are expressed is initially the same in both fishes and tetrapods. A second region of expression in the developing tetrapod limb, however, produces the extra skeletal elements that develop into foot bones. Thus, changes in the expression of these genes appear to have led to the evolution of walking legs from the paired fins of fishes.

Duplications of homeotic genes probably facilitated the origin of new body shapes in animals, including the evolution of vertebrates (animals with backbones) from invertebrates (animals without backbones). For example, consider the homeotic genes of a fruit fly (an invertebrate) and a mouse (a vertebrate). A fruit fly has a single cluster of homeotic genes that direct the development of major body parts, whereas a mouse has four clusters of these genes. Two duplications of these gene clusters appear to have occurred in the evolution of vertebrates from invertebrates. Mutations in these duplicated genes may then have led to the origin of novel vertebrate characteristics. For instance, some genes may have taken on new roles, such as directing the development of a backbone, jaws, and limbs.

Gills

▲ **Figure 14.12 Paedomorphosis.** The axolotl, a salamander, becomes an adult (shown here) and reproduces while retaining certain tadpole characteristics, including gills.

▼ **Figure 14.13 Comparison of human and chimpanzee skull development.** Starting with fetal skulls that are very similar (left), the differential growth rates of the bones making up the skulls produce adult heads with very different proportions. The grid lines will help you relate the fetal skulls to the adult skulls.

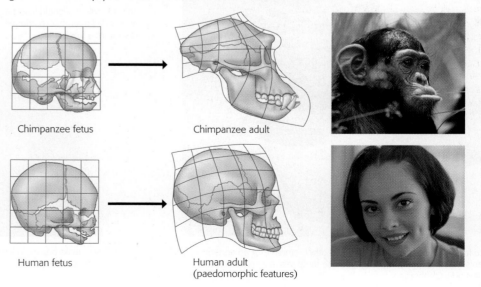

Chimpanzee fetus

Chimpanzee adult

Human fetus

Human adult (paedomorphic features)

Earth History and Macroevolution

Having examined how new species and evolutionary adaptations arise, we are ready to turn our attention to the history of biological diversity. Macroevolution is closely tied to the history of Earth.

Geologic Time and the Fossil Record

The fossil record, the sequence in which fossils appear in rock strata, is an archive of macroevolution. **Figure 14.14** surveys some of the diverse ways that organisms can fossilize. Sedimentary rocks are the richest sources

of fossils and provide a record of life on Earth in their strata (see Chapter 13). The fossils in each stratum of sedimentary rock are a local sample of the organisms that existed at the time the sediment was deposited.

By studying many sites over the past two centuries, geologists have established a **geologic time scale**, reflecting a consistent sequence of geologic periods (**Table 14.1**). The time line presented in Table 14.1 is separated into four broad divisions: the Precambrian (a general term for the time before about 540 million years ago), followed by the Paleozoic, Mesozoic, and Cenozoic eras. Each of these divisions represents a distinct age in the history of Earth and its life.

▼ Figure 14.14 **A gallery of fossils.**

Sedimentary fossils are formed when minerals seep into and replace organic matter. These petrified (stone) trees in Arizona's Petrified Forest National Park are about 190 million years old.

Sedimentary rocks are the richest hunting grounds for paleontologists, scientists who study fossils. This researcher is excavating a fossilized dinosaur skeleton from sandstone in Dinosaur National Monument, located in Utah and Colorado.

This 45-million-year-old insect is embedded in amber (hardened resin from a tree).

Trace fossils are footprints, burrows, and other remnants of an ancient organism's behavior. A dinosaur left these footprints in a creek bed in what is now Oklahoma.

These tusks belong to a whole 23,000-year-old mammoth, which scientists discovered in Siberian ice in 1999.

Table 14.1 The Geologic Time Scale

Geologic Time	Period	Epoch	Age (millions of years ago)	Some Important Events in the History of Life	Relative Time Span
Cenozoic era	Quaternary	Recent		Historical time	Cenozoic
			0.01		
		Pleistocene		Ice ages; humans appear	Mesozoic
			1.8		
	Tertiary	Pliocene		Origin of genus *Homo*	
			5		Paleozoic
		Miocene		Continued speciation of mammals and angiosperms	
			23		
		Oligocene		Origins of many primate groups, including apes	
			34		
		Eocene		Angiosperm dominance increases; origins of most living mammalian orders	
			56		
		Paleocene		Major speciation of mammals, birds, and pollinating insects	
			65		
Mesozoic era	Cretaceous			Flowering plants (angiosperms) appear; many groups of organisms, including most dinosaur lineages, become extinct at end of period (Cretaceous extinctions)	
			145		
	Jurassic			Gymnosperms continue as dominant plants; dinosaurs become dominant	
			200		
	Triassic			Cone-bearing plants (gymnosperms) dominate landscape; speciation of dinosaurs, early mammals, and birds	
			251		
Paleozoic era	Permian			Extinction of many marine and terrestrial organisms (Permian extinctions); speciation of reptiles; origins of mammal-like reptiles and most living orders of insects	
			299		
	Carboniferous			Extensive forests of vascular plants; first seed plants; origin of reptiles; amphibians become dominant	Pre-cambrian
			359		
	Devonian			Diversification of bony fishes; first amphibians and insects	
			416		
	Silurian			Early vascular plants dominate land	
			444		
	Ordovician			Marine algae are abundant; colonization of land by diverse fungi, plants, and animals	
			488		
	Cambrian			Origin of most living animal phyla (Cambrian explosion)	
			542		
Precambrian			600	Diverse algae and soft-bodied invertebrate animals appear	
			635	Oldest animal fossils	
			2,100	Oldest eukaryotic fossils	
			2,700	Oxygen begins accumulating in atmosphere	
			3,500	Oldest fossils known (prokaryotes)	
			4,600	Approximate time of origin of Earth	

The boundaries between eras are marked by mass extinctions, when many forms of life disappeared from the fossil record and were replaced by species that diversified from the survivors. For example, the beginning of the Cambrian period is delineated by a great diversity of fossilized animals that are absent in rocks of the late Precambrian. And most of the animals that lived during the late Precambrian became extinct at the end of that era.

Fossils are reliable chronological records only if we can determine their ages. The record of the rocks reveals only the *relative* ages of fossils and therefore the order in which groups of species evolved. However, the series of

sedimentary layers alone does not tell us the *absolute* ages of the embedded fossils. The rock strata are analogous to the layers of wallpaper you might peel from the walls of a very old house. You could determine the sequence in which the wallpapers had been applied, but not the year that each layer was added. Geologists use a variety of methods to determine the ages of rocks and the fossils they contain. The most common method is **radiometric dating** (Figure 14.15), which is based on the decay of radioactive isotopes (recall that isotopes are alternative forms of elements; see Table 2.1). The dates in the geologic time scale in Table 14.1 were established by radiometric dating. ☑

▼ Figure 14.15 **Radiometric dating.** Amounts of radioactive isotopes can be measured by the radiation they emit as they decompose to more stable atoms (see Chapter 2). Paleontologists use this clocklike decay to date fossils.

Radioactive decay of carbon-14.
Carbon-14 is a radioactive isotope with a half-life of 5,600 years. From the time an organism dies, it takes 5,600 years for half of the radioactive carbon-14 to decay; half of the remainder is present after another 5,600 years; and so on.

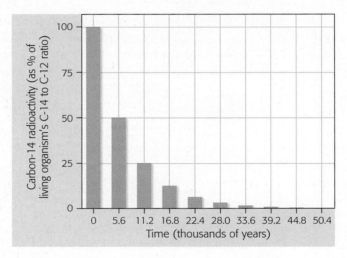

How carbon-14 dating is used to determine the vintage of a fossilized clam shell.

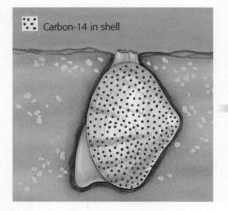

Carbon-14 in shell

❶ While an organism, in this case a clam, is alive, it assimilates the different isotopes of each element in proportions determined by their relative abundances in the environment. Carbon-14 is taken up in trace quantities, along with much larger quantities of the more common carbon-12.

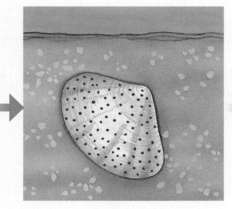

❷ After the clam dies, it is covered with sediment, and its shell eventually becomes consolidated into a layer of rock as the sediment is compressed. From the time the clam dies and ceases to assimilate carbon, the amount of carbon-14 relative to carbon-12 in the fossil declines due to radioactive decay.

❸ After the clam fossil is found, its age can be determined by measuring the ratio of the two isotopes to learn how many half-life reductions have occurred since it died. For example, if the ratio of carbon-14 to carbon-12 in this fossil clam was found to be 25% of that of a living organism, this fossil would be about 11,200 years old (see graph above).

Plate Tectonics and Macroevolution

The continents are not locked in place. They drift about Earth's surface like passengers on huge, irregularly shaped plates of crust floating on a flexible layer of hot, underlying material called the mantle. When two landmasses are embedded in different plates, their positions relative to each other change. For example, North America and Europe are presently drifting apart at a rate of about 2 cm per year. Many important geologic processes, including mountain building, volcanic activity, and earthquakes, occur at plate boundaries. California's infamous San Andreas Fault is at a border where two plates slide past each other (**Figure 14.16**).

Plate movements rearrange geography constantly, but two chapters in the continuing saga of continental drift had an especially strong influence on life. About 250 million years ago, near the end of the Paleozoic era, plate movements brought all the previously separated landmasses together into a supercontinent called Pangaea ("all land") (**Figure 14.17**). Imagine some of the possible effects on life. Species that had been evolving in isolation came together and competed. As the landmasses joined over millions of years, the total amount of shoreline was reduced. There is also evidence that the ocean basins increased in depth, lowering sea level and draining the shallow coastal seas. Then, as now, most marine species inhabited shallow waters, and the formation of Pangaea destroyed a considerable amount of that habitat. It was probably a long, traumatic period for terrestrial life as well. The continental interior, which had a drier and more erratic climate than coastal regions, increased in area substantially when the land came together. Changing ocean currents also undoubtedly affected land life as well as sea life. Thus,

the formation of Pangaea had a tremendous environmental impact that reshaped biological diversity by causing extinctions and providing new opportunities for the survivors, which diversified as a result of branching evolution.

The second dramatic chapter in the history of continental drift began about 180 million years ago, during the Mesozoic era. Pangaea started to break up, causing geographic isolation of colossal proportions. As the continents drifted apart, each became a separate evolutionary arena as climates changed and the organisms of the different biogeographic realms diverged.

The pattern of continental separations is the solution to many puzzles of biogeography. For example, paleontologists have discovered matching fossils of Mesozoic reptiles in Ghana (West Africa) and Brazil. These two parts of the world, now separated by 3,000 km of ocean, were contiguous during the early Mesozoic era. Plate tectonics also explains much about the current distribution of organisms. For example, the mammals of Australia contrast so sharply with those of the rest of the world because the plate on which Australia rests drifted away from other plates early in the evolutionary history of mammals, leaving marsupials free to evolve in isolation on that continent (see Figure 13.7). ☑

▶ **Figure 14.17 The history of plate tectonics.** The continents continue to drift, though not at a rate that's likely to cause any motion sickness for their passengers.

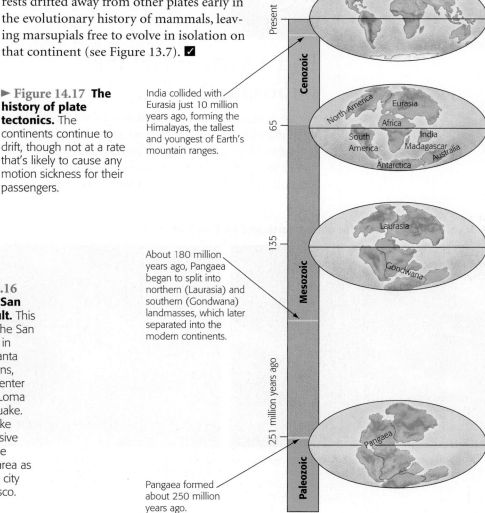

India collided with Eurasia just 10 million years ago, forming the Himalayas, the tallest and youngest of Earth's mountain ranges.

About 180 million years ago, Pangaea began to split into northern (Laurasia) and southern (Gondwana) landmasses, which later separated into the modern continents.

Pangaea formed about 250 million years ago.

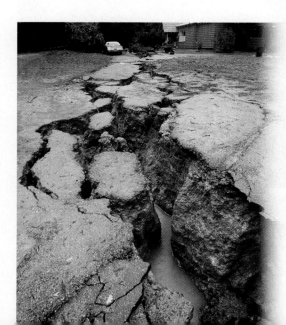

◀ **Figure 14.16 California's San Andreas fault.** This fissure is on the San Andreas fault in California's Santa Cruz Mountains, near the epicenter of the 1989 Loma Prieta earthquake. The earthquake caused extensive damage to the surrounding area as well as to the city of San Francisco.

Mass Extinctions and Explosive Diversifications of Life

As discussed in the Biology and Society section, the fossil record reveals that five mass extinctions have occurred over the last 500 million years. In each of these events, 50% or more of Earth's species died out. Of all the mass extinctions, those marking the ends of the Permian and Cretaceous periods have been the most intensively studied.

The Permian mass extinction, at about the time the merging continents formed Pangaea, claimed about 96% of marine species and took a tremendous toll on terrestrial life as well. Equally notable is the mass extinction at the end of the Cretaceous period. For 150 million years prior, dinosaurs dominated Earth's land and air, whereas mammals were few and small, resembling today's rodents. Then, about 65 million years ago, most of the dinosaurs became extinct, leaving behind only the descendants of one lineage, the birds. Remarkably, the massive die-off (which also included half of all other species) occurred in less than 10 million years—a brief period in geologic time.

But there is a flip side to the destruction. Each massive dip in species diversity has been followed by explosive diversification of certain survivors. Extinctions seem to have provided the surviving organisms with new environmental opportunities. For example, mammals existed for at least 75 million years before undergoing an explosive increase in diversity just after the Cretaceous period. Their rise to prominence was undoubtedly associated with the void left by the extinction of the dinosaurs. The world would be a very different place today if many dinosaur lineages had escaped the Cretaceous extinctions or if none of the mammals had survived.

THE PROCESS OF SCIENCE: Mass Extinctions

Did a Meteor Kill the Dinosaurs?

For decades, scientists have been debating the cause of the rapid dinosaur die-off that occurred 65 million years ago. Many **observations** provide clues. The fossil record shows that the climate had cooled and that shallow seas were receding from continental lowlands. It also shows that many plant species died out. Perhaps the most telling evidence was discovered by physicist Luis Alvarez and his geologist son Walter Alvarez, both of the University of California. In 1980, they found that rock deposited around 65 million years ago contains a thin layer of clay rich in iridium, an element very rare on Earth but common in meteors and other extraterrestrial material that occasionally falls to Earth. This discovery led the Alvarez team to ask the following **question**: Is

▼ Figure 14.18 **Trauma for planet Earth and its Cretaceous life.**

An artist's representation of the impact of an asteroid or comet.

The impact's immediate effect was most likely a cloud of hot vapor and debris that could have killed many of the plants and animals in North America within hours.

Chicxulub crater

The 65-million-year-old Chicxulub impact crater is located in the Caribbean Sea near the Yucatán Peninsula of Mexico. The horseshoe shape of the crater and the pattern of debris in sedimentary rocks indicate that an asteroid or comet struck at a low angle from the southeast.

the iridium layer the result of fallout from a huge cloud of dust that billowed into the atmosphere when a large meteor or asteroid hit Earth?

The father and son formed the **hypothesis** that the mass extinction 65 million years ago was caused by the impact of an extraterrestrial object. This hypothesis makes a clear **prediction**: A huge impact crater of the right age should be found somewhere on Earth's surface. (This is a good example of discovery science, which relies on verifiable observations rather than a direct **experiment**; see Chapter 1.) In 1981, two petroleum geologists found the **results** predicted by Alvarezes' hypothesis: the Chicxulub crater, located near Mexico's Yucatán Peninsula in the Caribbean Sea (**Figure 14.18**). This impact site, about 180 km wide (about 112 miles)

and dating from the predicted time, was created when a meteor or asteroid about 10 km in diameter (about 6 miles) slammed into Earth, releasing thousands of times more energy than is stored in the world's combined stockpile of nuclear weapons. Such a cloud could have blocked sunlight and disturbed climate severely for months, perhaps killing off many plant species and, later, the animals that depended on those plants for food.

Debate continues about whether this impact alone caused the dinosaurs to die out or whether other factors—such as continental movements or volcanic activity—also contributed. Most scientists agree, however, that the collision that created the Chicxulub crater could indeed have been a major factor in global climatic changes and mass extinctions.

Classifying the Diversity of Life

Systematics is a discipline of biology that focuses on classifying organisms and determining their evolutionary relationships. Systematics includes **taxonomy**, which is the identification, naming, and classification of species.

Some Basics of Taxonomy

Assigning scientific names to species is an essential part of systematics. Common names such as monkey, fly, and pea may work well in everyday communication, but they can be ambiguous because there are many species of each of these organisms. And some common names are misleading. For example, consider the following so-called "fishes": jellyfish (a cnidarian), crayfish (a crustacean), and silverfish (an insect).

Using an agreed-upon formal naming system eases communication among scientists, allows researchers to unambiguously identify an organism, and makes it easier to recognize when a new species is discovered. The formal taxonomic system used by biologists today dates back to Carolus Linnaeus (1707–1778), a Swedish physician and botanist (plant specialist). Linnaeus's system has two main characteristics: a two-part name for each species and a hierarchical classification of species into broader groups of organisms.

Naming Species

Linnaeus's system assigns to each species a two-part latinized name, or **binomial**. The first part of a binomial is the **genus** (plural, *genera*) to which the species belongs. The second part of a binomial is unique for each species within the genus. The two parts must be used together

to name a species. For example, the scientific name for the leopard is *Panthera pardus*. Notice that the first letter of the genus is capitalized and that the whole binomial is italicized and latinized. (You can name a bug you discover after a friend, but you must add the appropriate Latin ending.) The scientific name of our own species, *Homo sapiens*, which Linnaeus assigned in a show of optimism, means "wise man."

Hierarchical Classification

In addition to naming species, a major objective of systematics is to group species into an ordered hierarchy of categories. The first step of such a hierarchical classification is built into the binomial. We group species that are closely related into the same genus. For example, the leopard (*Panthera pardus*) belongs to a genus that also contains three other cats: the lion (*Panthera leo*), the tiger (*Panthera tigris*), and the jaguar (*Panthera onca*) (**Figure 14.19**). Grouping species is natural for us, at least in concept. We lump together several trees we know as oaks and distinguish them from several other species of trees we call maples. Indeed, oaks and maples belong to separate genera. Biology's taxonomic scheme formalizes our tendency to group related objects as a way of structuring our view of the world.

▼ **Figure 14.19 The four species within the genus *Panthera*.**

Leopard (*Panthera pardus*)

Tiger (*Panthera tigris*)

Lion (*Panthera leo*)

Jaguar (*Panthera onca*)

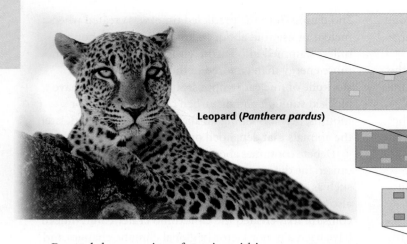

Leopard (*Panthera pardus*)

◀ **Figure 14.20**
Hierarchical classification. Taxonomy classifies species into increasingly comprehensive groups.

Species
Panthera pardus

Genus
Panthera

Family
Felidae

Order
Carnivora

Class
Mammalia

Phylum
Chordata

Kingdom
Animalia

Domain
Eukarya

Beyond the grouping of species within genera, taxonomy extends to progressively broader categories of classification. It places similar genera in the same **family**, puts families into **orders**, orders into **classes**, classes into **phyla** (singular, *phylum*), phyla into **kingdoms**, and kingdoms into **domains**. (See Chapter 1 to review the three domains of life.) **Figure 14.20** places the leopard in this taxonomic scheme of groups within groups. The resulting classification of a particular organism is somewhat like a postal address identifying a person in a particular apartment, in a building with many apartments, on a street with many apartment buildings, in a city with many streets, and so on. ☑

Classification and Phylogeny

Ever since Darwin, systematics has had a goal beyond simple organization: to have classification reflect evolutionary relationships. In other words, how an organism is named and classified should reflect its place within the evolutionary tree of life. Biologists use **phylogenetic trees** to depict hypotheses about the evolutionary history of species. These branching diagrams reflect the hierarchical classification of groups nested within more inclusive groups. The tree in **Figure 14.21** shows the classification of some carnivores and their probable evolutionary relationships. Note that each branch point represents the divergence of two lineages from a common ancestor.

Sorting Homology from Analogy

Homologous structures are one of the best sources of information about phylogenetic relationships. Recall from Chapter 13 that homologous structures in different species may vary in form and function but exhibit fundamental similarities because they evolved from the

▼ **Figure 14.21 The relationship of classification and phylogeny for some members of the order Carnivora.** The hierarchical classification is reflected in the finer and finer branching of the phylogenetic tree. Each branch point in the tree represents an ancestor common to species to the right of that branch point.

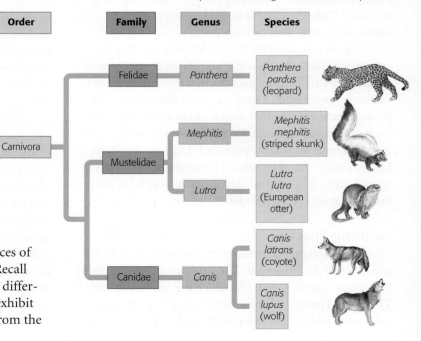

Order	Family	Genus	Species
	Felidae	*Panthera*	*Panthera pardus* (leopard)
Carnivora	Mustelidae	*Mephitis*	*Mephitis mephitis* (striped skunk)
		Lutra	*Lutra lutra* (European otter)
	Canidae	*Canis*	*Canis latrans* (coyote)
			Canis lupus (wolf)

same structure in a common ancestor. Among the vertebrates, for instance, the whale forelimb is adapted for steering in the water, whereas the bat wing is adapted for flight. Nonetheless, there are many basic similarities in the bones supporting these two structures, as you saw in Figure 13.8. The greater the number of homologous structures between two species, the more closely the species are related.

There are pitfalls in the search for homology: Not all likeness is inherited from a common ancestor. Species from different evolutionary branches may have certain structures that are superficially similar if natural selection has shaped analogous adaptations. This is called **convergent evolution**. Similarity due to convergence is called **analogy**, not homology. For example, the wings of insects and those of birds are analogous flight equipment: They evolved independently and are built from entirely different structures.

To develop phylogenetic trees and classify organisms according to evolutionary history, we must use only homologous similarities. This guideline is generally straightforward, but there can be complications. Adaptation can obscure homologies, and convergence can create misleading similarities. As we saw in Chapter 13, comparing the embryonic development of two species can often expose homology that is not apparent in the mature structures.

There is another clue to distinguishing homology from analogy: The more complex two similar structures are, the less likely it is they evolved independently. For example, compare the skulls of a human and a chimpanzee (see Figure 14.13). Although each is a fusion of many bones, they match almost perfectly, bone for bone. It is highly improbable that such complex structures matching in so many details could have separate origins. Most likely, the genes required to build these skulls were inherited from a common ancestor. ✔

Molecular Biology as a Tool in Systematics

If homology reflects common ancestry, then comparing the genes and gene products (proteins) of organisms gets right to the heart of their evolutionary relationships. Sequences of nucleotides in DNA are inherited, and they program corresponding sequences of amino acids in proteins. At the molecular level, the evolutionary divergence of species parallels the accumulation of differences in their genomes. The more recently two species have branched from a common ancestor, the more similar their DNA and amino acid sequences should be.

Today, the amino acid sequences for many proteins and the nucleotide sequences for a rapidly increasing number of genomes are in databases that are available on the Internet (see Chapter 12). These databases have fostered a boom in systematics as researchers use them to compare the hereditary information of different species in search of homology at its most basic level.

Molecular systematics provides a new way to test hypotheses about the phylogeny of species. The strongest support for any such hypothesis is agreement between molecular data and other means of tracing phylogeny, such as evaluating anatomical homology and analyzing the fossil record. And some fossils are preserved in such a way that DNA fragments can be extracted for comparison with living organisms (**Figure 14.22**).

▲ **Figure 14.22 Studying ancient DNA.** Some sedimentary fossils, such as this 40-million-year-old leaf, retain DNA and other organic material that scientists can analyze.

The Cladistic Revolution

Systematics entered a vigorous new era in the 1960s. Just as molecular methods became readily available for comparing species, computer technology that could crunch a new wealth of data helped usher in a new approach called cladistics.

Cladistics is the scientific search for clades (from the Greek word for "branch"). A **clade** consists of an ancestral species and all its descendants—a distinct branch in the tree of life. Identifying clades involves identifying homologies unique to a species or to a higher taxonomic cluster, such as a class or phylum (**Figure 14.23**).

Cladistics requires comparing an ingroup with an outgroup. The ingroup (for example, the three mammals in Figure 14.23) is the taxonomic cluster that is actually being analyzed. The outgroup (in Figure 14.23, the iguana, representing reptiles) is a species or group of species known to have diverged before the lineage that contains the groups being studied. By comparing members of the ingroup with each other and with the outgroup, we can determine what distinguishes the ingroup

☑**CHECKPOINT**

Our forearms and a bat's wings are derived from the same ancestral prototype; thus, they are _____. In contrast, the wings of a bat and the wings of a bee are derived from totally unrelated structures; thus, they are _____.

Answer: homologous; analogous

▶ **Figure 14.23
A simplified
example of
cladistics.**

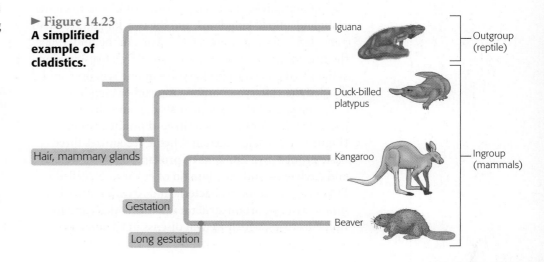

Iguana — Outgroup (reptile)

Duck-billed platypus

Hair, mammary glands

Kangaroo — Ingroup (mammals)

Gestation

Long gestation

Beaver

from the outgroup. All the mammals in the ingroup have hair and mammary glands. These were present in the ancestral mammal, but not in the outgroup. But gestation, the carrying of offspring in the uterus within the female parent, is absent from the duck-billed platypus (which lays eggs with a shell). From this absence we might infer that the duck-billed platypus represents an early branch point in the mammalian clade. Proceeding in this manner, we can construct a phylogenetic tree. Each branch represents the divergence of two groups from a common ancestor, with the emergence of a lineage possessing one or more new features. The sequence of branching represents the order in which they evolved and when groups last shared a common ancestor. In other words, cladistics focuses on the changes that define the branch points in evolution. Identifying clades makes it possible to construct classification schemes that reflect the branching pattern of evolution.

Cladistics has become the most widely used method in systematics. This approach clarifies evolutionary relationships that were not always apparent in other taxonomic classifications. For instance, biologists traditionally placed birds and reptiles in separate classes of vertebrates (class Aves and class Reptilia, respectively). This classification, however, is inconsistent with cladistics. An inventory of homologies indicates that birds and

crocodiles make up one clade, and lizards and snakes form another. If we go back as far as the ancestor that crocodiles share with lizards and snakes to make up a clade, then the class Reptilia must also include birds. The tree in **Figure 14.24** is thus more consistent with cladistics than with traditional classifications. ☑

▼ Figure 14.24 **How cladistics is shaking phylogenetic trees.** Strict application of cladistics sometimes produces phylogenetic trees that conflict with classical taxonomy.

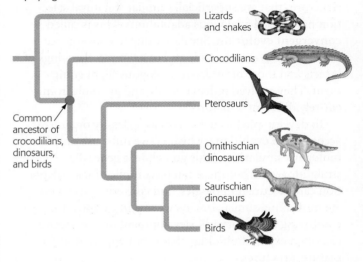

Classification: A Work in Progress

Phylogenetic trees are hypotheses about evolutionary history. Like all hypotheses, they are revised (or in some cases rejected) based on new evidence. Molecular systematics and cladistics are combining to remodel phylogenetic trees and challenge traditional classifications.

Linnaeus divided all known forms of life between the plant and animal kingdoms, and the two-kingdom system prevailed in biology for over 200 years. In the mid-1900s, the two-kingdom system was replaced by a five-kingdom system that placed all prokaryotes in one kingdom and divided the eukaryotes among four other kingdoms.

In the late 1900s, molecular studies and cladistics led to the development of a **three-domain system** (**Figure 14.25**). This current scheme recognizes three basic groups: two domains of prokaryotes—Bacteria and Archaea—and one domain of eukaryotes, called Eukarya. The domains Bacteria and Archaea differ in a number of important structural, biochemical, and functional features, which we will discuss in Chapter 15.

▲ Figure 14.25 **The three-domain classification system.** Molecular and cellular evidence supports the phylogenetic hypothesis that two lineages of prokaryotes, the domains Bacteria and Archaea, diverged very early in the history of life. Molecular evidence also suggests that domain Archaea is more closely related to domain Eukarya than to domain Bacteria.

The domain Eukarya is currently divided into kingdoms, but the exact number of kingdoms is still under debate. Biologists generally agree on the kingdoms Plantae, Fungi, and Animalia. These kingdoms consist of multicellular eukaryotes that differ in structure, development, and modes of nutrition. Plants make their own food by photosynthesis. Fungi live by decomposing the remains of other organisms and absorbing small organic molecules. Most animals live by ingesting food and digesting it within their bodies.

The remaining eukaryotes, the protists, include all those that do not fit the definition of plant, fungus, or animal—effectively, a taxonomic grab bag. Most protists are unicellular (amoebas, for example). But the protists also include certain large, multicellular organisms that are believed to be direct descendants of unicellular protists. For example, many biologists classify the seaweeds as protists because they are more closely related to some single-celled algae than they are to true plants.

It is important to understand that classifying Earth's diverse species is a work in progress as we learn more about organisms and their evolution. Charles Darwin envisioned the goals of modern systematics when he wrote in *The Origin of Species*, "Our classifications will come to be, as far as they can be so made, genealogies." ✔

✔**CHECKPOINT**

The current classification scheme favored by most biologists places life into _____ domains. The domain _____ includes humans in the kingdom called _____.

Answer: three; Eukarya; Animalia

EVOLUTION CONNECTION: Mass Extinctions

Rise of the Mammals

In this chapter, you've read about mass extinctions and their effects on the evolution of life on Earth. In the fossil record, each mass extinction was followed by a period of evolutionary change in which many new species arose whose adaptations allowed them to fill new habitats or community roles. The appearance of many new species followed each mass extinction as survivors became adapted to the many vacant ecological niches.

For example, fossil evidence indicates that the number of mammal species increased dramatically after the extinction of most of the dinosaurs around 65 million years ago (Figure 14.26). Although mammals originated nearly 180 million years ago, fossils older than 65 million years indicate that they were mostly small and not very diverse. Early mammals may have been eaten or outcompeted by the larger and more diverse dinosaurs. With the disappearance of most of the dinosaurs, mammals expanded greatly in both diversity and size, filling the ecological roles once occupied by dinosaurs. Had it not been for the dinosaur extinction, mammals may never have expanded their territories and become the predominant land animals. Therefore, we humans may owe our existence to the demise of older species. Through the process of evolution by natural selection, this pattern of death and renewal is repeated throughout the history of life on Earth.

▼ Figure 14.26 **The increase in mammalian species after the extinction of dinosaurs.** Although mammals originated nearly 150 million years ago, they did not begin to widely diverge until after the demise of the dinosaurs.

Monotremes (5 species, including platypuses and echidnas)

Marsupials (324 species, including kangaroos, opossums, and koalas)

Eutherians (placental mammals; 5,010 species)

American black bear

Chapter Review

 Go to the Study Area at **www.masteringbiology.com** for practice quizzes, eText,
BioFlix™ 3-D animations, MP3 Tutor Sessions, videos, current events, and more.

Macroevolution and the Diversity of Life

Microevolution refers to evolutionary change in the gene pool of a
population within a species. Macroevolution refers to evolutionary changes
evident at or above the species level, which are often seen in the fossil
record. It includes the appearance of new species, the origins of evolutionary
novelties, and the explosive diversification that follows some evolutionary
breakthroughs and mass extinctions.

The Origin of Species

What Is a Species?

According to the biological species concept, a species is a group of
populations whose members have the potential to interbreed in nature to
produce fertile offspring. The biological species concept is just one of several
possible ways to define species.

Reproductive Barriers between Species

Mechanisms of Speciation

When the gene pool of a population is severed from other gene pools of
the parent species, the splinter population can follow its own evolutionary
course.

Hybridization leading to polyploids is a common mechanism of sympatric
speciation in plants.

What Is the Tempo of Speciation?

According to the punctuated equilibria model, the time required for speciation
is relatively short compared with the overall duration of the species' existence.
This model accounts for the relative rarity of transitional fossils linking newer
species to older ones.

The Evolution of Biological Novelty

Adaptation of Old Structures for New Functions

An exaptation is a structure that evolves in one context and gradually
becomes adapted for other functions.

Evo-Devo: Development and Evolutionary Novelty

A subtle change in the genes that control a species' development can have
profound effects. In paedomorphosis, for example, the adult retains body
features that were strictly juvenile in ancestral species.

Earth History and Macroevolution

Geologic Time and the Fossil Record

Geologists have established a geologic time scale with four broad divisions:
Precambrian, Paleozoic, Mesozoic, and Cenozoic. The most common
method for determining the ages of fossils is radiometric dating.

Plate Tectonics and Macroevolution

About 250 million years ago, plate movements brought all the landmasses
together into the supercontinent Pangaea, causing extinctions and providing
new opportunities for the survivors to diversify. About 180 million years ago,
Pangaea began to break up, causing geographic isolation.

Mass Extinctions and Explosive Diversifications of Life

The fossil record reveals long, relatively stable periods punctuated by mass
extinctions followed by explosive diversification of certain survivors. For
example, during the Cretaceous extinctions, about 65 million years ago, the
world lost an enormous number of species, including most of the dinosaurs.
Mammals greatly increased in diversity after the Cretaceous period.

Classifying the Diversity of Life

Systematics, the study of biological diversity, includes taxonomy, which is
the identification, naming, and classification of species.

Some Basics of Taxonomy

Each species is assigned a two-part name. The first part is the genus, and
the second part is unique for each species within the genus. In the
taxonomic hierarchy, domain > kingdom > phylum > class > order >
family > genus > species.

Classification and Phylogeny

The goal of classification is to reflect phylogeny, which is the evolutionary
history of a species. Classification is based on the fossil record, homologous
structures, and comparisons of DNA and amino acid sequences. Homology
(similarity based on shared ancestry) must be distinguished from analogy

(similarity based on convergent evolution). Cladistics uses shared characteristics to group related organisms into clades—distinctive branches in the tree of life.

Classification: A Work in Progress

Biologists currently classify life into a three-domain system:

SELF-QUIZ

1. Contrast microevolution with macroevolution.

2. Bird guides once listed the myrtle warbler and Audubon's warbler as distinct species that lived side by side in parts of their ranges. However, recent books describe them as the eastern and western forms of a single species, the yellow-rumped warbler. Apparently, the two kinds of warblers
   a. live in the same areas.
   b. successfully interbreed.
   c. are almost identical in appearance.
   d. are merging to form a single species.

3. Identify each of the following reproductive barriers as prezygotic or postzygotic.
   a. One lilac species lives on acidic soil, another on basic soil.
   b. Mallard and pintail ducks mate at different times of year.
   c. Two species of leopard frogs have different mating calls.
   d. Hybrid offspring of two species of jimsonweed always die before reproducing.
   e. Pollen of one kind of pine tree cannot fertilize another kind.

4. Why is a small, isolated population more likely to undergo speciation than a large one?

5. Many species of plants and animals adapted to desert conditions probably did not arise there. Their success in living in deserts could be due to _____, structures that evolved in one context but became adapted for different functions.

6. Mass extinctions
   a. cut the number of species to the few survivors left today.
   b. resulted mainly from the separation of the continents.
   c. occurred regularly, about every million years.
   d. were followed by diversification of the survivors.

7. The animals and plants of India are almost completely different from the species in nearby Southeast Asia. Why might this be true?
   a. They have become separated by convergent evolution.
   b. The climates of the two regions are completely different.
   c. India is in the process of separating from the rest of Asia.
   d. India was a separate continent until relatively recently.

8. Place these levels of classification in order from least inclusive to most inclusive: class, domain, family, genus, kingdom, order, phylum, species.

9. A paleontologist estimates that when a particular rock formed, it contained 12 mg of the radioactive isotope potassium-40. The rock now contains 3 mg of potassium-40. The half-life of potassium-40 is 1.3 billion years. From this information, you can conclude that the rock is approximately _____ billion years old.

10. In the three-domain system, which two domains contain prokaryotic organisms?

Answers to the Self-Quiz questions can be found in Appendix D.

THE PROCESS OF SCIENCE

11. Imagine you are conducting fieldwork and discover two groups of mice living on opposite sides of a river. Assuming that you will not disturb the mice, design a study to determine whether these two groups belong to the same species. If you could capture some of the mice and bring them to the lab, how might that affect your experimental design?

BIOLOGY AND SOCIETY

12. Experts estimate that human activities cause the extinction of hundreds or thousands of species every year. The natural rate of extinction is thought to be a few species per year. As we continue to alter the global environment, especially by cutting down tropical rain forests, the resulting extinction will probably rival that at the end of the Cretaceous period. Most biologists are alarmed at this prospect. What are some reasons for their concern? Considering that life has endured numerous mass extinctions and has always bounced back, how is the present mass extinction different? What might be some consequences for the surviving species?

15 The Evolution of Microbial Life

Signs of life.
These present-day rocky outcrops closely resemble fossils of ancient prokaryotes.

CHAPTER CONTENTS

BIOLOGY AND SOCIETY: The Origin of Life

Can Life Be Created in the Lab?

In *The Origin of Species,* Charles Darwin described the emergence of new species as a consequence of natural selection acting on existing species. But if you take his argument back to its logical start, you arrive at the ultimate mystery: How did life first arise on Earth?

Some biologists seek insight into this question through experiments on artificial systems. While artificial life has not yet been created in the lab, several important steps toward that goal have been completed. In 2008, a team led by Craig Venter (famous for his role in sequencing the human genome—see Chapter 12) synthesized from scratch the entire genome of *Mycoplasma genitalium,* a species of bacteria found naturally in the human urinary tract. The researchers constructed DNA strands, about 6,000 bases each, and then connected them to make a complete genome. In earlier experiments, Venter's team had proved that they could transplant the complete genome of one species of *Mycoplasma* bacteria into another. In one fell swoop, the recipient organism took up the new genome and expressed its genes. By combining these two approaches, Venter's team hopes to create an artificial genome and then transplant it into a genome-free host cell, essentially "rebooting" that cell to express the artificial genome. If successful, a new, artificial life-form could result.

Why pursue this line of research? One reason is that it may help answer fundamental questions about the origins of life in nature. Additionally, Venter hopes to engineer artificial organisms that can be completely controlled—a bacterial strain that might, for example, clean up toxic wastes or generate biofuels but be unable to survive outside rigidly controlled conditions (so that it would not spread uncontrollably). But could the creation of artificial life have unintended consequences that might ultimately prove harmful? Are there certain experiments that just should not be performed?

The question, How did life begin? continues to fascinate scientists, though it may never be answered with certainty. Accordingly, this chapter starts with a discussion of some of the key events in the history of life on Earth. We'll then discuss the origins, structures, and diversity of microbes (a term generally used to describe microscopic organisms), starting with the prokaryotes and then moving on to the protists.

Major Episodes in the History of Life

This chapter is the first of three that survey Earth's life-forms and their evolution. This section helps to set the stage by giving a brief overview of the major events in the history of life on Earth.

Life began when Earth was young. The planet formed about 4.6 billion years ago, and its crust began to solidify about 4 billion years ago. A few hundred million years later, by 3.5 billion years ago, Earth was already inhabited by a diversity of organisms (**Figure 15.1**). Those earliest organisms were all **prokaryotes**, their cells lacking true nuclei. Within the next billion years, two distinct groups of prokaryotes—bacteria and archaea—diverged.

An oxygen revolution began about 2.7 billion years ago. Photosynthetic prokaryotes that split water molecules released oxygen gas, profoundly changing Earth's atmosphere (see the Evolution Connection section at the end of Chapter 6). The accumulation of O_2 in the atmosphere doomed many prokaryotic groups. Among the survivors, a diversity of metabolic modes evolved, including cellular respiration, which uses O_2 in extracting energy from food (see Chapter 6). All of this metabolic evolution occurred during the almost 2 billion years that prokaryotes had Earth to themselves.

The oldest eukaryotic fossils are about 2.1 billion years old. Recall from Chapter 4 that **eukaryotes** are composed of one or more cells that contain nuclei and many other organelles absent in prokaryotic cells. The eukaryotic cell evolved from a prokaryotic community, a host cell containing even smaller prokaryotes. The mitochondria of our cells and those of every other eukaryote are descendants of those smaller prokaryotes, as are the chloroplasts of plants and algae.

The origin of more complex cells launched a period of tremendous diversification of eukaryotic forms. These new organisms were the protists. Represented today by a great diversity of species, protists are mostly microscopic and unicellular. Examples you may recognize include algae, amoebas, and *Paramecium*.

The next great event in the evolution of life was multicellularity. As discussed in this chapter's Evolution Connection section, the first multicellular eukaryotes evolved at least 1.2 billion years ago as colonies of single-celled organisms. Their modern descendants

▼ **Figure 15.1 Some major episodes in the history of life.** The timing of events shown on this phylogenetic tree is based on fossil evidence and molecular analysis. (Review Figure 14.23 for a reminder of how to read and interpret phylogenetic trees.)

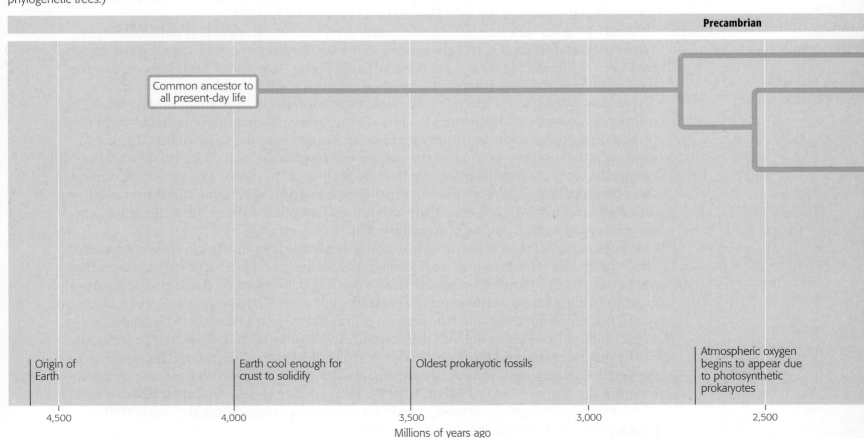

Precambrian

Common ancestor to all present-day life

Origin of Earth

Earth cool enough for crust to solidify

Oldest prokaryotic fossils

Atmospheric oxygen begins to appear due to photosynthetic prokaryotes

4,500   4,000   3,500   3,000   2,500

Millions of years ago

include multicellular protists, such as seaweeds. Other evolutionary branches stemming from the ancient protists gave rise to animals, fungi, and plants.

The greatest diversification of animals was the so-called Cambrian explosion. The Cambrian was the first period of the Paleozoic era, which began about 540 million years ago (see Table 14.1). The earliest animals lived in late-Precambrian seas, but they diversified extensively over a span of just 10 million years during the early Cambrian. In fact, all the major animal body plans—as well as all the major groups—had evolved by the end of that evolutionary eruption.

For over 85% of biological history—life's first 3 billion years—life was confined mostly to aquatic habitats. The colonization of land was a major milestone in the history of life. Plants and fungi together led the way about 500 million years ago. Plants transformed the landscape, creating new opportunities for all life-forms, especially herbivorous (plant-eating) animals and their predators.

The evolutionary venture onto land included vertebrate animals in the form of the first amphibians. These prototypes of today's frogs and salamanders descended from air-breathing fish with fleshy fins that could support the animal's weight on land. Further evolution by natural selection led to the appearance of reptiles and mammals. Among the mammals are the primates, the animal group that includes humans and their closest relatives, apes and monkeys.

Figure 15.2 uses the analogy of a clock ticking down from the origin of Earth 4.6 billion years ago to the present to summarize the major events in the history of life. As the figure makes clear, to trace the history of life on Earth, we must go back to the origin and diversification of microbes, starting with the prokaryotes.

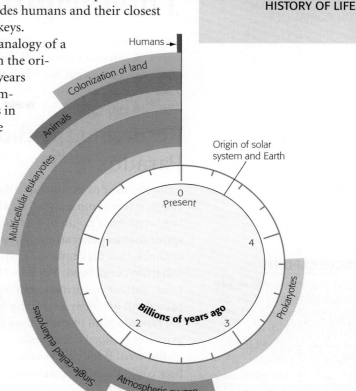

▶ Figure 15.2 **A clock analogy for the major events in the history of life on Earth.**

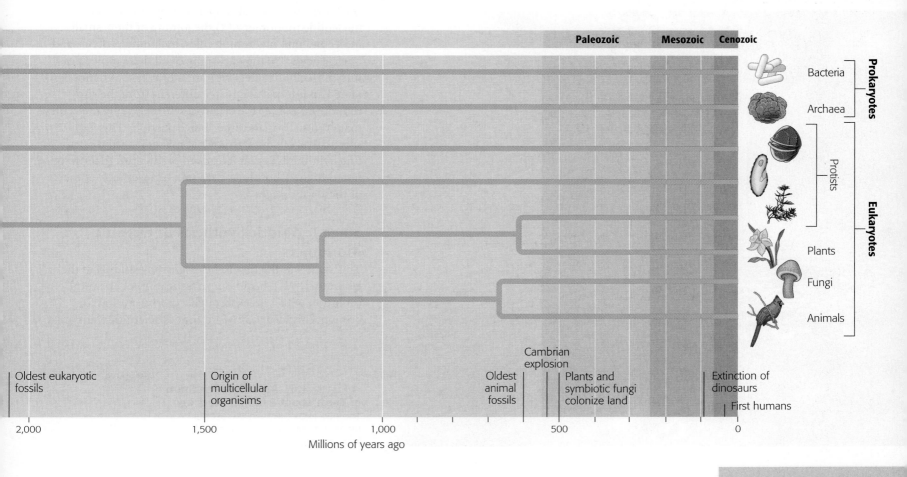

| | Paleozoic | Mesozoic | Cenozoic |

Prokaryotes
- Bacteria
- Archaea

Eukaryotes
- Protists
- Plants
- Fungi
- Animals

Oldest eukaryotic fossils

Origin of multicellular organisims

Oldest animal fossils

Cambrian explosion

Plants and symbiotic fungi colonize land

Extinction of dinosaurs

First humans

2,000   1,500   1,000   500   0

Millions of years ago

The Origin of Life

Although we'll never know for sure how life on Earth began, in this section we'll discuss hypotheses for the formation of the first life. Then we'll talk about its subsequent early evolution.

Resolving the Biogenesis Paradox

From the time of the ancient Greeks until well into the 1800s, it was commonly believed that life could regularly arise from nonliving matter, an idea called **spontaneous generation**. Many people believed, for instance, that flies could arise from rotting meat and fish from ocean mud. Then, in 1862, experiments by Louis Pasteur confirmed what many others had suspected: All life today, including microbes, arises only by the reproduction of preexisting life. This "life-from-life" principle is called **biogenesis**.

But wait! If life always arises from previous life, then how could the first organisms arise? Although there is no evidence that spontaneous generation occurs today,

it could have early in Earth's history, when conditions were very different. For instance, the O_2 in today's atmosphere is a corrosive agent that tends to disrupt chemical bonds, thereby preventing the formation of complex molecules—but O_2 wasn't present in the ancient atmosphere. And such energy sources as lightning, volcanic activity, and ultraviolet sunlight were all more intense on the early Earth than they are today. So life did not begin on a planet anything like the modern Earth, but on a young Earth that was a very different world (**Figure 15.3**). Most biologists now think that chemical and physical processes in Earth's primordial environment could have led to very simple cells through a sequence of stages. Debate abounds about the nature of those stages. ☑

A Four-Stage Hypothesis for the Origin of Life

According to one hypothesis for the origin of life, the first organisms were products of chemical evolution in four stages: (1) the abiotic (nonliving) synthesis of small organic molecules, such as amino acid and nucleotide monomers; (2) the joining of these small molecules into macromolecules, including proteins and nucleic acids; (3) the packaging of all these molecules into pre-cells, droplets with membranes that maintained an internal chemistry different from the surroundings; and (4) the origin of self-replicating molecules that eventually made inheritance possible. This is all speculative, of course, but what makes it a valid scientific hypothesis is that it leads to predictions that can be tested in the laboratory. Let's take a closer look at each of these four stages.

Stage 1: Abiotic Synthesis of Organic Monomers

This stage was the first to be extensively studied in the laboratory. We'll begin our investigation of chemical evolution with a breakthrough experiment that first brought this idea to the forefront of scientific thinking.

◀ Figure 15.3 **An artist's rendition of Earth about 3 billion years ago.** The pad-like objects in the scene represent colonies of prokaryotes known from the fossil record.

THE PROCESS OF SCIENCE: The Origin of Life

Can Biological Monomers Form Spontaneously?

In 1953, University of Chicago scientist Harold Urey and his 23-year-old graduate student Stanley Miller conducted what is now considered a classic experiment. They began with the **observation** that modern biological macromolecules (DNA, protein, carbohydrates, and so on) are all composed of elements (primarily oxygen, hydrogen, carbon, and nitrogen) that were present in abundance on the early Earth. This led to the **question**, Could biological molecules arise spontaneously under conditions like those on the early Earth? Miller and Urey began with the **hypothesis** that a closed system designed to simulate such conditions in the laboratory could produce biologically important organic molecules from inorganic ingredients.

Figure 15.4 shows the apparatus they created to test their hypothesis. A flask of warmed water simulated the primordial sea. An "atmosphere"—in the form of gases added to a reaction chamber—contained hydrogen gas (H_2), methane (CH_4), ammonia (NH_3), and water vapor (H_2O). To mimic the prevalent lightning of the early Earth, sparks were discharged into the chamber. A condenser cooled the atmosphere, causing water and any dissolved compounds to "rain" into the miniature "sea." Miller and Urey's **prediction** was that organic molecules would form and accumulate during this **experiment**.

Miller and Urey's **results** made front-page news. After the apparatus had run for a week, an abundance of organic molecules essential for life, including amino acids, the monomers of proteins, had collected in the "sea."

Since Miller and Urey's classic experiment, other scientists have repeated and extended the research, varying such conditions as the composition of the ancient "atmosphere" and "sea" (since Miller's "early atmosphere" was almost certainly incorrect in some ways). Laboratory analogs of the primeval Earth have produced all 20 amino acids and several sugars. These laboratory results support the concept of the abiotic synthesis of organic molecules on the early Earth.

▼ Figure 15.4 **The abiotic production of organic molecules: A laboratory simulation of early-Earth chemistry.**

Stanley Miller re-creating his 1953 experiment.

Miller and Urey's experiment

Stage 2: Abiotic Synthesis of Polymers

If the hypothesis of an abiotic origin of life is correct, then it should be possible to link organic monomers to form polymers such as proteins and nucleic acids without the help of enzymes and other cellular equipment. Researchers have brought about such polymerization by dripping solutions of organic monomers onto hot sand, clay, or rock. The heat vaporizes the water in the solutions and concentrates the monomers on the underlying material. Some of the monomers then spontaneously bond together to form polymers. On the early Earth, raindrops or waves may have splashed dilute solutions of organic monomers onto fresh lava or other hot rocks and then washed polypeptides and other polymers back into the sea. There, they could accumulate in great quantities (since no life existed to consume them). ☑

Stage 3: Formation of Pre-Cells

According to the four-stage hypothesis, a key step in the origin of life was the isolation of a collection of abiotically created molecules within a membrane. We'll call these molecular aggregates pre-cells—not really cells, but molecular packages with some of the properties of life. Within a confined space, certain combinations of molecules could be concentrated and interact more efficiently. Furthermore, the internal environment of a pre-cell could differ from its surroundings.

Laboratory experiments demonstrate that pre-cells could have formed spontaneously from abiotically produced organic compounds (see Figure 5.20). Such pre-cells produced in the laboratory display some lifelike properties: They have a selectively permeable surface, can grow by absorbing molecules from their surroundings, and swell or shrink when placed in solutions of different salt concentrations.

Stage 4: Origin of Self-Replicating Molecules

Life is defined partly by the process of inheritance, which is based on self-replicating molecules. Today's cells store their genetic information as DNA. They transcribe the information into RNA and then translate RNA messages into specific enzymes and other proteins (see Chapter 10). This mechanism of information flow probably emerged gradually through a series of small changes to much simpler processes.

What were the first genes like? One hypothesis is that they were short strands of RNA that replicated themselves without the assistance of proteins. In laboratory experiments, short RNA molecules can assemble spontaneously from nucleotide monomers in the absence of enzymes (**Figure 15.5**). The result is a population of RNA molecules, each with a random sequence of monomers.

Some of the molecules self-replicate, but their success at this reproduction varies. What happens can be described as molecular evolution: The RNA varieties that replicate fastest increase their frequency in the population.

In addition to the experimental evidence, there is another reason the idea of RNA genes in the primordial world is plausible. Cells actually have RNAs that can act as enzymes; they are called ribozymes. Perhaps early ribozymes catalyzed their own replication. That would help with the "chicken and egg" paradox of which came first, enzymes or genes. Maybe the "chicken and egg" came together in the same RNA molecules. The molecular biology of today may have been preceded by an ancient "RNA world." ☑

From Chemical Evolution to Darwinian Evolution

If pre-cells with self-replicating RNA (and later DNA) did form on the young Earth, they would be refined by natural selection. Mutations, errors in the copying of the "genes," would result in variation among the pre-cells. And the most successful of these pre-cells would grow, divide (reproduce), and continue to evolve. Of course, the gap between such pre-cells and even the simplest of modern cells is enormous. But with millions of years of incremental changes through natural selection, these molecular cooperatives could have become more and more cell-like. The point at which we stop calling them pre-cells and start calling them living cells is as fuzzy as our understanding of how life originated. But we do know that prokaryotes were already flourishing at least 3.5 billion years ago and that all branches of life arose from those ancient prokaryotes.

☑**CHECKPOINT**

What are ribozymes? Why are they a logical step in the formation of life?

Answer: A ribozyme is an RNA molecule that functions as an enzyme. Ribozymes can perform some of the functions of both DNA and protein.

▼ Figure 15.5 **Self-replication of RNA "genes."**

1 RNA monomers

2 Formation of short RNA polymers: simple "genes"

3 Assembly of a complementary RNA chain (pairing rules are G with C and A with U)

4 Complementary chain serves as template for making a copy of original "gene"

Original "gene"

Complementary RNA chain

Prokaryotes

The history of prokaryotic life is a success story spanning billions of years. Prokaryotes lived and evolved all alone on Earth for about 2 billion years. They have continued to adapt and flourish on an evolving Earth and in turn have helped to change the planet. In this section, you will study prokaryotic structure and function, diversity, effects on humans, and ecological significance.

Microbial Life

◄ **Figure 15.6 A window to early life?** An instrument on the research submarine *Alvin* samples the water around a hydrothermal vent more than 1.5 km below the ocean's surface. Prokaryotes that live near the vent use the emitted gases as an energy source. This environment, which is very dark, hot, and under high pressure, is among the most extreme in which life exists today.

They're Everywhere!

Today, prokaryotes are found wherever there is life. They thrive in habitats too cold, too hot, too salty, too acidic, or too alkaline for any eukaryote (**Figure 15.6**). Biologists have even discovered prokaryotes living on the walls of a gold mine 2 miles below Earth's surface. In addition, prokaryotes far outnumber eukaryotes. In fact, more prokaryotes are living in your mouth right now than the total number of people who have ever lived! The collective biological mass (biomass) of prokaryotes is at least ten times that of all eukaryotes.

Though individual prokaryotes are small organisms (**Figure 15.7**), they are giants in their collective impact on Earth and its life. In terms of their impact on humans, we hear most about the relatively few species that can cause serious illness. Bacterial infections are responsible for about half of all human diseases, including tuberculosis, cholera, many sexually transmissible diseases, and certain types of food poisoning. During the 1300s, Black Death—bubonic plague, a bacterial disease—spread across Europe, killing an estimated 25% of the human population.

However, prokaryotic life is much more than just a rogues' gallery. Benign or beneficial prokaryotes are far more common than harmful ones. For example, humans have long used bacteria to produce various types of food, converting milk to cheese, yogurt, and sour cream, for example. In addition, each of us harbors several hundred species of bacteria in and on our body. Bacteria in our intestines provide us with important vitamins, and others living in our mouth prevent harmful fungi from growing there.

The ecological significance of prokaryotes is hard to overstate. Prokaryotes recycle carbon and other vital chemical elements back and forth between organic matter and the soil and atmosphere. For example, some prokaryotes decompose dead organisms. Found in soil

► Figure 15.7 **Bacteria on the head of a pin.** The orange rods are individual bacteria, each about 5 μm long, on the head of a pin. Besides highlighting their tiny size, this micrograph will help you understand why a pin prick can cause infection.

Colorized SEM 605x

and at the bottom of lakes, rivers, and oceans, these decomposers return chemical elements to the environment in the form of inorganic compounds that can be used by plants, which in turn feed animals. If prokaryotic decomposers were to disappear, the chemical cycles that sustain life would come to a halt. All forms of eukaryotic life would also be doomed. In contrast, prokaryotic life would undoubtedly persist in the absence of eukaryotes, as it once did for 2 billion years.

The Structure and Function of Prokaryotes

Prokaryotes have a cellular organization fundamentally different from that of eukaryotes. Whereas eukaryotic cells have a membrane-enclosed nucleus and numerous other membrane-enclosed organelles, prokaryotic cells lack these structural features (see Chapter 4, especially Figures 4.4 and 4.5). And nearly all species of prokaryotes have cell walls exterior to their plasma membranes. In this section, we will discuss other features that are common to all prokaryotes, including aspects of their form, reproduction, and nutrition.

Prokaryotic Forms

Determining cell shape by microscopic examination is an important step in identifying a prokaryote (**Figure 15.8**). Spherical species are called **cocci** (singular, *coccus*). Cocci that occur in clusters are called staphylococci. Other cocci occur in chains; they are called streptococci. For example, the bacterium that causes strep throat in humans is a streptococcus. Rod-shaped prokaryotes are called **bacilli** (singular, *bacillus*). A third prokaryotic cell shape is spiral or curved. These include spirochetes, different species of which cause syphilis and Lyme disease.

Most prokaryotes are unicellular and very small, but there are exceptions to both of these generalizations. The cells of some species usually exist as groups of two or more cells, such as the streptococci already mentioned. Others form true colonies, which are permanent aggregates of identical cells (**Figure 15.9a**). And some species even exhibit a simple multicellular organization, with a division of labor among specialized types of cells (**Figure 15.9b**). Among unicellular species, moreover, there are some giants that actually dwarf most eukaryotic cells (**Figure 15.9c**).

▼ Figure 15.9 **A diversity of prokaryotic shapes and sizes.**

(a) Actinomycete. An actinomycete is a mass of branching chains of rod-shaped cells. These bacteria are common in soil, where they secrete antibiotics that inhibit the growth of other bacteria. Various antibiotic drugs, such as streptomycin, are obtained from actinomycetes.

Colorized SEM 4,800×

LM 400×

(b) Cyanobacteria. These photosynthetic cyanobacteria divide labor among their cells. The box highlights a cell that converts atmospheric nitrogen to ammonia, which can then be incorporated into amino acids and other organic compounds.

LM 18×

(c) Giant bacterium. The larger white blob in this photo is the marine bacterium *Thiomargarita namibiensis*. This prokaryotic cell is over 0.5 mm in diameter, about the size of the fruit fly's head below it.

▼ Figure 15.8 **Three common shapes of prokaryotic cells.**

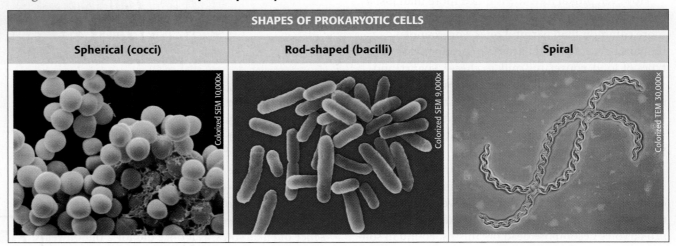

SHAPES OF PROKARYOTIC CELLS		
Spherical (cocci)	**Rod-shaped (bacilli)**	**Spiral**

Colorized SEM 10,000×   Colorized SEM 9,000×   Colorized TEM 30,000×

About half of all prokaryotic species are mobile. Many of those that travel have one or more flagella that propel the cells away from unfavorable places or toward more favorable places, such as nutrient-rich locales (**Figure 15.10**).

Prokaryotic Reproduction

Many prokaryotes can reproduce at a phenomenal rate if conditions are favorable. The cells copy their DNA almost continuously and divide again and again by the process called **binary fission**. Dividing by binary fission, a single cell becomes 2 cells, which then become 4, 8, 16, and so on. Some species can produce a new generation in only 20 minutes under optimal conditions. If reproduction continued unchecked at this rate, a single prokaryote could give rise to a colony outweighing Earth in only three days!

Fortunately, few prokaryotic populations can sustain this kind of exponential growth for long. Environments are usually limiting in resources such as food and space. Prokaryotes also produce metabolic waste products that may eventually pollute the colony's environment. Still, you can understand why certain bacteria can make you sick so soon after infection or why food can spoil so rapidly. Refrigeration retards food spoilage not because the cold kills bacteria, but because most microorganisms reproduce very slowly at such low temperatures.

Some prokaryotes can survive during very harsh conditions and for extended periods, even centuries, by forming specialized cells called endospores. An **endospore** is a thick-coated, protective cell produced within the prokaryotic cell when the prokaryote is exposed to unfavorable conditions (**Figure 15.11**). The endospore can survive all sorts of trauma, including lack of water and nutrients, extreme heat or cold, and most poisons. Not even boiling water kills most of these resistant cells. And when the environment becomes more hospitable, the endospore can absorb water and resume growth. To ensure that all cells, including endospores, are killed when laboratory equipment is sterilized, microbiologists use an appliance called an autoclave, a pressure cooker that uses high-pressure steam at a temperature of 121°C (250°F). The food-canning industry employs similar methods to kill endospores of dangerous soil bacteria such as *Clostridium botulinum*, which produces a toxin that causes the potentially fatal disease botulism. ☑

▼ **Figure 15.10 Prokaryotic flagella.** At the base of the prokaryotic flagellum is a motor and set of rings embedded in the plasma membrane and cell wall. This machinery actually spins like a wheel, rotating the filament of the flagellum.

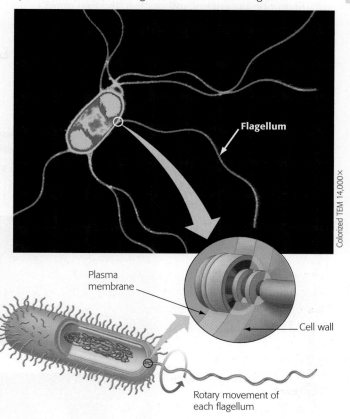

Colorized TEM 14,000×

Flagellum

Plasma membrane

Cell wall

Rotary movement of each flagellum

▼ **Figure 15.11 An endospore in an anthrax bacterium.** This prokaryote is *Bacillus anthracis*, the notorious bacterium that produces the deadly disease called anthrax in cattle, sheep, and humans. There are actually two cells here, one inside the other. The outer cell produced the specialized dormant inner cell, the endospore.

Endospore

Colorized SEM 45,000×

☑**CHECKPOINT**

1. Using a microscope, how could you distinguish the cocci that cause staph infections from those that cause strep throat?
2. Why do microbiologists autoclave their laboratory instruments and glassware rather than just washing them in very hot water?

Answers: 1. by the arrangement of the cell aggregates: grapelike clusters for staphylococcus and chains of cells for streptococcus 2. to kill bacterial endospores, which can survive boiling water

Prokaryotic Nutrition

When classifying diverse organisms, biologists often use the phrase "mode of nutrition" to describe how an organism obtains energy and carbon, the two main resources needed for synthesizing organic compounds. Species that obtain energy from light are called photo-trophs, while species that obtain energy from environmental chemicals are called chemotrophs. Species that obtain carbon from the inorganic compound carbon dioxide (CO_2) are called autotrophs, while species that obtain carbon from at least one organic nutrient—the sugar glucose, for instance—are called heterotrophs. We can combine energy source (phototroph versus chemotroph) and carbon source (autotroph versus heterotroph) to group all organisms according to the four major modes of nutrition shown in **Figure 15.12**. Two of these modes—photoautotrophs (such as most plants), and chemoheterotrophs (such as most animals)—are dominant among multicellular organisms. The other two modes are used only by certain prokaryotes. ☑

► Figure 15.12
Modes of nutrition.

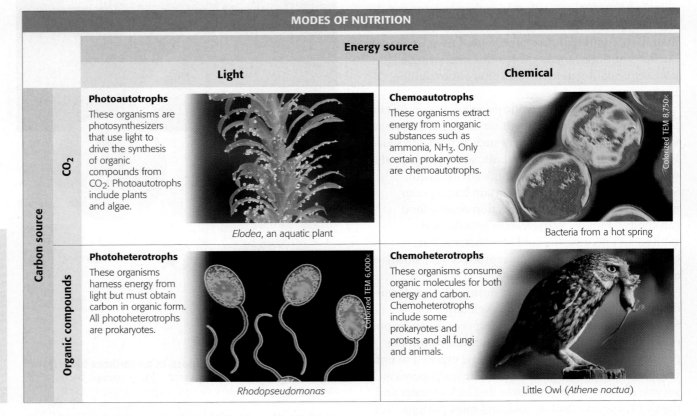

MODES OF NUTRITION

| | **Energy source** | |
	Light	**Chemical**
CO_2 (Carbon source)	**Photoautotrophs** These organisms are photosynthesizers that use light to drive the synthesis of organic compounds from CO_2. Photoautotrophs include plants and algae. *Elodea*, an aquatic plant	**Chemoautotrophs** These organisms extract energy from inorganic substances such as ammonia, NH_3. Only certain prokaryotes are chemoautotrophs. Colorized TEM 8,750× Bacteria from a hot spring
Organic compounds (Carbon source)	**Photoheterotrophs** These organisms harness energy from light but must obtain carbon in organic form. All photoheterotrophs are prokaryotes. Colorized TEM 6,000× *Rhodopseudomonas*	**Chemoheterotrophs** These organisms consume organic molecules for both energy and carbon. Chemoheterotrophs include some prokaryotes and protists and all fungi and animals. Little Owl (*Athene noctua*)

☑**CHECKPOINT**
A bacterium requires only water and the amino acid methionine to grow and lives in very deep caves where no light penetrates. Based on its mode of nutrition, this species would be classified as a _____.

Answer: *chemoheterotroph*

The Two Main Branches of Prokaryotic Evolution: Bacteria and Archaea

By comparing diverse prokaryotes at the molecular level, biologists have identified two major branches of prokaryotic evolution: **bacteria** and **archaea**. Thus, life is organized into three domains—**Bacteria**, **Archaea**, and **Eukarya** (review Figure 14.25). Although bacteria and archaea have prokaryotic cell organization in common, they differ in many structural and physiological characteristics. Some of these differences suggest that archaea are more closely related to eukaryotes than they are to bacteria. In this section, we'll focus on the special characteristics of archaea before turning our attention to bacteria.

The term *archaea* ("ancient") refers to the antiquity of this group's origin from the earliest cells. Even today, many species of archaea inhabit extreme environments, such as hot springs and salt ponds. Few other modern organisms (if any) can survive in some of these environments, which may resemble habitats on the early Earth.

Biologists refer to some archaea as "extremophiles," meaning "lovers of the extreme." There are extreme halophiles ("salt lovers"), archaea that thrive in such environments as Utah's Great Salt Lake, the Dead Sea, and seawater-evaporating ponds used to produce salt

(Figure 15.13a). There are also extreme thermophiles ("heat lovers") that live in very hot water (Figure 15.13b); some archaea even populate the deep-ocean vents that gush water hotter than 100°C (212°F), such as the one shown in Figure 15.6. Also among the archaea are the methanogens, which live in anaerobic (oxygen-free) environments and give off methane as a waste product. They are abundant in the mud at the bottom of lakes and swamps. You may have seen methane, also called marsh gas, bubbling up from a swamp. Great numbers of methanogens also inhabit the digestive tracts of animals. In humans, intestinal gas is largely the result of their metabolism. More importantly, methanogens aid digestion in cattle, deer, and other animals that depend heavily on cellulose for their nutrition. Normally, bloating does not occur in these animals because they regularly expel large volumes of gas produced by the methanogens. (And that may be more than you wanted to know about these gas-producing microbes!)

▼ Figure 15.13 **Archaeal "extremophiles."**

(a) **Salt-loving archaea.** This is an aerial photo of commercial salt-producing ponds at the edge of San Francisco Bay. The colors of the ponds result from dense growth of the harmless prokaryotes that thrive when the salinity of the water reaches five to eight times that of seawater.

(b) **Heat-loving archaea.** In this photo, you can see orange and yellow colonies of heat-loving prokaryotes growing near a Nevada geyser.

Bacteria and Humans

In this section, we'll consider some of the ways that bacteria interact with humans. These interactions can be for both good and ill.

Bacteria That Cause Disease

We are constantly exposed to bacteria, some of which are potentially harmful (**Figure 15.14**). Bacteria and other organisms that cause disease are called **pathogens**. We're healthy most of the time because our body's defenses check the growth of pathogens. Occasionally, the balance shifts in favor of a pathogen, and we become ill. Even some of the bacteria that are normal residents of the human body can make us sick when our defenses have been weakened by poor nutrition, medical treatment (especially cancer therapy), or a viral infection.

Most pathogenic bacteria cause disease by producing poisons. There are two classes of these poisons: exotoxins and endotoxins. **Exotoxins** are poisonous proteins secreted by bacterial cells. A single gram of the exotoxin that causes botulism could kill a million people. Another exotoxin producer is *Staphylococcus aureus* (abbreviated *S. aureus*). It is a common, usually harmless resident of our skin surface. However, if *S. aureus* enters the body through a cut or other wound or is swallowed in contaminated food, it can cause serious diseases collectively called staph infections. One type of *S. aureus* produces exotoxins that cause layers of skin to slough off ("flesh-eating disease"); another can cause vomiting and severe diarrhea; yet another can produce a potentially deadly disease called toxic shock syndrome.

In contrast to exotoxins, **endotoxins** are not cell secretions but are chemical components of the outer membrane of certain bacteria. All endotoxins induce the same general symptoms: fever, aches, and sometimes shock (a dangerous drop in blood pressure). The severity of symptoms varies with the host's condition and with the bacterium. Different species of *Salmonella*, for example, produce endotoxins that cause food poisoning and typhoid fever.

Colorized SEM 4,000x

Haemophilus influenzae

Cells of nasal lining

◄ Figure 15.14 **Bacteria that cause pneumonia.** *Haemophilus influenzae*, shown here on cells lining the interior of a human nose, causes pneumonia and other lung infections, which kill about 4 million people worldwide per year. Most victims are children in less industrialized countries, where malnutrition lowers resistance to all pathogens.

During the last century, following the discovery that "germs" cause disease, the incidence of bacterial infections declined, particularly in more industrialized nations. Sanitation is generally the most effective way to prevent bacterial disease. The installation of water treatment and sewage systems continues to be a public health priority throughout the world. Antibiotics have been discovered that can cure most bacterial diseases. However, resistance to widely used antibiotics has evolved in many of these pathogens (see Chapter 1).

In addition to sanitation and antibiotics, a third defense against bacterial disease is education. A case in point is Lyme disease, the most widespread pest-carried disease in the United States. The disease is caused by a spirochete bacterium carried by ticks (**Figure 15.15**). Disease-carrying ticks live on deer and field mice but also bite humans. Lyme disease usually starts as a red rash shaped like a bull's-eye around a tick bite. Antibiotics can cure the disease if administered within a month of exposure. If untreated, Lyme disease can cause debilitating arthritis, heart disease, and nervous system disorders. So far, the best defense is public education about avoiding tick bites and the importance of seeking treatment if a rash develops. The Centers for Disease Control and Prevention recommends that you avoid vegetation in tick-infested areas, wear light-colored clothing when walking through brush so that you can easily see any ticks, and use an appropriate insect repellent. ✔

☑CHECKPOINT

How can an exotoxin be harmful even after bacteria are killed?

Answer: Exotoxins are secreted poisons that can remain harmful even when the bacteria that secrete them are gone.

Bioterrorism

During the fall of 2001, five Americans died from the disease anthrax in a presumed terrorist attack (**Figure 15.16**). Unfortunately, this act of bioterrorism was hardly unprecedented; there is a long and ugly history of humans using organisms as weapons.

During the Middle Ages, the bacterium *Yersinia pestis* (the cause of bubonic plague) played a role in battle when armies hurled the bodies of plague victims into enemy ranks. Early conquerors, settlers, and warring armies in South and North America gave native peoples items purposely contaminated with infectious bacteria. In the 1930s, the Japanese government instituted a biowarfare program that killed tens of thousands of Chinese soldiers and civilians. Their weapon? The bacteria that cause plague, anthrax, and cholera. In 1984, members of a cult in Oregon contaminated restaurant salad bars with *Salmonella* bacteria; over 700 people became sick, and 45 were hospitalized. During the 1990s, another cult tried to start an anthrax epidemic in Tokyo; and in Iraq, the Iraqi army loaded missiles filled with harmful bacteria. Luckily, neither of these attempts resulted in casualties.

The United States opened its first biological weapons research facility in 1943. There, the military bred new strains of bacteria that cause such illnesses as anthrax and botulism. To "weaponize" these pathogens, researchers selected highly virulent strains and developed formulations for effective dispersal. But the practical difficulties of controlling such weapons—and a measure of moral repugnance—led President Richard Nixon to end the U.S. bioweapons program in 1969 and to order its products destroyed. In 1975, the United States signed the Biological Weapons Convention, pledging never to develop or store biological weapons. Eventually, 103 nations joined the ban, although not every signing nation has honored it.

▼ **Figure 15.15 Lyme disease, a bacterial disease transmitted by ticks.** The bacterium that causes Lyme disease (shown in micrograph at right) is carried from deer to humans by ticks.

16x

SEM 3,000x

"Bull's-eye" rash

Tick that carries the Lyme disease bacterium

Spirochete that causes Lyme disease

▼ **Figure 15.16 Cleaning up after a bioterrorist attack.** This 2001 photo shows hazardous-material workers spraying themselves after they had searched the Senate Office Building in Washington, DC, for anthrax spores.

The Ecological Impact of Prokaryotes

Pathogenic bacteria are in the minority among prokaryotes. Far more common are species that are essential to our well-being, either directly or indirectly. Let's turn our attention now to the vital role that prokaryotes play in sustaining the biosphere.

Prokaryotes and Chemical Recycling

Not too long ago, the atoms making up the organic molecules in your body were part of the inorganic compounds of soil, air, and water, as they will be again. Life depends on the recycling of chemical elements between the biological and physical components of ecosystems. Prokaryotes play essential roles in these chemical cycles. For example, all the nitrogen that plants use to make proteins and nucleic acids comes from prokaryotic metabolism in the soil. In turn, animals get their nitrogen compounds from plants.

Another vital function of prokaryotes, mentioned earlier in the chapter, is the breakdown of organic wastes and dead organisms. Prokaryotes decompose organic matter and, in the process, return elements to the environment in inorganic forms that can be used by other organisms. If it were not for such decomposers, carbon, nitrogen, and other elements essential to life would become locked in the organic molecules of corpses and waste products. You'll learn more about the role that prokaryotes play in chemical cycling in Chapter 20.

Prokaryotes and Bioremediation

People have put the metabolically diverse prokaryotes to work in cleaning up the environment. The use of organisms to remove pollutants from water, air, or soil is called **bioremediation**. One example of bioremediation is the use of prokaryotic decomposers to treat our sewage. Raw sewage is first passed through a series of screens and shredders, and solid matter settles out from the liquid waste. This solid matter, called sludge, is then gradually added to a culture of anaerobic prokaryotes, including both bacteria and archaea. The microbes decompose the organic matter in the sludge, converting it to material that can be used as landfill or fertilizer. The liquid wastes may then be passed through a trickling filter system consisting of a long horizontal bar that slowly rotates, spraying liquid wastes onto a bed of rocks (**Figure 15.17**). Aerobic prokaryotes and fungi growing on the rocks remove much of the organic matter from the liquid. The outflow from the rock bed is then sterilized and released back into the environment.

We are just beginning to explore the great potential that prokaryotes offer for bioremediation. Certain bacteria that occur naturally on ocean beaches can decompose petroleum and are useful in cleaning up oil spills (**Figure 15.18**). Genetically engineered bacteria may be able to degrade oil more rapidly than naturally occurring oil-eaters. Bacteria may also help us clean up old mining sites. The water that drains from mines is highly acidic and is also laced with heavy metals and other poisons. Bacteria called *Thiobacillus* thrive in the acidic waters that drain from mines. While obtaining energy by oxidizing sulfur or sulfur-containing compounds, the bacteria also accumulate metals from the mine waters. Unfortunately, their use in cleaning up mine wastes is limited because their metabolism also adds sulfuric acid to the water. If this problem is solved, perhaps through genetic engineering, *Thiobacillus* and other prokaryotes may help us overcome some environmental dilemmas that seem unsolvable today. One current research focus is a bacterium that tolerates radiation doses thousands of times stronger than those that would kill a person. This species may help clean up toxic dump sites that include radioactive waste.

The various modes of nutrition and metabolic pathways we find in organisms living today are all variations on themes that evolved in prokaryotes during their long reign as Earth's exclusive inhabitants. The subsequent breakthroughs in evolution were mostly structural, including the origin of the eukaryotic cell and the diversification of the protists. ✔

▲ Figure 15.17 **Putting microbes to work in sewage treatment facilities.** This is a trickling filter system, which uses bacteria, archaea, and fungi to treat liquid wastes after sludge is removed.

Labels: Rotating spray arm; Rock bed coated with aerobic prokaryotes and fungi; Liquid wastes; Outflow

✔CHECKPOINT

How do bacteria help restore the atmospheric CO_2 required by plants for photosynthesis?

Answer: By decomposing the organic molecules of dead organisms and organic refuse such as leaf litter, bacteria release carbon from the organic matter in the form of CO_2

▼ Figure 15.18 **Treatment of an oil spill in Alaska.** These workers are spraying fertilizers onto an oil-soaked beach. The fertilizers stimulate growth of naturally occurring bacteria that initiate the breakdown of the oil. This technique is the fastest and least expensive way yet devised to clean up spills on beaches.

Protists

Anton van Leeuwenhoek, an early Dutch microscopist, was the first person to describe the microbial world: "No more pleasant sight has met my eye than this of so many thousands of living creatures in one small drop of water," he wrote more than three centuries ago. It is a world every biology student should have the opportunity to rediscover by peering through a microscope into a droplet of pond water filled with diverse creatures called protists (**Figure 15.19**).

The term **protists** is a bit of a catch-all category that includes all eukaryotes that are not fungi, animals, or plants. It should therefore not be surprising that the protists are highly diverse. Most, but not all, are unicellular. Because their cells are eukaryotic, even the simplest protists are much more complex than any prokaryote.

The first eukaryotes to evolve from prokaryotic ancestors were protists. These primal eukaryotes were not only the predecessors of the great variety of modern protists; they were also ancestral to all other eukaryotes—plants, fungi, and animals. Two of the most significant chapters in the history of life—the origin of the eukaryotic cell and the subsequent emergence of multicellular eukaryotes—both occurred during the evolution of protists.

▼ Figure 15.19 **A diversity of protists in a drop of pond water.**

The Origin of Eukaryotic Cells

The fossil record indicates that eukaryotes evolved from prokaryotes around 2 billion years ago. One of biology's most engaging questions is how this happened—in particular, how the membrane-enclosed organelles of eukaryotic cells arose. A widely accepted theory is that eukaryotic cells evolved through a combination of two processes. In one process, the eukaryotic cell's endomembrane system—all of the membrane-enclosed organelles except mitochondria and chloroplasts (review Figure 4.18)—evolved from inward folds of the plasma membrane of a prokaryotic cell (**Figure 15.20a**).

A second, very different process, called endosymbiosis, generated mitochondria and chloroplasts. **Symbiosis** ("living together") is a close association between organisms of two or more species, and **endosymbiosis** refers to one species living inside another host species. Chloroplasts and mitochondria seem to have evolved from small symbiotic prokaryotes (symbionts) that established residence within other, larger host prokaryotes (**Figure 15.20b**). The ancestors of mitochondria may have been aerobic bacteria that were able to use oxygen to release large amounts of energy from organic molecules by cellular respiration. At some point, such a prokaryote might have been an internal parasite of a larger heterotroph, or an ancestral host cell may have ingested some of these aerobic cells for food. If some of the smaller cells were indigestible, they might have remained alive and continued to perform respiration in the host cell. In a similar way, photosynthetic bacteria ancestral to chloroplasts may have come to live inside a larger host cell. Because almost all eukaryotes have mitochondria but only some have chloroplasts, it is logical to suppose that mitochondria evolved first.

However the relationships began, it is not hard to imagine how a symbiosis between engulfed aerobic or photosynthetic cells and a larger host cell might have become mutually beneficial. In a world that was becoming increasingly aerobic, a cell that was itself an anaerobe would have benefited from aerobic endosymbionts that turned the oxygen to advantage. And a heterotrophic host could derive nourishment from photosynthetic endosymbionts. In the process of becoming more interdependent, the host and endosymbionts would have become a single organism, its parts inseparable.

Developed primarily by Lynn Margulis, of the University of Massachusetts, the endosymbiosis model is supported by extensive evidence. Present-day mitochondria and chloroplasts are similar to prokaryotic cells in a number of ways. For example, both types of organelles contain small amounts of DNA, RNA, and ribosomes that resemble prokaryotic versions more than eukaryotic ones. These components enable chloroplasts and mitochondria to exhibit some autonomy in their activities. The organelles transcribe and translate their DNA into polypeptides, contributing some of their own enzymes to the cell. They also replicate their own DNA and reproduce within the cell by a process resembling the binary fission of prokaryotes.

The origin of the eukaryotic cell made more complex organisms possible, and a vast variety of protists evolved. ☑

☑CHECKPOINT

Which organelles of eukaryotic cells probably descended from endosymbiotic bacteria?

Answer: mitochondria and chloroplasts

▼ Figure 15.20 **A two-stage hypothesis for the evolution of eukaryotes through endosymbiosis.**

(a) Origin of the endomembrane system

(b) Origin of mitochondria and chloroplasts

The Diversity of Protists

All protists are eukaryotes, but they are so diverse that it is hard to name any other general characteristics about them. Biologists used to classify all protists in a kingdom called Protista but now think that these organisms constitute several kingdoms within domain Eukarya. While our knowledge of the evolutionary relationships between these diverse groups remains incomplete, the term *protist* is still a convenient term to refer to eukaryotes that are not plants, animals, or fungi. This catch-all category will almost certainly be reorganized in the coming years.

Given their diversity, it is not surprising that protists vary in structure and function more than any other group of organisms. Because most protists are unicellular (although there are some colonial and multicellular species), they are justifiably considered the simplest eukaryotic organisms. But at the cellular level, many protists are exceedingly complex—among the most elaborate of all cells. We should expect this of organisms that must carry out, within the boundaries of a single cell, all the basic functions performed by the collective of specialized cells that make up the bodies of plants and animals. Each unicellular protist is not analogous to a single cell from a human, but is itself an organism as complete as any whole animal or plant.

The classification of protists remains a work in progress. Therefore, rather than discuss any hypothesis about protist phylogeny, let's instead look at four major categories of protists, grouped by lifestyle: protozoans, slime molds, unicellular algae, and seaweeds.

Protozoans

Protists that live primarily by ingesting food are called **protozoans** (Figure 15.21). Protozoans thrive in all types of aquatic environments. Most species eat bacteria or other protozoans, but some can absorb nutrients dissolved in the water. Protozoans that live as parasites in animals, though in the minority, cause some of the world's most harmful human diseases.

Flagellates are protozoans that move by means of one or more flagella. Most species are free-living (nonparasitic). However, this group also includes some nasty parasites that make humans sick. An example is *Giardia*, a flagellate that infects the human intestine and can cause abdominal cramps and severe diarrhea. People become infected mainly by drinking water contaminated with feces from infected animals. Another group of dangerous flagellates are the trypanosomes, including a species that causes sleeping sickness, a serious illness prevalent in tropical Africa and transmitted by the tsetse fly.

Amoebas are characterized by great flexibility in their body shape and the absence of permanent organelles for locomotion. Most species move and feed by means of **pseudopodia** (singular, *pseudopodium*), temporary extensions of the cell. Amoebas can assume virtually any shape as they creep over rocks, sticks, or mud at the bottom of a pond or ocean. One species of parasitic amoeba causes amoebic dysentery, responsible for up to 100,000 deaths worldwide every year. Other protozoans with pseudopodia include the **forams**, which also have shells.

Apicomplexans are all parasitic, and some cause serious human diseases. They are named for a structure at their apex (tip) that is specialized for penetrating host cells and tissues. This group of protozoans includes *Plasmodium*, the parasite that causes malaria. As part of the effort to combat malaria, scientists determined the complete sequence of the *Plasmodium* genome in 2002.

Ciliates are protozoans that use structures called cilia to move and feed. Nearly all ciliates are free-living (nonparasitic). The best-known example is the freshwater ciliate *Paramecium*. ✔

▼ Figure 15.21 **A diversity of protozoans.**

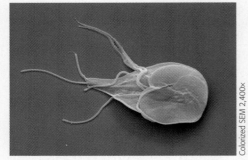

A flagellate: *Giardia*. This flagellated protozoan parasite can colonize and reproduce within the human intestine, causing disease.

Colorized SEM 2,400×

Another flagellate: trypanosomes. The squiggly forms among these human red blood cells are trypanosomes, parasitic flagellates that cause sleeping sickness, a debilitating disease common in parts of Africa.

Colorized SEM 5,440×

Food being ingested

Pseudopodium of amoeba

An amoeba. This amoeba is ingesting a smaller protozoan as food. The amoeba's pseudopodium arches around the prey and engulfs it into a food vacuole.

LM 13×

Slime Molds

These protists are more attractive than their name. Slime molds resemble fungi in appearance and lifestyle, but the similarities are due to convergent evolution; slime molds and fungi are not at all closely related. The weblike body of a slime mold, like that of a fungus, is an adaptation that increases exposure to the environment. This suits the role of these organisms as decomposers. The two main groups of these protists are plasmodial slime molds and cellular slime molds.

Plasmodial slime molds are named for the feeding stage in their life cycle, an amoeboid mass called a plasmodium (**Figure 15.22**)—not to be confused with *Plasmodium*, the apicomplexan parasite that causes malaria. You can find plasmodial slime molds among the leaf litter and other decaying material on a forest floor, and you won't need a microscope to see them. A plasmodium can measure several centimeters across, with its network of fine filaments taking in bacteria and bits of dead organic matter amoeboid style. Large as it is, the plasmodium is actually a single cell with many nuclei.

The study of **cellular slime molds** raise a question about what it means to be an individual organism, given the different forms that they can take during their successive life stages (**Figure 15.23**). The feeding stage of a cellular slime mold consists of ❶ solitary amoeboid cells. They function individually, using their pseudopodia to creep through organic matter and engulf bacteria. But when food is in short supply, the amoeboid cells swarm together to form ❷ a slug-like colony that moves and functions as a single unit. After a brief period of mobility, the colony ❸ extends a stalk and develops into a multicellular reproductive structure. ☑

▼ Figure 15.22 **A plasmodial slime mold.** The weblike form of the slime mold's feeding stage is an adaptation that enlarges the organism's surface area, increasing its contact with food, water, and oxygen.

▼ Figure 15.23 **Life stages of a cellular slime mold.**

❷ Slug-like colony

❶ Amoeboid cells

❸ Reproductive structure

A foram. A foram cell secretes a shell made of organic material hardened with calcium carbonate. The shells of fossilized forams are a major component of limestone.

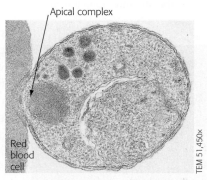

Apical complex

Red blood cell

An apicomplexan. *Plasmodium*, which causes malaria, enters red blood cells of its human host. The parasite feeds on the host cell from within, eventually destroying it.

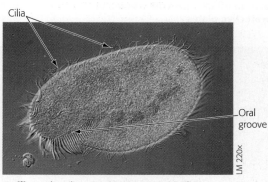

Cilia

Oral groove

A ciliate. The ciliate *Paramecium* uses its cilia to move through pond water. Cilia lining the oral groove keep a current of water containing food moving toward the cell "mouth."

Unicellular and Colonial Algae

Photosynthetic protists belong to an informal category called **algae** (singular, *alga*). Algae come in unicellular, colonial, and multicellular forms. Their chloroplasts support food chains in freshwater and marine ecosystems. Many unicellular algae are components of **plankton**, the communities of organisms, mostly microscopic, that drift or swim weakly near the surfaces of ponds, lakes, and oceans. We'll look at three groups of unicellular algae—dinoflagellates, diatoms, and green algae—and one type of colonial algae.

Dinoflagellates are abundant in the vast aquatic pastures of plankton. Each **dinoflagellate** species has a characteristic shape reinforced by external plates made of cellulose (**Figure 15.24a**). The beating of two flagella in perpendicular grooves produces a spinning movement. Dinoflagellate blooms—population explosions—sometimes cause warm coastal waters to turn pinkish orange, a phenomenon known as a red tide. Toxins produced by some red-tide dinoflagellates have caused massive fish kills, especially in the tropics, and are poisonous to humans as well.

Diatoms have glassy cell walls containing silica, the mineral used to make glass (**Figure 15.24b**). The cell wall consists of two halves that fit together like the bottom and lid of a shoe box. Diatoms store their food reserves in the form of an oil that provides buoyancy, keeping diatoms floating as plankton near the sunlit

CHECKPOINT

1. What metabolic process mainly distinguishes algae from protozoans?
2. Are seaweeds plants?

Answers: 1. photosynthesis 2. No, they are large, multicellular marine algae.

▼ Figure 15.24 **Unicellular and colonial algae.**

(a) A dinoflagellate, with its wall of protective plates

(b) A sample of diverse diatoms, which have glassy walls

(c) *Chlamydomonas*, a unicellular green alga with a pair of flagella

(d) *Volvox*, a colonial green alga

surface. Massive accumulations of fossilized diatoms make up thick sediments known as diatomaceous earth, which is mined for its use as a filtering material, an abrasive, and a natural insecticide.

Green algae are named for their grass-green chloroplasts. Unicellular green algae flourish in most freshwater lakes and ponds, as well as many home pools and aquariums. Some species are flagellated, such as *Chlamydomonas* (**Figure 15.24c**). The green algal group also includes colonial forms, such as *Volvox*, shown in **Figure 15.24d**. Each *Volvox* colony is a hollow ball of flagellated cells (the small green dots in the photo) that are very similar to certain unicellular green algae. The balls within the balls in Figure 15.24d are daughter colonies that will be released when the parent colonies rupture. Of all photosynthetic protists, green algae are the most closely related to plants. (We'll examine the evidence of this evolutionary relationship in the next chapter.)

Seaweeds

Defined as large, multicellular marine algae, **seaweeds** grow on rocky shores and just offshore beyond the zone of the pounding surf. Their cell walls have slimy and rubbery substances that cushion their bodies against the agitation of the waves. Some seaweeds are as large and complex as many plants. And though the word *seaweed* implies a plantlike appearance, the similarities between these algae and plants are a consequence of convergent evolution. In fact, the closest relatives of seaweeds are certain unicellular algae, which is why many biologists include seaweeds with the protists. Seaweeds are classified into three diferent groups, based partly on the types of pigments present in their chloroplasts: green algae, red algae, and brown algae (some of which are known as kelp) (**Figure 15.25**).

Many coastal people, particularly in Asia, harvest seaweeds for food. For example, in Japan and Korea, some seaweed species (such as brown algae called kombu) are ingredients in soups. Other seaweeds (such as red algae called nori) are used to wrap sushi. Marine algae are rich in iodine and other essential minerals. However, much of their organic material consists of unusual polysaccharides that humans cannot digest, which prevents seaweeds from becoming staple foods. They are ingested mostly for their rich tastes and unusual textures. The gel-forming substances in the cell walls of seaweeds are widely used as thickeners for such processed foods as puddings, ice cream, and salad dressing. The seaweed extract called agar provides the gel-forming base for the media microbiologists use to culture bacteria in petri dishes and for a type of gel commonly used to perform gel electrophoresis (as shown in Figure 12.17). ☑

▼ Figure 15.25 **The three major groups of seaweeds.**

Green algae. This sea lettuce is an edible species that inhabits the intertidal zone, where the land meets the ocean. In addition to seaweeds, green algae include unicellular and colonial species.

Red algae. These seaweeds are most abundant in the warm coastal waters of the tropics. Because their chloroplasts have special pigments that absorb the blue and green light that penetrates best through water, red algae can generally live in the deepest water.

Brown algae. This group includes the largest seaweeds, known as kelp, which grow as marine "forests" in relatively deep water beyond the intertidal zone.

EVOLUTION CONNECTION: The Origin of Life

The Origin of Multicellular Life

An orchestra can play a greater variety of musical compositions than a violin soloist can. Put simply, increased complexity makes more variations possible. Thus, the origin of the eukaryotic cell led to an evolutionary radiation of new forms of life. Unicellular protists, which are organized on the complex eukaryotic plan, are much more diverse in form than the simpler prokaryotes. The evolution of multicellular bodies crossed another threshold in structural organization.

Multicellular organisms are fundamentally different from unicellular ones. In a unicellular organism, all of life's activities are carried out by a single cell. In contrast, a multicellular organism has various specialized cells that perform different functions—such as feeding, waste disposal, gas exchange, and protection—and are dependent on each other.

The evolutionary links between unicellular and multicellular life were probably colonial forms, in which unicellular protists stuck together as loose federations of independent cells (**Figure 15.26**). The gradual transition from colonies to truly multicellular organisms involved the cells becoming increasingly interdependent as a division of labor evolved. We can see one level of specialization and cooperation in the colonial green alga *Volvox* (see Figure 15.24d). *Volvox* produces gametes (sperm and ova), which depend on nonreproductive (somatic) cells while developing. Cells in truly multicellular organisms are specialized for many more nonreproductive functions.

Multicellularity evolved many times among the ancestral stock of protists, leading to new waves of biological diversification. The diverse seaweeds are examples of their descendants, and so are plants, fungi, and animals. In the next chapter, we'll trace the long evolutionary movement of plants and fungi onto land.

▼ Figure 15.26 **A model for the evolution of multicellular organisms from unicellular protists.**

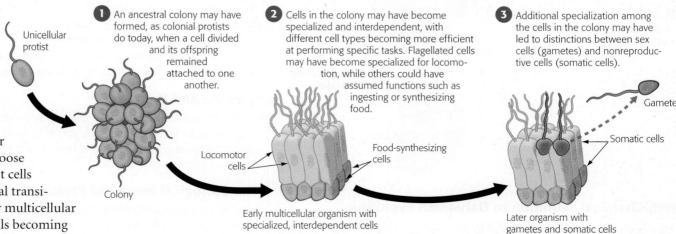

Unicellular protist

1 An ancestral colony may have formed, as colonial protists do today, when a cell divided and its offspring remained attached to one another.

Colony

2 Cells in the colony may have become specialized and interdependent, with different cell types becoming more efficient at performing specific tasks. Flagellated cells may have become specialized for locomotion, while others could have assumed functions such as ingesting or synthesizing food.

Locomotor cells

Food-synthesizing cells

Early multicellular organism with specialized, interdependent cells

3 Additional specialization among the cells in the colony may have led to distinctions between sex cells (gametes) and nonreproductive cells (somatic cells).

Gamete

Somatic cells

Later organism with gametes and somatic cells

Chapter Review

 Go to the Study Area at **www.masteringbiology.com** for practice quizzes, eText, BioFlix™ 3-D animations, MP3 Tutor Sessions, videos, current events, and more.

Major Episodes in the History of Life

Major episode	Millions of years ago
Plants and fungi colonize land	500
All major animal phyla established	530
First multicellular organisms	1,200
Oldest eukaryotic fossils	1,800
Accumulation of O_2 in atmosphere	2,400
Oldest prokaryotic fossils	3,500
Origin of Earth	4,600

The Origin of Life

Resolving the Biogenesis Paradox

All life today arises only by the reproduction of preexisting life. However, most biologists think it is possible that chemical and physical processes in Earth's primordial environment produced the first cells through a series of stages.

A Four-Stage Hypothesis for the Origin of Life

One scenario suggests that the first organisms were products of chemical evolution in four stages:

Inorganic compounds

1 **Abiotic synthesis of organic monomers**

Organic monomers

2 **Abiotic synthesis of polymers**

Polymer

3 **Formation of pre-cells**

Membrane-enclosed compartment

4 **Self-replicating molecules**

Complementary chain

From Chemical Evolution to Darwinian Evolution

Over millions of years, natural selection favored the most efficient pre-cells, which evolved into the first prokaryotic cells.

Prokaryotes

They're Everywhere!

Prokaryotes are found wherever there is life and greatly outnumber eukaryotes. Prokaryotes thrive in habitats where eukaryotes cannot live. A few prokaryotic species cause serious diseases, but most are either benign or beneficial to other forms of life.

The Structure and Function of Prokaryotes

Prokaryotic cells lack nuclei and other membrane-enclosed organelles. Most have cell walls. Some of the most common shapes of prokaryotes are:

| Spherical | Rod-shaped | Spiral |

About half of all prokaryotic species are mobile, most of these using flagella to move. Some prokaryotes can survive extended periods of harsh conditions by forming endospores. Many prokaryotes can reproduce by binary fission at high rates if conditions are favorable, but growth is usually restricted by limited resources.

Prokaryotes exhibit all four major modes of nutrition:

Nutritional Mode	Energy Source	Carbon Source
Photoautotroph	Sunlight	CO_2
Chemoautotroph	Inorganic chemicals	
Photoheterotroph	Sunlight	Organic compounds
Chemoheterotroph	Organic compounds	

The Two Main Branches of Prokaryotic Evolution: Bacteria and Archaea

The prokaryotic lineage includes domains Bacteria and Archaea. Many archaea are "extremophiles" capable of surviving under conditions (such as high heat or salt concentrations) that would kill other forms of life.

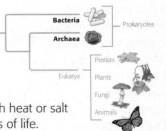

Bacteria and Humans

Most pathogenic bacteria cause disease by producing exotoxins or endotoxins. Sanitation, antibiotics, and education are the best defenses against bacterial disease. Humans have a long history of employing bacteria as bioterrorist agents.

The Ecological Impact of Prokaryotes

Prokaryotes help recycle chemical elements between the biological and physical components of ecosystems. Humans can use prokaryotes to remove pollutants from water, air, and soil in the process called bioremediation.

Protists

The Origin of Eukaryotic Cells

The nucleus and endomembrane system of eukaryotes probably evolved from infoldings of the plasma membrane of ancestral prokaryotes. Mitochondria and chloroplasts probably evolved from symbiotic prokaryotes that took up residence inside larger cells, a process called endosymbiosis.

The Diversity of Protists

Protists are unicellular eukaryotes and their closest multicellular relatives.

- Protozoans (including flagellates, amoebas, apicomplexans, and ciliates) primarily live in aquatic environments and ingest their food.
- Slime molds (including plasmodial slime molds and cellular slime molds) resemble fungi in appearance and lifestyle as decomposers, but are not at all closely related.
- Unicellular algae (including dinoflagellates, diatoms, and unicellular green algae) are photosynthetic protists that support food chains in freshwater and marine ecosystems.
- Seaweeds—which include green, red, and brown algae—are large, multicellular marine algae that grow on and near rocky shores.

SELF-QUIZ

1. Place these events in the history of life on Earth in the order that they occurred.
   a. colonization of land by animals
   b. colonization of land by plants and fungi
   c. diversification of animals (Cambrian explosion)
   d. origin of eukaryotes
   e. origin of humans
   f. origin of multicellular organisms
   g. origin of prokaryotes

2. Place the following steps in the origin of life in the order that they are hypothesized to have occurred.
   a. integration of abiotically produced molecules into membrane-enclosed pre-cells
   b. origin of the first molecules capable of self-replication
   c. abiotic joining of organic monomers into polymers
   d. abiotic synthesis of organic monomers
   e. natural selection among pre-cells

3. DNA replication relies on the enzyme DNA polymerase. Why does this suggest that the earliest genes were made from RNA?

4. Contrast exotoxins with endotoxins.

5. What is the difference between autotrophs and heterotrophs in terms of the source of their organic compounds?

6. The bacteria that cause tetanus can be killed only by prolonged heating at temperatures considerably above boiling. What does this suggest about tetanus bacteria?

7. To what nutritional classification do you belong? To what nutritional classification does a mushroom belong? (*Hint*: Review Figure 15.12.)

8. Why are protists especially important to biologists investigating the evolution of eukaryotic life?

9. Of the following, which describes protists most inclusively?
   a. multicellular eukaryotes
   b. protozoans
   c. eukaryotes that are not plants, fungi, or animals
   d. single-celled organisms closely related to bacteria

10. Which algal group is most closely related to plants?
    a. diatoms
    b. green algae
    c. dinoflagellates
    d. seaweeds

Answers to the Self-Quiz questions can be found in Appendix D.

THE PROCESS OF SCIENCE

11. Imagine you are on a team designing a moon base that will be self-contained and self-sustaining. Once supplied with building materials, equipment, and organisms from Earth, the base will be expected to function indefinitely. One of the members of your team has suggested that everything sent to the base be chemically treated or irradiated so that no bacteria of any kind are present. Do you think this is a good idea? Predict some of the consequences of eliminating all bacteria from an environment.

12. Your classmate says that organisms that require oxygen existed before photosynthetic organisms. Do you support this idea? Explain why or why not.

BIOLOGY AND SOCIETY

13. Many local newspapers publish a weekly list of restaurants that have been cited by inspectors for poor sanitation. Locate such a report and highlight the cases that are probably associated with food contamination by pathogenic prokaryotes.

14. What do you think should be done to prevent bioterrorism?

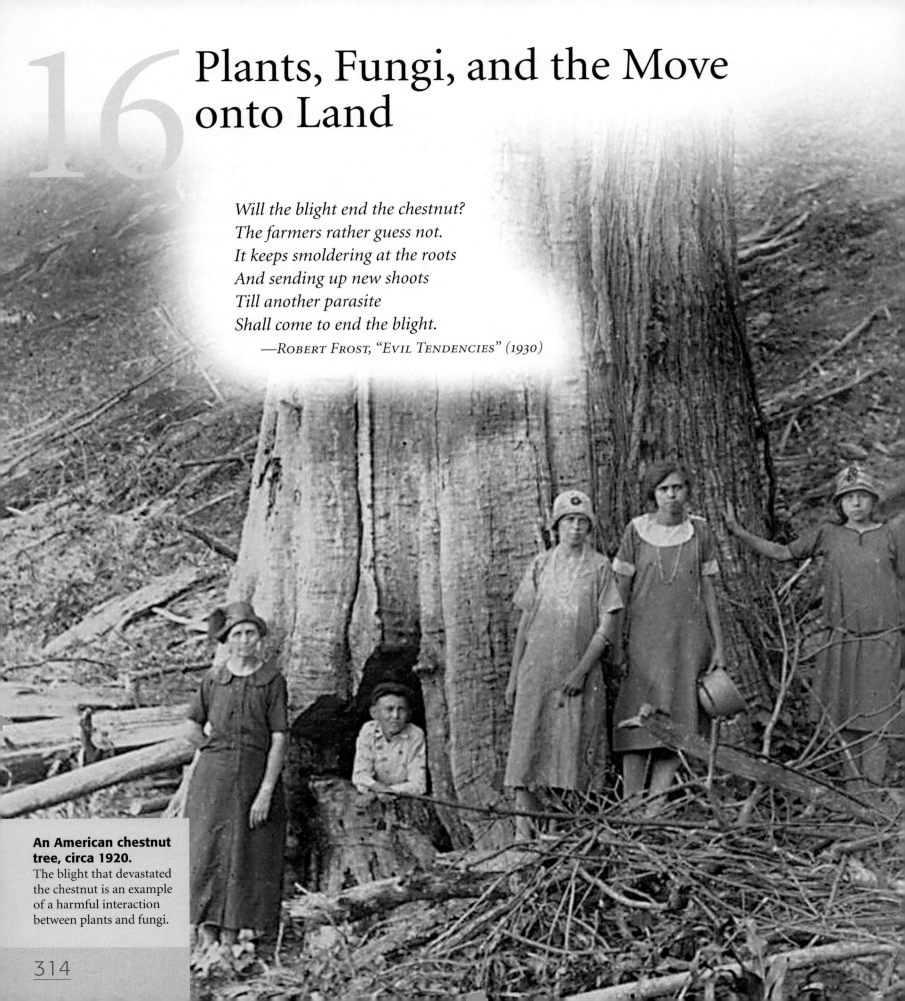

16 Plants, Fungi, and the Move onto Land

Will the blight end the chestnut?
The farmers rather guess not.
It keeps smoldering at the roots
And sending up new shoots
Till another parasite
Shall come to end the blight.

—ROBERT FROST, "EVIL TENDENCIES" (1930)

An American chestnut tree, circa 1920.
The blight that devastated the chestnut is an example of a harmful interaction between plants and fungi.

BIOLOGY AND SOCIETY: Plant-Fungus Symbiosis

Will the Blight End the Chestnut?

The forests of the eastern United States, from Maine to Georgia, were once dominated by the American chestnut tree (*Castanea dentate*). Prized for their rapid growth, huge size, rot-resistant wood (which made them ideal for log cabin foundations), and bountiful harvest of edible nuts (traditionally "roasting on an open fire" during the Thanksgiving-to-Christmas holiday season), American chestnuts were a mainstay of rural life.

Tragically, all this changed in just a few decades. Around 1900, an Asian fungus called *Cryphonectria parasitica* was accidentally introduced from China into North America. While many Asian trees had evolved defenses against the fungus, American trees had not. In just 25 years, blight caused by the fungus killed virtually every one of the estimated 3.5 billion adult American chestnut trees.

Despite the decimation of trees by the blight, the American chestnut is not extinct. You can still find small chestnut trees sprouting from old roots or stumps in many forests. Unfortunately, the blight fungus kills practically all the trees before they reach sexual maturity, so the trees are not propagating and restoring the population. However, forest researchers are working with the wealth of genetic material contained in the many surviving young sprouts to try to develop a blight-resistant strain through breeding and genetic engineering. Perhaps this magnificent tree will once again be a presence in American forests. But the question posed by Robert Frost in the opening line of his poem remains unanswered.

The chestnut blight is an example of **symbiosis**, a close association of two or more species. The harmful interaction between plant and fungus exemplified by the chestnut blight is an unusual case. As we explore the diversity of plants and fungi in this chapter, you will see that it is much more common for the members of these two kingdoms to benefit from each other's presence. Indeed, as we'll discuss first, aquatic plants probably never could have adapted to land without the aid of fungi.

Colonizing Land

Plants are terrestrial (land-dwelling) organisms. It is true that some, such as water lilies, have returned to the water, but they evolved from terrestrial ancestors (just as several species of aquatic mammals, such as whales, evolved from terrestrial mammals). In this section, we'll discuss some of the adaptations that allowed plants to move onto land.

Terrestrial Adaptations of Plants

What exactly is a plant? A **plant** is a multicellular eukaryote that makes organic molecules by photosynthesis (in other words, plants are photoautotrophs—see Figure 15.12). Photosynthesis distinguishes plants from the animal and fungal kingdoms. But what about large algae, including seaweeds, which we classified as protists in the preceding chapter? They, too, are multicellular, eukaryotic, and photosynthetic. What distinguishes plants from algae is a set of structural and reproductive terrestrial adaptations.

Structural Adaptations

Living on land poses different problems than living in water (**Figure 16.1**). In terrestrial habitats, the resources that a photosynthetic organism needs are found in two very different places. Light and carbon dioxide are mainly available in the air, while water and mineral nutrients are found mainly in the soil. Thus, the complex bodies of plants are specialized to take advantage of these two environments by having both aerial leaf-bearing organs called **shoots** and subterranean organs called **roots**.

Most plants have symbiotic fungi associated with their roots. These root-fungus combinations are called **mycorrhizae** ("fungus roots") (**Figure 16.2**). For their part, the fungi absorb water and essential minerals from the soil and provide these materials to the plant. The sugars produced by the plant nourish the fungi. Mycorrhizae are evident on some of the oldest plant fossils, suggesting that they are key adaptations that made it possible for plants to live on land.

Shoots also show structural adaptations to the terrestrial environment. Leaves are the main photosynthetic organs of most plants. Exchange of carbon dioxide (CO_2) and oxygen (O_2) between the atmosphere and the

▼ Figure 16.1 **Structural adaptations of algae and plants.**

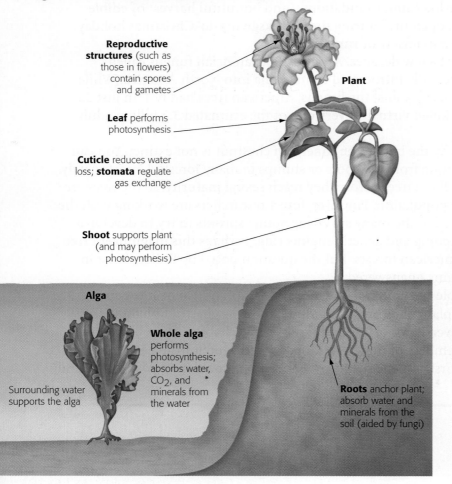

Reproductive **structures** (such as those in flowers) contain spores and gametes

Plant

Leaf performs photosynthesis

Cuticle reduces water loss; **stomata** regulate gas exchange

Shoot supports plant (and may perform photosynthesis)

Alga

Whole alga performs photosynthesis; absorbs water, CO_2, and minerals from the water

Surrounding water supports the alga

Roots anchor plant; absorb water and minerals from the soil (aided by fungi)

▼ Figure 16.2 **Mycorrhizae: symbiotic associations of fungi and roots.** The finely branched filaments of the fungus (white in the photo) provide an extensive surface area for absorption of water and minerals from the soil.

Roots

Fungus

Root surrounded by fungus

photosynthetic interior of a leaf occurs via **stomata** (singular, *stoma*), the microscopic pores found on a leaf's surface (see Figure 7.2). A waxy layer called the **cuticle** coats the leaves and other aerial parts of most plants, helping the plant body retain water. (You've probably noticed the waxy surface of some houseplant leaves.)

Differentiation of the plant body into shoot and root systems solved one problem but created new ones. For the shoot system to stand up straight in the air, it must have support. This is not a major problem in the water: Huge seaweeds do not need skeletons because the surrounding water buoys them. An important terrestrial adaptation of plants is **lignin**, a chemical that hardens cell walls. Imagine what would happen to you if your skeleton were to suddenly turn mushy. A tree would also collapse if it were not for its "skeleton," its framework of lignin-rich cell walls.

Specialization of the plant body into shoots and roots also introduced the problem of transporting vital materials between the distant organs. The terrestrial equipment of most plants includes **vascular tissue**, a system of tube-shaped cells that branch throughout the plant (**Figure 16.3**). The vascular tissue actually has two types of tissues specialized for transport: **xylem**, consisting of dead cells with tubular cavities for transporting water and minerals from roots to leaves; and **phloem**, consisting of living cells that distribute sugars from the leaves to the roots and other nonphotosynthetic parts of the plant.

Reproductive Adaptations

Adapting to land also required a new mode of reproduction. For algae, the surrounding water ensures that gametes (sperm and eggs) and developing offspring stay moist. The aquatic environment also provides a means of dispersing the gametes and offspring. Plants, however, must keep their gametes and developing offspring from drying out in the air. Plants (and some algae) produce their gametes in protective structures that are called **gametangia** (singular, *gametangium*). A gametangium has a jacket of protective cells surrounding a moist chamber where gametes can develop without dehydrating.

For most plants, sperm reach the eggs by traveling inside pollen grains, which are carried by wind or animals. The egg remains within tissues of the mother plant and is fertilized there. In plants, but not algae, the zygote (fertilized egg) develops into an embryo while still contained within the female parent, which protects the embryo and keeps it from dehydrating (**Figure 16.4**). Most plants rely on wind or animals, such as fruit-eating birds or mammals, to disperse their offspring, which are in the form of embryos contained in seeds. ☑

◄ Figure 16.3 **Network of vascular tissue in a leaf.** The vascular tissue of the leaf in the photograph is visible as yellow veins. This tissue delivers water and minerals absorbed by the roots and carries away the sugars produced in the leaves.

Phloem

Xylem

Vascular tissue

Oak leaf

▼ Figure 16.4 **The protected embryo of a plant.** Internal fertilization, with sperm and egg combining within a moist chamber on the mother plant, is an adaptation for living on land. The female parent continues to nurture and protect the plant embryo, which develops from the zygote.

LM 55x

Embryo

Maternal tissue

The Origin of Plants from Green Algae

The algal ancestors of plants carpeted moist fringes of lakes or coastal salt marshes over 500 million years ago. These shallow-water habitats were subject to occasional drying, and natural selection would have favored algae that could survive periodic droughts. Some species accumulated adaptations that enabled them to live permanently above the water line. A modern-day lineage of green algae, the **charophytes** (**Figure 16.5**), may resemble one of these early plant ancestors. Plants and present-day charophytes probably evolved from a common ancestor.

Adaptations making life on dry land possible had accumulated by about 475 million years ago, the age of the oldest known plant fossils. The evolutionary novelties of these first land plants opened the new frontier of a terrestrial habitat. Early plant life would have thrived in the new environment. Bright sunlight was abundant on land, the atmosphere had a wealth of carbon dioxide, and at first there were relatively few pathogens and plant-eating animals. The stage was set for an explosive diversification of plant life.

▼ Figure 16.5 **Two species of charophytes, the closest algal relatives of plants.**

LM 85x

LM 265x

Evolution of Plants

Plant Diversity

As we survey the diversity of modern plants, remember that the past is the key to the present. The history of the plant kingdom is a story of adaptation to diverse terrestrial habitats.

Bacteria

Archaea

Eukarya | Protists
Plants
Fungi
Animals

Highlights of Plant Evolution

The fossil record chronicles four major periods of plant evolution, which are also evident in the diversity of modern plants (**Figure 16.6**). Each stage is marked by the evolution of structures that opened new opportunities on land.

1 After plants originated from an algal ancestor approximately 475 million years ago, early diversification gave rise to nonvascular plants, including mosses, liverworts, and hornworts. These plants, called **bryophytes**, lack true roots and leaves. Bryophytes also lack lignin, the wall-hardening material that enables other plants to stand tall. Without lignified cell walls, bryophytes have weak upright support. The most familiar bryophytes are **mosses**. A mat of moss actually consists of many plants growing in a tight pack, holding one another up. Gametangia, which protect the gametes and embryos, are a terrestrial adaptation that originated in bryophytes.

2 The second period of plant evolution, begun about 425 million years ago, was the diversification of

▼ Figure 16.6 **Highlights of plant evolution.** This phylogenetic tree highlights the evolution of structures that allowed plants to move onto land; these structures still exist in modern plants. As we survey the diversity of plants, miniature versions of this tree will help you place each plant group in its evolutionary context.

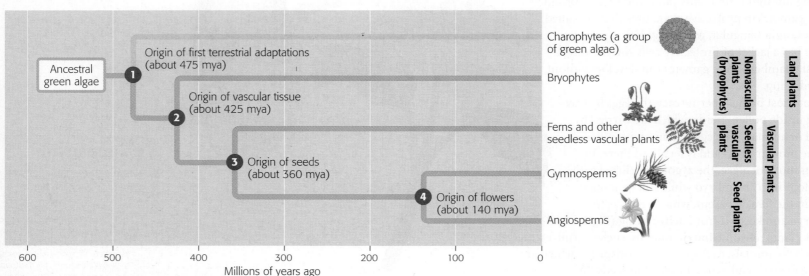

Ancestral green algae

1 Origin of first terrestrial adaptations (about 475 mya)

2 Origin of vascular tissue (about 425 mya)

3 Origin of seeds (about 360 mya)

4 Origin of flowers (about 140 mya)

Charophytes (a group of green algae)

Bryophytes

Ferns and other seedless vascular plants

Gymnosperms

Angiosperms

Nonvascular plants (bryophytes)

Seedless vascular plants

Seed plants

Land plants

Vascular plants

600   500   400   300   200   100   0
Millions of years ago

plants with vascular tissue. The presence of conducting tissues hardened with lignin allowed vascular plants to grow much taller, rising above the ground to achieve significant height. The earliest vascular plants lacked seeds. Today, this seedless condition is retained by **ferns** and a few other groups of vascular plants.

3 The third major period of plant evolution began with the origin of the seed about 360 million years ago. Seeds advanced the colonization of land by further protecting plant embryos from drying and other hazards. A **seed** consists of an embryo packaged along with a store of food within a protective covering. The seeds of early seed plants were not enclosed in any specialized chambers. These plants gave rise to the **gymnosperms** ("naked seeds"). Today, the most widespread and diverse gymnosperms are the **conifers**, consisting mainly of cone-bearing trees, such as pines.

4 The fourth major episode in the evolutionary history of plants was the emergence of flowering plants, or **angiosperms** ("contained seeds"), at least 140 million years ago. The **flower** is a complex reproductive structure that bears seeds within protective chambers called ovaries. This contrasts with the naked seeds of gymnosperms. The great majority of living plants—some 250,000 species—are angiosperms, including all our fruit and vegetable crops, grains, grasses, and most trees.

With these highlights as our framework, we are now ready to survey the four major groups of modern plants: bryophytes, ferns, gymnosperms, and angiosperms (**Figure 16.7**). ☑

▼ Figure 16.7 **The major groups of plants.**

PLANT DIVERSITY			
Bryophytes (nonvascular plants)	**Ferns** (seedless vascular plants)	**Gymnosperms** (naked-seed plants)	**Angiosperms** (flowering plants)

Bryophytes

Mosses, which are bryophytes, may sprawl as low mats over acres of land (**Figure 16.8**). Mosses display two of the key terrestrial adaptations that made the move onto land possible: (1) a waxy cuticle that helps prevent dehydration and (2) the retention of developing embryos within the mother plant's gametangium. However, mosses are not totally liberated from their ancestral aquatic habitat. Mosses need water to reproduce because their sperm need to swim to reach eggs within the female gametangium. (A film of rainwater or dew is enough moisture for the sperm to travel.) In addition,

Bryophytes

Ferns

Gymnosperms

Angiosperms

▼ Figure 16.8 **A peat moss bog in Scotland.** Mosses are bryophytes, which are nonvascular plants. Peat mosses, or sphagnum, carpet at least 3% of Earth's terrestrial surface. They are most commonly found in high northern latitudes. The ability of peat moss to absorb and retain water makes it an excellent addition to garden soil.

▼ **Figure 16.9 The two forms of a moss.** The feathery plant we generally know as a moss is the gametophyte. The stalk with the capsule at its tip is the sporophyte. This photo shows the capsule releasing its tiny spores.

Spores
Spore capsule
Sporophyte
Gametophytes

Alternation of Generations

▶ **Figure 16.10 Alternation of generations.** Plants have life cycles very different from ours. Each of us is a diploid individual; the only haploid stages in the human life cycle, as for nearly all animals, are sperm and eggs. By contrast, plants have alternating generations: Diploid (2n) individuals (sporophytes) and haploid (n) individuals (gametophytes) generate each other in the life cycle.

because most mosses have no vascular tissue to carry water from soil to aerial parts of the plant, they need to live in damp, shady places.

If you examine a mat of moss closely, you can see two distinct forms of the plant. The green, sponge-like plant that is the more obvious is called the **gametophyte**. Careful examination will reveal the other form of the moss, called a **sporophyte**, growing out of a gametophyte as a stalk with a capsule at its tip (**Figure 16.9**). The cells of the gametophyte are haploid (they have one set of chromosomes; see Chapter 8). In contrast, the sporophyte is made up of diploid cells (with two chromosome sets). These two different stages of the plant life cycle are named for the types of reproductive cells they produce. Gametophytes produce gametes (sperm and eggs), while sporophytes produce spores. As reproductive cells, **spores** differ from gametes in two ways: A spore can develop into a new organism without fusing with another cell (two gametes must fuse to form a zygote); and spores usually have tough coats that enable them to resist harsh environments (whereas gametes must stay moist).

The gametophyte and sporophyte are alternating generations that take turns producing each other. Gametophytes produce gametes that unite to form zygotes, which develop into new sporophytes. And sporophytes produce spores that give rise to new gametophytes. This type of life cycle, called **alternation of generations**, occurs only in plants and multicellular green algae (**Figure 16.10**). Among plants, mosses and other bryophytes are unique in having the gametophyte as the larger, more obvious plant. As we continue our survey of plants, we'll see an increasing dominance of the sporophyte as the more highly developed generation. ☑

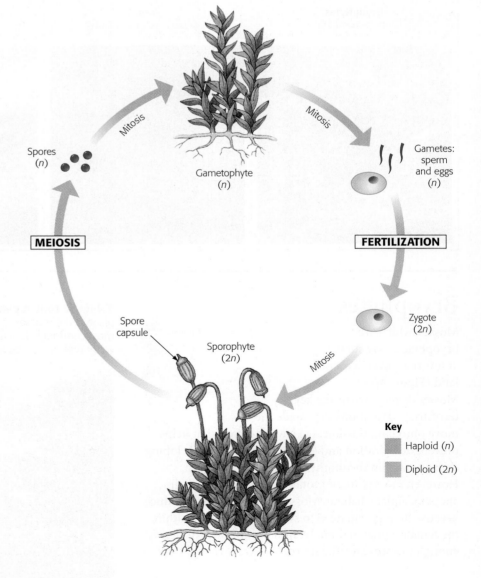

Spores (n)

Mitosis

Gametophyte (n)

Mitosis

Gametes: sperm and eggs (n)

MEIOSIS

FERTILIZATION

Spore capsule

Sporophyte (2n)

Mitosis

Zygote (2n)

Key
Haploid (n)
Diploid (2n)

Ferns

Ferns took terrestrial adaptation to the next level with the evolution of vascular tissue. Ferns are by far the most diverse seedless vascular plants, with more than 12,000 known species. However, the sperm of ferns, like those of mosses, have flagella and must swim through a film of water to fertilize eggs. Most ferns inhabit the tropics, although many species are found in temperate forests, such as many woodlands in the United States (**Figure 16.11**).

During the Carboniferous period, from about 360 to 300 million years ago, ancient ferns were part of a much greater diversity of seedless plants that formed vast, swampy tropical forests over much of what is now Eurasia and North America (**Figure 16.12**). As the plants died, they fell into stagnant wetlands and did not decay completely. Their remains formed thick organic deposits called peat. Later, seawater flooded the swamps, marine sediments covered the peat, and pressure and heat gradually converted the peat to coal. Coal is black sedimentary rock made up of fossilized plant material. Like coal, oil and natural gas also formed from the remains of long-dead organisms; thus, all three are known as **fossil fuels**. Since the Industrial Revolution, coal has been a crucial source of energy for human society. However, burning these fossil fuels releases CO_2 and other greenhouse gases into the atmosphere, contributing to global climate change. ☑

▼ Figure 16.11 **Ferns (seedless vascular plants).** The fern species in the foreground grows on the forest floor in the eastern United States. The fern generation familiar to us is the sporophyte generation. You would have to crawl on the forest floor and explore with careful hands and sharp eyes to find fern gametophytes (upper right), tiny plants growing on or just below the soil surface.

Spore capsule

"Fiddlehead" (young leaves ready to unfurl)

◄ Figure 16.12 **A "coal forest" of the Carboniferous period.** This painting, based on fossil evidence, reconstructs one of the great seedless forests. Most of the large trees with straight trunks are seedless vascular plants called lycophytes. On the left, the tree with numerous feathery branches is another type of seedless vascular plant called a horsetail. The plants near the base of the trees are ferns.

☑ CHECKPOINT

Why are ferns able to grow taller than mosses?

Answer: Vascular tissue hardened with lignin allows ferns to stand taller and transport nutrients farther.

Gymnosperms

"Coal forests" dominated the North American and Eurasian landscapes until near the end of the Carboniferous period. At that time, the global climate turned drier and colder, and the vast swamps began to disappear. This climatic change provided an opportunity for seed plants, which can complete their life cycles on dry land and withstand long, harsh winters. Of the earliest seed plants, the most successful were the gymnosperms, and several kinds grew along with the seedless plants in the Carboniferous swamps. Their descendants include the conifers, or cone-bearing plants. Let's take a closer look at the conifers before discussing the adaptations that allowed seed plants to dominate the land.

Conifers

Perhaps you have had the fun of hiking or skiing through a forest of conifers, the most common gymnosperms. Pines, firs, spruces, junipers, cedars, and redwoods are all conifers. A broad band of coniferous forests covers much of northern Eurasia and North America and extends southward in mountainous regions (**Figure 16.13**). Today, about 190 million acres of coniferous forests in the United States, mostly in the western states and Alaska, are designated national forests.

Conifers are among the tallest, largest, and oldest organisms on Earth. Coastal redwoods, native to the northern California coast, are the world's tallest trees. Three coastal redwoods discovered in 2006 are more than 110 m (370 ft)

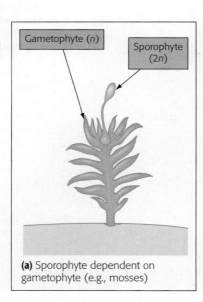

▼ Figure 16.13 **A coniferous forest in Banff National Park, in Alberta, Canada.** Coniferous forests are widespread in northern North America and Eurasia; conifers also grow in the Southern Hemisphere, though they are less numerous there.

tall. Some of the most massive organisms alive are the giant sequoias, relatives of redwoods that grow in the Sierra Nevada mountains of California. One, known as the General Sherman tree, is about 84 m (275 ft) high and weighs more than the combined weight of a dozen space shuttles. Bristlecone pines, another species of California conifer, are among the oldest organisms alive. One bristlecone, named Methuselah, is more than 4,600 years old; it was a young tree when humans invented writing.

Nearly all conifers are evergreens, meaning they retain leaves throughout the year. Even during winter, they perform a limited amount of photosynthesis on sunny days. And when spring comes, conifers already have fully developed leaves that can take advantage of the sunnier days. The needle-shaped leaves of pines and firs are also adapted to survive dry seasons. A thick cuticle covers the leaf, and the stomata are located in pits, further reducing water loss.

Coniferous forests are highly productive; you probably use products harvested from them every day. For example, conifers provide much of our lumber for building and wood pulp for paper production. What we call wood is actually an accumulation of vascular tissue with lignin, which gives the tree structural support.

Terrestrial Adaptations of Seed Plants

Compared to ferns, most gymnosperms have three additional adaptations that make survival in diverse terrestrial habitats possible: (1) further reduction of the gametophyte, (2) pollen, and (3) seeds.

The first adaptation is an even greater development of the diploid sporophyte compared to the haploid gametophyte generation (**Figure 16.14**). A pine tree or other conifer is actually a sporophyte with tiny

▶ Figure 16.14 **Three variations on alternation of generations in plants.**

Key

Haploid (n)

Diploid (2n)

(a) Sporophyte dependent on gametophyte (e.g., mosses)

(b) Large sporophyte and small, independent gametophyte (e.g., ferns)

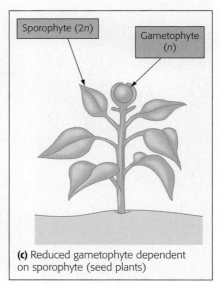

(c) Reduced gametophyte dependent on sporophyte (seed plants) gametophytes living in its cones (Figure 16.15). In contrast to bryophytes and ferns, gymnosperm gametophytes are totally dependent on and protected by the tissues of the parent sporophyte.

A second adaptation of seed plants to dry land came with the evolution of pollen. A **pollen grain** is actually the much-reduced male gametophyte; it houses cells that will develop into sperm. In the case of conifers, wind carries the pollen from male to female cones, where eggs develop within female gametophytes. This mechanism for sperm transfer contrasts with the swimming sperm of mosses and ferns. In seed plants, this use of tough, airborne pollen that carries sperm to egg is a terrestrial adaptation that led to even greater success and diversity of plants on land.

The third important terrestrial adaptation of seed plants is the seed itself. A seed consists of a plant embryo packaged along with a food supply within a protective coat. Seeds develop from structures called **ovules** (Figure 16.16). In conifers, the ovules are located on the scales of female cones. Conifers and other gymnosperms, lacking ovaries, bear their seeds "naked" on

the cone scales (though the seeds do have protective coats). Once released from the parent plant, the seed can remain dormant for days, months, or even years. Under favorable conditions, the seed can then **germinate**: Its embryo emerges through the seed coat as a seedling. Some seeds drop close to their parents, while others are carried far by the wind or animals. ☑

▼ Figure 16.16 **From ovule to seed.**

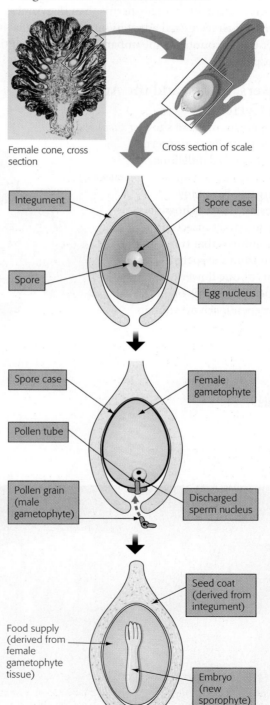

Female cone, cross section

Cross section of scale

Key

■ Haploid (*n*)

■ Diploid (2*n*)

(a) **Ovule.** The sporophyte produces spores within a tissue surrounded by a productive integument, which may be multilayered.

Integument

Spore case

Spore

Egg nucleus

(b) **Fertilized ovule.** The spore develops into a female gametophyte, which produces one or more eggs. When a pollen grain enters the ovule through a special pore in the integument, it discharges sperm cells that fertilize eggs.

Spore case

Female gametophyte

Pollen tube

Pollen grain (male gametophyte)

Discharged sperm nucleus

(c) **Seed.** Fertilization initiates the transformation of ovule to seed. The fertilized egg (zygote) develops into an embryo; the rest of the gametophyte forms a tissue that stockpiles food; and the integument of the ovule hardens to become the seed coat.

Seed coat (derived from integument)

Food supply (derived from female gametophyte tissue)

Embryo (new sporophyte)

▼ Figure 16.15 **A pine tree, the sporophyte, bearing two types of cones containing gametophytes.** Each scale of the female cone is actually a modified leaf that bears a structure called an ovule containing a female gametophyte. Male cones release clouds of millions of pollen grains, the male gametophytes. Some of these pollen grains land on female cones on trees of the same species. The sperm can fertilize eggs in the ovules of the female cones. The ovules eventually develop into seeds.

Scale

Ovule-producing cones; the scales contain female gametophytes

Pollen-producing cones; they produce male gametophytes

Ponderosa pine

Angiosperms

Angiosperms dominate
the modern landscape.
There are about 250,000
angiosperm species versus
about 700 species of gymnosperms. Several unique adaptations
account for the success of angiosperms. For example,
refinements in vascular tissue make water transport
even more efficient in angiosperms than in gymnosperms. Of all terrestrial adaptations, however, it is the
flower that accounts for the unparalleled success of the
angiosperms.

Flowers, Fruits, and the Angiosperm Life Cycle

No organism makes a showier display of its
sex life than the angiosperm.
From roses to dandelions,
flowers display a plant's
sex organs. For many
angiosperms, this showiness
helps to attract insects and
other animals that transfer
pollen from the sperm-bearing
organs of one flower to the
egg-bearing organs of another.
This dependence on animals

for pollen transfer targets the pollen to other plants of
the same species, rather than relying on uncertain winds
to blow the pollen around.

A flower is actually a short stem with four whorls of
modified leaves: sepals, petals, stamens, and carpels
(**Figure 16.17**). At the bottom of the flower are the
sepals, which are usually green. They enclose the flower
before it opens (think of the green "wrapping" on a
rosebud). Above the sepals are the **petals**, which are
often colorful and help to attract insects and other pollinators. The actual reproductive structures are multiple
stamens and one or more carpels. Each **stamen** consists
of a stalk—the **filament**—bearing a sac called an
anther, in which the pollen grains develop. The **carpel**
consists of a stalk—the **style**—with an ovary at the base
and a sticky tip known as the **stigma**,
which traps pollen. The **ovary** is a protective chamber containing one or
more ovules, in which the eggs
develop. As you can see in
Figure 16.18, the basic
structure of a flower can
exist in many beautiful
variations.

◄ Figure 16.17 **Structure of a flower.**

Bryophytes

Fems

Gymnosperms

Angiosperms

Petal

Stamen — Anther
Filament

Stigma
Style   — Carpel
Ovary

Ovule

Sepal

▼ Figure 16.18 **A diversity of flowers.**

Pansy

Bleeding heart

California poppy

Water lily

▼ Figure 16.19 **The angiosperm life cycle.**

Germinated pollen grain (male gametophyte) on stigma of carpel

Anther at tip of stamen

Pollen tube growing down style of carpel

Ovary (base of carpel)

Ovule

FERTILIZATION

Embryo sac (female gametophyte)

Endosperm

Egg

Zygote

Two sperm nuclei

Embryo (sporophyte)

Mature sporophyte plant with flowers

Sporophyte seedling

Seed

Germinating seed

Seed (develops from ovule)

Fruit (develops from ovary)

Key

Haploid (*n*)

Diploid (2*n*)

Figure 16.19 highlights key stages in the angiosperm life cycle. ❶ The flower we see is part of the sporophyte plant. As in gymnosperms, the pollen grain is the male gametophyte of angiosperms. The female gametophyte—the embryo sac—is located within an ovule, which in turn resides within a chamber of the ovary. ❷ A pollen grain that lands on the sticky stigma of a carpel extends a tube down to the ovule and ❸ deposits two sperm nuclei within the embryo sac. This **double fertilization** is an angiosperm characteristic. One sperm cell fertilizes an egg in the embryo sac. ❹ This produces a zygote, which ❺ develops into an embryo. The second sperm cell "fertilizes" another female gametophyte cell, which then develops into a nutrient-storing tissue called **endosperm**, which nourishes the embryo. Double fertilization therefore synchronizes the development of the embryo and food reserves within an ovule. ❻ The whole ovule develops into a seed. The seed's enclosure within an ovary is what distinguishes angiosperms from gymnosperms, which have a naked seed.

A **fruit** is the ripened ovary of a flower. As seeds are developing from ovules, the ovary wall thickens, forming the fruit that encloses the seeds. A pea pod is an example of a fruit, with seeds (mature ovules, the peas) encased in the ripened ovary (the pod). Fruits protect and help disperse seeds. As **Figure 16.20** demonstrates, many angiosperms depend on animals to disperse seeds. Conversely, most land animals, including humans, rely on angiosperms as a food source, directly or indirectly. ☑

Angiosperms and Agriculture

Whereas gymnosperms supply most of our lumber and paper, angiosperms supply nearly all of our food—as well as the food eaten by domesticated animals, such as cows and chickens. Over 90% of the plant kingdom is made up of angiosperms, including cereal grains such as wheat and corn, citrus and other fruit trees, garden vegetables, and cotton. Many types of garden produce—tomatoes, squash, strawberries, and oranges, to name just a few—are the edible fruits of plants we have domesticated. Fine hardwoods from flowering plants such as oak, cherry, and walnut trees supplement the lumber we get from conifers. We also grow angiosperms for fiber, medications, perfumes, and decoration.

Early humans probably collected wild seeds and fruits. Agriculture gradually developed as people began sowing seeds and cultivating plants to have a more dependable food source. And as they domesticated certain plants, people began to select those with improved yield and quality. Agriculture can thus be seen as yet another facet of the evolutionary relationship between plants and animals.

▼ Figure 16.20 **Fruits and seed dispersal.** Different types of fruits are adapted for different methods of dispersal.

Wind dispersal. Some angiosperms depend on wind for seed dispersal. Here, you can see seeds blowing off a plant called great reedmace.

Animal transportation. Some fruits are adapted to hitch free rides on animals. The cockleburs attached to the fur of this dog may be carried miles before opening and releasing seeds.

Animal ingestion. Many angiosperms produce fleshy, edible fruits that are attractive to animals as food. When a weasel eats a berry, it digests the fleshy part of the fruit, but most of the tough seeds pass unharmed through the animal's digestive tract. The weasel later deposits the seeds, along with a fertilizer supply, some distance from where it ate the fruit.

Plant Diversity as a Nonrenewable Resource

The exploding human population, with its demand for space and natural resources, is extinguishing plant species at an unprecedented rate. The problem is especially critical in the tropics, where more than half the human population lives and population growth is fastest. Tropical rain forests are being destroyed at a frightening pace (Figure 16.21). The most common cause of this destruction is large-scale slash-and-burn clearing of forest for agricultural use. Fifty million acres, an area about the size of the state of Washington, are cleared each year, a rate that could completely eliminate Earth's tropical forests within 25 years. As the forest disappears, so do thousands of plant species. Insects and other rain forest animals that depend on these plants are also vanishing. In all, researchers estimate that the destruction of habitat in the rain forest and other ecosystems is claiming hundreds of species each year. The toll is greatest in the tropics because that is where most species live; but environmental assault seems to be a general human tendency. Europeans eliminated most of their forests centuries ago, and habitat destruction is now endangering many species in North America. What is lost is irreplaceable—entire ecosystems that provide medicinal plants, food, timber, and clean water and air.

Many people have ethical concerns about contributing to the extinction of living forms. But there are also practical reasons to be concerned about the loss of plant diversity. As already mentioned, we depend on plants for thousands of products, including food, building materials, and medicines (Table 16.1). More than 120 prescription drugs are extracted from plants. However, researchers have investigated fewer than 5,000 of the 300,000 known plant species as potential sources of medicine. Pharmaceutical companies were led to most of these species by local peoples who use the plants in preparing their traditional medicines.

Scientists are now rallying to slow the loss of plant diversity, in part by offering less destructive ways for people to benefit from forests. The goal of such efforts is to encourage management practices that use forests as resources without damaging them. The solutions we propose must be economically realistic; people who live where there are tropical rain forests must be able to make a living. But if the only goal is profit for the short term, then we will continue to slash and burn until the forests are gone. We need to appreciate the rain forests and other ecosystems as living treasures that can regenerate only slowly. Only then will we learn to work with them in ways that preserve their biological diversity for the future.

Throughout our survey of plants in this chapter, we have seen how entangled the botanical world is with other terrestrial life. We switch our attention now to that other group of organisms that moved onto land with plants: the kingdom Fungi. ✓

✓ CHECKPOINT

In what way are forests renewable resources? In what way are they not?

Answer: Forests are renewable in the sense that new trees can grow where old growth has been removed by logging. Habitats that are permanently destroyed cannot be replaced, however, so forests must be harvested in a sustainable manner.

▼ Figure 16.21 **Clear-cutting of a tropical forest in Brazil.**

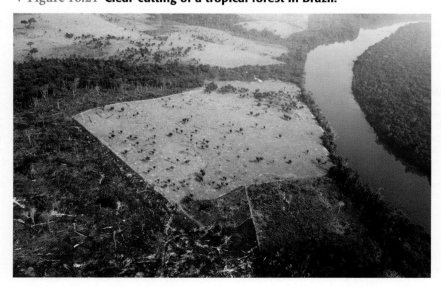

Table 16.1	A Sampling of Medicines Derived from Plants		
Compound	**Source**		**Example of Use**
Atropine	Belladonna plant		Pupil dilator in eye exams
Digitalin	Foxglove		Heart medication
Menthol	Eucalyptus tree		Ingredient in cough medicines
Morphine	Opium poppy		Pain reliever
Quinine	Quinine tree		Malaria preventive
Paclitaxel (Taxol)	Pacific yew		Ovarian cancer drug
Tubocurarine	Curare tree		Muscle relaxant during surgery
Vinblastine	Periwinkle		Leukemia drug

Source: Adapted from Randy Moore et al., Botany, 2nd ed. Dubuque, IA: Brown, 1998. Table 2.2, p. 37.

Fungi

The word *fungus* often evokes unpleasant images. Fungi rot timbers, spoil food, and afflict humans with athlete's foot and worse. However, ecosystems would collapse without fungi to decompose dead organisms, fallen leaves, feces, and other organic materials. Fungi recycle vital chemical elements back to the environment in forms that other organisms can assimilate. And you have already learned that nearly all plants have mycorrhizae, fungus-root associations that help plants absorb minerals and water from the soil. In addition to these ecological roles, fungi have been used by people in various ways for centuries. We eat fungi (mushrooms and truffles, for instance), culture fungi to produce antibiotics and other drugs, add them to dough to make bread rise, culture them in milk to produce a variety of cheeses, and use them to ferment beer and wine.

Fungi are eukaryotes, and most are multicellular, but many have body structures and modes of reproduction unlike those of any other organism (**Figure 16.22**). Molecular studies indicate that fungi and animals arose from a common ancestor around 1.5 billion years ago. The oldest undisputed fossils of fungi, however, are only about 460 million years old, perhaps because the ancestors of terrestrial fungi were microscopic and fossilized poorly. Despite appearances, a mushroom is more closely related to you than it is to any plant!

Biologists who study fungi have described over 100,000 species, and there may be as many as 1.5 million. Classifying fungi is an ongoing area of research. (One widely accepted phylogenetic tree divides the kingdom Fungi into five groups.) You are probably familiar with several kinds of fungi, including mushrooms, mold, and yeast. In this section, we'll discuss the characteristics common to all fungi and then survey their wide-ranging ecological impact. ✓

✓CHECKPOINT

Name three ways that we benefit from fungi in our environment.

Answer: Fungi help recycle nutrients by decomposing dead organisms; mycorrhizae help plants absorb water and nutrients; and some fungi serve us as food.

▼ Figure 16.22 **A gallery of diverse fungi.**

Orange fungi. These are the reproductive structures of a fungus that absorbs nutrients as it decomposes a fallen tree on a forest floor.

A "fairy ring." Some mushroom-producing fungi poke up "fairy rings," which can appear on a lawn overnight. A ring develops at the edge of the main body of the fungus, which consists of an underground mass of tiny filaments (hyphae) within the ring. As the underground fungal mass grows outward from its center, the diameter of the fairy ring produced at its expanding perimeter increases annually.

Mold. Molds grow rapidly on their food sources, which are often our food sources as well. The mold on this orange reproduces asexually by producing chains of microscopic spores (inset) that are dispersed via air currents.

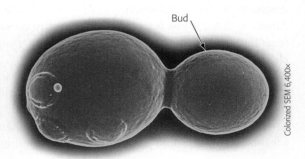
Budding yeast. Yeasts are unicellular fungi. This yeast cell is reproducing asexually by a process called budding.

Predatory fungus. This predatory fungus traps and feeds on tiny roundworms in the soil. The fungus is equipped with hoops that can constrict around a worm in less than a second.

Characteristics of Fungi

We'll begin our look at the structure and function of fungi with an overview of how fungi obtain nutrients.

Fungal Nutrition

Fungi are chemoheterotrophs (see Figure 15.12) that acquire their nutrients by **absorption**. In this mode of nutrition, small organic molecules are absorbed from the surrounding medium. A fungus digests food outside its body by secreting powerful digestive enzymes into the food. The enzymes decompose complex molecules to simpler compounds that the fungus can absorb. Fungi absorb nutrients from such nonliving organic material as fallen logs, animal corpses, and the wastes of live organisms.

Fungal Structure

The bodies of most fungi are constructed of threadlike filaments called **hyphae** (singular, *hypha*). Fungal hyphae are minute threads of cytoplasm surrounded by a plasma membrane and cell wall. The cell walls of fungi differ from the cellulose walls of plants. Fungal cell walls are usually built mainly of chitin, a strong but flexible polysaccharide that is also found in the external skeletons of insects. Most fungi have multicellular hyphae, which consist of chains of cells separated by cross-walls with pores. In many fungi, cell-to-cell channels allow ribosomes, mitochondria, and even nuclei to flow between cells.

Fungal hyphae branch repeatedly, forming an interwoven network called a **mycelium** (plural, *mycelia*), the feeding structure of the fungus (**Figure 16.23**). Fungal mycelia usually escape our notice because they are often subterranean, but they can be huge. In fact, scientists have discovered that the mycelium of one humongous fungus in Oregon is 5.5 kilometers (km)—that's 3.4 miles!—in diameter and spreads through 2,200 acres of forest. This fungus is at least 2,600 years old and weighs hundreds of tons, qualifying it as one of Earth's oldest and largest organisms.

A mycelium maximizes contact with its food source by mingling with the organic matter it is decomposing and absorbing. A bucketful of rich organic soil may contain as much as a kilometer of hyphae. A fungal mycelium grows rapidly, adding hyphae as it branches within its food. The great majority of fungi are nonmotile; they cannot run, swim, or fly in search of food. But the mycelium makes up for the lack of mobility by swiftly extending the tips of its hyphae into new territory. ☑

Fungal Reproduction

The mushroom in Figure 16.23 is actually made up of tightly packed hyphae. Mushrooms arise from an underground mycelium. While the mycelium obtains food from organic material via absorption, the function of the mushroom is reproduction. It must be above ground to disperse its spores on air currents.

Fungi typically reproduce by releasing haploid spores that are produced either sexually or asexually. The output of spores is mind-boggling. For example, puffballs, which are the reproductive structures of certain fungi, can spew clouds containing trillions of spores (see Figure 13.12). Easily carried by wind or water, spores germinate to produce mycelia if they land in a moist place where there is food. Spores thus function in dispersal and account for the wide geographic distribution of many species of fungi. The airborne spores of fungi have been found more than 160 km (100 miles) above Earth. Closer to home, try leaving a slice of bread out for a week and you will observe the furry mycelia that grow from the invisible spores raining down from the surrounding air.

Reproductive structure

Hyphae

Spore-producing structures

Mycelium

◀ **Figure 16.23 The fungal mycelium.** A mushroom consists of tightly packed hyphae that extend upward from a much more massive mycelium of hyphae growing underground. The photo shows a mycelium made up of the cottony threads that decompose organic litter.

☑**CHECKPOINT**

Describe how the structure of a fungal mycelium reflects its function.

Answer: The extensive network of hyphae puts a large surface area in contact with the food source.

The Ecological Impact of Fungi

Fungi have been major players in terrestrial communities ever since plants and fungi together moved onto land. Let's examine a few examples of how fungi continue to have an enormous ecological impact, including numerous interactions with humans.

Fungi as Decomposers

Fungi and bacteria are the principal decomposers that keep ecosystems stocked with the inorganic nutrients essential for plant growth. Without decomposers, carbon, nitrogen, and other elements would accumulate in nonliving organic matter. Plants and the animals they feed would starve because elements taken from the soil would not be returned (see Chapter 20).

Fungi are well adapted as decomposers of organic refuse. Their invasive hyphae enter the tissues and cells of dead organisms and digest polymers, including the cellulose of plant cell walls. A succession of fungi, in concert with bacteria and, in some environments, invertebrate animals, is responsible for the complete breakdown of organic litter. The air is so loaded with fungal spores that as soon as a leaf falls or an insect dies, it is covered with spores and soon after infiltrated by fungal hyphae.

We may applaud fungi that decompose forest litter or dung, but it's a different story when molds attack our fruit or our shower curtains. A significant amount of the world's fruit harvest is lost each year to fungal attack. And a wood-digesting fungus does not distinguish between a fallen oak limb and the oak planks of a boat. During the Revolutionary War, the British lost more ships to fungal rot than to enemy attack. What's more, soldiers stationed in the tropics during World War II watched as their tents, clothing, boots, and binoculars were destroyed by molds.

Parasitic Fungi

Parasitism is a relationship in which two species live in contact and one organism benefits while the other is harmed. Of the 100,000 known species of fungi, about 30% make their living as parasites. Parasitic fungi absorb nutrients from the cells or body fluids of living hosts.

About 50 species of fungi are known to be parasitic in humans and other animals. Among the diseases that fungi cause in humans are yeast infections of the lungs, some of which can be fatal, and vaginal yeast infections. Other fungal parasites produce a skin disease called ringworm, so named because it appears as circular red areas on the skin. Most commonly, ringworm fungus attacks the feet and causes intense itching and sometimes blisters. This condition, known as athlete's foot, is highly contagious but can be treated with fungicidal lotions and powders.

The great majority of fungal parasites infect plants. In some cases, fungi that infect plants have literally changed landscapes. In the Biology and Society section, you learned how a fungal blight devastated the American chestnut. A related fungus killed most American elm trees (**Figure 16.24a**). Fungi are also serious agricultural pests, and some of the fungi that attack food crops are toxic to humans. The seed heads of many kinds of grain and grasses, including rye, wheat, and oats, are sometimes infected with fungal growths called ergots (**Figure 16.24b**). Consumption of flour made from ergot-infested grain can cause hallucinations, temporary insanity, and death. In fact, lysergic acid, the raw material from which the hallucinogenic drug LSD is made, has been isolated from ergots. This fact may help explain a centuries-old mystery, as we'll see next.

(a) American elm trees killed by Dutch elm disease fungus

(b) Ergots

◄ Figure 16.24 **Parasitic fungi that cause plant disease. (a)** The parasitic fungus that causes Dutch elm disease evolved with European species of elm trees, and it is relatively harmless to them. But it has been deadly to American elms since it was accidentally introduced after World War I. **(b)** Ergots, parasitic fungi, are the dark structures on these rye seed heads.

Did a Fungus Lead to the Salem Witch Hunt?

In January 1692, eight young girls in the town of Salem, Massachusetts, began to act bizarrely. The girls suffered from incomprehensible speech, odd skin sensations, convulsions, and hallucinations. The worried community blamed the girls' symptoms on witchcraft and began to accuse one another. By the time the hysteria ended that autumn, more than 150 villagers had been accused of witchcraft, and 20 of them had been hanged. Finding the cause behind the "Salem witch hunt" has long intrigued historians.

In 1976, a University of California psychology graduate student offered a new explanation. She began with the **observation** that the symptoms reported by the girls were consistent with ergot poisoning (**Figure 16.25**). This led her to **question** whether an ergot outbreak could have been behind the witch hunt. The researcher tested her **hypothesis** by examining the historical records, making the **prediction** that facts consistent with ergot poisoning would be uncovered.

Her **results** were suggestive, though not conclusive. Agricultural records confirm that rye—the principal host for ergot—grew abundantly around Salem at that time and that the growing season of 1691 had been particularly warm and wet, conditions under which ergot thrives. This suggests that the rye crop consumed during the winter of 1691–1692 could easily have been contaminated. The summer of 1692, when the accusations began to die down, was dry, consistent with an ergot die-off. Most importantly, the reported symptoms appear consistent with those of ergot poisoning. These clues suggest (but do not prove) that the girls, and perhaps others in Salem, were in the grips of ergot-induced illness. Some historians dispute this idea, and other hypotheses have been proposed. Conclusive evidence may never be found, but this story reinforces the unifying thread of this chapter—the importance of the interaction of plants and fungi—and illustrates how the scientific method can be applied in a wide variety of academic disciplines.

▼ **Figure 16.25** **Ergot and the Salem witch hunt.** Ergot poisoning may have been the catalyst for the Salem witch hunt of 1692.

Commercial Uses of Fungi

It would not be fair to fungi to end our discussion with an account of diseases. In addition to their positive global impact as decomposers, fungi also have a number of practical uses for humans.

Most of us have eaten mushrooms, although we may not have realized that we were ingesting the reproductive extensions of subterranean fungi. Mushrooms are often cultivated commercially in artificial caves in which cow manure is piled—thus, you should be sure to wash your store-bought mushrooms thoroughly. Edible mushrooms also grow wild in fields, forests, and backyards, but so do poisonous ones. There are no simple rules to help the novice distinguish edible from deadly mushrooms. Only experts should dare to collect wild fungi for eating.

Mushrooms are only one of many fungi we eat (**Figure 16.26**). Truffles are fungi that grow in association with tree roots; they are highly prized by gourmets. And the distinctive flavors of certain kinds of cheeses (such as blue cheese) come from the fungi used to ripen them. Particularly important in food production are unicellular fungi, the yeasts. As discussed in Chapter 6, yeasts are used in baking, brewing, and winemaking.

▼ **Figure 16.26** **Fungi eaten by humans.**

Truffles (the fungal kind, not the chocolates). These are the reproductive structures of certain fungi that grow with tree roots as mycorrhizae. Truffles release strong odors that attract certain mammals, which excavate the fungi and disperse their spores. Truffle hunters use pigs or dogs to locate their prizes, which may command prices of hundreds of dollars per pound.

Blue cheese. The turquoise streaks in blue cheese are the mycelia of a specific fungus.

Chanterelle mushrooms. Gourmet mushrooms are highly prized by chefs for their earthy flavors and interesting textures.

☑**CHECKPOINT**

1. What is athlete's foot?
2. What do you think is the natural function of the antibiotics that fungi produce in their native environments?

Answers: 1. Athlete's foot is infection of the foot's skin with ringworm fungus. 2. The antibiotics block the growth of microorganisms, especially bacteria, that compete with the fungi for nutrients and other resources.

Fungi are medically valuable as well. Some fungi produce antibiotics that are used to treat bacterial diseases. In fact, the first antibiotic discovered was penicillin, made by the common mold *Penicillium* (**Figure 16.27**). In 2007, researchers discovered that several varieties of fungi that attack hazelnut trees produce a powerful anticancer drug.

As sources of medicines and food, as decomposers, and as partners with plants in mycorrhizae, fungi play vital roles in life on Earth. ☑

Penicillium   Zone of inhibited growth

Staphylococcus

▶ Figure 16.27 **Fungal production of an antibiotic.** Penicillin is made by the common mold *Penicillium*. In this petri dish, the clear area between the mold and the growing *Staphylococcus* bacteria is where the antibiotic produced by the *Penicillium* inhibits the growth of the bacteria.

EVOLUTION CONNECTION: Plant-Fungus Symbiosis

Mutually Beneficial Symbiosis

Discussing fungi in the same chapter as plants may seem to indicate that these two kingdoms are close relatives. Actually, as we've discussed, fungi are more closely related to animals than to plants. But the success of plants on land and the great diversity of fungi are interconnected; neither could have populated the land without the other.

Evolution is not just about the origin and adaptation of individual species. Relationships between species are also an evolutionary product. Symbiosis is the term used to describe ecological relationships between organisms of different species that are in direct physical contact.

Although the Biology and Society and Process of Science sections presented examples of harmful symbiotic relationships, there are many other examples where

symbiosis is mutually beneficial. Eukaryotic cells evolved from symbiosis among prokaryotes. And today, bacteria living in the roots of certain plants provide nitrogen compounds to their host and receive food in exchange. We have our own symbiotic bacteria that help keep our skin healthy and produce certain vitamins in our intestines. Particularly relevant to this chapter is the symbiotic association of fungi and plant roots—mycorrhizae—that made life's move onto land possible.

Lichens, symbiotic associations of fungi and algae, are striking examples of how two species can become so merged that the cooperative is essentially a new life-form. At a distance, it is easy to mistake lichens for mosses or other simple plants growing on rocks, rotting logs, trees, roofs, or gravestones (**Figure 16.28**). In fact, lichens are not mosses or any other kind of plant, nor are they even individual organisms. A lichen is a symbiotic association of millions of tiny algal cells embraced by a mesh of fungal hyphae. The photosynthetic algae feed the fungi. The fungal mycelium, in turn, provides a suitable habitat for the algae, helping the algae absorb and retain water and minerals. The mutualistic merger is so complete that lichens are actually named as species, as though they were individual organisms.

After protists and plants, fungi is the third group of eukaryotes that we have surveyed so far. Strong evidence suggests that they evolved from protist ancestors that also gave rise to the fourth and most diverse group of eukaryotes: the animals, the topic of the next chapter.

▼ Figure 16.28 **Lichens: symbiotic associations of fungi and algae.** Lichens generally grow very slowly, sometimes less than a millimeter per year. Some lichens are thousands of years old, rivaling the oldest plants as Earth's elders. The close relationship between the fungal and algal partners is evident in the microscopic blowup of a lichen.

Algal cell

Colorized SEM 1,400×

Fungal hyphae

Chapter Review

SUMMARY OF KEY CONCEPTS

 Go to the Study Area at **www.masteringbiology.com** for practice quizzes, eText, BioFlix™ 3-D animations, MP3 Tutor Sessions, videos, current events, and more.

Colonizing Land

Terrestrial Adaptations of Plants

Plants are multicellular photosynthetic eukaryotes with adaptations for living on land.

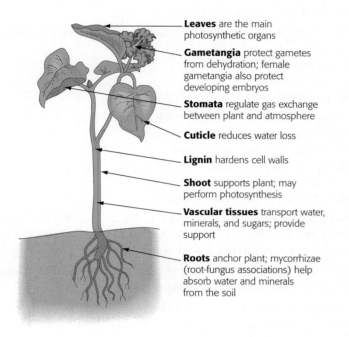

Leaves are the main photosynthetic organs

Gametangia protect gametes from dehydration; female gametangia also protect developing embryos

Stomata regulate gas exchange between plant and atmosphere

Cuticle reduces water loss

Lignin hardens cell walls

Shoot supports plant; may perform photosynthesis

Vascular tissues transport water, minerals, and sugars; provide support

Roots anchor plant; mycorrhizae (root-fungus associations) help absorb water and minerals from the soil

The Origin of Plants from Green Algae

Plants evolved from a group of multicellular green algae called charophytes.

Plant Diversity

Highlights of Plant Evolution

Four major periods of plant evolution are marked by terrestrial adaptations.

Origin of gametangia (protect gametes and embryos)	Origin of vascular tissue (conducts water and nutrients)	Origin of seeds (protect embryos from dessication and other hazards)	Origin of flowers (bear ovules within protective chambers called ovaries)

Bryophytes

 The most familiar bryophytes are mosses. Mosses display two key terrestrial adaptations: a waxy cuticle that prevents dehydration and the retention of developing embryos within the mother plant's gametangia. Mosses are most common in moist environments because their sperm must swim to the eggs and because they lack lignin in their cell walls and thus cannot stand tall. Bryophytes are unique among plants in having the gametophyte as the dominant generation in the life cycle.

Ferns

 Ferns are seedless plants that have vascular tissues but still use flagellated sperm to fertilize eggs. During the Carboniferous period, giant ferns were among the plants that decayed to thick deposits of organic matter, which were gradually converted to coal.

Gymnosperms

 A drier and colder global climate near the end of the Carboniferous period favored the evolution of the first seed plants. The most successful were the gymnosperms, represented by conifers. Needle-shaped leaves with thick cuticles and sunken stomata are adaptations to dry conditions. Conifers and most other gymnosperms have three additional terrestrial adaptations: (1) further reduction of the haploid gametophyte and greater development of the diploid sporophyte, (2) sperm-bearing pollen, which doesn't require water for transport, and (3) seeds, which consist of a plant embryo packaged along with a food supply inside a protective coat.

Angiosperms

Angiosperms supply nearly all our food and much of our fiber for textiles. The evolution of the flower and more efficient water transport help account for the success of the angiosperms. The dominant stage is a sporophyte with gametophytes in its flowers. The female gametophyte is located within an ovule, which in turn resides within a chamber of the ovary. Fertilization of an egg in the female gametophyte produces a zygote, which develops into an embryo. The whole ovule develops into a seed. The seed's enclosure within an ovary is what distinguishes angiosperms from gymnosperms, which have naked seeds. A fruit is the ripened ovary of a flower. Fruits protect and help disperse seeds. Angiosperms are a major food source for animals, while animals aid plants in pollination and seed dispersal. Agriculture constitutes a unique kind of evolutionary relationship among plants, humans, and other animals.

Plant Diversity as a Nonrenewable Resource

The exploding human population, with its demand for space and natural resources, is causing the extinction of plant species at an unprecedented rate. Scientists are researching how to slow this loss and help us learn to work with forests in ways that preserve their biological diversity.

Fungi

Characteristics of Fungi

Fungi are unicellular or multicellular eukaryotes; they are chemoheterotrophs that digest their food externally and absorb the nutrients from the environment. They are more closely related to animals than to plants. A fungus usually consists of a mass of threadlike hyphae, forming a mycelium. The cell walls of fungi are mainly composed of chitin. Although most fungi are nonmotile, a mycelium can grow very quickly, extending the tips of its hyphae into new territory. Mushrooms are reproductive structures that extend from the underground mycelium. Fungi reproduce and disperse by releasing spores that are produced either sexually or asexually.

Reproductive structure

Hyphae

Mycelium

The Ecological Impact of Fungi

Fungi and bacteria are the principal decomposers of ecosystems. Many molds destroy fruit, wood, and human-made materials. About 50 species of fungi are known to be parasites of humans and other animals. Fungi are also commercially important as food and in baking, beer and wine production, and the manufacture of antibiotics.

Evolution Connection: Mutually Beneficial Symbiosis

Lichens, in which algae are surrounded by fungal hyphae, are an example of a mutually beneficial symbiotic relationship.

SELF-QUIZ

1. Which of the following structures is common to all four major plant groups: vascular tissue, flowers, seeds, cuticle, pollen?

2. Angiosperms are distinguished from all other plants because only angiosperms have reproductive structures called _____.

3. Complete the following analogies:
   a. Gametophyte is to haploid as _____ is to diploid.
   b. _____ are to conifers as flowers are to _____.
   c. Ovule is to seed as ovary is to _____.

4. Under a microscope, a piece of a mushroom would look most like
   a. jelly.
   b. a tangle of string.
   c. grains of sand.
   d. a sponge.

5. During the Carboniferous period, the dominant plants, which later formed the great coal beds, were mainly
   a. mosses and other bryophytes.
   b. ferns and other seedless vascular plants.
   c. charophytes and other green algae.
   d. conifers and other gymnosperms.

6. You discover a new species of plant. Under the microscope, you find that it produces flagellated sperm. A genetic analysis shows that its dominant generation has diploid cells. What kind of plant do you have?

7. How does the evergreen nature of pines and other conifers adapt the plants for living where the growing season is very short?

8. Which of the following terms includes all others in the list? angiosperm, fern, vascular plant, gymnosperm, seed plant

9. Plant diversity is greatest in
   a. tropical forests.
   b. the temperate forests of Europe.
   c. deserts.
   d. the oceans.

10. What is a fruit?

11. Lichens are symbionts of photosynthetic _____ with _____.

12. Contrast the heterotrophic nutrition of a fungus with your own heterotrophic nutrition.

Answers to the Self-Quiz questions can be found in Appendix D.

THE PROCESS OF SCIENCE

13. In April 1986, an accident at a nuclear power plant in Chernobyl, Ukraine, scattered radioactive fallout for hundreds of miles. In assessing the biological effects of the radiation, researchers found mosses to be especially valuable as organisms for monitoring the damage. Radiation damages organisms by causing mutations. Explain why it is faster to observe the genetic effects of radiation on mosses than on other types of plants. Imagine that you are conducting tests shortly after a nuclear accident. Using potted moss plants as your experimental organisms, design an experiment to test the hypothesis that the frequency of mutations decreases with the organism's distance from the source of radiation.

14. You discover what you think may be one extremely large underground fungal mycelium living beneath your campus. How could you prove that it is, in fact, one individual organism spread across a very large area, as opposed to a group of separate organisms?

BIOLOGY AND SOCIETY

15. Why are tropical rain forests being destroyed at such an alarming rate? What kinds of social, technological, and economic factors are responsible? Most forests in more industrialized Northern Hemisphere countries have already been cut. Do the more industrialized nations have a right to pressure the less industrialized nations in the Southern Hemisphere to slow or stop the destruction of their forests? Defend your answer. What kinds of benefits, incentives, or programs might slow the assault on the rain forests?

16. Imagine you were charged with the task of managing a coniferous forest. How would you balance the need for productive use of the forest (to provide lumber, for example) with preservation of its diversity? What activities would you allow or prohibit in the forest (for example, snowmobiling, logging, hiking, mushroom harvesting, mining, camping, grazing, making campfires)? How would you defend your choices to the public?

A skull from one of Indonesia's "hobbit people."
Scientists are debating whether the skull is from an ancient human-like species.

BIOLOGY AND SOCIETY: Human Evolution

Rise of the Hobbit People

In 2003, Australian anthropologists digging on the Indonesian island of Flores stumbled upon bones—including a nearly complete female skeleton—of some highly unusual people. The find, dated to 12,000 to 18,000 years ago (very recent in terms of human evolution), had some very odd features: The bones belonged to adults who stood just over 3 feet tall with a brain less than one-third the size of modern humans (about the size of a chimpanzee's). Since that initial find, bones from about 12 similarly sized individuals have also been unearthed.

While there are many known humans of small stature, including some pygmies living on Flores, the tiny head and brain were unprecedented. The unusual bones were attributed to a previously unknown species named *Homo floresiensis* and nicknamed "hobbits." The discoverers speculated that a small-bodied, small-brained band of ancestral humans arrived on Flores millions of years ago from Africa and, within the isolated environment of the island, evolved into the even smaller *Homo floresiensis*. There is precedent for animal evolution of this type: Biologists have discovered island-bound dwarf populations of deer, elephants, and hippos. One hypothesis is that a lack of predators favors the evolution of smaller, more energy-efficient forms.

Other scientists are highly skeptical, arguing that the bones are from diseased members of *Homo sapiens*. Several human disorders can cause malformations similar to those found in the Flores bones. Are these bones from an isolated human-like species with a tiny head, or from diseased members of our own species? The debate rages on.

Homo sapiens and *Homo floresiensis* are just two of the over 1 million species of animals that have been named and described by biologists. This amazing diversity arose through hundreds of millions of years of evolution as natural selection shaped animal adaptations to Earth's many environments. In this chapter, we'll look at the 9 most abundant and widespread of the roughly 35 phyla (major groups) in the kingdom Animalia. We'll give special attention to the major milestones in animal evolution and conclude by reconnecting with the fascinating subject of human evolution.

The Origins of Animal Diversity

Animal life began in Precambrian seas with the evolution of multicellular creatures that ate other organisms. We are among their descendants.

What Is an Animal?

Animals are eukaryotic, multicellular, heterotrophic organisms that obtain nutrients by eating. This mode of nutrition contrasts animals with plants and other organisms that construct organic molecules through photosynthesis. It also contrasts with fungi, which obtain nutrients by absorption after digesting the food outside the body (see Chapter 16). Most animals digest their food within their bodies after ingesting other organisms, dead or alive, whole or by the piece (**Figure 17.1**).

Animal cells lack the cell walls that provide strong support in the bodies of plants and fungi. And most animals have muscle cells for movement and nerve cells that control the muscles. The most complex animals can use their muscular and nervous systems for many functions other than eating. Some species even use massive networks of nerve cells called brains to think.

Most animals are diploid and reproduce sexually; eggs and sperm are the only haploid cells. The life cycle of a sea star (**Figure 17.2**) includes basic stages found in most animal life cycles. ❶ Male and female adult animals make haploid gametes by meiosis, and ❷ an egg and a sperm fuse, producing a zygote. ❸ The zygote divides by mitosis, forming ❹ an

early embryonic stage called a **blastula**, which is usually a hollow ball of cells. ❺ In most animals, one side of the blastula folds inward, forming a stage called a **gastrula**. ❻ The gastrula develops into a saclike embryo with a two-layered wall and an opening at one end. After the gastrula stage, many animals develop directly into adults. ❼ Others, such as the sea star, develop into a **larva**, an immature individual that looks different from the adult animal (a tadpole, for another example, is a larval frog). ❽ The larva undergoes a major change of body form, called **metamorphosis**, in becoming an adult capable of reproducing sexually. ☑

▼ Figure 17.1 **Nutrition by ingestion, the animal way of life.** Few animals ingest a piece of food as large as the gazelle being eaten by this rock python. The snake will spend two weeks or more digesting its meal.

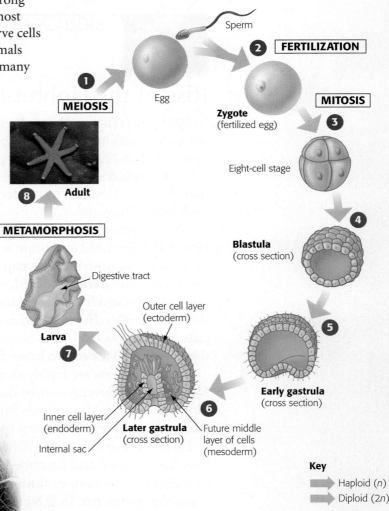

▼ Figure 17.2 **Life cycle of a sea star as an example of animal development.**

Sperm

FERTILIZATION

Egg

MEIOSIS

Zygote
(fertilized egg)

MITOSIS

Eight-cell stage

Adult

METAMORPHOSIS

Blastula
(cross section)

Digestive tract

Outer cell layer
(ectoderm)

Larva

Early gastrula
(cross section)

Inner cell layer
(endoderm)

Later gastrula
(cross section)

Future middle
layer of cells
(mesoderm)

Internal sac

Key

Haploid (*n*)

Diploid (2*n*)

Early Animals and the Cambrian Explosion

Animals probably evolved from a colonial, flag-ellated protist that lived in Precambrian seas about a billion years ago (Figure 17.3). By the late Precambrian, about 600–700 million years ago, a diversity of animals had evolved. At the beginning of the Cambrian period, 542 million years ago, animals underwent a relatively rapid diversification called "the Cambrian explosion." During a span of only about 15 million years, all the major animal body plans we see today evolved. Although many of the Cambrian animals seem bizarre, most zoologists now agree that Cambrian fossils can be classified as ancient representatives of modern animal phyla (Figure 17.4).

Why did the evolution of animal forms accel-erate so dramatically at that time? Hypotheses abound. One hypothesis emphasizes increasingly complex predator-prey relationships that led to diverse adaptations for feeding, motility, and protection. This hypothesis would help explain why most Cambrian ani-mals had shells or hard outer skeletons, in contrast with Precambrian animals, which were mostly soft-bodied. Another hypothesis focuses on the evolution of genes that control the development of animal form, such as the placement of body parts in embryos. The study of the re-lationship between evolution and development, called evo-devo, is a very active area of biological research. At least some "master control" genes are important in devel-opment of diverse animal phyla (see Chapter 11). Varia-tions in how, when, and where these genes are expressed in an embryo can produce some of the major differences in body form that distinguish the phyla. Perhaps such changes in gene expression were partly responsible for the relatively rapid diversification of animals during the early Cambrian period.

In the last half billion years, animal evolution has to a large degree merely generated variations of the animal forms that originated in the Cambrian seas. Continuing research will help test hypotheses about the Cambrian explosion. But even as the explosion becomes less myste-rious, it will seem no less wondrous. ✔

◄ Figure 17.4 **A Cambrian seascape.** This drawing is based on fossils (such as *Olenoides serratus*, a kind of trilobite, shown below) col-lected at a site called the Burgess Shale in British Columbia, Canada.

✔CHECKPOINT

Why is animal evolution dur-ing the early Cambrian re-ferred to as an "explosion"?

Answer: because a great diversity of animals evolved in a relatively short time span

▼ Figure 17.3 **One hypothesis for a sequence of stages in the origin of animals from a colonial protist.** With its specialized cells and a simple digestive compartment, the proto-animal shown at the end of the process could have fed on organic matter on the seafloor.

    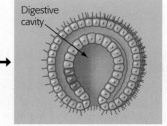

The earliest colonies may have consisted of only a few cells, all of which were flagellated and basically identical.	Some of the later colonies may have been hollow spheres of cells that ingested organic nutrients from the water.	Eventually, cells in the colony may have specialized, with some cells adapted for reproduction and others for locomotion and feeding.	A simple multicellular organism with cell layers may have evolved from a hollow colony, with cells on one side of the colony cupping inward.	A layered body plan would have enabled further division of labor, with cells specialized for locomotion, protection, reproduction, or feeding.

Reproductive cells

Digestive cavity

Animal Phylogeny

Historically, biologists have categorized animals by "body plan"—general features of body structure. Distinctions between body plans are used to help infer the evolutionary relationships between animal groups. More recently, a wealth of genetic data has allowed evolutionary biologists to modify and refine groups. **Figure 17.5** represents a set of hypotheses about the evolutionary relationships between nine major animal phyla based on structural and genetic similarities.

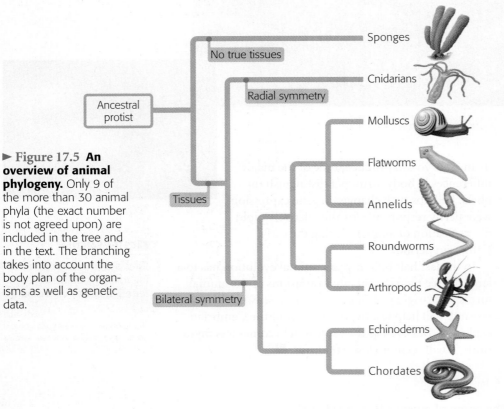

► Figure 17.5 **An overview of animal phylogeny.** Only 9 of the more than 30 animal phyla (the exact number is not agreed upon) are included in the tree and in the text. The branching takes into account the body plan of the organisms as well as genetic data.

A major branch point in animal evolution distinguishes sponges from all other animals based on structural complexity. Unlike more complex animals, sponges lack true tissues, groups of similar cells that perform a function (such as nervous tissue). A second major evolutionary split is based on body symmetry: radial versus bilateral (**Figure 17.6**). To understand this difference, imagine a pail and shovel. The pail has **radial symmetry**, identical all around a central axis. The shovel has **bilateral symmetry**, which means there's only one way to split it into two equal halves—right down the midline. A bilateral animal has a definite "head end" that first encounters food, danger, and other stimuli when traveling. In most bilateral animals, a nerve center in the form of

▼ Figure 17.6 **Body symmetry.**

Radial symmetry. Parts radiate from the center, so any slice through the central axis divides into mirror images.

Bilateral symmetry. Only one slice can divide left and right sides into mirror-image halves.

a brain is at the head end, near a concentration of sense organs such as eyes. Thus, bilateral symmetry is an adaptation that aids movement, such as crawling, burrowing, or swimming. Indeed, many radial animals are stationary, whereas most bilateral animals are mobile.

The evolution of body cavities also helped lead to more complex animals. A **body cavity** is a fluid-filled space separating the digestive tract from the outer body wall. The cavity enables the internal organs to grow and move independently of the outer body wall, and the fluid cushions them from injury. In soft-bodied animals such as earthworms, the fluid is under pressure and functions as a hydrostatic skeleton. Of the phyla shown in Figure 17.5, only sponges, cnidarians, and flatworms lack a body cavity.

Among animals with a body cavity, there are differences in how the cavity develops (**Figure 17.7**). In all cases, the cavity is at least partly lined by a middle layer of tissue, called mesoderm, which develops between the inner (endoderm) and outer (ectoderm) layers of the gastrula embryo. If the body cavity is not completely lined by tissue derived from mesoderm, it is called a **pseudocoelom**. Among the animal phyla discussed in this chapter, only the roundworms (nematodes) have a pseudocoelom. A true **coelom**, the type of body cavity humans and many other animals have, is completely lined by tissue derived from mesoderm. (Despite its name, a pseudocoelom functions just like a coelom.)

With the overview of animal evolution in Figure 17.5 as our guide, we're ready to take a closer look at the nine most numerous animal phyla. ☑

▼ Figure 17.7 **Body plans of bilateral animals.** The various organ systems of these animals develop from the three tissue layers that form in the embryo.

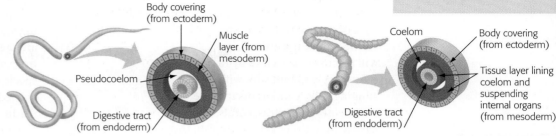

(a) No body cavity: for example, flatworm

Body covering (from ectoderm)
Tissue-filled region (from mesoderm)
Digestive tract (from endoderm)

(b) Pseudocoelom: a body cavity only partially lined by the mesoderm, the middle tissue layer; for example, roundworm

Body covering (from ectoderm)
Muscle layer (from mesoderm)
Pseudocoelom
Digestive tract (from endoderm)

(c) True coelom: a fluid-filled body cavity completely lined by mesoderm; for example, annelid

Coelom
Body covering (from ectoderm)
Tissue layer lining coelom and suspending internal organs (from mesoderm)
Digestive tract (from endoderm)

Major Invertebrate Phyla

Living as we do on land, our sense of animal diversity is biased in favor of vertebrates, animals with a backbone, such as amphibians, reptiles, and mammals. However, vertebrates make up less than 5% of all animal species. If we were to sample the animals in an aquatic habitat, such as a pond, tide pool, or coral reef, or if we were to consider the millions of insects that share our terrestrial world, we would find ourselves in the realm of **invertebrates**, animals without backbones. We give special attention to the vertebrates simply because we humans are among the backboned ones. However, by exploring the other 95% of the animal kingdom—the invertebrates—we'll discover an astonishing diversity of beautiful creatures that too often escape our notice.

cavity and then flows out of the sponge through a larger opening **(Figure 17.8)**. Cells called choanocytes have flagella that sweep water through the sponge's porous body. Choanocytes trap bacteria and other food particles in mucus and then engulf the food by phagocytosis (see Chapter 4). Cells called amoebocytes pick up food from the choanocytes, digest it, and carry the nutrients to other cells. Amoebocytes also manufacture the fibers that make up a sponge's skeleton. In some sponges, these fibers are sharp and spur-like. Other sponges have softer, more flexible skeletons; these pliant, honeycombed skeletons are often used as natural sponges in the bath or to wash cars. ✔

✔CHECKPOINT

In what fundamental way does the structure of a sponge differ from that of all other animals?

Answer: A sponge has no true tissues.

Sponges

Sponges were once grouped into a single phlyum Porifera. However, genetic evidence now suggests that there are multiple phyla. Sponges are stationary animals that appear so sedate that the ancient Greeks believed them to be plants. The simplest of all animals, sponges probably evolved very early from colonial protists. Sponges range in height from about 1 cm to 2 m. They have no nerves or muscles, but their individual cells can sense and react to changes in the environment. The cell layers of sponges are loose federations of cells, not really tissues, because the cells are relatively unspecialized. Of the 9,000 or so species of sponges, only about 100 live in fresh water; the rest are marine.

The body of a sponge resembles a sac perforated with holes. Water is drawn through the pores into a central

Sponges
Cnidarians
Molluscs
Flatworms
Annelids
Roundworms
Arthropods
Echinoderms
Chordates

▼ Figure 17.8 **Anatomy of a sponge.**

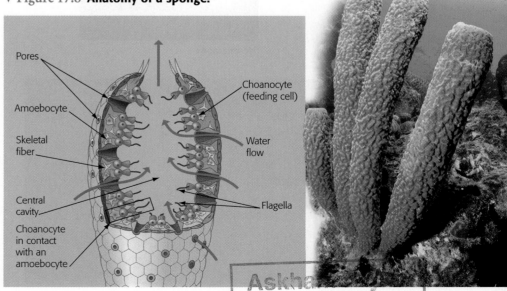

Pores
Amoebocyte
Skeletal fiber
Central cavity
Choanocyte in contact with an amoebocyte
Choanocyte (feeding cell)
Water flow
Flagella

Cnidarians

Cnidarians (phylum Cnidaria) are characterized by the presence of body tissues—as are all the remaining animals we will discuss—as well as by radial symmetry and tentacles with stinging cells. Cnidarians include sea anemones, hydras, corals, and jellies (sometimes called jellyfish, though they are not fish). Most of the 10,000 cnidarian species are marine.

The basic body plan of a cnidarian is a sac with a central digestive compartment, the **gastrovascular cavity**. A single opening to this cavity functions as both mouth and anus. This basic body plan has two variations: the stationary **polyp** and the floating **medusa** (Figure 17.9). Polyps adhere to larger objects and extend their tentacles,

waiting for prey. Examples of the polyp body plan are hydras, sea anemones, and corals. A medusa (plural, *medusae*) is a flattened, mouth-down version of the polyp. It moves freely by a combination of passive drifting and contractions of its bell-shaped body. Some jellies are medusae with tentacles over 100 m long dangling from an umbrella up to 2 m in diameter. There are some species of cnidarians that live only as polyps, others only as medusae, and still others that pass through both a medusa stage and a polyp stage in their life cycle.

Cnidarians are carnivores that use tentacles arranged in a ring around the mouth to capture prey and push the food into the gastrovascular cavity, where digestion begins. The undigested remains are eliminated through the mouth/anus. The tentacles are armed with batteries of cnidocytes ("stinging cells") that function in defense and in the capture of prey (Figure 17.10). The phylum Cnidaria is named for these stinging cells. ☑

▼ **Figure 17.9 Polyp and medusa forms of cnidarians.** Note that cnidarians have two tissue layers, distinguished in the diagrams by blue and yellow. The gastrovascular cavity has only one opening, which functions as both mouth and anus.

Polyp form

Coral

Sea anemone

Hydra

▼ **Figure 17.10 Cnidocyte action.** When a trigger on a tentacle is stimulated by touch, a fine thread shoots out from a capsule. Some cnidocyte threads entangle prey, while others puncture the prey and inject a poison.

Medusa form

Jelly

Molluscs

Snails and slugs, oysters and clams, and octopuses and squids are all **molluscs** (phylum Mollusca). Molluscs are soft-bodied animals, but most are protected by a hard shell. Slugs, squids, and octopuses have reduced shells, most of which are internal, or they have lost their shells completely during their evolution. Many molluscs feed by using a file-like organ called a **radula** to scrape up food. Garden snails use their radulas like tiny saws to cut pieces out of leaves.

All molluscs have a similar body plan (**Figure 17.11**). The body has three main parts: a muscular foot, usually used for movement; a visceral mass containing most of the internal organs; and a fold of tissue called the mantle. The **mantle** drapes over the visceral mass and secretes the shell if one is present.

There are 93,000 known species of molluscs, with most being marine animals. The three major groups of molluscs are gastropods (including snails and slugs), bivalves (including clams and oysters), and cephalopods (including squids and octopuses) (**Figure 17.12**).

Most **gastropods** are protected by a single spiraled shell into which the animal can retreat when threatened. Many have a distinct head with eyes at the tips of tentacles (think of a garden snail). Marine, freshwater, and terrestrial gastropods make up about three-quarters of the living mollusc species.

The **bivalves**, including clams, oysters, mussels, and scallops, have shells divided into two halves hinged together. None of the bivalves have a radula. There are both marine and freshwater species, with most being sedentary, using their muscular foot for digging and anchoring in sand or mud.

Cephalopods are all marine animals and generally differ from gastropods and sedentary bivalves in that their bodies are fast and agile. A few have large, heavy shells, but in most the shell is small and internal (as in squids) or missing (as in octopuses). Cephalopods have large brains and sophisticated sense organs, which contribute to their success as mobile predators. They use beak-like jaws and a radula to crush or rip prey apart. The mouth is at the base of the foot, which is drawn out into several long tentacles for catching and holding prey. Cephalopods include the largest invertebrates: the giant squid (averaging about 10 m in length) and the closely related colossal squid (up to 13 m). Very little is known about these majestic creatures. In fact, until 2004, none was ever seen alive. In that year, Japanese zoologists captured the first video footage of a giant squid in its natural habitat. ✔

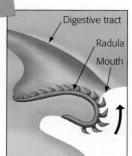

◄ Figure 17.11 The general body plan of a mollusc. Note the body cavity (a true coelom, though a small one) and the complete digestive tract, with both mouth and anus.

Visceral mass — Coelom — Kidney — Heart — Reproductive organs — Digestive tract — Shell — **Mantle** — Mantle cavity — Anus — Gill — **Radula** — Mouth — **Foot** — Nerve cords

Digestive tract — Radula — Mouth

☑CHECKPOINT

Classify these molluscs: A garden snail is an example of a _____; a clam is an example of a _____; a squid is an example of a _____.

Answer: gastropod; bivalve; cephalopod

▼ Figure 17.12 **Mollusc diversity.**

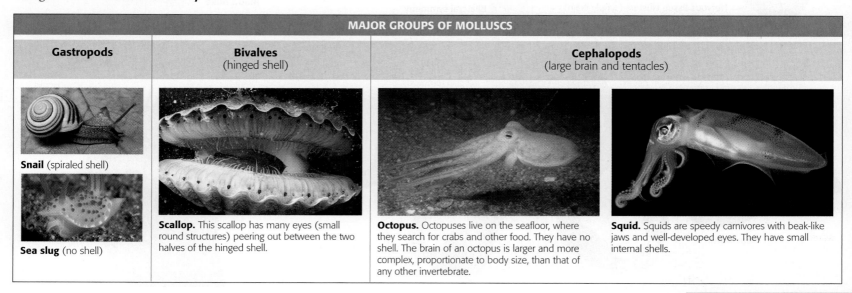

Gastropods	Bivalves (hinged shell)	Cephalopods (large brain and tentacles)

MAJOR GROUPS OF MOLLUSCS

Snail (spiraled shell)

Sea slug (no shell)

Scallop. This scallop has many eyes (small round structures) peering out between the two halves of the hinged shell.

Octopus. Octopuses live on the seafloor, where they search for crabs and other food. They have no shell. The brain of an octopus is larger and more complex, proportionate to body size, than that of any other invertebrate.

Squid. Squids are speedy carnivores with beak-like jaws and well-developed eyes. They have small internal shells.

Sponges — Cnidarians — **Molluscs** — Flatworms — Annelids — Roundworms — Arthropods — Echinoderms — Chordates

Flatworms

Flatworms (phylum Platy-helminthes) are the simplest animals with bilateral symmetry. True to their name, these worms are ribbonlike and range from about 1 mm to about 20 m in length. Most flatworms have a gastrovascular cavity with a single opening. There are about 20,000 species of flatworms living in marine, freshwater, and damp terrestrial habitats (**Figure 17.13**).

The gastrovascular cavity of free-living flatworms called planarians is highly branched, providing an extensive surface area for the absorption of nutrients. When the animal feeds, a muscular tube projects through the mouth and sucks food in. Planarians live on the undersurfaces of rocks in freshwater ponds and streams.

Parasitic flatworms called blood flukes (*Schistosoma mansoni*) are a major health problem in the tropics. These worms have suckers that attach to the inside of the blood vessels near the human host's intestines. Infection by these flatworms causes a long-lasting disease called schistosomiasis or blood fluke disease, with such symptoms as severe abdominal pain, anemia, and dysentery. About 250 million people in 70 countries suffer from blood fluke disease each year.

Tapeworms parasitize many vertebrates, including humans. Most tapeworms have a very long, ribbonlike body with repeated parts. They differ from other flatworms in not having a gastrovascular cavity. Living in partially digested food in the intestines of their hosts, tapeworms simply absorb nutrients across their body surface. The head of a tapeworm is armed with suckers and hooks that lock the worm to the intestinal lining of the host. Behind the head is a long ribbon of units that are little more than sacs of sex organs. At the back of the worm, mature units containing thousands of eggs break off and leave the host's body with the feces. Humans can become infected with tapeworms by eating undercooked beef, pork, or fish containing tapeworm larvae. The larvae are microscopic, but the adults can reach lengths of 20 m in the human intestine. Such large tapeworms can cause intestinal blockage and rob enough nutrients from the human host to cause nutritional deficiencies. Fortunately, an orally administered drug can kill the adult worms. ☑

Sponges
Cnidarians
Molluscs
Flatworms
Annelids
Roundworms
Arthropods
Echinoderms
Chordates

☑**CHECKPOINT**

Flatworms are the simplest animals to display a body plan that is _____.

Answer: bilaterally symmetrical

▼ Figure 17.13 **Flatworm diversity.**

Digestive tract (gastrovascular cavity)

Nerve cords

Mouth

Eyespots (detect light)

Nervous tissue clusters (simple brain)

Planarian

Bilateral symmetry

Blood fluke

Reproductive unit with skin removed

Head

Hooks

Suckers

Tapeworm

Annelids

Annelids (phylum Annelida) are worms that have **body segmentation**, which is the subdivision of the body along its length into a series of repeated parts called segments. In annelids, the segments look like a set of fused rings. There are about 16,000 annelid species, ranging in length from less than 1 mm to the 3-m giant Australian earthworm. Annelids live in damp soil, the sea, and most freshwater habitats. There are three main groups: earthworms, polychaetes, and leeches (**Figure 17.14**).

Annelids exhibit two innovations shared by all other bilateral animals except flatworms. One is a **complete digestive tract**, which is a digestive tube with two openings: a mouth and an anus. A complete digestive tract can process food and absorb nutrients as a meal moves in one direction from one specialized digestive organ to the next. In humans, for example, the mouth, stomach, and intestines act as digestive organs. A second innovation is a body cavity, which in this case is a coelom (a body cavity completely lined by mesoderm-derived tissue; see Figure 17.7c).

Earthworms, like all annelids, are segmented externally and internally (**Figure 17.15**). Many of the internal structures are repeated, segment by segment. The coelom (body cavity) is partitioned by walls (only two segment walls are fully shown here). The nervous system (yellow in the figure) and waste-disposal organs (green) are repeated in each segment. The digestive tract, however, is not segmented; it passes through the

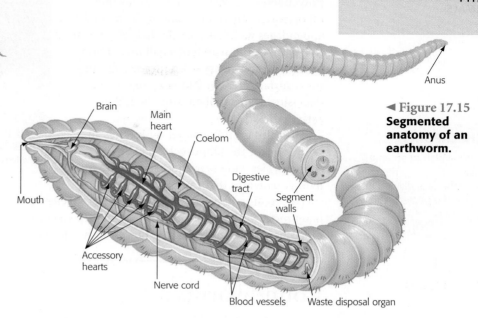

◀ **Figure 17.15 Segmented anatomy of an earthworm.**

segment walls from the mouth to the anus. Segmental blood vessels include one main heart and five pairs of accessory hearts.

Earthworms eat their way through the soil, extracting nutrients as the soil passes through the digestive tract. Undigested material, mixed with mucus secreted into the digestive tract, is eliminated as castings through the anus. Farmers and gardeners value earthworms because the animals till the earth, and the castings improve the texture of the soil. Charles Darwin estimated that each acre of British farmland contained about 50,000 earthworms that produced 18 tons of castings per year.

▼ Figure 17.14 **Annelid diversity.**

MAJOR GROUPS OF ANNELIDS		
Earthworms	**Polychaetes**	**Leeches**
This giant Australian earthworm is bigger than many snakes. Perhaps you've slipped on slimy worms, but imagine actually tripping over one!	Polychaetes have segmental appendages that function in movement and as gills. This species is called a Christmas tree worm. The feathery, spiral structures are a pair of gills from a single worm.	The European freshwater leech *Hirudo medicinalis* can be used to repair tissue damage after reattachment surgery and other human trauma.

In contrast to earthworms, most **polychaetes** are marine, mainly crawling on or burrowing in the seafloor. Segmental appendages with hard bristles help the worm wriggle about in search of small invertebrates to eat. The appendages also increase the animal's surface area for taking up oxygen and disposing of metabolic wastes, including carbon dioxide.

The third group of annelids, **leeches**, are notorious for the bloodsucking habits of some species. However, most species are free-living carnivores that eat small invertebrates such as snails and insects. A few terrestrial species inhabit moist vegetation in the tropics, but the majority of leeches live in fresh water. A European freshwater

Medicinal leech

species called *Hirudo medicinalis* is used for the treatment of circulatory complications. Most commonly, leeches are applied after reconstructive microsurgery in which limbs or digits are reattached. Because arteries (which transport blood into a reattached area) are easier to reconnect than veins (which transport blood out), blood can pool in the reattached area and stagnate, starving the healing tissue of oxygen. Medicinal leeches have razor-like jaws with hundreds of tiny teeth that cut through the skin. They secrete saliva containing an anesthetic and an anticoagulant into the wound. The anesthetic makes the bite virtually painless, and the anticoagulant prevents clotting as the leech drains excess blood from the wound. ✓

Roundworms

Roundworms (also called **nematodes**, members of the phylum Nematoda) get their common name from their cylindrical body, which is usually tapered at both ends **(Figure 17.16)**. Roundworms are among the most diverse (in species number) and widespread of all animals. About 25,000 species of roundworms are known, and perhaps ten times that

Sponges
Cnidarians
Molluscs
Flatworms
Annelids
Roundworms
Arthropods
Echinoderms
Chordates

number actually exist. Roundworms range in length from about a millimeter to a meter. They are found in most aquatic habitats, in wet soil, and as parasites in the body fluids and tissues of plants and animals.

Free-living roundworms in the soil are important decomposers. A single rotting apple may contain 90,000 nematodes; an acre of topsoil contains billions. Other species of nematodes are major agricultural pests that attack the roots of plants. At least 50 parasitic roundworm species infect humans, including pinworms, hookworms, and the parasite that causes trichinosis. ✓

▼ Figure 17.16 **Roundworm diversity.**

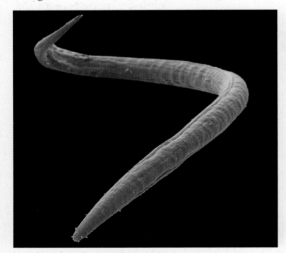

(a) A free-living roundworm. This species has the classic roundworm shape: cylindrical with tapered ends. The ridges indicate muscles that run the length of the body.

(b) Parasitic roundworms in pork. The potentially fatal disease trichinosis is caused by eating undercooked pork infected with *Trichinella* roundworms. The worms (shown here in pork tissue) burrow into the human intestine and then invade muscle tissue.

(c) Canine heart infected with parasitic roundworms. Heartworms are spread by mosquitoes that transmit fertilized eggs from one infected host to another. Regular doses of medication can protect dogs from heartworms.

Arthropods

Arthropods (phylum Arthropoda) are named for their jointed appendages. Crustaceans (such as crabs and lobsters), arachnids (such as spiders and scorpions), and insects (such as grasshoppers and moths) are examples of arthropods (**Figure 17.17**). Zoologists estimate that the total arthropod population numbers about a billion billion (10^{18}) individuals—that's about 150 million arthropods for each person! Researchers have identified over a million arthropod species, mostly insects. In fact, two out of every three species of life that have been scientifically described are arthropods. And arthropods are represented in nearly all habitats of the biosphere. In species diversity, distribution, and sheer numbers, arthropods must be regarded as the most successful animal phylum.

Sponges
Cnidarians
Molluscs
Flatworms
Annelids
Roundworms
Arthropods
Echinoderms
Chordates

◄ **Figure 17.17 Arthropod diversity.**

MAJOR GROUPS OF ARTHROPODS
Arachnids
Crustaceans
Millipedes and Centipedes
Insects

General Characteristics of Arthropods

Arthropods are segmented animals. In contrast with the repeating similar segments of annelids, however, arthropod segments and their appendages have become specialized for a great variety of functions. This evolutionary flexibility contributed to the great diversification of arthropods. Specialization of segments (or of fused groups of segments) provides for an efficient division of labor among body regions. For example, the appendages of different segments are variously adapted for walking, feeding, sensory reception, swimming, and defense (**Figure 17.18**).

The body of an arthropod is completely covered by an **exoskeleton**, an external skeleton. This coat is constructed from layers of protein and a polysaccharide called chitin. The exoskeleton can be a thick, hard armor over some parts of the body (such as the head), yet be paper-thin and flexible in other locations (such as the joints). The exoskeleton protects the animal and provides points of attachment for the muscles that move the appendages. There are, of course, advantages to wearing hard parts on the outside. Our own skeleton is interior to most of our soft tissues, an arrangement that doesn't provide much protection from injury. But our skeleton does offer the advantage of being able to grow along with the rest of our body. In contrast, a growing arthropod must occasionally shed its old exoskeleton and secrete a larger one. This process, called molting, leaves the animal temporarily vulnerable to predators and other dangers. The next five pages explore the major groups of arthropods. ☑

☑CHECKPOINT

What is the primary difference between your skeleton and a crab's skeleton?

Answer: Your skeleton is interior, whereas a crab has an exterior skeleton (an exoskeleton).

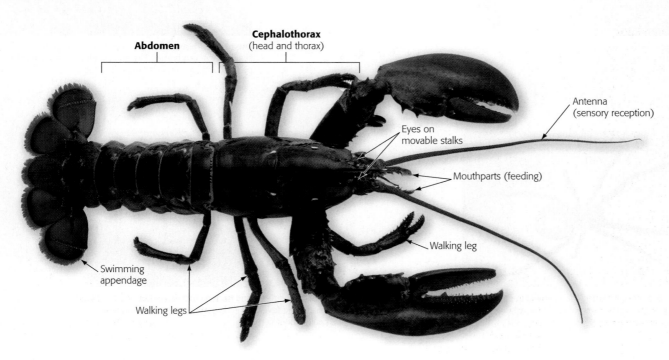

Abdomen   **Cephalothorax** (head and thorax)

Eyes on movable stalks

Antenna (sensory reception)

Mouthparts (feeding)

Walking leg

Swimming appendage

Walking legs

◄ **Figure 17.18 Anatomy of a lobster, a crustacean.** The whole body, including the appendages, is covered by an exoskeleton. The body is segmented, but this characteristic is obvious only in the abdomen.

Arachnids

Most **arachnids** live on land. Scorpions, spiders, ticks, and mites are examples (**Figure 17.19**). Arachnids usually have four pairs of walking legs and a specialized pair of feeding appendages. In spiders, these feeding appendages are fang-like and equipped with poison glands. As a spider uses these appendages to immobilize and dismantle its prey, it spills digestive juices onto the torn tissues and sucks up its liquid meal.

▼ Figure 17.19 **Arachnid characteristics and diversity.**

Two feeding appendages

Leg (four pairs)

Scorpion. Scorpions have a pair of large pincers that function in defense and food capture. The tip of the tail bears a poisonous stinger. Scorpions sting people only when prodded or stepped on.

Spider. Like most spiders, including the tarantula in the large photo, this black widow spins a web of liquid silk, which solidifies as it comes out of specialized glands. A black widow's venom can kill small prey but is rarely fatal to humans.

Dust mite. This microscopic house dust mite is a ubiquitous scavenger in our homes. Dust mites are harmless except to people who are allergic to the mites' feces.

Wood tick. Wood ticks carry bacteria that cause Rocky Mountain spotted fever. Lyme disease is caused by a separate species called deer ticks.

Crustaceans

Crustaceans are nearly all aquatic. Crabs, lobsters, crayfish, shrimps, and barnacles are all crustaceans (**Figure 17.20**). They all exhibit the arthropod hallmark of multiple pairs of specialized appendages. One group of crustaceans, the isopods, is represented on land by pill bugs, which you can find on the undersides of moist leaves and other organic debris.

▼ Figure 17.20 **Crustacean characteristics and diversity.**

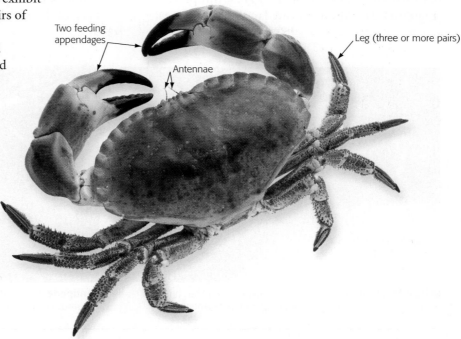

Two feeding appendages

Antennae

Leg (three or more pairs)

Crab. Ghost crabs are common along shorelines throughout the world. They scurry along the surf's edge and then quickly bury themselves in sand.

Shrimp. Naturally found in Pacific waters from Africa to Asia, giant prawns are widely cultivated as food.

Pill bug. Commonly found in moist locations with decaying leaves, such as under logs, pill bugs get their name from their tendency to roll up into a tight ball when they sense danger.

Crayfish. The Miami cave crayfish, an endangered crustacean, was only recently discovered living in underground Florida aquifers.

Barnacles. Barnacles are stationary crustaceans with exoskeletons hardened into shells by calcium carbonate (lime). The jointed appendages projecting from the shell capture small plankton.

Millipedes and Centipedes

Millipedes and **centipedes** have similar segments over most of the body and superficially resemble annelids, but their jointed legs reveal they are arthropods (**Figure 17.21**). Millipedes are landlubbers that eat decaying plant matter. They have two pairs of short legs per body segment. Centipedes are terrestrial carnivores, with a pair of poison claws used in defense and to paralyze prey, such as cockroaches and flies. Each of their body segments bears a single pair of legs.

► Figure 17.21 **Millipedes and centipedes.**

Two pairs of legs per segment

Millipede. Like most millipedes, this blue millipede has an elongated body with two pairs of legs per trunk segment.

One pair of legs per segment

Centipede. Centipedes can be found in dirt and leaf litter. Their venomous claws can harm cockroaches and spiders, but not humans.

Insect Anatomy

Like the grasshopper in **Figure 17.22**, most **insects** have a three-part body: head, thorax, and abdomen. The head usually bears a pair of sensory antennae and a pair of eyes. The mouthparts of insects are adapted for particular kinds of eating—for example, for biting and chewing plant material in grasshoppers; for lapping up fluids in houseflies; and for piercing skin and sucking blood in mosquitoes. Most adult insects have three pairs of legs and one or two pairs of wings, all extending from the thorax.

Flight is obviously one key to the great success of insects. An animal that can fly can escape many predators, find food and mates, and disperse to new habitats much faster than an animal that must crawl on the ground. Because their wings are extensions of the exoskeleton and not true appendages, insects can fly without sacrificing legs. By contrast, the flying vertebrates—birds and bats—have one of their two pairs of legs modified for wings, which explains why these vertebrates are generally not very swift on the ground.

▼ Figure 17.22 **Anatomy of a grasshopper.**

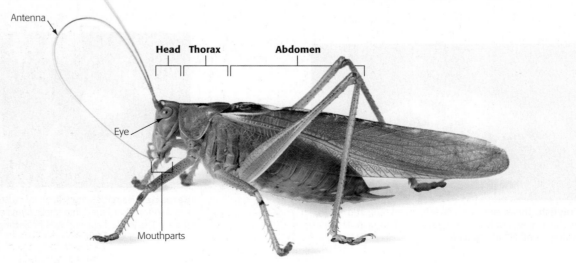

Antenna

Head   Thorax   Abdomen

Eye

Mouthparts

Insect Diversity

In species diversity, insects outnumber all other forms of life combined (Figure 17.23). They live in almost every terrestrial habitat and in fresh water, and flying insects fill the air. Insects are rare in the seas, where crustaceans are the dominant arthropods. The oldest insect fossils date back to about 400 million years ago. Later, the evolution of flight sparked an explosion in insect variety.

▼ Figure 17.23 **Insect diversity.**

Banded Orange Heliconian. The range of this butterfly extends from Brazil to Kansas, where it can be seen feeding on the nectar of flowers in summertime.

Giraffe weevil. This species is found in Madagascar. Only males have an extended neck.

Peacock katydid. The eyespots on the wings of this katydid may startle predators.

Leaf roller. Unlike most moths, the Costa Rican leaf roller moth is active during the daytime.

Praying mantis. There are over 2,000 species of praying mantises throughout the world.

Yellow jacket wasp. Common in North America, yellow jackets are important insect predators.

Leaf beetle. This beetle belongs to the family of leaf beetles, the most commonly encountered beetles in North America.

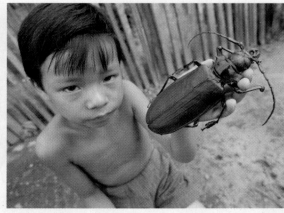

Longhorn beetle. The longhorn or Titan beetle is the largest known beetle in the Amazon rain forest and among the largest insects in the world.

☑**CHECKPOINT**

Which major arthropod group is mainly aquatic? Which is the most numerous?

Answer: crustaceans; insects

Many insects undergo metamorphosis in their development. In the case of grasshoppers and some other insect groups, the young resemble adults but are smaller and have different body proportions. The animal goes through a series of molts, each time looking more like an adult, until it reaches full size. In other cases, insects have distinctive larval stages specialized for eating and growing that are known by such names as maggots (fly larvae) or caterpillars (larvae of moths and butterflies). The larval stage looks entirely different from the adult stage, which is specialized for dispersal and reproduction. Metamorphosis from the larva to the adult occurs during a pupal stage (**Figure 17.24**).

Doctors can use maggots—specifically, juvenile blowflies (*Phaenicia sericata*)—to cleanse infected wounds. This practice is becoming increasingly common for the treatment of bedsores, burns, surgical wounds, and traumatic injuries. It is safe and effective because the enzymes secreted by the immature flies break down only dead or dying tissue, leaving healthy tissue unharmed.

Animals so numerous, diverse, and widespread as insects are bound to affect the lives of all other terrestrial organisms, including humans, in many ways. On one hand, we depend on bees, flies, and other insects to pollinate our crops and orchards. On the other hand, insects are carriers of the microbes that cause many human diseases, such as malaria and West Nile disease. Insects also compete with humans for food by eating our field crops. Trying to minimize their losses, farmers in the United States spend billions of dollars each year on pesticides, spraying crops with massive doses of insecticide poisons. But try as they might, not even humans have significantly challenged the preeminence of insects and their arthropod kin. As Cornell University's Thomas Eisner puts it: "Bugs are not going to inherit the Earth. They own it now. So we might as well make peace with the landlord." ☑

▼ Figure 17.24 **Metamorphosis of a monarch butterfly.**

The **larva (caterpillar)** spends its time eating and growing, molting as it grows.

After several molts, the larva becomes a **pupa** encased in a cocoon.

Within the pupa, the larval organs break down and adult organs develop from cells that were dormant in the larva.

Finally, the **adult** emerges from the cocoon.

The butterfly flies off and reproduces, nourished mainly by calories stored when it was a caterpillar.

Echinoderms

The **echinoderms** (phylum Echinodermata) are named for their spiny surfaces (*echin* is Greek for "spiny"). Among the echinoderms are sea stars, sea urchins, sea cucumbers, and sand dollars (**Figure 17.25**).

Echinoderms include about 7,000 species, all of them marine. Most move slowly (if at all). Echinoderms lack body segments, and most have radial symmetry as adults. Both the external and the internal parts of a sea star, for instance, radiate from the center like spokes of a wheel. In contrast to the adult, the larval stage of echinoderms is bilaterally symmetrical. This supports other evidence that echinoderms are not closely related to other radial animals, such as cnidarians, that never show bilateral symmetry. Most echinoderms have an **endoskeleton** (interior skeleton) constructed from hard plates just beneath the skin. Bumps and spines of this endoskeleton account for the animal's rough or prickly surface. Unique to echinoderms is the **water vascular system**, a network of water-filled canals that circulate water throughout the echinoderm's body, facilitating gas exchange (the entry of O_2 and the removal of CO_2) and waste disposal. The water vascular system also branches into extensions called tube feet. A sea star or sea urchin pulls itself slowly over the seafloor using its suction-cup-like tube feet. Sea stars also use their tube feet to grip prey during feeding.

Looking at sea stars and other adult echinoderms, you may think they have little in common with humans and other vertebrates. But as shown by the phylogenetic tree at the left, echinoderms share an evolutionary branch with chordates, the phylum that includes vertebrates. Analysis of embryonic development can differentiate the echinoderms and chordates from the evolutionary branch that includes molluscs, flatworms, annelids, roundworms, and arthropods. With this context in mind, we're now ready to make the transition from invertebrates to vertebrates. ☑

▼ **Figure 17.25 Echinoderm diversity.**

Sea star. When a sea star encounters an oyster or clam, it grips the mollusc's shell with its tube feet (see inset) and positions its mouth next to the narrow opening between the two halves of the prey's shell. The sea star then pushes its stomach out through its mouth and the crack in the mollusc's shell.

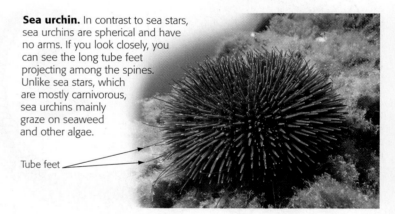

Sea urchin. In contrast to sea stars, sea urchins are spherical and have no arms. If you look closely, you can see the long tube feet projecting among the spines. Unlike sea stars, which are mostly carnivorous, sea urchins mainly graze on seaweed and other algae.

Tube feet

Sea cucumber. On casual inspection, this California sea cucumber does not look much like other echinoderms. However, a closer look reveals many echinoderm traits, including five rows of tube feet.

Sand dollar. Live sand dollars have a skin of movable spines covering a rigid skeleton. A set of five pores (arranged in a star pattern) allows seawater to be drawn into the sand dollar's body.

Vertebrate Evolution and Diversity

Most of us are curious about our family ancestry. Biologists are also interested in the larger question of tracing human ancestry within the animal kingdom. In this section, we trace the evolution of the vertebrates, the group that includes humans and their closest relatives. All vertebrates have endoskeletons, a characteristic shared with most echinoderms. However, vertebrate endoskeletons are unique in having a cranium (skull) and a backbone, a series of bones called vertebrae (singular, vertebra), for which the group is named (**Figure 17.26**). Our first step in tracing the vertebrate lineage is to determine where vertebrates fit in the animal kingdom.

▼ **Figure 17.26 A vertebrate endoskeleton.** This snake skeleton, like those of all vertebrates, has a cranium (skull) and a backbone consisting of vertebrae.

Cranium
(protects brain)

Vertebra

Characteristics of Chordates

The last phylum in our survey of the animal kingdom is the phylum Chordata. **Chordates** share four key features that appear in the embryo and sometimes in the adult (**Figure 17.27**). These four chordate hallmarks are (1) a **dorsal, hollow nerve cord**; (2) a **notochord**, which is a flexible, longitudinal rod located between the digestive tract and the nerve cord; (3) **pharyngeal slits**, which are grooves in the pharynx, the region of the digestive tube just behind the mouth; and (4) a **post-anal tail**, which is a tail to the rear of the anus. Though these chordate characteristics are often difficult to recognize in the adult animal, they are always present in chordate embryos. For example, the notochord, for which our phylum is named, persists in adult humans only in the form of the cartilage disks that function as cushions between the vertebrae. (Back injuries described as "ruptured disks" or "slipped disks" refer to these structures.)

Body segmentation is another chordate characteristic. Chordate segmentation is apparent in the backbone of vertebrates (see Figure 17.26) and is also evident in the segmental muscles of all chordates (see the chevron-shaped—>>>>—muscles in the lancelet in Figure 17.28). Segmental musculature is not so obvious in adult humans unless one is motivated enough to sculpt those "washboard abs."

Three groups of chordates are invertebrates. Two of these groups, **tunicates** and **lancelets**, have no cranium (**Figure 17.28**). The third group— hagfishes—have a cranium. (Hagfishes will be described shortly.) All other

Sponges
Cnidarians
Molluscs
Flatworms
Annelids
Roundworms
Arthropods
Echinoderms
Chordates

▼ Figure 17.27 **Chordate characteristics.**

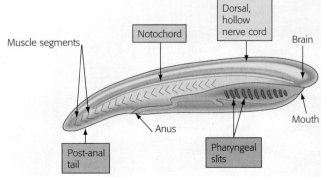

Muscle segments

Notochord

Dorsal, hollow nerve cord

Brain

Anus

Mouth

Post-anal tail

Pharyngeal slits

chordates are **vertebrates**, which retain the basic chordate characteristics but have additional features that are unique—including, of course, the backbone. **Figure 17.29** is an overview of chordate and vertebrate evolution that will provide a context for our survey. ☑

▶ Figure 17.28 **Chordates that have no cranium or vertebrae.**

Lancelet. This marine invertebrate owes its name to its bladelike shape. Only a few centimeters long, lancelets wiggle backward into the sand, leaving their mouth exposed, and filter tiny food particles from the seawater.

Tunicates. The tunicate, or sea squirt, is a stationary animal that filters food from the water. These pastel sea squirts get their nickname from their coloration and the fact that they can quickly expel water to startle intruders.

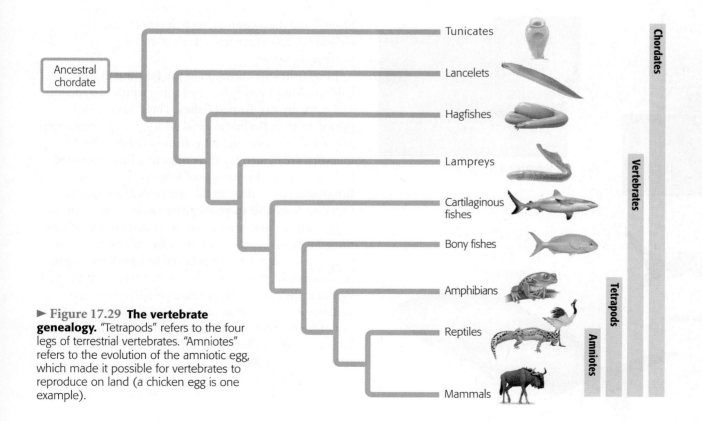

▶ Figure 17.29 **The vertebrate genealogy.** "Tetrapods" refers to the four legs of terrestrial vertebrates. "Amniotes" refers to the evolution of the amniotic egg, which made it possible for vertebrates to reproduce on land (a chicken egg is one example).

Fishes

The first vertebrates were aquatic and probably evolved during the early Cambrian period about 542 million years ago. In contrast with most other vertebrates, they lacked jaws, hinged bone structures that work the mouth.

Two types of jawless fishes survive today: hagfishes and lampreys. As noted earlier, hagfishes are invertebrates but have a cranium. Present-day hagfishes scavenge dead or dying animals on the cold, dark seafloor. When threatened, a hagfish exudes an enormous amount of slime from special glands on the sides of its body (**Figure 17.30a**). Recently, hagfishes have become endangered because their skin is used to make "eel-skin" belts, purses, and boots. Unlike hagfishes, lampreys are vertebrates. Some lampreys are parasites that use their jawless mouths as suckers to attach to the sides of large fish and draw blood (**Figure 17.30b**).

We know from the fossil record that the first jawed vertebrates were fishes that evolved about 470 million years ago. They had two pairs of fins, making them agile swimmers. Some early fishes were active predators up to 10 m in length that could chase prey and bite off chunks of flesh. Even today, most fishes are carnivores.

Cartilaginous fishes, such as sharks and rays, have a flexible skeleton made of cartilage (**Figure 17.30c**). Most sharks are adept predators because they are fast swimmers with streamlined bodies, acute senses, and powerful jaws. A shark does not have keen eyesight, but its sense of smell is very sharp. In addition, special electrosensors on the head can detect minute electrical fields produced by muscle contractions in nearby animals. Sharks also have a **lateral line system**, a row of sensory organs running along each side of the body. Sensitive to changes in water pressure, the lateral line system enables a shark to detect minor vibrations caused by animals swimming in its neighborhood. There are about 750 living species of cartilaginous fishes, nearly all of them marine.

Most fishes—about 27,000 species—are **bony fishes** with skeletons reinforced by calcium (**Figure 17.30d**). They also have a lateral line system, a keen sense of smell, and excellent eyesight. On each side of the head, a protective flap called the **operculum** (plural, *opercula*) covers a chamber housing the gills, feathery external organs that extract oxygen from water. Movement of the operculum allows the fish to breathe without swimming. By contrast, sharks lack opercula and must swim to pass water over their gills. The need to move water over the gills is why a shark must keep moving to stay alive. Also unlike sharks, bony fishes have an organ that helps keep them buoyant—the **swim bladder**, a gas-filled sac. Thus, many bony fishes can conserve energy by remaining almost motionless, in contrast to sharks, which sink if they stop swimming. Some bony fishes have a connection between the swim bladder and the digestive tract that enables them to gulp air and extract oxygen from it when the oxygen level in the water gets too low.

Bony fishes are common in the seas and in freshwater habitats. Most bony fishes, including trout, bass, perch, and tuna, are **ray-finned fishes**. Their fins are supported by thin, flexible skeletal rays. A second evolutionary branch includes the **lobe-finned fishes**. The lobe-fins are named for their muscular fins supported by stout bones. One lineage of lobe-finned fishes, **lungfishes**, lives today in the Southern Hemisphere. Lungfishes inhabit stagnant ponds and swamps, surfacing to gulp air into their lungs. A second lineage of lobe-finned fishes is represented today by the coelacanth, a deep-sea dweller that was believed to have been extinct for millions of years before being "rediscovered" in 1938. Coelacanths may use their fins to waddle along the seafloor. A third lineage of lobe-finned fishes migrated out of fresh water and adapted to life on land, playing a key role in the evolution of amphibians, the first terrestrial vertebrates. ☑

▼ Figure 17.30 **Fish diversity.**

(a) Hagfish (inset: slime)

(b) Lamprey (inset: mouth)

(c) Shark, a cartilaginous fish

Lateral line

Operculum

(d) Bony fish

Amphibians

In Greek, the word *amphibios* means "living a double life." Most **amphibians** exhibit a mixture of aquatic and terrestrial adaptations. Most species are tied to water because their eggs, lacking shells, dry out quickly in the air. A frog may spend much of its time on land, but it lays its eggs in water. An egg develops into a larva called a tadpole, a legless, aquatic algae-eater with gills, a lateral line system resembling that of fishes, and a long-finned tail. In changing into a frog, the tadpole undergoes a radical metamorphosis (**Figure 17.31a**). When a young frog crawls onto shore and begins life as a terrestrial insect-eater, it has four legs, air-breathing lungs instead of gills, external eardrums, and no lateral line system. But even as adults, amphibians are most abundant in damp habitats, such as swamps and rain forests. This is partly because amphibians depend on their moist skin to supplement lung function in exchanging gases with the environment. Thus, even those frogs that are adapted to relatively dry habitats spend much of their time in humid burrows or under piles of moist leaves. The amphibians of today, including frogs and salamanders, account for about 12% of all living vertebrates, or about 6,000 species (**Figure 17.31b**).

Amphibians were the first vertebrates to colonize land. They descended from fishes that had lungs and fins with muscles and skeletal supports strong enough to enable some movement, however clumsy, on land (**Figure 17.32**). The fossil record chronicles the evolution of four-limbed amphibians from fishlike ancestors. Terrestrial vertebrates—amphibians, reptiles, and mammals—are collectively called **tetrapods**, which means "four feet." ☑

Tunicates
Lancelets
Hagfishes
Lampreys
Cartilaginous fishes
Bony fishes
Amphibians
Reptiles
Mammals

☑CHECKPOINT

Amphibians were the first _____, four-footed terrestrial vertebrates.

Answer: tetrapods

▼ Figure 17.31 **Amphibian diversity.**

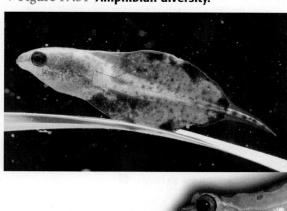

(a) Tadpole and adult golden palm tree frog

Red-eyed tree frog

Texas barred tiger salamander

(b) Frogs and salamanders: the two major groups of amphibians

▼ Figure 17.32 **The origin of tetrapods.**

Lobe-finned fish. Fossils of some lobe-finned fishes have skeletal supports extending into their fins.

Early amphibian. Fossils of early amphibians have limb skeletons that probably functioned in helping them move on land.

Reptiles

Reptiles (including birds) and mammals are **amniotes**. The evolution of amniotes from an amphibian ancestor included many adaptations for living on land. The adaptation that gives the group its name is the **amniotic egg**, a fluid-filled egg inside of which the embryo develops. The amniotic egg functions as a self-contained "pond" that enables amniotes to complete their life cycle on land.

The **reptiles** include snakes, lizards, turtles, crocodiles, alligators, and birds, along with a number of extinct groups, including most of the dinosaurs **(Figure 17.33)**. The bull snake in Figure 17.33 displays two reptilian adaptations to living on land: scaled waterproof skin, preventing dehydration in dry air, and amniotic eggs with shells, providing a watery, nutritious internal environment where the embryo can develop. These adaptations allowed reptiles to break their ancestral ties to aquatic habitats. Reptiles cannot breathe through their dry skin and so obtain most of their oxygen through their lungs.

*(Cladogram: Tunicates, Lancelets, Hagfishes, Lampreys, Cartilaginous fishes, Bony fishes, Amphibians, **Reptiles**, Mammals)*

Nonbird Reptiles

Nonbird reptiles are sometimes referred to as "cold-blooded" animals because they do not use their metabolism extensively to control body temperature. Reptiles do regulate body temperature, but largely through behavioral adaptations. For example, many lizards regulate their internal temperature by basking in the sun when the air is cool and seeking shade when the air is too warm. Because lizards and other nonbird reptiles absorb external heat rather than generating much of their own, they are said to be **ectotherms**, a term more accurate than "cold-blooded." By heating directly with solar energy rather than through the metabolic breakdown of food, a nonbird reptile can survive on less than 10% of the calories required by a mammal of equivalent size.

As successful as reptiles are today, they were far more widespread, numerous, and diverse during the Mesozoic era, which is sometimes known as the "age of reptiles." Reptiles diversified extensively during that era, producing a dynasty that lasted until about 65 million years ago. Dinosaurs, the most diverse reptile group, included the largest animals ever to inhabit land. Some were gentle giants that lumbered about while browsing vegetation. Others were voracious carnivores that chased their larger prey on two legs.

The age of reptiles began to fade about 70 million years ago. Around that time, the global climate became cooler and more variable (see Chapter 14). This was a period of mass extinctions that claimed all the dinosaurs by about 65 million years ago, except for one lineage. That lone surviving lineage is represented today by the reptilian group we know as birds. ✔

▼ **Figure 17.33 Reptile diversity.**

Snake. Like all snakes, bull snakes lay eggs. Bull snakes are commonly found in North American prairies, where their coloration helps them blend into the background.

Turtle. A fully grown Galápagos tortoise can weigh more than 600 pounds and live for more than 150 years.

Lizard. The leopard gecko, a popular pet, is native to the deserts of the Middle East.

Birds

Although **birds** were previously placed in their own class—class Aves—recent genetic evidence shows that they are reptiles, having evolved from a lineage of small, two-legged dinosaurs called theropods. But today birds look quite different from reptiles because of their feathers and other distinctive flight equipment. Almost all of the 10,000 living bird species are airborne. The few flightless species, including the ostrich and the penguin, evolved from flying ancestors.

Appreciating the avian world is all about understanding flight. Almost every element of bird anatomy is modified in some way that enhances flight. The bones have a honeycombed structure that makes them strong but light (the wings of airplanes have the same basic construction). For example, a huge seagoing species called the frigate bird has a wingspan of more than 2 m, but its whole skeleton weighs only about 113 g (4 oz). Another adaptation that reduces the weight of birds is the absence of some internal organs found in other vertebrates. Female birds, for instance, have only one ovary instead of a pair. Also, today's birds are toothless, an adaptation that trims the weight of the head (preventing uncontrolled nosedives). Birds do not chew food in the mouth but grind it in the gizzard, a chamber of the digestive tract near the stomach.

Flying requires a great expenditure of energy and an active metabolism. Unlike other reptiles, birds are **endotherms**, meaning they use their own metabolic heat to maintain a warm, constant body temperature.

A bird's most obvious flight equipment is its wings. Bird wings are airfoils that illustrate the same principles of aerodynamics as the wings of an airplane (**Figure 17.34**). A bird's flight motors are its powerful breast muscles, which are anchored to a keel-like breastbone. It is mainly these flight muscles that we call "white meat" on poultry. Some birds, such as eagles and hawks, have wings adapted for soaring on air currents and flap their wings only occasionally. Other birds, including hummingbirds, excel at maneuvering but must flap continuously to stay aloft. Feathers are made of the protein that forms the scales of reptiles. Feathers may have functioned first as insulation, helping birds retain body heat, only later being adapted as flight gear. ✓

☑CHECKPOINT

Birds differ from other reptiles in their main source of body heat, with birds being _____ and other reptiles being _____.

Answer: endotherms; ectotherms

► Figure 17.34 **The aerodynamics of a bald eagle in flight.** Both birds and airplanes owe their "lift" to changes in air pressure caused by the shape of their wings.

Lower air pressure

Higher air pressure

Airfoil

Crocodile. Found throughout central and southern Africa, the Nile crocodile can grow up to 20 feet long and weigh nearly 2,000 pounds.

Birds. These large birds, called red-crowned cranes, are highly endangered, with fewer than 2,000 left in the wild, mostly in China.

Dinosaur. This *Herrarasaurus* skeleton is from a carnivorous biped discovered in Argentina.

Mammals

There are two major lineages of amniotes: one that led to the reptiles and one that produced the mammals. The first **mammals** arose about 200 million years ago and were probably small, nocturnal insect-eaters. Mammals became much more diverse after the downfall of the dinosaurs. Most mammals are terrestrial. These include nearly 1,000 species of winged mammals, the bats. And about 80 species of dolphins, porpoises, and whales are totally aquatic. The blue whale, an endangered mammal that grows to lengths of nearly 30 m, is the largest animal that has ever lived. Two features—hair and mammary glands that produce milk that nourishes the young—are mammalian hallmarks. The main function of hair is to insulate the body and help maintain a warm, constant internal temperature; mammals, like birds, are endotherms.

The major groups of mammals are monotremes, marsupials, and eutherians (**Figure 17.35**). The duck-billed platypus and the echidna are the only existing species of **monotremes**, egg-laying mammals. The platypus lives along rivers in eastern Australia and on the nearby island of Tasmania. The female usually lays two eggs and incubates them in a leaf nest. After hatching, the young nurse by licking up milk secreted onto the mother's fur.

Most mammals are born rather than hatched. During pregnancy in marsupials and eutherians, the embryos are nurtured inside the mother by an organ called the **placenta**. Consisting of both embryonic and maternal tissues, the placenta joins the embryo to the mother within the uterus. The embryo is nurtured by maternal blood that flows close to the embryonic blood system in the placenta.

Marsupials, the so-called pouched mammals, include kangaroos, koalas, and opossums. These mammals have a brief pregnancy and give birth to tiny embryonic offspring that complete development while attached to the mother's nipples. The nursing young are usually housed in an external pouch on the mother's abdomen. Nearly all marsupials live in Australia, New Zealand, and North and South America. Australia has been a marsupial sanctuary for much of the past 60 million years. Australian marsupials have diversified extensively, filling terrestrial habitats that on the other continents are occupied by eutherian mammals.

Eutherians are also called **placental mammals** because their placentas provide a more intimate and longer-lasting association between the mother and her developing young than do marsupial placentas. Eutherians make up almost 95% of the 5,300 species of living mammals. Dogs, cats, cows, rodents, rabbits, bats, and whales are all examples of eutherian mammals. One of the eutherian groups is the primates, which include monkeys, apes, and humans. ✔

Tunicates
Lancelets
Hagfishes
Lampreys
Cartilaginous fishes
Bony fishes
Amphibians
Reptiles
Mammals

✔**CHECKPOINT**

What are two hallmarks of mammals?

Answer: hair and mammary glands

▼ **Figure 17.35 Mammalian diversity.**

MAJOR GROUPS OF MAMMALS

Monotremes
(hatched from eggs)

Marsupials
(embryonic at birth)

Eutherians
(fully developed at birth)

Monotremes, such as this echidna, are the only mammals that lay eggs (inset).

The young of marsupials are born very early in their development. The newborn kangaroo (inset) will finish its growth while nursing from a nipple in its mother's pouch.

In eutherians (placental mammals), young develop within the uterus of the mother. There they are nurtured by the flow of blood through the dense network of vessels in the placenta. This newborn wildebeest is coated by remnants of the placenta.

The Human Ancestry

We have now traced animal phylogeny to the **primates**, the mammalian group that includes *Homo sapiens* and its closest kin. To understand what that means, we must follow our ancestry back to the trees, where some of our most treasured traits originated.

The Evolution of Primates

Primate evolution provides a context for understanding human origins. The fossil record supports the hypothesis that primates evolved from insect-eating mammals during the late Cretaceous period, about 65 million years ago. Those early primates were small, arboreal (tree-dwelling) mammals. Thus, primates were first distinguished by characteristics that were shaped, through natural selection, by the demands of living in the trees. For example, primates have limber shoulder joints, which make it possible to swing from branch to branch. The agile hands of primates can hang on to branches and manipulate food. Nails have replaced claws in many primate species, and the fingers are very sensitive. The eyes of primates are close together on the front of the face. The overlapping fields of vision of the two eyes enhance depth perception, an obvious advantage when swinging in trees. Excellent eye-hand coordination is also important for arboreal maneuvering. Parental care is essential for young animals in the trees. Mammals devote more energy to caring for their young than most other vertebrates, and primates are among the most attentive parents of all mammals. Most primates have single births and nurture their offspring for a long time. Although humans do not live in trees, we retain in modified form many traits that originated there.

Taxonomists divide the primates into three main groups (**Figure 17.36**). The first includes lorises, pottos, and lemurs. These primates live in Madagascar, Africa, and southern Asia. Tarsiers, small nocturnal tree-dwellers found only in Southeast Asia, form the second group of primates. The third group of primates, **anthropoids**, includes monkeys, apes, and humans. All monkeys in the New World (the Americas) are arboreal and are distinguished by prehensile (grasping) tails that function as an extra appendage for swinging. (If you see a monkey in a zoo swinging by its tail, you know it's from the New World.) Although some Old World (African and Asian) monkeys are also arboreal, their tails are not prehensile. And many Old World monkeys, including baboons, macaques, and mandrills,

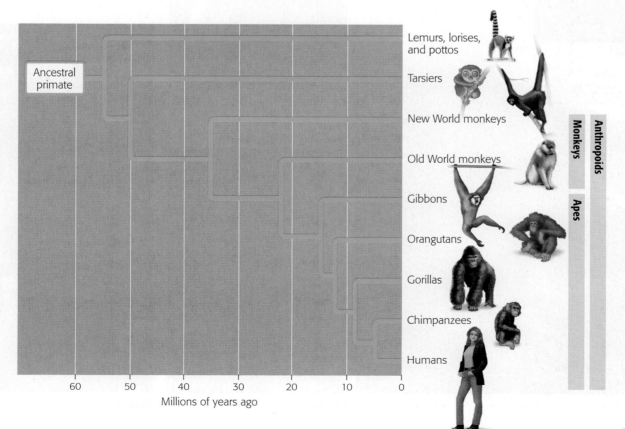

Ancestral primate

Lemurs, lorises, and pottos

Tarsiers

New World monkeys

Old World monkeys

Gibbons

Orangutans

Gorillas

Chimpanzees

Humans

Monkeys

Apes

Anthropoids

60 50 40 30 20 10 0

Millions of years ago

◄ Figure 17.36 **Primate phylogeny.**

are mainly ground-dwellers. Anthropoids also have a fully opposable thumb; that is, they can touch the tips of all four fingers with their thumb.

Our closest anthropoid relatives are the nonhuman apes: gibbons, orangutans, gorillas, and chimpanzees. They live only in tropical regions of the Old World. Except for some gibbons, apes are larger than monkeys, with relatively long arms, short legs, and no tail. Although all apes are capable of living in trees, only gibbons and orangutans are primarily arboreal. Gorillas and chimpanzees are highly social. Apes have larger brains proportionate to body size than monkeys, and their behavior is more adaptable. And, of course, the apes include humans. **Figure 17.37** shows examples of primates.

▼ Figure 17.37 **Primate diversity.**

Ring-tailed lemur

Tarsier

Patas monkey (Old World monkey)

Black spider monkey
(New World monkey)

Gibbon (ape)

Orangutan (ape)

Gorilla (ape)

Chimpanzee (ape)

Human

The Emergence of Humankind

Humanity is one very young twig on the tree of life. In the continuum of life spanning 3.5 billion years, the fossil record and molecular systematics indicate that humans and chimpanzees have shared a common African ancestry for all but the last 5–7 million years (see Figure 17.36). Put another way, if we compressed the history of life to a year, the human branch has existed for only 18 hours.

Some Common Misconceptions

Certain misconceptions about human evolution persist in the minds of many, long after these myths have been debunked by the fossil evidence. Let's first dispose of the myth that our ancestors were chimpanzees or any other modern ape. Chimpanzees and humans represent two divergent branches of the anthropoid tree that evolved from a common, less specialized ancestor. Chimps are not our parent species, but more like our phylogenetic siblings or cousins.

Another misconception envisions human evolution as a ladder with a series of steps leading directly from an ancestral anthropoid to *Homo sapiens*. This is often illustrated as a parade of fossil **hominins** (members of the human family) becoming progressively more modern as they march across the page. If human evolution is a parade, then it is a disorderly one, with many splinter groups having traveled down dead ends. At times in hominin history, several human species coexisted (**Figure 17.38**). Human phylogeny is more like a multi-branched bush than a ladder, with our species being the tip of the only twig that still lives.

One more myth is that various human characteristics, such as upright posture and an enlarged brain, evolved in unison. A popular image is of early humans as half-stooped, half-witted cave-dwellers. In fact, the fossil record reveals that different human features evolved at different rates, with erect posture, or bipedalism, leading the way. Our pedigree includes ancestors who walked upright but had ape-sized brains.

After dismissing some of the folklore on human evolution, however, we must admit that many questions about our ancestry have not yet been resolved.

Human Evolution

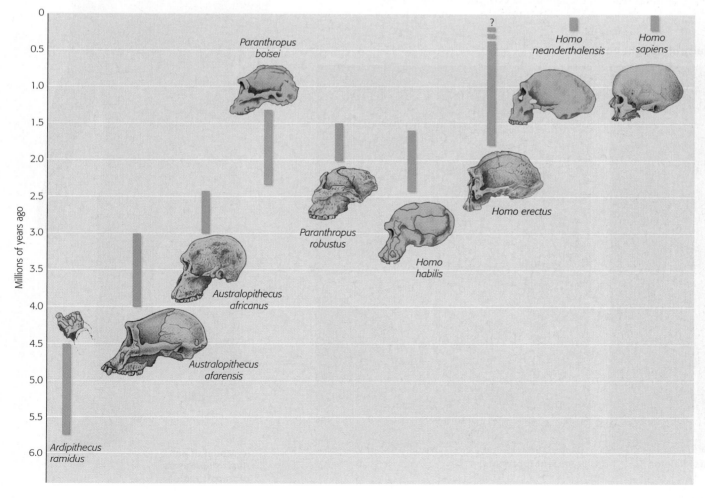

◀ **Figure 17.38 A time line of human evolution.** Notice that there have been times when two or more hominin species coexisted. The skulls are all drawn to the same scale so you can compare the sizes of craniums and hence brains.

Australopithecus and the Antiquity of Bipedalism

Before there was the genus *Homo*, several hominin species of the genus *Australopithecus* walked the African savanna. Fossil evidence indicates that bipedalism first arose in *Australopithecus afarensis* at least 4 million years ago **(Figure 17.39)**.

One of the most complete fossil skeletons of the species *Australopithecus afarensis* dates to about 3.2 million years ago in East Africa. Nicknamed Lucy by her discoverers, the individual was a female, only about 3 feet tall and with a head about the size of a softball. Lucy and her kind lived in savanna areas and may have subsisted on nuts and seeds, bird eggs, and whatever animals they could catch or scavenge from kills made by more efficient predators such as large cats and dogs.

Homo habilis and the Evolution of Inventive Minds

Enlargement of the human brain is first evident in fossils from East Africa dating to about 2.4 million years ago. Thus, the fundamental human trait of an enlarged brain evolved a few million years after the other major human trait—bipedalism. As evolutionary biologist Stephen Jay Gould put it, "Mankind stood up first and got smart later."

Anthropologists have found skulls with brain capacities intermediate in size between those of the latest *Australopithecus* species and those of *Homo sapiens*. Simple handmade stone tools are sometimes found with the larger-brained fossils, which have been dubbed *Homo habilis* ("handy man"). After walking upright for about 2 million years, humans were finally beginning to use their manual dexterity and big brains to invent tools that enhanced their hunting, gathering, and scavenging on the African savanna.

▼ Figure 17.39 **The antiquity of upright posture.**

(a) *Australopithecus afarensis* **skeleton**

(b) **Ancient footprints**

(c) **Model of an** *Australopithecus afarensis* **male**

Homo erectus and the Global Dispersal of Humanity

The first species to extend humanity's range from Africa to other continents was *Homo erectus*, perhaps a descendant of *Homo habilis*. Skeletons of *Homo erectus* dating to 1.8 million years ago found in the former Soviet republic of Georgia represent the oldest known fossils of hominins outside Africa. *Homo erectus* migrated to populate many regions of Asia and Europe, eventually moving as far as Indonesia. Remember the "hobbits" (*Homo floresiensis*) from the Biology and Society section? Some anthropologists speculate that they may actually be a long-isolated group of *Homo erectus*.

Homo erectus was taller than *Homo habilis* and had a larger brain capacity. During the 1.5 million years the species existed, the *Homo erectus* brain increased to as large as 1,200 cm³, comparable to modern humans. Intelligence enabled humans to continue succeeding in Africa and also to survive in the colder climates of the north. *Homo erectus* resided in huts or caves, built fires, made clothes from animal skins, and designed stone tools. In anatomical and physiological adaptations, *Homo erectus* was poorly equipped for life outside the tropics, but made up for the deficiencies with cleverness and social cooperation.

Some African, Asian, European, and Australasian (from Indonesia, New Guinea, and Australia) populations of *Homo erectus* gave rise to regionally diverse descendants that had even larger brains. Among these descendants of *Homo erectus* was *Homo neanderthalensis* (the Neanderthals), who lived in Europe, the Middle East, and Asia from about 200,000 years ago to about 30,000 years ago. (They are named Neanderthals because their fossils were first found in the Neander Valley of Germany.) Compared with us, Neanderthals had slightly heavier browridges and less pronounced chins, but their brains were similar to ours. Neanderthals were skilled toolmakers, and they participated in burials and other rituals that required abstract thought. From ancient bones, scientists can tell the shape of Neanderthal bodies. But given their prehistoric extinction, is it possible to infer their coloration? ✔

THE PROCESS OF SCIENCE: Human Evolution

What Did Neanderthals Look Like?

To try to put a face to Neanderthals, an international team of geneticists began with the **observation** that a gene called *mc1r* has a large effect on the hair and skin pigmentation of modern humans. This led the scientists to **question** what form of the gene would be found in Neanderthals. They formed the **hypothesis** that determining the DNA sequence of this gene would lend insight into the physical appearance of Neanderthals.

Their **experiment** used the polymerase chain reaction (see Chapter 12) to sequence the *mc1r* pigment gene in DNA isolated from Neanderthal bones. Their **results** were surprising: The Neanderthal gene contained a mutation not found in any modern humans. Next, the scientists genetically engineered modern human cells to carry the Neanderthal version of *mc1r*. They found that the mutation impaired the activity of a plasma membrane receptor protein that helps regulate the balance between skin pigments. Anyone with two identical alleles for this mutation (estimated to include at least 1 in 100 Neanderthals and possibly many more) would have red hair and pale skin (**Figure 17.40**).

Interestingly, this coloration is seen in modern Europeans who live in the Neanderthal's old territory. In less sunny regions, pale skin facilitates the production of the essential nutrient vitamin D. Thus, the Neanderthals and *Homo sapiens* who coexisted in Europe may have looked more alike than previously realized.

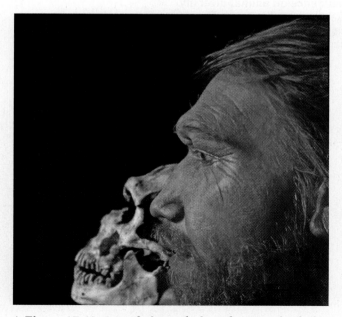

▲ Figure 17.40 **An artist's rendering of a Neanderthal.**

The Origin and Dispersal of *Homo sapiens*

The oldest known fossils of our own species, *Homo sapiens*, were discovered in Ethiopia and date from 160,000 to 195,000 years ago. These early humans lacked the heavy browridges of *Homo erectus* and *Homo neanderthalensis* and were more slender, suggesting that they belong to a distinct lineage. The Ethiopian fossils support molecular evidence about the origin of humans: DNA studies strongly suggest that all living humans can trace their ancestry back to a single African *Homo sapiens* woman who lived 160,000 to 200,000 years ago.

Fossil evidence suggests that our species emerged from Africa in one or more waves, spreading first into Asia and then to Europe and Australia. The oldest fossils of *Homo sapiens* outside Africa date back about 50,000 years. The date of the first arrival of humans in the New World is uncertain, although the generally accepted evidence suggests a minimum of 15,000 years ago. ✓

Cultural Evolution

Certain uniquely human traits have allowed for the development of human societies. The primate brain continues to grow after birth, and the period of growth is longer for a human than for any other primate. The extended period of human development also lengthens the time parents care for their offspring, which contributes to the child's ability to benefit from the experiences of earlier generations. This is the basis of **culture**—social transmission of accumulated knowledge, customs,

beliefs, and art over generations. The major means of this transmission is language, spoken and written.

There have been three major stages of cultural evolution. The first stage began with nomads who hunted and gathered food on the African grasslands 2 million years ago. They made tools, organized communal activities, and divided labor. Beautiful ancient art, such as the 30,000-year-old cave paintings shown in **Figure 17.41**, is just one example of our cultural roots in early societies. The second main stage of cultural evolution came with the development of agriculture in Africa, Eurasia, and the Americas about 10,000 to 15,000 years ago. Along with agriculture came permanent settlements and the first cities. The third major stage in our cultural evolution was the Industrial Revolution, which began in the 1700s. Since then, the development of new technology has escalated exponentially; the span of time from the flight of the Wright brothers to Neil Armstrong's walk on the moon was less than an average American's life span. And it took less than a decade for the Internet to transform commerce, communication, and education.

Through all this cultural evolution, from simple hunter-gatherers to high-tech societies, we have not changed biologically in any significant way. We are probably no more intelligent than our cave-dwelling ancestors. The know-how to build skyscrapers, computers, and spacecraft is stored not in our genes but in the cumulative product of hundreds of generations of human experience, passed along by parents, teachers, books, and electronic media.

✓**CHECKPOINT**
1. Humans first evolved on which continent?
2. Is it possible that a *Homo sapiens* individual ever met a Neanderthal?

Answers: **1.** *Africa* **2.** *Yes—Neanderthals lived in Europe and Asia until about 30,000 years ago, and Homo sapiens first migrated out of Africa about 50,000 years ago.*

▼ **Figure 17.41 Art history goes way back—and so does our fascination with and dependence on animal diversity.** Early artists created these remarkable images beginning about 30,000 years ago. Three cave explorers found this prehistoric art gallery in a cavern in southern France.

Nothing has had a greater impact on life on Earth than one species: *Homo sapiens*. The global consequences of human evolution have been enormous (Figure 17.42). Cultural evolution made *Homo sapiens* a new force in the history of life—a species that could defy its physical limitations and shortcut biological evolution. We do not have to wait to adapt to an environment through natural selection; we simply change the environment to meet our needs.

We are the most numerous and widespread of all large animals, and wherever we go we bring environmental change faster than many species can adapt; the rate of extinctions during the 1900s was 50 times greater than the average for the past 100,000 years. In the next unit, on ecology, we'll examine the interactions of humans—as well as other species—with the environment.

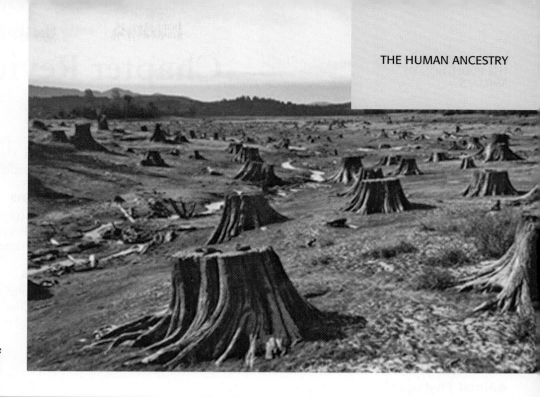

▶ Figure 17.42 **A clear-cut forest, an example of environmental change caused by *Homo sapiens*.**

EVOLUTION CONNECTION: Human Evolution

Recent Human Evolution

In the Biology and Society section, you learned about the discovery of a possible new human species called *Homo floresiensis*. And in the Process of Science section, you read about a genetic study involving our closest hominin relative, *Homo neanderthalensis*. Such investigations point out how evolution has shaped our own species after our split with the chimpanzees.

Biologists have identified a number of genes that appear to be undergoing rapid evolution in humans. Among them are genes involved in defense against malaria and tuberculosis and at least one gene regulating brain size. Another rapidly evolving human gene is called *FOXP2*. Several lines of evidence suggest that *FOXP2* functions in speech, a characteristic that obviously distinguishes humans from chimpanzees (Figure 17.43). For example, humans with mutations in this gene generally have severe speech impairment. Additionally, there is evidence that the *FOXP2* gene is expressed in the brains of zebra finches and canaries at the time when these songbirds are learning their songs. But the human version of *FOXP2* is unique, perhaps partially explaining our unique linguistic abilities. Further research has determined that the human form of *FOXP2* likely arose within the last 100,000 years, a time period that matches the emergence of *Homo sapiens*. Interestingly, the experiment described in the Process of Science section found that Neanderthals had the same version of *FOXP2* that we do.

▲ Figure 17.43 **Our closest primate relative: the chimpanzee.** The ability to speak is one trait that clearly differentiates humans from chimpanzees.

Chapter Review

SUMMARY OF KEY CONCEPTS

The Origins of Animal Diversity

What Is an Animal?

Animals are eukaryotic, multicellular, heterotrophic organisms that obtain nutrients by ingestion. Most animals reproduce sexually and develop from a zygote to a blastula to a gastrula. After the gastrula stage, some develop directly into adults, whereas others pass through a larval stage.

Early Animals and the Cambrian Explosion

Animals probably evolved from a colonial, flagellated protist more than 700 million years ago. At the beginning of the Cambrian period, 542 million years ago, animal diversity increased rapidly.

Animal Phylogeny

Major branches of animal evolution are defined by two key evolutionary differences: the presence or absence of tissues and radial versus bilateral body symmetry. A body cavity (coelom) at least partly lined by mesoderm evolved in a number of later branches.

Radial symmetry   **Bilateral symmetry**

Major Invertebrate Phyla

This tree shows the eight major invertebrate phyla, as well as chordates, which include a few invertebrates.

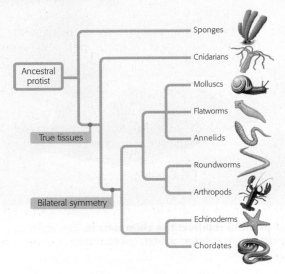

Sponges

Sponges (multiple phyla) are stationary animals with porous bodies but no true tissues. They feed by drawing water through pores in the sides of the body and trapping food particles in mucus.

Cnidarians

Cnidarians (phylum Cnidaria) have radial symmetry, a gastrovascular cavity with a single opening, and tentacles with stinging cnidocytes. The body is either a stationary polyp or a floating medusa.

Molluscs

Molluscs (phylum Mollusca) are soft-bodied animals often protected by a hard shell. The body has three main parts: a muscular foot, a visceral mass, and a fold of tissue called the mantle.

MOLLUSCS		
Gastropods	**Bivalves**	**Cephalopods**

Flatworms

Flatworms (phylum Platyhelminthes) are the simplest bilateral animals. They may be free-living (such as planarians) or parasitic (such as tapeworms).

Annelids

Annelids (phylum Annelida) are segmented worms with complete digestive tracts. They may be free-living or parasitic.

Roundworms

Roundworms, also called nematodes (phylum Nematoda), are unsegmented and cylindrical with tapered ends. They may be free-living or parasitic.

Arthropods

Arthropods (phylum Arthropoda) are segmented animals with an exoskeleton and specialized, jointed appendages.

Antenna

Head Thorax   Abdomen

Echinoderms

Echinoderms (phylum Echinodermata) are stationary or slow-moving marine animals that lack body segments and possess a unique water vascular system. Bilaterally symmetrical larvae usually change to radially symmetrical adults. Echinoderms have a bumpy endoskeleton.

Vertebrate Evolution and Diversity

Characteristics of Chordates

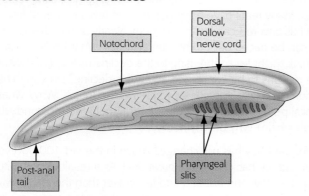

Tunicates and lancelets are invertebrate chordates. The vast majority of chordates are vertebrates, possessing a cranium and backbone.

Fishes

Hagfishes have skulls but lack vertebrae. Lampreys are jawless vertebrates. Cartilaginous fishes, such as sharks, are mostly predators with powerful jaws and a flexible skeleton made of cartilage. Bony fishes have a stiff skeleton reinforced by calcium. Bony fishes are further classified into ray-finned fishes and lobe-finned fishes (including lungfishes).

Amphibians

Amphibians are tetrapod vertebrates that usually deposit their eggs (lacking shells) in water. Aquatic larvae typically undergo a radical metamorphosis into the adult stage. Their moist skin requires that amphibians spend much of their adult life in humid environments.

Reptiles

Reptiles are amniotes, vertebrates that develop in a fluid-filled egg enclosed by a shell. Reptiles include terrestrial ectotherms with lungs and waterproof skin covered by scales. Scales and amniotic eggs enhanced reproduction on land. Birds are endothermic reptiles with wings, feathers, and other adaptations for flight.

Mammals

Mammals are endothermic vertebrates with hair and mammary glands. There are three major groups: Monotremes lay eggs; marsupials use a placenta but give birth to tiny embryonic offspring that usually complete development while attached to nipples inside the mother's pouch; and eutherians, or placental mammals, use their placenta in a longer-lasting association between the mother and her developing young.

MAMMALS		
Monotremes	**Marsupials**	**Eutherians**

The Human Ancestry

The Evolution of Primates

The first primates were small, arboreal mammals that evolved from insect-eating mammals about 65 million years ago. Anthropoids consist of New World monkeys (with prehensile tails), Old World monkeys (without prehensile tails), apes, and humans.

The Emergence of Humankind

Chimpanzees and humans evolved from a common ancestor about 5–7 million years ago. Upright posture evolved in several hominin species of the genus *Australopithecus* at least 4 million years ago. Enlargement of the human brain in *Homo habilis* came later, about 2.4 million years ago. *Homo erectus* was the first species to extend humanity's range from its birthplace in Africa to other continents. *Homo erectus* gave rise to regionally diverse descendants, such as the Neanderthals (*H. neanderthalensis*). Current data indicate a relatively recent dispersal of modern Africans that gave rise to today's human diversity. Human cultural evolution began in Africa with wandering hunter-gatherers, progressed to the development of agriculture and the Industrial Revolution, and continues today with accelerating technological change.

SELF-QUIZ

1. Bilateral symmetry in the animal kingdom is best correlated with
   a. an ability to sense equally in all directions.
   b. the presence of a skeleton.
   c. motility and active predation and escape.
   d. development of a true coelom.
2. The cavity between your outer body wall and your digestive tract that is fully lined by mesoderm is an example of a true _____. Roundworms have a _____, a body cavity not completely lined by mesoderm.

3. Identify which of the following categories includes all others in the list: arthropod, arachnid, insect, butterfly, crustacean, millipede.

4. The oldest group of tetrapods is the _____.

5. Reptiles are much more extensively adapted to life on land than amphibians because reptiles
   a. have a complete digestive tract.
   b. lay eggs that are enclosed in shells.
   c. are endothermic.
   d. go through a larval stage.

6. What is the name of the phylum to which humans belong? For what anatomical structure is the phylum named? Where in your body is a derivative of this anatomical structure found?

7. Fossils suggest that the first major trait distinguishing human primates from other primates was _____.

8. Which of the following types of animals is not included in the human ancestry? (*Hint*: See Figure 17.29.)
   a. a bird
   b. a bony fish
   c. an amphibian
   d. a primate

9. Put the following list of species in order, from the oldest to the most recent: *Homo erectus*, *Australopithecus* species, *Homo habilis*, *Homo sapiens*.

10. Match each of the following animals to its phylum:
   a. human   1. Echinodermata
   b. leech   2. Arthropoda
   c. sea star   3. Cnidaria
   d. lobster   4. Chordata
   e. sea anemone   5. Annelida

Answers to the Self-Quiz questions can be found in Appendix D.

THE PROCESS OF SCIENCE

11. Imagine that you are a marine biologist. As part of your exploration, you dredge up an unknown animal from the seafloor. Describe some of the characteristics you should look at to determine the phylum to which the creature should be assigned.

12. Many people describe themselves as vegetarians. Strictly speaking, vegetarians eat only plant products. Most vegetarians are not, in fact, that strict. Interview acquaintances who describe themselves as vegetarians and determine which taxonomic groups they avoid eating (see Figures 17.5 and 17.29). Try to generalize about their diet. For example, do they avoid eating vertebrates but eat some invertebrates? Do they avoid only birds and mammals? Do they eat dairy products or eggs?

13. Some researchers think that drying and cooling of the climate caused expansion of the African savanna and that this environment favored upright walking in early humans. Why might bipedalism be advantageous in the savanna? How might an erect posture relate to the evolution of a larger brain?

BIOLOGY AND SOCIETY

14. Coral reefs harbor a greater diversity of animals than any other environment in the sea. Australia's Great Barrier Reef has been protected as a marine reserve and is a mecca for scientists and nature enthusiasts. Elsewhere, such as in Indonesia and the Philippines, coral reefs are in danger. Many reefs have been depleted of fish, and runoff from the shore has covered coral with sediment. Nearly all the changes in the reefs can be traced back to human activities. What kinds of activities do you think might be contributing to the decline of the reefs? What are some reasons to be concerned about this decline? Do you think the situation is likely to improve or worsen in the future? Why? What might the local people do to halt the decline? Should the more industrialized countries help? Why or why not?

15. The human body has not changed much in the last 100,000 years, but human culture has changed a great deal. As a result of our culture, we change the environment at a rate far greater than the rate at which many species, including our own, can evolve. What evidence of rapid environmental change do you see regularly? What aspects of human culture are responsible for these changes? Do you see any evidence of a decrease in the rate of human-caused environmental changes?

Unit 4
Ecology

Chapter 18: **An Introduction to Ecology and the Biosphere**

Chapter Thread: **Global Climate Change**

Chapter 19: **Population Ecology**

Chapter Thread: **Invasive Species**

Chapter 20: **Communities and Ecosystems**

Chapter Thread: **The Loss of Biodiversity**

18 An Introduction to Ecology and the Biosphere

Polar bear on arctic ice.
Global climate change is shrinking the permanent sea ice that polar bears use as hunting platforms.

CHAPTER CONTENTS
■■■ Chapter Thread: **GLOBAL CLIMATE CHANGE**

BIOLOGY AND SOCIETY: Global Climate Change

Penguins and Polar Bears in Peril

The scientific debate is over. The great majority of scientists now agree that the global climate is changing. In the past few years, climate change has also become a hot topic for the general public. Articles and documentaries appear regularly, and it seems as if every magazine on the newsstand—including such unlikely publications as *Elle* and *Sports Illustrated*—has featured at least one cover story on climate change. What do we know about climate change now, and what can we expect for the future?

Average global temperatures have risen 0.8°C (about 1.4°F) over the past century, mostly over the last 30 years. The northernmost regions of the Northern Hemisphere and the Antarctic Peninsula have heated up the most. In parts of Alaska, for example, winter temperatures have risen by 5–6°F. The permanent arctic sea ice is shrinking; each summer brings thinner ice and more open water. Polar bears, which stalk their prey on ice and need to store up body fat for the warmer months when there is no ice, are showing signs of starvation as their winter hunting grounds melt away. At the other end of the planet, diminishing sea ice near the Antarctic Peninsula limits the access of Adélie penguins to their food supply, and spring blizzards of unprecedented frequency and severity are taking a heavy toll on their eggs and chicks. But the warming trend isn't the whole story of climate change. Precipitation patterns have also changed, bringing longer and more intense drought to some regions. In other areas, a greater proportion of the total rainfall is falling in torrential downpours that cause flooding.

Any predictions that scientists make now about the future ecological impact of global climate change are based on incomplete information. Much remains to be discovered about species diversity and about the complex interactions of organisms with each other and with their environments. There is overwhelming evidence that human enterprises are responsible for the changes that are occurring. How we respond to this crisis will determine whether circumstances improve or worsen. And the process begins with understanding the basic concepts of ecology, which we start to explore in this chapter.

An Overview of Ecology

In your study of biology so far, you have learned about the diversity of life on Earth and about the molecular and cellular structures and processes that make life tick. **Ecology**, the scientific study of the interactions between organisms and their environments, offers a different perspective on life—biology from the skin out, so to speak.

Humans have always had an interest in other organisms and their environments. As hunters and gatherers, prehistoric people had to learn where and when game and edible plants could be found in greatest abundance. Naturalists, from Aristotle to Darwin and beyond, made the process of observing and describing organisms in their natural habitats an end in itself rather than simply a means of survival. Extraordinary insight can still be gained from this discovery-based approach of watching nature and recording its structure and processes (**Figure 18.1**). As you might expect, field research, whether discovery-based natural history or hypothesis-driven science, is fundamental to ecology. But ecologists also test hypotheses using laboratory experiments, where conditions can be simplified and controlled. And some ecologists take a theoretical approach, devising mathematical and computer models, which enable them to simulate large-scale experiments that are impossible to conduct in the field.

▼ Figure 18.1 **Discovery science in a rain forest canopy.**

(a) Establishing a canopy research station. This hot-air dirigible is placing a giant rubber raft on the treetops of a tropical rain forest in French Guiana, in northeastern South America.

(b) Studying the canopy. Living and working on this field station 30 m above the forest floor, an international research team studies the canopy environment. French biologist Pierre Grard and two other scientists are cataloging plants and the insects that feed on those plants.

Ecology and Environmentalism

Technological innovations have enabled humans to colonize just about every environment on Earth. Even so, our survival depends on Earth's resources, which have been profoundly altered by human activities (**Figure 18.2**). Global climate change is just one of the many environmental issues that have stirred public concern in recent decades. Some of our industrial and agricultural practices have contaminated the air, soil, and water. Our relentless quest for land and other resources has endangered a lengthy list of plant and animal species and has even driven some to extinction.

The science of ecology can provide the understanding needed to solve environmental problems. But these problems cannot be solved by ecologists alone, because they require making decisions based on values and ethics. On a personal level, each of us makes daily choices that affect our ecological impact. And legislators and corporations, motivated by environmentally aware voters and consumers, must address questions that have wider implications: How should land use be regulated? Should we try to save all species or just certain ones? What alternatives to environmentally destructive practices can be developed? How can we balance environmental impact with economic needs?

▼ Figure 18.2 **Human impact on the environment.** A man paddles a canoe through a trash-clogged waterway in Manila, capital of the Philippines.

A Hierarchy of Interactions

Many different factors can potentially affect an organism's interaction with the environment. **Biotic factors**—all of the organisms in the area—make up the living component of the environment. Other organisms may compete with an individual for food and other resources, prey upon it, or change its physical and chemical environment. **Abiotic factors** make up the environment's nonliving component and include chemical and physical factors, such as temperature, light, water, minerals, and air. An organism's **habitat**, the specific environment it lives in, includes the biotic and abiotic factors of its surroundings.

When we study the interactions between organisms and their environments, it is convenient to divide ecology into four increasingly comprehensive levels: organismal ecology, population ecology, community ecology, and ecosystem ecology.

Organismal ecology is concerned with the evolutionary adaptations that enable individual organisms to meet the challenges posed by their abiotic environments. The distribution of organisms is limited by the abiotic conditions they can tolerate. For example, amphibians such as the salamander in **Figure 18.3a** are generally restricted to moist environments because their skin does not prevent dehydration (in fact, it plays a significant role in exchanging gases with the environment). You'll encounter other examples of organismal ecology throughout this chapter.

The next level of organization in ecology is the **population**, a group of individuals of the same species living in a particular geographic area. **Population ecology** concentrates mainly on factors that affect population density and growth (**Figure 18.3b**). Chapter 19 focuses on population ecology.

A **community** consists of all the organisms that inhabit a particular area; it is an assemblage of populations of different species. Questions in **community ecology** focus on how interactions between species, such as predation and competition, affect community structure and organization (**Figure 18.3c**).

An **ecosystem** includes all the abiotic factors in addition to the community of species in a certain area. For example, a savanna ecosystem includes not only the organisms, such as diverse plants and animals, but also the soil, water sources, sunlight, and other abiotic factors of the environment. In **ecosystem ecology**, questions concern energy flow and the cycling of chemicals among the various biotic and abiotic factors (**Figure 18.3d**). You'll learn about community and ecosystem ecology in Chapter 20.

The **biosphere** is the global ecosystem—the sum of all the planet's ecosystems, or all of life and where it

▼ Figure 18.3 **Examples of questions at different levels of ecology.**

Ecological Hierarchy

(a) Organismal ecology. What range of temperatures can a red salamander tolerate?

(b) Population ecology. What factors limit the number of striped mice that can inhabit a particular area?

(c) Community ecology. How do predators such as this beech marten affect the diversity of rodents in a community?

(d) Ecosystem ecology. What processes recycle vital chemical elements such as nitrogen within a savanna ecosystem in Africa?

lives. The most complex level in ecology, the biosphere includes the atmosphere to an altitude of several kilometers, the land down to water-bearing rocks about 1,500 m deep, lakes and streams, caves, and the oceans to a depth of several kilometers. Isolated in space, the biosphere is self-contained, or closed, except that its photosynthetic producers derive energy from sunlight and it loses heat to space. ☑

☑CHECKPOINT

What does the ecosystem level of classification have in common with the community level of classification? What does the ecosystem level include that the community level does not?

Answer: all the biotic factors of the area; the abiotic factors of the area

Living in Earth's Diverse Environments

Whether you have seen the world by traveling or through television and movies, you have probably noticed that there are striking regional patterns in the distribution of life. For example, some terrestrial areas, such as the tropical forests of South America and Africa, are home to plentiful plant life, whereas other areas, such as deserts, are relatively barren. Coral reefs are alive with vibrantly colored organisms; other parts of the ocean appear empty by comparison.

The distribution of life varies on a local scale, too. In the aerial view of an Alaskan wilderness in **Figure 18.4**, we can see a mixture of forest, small lakes, a meandering river, and open meadows. Within these different environments, variation occurs on a smaller scale. For example, we would find that each lake has several different habitats, and each habitat has a characteristic community of organisms.

▼ **Figure 18.4 Local variation of the environment in an Alaskan wilderness.**

Abiotic Factors of the Biosphere

Patterns in the distribution of life mainly reflect differences in the abiotic factors of the environment. In this section, we look at several major abiotic factors that influence where organisms live.

Energy Source

All organisms require a usable source of energy to live. Solar energy from sunlight, captured by chlorophyll during the process of photosynthesis, powers most ecosystems. In the image shown in **Figure 18.5**, colors are keyed to the relative abundance of chlorophyll. Green areas on land indicate high densities of plant life. Orange areas on land, including the Sahara region of Africa and much of the western United States, are much less productive. Green regions of the ocean contain an abundance of algae and photosynthetic bacteria compared to darker regions.

Lack of sunlight is seldom the most important factor limiting plant growth for terrestrial ecosystems (although shading by trees does create intense competition for light among plants growing on forest floors). In many aquatic environments, however, light cannot penetrate beyond certain depths. As a result, most photosynthesis in a body of water occurs near the surface. Surprisingly, life also thrives in environments that are completely dark. A mile or more below the ocean's

surface lies the unique world of hydrothermal vents, sites near the adjoining edges of giant plates of Earth's crust where molten rock and hot gases surge upward from Earth's interior. Towering chimneys, sometimes as high as 30 m, emit scalding water and hot gases. These are ecosystems powered by chemoautotrophic bacteria (see Figure 15.12) that derive energy from the oxidation of inorganic chemicals such as hydrogen sulfide (**Figure 18.6**). Bacteria with similar metabolic talents support communities of cave-dwelling organisms.

▼ **Figure 18.5**
Distribution of life in the biosphere. In this image of Earth, colors are keyed to the relative abundance of chlorophyll, which correlates with the regional densities of photosynthetic organisms.

▼ **Figure 18.6 A deep-sea hydrothermal vent.** "Black smokers" west of Vancouver Island spew plumes of hot gases from Earth's interior. Giant tube worms (inset), annelids that may grow to 2 m long, are members of the vent community.

Temperature

Temperature is an important abiotic factor because of its effect on metabolism. Few organisms can maintain a sufficiently active metabolism at temperatures close to 0°C, and temperatures above 45°C destroy the enzymes of most organisms. Still, extraordinary adaptations enable some species to live outside this temperature range. For example, bacteria and archaea living in hot springs have enzymes that function optimally at extremely high temperatures (**Figure 18.7**). At the other extreme, communities consisting of bacteria, algae, and small invertebrates inhabit sea ice at both poles.

▲ Figure 18.7 **Home of hot prokaryotes.** "Heat-loving" archaea thrive in this pool at Waiotapu geothermal area in the volcanic zone of New Zealand's North Island, where temperatures may exceed 80°C (176°F).

Water

Water is essential to all life. For terrestrial organisms, the primary threat is drying out in the air. Many land species have watertight coverings that reduce water loss. A waxy coating on the leaves and other aerial parts of most plants helps prevent dehydration. Aquatic organisms are surrounded by water, but they face problems of water balance if their own solute concentration does not match that of their surroundings (see Figure 5.14).

Nutrients

The distribution and abundance of photosynthetic organisms, including plants, algae, and photosynthetic bacteria, depend on the availability of inorganic nutrients such as compounds of nitrogen and phosphorus. Plants obtain these nutrients from the soil. Soil structure, pH, and nutrient content often play major roles in determining the distribution of plants. In many aquatic ecosystems, low levels of nitrogen and phosphorus limit the growth of algae and photosynthetic bacteria.

Other Aquatic Factors

Several abiotic factors are important in aquatic, but not terrestrial, ecosystems. While terrestrial organisms have a plentiful supply of oxygen from the air, aquatic organisms must depend on oxygen dissolved in water. This is a critical factor for many species of fish. Cold, fast-moving water has a higher oxygen content than warm or stagnant water (**Figure 18.8**). Salinity (saltiness), currents, and tides also play a role in many aquatic ecosystems.

Other Terrestrial Factors

Some abiotic factors affect terrestrial, but not aquatic, ecosystems. For example, wind is often an important abiotic factor on land. Wind increases an organism's rate of water loss by evaporation. The resulting increase in evaporative cooling can be advantageous on a hot summer day, but it can cause dangerous wind chill in the winter. In some ecosystems, frequent occurrences of natural disturbances such as storms or fire play a role in the distribution of organisms. ☑

▶ Figure 18.8 **Trout stream in Aspen, Colorado.** Fish such as trout and salmon need the higher oxygen content found in cold, fast-moving water.

The Evolutionary Adaptations of Organisms

The ability of organisms to live in Earth's diverse environments demonstrates the close relationship between the fields of ecology and evolutionary biology. Charles Darwin was an ecologist (although he predated the word *ecology*). It was the geographic distribution of organisms and their exquisite adaptations to specific environments that provided Darwin with evidence for evolution. Evolutionary adaptation via natural selection results from the interactions of organisms with their environments, which brings us back to our definition of ecology. Thus, events that occur in the short term, during the course of an individual's lifetime, may translate into effects over the longer scale of evolutionary time. For example, because the availability of water affects a plant's growth and ultimately its reproductive success, precipitation has an impact on the gene pool of a plant population. After a period of below-average rainfall, drought-resistant individuals may be more prevalent in a plant population. We'll look at some evolutionary responses to biotic interactions in Chapter 20. ✓

Adjusting to Environmental Variability

The abiotic factors in a habitat may vary from year to year, seasonally, or over the course of a day. An individual's abilities to adjust to environmental changes that occur during its lifetime are themselves adaptations refined by natural selection. For instance, if you see a bird on a cold day, it may look unusually fluffy (**Figure 18.9**). Small muscles in the skin raise the bird's feathers, a physiological response that traps insulating pockets of air. Some species of birds adjust to seasonal cold by growing heavier feathers. And some bird species respond to the onset of cold weather by migrating to warmer regions—a behavioral response. Note that these responses occur during the lifetime of an individual, so they do not qualify as evolution, which is change in a population over time.

▼ **Figure 18.9** **A gray-headed chickadee demonstrating its physiological response to cold weather.**

Physiological Responses

Like birds, mammals can create a temporary layer of insulation on cold days by contracting skin muscles, in this case attached to hairs. (Our own muscles do this, too, but we just get "goose bumps" instead of a furry insulation.) The blood vessels in the skin also constrict, which slows the loss of body heat. In both cases, the adjustment occurs in just seconds.

A gradual, though still reversible, physiological adjustment that occurs in response to an environmental change is called **acclimation**. For example, suppose you moved from Boston, which is essentially at sea level, to the mile-high city of Denver. One physiological response to the lower oxygen supply in your new environment would be a gradual increase in the number of your red blood cells, which transport O_2 from your lungs to other parts of your body. Acclimation can take days or weeks. This is why high-altitude climbers, such as those attempting to scale Mount Everest, need extended stays at a high-elevation base camp before proceeding to the summit.

The ability to acclimate is generally related to the range of environmental conditions a species naturally experiences. Species that live in very warm climates, for example, usually cannot acclimate to extreme cold. Among vertebrates, birds and mammals can generally tolerate the greatest temperature extremes because, as endotherms, they use their metabolism to regulate internal temperature. In contrast, ectothermic reptiles can only tolerate a more limited range of temperatures (**Figure 18.10**).

▼ **Figure 18.10** **The number of lizard species in different regions of the contiguous United States.** Notice that there are fewer and fewer lizard species in more northern regions. This reflects lizards' ectothermic physiology, which depends on environmental heat for keeping the body warm enough for the animal to be active.

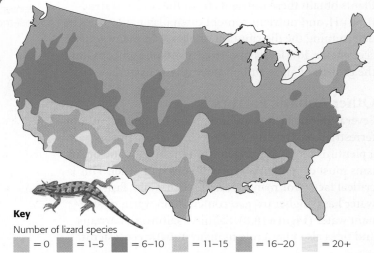

Key

Number of lizard species

■ = 0   ■ = 1–5   ■ = 6–10   ■ = 11–15   ■ = 16–20   ■ = 20+

Anatomical Responses

Many organisms respond to environmental challenge with some type of change in body shape or structure. When the change is reversible, the response is an example of acclimation. Many mammals, for example, grow a heavier coat of fur in response to cold, and shed it when summer comes. Sometimes fur or feather color changes seasonally as well, camouflaging the animal against winter snow and summer vegetation (Figure 18.11).

Other anatomical changes are irreversible over the lifetime of an individual. Environmental variation can affect growth and development so much that there may be remarkable differences in body shape within a population. You can see an example in Figure 18.12, which shows the "flagging" that wind causes in certain trees. In general, plants are more anatomically changeable than animals. Rooted and unable to move to a better location, plants rely entirely on their anatomical and physiological responses to survive environmental fluctuations.

▲ Figure 18.11 **The arctic fox in winter and summer coats.**

▼ Figure 18.12 **Wind as an abiotic factor that shapes trees.** The mechanical disturbance of the prevailing wind hinders limb growth on the windward side of these fir trees on Oregon's Mount Hood, while limbs on the other side grow normally. This anatomical response is an evolutionary adaptation that reduces the number of limbs that are broken during strong winds.

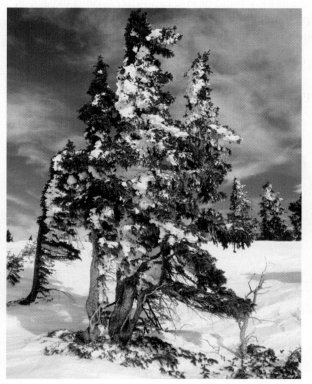

Behavioral Responses

In contrast to plants, most animals can respond to an unfavorable change in the environment by moving to a new location. Such movement may be fairly localized. For example, many desert ectotherms, including reptiles, maintain a reasonably constant body temperature by shuttling between sun and shade. Some animals are capable of migrating great distances in response to such environmental cues as the changing seasons. Many migratory birds overwinter in Central and South America, returning to northern latitudes to breed during summer. And humans, with their large brains and available technology, have an especially rich range of behavioral responses available to them (Figure 18.13). ☑

▼ Figure 18.13 **Behavioral responses have expanded the geographic range of humans.** Dressing for the weather is a thermoregulatory behavior unique to humans.

☑**CHECKPOINT**

What is acclimation?

Answer: a gradual, reversible change in anatomy or physiology in response to an environmental change

Biomes

The abiotic factors you learned about in the previous section are largely responsible for the distribution of life on Earth. (You'll learn about the role of biotic factors in species distribution in Chapter 20.) Using various combinations of these factors, ecologists have categorized Earth's environments into biomes. A **biome** is a major terrestrial or aquatic life zone, characterized by vegetation type in terrestrial biomes or the physical environment in aquatic biomes. In this section, we'll briefly survey the aquatic biomes, followed by the terrestrial biomes.

Aquatic biomes, which occupy roughly 75% of Earth's surface, are determined by their salinity and other physical factors. Freshwater biomes (lakes, streams and rivers, and wetlands) typically have a salt concentration of less than 1%. The salt concentrations of marine biomes (oceans, intertidal zones, coral reefs, and estuaries) are generally around 3%.

Freshwater Biomes

Freshwater biomes cover less than 1% of Earth, and they contain a mere 0.01% of its water. But they harbor a disproportionate share of biodiversity—an estimated 6% of all described species. Moreover, we depend on freshwater biomes for drinking water, crop irrigation, sanitation, and industry.

Freshwater biomes fall into two broad groups: standing water, which includes lakes and ponds, and flowing water, such as rivers and streams. The difference in water movement results in profound differences in ecosystem structure.

Lakes and Ponds

Standing bodies of water range from small ponds only a few square meters in area to large lakes, such as North America's Great Lakes, that are thousands of square kilometers (Figure 18.14).

◀ Figure 18.14 **Satellite view of the Great Lakes.**

In lakes and large ponds, the communities of plants, algae, and animals are distributed according to the depth of the water and its distance from shore (Figure 18.15). Shallow water near shore and the upper layer of water away from shore make up the **photic zone**, so named because light is available for photosynthesis. Microscopic algae and cyanobacteria grow in the photic zone, joined by rooted plants and floating plants such as water lilies in the photic area near shore. If a lake or pond is deep enough or murky enough, it has an **aphotic zone**, where light levels are too low to support photosynthesis.

At the bottom of all aquatic biomes is the **benthic realm**. Made up of sand and organic and inorganic sediments, the benthic realm is occupied by communities of organisms that are collectively called benthos. (If you have ever waded barefoot into a pond or lake, you have felt the benthos squish between your toes.) Dead material that "rains" down from the productive surface waters of the photic zone is a major source of food for the benthos.

The mineral nutrients nitrogen and phosphorus typically regulate the amount of phytoplankton growth in a lake or pond. Many lakes and ponds are affected by large inputs of nitrogen and phosphorus from sewage and runoff from fertilized lawns and farms. These nutrients often produce heavy growth of algae, which reduces light penetration. When the algae die and decompose, a pond or lake can suffer serious oxygen depletion, killing fish that are adapted to high-oxygen conditions. ☑

▼ Figure 18.15 **Zones in a lake.**

▲ Figure 18.16 **A stream in the Appalachian Mountains.**

Rivers and Streams

Rivers and streams, which are bodies of flowing water, generally support quite different communities of organisms than lakes and ponds (**Figure 18.16**). A river or stream changes greatly between its source (perhaps a spring or snowmelt in the mountains) and the point at which it empties into a lake or the ocean. Near a source, the water is usually cold, low in nutrients, and clear. The channel is often narrow, with a swift current that does not allow much silt to accumulate on the bottom. The current also inhibits the growth of phytoplankton; most of the organisms found here are supported by the photosynthesis of algae attached to rocks or by organic material (such as leaves) carried into the stream from the surrounding land. The most abundant benthic animals are usually insects that eat algae, leaves, or one another. Trout are often the predominant fishes, locating their food, including insects, mainly by sight in the clear water.

Downstream, a river or stream typically widens and slows. There the water is usually warmer and may be murkier because of sediments and phytoplankton suspended in it. Worms and insects that burrow into mud are often abundant, as are waterfowl, frogs, and catfish and other fishes that find food more by scent and taste than by sight.

Many streams and rivers have been affected by pollution from human activities. We have also altered rivers by constructing dams used for flood control, to provide reservoirs of drinking water, or to generate hydroelectric power. In many cases, dams have completely changed the downstream ecosystems, altering the rate and volume of water flow and affecting fish and invertebrate populations (**Figure 18.17**).

▲ Figure 18.17 **Damming the Columbia River Basin.** This map shows only the largest of the 250 dams that have altered freshwater ecosystems throughout the Pacific Northwest. These great concrete obstacles make it difficult for salmon to swim upriver to their breeding streams, though many dams now have "fish ladders" that provide detours (inset).

Wetlands

A **wetland** is a transitional biome between an aquatic ecosystem and a terrestrial one. Freshwater wetlands include swamps, bogs, and marshes (**Figure 18.18**). Covered with water either permanently or periodically, wetlands support the growth of aquatic plants and are among the richest of biomes in terms of species diversity. They provide water storage areas that reduce flooding and improve water quality by trapping pollutants such as metals and organic compounds in their sediments. The recognition of their ecological and economic value has led to government and private efforts to protect and restore wetlands.

▶ Figure 18.18 **A marsh in the Pocono Mountains, Pennsylvania.**

Marine Biomes

Gazing out over a vast ocean, you might think that it is the most uniform environment on Earth. But marine habitats can be as different as night and day. The deepest ocean, where hydrothermal vents are located, is perpetually dark. In contrast, the vivid coral reefs nearer the surface are utterly dependent on sunlight. Habitats near shore are different from those in mid-ocean, and the seafloor hosts different communities than the open waters.

As in freshwater biomes, the seafloor is known as the benthic realm (**Figure 18.19**). The **pelagic realm** of the oceans includes all open water. In shallow areas, such as the submerged parts of continents, called continental shelves, the photic zone includes both pelagic and benthic regions. In these sunlit areas, photosynthesis by **phytoplankton** (photosynthetic algae and bacteria) and multicellular algae provides energy for a diverse community of animals. Sponges, burrowing worms, clams, sea anemones, crabs, and echinoderms inhabit the benthos. **Zooplankton** (free-floating animals, including many microscopic ones), fishes, marine mammals, and many other types of animals are abundant in the pelagic photic zone.

The **coral reef** biome occurs in the photic zone of warm tropical waters in scattered locations around the globe (**Figure 18.20**). A coral reef is built up slowly by successive generations of coral animals—a diverse group of cnidarians that secrete a hard external skeleton—and by multicellular algae encrusted with limestone. Unicellular algae live within the coral's cells, providing the coral with food. The physical structure and productivity of coral reefs support a huge variety of invertebrates and fishes.

The photic zone extends down a maximum of 200 m in the ocean. Although there is not enough light for photosynthesis between 200 and 1,000 m, some light does reach these depths of the aphotic zone. This dimly lit world, sometimes called the twilight zone, is dominated by a fascinating variety of small fishes and crustaceans. Food sinking from the photic zone provides some sustenance for these animals. In addition, many of them migrate to the surface at night to feed. Some fishes in the twilight zone have enlarged eyes, enabling them to see in the very dim light, and light-emitting organs that attract mates and prey.

Below 1,000 m, the ocean is completely and permanently dark. Adaptation to this environment has produced many bizarre-looking creatures. Most of the benthic organisms here are deposit feeders, animals that consume dead organic material on the seafloor. Crustaceans, polychaete worms, sea anemones, and echinoderms such as sea cucumbers, sea stars, and sea

▼ Figure 18.19 **Ocean life.** (Zone depths and organisms not drawn to scale.)

▲ Figure 18.20 **A coral reef in the Red Sea off the coast of Egypt.**

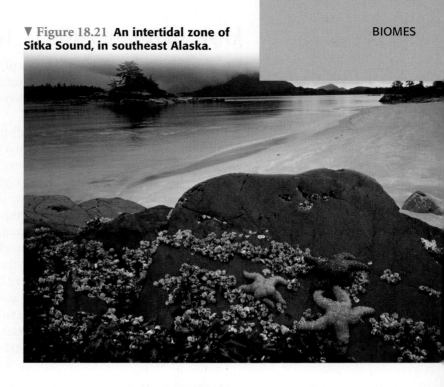

▼ Figure 18.21 **An intertidal zone of Sitka Sound, in southeast Alaska.**

urchins are common. Food is scarce, however. The density of animals is low except at hydrothermal vents, the prokaryote-powered ecosystems mentioned earlier (see Figure 18.6).

The marine environment also includes distinctive biomes where the ocean interfaces with land or with fresh water. In the **intertidal zone**, where the ocean meets land, the shore is pounded by waves during high tide and exposed to the sun and drying winds during low tide (**Figure 18.21**). The rocky intertidal zone is home to many sedentary organisms, such as algae, barnacles, and mussels, which attach to rocks and are thus prevented from being washed away. On sandy beaches, suspension-feeding worms, clams, and predatory crustaceans bury themselves in the ground.

Figure 18.22 shows an **estuary**, a transition area between a river and the ocean. The saltiness of estuaries ranges from nearly that of fresh water to that of the ocean. With their waters enriched by nutrients from the river, estuaries, like freshwater wetlands, are among the most productive areas on Earth. Oysters, crabs, and many fishes live in estuaries or reproduce in them. Estuaries are also crucial nesting and feeding areas for waterfowl. Mudflats and salt marshes are extensive coastal wetlands that often border estuaries.

For centuries, people viewed the ocean as a limitless resource, harvesting its bounty and using it as a dumping ground for wastes. These practices caused little damage when Earth's human population was small, but their negative impact is now great. From worldwide declines in commercial fish species to dying coral reefs, to beaches closed by pollution, danger signs abound. Because of their proximity to land, estuaries are especially vulnerable. Many have been completely replaced by development on landfill. Other threats include nutrient pollution, contamination by pathogens or toxic

chemicals, alteration of freshwater inflow, and introduction of non-native species. Coral reefs have suffered from many of the same problems. Excessive fishing has also upset the species balance in some reef communities and greatly reduced diversity. The widespread, massive deaths of reef-building corals in some regions have been attributed to rising sea surface temperatures; ocean acidification is also a serious threat. A 2008 assessment of coral reefs found that 19% had already fallen victim to a combination of factors and warned that most of the remaining reefs could disappear by 2040 if current trends continue. ☑

▼ Figure 18.22 **An estuary on the edge of Chesapeake Bay, Maryland.**

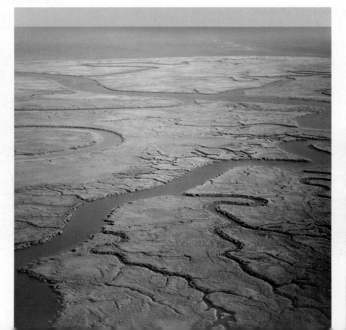

How Climate Affects Terrestrial Biome Distribution

Terrestrial biomes are determined primarily by climate, especially temperature and rainfall. Before we survey these biomes, let's look at the broad patterns of global climate that help explain their locations.

Earth's global climate patterns are largely the result of the input of radiant energy from the sun and the planet's movement in space. Because of its curvature, Earth receives an uneven distribution of solar energy (**Figure 18.23**). The equator receives the greatest intensity of solar radiation. As it is heated by the direct rays of the sun, air at the equator rises. It then cools, forms clouds, and drops rain (**Figure 18.24**). This largely explains why rain forests are concentrated in the **tropics**—the region from the Tropic of Cancer to the Tropic of Capricorn.

After losing moisture over equatorial zones, dry high-altitude air masses spread away from the equator until they cool and descend again at latitudes of about 30° north and south. Many of the world's great deserts—the Sahara in North Africa and the Arabian on the Arabian Peninsula, for example—are centered at these latitudes because of the dry air they receive.

Latitudes between the tropics and the Arctic Circle in the north and the Antarctic Circle in the south are called **temperate zones**. Generally, these regions have milder climates than the tropics or the polar regions. Notice in Figure 18.24 that some of the descending dry air heads into the latitudes above 30°. At first these air masses pick up moisture, but they tend to drop it as they cool at higher latitudes. This is why the north and south temperate zones tend to be relatively wet. Coniferous forests dominate the landscape at the wet but cool latitudes around 60° north.

Proximity to large bodies of water and the presence of landforms such as mountain ranges also affect climate. Oceans and large lakes moderate climate by absorbing heat when the air is warm and releasing heat to cold air. Mountains affect climate in two major ways. First, air temperature drops as elevation increases. As a result, driving up a tall mountain offers a quick tour of several biomes. **Figure 18.25** shows the scenery you might encounter on a journey from the scorching lowlands of the Sonoran desert to a cool coniferous forest at an elevation of 11,000 feet above sea level.

▼ Figure 18.23 **Uneven heating of Earth.**

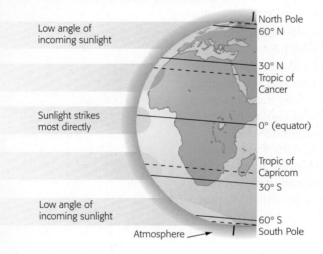

▼ Figure 18.24 **How uneven heating of Earth produces various climates.**

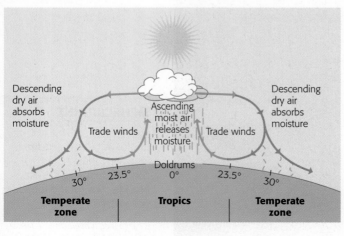

▶ Figure 18.25 **Effect of altitude on vegetation.** The zones shown are typical of the Sonoran Desert region in southwestern North America.

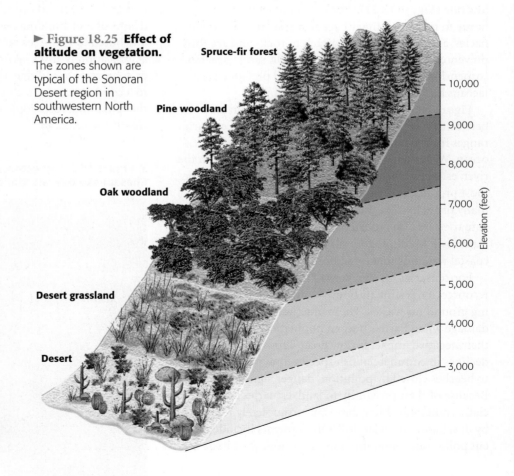

Second, mountains can block the flow of cool, moist air from a coast, causing radically different climates on opposite sides of a mountain range. In the example shown in **Figure 18.26**, moist air moves in off the Pacific Ocean and encounters the Coast Range in the state of Washington. Air flows upward, cools at higher altitudes, and drops a large amount of rainfall. The biological community in this wet region is a temperate rain forest. Precipitation increases again farther inland as the air moves up and over the Cascade Range. On the eastern side of the Cascades, there is little precipitation in the area known as a rain shadow. As a result, eastern Washington is very arid, almost qualifying as a desert. ☑

▼ Figure 18.26 **How mountains affect rainfall.**

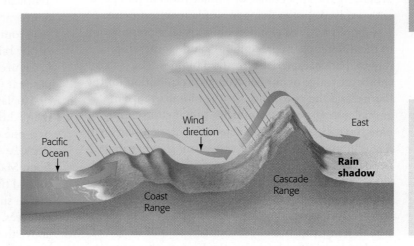

☑CHECKPOINT

Why is there so much rainfall in the tropics?

Answer: Air at the equator rises as it is warmed by direct sunlight. As the air rises, it cools. This causes cloud formation and rainfall because cool air holds less moisture than warm air.

Terrestrial Biomes

Terrestrial ecosystems are grouped into biomes primarily on the basis of their vegetation type (**Figure 18.27**). By providing food, shelter, nesting sites, and much of the organic material for the decomposers that recycle mineral nutrients, plants build the foundation for the communities of animals and other organisms typical of each biome. The geographic distribution of plants, and thus of biomes, largely depends on climate, with temperature and rainfall often the key factors determining the kind of biome that exists in a particular region. If the climate in two geographically separate areas is similar, the same type of biome may occur in both. Coniferous forests,

for instance, extend in a broad band across North America, Europe, and Asia. There is local variation within each biome, giving the vegetation a patchy, rather than a uniform, appearance. For example, in northern coniferous forests, snowfall may break branches and small trees, causing openings where broadleaf trees such as aspen and birch can grow. Local storms and fires also create openings in many biomes.

Each biome is characterized by a type of biological community, rather than an assemblage of particular species. For example, the groups of species living in the deserts of southwestern North America and in the Sahara Desert of Africa are different, but both groups are adapted to desert conditions. Organisms in widely

► Figure 18.27 **Map of the major terrestrial biomes.** Although this map has sharp boundaries, biomes actually grade into one another. We'll use smaller versions of this map, highlighted by color coding, during our closer look at the terrestrial biomes in the next several pages.

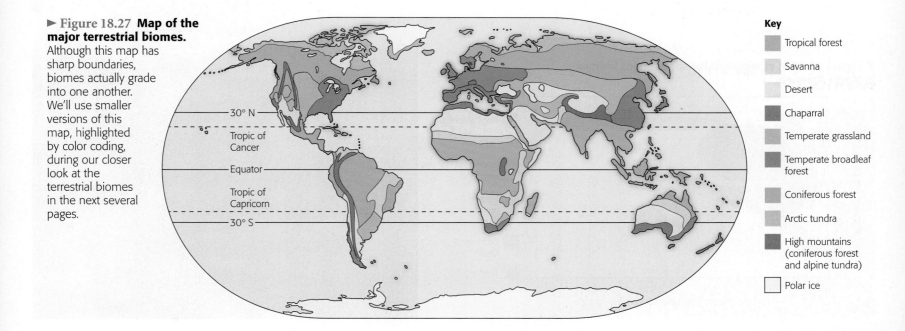

Key

- Tropical forest
- Savanna
- Desert
- Chaparral
- Temperate grassland
- Temperate broadleaf forest
- Coniferous forest
- Arctic tundra
- High mountains (coniferous forest and alpine tundra)
- Polar ice

separated biomes may look alike because of convergent evolution, the appearance of similar traits in independently evolved species living in similar environments.

Figure 18.28 is a climograph, a visual representation of the differences in precipitation and temperature ranges that characterize terrestrial biomes. The x-axis shows the range of annual average precipitation for the biome, and the y-axis displays the biome's range of annual average temperature. By studying the climograph plots, we can see how these abiotic factors compare in different biomes. For example, organisms that live in temperate broadleaf forests must have adaptations that allow them to cope with a wide range of temperatures, while organisms in tropical forests experience very little temperature variation. Although the range of precipitation in temperate broadleaf forests is similar to that of northern coniferous forests, the lower range of temperatures in northern coniferous forests reveals a significant difference in the abiotic environments of these two biomes.

Today, concern about global warming is generating intense interest in the effect of climate on vegetation patterns. Using powerful new tools, such as satellite imagery, scientists are documenting latitudinal shifts in biome borders, decreases in snow and ice coverage, and changes in the length of the growing season. At the same time, many natural biomes have been fragmented and altered by human activity. We'll discuss both of these issues after we survey the major terrestrial biomes, beginning near the equator and generally approaching the poles.

☑CHECKPOINT

Why are climbing plants common in tropical rain forests?

Answer: Climbing is a plant adaptation for reaching sunlight in a closed canopy, where little sunlight reaches the forest floor.

Tropical Forest

Tropical forests occur in equatorial areas where the temperature is warm and days are 11–12 hours long year-round. The type of vegetation is determined primarily by rainfall. Tropical rain forests, like the one shown in **Figure 18.29**, receive 200–400 cm (79–157 in.) of rain per year.

The layered structure of tropical rain forests provides many different habitats. Treetops form a closed canopy over one or two layers of smaller trees and a shrub understory. Few plants grow in the deep shade of the forest floor. Many trees are covered by woody vines growing toward the light. Other plants, such as orchids, gain access to sunlight by growing on the branches or trunks of tall trees. Scattered trees reach full sunlight by towering above the canopy. Many of the animals also dwell in trees, where food is abundant. Monkeys, birds, insects, snakes, bats, and frogs find food and shelter many meters above the ground.

Rainfall is less plentiful in other tropical forests. Tropical dry forests predominate in lowland areas that have a prolonged dry season or scarce rainfall at any time. The plants found there are a mixture of thorny shrubs and trees and succulents. In regions with distinct wet and dry seasons, tropical deciduous trees are common. ☑

▼ **Figure 18.29 Tropical rain forest in Borneo.**

▼ **Figure 18.28 A climograph for some major biomes in North America.**

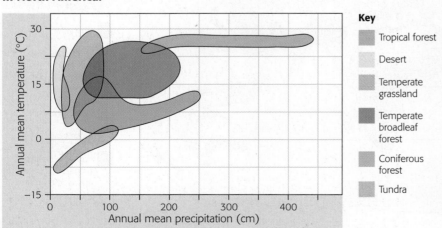

Key

Tropical forest

Desert

Temperate grassland

Temperate broadleaf forest

Coniferous forest

Tundra

Savanna

Savannas, such as the one shown in **Figure 18.30**, are dominated by grasses and scattered trees. The temperature is warm year-round. Rainfall averages 30–50 cm (roughly 12–20 in.) per year, with dramatic seasonal variation.

Fire, caused by lightning or human activity, is an important abiotic factor in the savanna. The grasses survive burning because the growing points of their shoots are below ground. Other plants have seeds that sprout rapidly after a fire. Poor soil and lack of moisture, along with fire and grazing animals, prevent the establishment of most trees. The luxuriant growth of grasses and small broadleaf plants during the rainy season provides a rich food source for plant-eating animals.

Many of the world's large grazing mammals and their predators inhabit savannas. African savannas are home to zebras and many species of antelope, as well as to lions and cheetahs. Several species of kangaroo are the dominant grazers of Australian savannas. Oddly, though, the large grazers are not the dominant plant-eaters in savannas. That distinction belongs to insects, especially ants and termites. Other animals include burrowers such as mice, moles, gophers, and ground squirrels.

Desert

Deserts are the driest of all biomes, characterized by low and unpredictable rainfall—less than 30 cm (about 12 in.) per year. Some deserts are very hot, with daytime soil surface temperatures above 60°C (140°F) and large daily temperature fluctuations. Other deserts, such as those west of the Rocky Mountains and the Gobi Desert, spanning northern China and southern Mongolia, are relatively cold. Air temperatures in cold deserts may fall below −30°C (−22°F).

Desert vegetation typically includes water-storing plants such as cacti and deeply rooted shrubs. Various snakes, lizards, and seed-eating rodents are common inhabitants. Arthropods such as scorpions and insects also thrive in the desert. Evolutionary adaptations of desert plants and animals include a remarkable array of mechanisms that conserve water. For example, the "pleated" stem of saguaro cacti (**Figure 18.31**) enables the plants to expand when they absorb water during wet periods. Some desert mice *never* drink, deriving all their water from the metabolic breakdown of the seeds they eat. Protective adaptations that deter feeding by mammals and insects, such as spines on cacti and poisons in the leaves of shrubs, are common in desert plants. ☑

☑CHECKPOINT

1. How does the savanna climate vary seasonally?
2. What abiotic factor characterizes deserts?

Answers: **1.** *Temperature stays about the same year-round, but rainfall varies dramatically.* **2.** *Rainfall is low and unpredictable.*

▼ Figure 18.30 **Savanna in Kenya.**

▼ Figure 18.31 **Sonoran desert.**

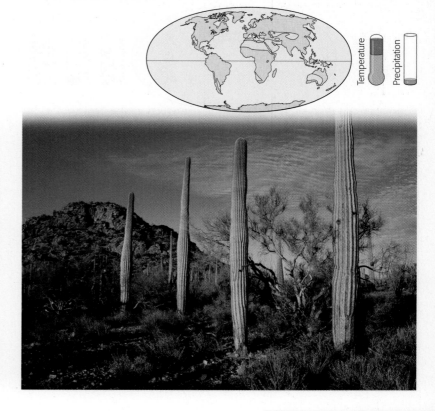

Chaparral

The climate that supports **chaparral** vegetation results mainly from cool ocean currents circulating offshore, producing mild, rainy winters and hot, dry summers. As a result, this biome is limited to small coastal areas, some in California (**Figure 18.32**). The largest region of chaparral surrounds the Mediterranean Sea; in fact, Mediterranean is another name for this biome. Dense, spiny, evergreen shrubs dominate chaparral. Annual plants are also common during the wet winter and spring months. Animals characteristic of the chaparral are deer, fruit-eating birds, seed-eating rodents, and lizards and snakes.

Chaparral vegetation is adapted to periodic fires caused by lightning. Many plants contain flammable chemicals and burn fiercely, especially where dead brush has accumulated. After a fire, shrubs use food reserves stored in the surviving roots to support rapid shoot regeneration. Some chaparral plants produce seeds that will germinate only after a hot fire. The ashes of burned vegetation fertilize the soil with mineral nutrients, promoting regrowth of the plant community. Houses do not fare as well. The firestorms that race through the densely populated canyons of Southern California can be devastating to the human inhabitants.

Temperate Grassland

Temperate grasslands have some of the characteristics of tropical savannas, but they are mostly treeless, except along rivers or streams, and are found in regions of relatively cold winter temperatures. Rainfall, averaging between 25 and 75 cm per year (approximately 10–30 in.), with frequent severe droughts, is too low to support forest growth. Periodic fires and grazing by large mammals also prevent invasion by woody plants. These grazers include the bison and pronghorn in North America, the wild horses and sheep of the Asian steppes, and kangaroos in Australia. As in the savanna, however, the dominant plant-eaters are invertebrates, especially grasshoppers and soil-dwelling nematodes.

Without trees, many birds nest on the ground. Many small mammals, such as rabbits, voles, ground squirrels, prairie dogs, and pocket gophers, dig burrows to escape predators. Temperate grasslands like the one shown in **Figure 18.33** once covered much of central North America.

Because grassland soil is both deep and rich in nutrients, these habitats provide fertile land for agriculture. Most grassland in the United States has been converted to cropland or pasture, and very little natural prairie exists today. ✔

▼ Figure 18.32 **Chaparral in California.**

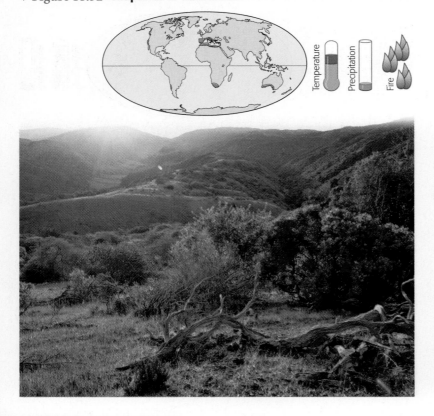

▼ Figure 18.33 **Temperate grassland in South Dakota.**

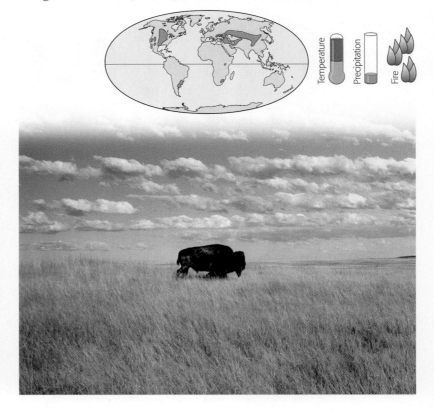

Temperate Broadleaf Forest

Temperate broadleaf forests occur throughout midlatitudes where there is sufficient moisture to support the growth of large trees. Annual precipitation is relatively high, ranging from 75–150 cm (30–60 in.), and typically distributed evenly around the year. Annual temperature varies over a wide range, with hot summers and cold winters. In the Northern Hemisphere, dense stands of deciduous trees are trademarks of temperate forests, such as the one pictured in **Figure 18.34**. Deciduous trees drop their leaves before winter, when temperatures are too low for effective photosynthesis and water lost by evaporation is not easily replaced from frozen soil.

Numerous invertebrates live in the soil and the thick layer of leaf litter that accumulates on the forest floor. Some vertebrates, such as mice, shrews, and ground squirrels, burrow for shelter and food, while others, including many species of birds, live in the trees. Predators include bobcats, foxes, black bears, and mountain lions. Many mammals that inhabit these forests enter a dormant winter state called hibernation, and some bird species migrate to warmer climates.

Virtually all the original temperate broadleaf forests in North America were destroyed by logging or cleared for agriculture or development. These forests tend to recover after disturbance, however, and today we see deciduous trees growing in undeveloped areas over much of their former range.

Coniferous Forest

Cone-bearing evergreen trees such as pine, spruce, fir, and hemlock dominate **coniferous forests** in the Northern Hemisphere. (Other kinds of conifers grow in parts of South America, Africa, and Australia.) The northern coniferous forest, or **taiga** (**Figure 18.35**), is the largest terrestrial biome on Earth, stretching in a broad band across North America and Asia south of the Arctic Circle. Taiga is also found at cool, high elevations in more temperate latitudes, for example, in much of the mountainous region of western North America. The taiga is characterized by long, snowy winters and short, wet summers that are sometimes warm. The slow decomposition of conifer needles in the thin, acidic soil makes few nutrients available for plant growth. The conical shape of many conifers prevents too much snow from accumulating on their branches and breaking them. Animals of the taiga include moose, elk, hares, bears, wolves, grouse, and migratory birds. The Asian taiga is home to the dwindling number of Siberian tigers that remain in the wild.

The **temperate rain forests** of coastal North America (from Alaska to Oregon) are also coniferous forests. Warm, moist air from the Pacific Ocean supports this unique biome, which, like most coniferous forests, is dominated by a few tree species, typically hemlock, Douglas fir, and redwood. These forests are heavily logged, and the old-growth stands of trees are rapidly disappearing. ☑

☑CHECKPOINT

1. How does the loss of leaves function as an adaptation of deciduous trees to cold winters?
2. What type of trees are characteristic of the taiga?

Answers: 1. by reducing loss of water from the trees when water cannot be replaced because of frozen soil 2. conifers such as pine, spruce, fir, and hemlock

▼ **Figure 18.34**
Temperate broadleaf forest in Great Smokies National Park, North Carolina.

▼ **Figure 18.35**
Northern coniferous forest in Finland, with the sky lit by the northern lights.

Tundra

Tundra covers expansive areas of the Arctic between the taiga and polar ice. **Permafrost** (permanently frozen subsoil), bitterly cold temperatures, and high winds are responsible for the absence of trees and other tall plants in the arctic tundra shown in **Figure 18.36**. The arctic tundra receives very little annual precipitation. However, water cannot penetrate the underlying permafrost, so melted snow and ice accumulate in pools on the shallow topsoil during the short summer.

Tundra vegetation includes small shrubs, grasses, mosses, and lichens. When summer arrives, annual plants grow quickly and flower in a rapid burst. Caribou, musk oxen, wolves, and small rodents called lemmings are among the mammals found in the arctic tundra. Many migratory animals use the tundra as a summer breeding ground. During the brief but productive warm season, the marshy ground supports the aquatic larvae of insects, providing food for migratory waterfowl, and clouds of mosquitoes often fill the tundra air.

High winds and cold temperatures create plant communities called alpine tundra on very high mountaintops at all latitudes, including the tropics. Although these communities are similar to arctic tundra, there is no permafrost beneath alpine tundra.

Polar Ice

Polar ice covers the land at high latitudes north of the arctic tundra in the Northern Hemisphere and in Antarctica in the Southern Hemisphere (**Figure 18.37**). The temperature in these regions is extremely cold year-round, and precipitation is very low. Only a small portion of these landmasses is free of ice or snow, even during the summer. Nevertheless, small plants, such as mosses and lichens, eke out a living, and invertebrates such as nematodes, mites, and wingless insects called springtails inhabit the frigid soil. Nearby sea ice provides feeding platforms for large animals such as polar bears (in the Northern Hemisphere), penguins (in the Southern Hemisphere), and seals. Seals, penguins, and other marine birds visit the land to rest and breed. The polar marine biome provides the food that sustains these birds and mammals. In the Antarctic, penguins feed at sea, eating a variety of fish, squid, and small shrimplike crustaceans known as krill. Antarctic krill, an important food source for many species of fish, seals, squid, seabirds, and filter-feeding whales as well as penguins, depend on sea ice for breeding and as a refuge from predators. As the amount and duration of sea ice decline as a consequence of global climate change, krill habitat is shrinking. ✔

▼ Figure 18.36 **Arctic tundra in central Alaska.**

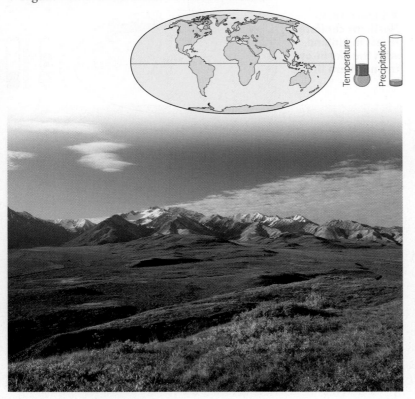

▼ Figure 18.37 **Polar ice in Antarctica.**

The Water Cycle

Biomes are not self-contained units. Rather, all parts of the biosphere are linked by the global water cycle, illustrated in **Figure 18.38**, and by nutrient cycles, which you will learn about in Chapter 20. Consequently, events in one biome may reverberate throughout the biosphere.

As you learned earlier in this chapter, water and air move in global patterns driven by solar energy. Precipitation and evaporation continuously move water between the land, oceans, and the atmosphere. Water also evaporates from plants, which pull it from the soil in a process called transpiration.

Over the oceans, evaporation exceeds precipitation. The result is a net movement of water vapor to clouds that are carried by winds from the oceans across the land. On land, precipitation exceeds evaporation and transpiration. Excess precipitation forms systems of surface water (such as lakes and rivers) and groundwater, all of which flow back to the sea, completing the water cycle.

Just as the water draining from your shower carries dead skin cells from your body along with the day's grime, the water washing over and through the ground carries traces of the land and its history. For example, water flowing from land to the sea carries with it silt (fine soil particles) and chemicals such as fertilizers and pesticides.

Erosion from coastal development has caused silt to muddy the waters of some coral reefs, dimming the light available to the photosynthetic algae that power the reef community. Chemicals in surface water may travel hundreds of miles by stream and river to the ocean, where currents then carry them even farther from their point of origin. For instance, traces of pesticides and chemicals from industrial wastes have been found in marine mammals in the Arctic and in deep-sea octopuses and squid. Airborne pollutants such as nitrogen oxides and sulfur oxides, which combine with water to form acid precipitation, are also distributed by the water cycle.

Human activity also affects the global water cycle itself in a number of important ways. One of the main sources of atmospheric water is transpiration from the dense vegetation making up tropical rain forests. The destruction of these forests changes the amount of water vapor in the air. This, in turn, will likely alter local, and perhaps global, weather patterns. Pumping large amounts of groundwater to the surface for irrigation affects the water cycle, too. This practice can increase the rate of evaporation over land and may deplete groundwater supplies. We'll consider some of these environmental impacts in the next section. ☑

▼ Figure 18.38 **The global water cycle.**

☑CHECKPOINT

What is the main way that living organisms contribute to the water cycle?

Answer: *Plants move water from the ground to the air via transpiration.*

Human Impact on Biomes

For hundreds of years, people have been using increasingly effective technologies to capture or produce food, to extract resources from the environment, and to build cities. It is now clear that the environmental costs of these enterprises are staggering. In this section, you'll see some examples of how human activities are affecting forest and freshwater resources. Throughout the remainder of this unit, you'll learn about the role of ecological knowledge in achieving **sustainability**, the goal of developing, managing, and conserving Earth's resources in ways that meet the needs of people today without compromising the ability of future generations to meet their needs.

Forests

The map in Figure 18.27 shows the terrestrial biomes that would be expected to flourish under the prevailing climatic conditions. However, about three-quarters of Earth's land surface has been altered by thousands of years of human occupation. Most of the land that we've appropriated is used for agriculture; another hefty chunk is covered by the asphalt and concrete of development. Changes in vegetation are especially dramatic in regions like tropical forests that escaped large-scale human intervention until recently. Satellite photos of a small area in Brazil show how thoroughly a landscape can be altered in a short amount of time (Figure 18.39).

Every year, more and more forested land is cleared for agriculture (Figure 18.40). You might think that this land is needed to feed new mouths as the human population continues to grow, but that's not entirely the case. Unsustainable agricultural practices have degraded much of the world's cropland so severely that it is unusable. Researchers estimate that replacing worn-out farmland accounts for up to 80% of the deforestation occurring today. Forests are also being lost to logging, mining, and air pollution, problems that are hitting coniferous forests especially hard. (As we mentioned in the previous section, most temperate broadleaf forests were replaced by human enterprises long ago.) Land that hasn't been directly converted to food production and living space also bears the imprint of our presence. Roads penetrate regions that are otherwise unaltered, bringing pollution to the wilderness, providing avenues for new diseases to emerge, and slicing vast tracts of biome into segments that are too small to support a full array of species.

Land uses that provide resources such as food, fuel, and shelter are clearly beneficial to us. But natural ecosystems also provide services that support the human population—purification of air and water, nutrient cycling, and recreation, to name just a few. We'll return to the topic of ecosystem services in Chapter 20.

Fresh Water

The impact of human activities on freshwater ecosystems may pose an even greater threat to life on Earth—including ourselves—than the damage to terrestrial ecosystems. Freshwater ecosystems are being polluted by large amounts of nitrogen and phosphorus compounds

▼ Figure 18.39 **Satellite photos of the Rondonia area of the Brazilian rain forest.**

1975. In 1975, the forest in this remote region was virtually intact.

2001. Same area in 2001, after a paved highway through the region brought loggers and farmers. The "fishbone" pattern marks the new network of roads carved through the forest.

▼ Figure 18.40 **Deforestation by cutting and burning trees in the Amazon basin, Brazil.**

that run off from heavily fertilized farms or from livestock feedlots. A wide variety of other pollutants, such as industrial wastes, also contaminate freshwater habitats, drinking water, and groundwater. Some regions of the world face dire shortages of water as a result of the overuse of groundwater for irrigation, extended droughts (partially caused by global climate change), or poor water management practices.

Las Vegas, the population center of Clark County, Nevada, is one example of a city whose water resources are increasingly stressed by drought and overuse. **Figure 18.41a** is a satellite photo of Las Vegas in 1973, when the population of Clark County was 319,400. **Figure 18.41b** shows the same area less than 30 years later, when the population had swelled to 1,620,748. In contrast to the disappearance of greenery in the photos of Brazilian rain forest, the mark of human activities in Figure 18.41b is the notable expansion of greenery—the result of watering lawns and golf courses. Las Vegas is situated in a high valley in the Mojave Desert. Where does it get the water to turn barren desert into green fields?

Las Vegas taps underground aquifers for some water, but its main water supply is Lake Mead. Lake Mead is an enormous reservoir formed by the Hoover Dam on the Colorado River, which in turn receives almost all of its water from snowmelt in the Rocky Mountains. With decreased annual snowfall, attributable largely to global warming, the flow of the Colorado has greatly diminished. The water level in Lake Mead has dropped drastically (**Figure 18.42**), and parched cities and farms farther downstream are pleading for more water.

It is clear that even with restrictions on water use, Las Vegas must look elsewhere to supply its needs.

Among other options, Las Vegas is eyeing the abundant supply of groundwater in the northern end of the valley where it lies. Although sparsely populated, that area is home to many ranchers whose livelihoods depend on the groundwater. It is also home to numerous endangered species. Not surprisingly, environmentalists and residents of the north valley are resisting efforts to pipe its groundwater to Las Vegas.

Nevada is just one of many places where the hard realities of climate change are beginning to affect daily life. Battles over water resources are shaping up throughout the arid West and Southwest of the United States, where changing precipitation patterns due to global warming are projected to continue the drought for many years to come. Record drought also hit the southeastern United States in 2006, draining reservoirs, threatening water supplies, and providing fuel for massive wildfires. Conditions there, however, have begun to improve.

While policymakers are dealing with current crises and planning how to manage resources in the future, researchers are seeking methods of sustainable agriculture and water use. Basic ecological research is an essential component of ensuring that sufficient food and water will be available for people now—and for the generations to come. You'll learn more about the goal of sustainability in the next two chapters. But first, let's take a closer look at a major threat to sustainability: global climate change. ☑

▼ **Figure 18.42 Low water level in Lake Mead.** The white "bathtub ring" is caused by mineral deposits on rocks that were once submerged.

▼ Figure 18.41 **Satellite photos of Las Vegas, Nevada.**

(a) May 1973

(b) May 2000

Global Climate Change

Rising concentrations of carbon dioxide (CO_2) and certain other gases in the atmosphere are changing global climate patterns. This was the overarching conclusion of heat. Some of these so-called **greenhouse gases** are natural, including CO_2, water vapor, and methane. Others, such as chlorofluorocarbons (CFCs, found in some

THE PROCESS OF SCIENCE: Global Climate Change

How Does Climate Change Affect Species Distribution?

As you've learned in this chapter, abiotic factors of the environment are fundamental determinants of where organisms live. It stands to reason that changes in temperature and precipitation patterns will have a significant effect on the distribution of life. With rising temperatures, the ranges of many species have already shifted toward the poles or to higher elevations. For example, shifts in the ranges of many bird species have been reported; the Inuit peoples living north of the Arctic Circle have sighted birds such as robins in the region for the first time.

Let's examine how a team of ecologists investigated the impact of climate change on European butterflies.

Using the **observation** that the average temperature in Europe has risen 0.8°C and that butterflies are sensitive to temperature change, the researchers asked the **question**: Have the ranges of butterflies changed in response to the temperature changes? This question led to the **hypothesis** that butterfly range boundaries are shifting in line with the warming trend. The researchers **predicted** that butterfly species will be establishing new populations to the north of their former ranges, and populations at the southern edges of their ranges will become extinct. The **experiment** involved analyzing historical data on the ranges of 35 species of butterflies in Europe. The **results** showed that more than 60% of the species have pushed their northern range boundaries poleward over the last century, some by as much as 150 miles. The southern boundaries have simultaneously contracted for some species, but not for others.

Figure 18.47 shows the range shift for the species *Argynnis paphia*.

While some organisms have the dispersal ability and the room for northward population shifts, species that live on mountaintops or in polar regions have nowhere to go. For example, researchers in Costa Rica have reported the disappearance of 20 species of frogs and toads as warmer Pacific Ocean temperatures reduce the dry-season mists in their mountain habitats. And as we mentioned in the Biology and Society section, animals at both poles are also at risk.

▶ **Figure 18.47 Northward shift of *Argynnis paphia*.** On the map, orange represents the butterfly's range in 1970; its 1997 range is shown in light green.

Sweden
Norway
Finland
Estonia
Russia
Lativia
Denmark
Lithuania

Europe

Africa

Argynnis paphia (silver-washed fritillary butterfly)

Effects of Climate Change on Ecosystems

The sentiment expressed by the poet John Donne that "no man is an island" is equally true for every other species in nature: Every species needs others to survive. Climate change is knocking some of these interactions out of sync. In many plants and animals, life cycle events are triggered by warming temperatures. Across the Northern Hemisphere, the warm temperatures of spring are arriving earlier. Satellite images show earlier greening of the landscape, and flowering occurs sooner. A variety of species, including birds and frogs, have begun their breeding seasons earlier. But for other species, the environmental cue that spring has arrived is daylength, which is not affected by climate change. Consequently, plants may bloom before pollinators have emerged, or eggs may hatch before a dependable food source is available.

The combined effects of climate change on forest ecosystems in western North America have spawned catastrophic wildfire seasons (**Figure 18.48**). In these regions, spring snowmelt in the mountains releases water into streams that sustain forest moisture levels over the summer dry season. With the earlier arrival of spring, snowmelt begins earlier and dwindles away before the dry season ends. As a result, the fire season is lasting longer. Meanwhile, bark beetles, which bore into conifers to lay their eggs, have benefited from global warming. Healthy trees can fight off the pests, but drought-stressed trees are too weak to resist (**Figure 18.49**). As a bonus for the beetles, the warmer weather allows them to reproduce twice a year rather than once. In turn, vast numbers of dead trees add fuel to a fire. Wildfires burn longer, and the number of acres burned has increased dramatically.

The map of terrestrial biomes (see Figure 18.27), which is primarily determined by temperature and rainfall, is also changing. Melting permafrost is shifting the boundary of the tundra northward as shrubs and conifers are able to stretch their ranges into the previously frozen ground. Prolonged droughts are extending the boundaries of deserts. Scientists also predict that great expanses of the Amazonian rain forest will gradually become savanna as increased temperatures dry out the soil.

Global climate change has significant consequences for humans, too, as changing temperature and precipitation patterns affect food production, the availability of fresh water, and the structural integrity of buildings and roads. All projections point to an even greater impact in the future. Unlike other species, however, humans can take action to reduce greenhouse gas emissions and maybe even reverse the warming trend. ✓

▼ Figure 18.49 **Pines in Southern California infested by bark beetles.** Red or light-colored foliage indicates dead or dying trees. Green trees are still healthy.

▼ Figure 18.48 **A forest fire near South Lake Tahoe, California, June 2007.**

✓CHECKPOINT

Why does the earlier arrival of spring temperatures affect life cycle events for some organisms?

Answer: In some organisms, life cycle events are triggered by warm temperatures.

Looking to Our Future

Emissions of greenhouse gases are accelerating. From 2000 to 2005, global CO_2 emissions increased *four times faster* than in the preceding ten-year span. At this rate, further climate change is inevitable. With effort and ingenuity, however, we may be able to begin reducing emissions. International negotiations are under way on a new climate change treaty to succeed the 1997 Kyoto Protocol, which required industrialized nations to reduce their greenhouse gas emissions.

Given the vast scope and complexity of the problem, you might think that the actions of a single individual would have little impact on greenhouse gas emissions. But it was the collective activities of individuals that caused—and are still causing—emissions to rise. The amount of greenhouse gas emitted as a result of the actions of a single individual is that person's **carbon footprint** (from the fact that the most important greenhouse gas is CO_2). A carbon footprint can be estimated using a set of rough calculations; several different calculators are available online.

Home energy use is one major contributor to the carbon footprint. It's easy to reduce your energy use by turning off the lights, TV, and other electronics when they aren't in use. You can also switch to energy-efficient lightbulbs. In addition, you can unplug items that consume electricity even when they are not in use; examples are cell phone chargers, videogame consoles, and printers.

Transportation is another significant part of the carbon footprint. If you have a car, keep it well maintained, consolidate trips, share rides with friends, and use alternative means of transportation whenever possible.

Manufactured goods are a third category in the carbon footprint. Every item you purchase generated its own carbon footprint in the process of going from raw materials to store shelves. You can reduce your carbon emissions by not buying unnecessary goods and by recycling—or better yet, reusing—items instead of putting them in the trash (**Figure 18.50**). Landfills are the largest human-related source of methane, a greenhouse gas more potent than CO_2. The methane is released by bacteria that decompose (break down) landfill waste.

Changes in your eating habits can also shrink your carbon footprint. Like landfills, the digestive system of cattle depends on methane-producing bacteria. The methane released by cattle and by the bacteria that decompose their manure accounts for about 20% of methane emissions in the United States. Thus, replacing beef and dairy products in your diet with fish, chicken, eggs, and vegetables reduces your carbon footprint. In addition, eating locally grown fresh foods may lower the greenhouse gas emissions that result from food processing and transportation (**Figure 18.51**). Many websites offer additional suggestions for reducing your carbon footprint. ☑

☑**CHECKPOINT**

What is a personal carbon footprint?

Answer: the amount of carbon that a person is responsible for emitting

▼ **Figure 18.50 Turning trash into treasure.** If you have things that you no longer need, don't trash them when you move out. Environmentally conscious students at Davidson College in North Carolina collect discarded items for local charities.

▼ **Figure 18.51 Eating locally grown foods may reduce your carbon footprint—and they taste good, too!**

Climate Change as an Agent of Natural Selection

Environmental change has always been a part of life; in fact, it is a key ingredient of evolutionary change. Will evolutionary adaptation counteract the negative effects of climate change on organisms? Researchers have documented microevolutionary shifts in a few populations, including red squirrels, a few bird species, and a tiny mosquito (**Figure 18.52a**). It appears that some populations, especially those with high genetic variability and short life spans, may avoid extinction by means of evolutionary adaptation. However, evolutionary adaptation is unlikely to save long-lived species such as polar bears and penguins (**Figure 18.52b**) that are experiencing rapid habitat loss. The rate of climate change is incredibly fast compared to major climate shifts in evolutionary history. If climate change continues on its present course, thousands of species—the Intergovernmental Panel on Climate Change estimates as many as 30% of plants and animals—will face extinction by midcentury.

▼ Figure 18.52 **Which species will survive climate change?**

(a) Pitcher plant mosquito. The pitcher plant mosquito, pictured inside the carnivorous plant for which it is named, may be able to evolve quickly enough.

(b) Adélie penguin. Adélie penguins, which live on the rapidly warming Antarctic Peninsula, are not likely to make it.

Chapter Review

SUMMARY OF KEY CONCEPTS

 Go to the Study Area at **www.masteringbiology.com** for practice quizzes, eText, BioFlix™ 3-D animations, MP3 Tutor Sessions, videos, current events, and more.

An Overview of Ecology

Ecology is the scientific study of interactions between organisms and their environments. The environment includes abiotic (nonliving) and biotic (living) components. Ecologists use observation, experiments, and computer models to test hypothetical explanations of these interactions.

Ecology and Environmentalism

Human activities have had an impact on all parts of the biosphere. Ecology provides the basis for understanding and addressing these environmental problems.

A Hierarchy of Interactions

Ecologists study interactions at four increasingly complex levels.

Organismal ecology (individual) → Population ecology (group of individuals) → Community ecology (all organisms in a particular area) → Ecosystem ecology (all organisms and abiotic factors)

Living in Earth's Diverse Environments

The biosphere is an environmental patchwork in which abiotic factors affect the distribution and abundance of organisms.

Abiotic Factors of the Biosphere

These include the availability of sunlight, water, nutrients, and temperature. In aquatic habitats, dissolved oxygen, salinity, current, and tides are also important. Additional factors in terrestrial environments include wind and fire.

The Evolutionary Adaptations of Organisms

Adaptation via natural selection results from the interactions of organisms with their environments.

Adjusting to Environmental Variability

Organisms also have adaptations that enable them to cope with environmental variability, including physiological, behavioral, and anatomical responses to changing conditions.

Biomes

A biome is a major terrestrial or aquatic life zone.

Freshwater Biomes

Freshwater biomes include lakes, ponds, rivers, streams, and wetlands. Lakes vary, depending on depth, with regard to light penetration, temperature, nutrients, oxygen levels, and community structure. Rivers change greatly from their source to the point at which they empty into a lake or an ocean.

Marine Biomes

Marine life is distributed into distinct realms (benthic and pelagic) and zones (photic, aphotic, and intertidal) according to the depth of the water, degree of light penetration, distance from shore, and open water versus bottom. The coral reef biome, which occurs in warm tropical waters above the continental shelf, has an abundance of biological diversity. A biome found near hydrothermal vents in the deep ocean is powered by chemical energy from Earth's interior instead of sunlight. Estuaries, located where a freshwater river or stream merges with the ocean, are one of the most biologically productive environments on Earth.

How Climate Affects Terrestrial Biome Distribution

The geographic distribution of terrestrial biomes is based mainly on regional variations in climate. Climate is largely determined by the uneven distribution of solar energy on Earth. Proximity to large bodies of water and the presence of landforms such as mountains also affect climate.

Terrestrial Biomes

Most biomes are named for major physical or climatic features and for their predominant vegetation. The major terrestrial biomes include tropical forest, savanna, desert, chaparral, temperate grassland, temperate broadleaf forest, coniferous forest, tundra, and polar ice. If the climate in two geographically separate areas is similar, the same type of biome may occur in both.

The Water Cycle

The global water cycle links aquatic and terrestrial biomes. Human activities are disrupting the water cycle.

Human Impact on Biomes

Land use by humans has altered vast tracts of forest and degraded the services provided by natural ecosystems. Unsustainable agricultural practices have depleted cropland fertility. Human activities have polluted freshwater ecosystems, which are vital for life. Agriculture, population growth, drought, and declining snowfall are all factors in the rapid depletion of freshwater resources in some regions.

Global Climate Change

The Greenhouse Effect and Global Warming

So-called greenhouse gases, including CO_2 and methane, increase the amount of heat retained in Earth's atmosphere. The accumulation of these gases has caused increases in the average global temperature.

The Accumulation of Greenhouse Gases

Human activities, especially the burning of fossil fuels, are responsible for the rise in greenhouse gases over the past century. Release of CO_2 has exceeded the amount that can be absorbed by natural processes.

Effects of Climate Change on Ecosystems

Climate change is disrupting interactions between species. Devastating wildfires are among the effects of climate change in certain ecosystems. Climate change is also shifting biome boundaries.

Looking to Our Future

Each person has a carbon footprint—that person's responsibility for a portion of global greenhouse gas emissions. We can take action to reduce our carbon footprints.

SELF-QUIZ

1. Place these levels of ecological study in order, from the least to the most comprehensive: community ecology, ecosystem ecology, organismal ecology, population ecology.

2. Name several abiotic factors that might affect the community of organisms living inside a home fish tank.

3. The formation of goose bumps on your skin in cold weather is an example of a (an) _____ response, while seasonal migration is an example of a (an) _____ response.

4. Which of the following sea creatures might be described as a pelagic animal of the aphotic zone?
   a. a coral reef fish
   b. a giant clam near a deep-sea hydrothermal vent
   c. an intertidal snail
   d. a deep-sea squid
   e. a species of phytoplankton

5. Identify the following biomes on the climograph below: tundra, coniferous forest, desert, grassland, temperate broadleaf forest, and tropical rain forest.

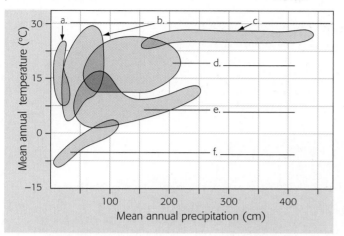

6. We are on a coastal hillside on a hot, dry summer day among evergreen shrubs that are adapted to fire. We are most likely standing in a _____ biome.

7. What three abiotic factors account for the rarity of trees in arctic tundra?

8. What human activity is responsible for the greatest amount of deforestation?

9. What is the greenhouse effect? How is the greenhouse effect related to global warming?

10. The recent increase in atmospheric CO_2 concentration is mainly a result of an increase in
   a. primary productivity.
   b. the absorption of heat radiating from Earth.
   c. the burning of fossil fuels and wood.
   d. cellular respiration by the increasing human population.

11. What populations of organisms are most likely to survive climate change via evolutionary adaptation?

Answers to the Self-Quiz questions can be found in Appendix D.

THE PROCESS OF SCIENCE

12. Design a laboratory experiment to measure the effect of water temperature on the population growth of a certain phytoplankton species from a pond.

13. Some people are not convinced that human-induced global climate change is a real phenomenon. Using your knowledge of the scientific process from Chapter 1 and the information from this chapter, develop arguments you could use to explain the scientific basis for saying that global climate change is truly occurring and that humans are responsible for it.

BIOLOGY AND SOCIETY

14. Near Lawrence, Kansas, there was a rare patch of the original North American temperate grassland that had never been plowed. It was home to numerous native grasses, annual plants, and grassland animals. Among the species present were two endangered plants. Environmental activists thought the area should be set aside as a nature preserve, and they started to raise money to save the patch of land. In 1990, the owner of the land plowed it, stating that there are no federal laws protecting endangered plants on private grasslands and that he did not want to be told what he could do with his property. What issues and values are in conflict in this situation? How could this story have had a more satisfactory ending for all concerned? What would you have done if you were an environmental activist? If you were the farmer?

15. Land clearing for agricultural use is responsible for most of the deforestation that is currently taking place. Agriculture also pumps enormous amounts of groundwater for irrigation and pollutes surface waters with nitrogen via runoff. But agriculture also supplies the world's food. Are the environmental consequences of agriculture simply the price we must pay to eat? What are the long-term costs? Suggest some ways to alleviate the environmental impact of agriculture.

16. Research your country's per capita (per person) carbon emissions. Calculate your own carbon footprint and compare it with the average for your country. Make a list of actions you are willing to take to reduce your personal carbon footprint. What actions could you take to persuade others to reduce their carbon footprints?

17. In the summer of 2007, unprecedented drought conditions brought the city of Atlanta, Georgia, within weeks of running out of water. Most of Atlanta's water comes from Lake Lanier, a reservoir that the Army Corps of Engineers created by damming the Chattahoochee River. With Lake Lanier drying up from lack of rainfall, Georgia petitioned the Corps, which manages the dam, to reduce the amount of water released downstream. The Corps refused, citing its obligation under the Endangered Species Act to protect the habitats of a species of sturgeon (a fish) and two species of mussel (a mollusc). Objections were also raised by Alabama and Florida, where hundreds of towns, recreational facilities, and power plants depend on the water released downstream. Some people thought that Atlanta authorities had brought the water shortage on themselves by allowing developers to build without considering whether adequate water was available. Florida also argued that reduction of freshwater inflows would harm its oyster fisheries. How would you prioritize the competing claims on the water from Lake Lanier? Who should allocate scarce water resources? How can cities and states plan more wisely for future shortages?

Population Ecology

Rabbit plague.
Methods to control the
rabbit population
included driving the
rabbits into pens to be
killed.

BIOLOGY AND SOCIETY: Invasive Species

Multiplying Like Rabbits

In 1859, 12 pairs of European rabbits were released on a ranch in southern Australia by an Englishman who wanted to hunt familiar game. The animals quickly became a nuisance. In 1865, 20,000 rabbits were killed on that ranch alone. By 1900, several hundred *million* rabbits were distributed over most of the continent. The rabbit invasion was a catastrophe in several ways. Their activities destroyed farm and grazing land by eating vegetation down to the roots. Especially in arid regions, the loss of plant cover led to soil erosion. Extensive underground rabbit burrows made grazing treacherous for cattle and sheep. The rabbits also competed directly with native marsupials, such as kangaroos and bilbies, for food and living space.

The rabbits were soon joined by another sportsman's favorite game animal, the European red fox. This adaptable predator also thrived in Australia, thanks to the seemingly endless supply of rabbits. The fox isn't a fussy eater, however. As they followed the spread of rabbits across the continent, red foxes also ate several species of native birds and small mammals to extinction, and they continue to threaten other native species today.

Rabbits, foxes, and a long list of other non-native animal and plant species are still entrenched in Australia, damaging the environment and costing hundreds of millions of dollars in economic losses. Eradication is impossible; the best that can be done is to minimize the impact of these species. Nor is Australia alone with this problem. For as long as humans have traveled the globe, they have carried—intentionally or accidentally—thousands of species to new habitats. Like the European rabbits and foxes, many of these non-native species have established populations that spread far and wide, leaving environmental havoc in their wake. As you learned at the end of Chapter 17, humans have also multiplied and spread far from our point of origin, and we have radically changed our environment. As you explore population ecology in this chapter, you'll learn about trends in human population growth and other applications of this area of ecological research. Along the way, you'll find out why those rabbits multiplied so freely and whether they have yet been brought under control.

An Overview of Population Ecology

In Chapter 18, you learned about the factors that characterize an environment, how organisms respond to those factors, and how environmental factors determine the distribution of organisms in biomes. In this chapter, we turn our attention to another of the four hierarchical levels of ecology: **population ecology**, the study of factors that affect population density and growth.

Ecologists generally define a **population** as a group of individuals of a single species that occupy the same general area. Members of a population are likely to interact and breed with one another. They rely on the same resources and are influenced by the same environmental factors. Population ecology focuses on the factors that influence a population's density (number of individuals per unit area or volume), structure (such as relative numbers of individuals of different ages), size (number of individuals), and growth rate (rate of change in population size) (**Figure 19.1**).

Population ecology also plays a key role in applied research. For example, it provides critical information for identifying and saving endangered species. Population ecology is being used to develop sustainable fisheries throughout the world and to manage wildlife

populations. Studying the population ecology of pests and pathogens provides insight into controlling how they spread. Population ecologists also study human population growth, one of the most critical environmental issues of our time.

Let's consider what a snapshot of a population might look like. The first question is, Which individuals are included in this population? A population's geographic boundaries may be natural, as with certain species of largemouth bass in an isolated lake. But ecologists often define a population's boundaries in more arbitrary ways that fit their research questions. For example, an ecologist studying the contribution of asexual reproduction to the population growth of sea anemones might define a population as all the anemones of one species in a particular tide pool. Another researcher studying the effects of hunting on deer might define a population as all the deer within a particular state. Yet another researcher, attempting to determine which segment of the human population will be most affected by the AIDS epidemic, might study the HIV infection rate of the human population in one nation or throughout the world.

▼ Figure 19.1 **An ecologist studying a population of black-browed albatross in the Falkland Islands, east of the tip of South America.**

Population Density

Our snapshot of a population would include **population density**, the number of individuals of a species per unit area or volume of the habitat: the number of largemouth bass per cubic kilometer (km^3) of a lake, for example, or the number of oak trees per square kilometer (km^2) in a forest, or the number of nematodes per cubic meter (m^3) in the forest's soil. In rare cases, an ecologist can actually count all the individuals within the boundaries of the population. For example, we could count the total number of oak trees (say, 200) in a forest covering 50 km^2. The population density would be the total number of trees divided by the area, or 4 trees per square kilometer ($4/km^2$).

In most cases, however, it is impractical or impossible to count all individuals in a population. Instead, ecologists use a variety of sampling techniques to estimate population density. For example, they might estimate the density of alligators in the Florida Everglades based on a count of individuals in a few sample plots of 1 km^2 each. Generally speaking, the larger the number and size of sample plots, the more accurate the estimates. Population densities may also be estimated not by counts of organisms but by indirect indicators, such as number of bird nests or rodent burrows (**Figure 19.2**).

Keep in mind that population density is not a constant number. It changes when individuals are born or die and when new individuals enter the population (immigration) or leave it (emigration).

Population Age Structure

The **age structure** of a population—the distribution of individuals in different age-groups—reveals information that is not apparent from population density. For instance, age structure can provide insight into the history of a population's survival or reproductive success and how it relates to environmental factors. **Figure 19.3** shows the age structure of males in a population of cactus finches on the Galápagos Islands in 1987. (See Figure 13.11 for other examples of Galápagos finches.) Four-year-old birds, born in 1983, made up almost half the population, while no 2- or 3-year-olds were present at all. Why was there such dramatic variation? For their food, cactus finches depend on vegetation, which in turn depends on rainfall. The 1983 baby boom resulted from unusually wet weather that produced abundant food. Severe droughts in 1984 and 1985 limited the food supply, preventing reproduction and causing many deaths. As we'll see later in this chapter, age structure is also a useful tool for predicting future changes in a population. ✔

▼ Figure 19.2 **An indirect census of a prairie dog population.** We could get a rough estimate of the number of prairie dogs in this colony in South Dakota by counting the number of mounds constructed by the rodents and then multiplying by the number of animals using a typical burrow.

▼ Figure 19.3 **Age structure for the males in a population of large cactus finches (inset) on one of the Galápagos Islands in 1987.**

Life Tables and Survivorship Curves

Life tables track survivorship, the chance of an individual in a given population surviving to various ages. The life insurance industry uses life tables to predict how long, on average, a person of a given age will live. Starting with a population of 100,000 people, Table 19.1 shows the number who are expected to be alive at the beginning of each age interval, based on death rates in 2004. For example, 93,735 out of 100,000 people are expected to live to age 50. The chance of surviving to age 60, shown in the last column of the same row, is 0.939; about 94% of 50-year-olds will reach the age of 60. The chance of surviving to age 90, however, is only 0.412. Population ecologists have adopted this technique and constructed life tables to help them understand the structure and dynamics of various plant and animal species. By identifying the most vulnerable stage of the life cycle, life table data may also help conservationists develop effective measures to protect species whose populations are declining.

Ecologists represent life table data graphically in a **survivorship curve**, a plot of the number of individuals still alive at each age in the maximum life span (Figure 19.4). By using a percentage scale instead of actual ages on the *x*-axis, we can compare species with widely varying life spans on the same graph. The curve for the human population (red) shows that most people survive to the older age intervals. Ecologists refer to the shape of this curve as Type I survivorship. Species that

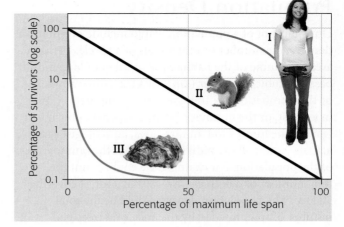

▲ Figure 19.4 **Three idealized types of survivorship curves.**

exhibit a Type I curve—humans and many other large mammals—usually produce few offspring but give them good care, increasing the likelihood that they will survive to maturity.

In contrast, a Type III curve (blue) indicates low survivorship for the very young, followed by a period when survivorship is high for those few individuals who live to a certain age. Species with this type of survivorship curve usually produce very large numbers of offspring but provide little or no care for them. Some species of fish, for example, produce millions of eggs at a time, but most of these offspring die as larvae from predation or other causes. Many invertebrates, such as oysters, also have Type III survivorship curves.

A Type II curve (black) is intermediate, with survivorship constant over the life span. That is, individuals are no more vulnerable at one stage of the life cycle than another. This type of survivorship has been observed in some invertebrates, lizards, and rodents. ☑

Life History Traits as Evolutionary Adaptations

A population's pattern of survivorship is a key feature of its **life history**, the set of traits that affect an organism's schedule of reproduction and survival. Let's take a closer look now at how natural selection affects the reproductive patterns that evolve in populations.

Key life history traits include the age at first reproduction, the frequency of reproduction, the number of offspring, and the amount of parental care given. For a given population in a particular environment, natural selection will favor the combination of life history traits that maximizes an individual's output of viable, fertile

Table 19.1	Life Table for the U.S. Population in 2004		
	Number Living at Start of Age Interval	Number Dying During Interval	Chance of Surviving Interval
Age Interval	(*N*)	(*D*)	1 − (*D/N*)
0−10	100,000	871	0.991
10−20	99,129	419	0.996
20−30	98,709	933	0.991
30−40	97,776	1,259	0.987
40−50	96,517	2,781	0.971
50−60	93,735	5,697	0.939
60−70	88,038	11,847	0.865
70−80	76,191	22,267	0.708
80−90	53,925	31,706	0.412
90+	22,219	22,219	0.000

☑CHECKPOINT

Sea turtles lay their eggs in nests on sandy beaches. Although a single nest may contain as many as 200 eggs, only a small proportion of the hatchlings survive the journey from the beach to the open ocean. Once sea turtles have matured, however, the mortality rate is low. What type of survivorship do sea turtles exhibit?

Answer: Type III

offspring. In other words, life history traits, like anatomical features, are shaped by evolutionary adaptation.

Natural selection cannot maximize all life history traits simultaneously because an organism has limited time, energy, and nutrients. For example, an organism that gives birth to a large number of offspring will not be able to provide a great deal of parental care. Consequently, the combination of life history traits represents trade-offs that balance the demands of reproduction and survival. Because selection pressures vary, life histories are very diverse. Nevertheless, ecologists have observed some patterns that are useful for understanding how natural selection influences life history characteristics.

One life history pattern is typified by small-bodied, short-lived animals (for example, insects and small rodents) that develop and reach sexual maturity rapidly, have a large number of offspring, and offer little or no parental care. In plants, "parental care" can be measured by the amount of nutritional material stocked in each seed. Many small, nonwoody plants (dandelions, for example) produce thousands of tiny seeds. Such organisms have an **opportunistic life history**, one that enables the plant or animal to take immediate advantage of favorable conditions. In general, populations with this life history pattern exhibit a Type III survivorship curve.

In contrast, some organisms have an **equilibrial life history**, a pattern of developing and reaching sexual maturity slowly and producing few, well-cared-for offspring. Organisms that have an equilibrial life history are typically larger-bodied, longer-lived species (for example, bears and elephants). Populations with this life history pattern exhibit a Type I survivorship curve. Plants with comparable life history traits include certain trees. For example, coconut palms produce relatively few seeds that are well stocked with nutrient-rich material. Table 19.2 compares key traits of opportunistic and equilibrial life history patterns.

What accounts for the differences in life history patterns?

Some ecologists hypothesize that the potential survival rate of the offspring and the likelihood that the adult will live to reproduce again are the critical factors. In a harsh, unpredictable environment, where the adult may have just one good shot at reproduction, it may be an advantage to invest in quantity rather than quality. On the other hand, in an environment where favorable conditions are more dependable, an adult is more likely to survive to reproduce again. Seeds are more likely to fall on fertile ground and newly emerged animals are more likely to survive to adulthood. In that case, it may be more advantageous for the adult to invest its energy in producing a few well-cared-for offspring at a time.

Of course, there is much more diversity in life history patterns than the two extremes described here. Nevertheless, the contrasting patterns are useful for understanding the interactions between life history traits and our next topic, population growth. ☑

☑CHECKPOINT

How does the term *opportunistic* capture the key characteristics of that life history pattern?

Answer: An opportunistic life history is characterized by an ability to produce a large number of offspring very rapidly when the environment affords a temporary opportunity for a burst of reproduction.

Dandelions have an opportunistic life history.

Elephants have an equilibrial life history.

Table 19.2	Some Life History Characteristics of Opportunistic and Equilibrial Populations	
Characteristic	**Opportunistic Populations (such as many wildflowers)**	**Equilibrial Populations (such as many large mammals)**
Climate	Relatively unpredictable	Relatively predictable
Maturation time	Short	Long
Life span	Short	Long
Death rate	Often high	Usually low
Number of offspring per reproductive episode	Many	Few
Number of reproductions per lifetime	Usually one	Often several
Timing of first reproduction	Early in life	Later in life
Size of offspring or eggs	Small	Large
Parental care	Little or none	Often extensive

Population Growth Models

BioFlix ™

Population Ecology

Population size fluctuates as new individuals are born or immigrate into an area and others die or emigrate out of an area. Some populations—for example, trees in a mature forest—are relatively constant over time. Other populations change rapidly, even explosively. Consider a single bacterium that divides every 20 minutes. There would be two bacteria after 20 minutes, four after 40 minutes, eight after 60 minutes, and so on. In just 12 hours, the population would approach 70 billion cells. If reproduction continued for a day and a half—a mere 36 hours—there would be enough bacteria to form a layer a foot deep over the entire Earth. Population ecologists use idealized models to investigate how the size of a particular population may change over time under different conditions. We'll describe two simple mathematical models that illustrate fundamental concepts of population growth.

The Exponential Population Growth Model: The Ideal of an Unlimited Environment

Exponential population growth describes the expansion of a population in an ideal, unlimited environment. In the exponential model, the population size of each new generation is calculated by multiplying the current population size by a constant factor that represents the number of births minus the number of deaths. Let's look at how such a population grows. In **Figure 19.5**, we begin with a population of 20 rabbits, indicated on the *y*-axis. Each month, there are more rabbit births than deaths; as a result, the population size increases each month.

Notice in Figure 19.5 that each increase is larger than the previous one—the *rate* of population growth changes with population size. The increasing rate of population growth produces a J-shaped curve that is typical of exponential growth. The slope of the curve represents the growth rate of the population. At the outset, when the population is small, the curve is almost flat—over the first 4 months, the population increases by a total of only 37 individuals, an average of 9.25 births per month. By the end of 7 months, 105 rabbits have been added to the population and the growth rate has increased to an average of 15 births per month. The largest growth is seen in the period from 10 to 12 months, when an average of 85 rabbits are born each month. Thus, the exponential growth model explains how a few dozen rabbits can multiply into millions and overrun a continent.

Exponential population growth is common in certain situations. For example, a disturbance such as a fire, flood, hurricane, drought, or cold snap may suddenly reduce the size of a population. Organisms that have opportunistic life history patterns can rapidly take advantage of the lack of competition and quickly recolonize the habitat by exponential population growth. Human activity can also be a major cause of disturbance, and plants and animals with opportunistic life history traits commonly occupy road cuts, freshly cleared fields and woodlots, and poorly maintained lawns. However, no natural environment can sustain exponential growth indefinitely. ☑

▼ Figure 19.5 **Exponential growth of a rabbit population.**

The Logistic Population Growth Model: The Reality of a Limited Environment

Most natural environments do not have an unlimited supply of the resources needed to sustain population growth. Environmental factors that hold population growth in check are called **limiting factors**. Limiting factors ultimately restrict the number of individuals that can occupy a habitat. Ecologists define **carrying capacity** as the maximum population size that a particular environment can sustain. In **logistic population growth**, the growth rate decreases as the population size approaches carrying capacity. When the population is at carrying capacity, the growth rate is zero.

You can see the effect of limiting factors in the graph in **Figure 19.6**, which shows the growth of a population of fur seals on St. Paul Island, off the coast of Alaska. (For simplicity, only the mated bulls were counted. Each has a harem of females, as shown in the photograph.) Before 1925, the seal population on the island remained low—between 1,000 and 4,500 mated bulls—because of uncontrolled hunting. After hunting was controlled, the population increased rapidly until about 1935, when it began to level off and started fluctuating around a population size of about 10,000 bull seals—the carrying capacity for St. Paul Island. In this instance, the main limiting factor was the amount of space suitable for breeding territories.

The carrying capacity for a population varies, depending on the species and the resources available in the habitat. For example, carrying capacity might be considerably less than 10,000 for a fur seal population on a smaller island with fewer breeding sites. Even in one location, it is not a fixed number. Organisms interact with other organisms in their communities, including predators, pathogens, and food sources, and these interactions may affect carrying capacity. Changes in abiotic factors may also increase or decrease carrying capacity. In any case, the concept of carrying capacity expresses an essential fact of nature: Resources are finite.

Ecologists hypothesize that selection for organisms exhibiting equilibrial life history patterns occurs in environments where the population size is at or near carrying capacity. Because competition for resources is keen under these circumstances, organisms gain an advantage by allocating energy to their own survival and to the survival of their descendants.

Figure 19.7 compares logistic growth (blue) with exponential growth (red). As you can see, the logistic curve is J-shaped at first, but gradually levels off to resemble an S shape as carrying capacity is reached. Both the logistic model and the exponential model are theoretical ideals of population growth. No natural population fits either one perfectly. However, these models are useful starting points for studying population growth. Ecologists use them to predict how populations will grow in certain environments and as a basis for constructing more complex models. ✔

▼ Figure 19.6 **Logistic growth of a seal population.**

✔CHECKPOINT

If a population's changing growth rate approximates the logistic model, what causes the population's size to level off?

Answer: finite resources

◄ Figure 19.7 **Comparison of exponential and logistic growth.**

Regulation of Population Growth

Now let's take a closer look at how population growth is regulated in nature. What stops a population from continuing to increase after reaching carrying capacity?

Density-Dependent Factors

The major biological assumption of the logistic model is that increasing population density reduces the resources available for individual organisms, ultimately limiting population growth. The logistic model is a description of **intraspecific competition**—competition between individuals of the same species for the same

limited resources. As population size increases, competition becomes more intense, and the growth rate declines in proportion to the intensity of competition. Thus, the growth rate of the population depends on population density. A **density-dependent factor** is a population-limiting factor whose effects intensify as the population increases in density. For example, as a limited food supply is divided among more and more individuals, birth rates may decline. **Figure 19.8a** shows the results of a bird population study where such a decline occurred. Clutch size (the number of eggs a female bird lays at one time) declined as the population density—the number of competitors—increased.

Density-dependent factors often depress a population's growth by increasing the death rate. In a laboratory experiment with flour beetles, for example, survivorship declined with increasing population density (**Figure 19.8b**). In a natural setting, plants that are growing close together may experience increased mortality. In an animal population, the death rate may climb as a result of increased disease transmission under crowded conditions or the accumulation of toxic waste products. Predation may also be an important cause of density-dependent mortality. A predator may concentrate on and capture more of a particular kind of prey as that prey becomes abundant.

A limited resource may be something other than food or nutrients. In many vertebrates that defend a territory, the availability of space may limit reproduction. For instance, the number of nesting sites on rocky islands may limit the population size of oceanic birds such as gannets, which maintain breeding territories (**Figure 19.9**).

▼ Figure 19.8 **Density-dependent regulation of population growth.**

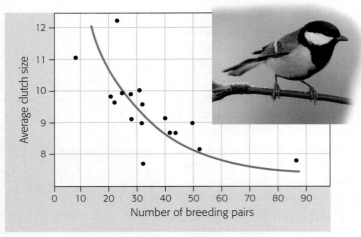

(a) **Decreasing birth rate with increasing density in a population of great tits**

(b) **Decreasing survival rates with increasing density in a population of flour beetles**

▼ Figure 19.9 **Space as a limiting resource in a population of gannets.**

Density-Independent Factors

In many natural populations, abiotic factors such as weather may limit or reduce population size well before other limiting factors become important. A population-limiting factor whose intensity is unrelated to population density is called a **density-independent factor**. If we look at the growth curve of such a population, we see something like exponential growth followed by a rapid decline rather than a leveling off. **Figure 19.10** shows this effect for a population of aphids, insects that feed on the sugary sap of plants. These and many other insects undergo virtually exponential growth in the spring and then rapid die-off when the weather turns hot and dry in the summer. A few individuals may remain, allowing population growth to resume if favorable conditions return. In some populations of insects—many mosquitoes and grasshoppers, for instance—the adults die off entirely, leaving behind eggs that will initiate population growth the following year. In addition to seasonal changes in the weather, environmental disturbances, such as fire, floods, and storms, can affect a population's size regardless of its density.

Over the long term, most populations are probably regulated by a complex interaction of density-dependent and density-independent factors. Although some populations remain fairly stable in size and are presumably close to a carrying capacity that is determined by biotic factors such as competition or predation, most populations for which we have long-term data do fluctuate.

Population Cycles

Some populations of insects, birds, and mammals undergo dramatic fluctuations in density with remarkable regularity. "Booms" characterized by rapid exponential growth are followed by "busts," during which the population falls back to a minimal level. A striking example is the boom-and-bust growth cycles of lemming populations that occur every three to four years. (Lemmings are small rodents that live in the tundra.) Some researchers hypothesize that natural changes in the lemmings' food supply may be the underlying cause. Another hypothesis is that stress from crowding during the "boom" may cause the "bust" by reducing reproduction via physiological mechanisms.

Figure 19.11 illustrates another example—the cycles of snowshoe hare and lynx. The lynx is one of the main predators of the snowshoe hare in the far northern forests of Canada and Alaska. About every ten years, both hare and lynx populations show a rapid increase followed by a sharp decline. What causes these boom-and-bust cycles? Since ups and downs in the two populations seem to almost match each other on the graph, does this mean that changes in one directly affect the other? For the hare cycles, there are three main hypotheses. First, cycles may be caused by winter food shortages that result from overgrazing. Second, cycles may be due to predator-prey interactions. Many predators other than lynx, such as coyotes, foxes, and great-horned owls, eat hares, and together these predators might overexploit their prey. Third, cycles may be affected by a combination of food resource limitation and excessive predation. Recent field studies support the hypothesis that the ten-year cycles of the snowshoe hare are largely driven by excessive predation, but are also influenced by fluctuations in the hare's food supplies. Long-term studies are the key to unraveling the complex causes of such population cycles. ☑

☑**CHECKPOINT**

List some density-dependent factors that limit population growth.

Answer: food and nutrient limitations, insufficient space for territories or nests, increase in disease and predation, accumulation of toxins

▼ Figure 19.10 **Weather change as a density-independent factor limiting growth of an aphid population.**

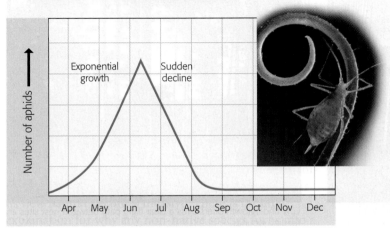

▼ Figure 19.11 **Population cycles of the snowshoe hare and the lynx.**

cheatgrass seeds mature in early summer, the plants become extremely dry and flammable, creating abundant fuel that is easily ignited by lightning or a stray spark. Cheatgrass fires are more intense and occur much more frequently than the fires that native plants have evolved to tolerate. After a few fire cycles, the native plants are gone, robbing more than 150 species of birds and mammals of the food and shelter they derive from sagebrush. Global climate change is also hastening the transition of rangeland into fields of cheatgrass. Studies have shown that cheatgrass responds to increased CO_2 levels by growing faster and accumulating more tissue, which in turn becomes more fuel for the fires that extend its domain.

Many invasive animals also deprive native species of resources. For example, European starlings (**Figure 19.15**) aggressively displace woodpeckers, bluebirds, and swallows from their nesting sites, even pulling out nestlings to make room for their own offspring. These destructive birds were first released in New York in 1890. From the original introduction of 100 or so birds, it took just 50 years for starlings to spread throughout North America and reach an estimated population of over 200 million. Flocks numbering in the tens or hundreds of thousands devour grain and fruits from fields, feedlots, and orchards. In urban areas, the acidic droppings from dense starling populations are corrosive to buildings, and foul smelling, too. ✔

✔CHECKPOINT

What distinguishes invasive species from organisms that are introduced to non-native habitats but do not become invasive?

Answer: Invasive species spread far from where they were introduced, and they cause environmental or economic damage.

▼ Figure 19.15 **A large flock of starlings (inset) settling into a tree to roost for the night.**

Biological Control of Pests

The absence of biotic factors such as pathogens, predators, or herbivores that suppress population growth rate by increasing mortality may also contribute to the success of invasive species. Accordingly, efforts to eliminate or control these troublesome organisms often focus on **biological control**, the intentional release of a natural enemy to attack a pest population. Agricultural researchers have long been interested in identifying potential biological agents to control insects, weeds, and other organisms that reduce crop yield.

It was biological control that finally brought the Australian rabbit population down to manageable levels. After fruitless attempts to stem the tide by poisoning, destroying rabbit colonies, and even herding rabbits into corrals and shooting hundreds at a time, the Australian government introduced a lethal virus into the rabbits' environment. The campaign, begun in 1950, was initially successful at driving down the population size. However, following the familiar pattern of antibiotic resistance in bacteria and pesticide resistance in insects, natural selection favored rabbits that were resistant to the deadly virus. As the percentage of resistant rabbits in the population increased, the population size quickly rebounded. The virus evolved, too. Genetic strains of the virus that were not fatal or that killed their hosts more slowly left more descendants. Evolutionary changes such as these, in which an adaptation of one species leads to a counteradaptation in a second species, are known as **coevolution**. Over the course of many generations, the rabbit population became more resistant to the disease and the virus became less lethal. The government managed to stave off a complete resurgence of the rabbit population by introducing new viral strains, but in 1995, they had to switch to a different pathogen to maintain control.

Coevolution is just one potential pitfall of biological control. Among other potential problems, an imported control agent may turn out to be as invasive as its target. Thus, rigorous research is needed to assess the safety and efficacy of potential biological control agents.

THE PROCESS OF SCIENCE: Invasive Species

Can Biological Control Defeat Kudzu?

Let's take a brief look at the search for a biological agent to control kudzu, an invasive vine known as "the plant that ate the South" (**Figure 19.16**). In the 1930s, the U.S. Department of Agriculture distributed plantings of this Asian import to help control erosion along road cuts and irrigation canals. Today, kudzu covers an estimated 12,000 square miles. With its formidable growth rate of up to a foot per day, kudzu climbs over forest trees and blankets the ground in dense greenery. The shoots die back in winter but quickly regenerate from the roots in the spring. Cold winters have limited its acquisition of new territory—the roots don't survive being frozen. However, as global climate change brings warmer winters, kudzu is advancing farther north.

Unlike many invasive species, kudzu does have natural enemies in the United States, but it easily outgrows the damage they inflict. Researchers are investigating the possibility that one of these native pathogens or insects

could be manipulated to provide effective control. Several possibilities have already been tested and discarded. For example, experiments showed that larvae of a species of moth have a prodigious appetite for kudzu. Further investigation, however, proved that the larvae actually preferred soybeans—an important crop species closely related to kudzu. At present, a fungal pathogen called *Myrothecium verrucaria* appears to be a promising candidate.

Researchers chose to test *M. verrucaria* because of **observations** that it causes severe disease in other weeds belonging to the same family as kudzu. Preliminary tests in a greenhouse, in controlled environment chambers, and in small outdoor plantings established that *M. verrucaria* kills kudzu when a high enough concentration of *M. verrucaria* spores is sprayed on the plants along with a "wetting agent" (a soap-like substance that reduces the surface tension of water). These findings led them to ask the **question**, Will *M. verrucaria* treatment work on an established stand of kudzu in a natural setting? Their **hypothesis** was that the *M. verrucaria* treatment that was most effective in the small outdoor plantings would also be most effective in a natural setting. Their **prediction** was that the greatest kudzu mortality would result from the treatment that sprayed the highest concentration of spores in combination with a wetting agent. The **results** of this field experiment, which are shown in **Figure 19.17**, support the hypothesis. However, these experiments are only the initial steps toward biological control of kudzu. A great deal of research is needed to ensure that the method is safe, effective, and practical.

▼ Figure 19.16 **Kudzu (*Pueraria lobata*).** Unseen in this photo are the plants that have died for lack of sunlight beneath the large kudzu leaves.

▼ Figure 19.17 **Biological control of a natural infestation of kudzu with the fungus *Myrothecium verrucaria*.**

Treatment	Mortality (%)
Spores (2×10^6/ml) + water	~7
Spores (2×10^7/ml) + water	~20
Spores (2×10^6/ml) + wetting agent	~37
Spores (2×10^7/ml) + wetting agent	~100
Wetting agent only	~5
Untreated	0

Integrated Pest Management

In contrast to enterprises such as fisheries, which harvest resources from natural ecosystems, agricultural operations create their own highly managed ecosystems. A typical crop population consists of genetically similar individuals (a monoculture) planted in close proximity to each other—a banquet laid out for the many plant-eating animals and pathogenic bacteria, viruses, and fungi in the community. The tilled, fertile ground nurtures weeds as well as crops. Thus, farmers wage an eternal war against pests that compete with their crop for soil minerals, water, and light; that siphon nutrients from the growing plants; or that consume their leaves, roots, fruits, or seeds. At home, you may be engaged in combat against pests on a smaller scale as you attempt to eradicate the weeds, insects, fungi, and bacteria that attack your lawn and garden or the mosquitoes that make your summer evenings miserable.

Like invasive species, most crop pests have an opportunistic life history pattern that enables them to rapidly take advantage of a favorable habitat. The history of agriculture abounds with examples of devastating pest outbreaks. For example, in the 1840s, a protist that causes a disease called potato rot wiped out the entire potato crop of Ireland in just a few years, causing the "Great Famine." Plant-eating insects can cause massive damage as well (**Figure 19.18**). When synthetic herbicides and insecticides such as DDT were developed in the 1940s, they quickly became the method of choice in agriculture. However, chemical solutions to pest problems bring numerous problems of their own. These chemicals are pollutants that can be carried great distances by air or water currents. And as you saw in Figure 13.14, natural selection may result in populations that are not affected by a pesticide. Furthermore, most insecticides kill both the pest and their natural predators. Because prey species often have a higher reproductive rate than predators, pest populations rapidly rebound before their predators can reproduce. There may be other unintended damage as well, such as killing pollinators that are essential for both agricultural and natural ecosystems.

Integrated pest management (IPM) uses a combination of biological, chemical, and cultural methods for sustainable control of agricultural pests. Researchers are also investigating IPM approaches to invasive species. IPM relies on knowledge of the population ecology of the pest and its associated predators and parasites, as well as plant growth dynamics. In contrast to traditional methods of pest control, IPM advocates tolerating a low level of pests rather than attempting total eradication. Thus, many pest control measures are aimed at lowering the habitat's carrying capacity for the pest population by using pest-resistant varieties and mixed-species plantings and rotating crops to deprive the pest of a dependable food source. Biological control is also used when possible. Pesticides are applied when necessary, but adherence to the principles of IPM prevents the overuse of chemicals. ☑

▼ Figure 19.18 **Plants infested with Japanese beetles (left) and grasshoppers (right).**

Human Population Growth

Now that we have examined the regulation of population growth in other organisms, what about our own species? Let's begin by looking at the history of the human population and then consider some current and future trends in population growth.

The History of Human Population Growth

In the time it takes to read this sentence, approximately 26 babies will be born somewhere in the world and 10 people will die. An imbalance between births and deaths is the cause of population growth (or decline), and as the line graph in **Figure 19.19** shows, the human population is expected to continue increasing for at least the next several decades. The bar graph in Figure 19.19 tells a different part of the story. The number of people added to the population each year has been declining since the 1980s. How do we explain these patterns of human population growth?

Let's begin with the rise in world population from approximately 480 million people in 1500 to the current population of more than 6.7 billion. In the exponential population growth model introduced earlier in this chapter, we assumed that the *rate* of the population growth was constant—births and deaths were roughly equal. As a result, population growth depended only on the size of the existing population. Throughout most of human history, this assumption held true. Although parents had many children, mortality was also high. As a result, human population growth was initially very slow. (If we extended the *x*-axis of Figure 19.19 back in

time to year 1, when the population was roughly 300 million, the line would be almost flat for 1,500 years.) The 1 billion mark was not reached until the early 1800s. As economic development in Europe and the United States led to advances in nutrition and sanitation, and later, medical care, humans took control of their population's growth rate. At first, the death rate decreased while the birth rate remained the same. The net rate of increase rose, and population growth began to pick up steam by the beginning of the 1900s. By midcentury, improvements in nutrition, sanitation, and health care had spread to the developing world, spurring growth at a breakneck pace as birth rates far outstripped death rates.

As the world population skyrocketed from 2 billion in 1927 to 3 billion just 33 years later, some scientists became alarmed. They feared that Earth's carrying capacity would be reached and that density-dependent factors would maintain that population size through human suffering and death. But the overall growth rate peaked in 1962. In the more developed nations, advanced medical care continued to improve survivorship, but effective contraceptives held down the birth rate. As a result, the overall growth rate of the world's population began a downward trend as the difference between birth rate and death rate decreased.

Because economic development has occurred at different times in different regions, worldwide population growth rates reflect a mosaic of the changes occurring in different countries. In the most developed nations, the overall growth rates are near zero (**Table 19.3**). In the developing world, on the other hand, death rates have dropped, but high birth rates persist. As a result, these populations are growing rapidly. Of the nearly 82 million people added to the world each year, almost 80 million are in developing nations. ☑

▼ Figure 19.19 **Five centuries of human population growth, projected to 2050.**

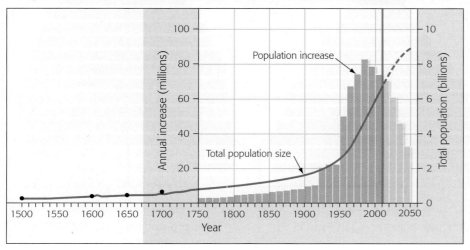

Table 19.3	Population Trends in 2008		
Population	**Birth Rate per 1,000**	**Death Rate per 1,000**	**Growth Rate (%)**
World	21	8	1.2
More developed countries	12	10	0.2
Less developed countries	23	8	1.5

Age Structures

Age structures, which were introduced at the beginning of this chapter, are helpful for predicting a population's future growth. **Figure 19.20** shows the estimated and projected age structures of Mexico's population in 1985, 2010, and 2035. In these diagrams, the area to the left of each vertical line represents the number of males in each age-group; females are represented on the right side of the line. The three different colors represent the portion of the population in their prereproductive years (0–14), prime reproductive years (15–44), and postreproductive years (45 and older). Within each of these broader groups, each horizontal bar represents the population in a 5-year age-group.

In 1985, each age-group was larger than the one above it, indicating a high birth rate. The pyramidal shape of this age structure is typical of a population that is growing rapidly. In 2010, the population's growth rate is lower—notice that the youngest (bottommost) three age-groups are roughly the same size. However, the population continues to be affected by its earlier expansion. This situation, which results from the increased proportion of women of childbearing age in the population, is known as **population momentum**. Girls 0–14 years old in the 1985 age structure (outlined in pink) are in their reproductive prime in 2010, and girls who are 0–14 years old in 2010 (outlined in blue) will carry the legacy of rapid growth forward to 2035. Putting the brakes on a rapidly expanding population is like stopping a freight train; the actual event takes place

long after the decision to do it was made. Even when fertility is reduced to replacement rate (an average of two children per female), the total population size will continue to increase for several decades. Thus, the percentage of individuals under the age of 15 gives a rough idea of future growth. In the developing countries, about 30% of the population is in this age-group. In contrast, roughly 17% of the population of developed nations is under the age of 15. Population momentum also explains why the total population size in Figure 19.19 continues to increase even though fewer people are added to the population each year.

Age structure diagrams may also indicate social conditions. For instance, an expanding population has an increasing need for schools, employment, and infrastructure. A large elderly population requires that extensive resources be allotted to health care. Let's look at trends in the age structure of the United States from 1983 to 2033 (**Figure 19.21**). The noticeable bulge in the 1983 population (yellow screen) corresponds to the "baby boom" that lasted for about two decades after World War II ended in 1945. The large number of children swelled school enrollments, prompting construction of new schools and creating a demand for teachers. On the other hand, graduates who were born near the end of the boom faced stiff competition for jobs. Because they make up such a large segment of the population, boomers have had an enormous influence on social, economic, and political trends. They also

▼ Figure 19.20 **Population momentum in Mexico.**

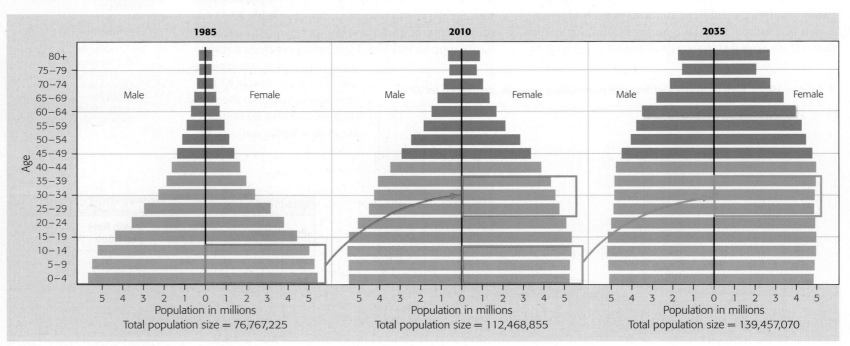

▼ Figure 19.21 **Age structures for the United States in 1983, 2008, and 2033 (projected).**

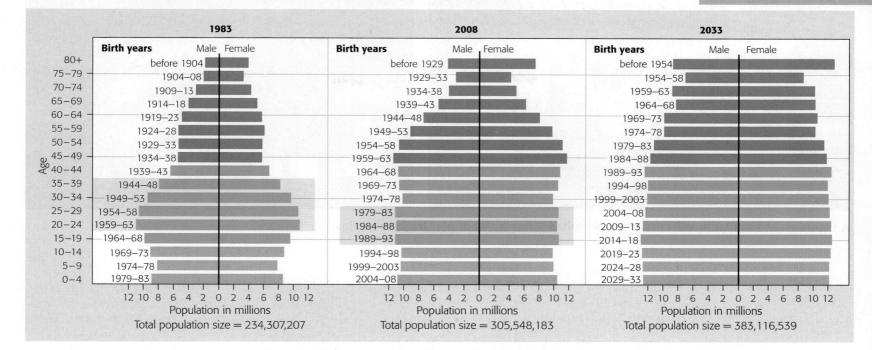

produced a boomlet of their own, seen in the 0–4 age group in 1983 and the bump (pink screen) in the 2008 age structure.

Where are the baby boomers now? The leading edge is reaching retirement age and will soon place pressure on programs such as Medicare and Social Security. In 2008, 60% of the U.S. population was between 20 and 64, the ages most likely to be in the workforce, and 13% was over 65. In 2033, the percentages are projected to be 54 and 19.7, respectively. In part, the increase in the elderly population results from people living longer. The percentage of the population over 80, which was 2.4% in 1983, is projected to rise to 5.6% in 2033 and 8%—close to 34 million people—in 2050. ☑

Our Ecological Footprint

Figure 19.19 shows that the world's population is growing exponentially, but at a slower rate than it did in the last century. The rate of increase, as well as population momentum, predicts that the populations of most developing nations will continue to increase for the foreseeable future. The U.S. Census Bureau projects a global population of 8 billion by 2025 and 9.5 billion by the middle of the century. But these numbers are only part of the story. Trillions of bacteria can live in a petri dish *if* they have sufficient resources. What is Earth's carrying capacity for the human population? Are there sufficient resources to sustain 8 or 9 billion people?

To accommodate all the people expected to live on our planet in the coming decades and improve the diets of those who are currently malnourished or undernourished, world food production must increase dramatically. As you learned in Chapter 18, cropland is being depleted and our current levels of water use are not sustainable. Overgrazing by the world's growing herds of livestock is turning vast areas of grassland into desert. And as additional natural resources are commandeered to support the expanding human population, many thousands of other species are expected to become extinct.

An **ecological footprint** is one approach to understanding resource availability and usage. The footprint is an estimate of the amount of land required to provide the raw materials an individual or a population consumes, including food, fuel, water, housing, and waste disposal. When the total area of ecologically productive land on Earth is divided by the global population, we each have a share of about 2.1 hectares (1 ha = 2.47 acres). In 2005, the average ecological footprint for the world's population was 2.7 ha. We have already overshot the planet's capacity to sustain us.

The ecological footprint of the United States (9.4 ha per person) is almost twice what its own land and

▲ Figure 19.22 **Families in India (left) and the United States (right) display their possessions.**

resources can support (5 ha per person); in other words, it has an enormous ecological deficit. Looking at **Figure 19.22**, it is not difficult to understand why. Compared to a family in rural India, Americans have an abundance of possessions. We also consume a disproportionate amount of food and fuel. By this measure, the ecological impact of affluent nations such as the United States is potentially as damaging as the unrestrained population growth in the developing world. So the problem is not just overpopulation, but overconsumption.

Figure 19.23 is a graphic representation of the disparity in consumption throughout the world. Countries are drawn in proportion to the amount of resources they consume. Regions depicted in red and pink have large ecological deficits—the average personal footprint is far greater than the global average.

The world's richest countries, with 20% of the global population, use 86% of the world's resources, leaving just 14% of global resources—energy, food, water, and other essentials—for the other 80% of the world's population to share. Indeed, the poorest 20% of the population accounts for just 1.3% of resource consumption. Some researchers estimate that providing everyone with the same standard of living as the United States would require the resources of three more planet Earths.

If you would like to learn about your personal resource consumption, several versions of online quizzes can provide a rough estimate of your ecological footprint. Like the carbon footprint calculators described in Chapter 18, these tools are useful for learning how to reduce your environmental impact. ☑

☑CHECKPOINT

How does an individual's large ecological footprint affect Earth's carrying capacity?

Answer: The more resources required to sustain an individual, the lower Earth's carrying capacity will be. (That is, Earth can sustain fewer people if each of those people consumes a large share of available resources.)

▼ Figure 19.23 **World map with area corresponding to ecological footprint.**

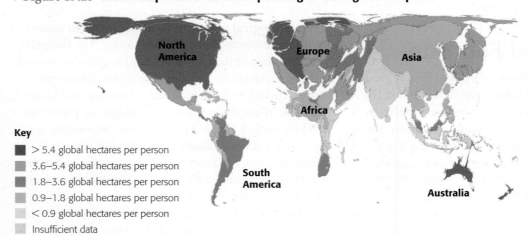

North America

Europe

Asia

Africa

South America

Australia

Key

■ > 5.4 global hectares per person
■ 3.6–5.4 global hectares per person
■ 1.8–3.6 global hectares per person
■ 0.9–1.8 global hectares per person
■ < 0.9 global hectares per person
■ Insufficient data

Humans as an Invasive Species

The magnificent pronghorn antelope (*Antilocapra americana*) is the descendant of ancestors that roamed the open plains and shrub deserts of North America millions of years ago (**Figure 19.24**). With strides that cover 6 m (20 ft) or more at its top speed of 97 km/h (60 mph), it is easily the fastest mammal on the continent. The pronghorn's speed is more than a match for its major predator, the wolf, which typically takes adults that have been weakened by age or illness. What selection pressure promoted such extravagant speed? Ecologists hypothesize that the pronghorn's ancestors were running from the now-extinct American cheetah, a fleet-footed predator that bore some similarities to the more familiar African cheetah.

Cheetahs were not the only danger in the pronghorn's environment. During the Pleistocene epoch, which lasted from 1.8 million to 10,000 years ago, North America was also home to an intimidating list of other predators: lions, jaguars, saber-toothed cats with canines up to 7 inches long, and towering short-faced bears, which stood 11 feet tall and weighed $\frac{3}{4}$ of a ton. There were plenty of potential prey for these fearsome predators, including massive ground sloths, bison with horns that spread 10 feet, elephant-like mammoths, a variety of horses and camels, and several species of pronghorns.

Of all these species of large mammals, only *Antilocapra americana* remained at the end of the Pleistocene. The others went extinct during a relatively brief period of time that coincided with the spread of humans throughout North America. Although the cause of the extinctions has been hotly disputed, many scientists think that the human invasion, combined with climate change at the end of the last ice age, were responsible. Taken together, changes in the biotic and abiotic environments happened too rapidly for an evolutionary response by these large mammals.

The role of humans in the Pleistocene extinctions was merely a preview of things to come. The human population continues to increase, colonizing almost every corner of the globe. Like other invasive species, we change the environment of the other organisms that share our habitats. As the scope and speed of human-induced environmental changes increase, extinctions are occurring at an accelerating pace. This rapid loss of biodiversity will be our unifying thread in the next chapter.

▼ **Figure 19.24**
A pronghorn antelope racing across the North American plains.

Chapter Review

SUMMARY OF KEY CONCEPTS

 Go to the Study Area at **www.masteringbiology.com** for practice quizzes, eText, BioFlix™ 3-D animations, MP3 Tutor Sessions, videos, current events, and more.

An Overview of Population Ecology

A population consists of members of a species living in the same place at the same time. Population ecology focuses on the factors that influence a population's size, growth rate, density, and structure.

Population Density

Population density, the number of individuals of a species per unit area or volume, can be estimated by a variety of sampling techniques.

Population Age Structure

A graph showing the distribution of individuals in different age-groups often provides useful information about the population.

Life Tables and Survivorship Curves

A population's pattern of mortality is a key feature of life history. A life table tracks survivorship and mortality in a population. Survivorship curves can be classified into three general types, depending on the rate of mortality over the entire life span.

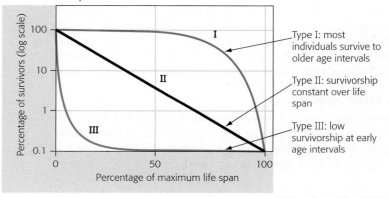

Type I: most individuals survive to older age intervals

Type II: survivorship constant over life span

Type III: low survivorship at early age intervals

421

Life History Traits as Evolutionary Adaptations

Life history traits are shaped by evolutionary adaptation; they may vary within a species and may change as the environmental context changes. Most populations probably fall between the extreme opportunistic life histories (reach sexual maturity rapidly; produce many offspring; little or no parental care) of many insects and the equilibrial life histories (develop slowly and produce few, well-cared-for offspring) of many larger-bodied species.

Population Growth Models

The Exponential Population Growth Model: The Ideal of an Unlimited Environment

Exponential population growth is the accelerating increase that occurs when growth is unlimited. The exponential model predicts that the larger a population becomes, the faster it grows. Exponential growth in nature is generally a short-lived consequence of organisms being introduced to a new or underexploited environment.

The Logistic Population Growth Model: The Reality of a Limited Environment

Logistic population growth occurs when growth is slowed by limiting factors. The logistic model predicts that a population's growth rate will be small when the population size is either small or large, and highest when the population is at an intermediate level relative to the carrying capacity.

Regulation of Population Growth

Over the long term, most population growth is limited by a mixture of density-independent and density-dependent factors. Density-dependent factors intensify as a population increases in density, increasing the death rate, decreasing the birth rate, or both. Density-independent factors affect the same percentage of individuals regardless of population size. Some populations have regular boom-and-bust cycles.

Applications of Population Ecology

Conservation of Endangered Species

Endangered and threatened species are characterized by very small population sizes. One approach to conservation is identifying and attempting to supply the critical combination of habitat factors needed by the population.

Sustainable Resource Management

According to the logistic growth model, a population's highest growth rate occurs when the population size is half the carrying capacity. For some populations, harvesting at this level produces sustainable yields. However, some populations don't meet the assumptions of the logistic model. Thorough scientific knowledge is needed to estimate yields that are truly sustainable.

Invasive Species

Invasive species are non-native organisms that spread far beyond their original point of introduction and cause environmental and economic damage. Typically, invasive species have an opportunistic life history pattern, and many are superior to native species at obtaining resources.

Biological Control of Pests

Biological control, the intentional release of a natural enemy to attack a pest population, is sometimes effective against invasive species. However, coevolution may eventually lessen the effectiveness of biological control. Researchers are currently investigating whether a fungal pathogen could be used to control kudzu, an invasive plant.

Integrated Pest Management

Crop scientists have developed integrated pest management (IPM) strategies, combinations of biological, chemical, and cultural methods, to deal with agricultural pests. IPM may prove useful at combating invasive species, too.

Human Population Growth

The History of Human Population Growth

The human population grew rapidly during the 1900s and currently stands at more than 6.7 billion. A shift from high birth and death rates to low birth and death rates has lowered the rate of growth in more developed countries. In developing nations, death rates have dropped, but birth rates are still high.

Age Structures

The age structure of a population affects its future growth. The wide base of the age structure of Mexico in 1985—the 0–14 age-group—predicts continued population growth in the next generation. Population momentum is the continued growth that occurs (despite fertility having been reduced to replacement rate) as a result of girls in the 0–14 age-group of a previously expanding population reaching their childbearing years. Age structures may also indicate social and economic trends, as in the age structure on the right below.

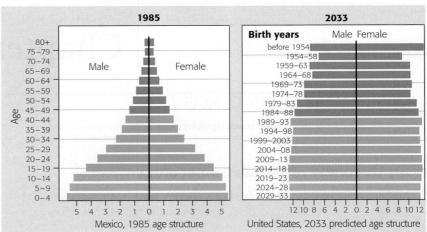

Mexico, 1985 age structure   United States, 2033 predicted age structure

Our Ecological Footprint

The ecological footprint represents the amount of land per person needed to support a nation's resource needs. The carrying capacity of the world may already be smaller than the population's ecological footprint. There is a huge disparity between resource consumption in more developed and less developed nations.

SELF-QUIZ

1. What two values would you need to know to figure out the human population density of your community?

2. If members of a species produce a large number of offspring but provide minimal parental care, then a Type _____ survivorship curve is expected. In contrast, if members of a species produce few offspring and provide them with long-standing care, then a Type _____ survivorship curve is expected.

3. Label the following characteristics as typical of opportunistic (O) or equilibrial (E) life history patterns:
   _____ produce many offspring per reproductive episode
   _____ typical of habitats that have unpredictable climate or frequent disturbances
   _____ life span typically long
   _____ maturation time short
   _____ often provide extensive care to offspring

4. Use this graph of the idealized exponential and logistic growth curves to complete the following.

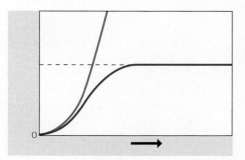

   a. Label the axes and curves on the graph.
   b. What does the dotted line represent?
   c. For each curve, indicate and explain where population growth is the most rapid.
   d. Which of these curves better represents global human population growth?

5. Which of the following shows the effects of a density-dependent limiting factor?
   a. A forest fire kills all the pine trees in a patch of forest.
   b. Early rainfall triggers the explosion of a locust population.
   c. Drought decimates a wheat crop.
   d. Rabbits multiply, and their food supply begins to dwindle.

6. Which life history pattern is typical of invasive species?

7. Skyrocketing growth of the human population since the beginning of the Industrial Revolution appears to be mainly a result of
   a. migration to thinly settled regions of the globe.
   b. better nutrition boosting the birth rate.
   c. a drop in the death rate due to better nutrition and health care.
   d. the concentration of humans in cities.

8. How does the age structure of the human population explain the current surplus in the Social Security fund?

9. According to the ecological footprint study produced in 2005,
   a. the carrying capacity of the world is 10 billion.
   b. the carrying capacity of the world would increase if all people ate more meat.
   c. the current demand on global resources by more developed countries is less than the resources available in those countries.
   d. the United States has a larger ecological footprint than its own resources can provide.
   e. nations with the largest ecological footprints have the fastest population growth rates.

Answers to the Self-Quiz questions can be found in Appendix D.

THE PROCESS OF SCIENCE

10. If researchers establish that *Myrothecium verrucaria* is an effective biological control agent against kudzu, they must then demonstrate that the pathogen will not harm desirable species such as soybeans (a close relative of kudzu). Describe experiments that could fulfill this purpose.

11. Propose a hypothesis to explain why North American mammals such as cheetahs, short-faced bears, and ground sloths did not survive the changes in their environment through evolutionary adaptation.

BIOLOGY AND SOCIETY

12. The mountain gorilla, spotted owl, giant panda, snow leopard, and grizzly bear are all endangered by human encroachment on their environments. Another thing these animals have in common is that they all have equilibrial life history traits. Why might they be more easily endangered than animals with opportunistic life history traits? What general type of survivorship curve would you expect these species to exhibit? Explain your answer.

13. Many people regard the rapid population growth of less developed countries as our most serious environmental problem. Others think that resource consumption in more developed countries is actually a greater threat to the environment. What kinds of problems result from these two issues? Which do you think is the greater environmental threat?

20 Communities and Ecosystems

Enjoying nature.
A kayaker and humpback whale share the waters of the Inside Passage in southeast Alaska.

CHAPTER CONTENTS

BIOLOGY AND SOCIETY: The Loss of Biodiversity

Does Biodiversity Matter?

As the human population has expanded, hundreds of species have become extinct and thousands more are endangered or threatened with extinction. These changes represent a loss in biological diversity, or *biodiversity*. Biodiversity loss goes hand in hand with the disappearance of natural ecosystems. Only about a quarter of Earth's land surfaces remain untouched by human alterations, and our activities have had an enormous impact on aquatic biomes as well. Most of us are only aware of the extent of biodiversity loss when it is brought to our attention by media headlines or biology textbooks. And frankly, it may be hard for us to imagine why it matters. In industrialized countries, clean water flows from our taps; food, clothing, and other necessities are readily available; and wastes vanish into plumbing fixtures or landfills. Life is good.

Most people appreciate the direct benefits provided by certain ecosystems. For example, you are probably aware that we use resources, such as water, wood, and fish and shellfish, that come from natural or near-natural ecosystems. You may know that 25% of all prescription drugs contain substances derived from plants. And many people enjoy activities such as hiking or whitewater rafting in pristine ecosystems, while others enjoy nature in less strenuous ways. But human well-being also depends on less obvious services provided by ecosystems whose biodiversity is intact. Healthy ecosystems purify air and water, decompose wastes, and recycle nutrients. Wetlands buffer coastal populations against hurricanes, reduce the impact of flooding rivers, and filter pollutants. Natural vegetation helps retain fertile soil and prevent landslides. Other ecosystem services include the control of agricultural pests by natural predators and pollination of crops. Some scientists have attempted to assign an economic value to these benefits, arriving at an average annual value of ecosystem services at $33 trillion. In contrast, the global gross national product for the same year was $18 trillion. Although rough, these estimates make the important point that we cannot afford to take biodiversity for granted.

In this chapter, we'll examine the interactions among organisms and how those relationships determine the features of communities. On a larger scale, we'll explore the dynamics of ecosystems. Finally, we'll consider how scientists are working to save biodiversity.

The Loss of Biodiversity

☑**CHECKPOINT**

How does the loss of genetic diversity endanger a population?

Answer: A population with decreased genetic diversity has less ability to evolve in response to environmental change.

As mentioned previously, **biodiversity** is short for biological diversity, the variety of living things that you learned about in Unit 3. It includes genetic diversity, species diversity, and ecosystem diversity. Thus, the loss of biodiversity encompasses more than just the fate of individual species.

Genetic Diversity

When species are lost, so are their unique genes. The genetic diversity within populations of a species is the raw material that makes microevolution and adaptation to the environment possible. If local populations are lost and the number of individuals in a species declines, so do the genetic resources for that species. Severe reduction in genetic variation threatens the survival of a species. The enormous genetic diversity of all the organisms on Earth has great potential benefit for people, too. Many researchers and biotechnology leaders are enthusiastic about the potential that genetic "bioprospecting" holds for future development of new medicines, industrial chemicals, and other products. Bioprospecting may also hold the key to the world's food supply. For example, researchers are currently scrambling to stop the spread of a deadly new strain of wheat stem rust, a fungal pathogen that has devastated harvests in eastern Africa and central Asia. At least 75% of the wheat varieties planted worldwide are susceptible to this pathogen, but researchers hope to find a resistance gene in the wild relatives of wheat (**Figure 20.1**).

Species Diversity

In view of the damage we are doing to the biosphere, ecologists believe that we are pushing species toward extinction at an alarming rate. The present rate of species loss may be as much as 1,000 times higher than at any time in the past 100,000 years. Several researchers estimate that at the current rate of destruction, over half of all currently living plant and animal species will be gone by the end of this century. **Figure 20.2** shows two recent victims. Here are some examples of where things stand:

- Approximately 12% of the 9,934 known bird species and 20% of the 5,416 known mammalian species in the world are threatened with extinction.

- About 20% of the known freshwater fishes in the world have either become extinct during human history or are seriously threatened.

- Roughly 32% of all known amphibian species are either near extinction or endangered.

- Of the approximately 20,000 known plant species in the United States, 200 species have become extinct since dependable records have been kept, and 730 species are endangered or threatened. ☑

▼ **Figure 20.2 Recent additions to the list of human-caused extinctions.**

A Chinese river dolphin. Formerly a resident of the Yangtze River, the Chinese river dolphin was the first large animal to become extinct since the 1950s.

Golden toads. These small frogs, unique to high-altitude Costa Rican cloud forests, have not been seen since 1989. The males shown here are gathered near a puddle to wait for mates.

Ecosystem Diversity

Ecosystem diversity is the third component of biological diversity. Recall from Chapter 18 that an ecosystem includes both the organisms and the abiotic factors in a particular area. Because of the network of interactions among populations of different species within an ecosystem, the local extinction of one species can have a negative effect on the entire ecosystem. In addition, **ecosystem services**, functions performed by an ecosystem that directly or indirectly benefit people, have been lost. These vital services include air and water purification, climate regulation, and erosion control. In Figure 18.39, you saw an all-too-common example of ecosystem damage and loss—the destruction of tropical forests to make room for and support the growing human population. Like tropical forests, coral reefs are rich in species diversity (**Figure 20.3**). An estimated 20% of the world's coral reefs have been destroyed by human activities, and 24% are in imminent danger of collapse. Scientists predict that another 26% of coral reefs will succumb in the next few decades if they are not protected. ☑

Causes of Declining Biodiversity

Ecologists have identified four main factors responsible for the loss of biodiversity: habitat destruction and fragmentation, invasive species, overexploitation, and pollution. The ever-expanding size and dominance of the human population are at the root of all four factors.

Habitat Destruction

The massive destruction and fragmentation of habitats caused by agriculture, urban development, forestry, and mining pose the single greatest threat to biodiversity (**Figure 20.4**). According to the International Union for the Conservation of Nature, which compiles information on the conservation status of species worldwide, habitat destruction is implicated in the decline of 73% of the species in modern history that have become extinct, endangered, vulnerable, or rare. We'll take a closer look at the consequences of habitat fragmentation later in this chapter.

Invasive Species

Ranking second behind habitat loss as a cause of biodiversity loss is the introduction of invasive species. In Chapter 19, we saw how uncontrolled population growth of human-introduced species to non-native habitats has caused havoc when the introduced species have competed with, preyed on, or parasitized native species. The lack of interactions with other species that could keep the newcomers in check is often a key factor in a non-native species becoming invasive.

▼ Figure 20.4 **Habitat destruction.** In a controversial method known as mountaintop removal, mining companies blast the tops off of mountains and then scoop out coal. The earth removed from the mountain is dumped into a neighboring valley.

▼ Figure 20.3 **Coral reef, a colorful display of biodiversity.**

Overexploitation

The marine fisheries we talked about in Chapter 19 demonstrate how people can overexploit wildlife by harvesting at rates that exceed the ability of populations to rebound **(Figure 20.5)**. Tigers, the American bison, and Galápagos tortoises are among the terrestrial species whose numbers have been drastically reduced by excessive commercial harvesting, poaching, or sport hunting. Overharvesting also threatens some plants, including rare trees such as mahogany and rosewood that produce valuable wood.

Pollution

Acid precipitation is a threat to forest and aquatic ecosystems. Aquatic ecosystems also may be polluted by toxic chemicals, nutrients, or other contaminants. Air or water pollution is a contributing factor in declining populations of hundreds of species worldwide. ☑

☑CHECKPOINT

What are the four main causes of declining biodiversity?

Answer: habitat destruction, invasive species, overexploitation, and pollution

▲ Figure 20.5 **Bluefin tuna ready for sale.** Bluefin is the tastiest tuna for sushi and sashimi, but the surging popularity of those dishes has left the species plummeting toward extinction—and the price skyrocketing. A single fish can fetch more than $100,000.

Community Ecology

On your next walk through a field or woodland, or even across campus or your own backyard, observe the variety of species present. You may see birds in trees, butterflies on flowers, dandelions in the grass of a lawn, or lizards darting for cover as you approach. Each of these organisms interacts with other organisms as it goes about looking for food, nesting sites, living space, or shelter. An organism's biotic environment includes not just individuals from its own population, but also populations of other species living in the same area. Such an assemblage of species living close enough together for potential interaction is called a **community**. In **Figure 20.6**, the lion, the zebra, the hyena, the vultures, the plants, and the unseen microbes are all members of an ecological community in Kenya.

Interspecific Interactions

Our study of communities begins with **interspecific interactions**—that is, interactions between species. Interspecific interactions can be classified according to the effect on the populations concerned, which may be helpful (+) or harmful (−). In some cases, two populations in a community vie for a resource such as food or space. The effect of this interaction is generally negative for both species (−/−)—neither species has access to the full range of resources offered by the habitat. On the other hand, some interspecific interactions benefit both parties (+/+). For example, the interactions between flowers and their pollinators are mutually beneficial. In a third type of interspecific interaction, one species exploits another species as a source of food. The effect of this interaction is clearly beneficial to one population and harmful to the other (+/−). In the next several pages, you will learn more about these interspecific interactions and how they affect communities. We will also look at some examples of interspecific interactions as powerful agents of natural selection.

▼ Figure 20.6 **Diverse species interacting in a Kenyan savanna community.**

Interspecific Competition (−/−)

In the logistic model of population growth (see Figure 19.6), increasing population density reduces the amount of resources available for each individual. Intraspecific (within-species) competition for limited resources ultimately limits population growth. In **interspecific competition** (between-species competition), the population growth of a species may be limited by the population densities of competing species as well as by the density of its own population.

What determines whether populations in a community compete with each other? Each species has an **ecological niche**, defined as the sum of its use of the biotic and abiotic resources in its environment. For example, the ecological niche of a small bird called the Virginia's warbler (**Figure 20.7a**) includes its nest sites and nest-building materials, the insects it eats, and climatic conditions such as the amount of precipitation and the temperature and humidity that enable it to survive. In other words, the ecological niche encompasses everything the Virginia's warbler needs for its existence. The ecological niche of the orange-crowned warbler (**Figure 20.7b**) includes some of the same resources used by the Virginia's warbler. Thus, when these two species inhabit the same area, they are competitors.

Ecologists investigated the effects of interspecific competition between populations of these two birds in a community in central Arizona. When they removed either Virginia's warblers or orange-crowned warblers from the study site, members of the remaining species were significantly more successful in raising their offspring. Thus, interspecific competition had a direct, negative effect on reproductive fitness.

If the ecological niches of two species are too similar, they cannot coexist in the same place. Ecologists call this the **competitive exclusion principle**, a concept introduced by Russian ecologist G. F. Gause, who demonstrated this effect with an elegant series of experiments. Gause used two closely related species of protists, *Paramecium caudatum* and *P. aurelia*. First, he established the carrying capacity for each species separately under the conditions used to culture (grow) them in the laboratory (**Figure 20.8**, top graph). Then he cultured the two species in the same habitat. Within two weeks, the *P. caudatum* population had crashed (bottom graph). Gause concluded that the requirements of these two species were so similar that the superior competitor—in this case, *P. aurelia*—deprived *P. caudatum* of essential resources. ☑

▼ **Figure 20.8** **Competitive exclusion in laboratory populations of *Paramecium*.**

Paramecium aurelia

Paramecium caudatum

▼ **Figure 20.7** **Species that use similar resources.**

(a) Virginia's warbler

(b) Orange-crowned warbler

☑CHECKPOINT

In another experiment, Gause found that *Paramecium caudatum* and *P. bursaria* could coexist in the same habitat, but both populations were smaller than when each species occupied the habitat alone. How would you explain these results?

Answer: Competition between P. caudatum and P. bursaria reduces the resources available to each population—some aspects of their ecological niches are similar. However, neither species outcompetes the other for essential resources.

Mutualism (+/+)

In **mutualism**, both species benefit from an interaction. For example, symbiotic root-fungus associations called mycorrhizae (see Figure 16.2) are mutualistic. Mutualism can also occur between species that are not symbiotic, such as flowers and their pollinators. Coral reef ecosystems depend on the mutualism between certain species of coral animals and unicellular algae. Reefs are constructed by successive generations of colonial corals that secrete an external calcium carbonate skeleton. The formation of the skeleton must outpace erosion and competition for space from fast-growing seaweeds. Massive coral reefs are made possible by the millions of algae that live in the cells of each coral polyp (**Figure 20.9**). In turn, the reef provides the food, shelter, and living space that support the splendid diversity of the reef community. The sugars that the algae produce by photosynthesis provide at least half of the energy used by the coral animals. In return, the algae gain a secure shelter that allows access to light. They also use the coral's waste products, including CO_2 and ammonia, a valuable source of nitrogen.

▲ Figure 20.9 **Mutualism.** Coral polyps are inhabited by unicellular algae.

Predation (+/−)

Predation refers to an interaction in which one species (the predator) kills and eats another (the prey). Because predation has such a negative impact on the reproductive success of the prey, numerous adaptations for predator avoidance have evolved in prey populations through natural selection. For example, some prey species, like the pronghorn antelope, run fast enough to escape their predators (see the Evolution Connection section in Chapter 19). Others, like rabbits, flee into shelters. Still other prey species rely on mechanical defenses, such as the porcupine's sharp quills or the hard shells of clams and oysters.

Adaptive coloration is a type of defense that has evolved in many species of animals. Camouflage, called **cryptic coloration**, makes potential prey difficult to spot against its background (**Figure 20.10**). **Warning coloration**, bright patterns of yellow, red, or orange in combination with black, often marks animals with effective chemical defenses. Predators learn to associate these color patterns with undesirable consequences, such as a noxious taste or painful sting, and avoid potential prey with similar markings. The vivid colors of the poison dart frog (**Figure 20.11**), an inhabitant of Costa Rican rain forests, warn of noxious chemicals in the frog's skin.

A prey species may also gain significant protection through mimicry, a "copycat" adaptation in which one species looks like another. For example, the pattern of alternating red, yellow, and black rings of the harmless scarlet king snake resembles the bold color pattern of the venomous eastern coral snake (**Figure 20.12**). Some insects have combined protective coloration with adaptations of body structures in elaborate disguises. For instance, there are insects that resemble twigs, leaves, and bird droppings. Some

▼ Figure 20.10 **Cryptic coloration.** Camouflage conceals the Argentine horned frog from predators.

► Figure 20.11 **Warning coloration of a poison dart frog.**

▲ Figure 20.12 **Mimicry in snakes.** The color pattern of the nonvenomous scarlet king snake (left) is similar to that of the venomous eastern coral snake (right).

even do a passable imitation of a vertebrate. For example, the colors on the dorsal side of a hawk moth caterpillar are an effective camouflage, but when disturbed, the caterpillar flips over to reveal the snakelike eyes of its ventral side (**Figure 20.13**). Eyespots that resemble vertebrate eyes are common in several groups of moths and butterflies. A flash of these large "eyes" startles would-be predators. In other species, an eyespot may deflect a predator's attack away from vital body parts.

Herbivory (+/−)

Herbivory is the consumption of plant parts or algae by an animal. Although herbivory is not usually fatal, a plant whose body parts have been partially eaten by an animal must expend energy to replace the loss. Consequently, numerous defenses against herbivores have evolved in plants. Spines and thorns are obvious antiherbivore devices, as anyone who has plucked a rose from a thorny rosebush or brushed against a spiky cactus knows. Chemical toxins are also very common in plants. Like the chemical defenses of animals, toxins in plants

are distasteful, and herbivores learn to avoid them. Among such chemical weapons are the poison strychnine, produced by a tropical vine called *Strychnos toxifera*; morphine, from the opium poppy; nicotine, produced by the tobacco plant; mescaline, from peyote cactus; and tannins, from a variety of plant species. Other defensive compounds that are not toxic to humans but may be distasteful to herbivores are responsible for the familiar flavors of peppermint, cloves, and cinnamon (**Figure 20.14**). Some plants even produce chemicals that cause abnormal development in insects that eat them. Chemical companies have taken advantage of the poisonous properties of certain plants to produce pesticides. For example, nicotine is used as an insecticide. ☑

☑CHECKPOINT
People find most bitter-tasting foods objectionable. Why do you suppose we have taste receptors for bitter-tasting chemicals?

Answer: Taste receptors sensitive to bitter chemicals presumably enabled the ancestors of humans to identify potentially toxic plants when they foraged for food and thus to survive longer.

▼ Figure 20.14 **Flavorful plants.**

Peppermint. Parts of the peppermint plant yield a pungent oil.

Cloves. The cloves used in cooking are the flower buds of this plant.

▼ Figure 20.13 **An insect mimicking a snake.** When disturbed, the hawk moth larva flips over (left) and resembles a snake (right).

Cinnamon. Cinnamon comes from the inner bark of this tree.

Parasites and Pathogens (+/−)

Both plants and animals may be victimized by parasites
or pathogens. These interactions are beneficial to one
species (the parasite or pathogen) and harmful to the
other, known as the host. A **parasite** lives on or in a
host from which it obtains nourishment. Pathogens are
disease-causing bacteria, viruses, fungi, or protists that
can be thought of as microscopic parasites. You may
recall learning about several invertebrate parasites in
Chapter 17, including flukes and tapeworms and a variety of roundworms, which live inside a host organism's
body. External parasites include arthropods such as
ticks, lice, mites, and mosquitoes, which attach to their
victims temporarily to feed on blood or other body
fluids. Plants are also attacked by parasites, including
roundworms and aphids, tiny insects that tap into the
phloem to suck plant sap (see Figure 19.10). In any parasite population, reproductive success is greatest for
individuals that are best at locating and feeding on their
hosts. For example, some aquatic leeches first locate a
host by detecting movement in the water and then confirm its identity based on the host's body temperature
and chemical cues on its skin.

Non-native pathogens, whose impact can be rapid
and dramatic, have provided some opportunities to
investigate the effects of pathogens on communities. In
one example, ecologists studied the consequences of the
epidemic of chestnut blight. The loss of chestnuts, massive canopy trees that once dominated many forest communities in North America, had a significant impact on
community composition and structure. Trees such as
oaks and hickories that had formerly competed with
chestnuts became more numerous; overall, the diversity
of tree species increased. Dead chestnut trees also
furnished niches for other organisms, such as insects,
cavity-nesting birds, and, eventually, decomposers. ☑

Trophic Structure

Now that we have looked at how populations in a community interact with one another, let's consider the community as a whole. The feeding relationships among the
various species in a community are referred to as its
trophic structure. A community's trophic structure
determines the passage of energy and nutrients from
plants and other photosynthetic organisms to herbivores and then to predators. The sequence of food
transfer between trophic levels is called a **food chain**.

Figure 20.15 compares a terrestrial food chain and
an aquatic food chain. Starting at the bottom, the
trophic level that supports all others consists of autotrophs, which ecologists call **producers**. Photosynthetic producers use light energy to power the
synthesis of organic compounds. Plants are the main
producers on land. In water, the producers are mainly
photosynthetic protists and cyanobacteria, collectively
called phytoplankton. Multicellular algae and aquatic
plants are also important producers in shallow waters.
In a few communities, such as those surrounding
hydrothermal vents, the producers are chemosynthetic
prokaryotes.

All organisms in trophic levels above the producers
are heterotrophs, or **consumers**, and all consumers
depend directly or indirectly on the output of producers. **Herbivores**, which eat plants, algae, or phytoplankton, are the **primary consumers**. Primary consumers on
land include grasshoppers and many other insects,
snails, and certain vertebrates, such as grazing mammals
and birds that eat seeds and fruits. In aquatic environments, primary consumers include a variety of

▼ Figure 20.15 **Examples of food chains.** The arrows
trace the transfer of food from producers through various
levels of consumers in terrestrial and aquatic ecosystems.

Carnivore — Quaternary consumers — Carnivore

Carnivore — Tertiary consumers — Carnivore

Carnivore — Secondary consumers — Carnivore

Herbivore — Primary consumers — Zooplankton

Plant — Producers — Phytoplankton

A terrestrial food chain   **An aquatic food chain**

zooplankton (mainly protists and microscopic animals such as small shrimps) that eat phytoplankton.

Above the primary consumers, the trophic levels are made up of **carnivores**, which eat the consumers from the level below. On land, **secondary consumers** include many small mammals, such as the mouse shown in Figure 20.15 eating an herbivorous insect, and a great variety of birds, frogs, and spiders, as well as lions and other large carnivores that eat grazers. In aquatic ecosystems, secondary consumers are mainly small fishes that eat zooplankton. Higher trophic levels include **tertiary consumers**, such as snakes that eat mice and other secondary consumers. Most communities have secondary and tertiary consumers. As the figure indicates, some also have a higher level, **quaternary consumers**. These include hawks in terrestrial ecosystems and killer whales in the marine environment.

Figure 20.15 shows only those consumers that eat living organisms. Some consumers derive their energy from **detritus**, the dead material left by all trophic levels, including animal wastes, plant litter, and dead organisms. **Detritivores**, which are often called scavengers, consume detritus. They may occupy any consumer level, depending on the source of the detritus. A great variety of animals, including earthworms, many rodents and insects, and vultures, are detritivores. Detritivores in aquatic communities include catfish, crabs, and crayfish. **Decomposers**, consisting of prokaryotes, fungi, and protists, secrete enzymes that digest molecules in organic material and convert them to inorganic forms (**Figure 20.16**). Enormous numbers of microscopic decomposers in the soil and in the mud at the bottom of lakes and oceans break down most of the community's organic materials to inorganic compounds that plants or phytoplankton can use. Decomposition of organic materials is essential for all communities and, indeed, for the continuation of life on Earth.

▼ Figure 20.16 **Fungi decomposing a dead log.**

Biological Magnification

As you learned in Chapter 18, toxic human-made chemicals, including pesticides and industrial wastes, frequently find their way into aquatic ecosystems. Organisms don't metabolize these chemicals; when these chemicals are consumed, they remain in the body. **Figure 20.17** shows the fate of chemicals called PCBs (organic compounds used in electrical equipment until 1977) in a Great Lakes food chain. Zooplankton feed on phytoplankton that have been contaminated by PCBs from the water. Because each individual consumes many phytoplankton, the concentration of PCBs is higher in zooplankton than in phytoplankton. For the same reason, the concentration of PCBs increases in the individuals that occupy each successive trophic level. This accumulation of toxins in the tissues of consumers in a food chain is called **biological magnification**. Thus, the top-level predators—herring gulls in Figure 20.17—have the highest concentrations of PCBs in the food chain and are the organisms most severely affected by any toxic compounds in the environment. In the research results shown in Figure 20.17, the concentration of PCBs in herring gulls eggs in the Great Lakes food chain was almost 5,000 times higher than that measured in phytoplankton. Fewer of these contaminated eggs hatched, resulting in a decline in the reproductive success of herring gulls.

▼ Figure 20.17 **Biological magnification of PCBs in a Great Lakes food chain in the early 1960s.**

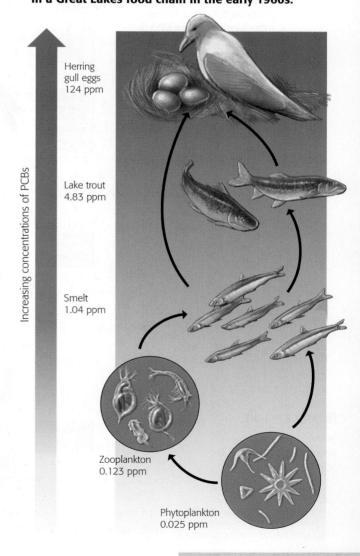

Increasing concentrations of PCBs

Herring gull eggs 124 ppm

Lake trout 4.83 ppm

Smelt 1.04 ppm

Zooplankton 0.123 ppm

Phytoplankton 0.025 ppm

Food Webs

Actually, few, if any, communities are so simple that they are characterized by a single unbranched food chain. Several types of primary consumers usually feed on the same plant species, and one species of primary consumer may eat several different plants. Such branching of food chains occurs at the other trophic levels as well. For example, adult frogs, which are secondary consumers, eat several insect species that may also be eaten by various birds. In addition, some consumers feed at several different trophic levels. An owl, for instance, may eat mice, which are mainly primary consumers that may also eat certain invertebrates. But an owl may also feed on snakes, which are strictly carnivorous. **Omnivores**, including humans, eat producers as well as consumers of different levels. Thus, the feeding relationships in a community are usually woven into elaborate **food webs**.

Figure 20.18 shows a simplified food web. An actual food web would involve many more organisms at each trophic level, and most of the animals would have a more diverse diet than shown in the figure. Note that some species feed at more than one trophic level and that the same species may be eaten by more than one consumer. ☑

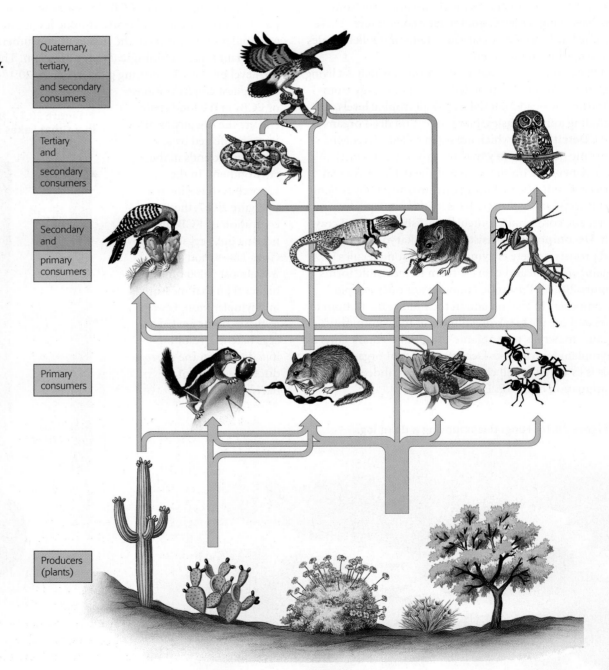

► **Figure 20.18 A simplified food web for a Sonoran desert community.** As in the food chains of Figure 20.15, the arrows in this web indicate "who eats whom," the direction of nutrient transfers. We also continue the color coding introduced in Figure 20.15 for the trophic levels and food transfers.

Quaternary, tertiary, and secondary consumers

Tertiary and secondary consumers

Secondary and primary consumers

Primary consumers

Producers (plants)

Species Diversity in Communities

The **species diversity** of a community—the variety of species that make up the community—has two components. The first component is **species richness**, or the number of different species in the community. The other component is the **relative abundance** of the different species, the proportional representation of a species in a community. To understand why both components are important for describing species diversity, imagine walking through the woodlands shown in **Figure 20.19**. On the path through woodland A, you would pass by four different species of trees, but most of the trees you encounter would be the same species. Now imagine walking on a path through woodland B. You would see the same four species of trees that you saw in woodland A—the species richness of the two woodlands is the same. However, woodland B might seem more diverse to you because no single species predominates. As **Figure 20.20** shows, the relative abundance of one species in woodland A is much higher than the relative abundances of the other three species. In woodland B, all four species are equally abundant. As a result, species diversity is greater in woodland B. Because plants provide food and shelter for many animals, a diverse plant community promotes animal diversity.

Although the abundance of a dominant species such as a forest tree can have an impact on the diversity of other species in the community, a nondominant species may also exert control over community composition. A **keystone species** is a species whose impact on its community is much larger than its total mass or abundance indicates. The term *keystone species* was derived from the wedge-shaped stone at the top of an arch that locks the other pieces in place. If the keystone is removed, the arch collapses. A keystone species occupies an ecological niche that holds the rest of its community in place.

To investigate the role of a potential keystone species in a community, ecologists compare diversity when the species is present or absent. Experiments by Robert Paine in the 1960s were among the first to provide evidence of the keystone species effect. Paine manually removed a predator, a sea star of the genus *Pisaster* (**Figure 20.21**), from experimental areas within the intertidal zone of the Washington coast. The result was that *Pisaster*'s main prey, a mussel, outcompeted many of the other shoreline organisms (algae, barnacles, and snails, for instance) for the important resource of space on the rocks. The number of different organisms present in experimental areas dropped from more than 15 species to fewer than 5 species.

Ecologists have identified other species that play a key role in ecosystem structure. For instance, the decline of sea otters off the western coast of Alaska allowed populations of sea urchins, their main prey, to increase. The abundance of urchins, which consume seaweeds such as kelp, has resulted in the loss of many of the kelp "forests" (see Figure 15.25) and the diversity of marine life that they support. In many ecosystems, however, ecologists are just beginning to understand the complex relationships among species; the value of an individual species may not be apparent until it is gone. ✓

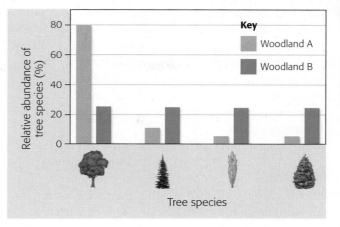

◀ **Figure 20.20**
Relative abundance of tree species in woodlands A and B.

▼ **Figure 20.21** **A *Pisaster* sea star eating its favorite food, a mussel.**

▼ **Figure 20.19** **Which woodland is more diverse?**

Woodland A

Woodland B

Disturbances in Communities

Most communities are constantly changing in response to disturbances. **Disturbances** are episodes that damage biological communities, at least temporarily, by destroying organisms and altering the availability of resources such as mineral nutrients and water. Examples of natural disturbances are storms, fire, floods, and droughts, but humans are by far the most significant agents of disturbance today.

Understanding the effects of disturbance is especially important because of the potential impact on the human population. For instance, one consequence of human-caused disturbance is the emergence of previously unknown infectious diseases. Three-quarters of emerging diseases have jumped to humans from another vertebrate species. In many cases, people come into contact with the pathogens through activities such as clearing land for agriculture, road building, or hunting in previously isolated ecosystems. HIV, which may have been transmitted to humans from the blood of primates butchered for food, is probably the best-known example. Others include potentially fatal hemorrhagic fevers, such as Ebola. Habitat destruction may also cause pathogen-carrying animals to venture closer to human dwellings in search of food.

Small-scale natural disturbances often have positive effects on a biological community. For example, when a large tree falls in a windstorm, it creates new habitats

(Figure 20.22). More light may now reach the forest floor, giving small seedlings the opportunity to grow; or the depression left by its roots may fill with water and be used as egg-laying sites by frogs, salamanders, and numerous insects. ☑

Ecological Succession

Communities change drastically following a severe disturbance that strips away vegetation and even soil. The disturbed area may be colonized by a variety of species, which are gradually replaced by a succession of other species, in a process called **ecological succession**.

When ecological succession begins in a virtually lifeless area with no soil, it is called **primary succession** (Figure 20.23). Examples of such areas are cooled lava flows on volcanic islands and the rubble left by a retreating glacier. Often the only life-forms initially present are autotrophic bacteria. Lichens and mosses, which grow from windblown spores, are commonly the first multicellular photosynthesizers to colonize the area. Soil develops gradually as rocks weather and organic matter accumulates from the decomposed remains of the early colonizers. Lichens and mosses are eventually overgrown by grasses and shrubs that sprout from seeds blown in from nearby areas or carried in by animals. Finally, the area is colonized by plants that become the community's prevalent form of vegetation. Primary succession can take hundreds or thousands of years.

☑CHECKPOINT

Why might the effect of a small-scale natural disturbance, such as a forest tree felled by wind, be considered positive?

Answer: A fallen tree alters abiotic factors such as water and sunlight in a small patch of the forest and provides new habitats for other organisms. The loss of one tree is not a major disruption to the forest community as a whole (unless it is a rare tree).

▼ Figure 20.22 **A small-scale disturbance.** When this tree fell during a windstorm, its root system and the surrounding soil uplifted, resulting in a depression that filled with water. The dead tree, the root mound, and the water-filled depression are new habitats.

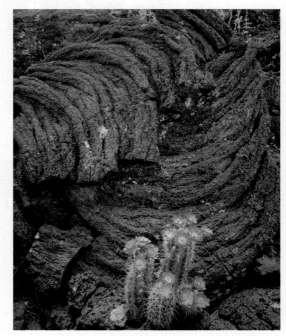

▼ Figure 20.23 **Primary succession under way on a lava flow in Valley of Fires State Park, New Mexico.**

Secondary succession occurs where a disturbance has destroyed an existing community but left the soil intact. For example, secondary succession occurs as areas recover from floods or fires (**Figure 20.24**). Disturbances that lead to secondary succession are also caused by human activities. Even before colonial times, humans were clearing the temperate deciduous forests of eastern North America for agriculture and settlements. Some of this land was later abandoned as the soil was depleted of its chemical nutrients or the residents moved west to new territories. Whenever human intervention stops, secondary succession begins. ☑

▼ Figure 20.24 **Secondary succession after a fire.**

☑**CHECKPOINT**
What is the main abiotic factor that distinguishes primary from secondary succession?

Answer: absence of soil (primary succession) versus presence of soil (secondary succession) at the onset of succession

Ecosystem Ecology

Energy Flow in Ecosystems

In addition to the community of species in a given area, an **ecosystem** includes all the abiotic factors, such as energy, soil characteristics, and water, described in Chapter 18. Perhaps you have seen or even made a terrarium like the one in **Figure 20.25**. Such a microcosm exhibits the two major processes that sustain all ecosystems: energy flow and chemical cycling. **Energy flow** is the passage of energy through the components of the ecosystem. **Chemical cycling** is the use and reuse of chemical elements such as carbon and nitrogen within the ecosystem.

Energy enters the terrarium in the form of sunlight (yellow arrows). Plants (producers) convert light energy to chemical energy through the process of photosynthesis. Animals (consumers) take in some of this chemical energy in the form of organic compounds when they eat the plants. Detritivores and decomposers in the soil obtain chemical energy when they feed on the dead remains of plants and animals. Every use of chemical energy by organisms involves a loss of some energy to the surroundings in the form of heat (red arrows). Because so much of the energy captured by photosynthesis is lost as heat, this ecosystem would run out of energy if it were not powered by a continuous inflow of energy from the sun.

In contrast to energy flow, chemical cycling (blue arrows in Figure 20.25) involves the transfer of materials within the ecosystem. While most ecosystems have a constant input of energy from sunlight or another source, the supply of the chemical elements used to construct molecules is limited. Chemical elements such as carbon and nitrogen are cycled between the abiotic component of the ecosystem, including the air, water, and soil, and the biotic component of the ecosystem (the community). Plants acquire these elements in inorganic

form from the air and soil and fix them into organic molecules. Animals, such as the snail in Figure 20.25, consume some of these organic molecules. When the plants and animals become detritus, decomposers return most of the elements to the soil and air in inorganic form. Some elements are also returned to the air and soil as the by-products of plant and animal metabolism.

In summary, both energy flow and chemical cycling involve the transfer of substances through the trophic levels of the ecosystem. However, energy flows through, and ultimately out of, ecosystems, whereas chemicals are recycled within and between ecosystems.

▼ Figure 20.25 **A terrarium ecosystem.** Though it is small and artificial, this sealed terrarium illustrates the two major ecosystem processes: energy flow and chemical cycling.

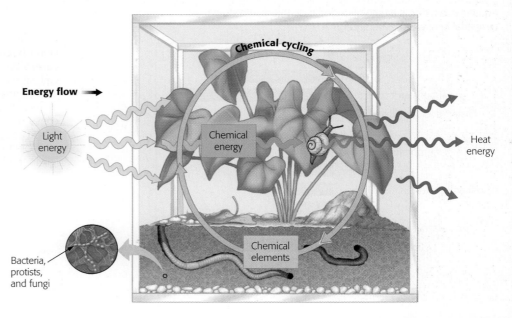

Energy Flow in Ecosystems

All organisms require energy for growth, maintenance, reproduction, and, in many species, locomotion. In this section, we take a closer look at energy flow through ecosystems. Along the way, we'll answer two key questions: What limits the length of food chains? and How do lessons about energy flow apply to people's use of resources?

Primary Production and the Energy Budgets of Ecosystems

Each day, Earth receives about 10^{19} kcal of solar energy, the energy equivalent of about 100 million atomic bombs. Most of this energy is absorbed, scattered, or reflected by the atmosphere or by Earth's surface. Of the visible light that reaches plants, algae, and cyanobacteria, only about 1% is converted to chemical energy by photosynthesis.

Ecologists call the amount, or mass, of living organic material in an ecosystem the **biomass**. The rate at which an ecosystem's producers convert solar energy to the chemical energy stored in biomass is called **primary production**. The primary production of the entire biosphere is roughly 165 billion tons of biomass per year.

Different ecosystems vary considerably in their primary production (**Figure 20.26**) as well as in their contribution to the total production of the biosphere. Tropical rain forests are among the most productive terrestrial ecosystems and contribute a large portion of the planet's overall production of biomass. Coral reefs also

have very high production, but their contribution to global production is small because they cover such a small area. Interestingly, even though the open ocean has very low production, it contributes the most to Earth's total net primary production because of its huge size—it covers 65% of Earth's surface area. Whatever the ecosystem, primary production sets the spending limit for the energy budget of the entire ecosystem because consumers must acquire their organic fuels from producers. Now let's see how this energy budget is divided among the different trophic levels in an ecosystem's food web. ☑

Ecological Pyramids

When energy flows as organic matter through the trophic levels of an ecosystem, much of it is lost at each link in the food chain. Consider the transfer of organic matter from plants (producers) to herbivores (primary consumers). In most ecosystems, herbivores manage to eat only a fraction of the plant material produced, and they can't digest all of what they do consume. For example, a caterpillar feeding on leaves passes about half the energy in the leaves as feces (**Figure 20.27**). Another 35% of the energy is expended in cellular respiration. Only about 15% of the energy in the caterpillar's food is transformed into caterpillar biomass. Only this biomass (and the energy it contains) is available to the consumer that eats the caterpillar.

Figure 20.28, called a **pyramid of production**, illustrates the cumulative loss of energy with each transfer in a food chain. Each tier of the pyramid represents all of the organisms in one trophic level, and the width of each tier indicates how much of the chemical energy of the tier below is actually incorporated into the organic matter of that trophic level. Note that producers convert

☑CHECKPOINT

Deserts and semidesert scrub cover about the same amount of surface area as tropical rain forests but contribute less than 1% of Earth's primary production, while rain forests contribute 22%. Explain this difference.

Answer: The primary production of rain forests is over 20 times greater than that of deserts and semidesert scrub ecosystems.

▼ **Figure 20.26 Primary production of different ecosystems.**
Primary production is the amount of biomass created by the producers of an ecosystem over a unit of time, in this case a year. Aquatic ecosystems are color-coded blue in these histograms; terrestrial ecosystems are green.

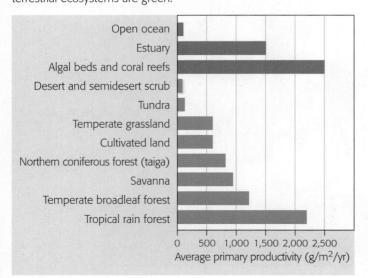

► **Figure 20.27**
What becomes of a caterpillar's food?
Only about 15% of the calories of plant material this herbivore consumes will be stored as biomass available to the next link in the food chain.

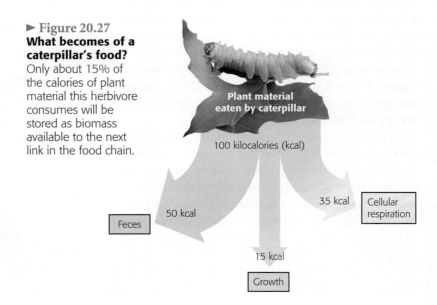

▼ Figure 20.28 **An idealized pyramid of production.**

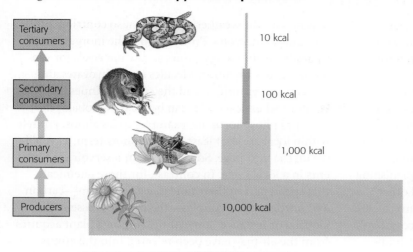

only about 1% of the energy in the sunlight available to them to primary production. In this generalized pyramid, 10% of the energy available at each trophic level becomes incorporated into the next higher level. Actual efficiencies of energy transfer are usually in the range of 5–20%. In other words, 80–95% of the energy at one trophic level never reaches the next.

An important implication of this stepwise decline of energy in a trophic structure is that the amount of energy available to top-level consumers is small compared with that available to lower-level consumers. Only a tiny fraction of the energy stored by photosynthesis flows through a food chain to a tertiary consumer, such as a snake feeding on a mouse. This explains why top-level consumers such as lions and hawks require so much geographic territory; it takes a lot of vegetation to support trophic levels so many steps removed from photosynthetic production. You can also understand why most food chains are limited to three to five levels;

there is simply not enough energy at the very top of an ecological pyramid to support another trophic level. There are, for example, no nonhuman predators of lions, eagles, and killer whales; the biomass in populations of these top-level consumers is insufficient to support yet another trophic level with a reliable source of nutrition.

Ecosystem Energetics and Human Resource Use

The dynamics of energy flow apply to the human population as much as to other organisms. As omnivores, we eat both plant material and meat. When we eat grain or fruit and vegetables, we are primary consumers; when we eat beef or other meat from herbivores, we are secondary consumers. When we eat fish such as trout and salmon (which eat insects and other small animals), we are tertiary or quaternary consumers.

The two production pyramids in **Figure 20.29** are based on the same generalized model used to construct Figure 20.28. The pyramid on the left shows energy flow from primary producers (represented by corn) to human vegetarians. In the pyramid on the right, the primary producers are eaten by cattle; people then eat the beef, making them secondary consumers. The human population has about ten times more energy available to it when people eat corn than when they process the same amount of corn through another trophic level and eat corn-fed beef.

Eating meat of any kind is both economically and environmentally expensive. Producing meat for human consumption usually requires that more land be used for growing grain, which also means greater use of fertilizers and pesticides and more water for irrigation. ☑

▼ Figure 20.29 **Food energy available to the human population at different trophic levels.**

☑**CHECKPOINT**

What is the role of abiotic reservoirs in biogeochemical cycles? What are the three major abiotic reservoirs?

Answer: Abiotic reservoirs are sources of the nutrients used by producers for synthesizing organic molecules. The atmosphere, soil, and water are the three major abiotic reservoirs.

▼ **Figure 20.30**
Plant growth on fallen tree. In the temperate rain forest of Olympic National Park, in Washington State, plants—including other trees—quickly take advantage of the mineral nutrients supplied by decomposing "nurse logs."

Chemical Cycling in Ecosystems

The sun (or in some cases Earth's interior) supplies ecosystems with a continual input of energy, but aside from an occasional meteorite, there are no extraterrestrial sources of chemical elements. Life, therefore, depends on the recycling of chemicals. While an organism is alive, much of its chemical stock changes continuously, as nutrients are acquired and waste products are released. Atoms present in the complex molecules of an organism at the time of its death are returned to the environment by the action of decomposers, replenishing the pool of inorganic nutrients that plants and other producers use to build new organic matter (**Figure 20.30**). In a sense, each living thing only borrows an ecosystem's chemical elements, returning what is left in its body after it dies. Let's take a closer look at how chemicals cycle between organisms and the abiotic components of ecosystems.

The General Scheme of Chemical Cycling

Because chemical cycles in an ecosystem involve both biotic and abiotic components, they are called **biogeochemical cycles**. **Figure 20.31** is a general scheme for the cycling of a nutrient within an ecosystem. Note that the cycle has an **abiotic reservoir** (white box) where a chemical accumulates or is stockpiled outside of living organisms. The atmosphere, for example, is an abiotic reservoir for carbon. Phosphorus, on the other hand, is available only from the soil. The water of aquatic ecosystems contains dissolved carbon, nitrogen, and phosphorus compounds.

Let's trace our way around our general biogeochemical cycle. ❶ Producers incorporate chemicals from the abiotic reservoir into organic compounds. ❷ Consumers feed on the producers, incorporating some of the chemicals into their own bodies. ❸ Both producers and consumers release some chemicals back to the environment in waste products. ❹ Decomposers play a central role by breaking down the complex organic molecules in detritus such as plant litter, animal wastes, and dead organisms. The products of this metabolism are inorganic molecules that replenish the abiotic reservoirs. Geologic processes such as

erosion and the weathering of rock also contribute to the abiotic reservoirs. Producers use the inorganic molecules from abiotic reservoirs as raw materials for synthesizing new organic molecules (carbohydrates and proteins, for example), and the cycle continues.

Biogeochemical cycles can be local or global. The chemical phosphorus, for example, cycles almost entirely within local areas, at least over the short term, as you'll see on the next page. Soil is the main reservoir for nutrients in a local cycle. In contrast, for those chemicals that spend part of their time in gaseous form—carbon and nitrogen are examples—the cycling is essentially global. For instance, some of the carbon a plant acquires from the air may have been released into the atmosphere by the respiration of a plant or animal on another continent.

Now let's examine three important biogeochemical cycles more closely: the cycles for carbon, phosphorus, and nitrogen. As you study the cycles, look for the four basic steps we described, as well as the geologic processes that may move chemicals around and between ecosystems. In all the diagrams, the main abiotic reservoirs appear in white boxes. ☑

▼ **Figure 20.31 General scheme for biogeochemical cycles.**

The Carbon Cycle

You saw a simple version of the carbon cycle in Figure 18.46, where you learned that the abiotic reservoirs of carbon include the atmosphere, fossil fuels, and dissolved carbon compounds in the oceans. The reciprocal metabolic processes of photosynthesis and cellular respiration, also familiar from previous chapters, are mainly responsible for the cycling of carbon between the biotic and abiotic worlds (**Figure 20.32**). **1** Photosynthesis removes CO_2 from the atmosphere and incorporates it into organic molecules, which are **2** passed along the food chain by consumers. **3** Cellular respiration returns CO_2 to the atmosphere. **4** Decomposers break down the carbon compounds in detritus; that carbon, too, is eventually released as CO_2. On a global scale, the return of CO_2 to the atmosphere by respiration closely balances its removal by photosynthesis. As you saw in Chapter 18, however, increasing levels of CO_2 caused by **5** the burning of wood and fossil fuels (coal and petroleum) is contributing to global climate change.

The Phosphorus Cycle

Organisms require phosphorus as an ingredient of nucleic acids, phospholipids, and ATP and (in vertebrates) as a mineral component of bones and teeth. In contrast to the carbon cycle and the other major biogeochemical cycles, the phosphorus cycle does not have an atmospheric component. Rocks are the only source of phosphorus for terrestrial ecosystems; in fact, rocks that have high phosphorus content are mined for fertilizer.

At the center of **Figure 20.33**, **1** the weathering (breakdown) of rock gradually adds inorganic phosphate (PO_4^{3-}) to the soil. **2** Plants absorb dissolved phosphate from the soil and assimilate it by building the phosphorus atoms into organic compounds. **3** Consumers obtain phosphorus in organic form from plants. **4** Phosphates are returned to the soil by the action of decomposers on animal waste and the remains of dead plants and animals. **5** Some of the phosphates drain from terrestrial ecosystems into the sea, where they may settle and eventually become part of new rocks. Phosphorus removed from the cycle in this way will not be available to living organisms until **6** geologic processes uplift the rocks and expose them to weathering.

Because weathering is generally a slow process, the amount of phosphates available to plants in natural ecosystems is often quite low and may be a limiting factor. Farmers and gardeners often use crushed phosphate rock, bone meal (finely ground bones from slaughtered livestock or fish), or guano to add phosphorus to the soil. Guano, the droppings of sea birds and bats, is mined from densely populated colonies or caves, where meters-deep deposits have accumulated.

▼ Figure 20.32 **The carbon cycle.**

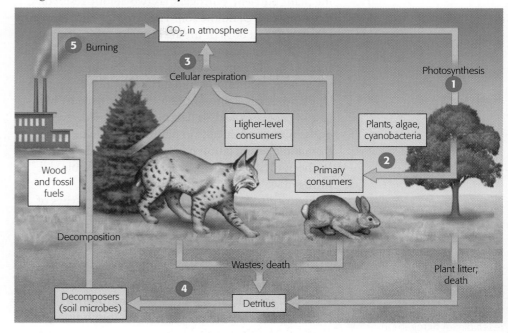

▼ Figure 20.33 **The phosphorus cycle.**

The Nitrogen Cycle

As an ingredient of proteins and nucleic acids, nitrogen is essential to the structure and functioning of all organisms. In particular, it is a crucial and often limiting plant nutrient. Nitrogen has two abiotic reservoirs, the atmosphere and the soil. The atmospheric reservoir is huge; almost 80% of the atmosphere is nitrogen gas (N_2). However, plants cannot assimilate nitrogen in the form of N_2. The process of **nitrogen fixation** converts gaseous N_2 to ammonia and nitrates, which can be used by plants. A small amount of nitrogen is fixed by high-energy processes such as lightning strikes. However, most of the nitrogen available in natural ecosystems comes from biological fixation performed by certain bacteria. Without these organisms, the reservoir of usable soil nitrogen would be extremely limited.

Figure 20.34 illustrates the actions of two types of nitrogen-fixing bacteria. ❶ Some bacteria live symbiotically in the roots of certain species of plants, supplying their hosts with a direct source of usable nitrogen. The largest group of plants with this mutualistic relationship is the legumes, a family that includes peanuts, soybeans, and kudzu. ❷ Free-living nitrogen-fixing bacteria in soil or water convert N_2 to ammonium (NH_4^+). The N_2 is available in air pockets in the soil and dissolved in water.

❸ After nitrogen is "fixed," some of the NH_4^+ is taken up and used by plants. ❹ Nitrifying bacteria in the soil also convert some of the NH_4^+ to nitrate (NO_3^-), ❺ which is more readily assimilated by plants. Plants use the nitrogen they assimilate to synthesize molecules such as amino acids, which are then incorporated into proteins.

❻ When an herbivore (represented here by a rabbit) eats a plant, it digests the proteins into amino acids and then uses the amino acids to build the proteins it needs. Higher-order consumers get nitrogen from the nitrogen-containing organic molecules of their prey. Also, nitrogen-containing waste products are formed during protein metabolism; consumers excrete some nitrogen as well as incorporating it into their body tissues. The urine that rabbits and other mammals excrete contains urea, a nitrogen-containing substance that is widely used as fertilizer.

Organisms that are not consumed eventually die and become detritus, which is decomposed by bacteria and fungi. ❼ Decomposition releases NH_4^+ from organic compounds back into the soil, replenishing the soil reservoir. Under low-oxygen conditions, however, ❽ soil bacteria known as denitrifying bacteria strip the oxygen atoms from NO_3^-, releasing N_2 back into the atmosphere and depleting the soil reservoir of usable nitrogen.

Although not shown in the figure, some NH_4^+ and NO_3^- are made in the atmosphere by chemical reactions involving N_2 and ammonia gas (NH_3). The ions produced by these chemical reactions reach the soil in precipitation and dust, which are a crucial source of nitrogen for plants in some ecosystems. ☑

☑CHECKPOINT

What are the abiotic reservoirs of nitrogen? In what form does nitrogen occur in each reservoir?

Answer: atmosphere: N_2; soil: NH_4^+ and NO_3^-

▼ Figure 20.34 **The nitrogen cycle.**

Nutrient Pollution

Low nutrient levels, especially of phosphorus and nitrogen, often limit the growth of algae and cyanobacteria in aquatic ecosystems. Nutrient pollution occurs when human activities add excess amounts of these chemicals to aquatic ecosystems.

In many areas, phosphate pollution comes from agricultural fertilizers. Phosphates are also a common ingredient in pesticides. Other major sources of phosphates include outflow from sewage treatment facilities and runoff of animal waste from livestock feedlots (where hundreds of animals are penned together). Phosphate pollution of lakes and rivers results in heavy growth of algae and cyanobacteria (**Figure 20.35**). Microbes consume a great deal of oxygen as they decompose the extra biomass, a process that depletes the water of oxygen. These changes lead to reduced diversity of aquatic species and a much less appealing body of water.

Sewage treatment facilities may discharge large amounts of dissolved inorganic nitrogen compounds into rivers or streams when extreme conditions (such as unusual storms) or malfunctioning equipment prevent them from meeting water quality standards. Agricultural sources of nitrogen include feedlots and the large amounts of inorganic nitrogen fertilizers that are routinely applied to crops. Lawns and golf courses also receive sizable applications of nitrogen fertilizers. Crop and lawn plants take up some of the nitrogen compounds,

and denitrifiers convert some to atmospheric N_2, but nitrate is not bound tightly by soil particles and is easily washed out of the soil by rain or irrigation. As a result, chemical fertilizers often exceed the soil's natural recycling capacity.

In an example of how far-reaching this problem can be, nitrogen runoff from Midwestern farm fields has been linked to an annual summer "dead zone" in the Gulf of Mexico (**Figure 20.36**). Vast algal blooms extend outward from where the Mississippi River deposits its nutrient-laden waters. As the algae die, decomposition of the huge quantities of biomass diminishes the supply of dissolved oxygen over an area that ranges from 13,000 km^2 (roughly the area of Connecticut) to 22,000 km^2 (about the area of Michigan). Oxygen depletion disrupts benthic communities, displacing fish and invertebrates that can move and killing organisms that are attached to the substrate. More than 400 recurring and permanent coastal dead zones totaling approximately 245,000 km^2 have been documented worldwide. ✓

✓CHECKPOINT

How does the excessive addition of mineral nutrients to a pond eventually result in the loss of most fish in the pond?

Answer: Overfertilization by nutrient pollution initially causes population explosions of algae and the organisms that feed on them. The respiration of so much life, especially of the microbes decomposing all the organic refuse, consumes most of the lake's oxygen, which the fish require.

▼ Figure 20.35 **Algal growth resulting from nutrient pollution.** The flat green area in this photo is not a lawn, but rather the surface of a polluted pond.

▼ Figure 20.36 **The Gulf of Mexico dead zone.**

Mississippi River

Gulf of Mexico

Light blue lines represent rivers draining into the Mississippi River (shown in dark blue). Nitrogen runoff carried by these rivers ends up in the Gulf of Mexico. In the images below, red and orange indicate high concentrations of phytoplankton. Bacteria feeding on dead phytoplankton deplete the water of oxygen, creating a "dead zone."

Gulf of Mexico

Summer

Gulf of Mexico

Winter

Conservation and Restoration Biology

As we have seen in this unit, many of the environmental problems facing us today have been caused by human enterprises. But the science of ecology is not just useful for telling us how things have gone wrong. Ecological research is also the foundation for finding solutions to these problems and for reversing the negative consequences of ecosystem alteration. Thus, we end the ecology unit by highlighting the beneficial applications of ecological research.

Conservation biology is a goal-oriented science that seeks to understand and counter the loss of biodiversity. Conservation biologists recognize that biodiversity can be sustained only if the evolutionary mechanisms that have given rise to species and communities of organisms continue to operate. Thus, the goal is not simply to preserve individual species but to sustain ecosystems, where natural selection can continue to function, and to maintain the genetic variability on which natural selection acts. The new and expanding field of **restoration ecology** uses ecological principles to develop methods of returning degraded areas to their natural state.

Biodiversity "Hot Spots"

Conservation biologists are applying their understanding of population, community, and ecosystem dynamics in establishing parks, wilderness areas, and other legally protected nature reserves. Choosing locations for protection often focuses on **biodiversity hot spots**. These relatively small areas have a large number of endangered and threatened species and an exceptional concentration of **endemic species**, species that are found nowhere else. Together, the "hottest" of Earth's biodiversity hot spots, shown in **Figure 20.37**, total less than 1.5% of Earth's land surface but are home to a third of all species of plants and vertebrates. For example, all lemurs are endemic to Madagascar, a large island off the eastern coast of Africa that is home to more than 50 species of lemurs. In fact, almost all of the mammals, reptiles, amphibians, and plants that inhabit Madagascar are endemic. There are also hot spots in aquatic ecosystems, such as certain river systems and coral reefs. Because biodiversity hot spots can also be hot spots of extinction, they rank high on the list of areas demanding strong global conservation efforts.

Concentrations of species provide an opportunity to protect many species in very limited areas. However, the "hot spot" designation tends to favor the most noticeable organisms, especially vertebrates and plants. Invertebrates and microorganisms are often overlooked. Furthermore, species endangerment is a global problem, and focusing on hot spots should not detract from efforts to conserve habitats and species diversity in other areas. Finally, even the protection of a nature reserve does not shield organisms from the effects of climate change or other threats, such as invasive species or infectious disease. The golden toad you saw in Figure 20.2 went extinct despite living in a protected reserve—a victim of changing weather patterns, airborne pollution, and disease. To stem the tide of biodiversity loss, we will have to address environmental problems globally as well as locally. ☑

▼ Figure 20.37 **Earth's terrestrial biodiversity hot spots (purple).**

Equator

Conservation at the Ecosystem Level

Most conservation efforts in the past have focused on saving individual species, and this work continues. (You have already learned about one example, the red-cockaded woodpecker; see Figure 19.12.) More and more, however, conservation biology aims at sustaining the biodiversity of entire communities and ecosystems. On an even broader scale, conservation biology considers the biodiversity of whole landscapes. Ecologically, a **landscape** is a regional assemblage of interacting ecosystems, such as an area with forest, adjacent fields, wetlands, streams, and streamside habitats. **Landscape ecology** is the application of ecological principles to the study of land-use patterns. Its goal is to make ecosystem conservation a functional part of the planning for land use.

▼ **Figure 20.38 Edges between ecosystems within a landscape.**

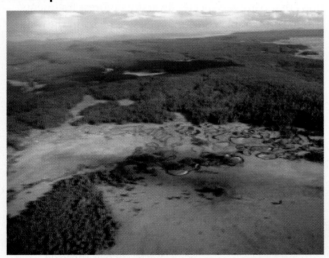

Natural edges. Grasslands border forest ecosystems in Yellowstone National Park.

Edges created by human activity. Pronounced edges (roads) surround clear-cuts in this photograph of a heavily logged rain forest in Malaysia.

Edges between ecosystems are prominent features of landscapes, whether natural or altered by people (**Figure 20.38**). Such edges have their own sets of physical conditions—such as soil type and surface features—that differ from the ecosystems on either side of them. Edges also may have their own type and amount of disturbance. For instance, the edge of a forest often has more blown-down trees than a forest interior because the edge is less protected from strong winds. Because of their specific physical features, edges also have their own communities of organisms. Some organisms thrive in edges because they require resources found only there. For instance, whitetail deer browse on woody shrubs found in edge areas between woods and fields, and their populations often expand when forests are logged or interrupted by development.

Edges can have both positive and negative effects on biodiversity. A recent study in a tropical rain forest in western Africa indicated that natural edge communities are important sites of speciation. On the other hand, landscapes where human activities have produced edges often have fewer species.

Another important landscape feature, especially where habitats have been severely fragmented, is the **movement corridor**, a narrow strip or series of small clumps of suitable habitat connecting otherwise isolated patches. In places where there is extremely heavy human impact, artificial corridors are sometime constructed (**Figure 20.39**). Corridors can promote dispersal and help sustain populations, and they are especially important to species that migrate between different habitats seasonally. But a corridor can also be harmful—as, for example, in the spread of disease, especially among small subpopulations in closely situated habitat patches. ☑

► **Figure 20.39 An artificial corridor.** This bridge over a road provides an artificial corridor for animals in Banff National Park, Canada.

How Does Tropical Forest Fragmentation Affect Biodiversity?

Long-term studies are essential for learning how we might best conserve biodiversity and other natural resources. One site for such studies is the Biological Dynamics of Forest Fragmentation Project (BDFFP), a 1,000-km^2 ecological "laboratory" located deep in the Amazonian forest of Brazil. When the project was begun in 1979, laws required landowners who cleared forest for ranching or other agricultural operations to leave scattered tracts of untouched forest. Biologists recruited some of these landowners to create reserves in isolated fragments of 1 ha (roughly 2.5 acres), 10 ha, and 100 ha. **Figure 20.40** shows some of these forest "islands." Before each area was isolated from the main forest, a small army of specialists inventoried the organisms present and measured the trees.

Hundreds of researchers have used the BDFFP sites to investigate the effects of forest fragmentation on all levels of ecological study. The initial **observations** for these investigations were gleaned from the results of other ecological studies—for example, the effects of

fragmentation in temperate forests or differences in the biodiversity of small islands compared to mainland ecosystems. These observations led many researchers to ask the **question**, How does fragmentation of tropical forests affect species diversity within the fragments? Based on previous research on numerous species, a reasonable **hypothesis** might be: Species diversity declines with the size of the forest fragment. An ecologist studying large predators such as jaguars and pumas, which require large hunting territories, might therefore make the **prediction** that predators will only be found in the largest areas. The BDFFP is unique because it allows researchers to test their predictions with new observations that compare species diversity in forest fragments with a comparable control: species diversity in the same area when it was intact. An undisturbed area of 25,400 acres is also available for comparison with fragmented areas. In addition, data can be collected over a period of years or even decades.

Since the BDFFP was established, scientists have studied many different groups of plants and animals. In general, the **results** have shown that fragmentation of forest into smaller pieces leads to a decline in species diversity. Species richness decreases as a result of local extinctions of many species of large mammals, insects, and insectivorous birds. The population density of remaining species often declines. Researchers also documented edge effects, such as those described in the previous section. Changes in abiotic factors along the fragment edges, including increased wind disturbance, higher temperature, and decreased soil moisture, played a role in community alterations. For example, ecologists found that tree mortality was higher than normal at fragment edges. They also observed changes in the composition of communities of invertebrates that inhabit the soil and leaf litter.

▼ Figure 20.40 **Fragments of forest in the Amazon that were created as part of the Biological Dynamics of Forest Fragmentation Project.** The patch of forest on the right is one hectare.

Restoring Ecosystems

One of the major strategies in restoration ecology is **bioremediation**, the use of living organisms to detoxify polluted ecosystems. For example, bacteria have been used to clean up old mining sites and oil spills (see Figure 15.18). Researchers are also investigating the potential of using plants to remove toxic substances such as heavy metals and organic pollutants (for example, PCBs) from contaminated soil (**Figure 20.41**).

Some restoration projects have the broader goal of returning ecosystems to their natural state, which may involve replanting vegetation, fencing out non-native animals, or removing dams that restrict water flow. Hundreds of restoration projects are currently under way in the United States. One of the most ambitious endeavors is the Kissimmee River Restoration Project in south central Florida.

The Kissimmee River was once a meandering shallow river that wound its way from Lake Kissimmee southward into Lake Okeechobee. Periodic flooding of the river covered a wide floodplain during about half of the year, creating wetlands that provided habitat for large numbers of birds, fishes, and invertebrates. And as the floods deposited the river's load of nutrient-rich silt on the floodplain, they boosted soil fertility and maintained the water quality of the river.

Between 1962 and 1971, the U.S. Army Corps of Engineers converted the 166-km wandering river to a straight canal 9 m deep, 100 m wide, and 90 km long. This project, designed to allow development on the floodplain, drained approximately 31,000 acres of wetlands, with significant negative impacts on fish and wetland bird populations. Without the marshes to help filter and reduce agricultural runoff, the river transported phosphates and other excess nutrients from Lake Okeechobee to the Everglades ecosystem to the south.

The restoration project involves removing water control structures such as dams, reservoirs, and channel modifications and filling in about 35 km of the canal (**Figure 20.42**). The first phase of the project was completed in 2004. The photo shows a section of the Kissimmee canal that has been plugged, diverting flow into the remnant river channels. Birds and other wildlife have returned in unexpected numbers to the 11,000 acres of wetlands that have been restored. The marshes are filled with native vegetation, and game fishes again swim in the river channels. ✓

☑CHECKPOINT

The water in the Kissimmee River eventually flows into the Everglades. How will the Kissimmee River Restoration Project affect water quality in the Everglades ecosystem?

Answer: Wetlands filter agricultural runoff, which prevents nutrient pollution from flowing downstream. By restoring this ecosystem service, the project will improve water quality in the Everglades.

▼ **Figure 20.41 Bioremediation using plants.** A researcher from the U.S. Department of Agriculture investigates the use of canola plants to reduce toxic levels of selenium in contaminated soil.

▼ **Figure 20.42 The Kissimmee River Restoration Project.**

Lake Kissimmee

Widened canal

Water control structure remaining

Water control structure removed

River channel restored

Canal backfilled

Phase 1 completed

Water control structure to be removed in phase 2

Former canal

Kissimmee River

Kissimmee River Floodplain

Detail
ATLANTIC OCEAN

FLORIDA

GULF OF MEXICO

0   10
Miles

Lake Okeechobee

The Goal of Sustainable Development

The demand for the "provisioning" services of ecosystems, such as food, wood, and water, is increasing as the world population grows and becomes more affluent. Although these demands are currently being met, they are satisfied at the expense of other critical ecosystem services, such as climate regulation and protection against natural disasters. Clearly, we have set ourselves and the rest of the biosphere on a precarious path into the future. How can we best manage Earth's resources to ensure that all generations inherit an adequate supply of natural and economic resources and a relatively stable environment?

Many nations, scientific societies, and private foundations have embraced the concept of **sustainable development**. The Ecological Society of America, the world's largest organization of ecologists, endorses a research agenda called the Sustainable Biosphere Initiative. The goal of this initiative is to acquire the ecological information necessary for the responsible development, management, and conservation of Earth's resources. The research agenda includes the search for ways to sustain the productivity of natural and artificial ecosystems and studies of the relationship between

biological diversity, global climate change, and ecological processes.

Sustainable development depends on more than continued research and application of ecological knowledge. It also requires that we connect the life sciences with the social sciences, economics, and humanities. Conservation of biodiversity is only one side of sustainable development; the other key factor is improving the human condition. Public education and the political commitment and cooperation of nations are essential to the success of this endeavor.

An awareness of our unique ability to alter the biosphere and jeopardize the existence of other species, as well as our own, may help us choose a path toward a sustainable future. The risk of a world without adequate natural resources for all its people is not a vision of the distant future. It is a prospect for your children's lifetime, or perhaps even your own. But although the current state of the biosphere is grim, the situation is far from hopeless. Now is the time to aggressively pursue greater knowledge about the diversity of life on our planet and to work toward long-term sustainability. ✓

EVOLUTION CONNECTION: The Loss of Biodiversity

Biophilia and an Environmental Ethic

For millions of years, the diversity of life has flourished via evolutionary adaptation in response to environmental change. In the Evolution Connections sections in Chapters 18 and 19, however, we pointed out that for many species, the pace of evolution can't match the breakneck speed at which humans are changing the environment. Perhaps those species are doomed to extinction. On the other hand, perhaps they can be saved by one human characteristic that works in their favor: biophilia.

Biophilia, which literally means "love of life," is a term that Edward O. Wilson, one of the world's foremost experts on biodiversity and conservation, uses for the human desire to affiliate with other life in its many forms (**Figure 20.43**). People develop close relationships with pets, nurture houseplants, invite

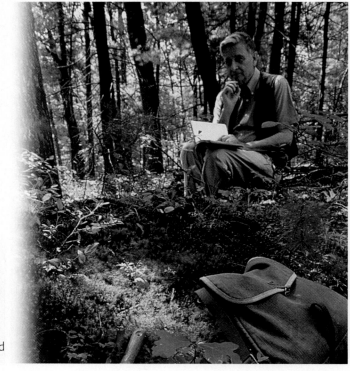

▶ Figure 20.43 **Doing what comes naturally.** Edward O. Wilson has helped teach scientists and the general public a greater respect for Earth's biodiversity. This photograph finds biophiliac Wilson in the woods near Massachusetts's Walden Pond, a landscape immortalized by another great naturalist and writer, Henry David Thoreau.

avian visitors with backyard feeders, and flock to zoos, gardens, and nature parks (**Figure 20.44**). Our attraction to pristine landscapes with clean water and lush vegetation is also testimony to our biophilia. Wilson proposes that our biophilia is innate, an evolutionary product of natural selection acting on a brainy species whose survival depended on a close connection to the environment and a practical appreciation of plants and animals. We evolved in natural environments rich in biodiversity, and we still have an affinity for such settings. Our behavior reflects remnants of our ancestral attachment to nature and the diversity of life.

It will come as no surprise that many biologists have embraced the concept of biophilia. After all, these are people who have turned their passion for nature into careers. But biophilia strikes a chord with biologists for another reason. If biophilia is evolutionarily embedded in our genomes, then there is hope that we can become better custodians of the biosphere. If we all pay more attention to our biophilia, a new environmental ethic could catch on among individuals and societies. And that ethic is a resolve never to knowingly allow a single species to become extinct as a result of our actions or any ecosystem to be destroyed as long as there are reasonable ways to prevent it. Yes, we should be motivated to preserve biodiversity because we depend on it for food, medicine, building materials, fertile soil, flood control, habitable climate, drinkable water, and breathable air. But maybe we can also work harder to prevent the extinction of other forms of life just because it is the ethical thing for us to do. Again, Wilson sounds the call: "Right now, we're pushing the species of the world through a bottleneck. We've got to make it a major moral principle to get as many of them through this as possible. It's the challenge now and for the rest of the century. And there's one good thing about our species: We like a challenge!"

Biophilia is a fitting capstone for this unit. Modern biology is the scientific extension of our human tendency to feel connected to and curious about all forms of life. We are most likely to save what we appreciate, and we are most likely to appreciate what we understand. We hope that our discussion of biodiversity has deepened your biophilia and broadened your education.

▼ **Figure 20.44 Biophilia.** Whether we seek other organisms in their own habitats or invite them into ours, we clearly find pleasure in the diversity of life.

Chapter Review

SUMMARY OF KEY CONCEPTS

 Go to the Study Area at **www.masteringbiology.com** for practice quizzes, eText, BioFlix™ 3-D animations, MP3 Tutor Sessions, videos, current events, and more.

The Loss of Biodiversity

THE COMPONENTS OF BIODIVERSITY		
Genetic Diversity	**Species Diversity**	**Ecosystem Diversity**
Loss of genetic diversity threatens the survival of a species and eliminates potential benefits to people.	The current rate of species extinctions is extremely high compared to the rate of natural extinctions over the past 100,000 years.	Destruction of ecosystems results in the loss of essential ecosystem services.

Causes of Declining Biodiversity

Habitat destruction is the leading cause of extinctions. Invasive species, overexploitation, and pollution are also significant factors.

Community Ecology

Interspecific Interactions

Populations in a community interact in a variety of ways that can be generally categorized as being beneficial (**+**) or harmful (**−**) to the populations. Because **+/−** interactions (exploitation of one species by another species) may have such a negative impact on the individual that is harmed, defensive evolutionary adaptations are common.

Trophic Structure

The trophic structure of a community defines the feeding relationships among organisms. These relationships are sometimes organized into food chains or food webs. Toxins may accumulate by the process of biological magnification as they are passed up a food chain to the top predators.

Increasing PCB concentration

Species Diversity in Communities

Diversity within a community includes species richness and relative abundance of different species. A keystone species is a species that has a great impact on the composition of the community despite relatively low abundance or biomass.

Disturbances in Communities

Disturbances are episodes that damage communities, at least temporarily, by destroying organisms or altering the availability of resources such as mineral nutrients and water. Humans are the most significant cause of disturbances today.

Ecological Succession

The sequence of changes in a community after a disturbance is called ecological succession. Primary succession occurs where a community arises in a virtually lifeless area with no soil. Secondary succession occurs where a disturbance has destroyed an existing community but left the soil intact.

INTERACTIONS BETWEEN SPECIES IN A COMMUNITY					
Interspecific Interaction	**Effect on Species 1**	**Effect on Species 2**	**Interspecific Interaction**	**Effect on Species 1**	**Effect on Species 2**
Competition	−	−	Exploitation		
			Predation	+	−
Mutualism	+	+	Herbivory	+	−
			Parasites and Pathogens	+	−

Ecosystem Ecology

Energy Flow in Ecosystems

An ecosystem is a biological community and the abiotic factors with which the community interacts. Energy must flow continuously through an ecosystem, from producers to consumers and decomposers. Chemical elements can be recycled between an ecosystem's living community and the abiotic environment. Trophic relationships determine an ecosystem's routes of energy flow and chemical cycling.

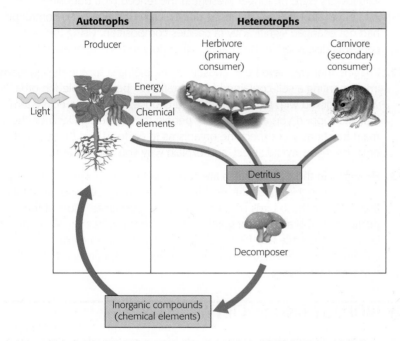

Primary production is the rate at which plants and other producers build biomass. Ecosystems vary considerably in their productivity. Primary production sets the spending limit for the energy budget of the entire ecosystem because consumers must acquire their organic fuels from producers.

In a food chain, only about 10% of the biomass at one trophic level is available to the next, resulting in a pyramid of production:

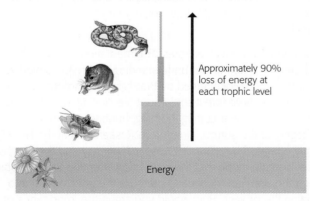

When people eat producers instead of consumers, less photosynthetic production is required, which reduces the impact to the environment.

Chemical Cycling in Ecosystems

Biogeochemical cycles involve biotic and abiotic components. Each circuit has an abiotic reservoir through which the chemical cycles. Some chemical elements require "processing" by certain microorganisms before they are available to plants as inorganic nutrients. A chemical's specific route through an ecosystem varies with the element and the trophic structure of the ecosystem. Phosphorus is not very mobile and is cycled locally. Carbon and nitrogen spend part of their time in gaseous form and are cycled globally. Runoff of nitrogen and phosphorus, especially from agricultural land, causes algal blooms in aquatic ecosystems, lowering water quality and sometimes depleting the water of oxygen.

Conservation and Restoration Biology

Biodiversity "Hot Spots"

Conservation biology is a goal-oriented science that seeks to counter the loss of biodiversity. The front lines for conservation biology are biodiversity "hot spots," relatively small geographic areas that are especially rich in endangered species.

Conservation at the Ecosystem Level

Increasingly, conservation biology aims at sustaining the biodiversity of entire communities, ecosystems, and landscapes. Edges between ecosystems are prominent features of landscapes, with positive and negative effects on biodiversity. Corridors can promote dispersal and help sustain populations.

Restoring Ecosystems

In some cases, toxic substances such as heavy metals can be removed from an ecosystem by microbes or plants. Ecologists are working to revitalize some ecosystems by planting native vegetation, removing barriers to wildlife, and other means. The Kissimmee River Restoration Project is an attempt to undo the ecological damage done when the river was engineered into straight channels.

The Goal of Sustainable Development

Balancing human needs with the health of the biosphere, sustainable development has the goal of long-term prosperity of human societies and the ecosystems that support them.

Evolution Connection: Biophilia and an Environmental Ethic

Biophilia is a term used by biologist E. O. Wilson to describe the innate affinity humans have for nature and for other organisms. Wilson—and many other biologists—hope that our biophilia will help us develop an environmental ethic of preserving biodiversity.

SELF-QUIZ

1. Currently, the number one cause of biodiversity loss is _____.

2. According to the concept of competitive exclusion,
   a. two species cannot coexist in the same habitat.
   b. extinction or emigration is the only possible result of competitive interactions.
   c. intraspecific competition results in the success of the best-adapted individuals.
   d. two species cannot share the same niche in a community.

3. The concept of trophic structure of a community emphasizes the
   a. prevalent form of vegetation.
   b. keystone species concept.
   c. feeding relationships within a community.
   d. species richness of the community.

4. Match each organism with its trophic level (you may choose a level more than once).
   a. alga   1. decomposer
   b. grasshopper   2. producer
   c. zooplankton   3. tertiary consumer
   d. eagle   4. secondary consumer
   e. fungus   5. primary consumer

5. Why are the top predators in food chains most severely affected by pesticides such as DDT?

6. Over a period of many years, grass grows on a sand dune, then shrubs grow, and then eventually trees grow. This is an example of ecological _____.

7. According to the pyramid of production, why is eating grain-fed beef a relatively inefficient means of obtaining the energy trapped by photosynthesis?

8. Local conditions, such as heavy rainfall or the removal of plants, may limit the amount of nitrogen, phosphorus, or calcium available to a particular terrestrial ecosystem, but the amount of carbon available to the ecosystem is seldom a problem. Why?

9. A _____ is a local grouping of interacting ecosystems with several adjacent habitats.

10. Movement corridors are
    a. the routes taken by migratory animals.
    b. strips or clumps of habitat that connect isolated fragments.
    c. landscapes that include several different ecosystems.
    d. edges or boundaries between ecosystems.
    e. buffer zones that protect the long-term viability of protected areas.

Answers to the Self-Quiz questions can be found in Appendix D.

THE PROCESS OF SCIENCE

11. An ecologist studying desert plants performed the following experiment. She staked out two identical plots that included a few sagebrush plants and numerous small annual wildflowers. She found the same five wildflower species in similar numbers in both plots. Then she enclosed one of the plots with a fence to keep out kangaroo rats, the most common herbivores in the area. After two years, four species of wildflowers were no longer present in the fenced plot, but one wildflower species had increased dramatically. The unfenced control plot had not changed significantly in species composition. Using the concepts discussed in the chapter, what do you think happened?

12. Imagine that you have been chosen as the biologist for the design team implementing a self-contained space station to be assembled in orbit. It will be stocked with the organisms you choose, creating an ecosystem that will support you and five other people for two years. Describe the main functions you expect the organisms to perform. List the types of organisms you would select, and explain why you chose them.

13. Biologists in the United States are concerned that populations of many migratory songbirds, such as warblers, are declining. Evidence suggests that some of these birds might be victims of pesticides. Most of the pesticides implicated in songbird mortality have not been used in the United States since the 1970s. Suggest a hypothesis to explain the current decline in songbird numbers. Design an experiment that could test your hypothesis.

BIOLOGY AND SOCIETY

14. Some organizations are starting to envision a sustainable society—one in which each generation inherits sufficient natural and economic resources and a relatively stable environment. The Worldwatch Institute, an environmental policy organization, estimates that we must reach sustainability by the year 2030 to avoid economic and environmental collapse. In what ways is our current system not sustainable? What might we do to work toward sustainability, and what are the major roadblocks to achieving it? How would your life be different in a sustainable society?

15. The Biological Dynamics of Forest Fragmentation Project, which has contributed so much to our understanding of threats to biodiversity, is itself endangered. Encouraged by Brazilian government agencies, urban sprawl and intensive forest settlement are closing in on the study site. Activities such as clear-cutting, burning, hunting, and logging threaten the integrity of the surrounding forest. Researchers at the BDFFP, which is jointly operated by a Brazilian research agency and the Smithsonian Tropical Institute, hope that they can bring attention to the problem through the Brazilian media and pressure the government into protecting the project. How would you argue the importance of protecting the BDFFP against ecologically destructive activities in a letter to a newspaper editor or to a government official?

Unit 5
Animal Structure and Function

Chapter 21: **Unifying Concepts of Animal Structure and Function**

Chapter Thread: **Thermoregulation**

Chapter 22: **Nutrition and Digestion**

Chapter Thread: **Obesity**

Chapter 23: **Circulation and Respiration**

Chapter Thread: **Lifesaving Procedures**

Chapter 24: **The Body's Defenses**

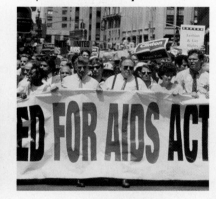

Chapter Thread: **HIV and AIDS**

Chapter 25: **Hormones**

Chapter Thread: **Sex Hormones**

Chapter 26: **Reproduction and Development**

Chapter Thread: **Reproductive Technologies**

Chapter 27: **Nervous, Sensory, and Motor Systems**

Chapter Thread: **Extrahuman Senses**

Animal Structure

The Structural Organization of Animals

Life is characterized by a hierarchy of organization. In animals, individual cells are grouped into tissues, which are in turn grouped into organs, which participate in organ systems, which together make up the entire organism (**Figure 21.1**). Parts of the body at each level of the hierarchy act together to perform the functions of life, such as regulating an animal's internal environment.

For example, the regulation of body temperature within an athlete's body requires cooperation of several levels of body organization. The brain, an organ, sends signals via the nervous system that trigger specific changes in tissues and individual cells. The result is a change within the whole body. ☑

Form Fits Function

Which is the better tool: a hammer or a screwdriver? The answer, of course, depends on what you're trying to do. Given a choice of tools, you would not use a hammer to loosen a screw or a screwdriver to pound a nail. How a device works is correlated with its structure: Form fits function. A chair, for example, must have a surface to sit on and a stable base to support that surface.

▼ Figure 21.1 **Structural hierarchy in a human.**

Cellular level: Muscle cell

Tissue level: Cardiac muscle

Organ level: Heart

Organ system level: Circulatory system

Organism level: Multiple organ systems functioning together

These structural elements are dictated by the task that the chair must perform. The same principle applies to life at its many structural levels, from cells to organisms. Analyzing a biological structure gives us clues about what it does and how it works. Conversely, knowing the function of a structure provides insight about its construction. In exploring life at its many levels, we will discover functional elegance at every turn (Figure 21.2).

When discussing form and function, biologists distinguish anatomy from physiology. **Anatomy** is the study of the structure, or form, of an organism and its parts. **Physiology** is the study of the function of an organism's structural equipment. For example, an anatomist might study the arrangement of blood vessels near the surface of the skin, whereas a physiologist might study how those blood vessels dilate in response to a sudden rise in temperature. Despite their different approaches, the two disciplines serve the same purpose: to better understand the connections between form and function.

The correlation of form and function is an important fundamental principle of biology that is evident at all levels of life's hierarchy; it is a principle that will guide us throughout our study of animals. But this rule of "design" does not mean that such biological tools as a bird's wings, a fish's gills, or a mammal's teeth are products of purposeful invention. Here the analogy to household tools such as hammers and screwdrivers fails because those tools were designed with specific goals. It is natural selection that refines biological equipment. It does so by screening for the most effective variations among individuals of a population—those variations that are most advantageous in the local environment. Generations of selecting for what works best in a particular environment will fit form to function without any goal-oriented plan. Thus, the form-function principle is just another facet of biology's unifying theme: evolution. ☑

Tissues

As you learned in Chapter 4, the cell is the basic unit of all living organisms. In most multicellular animals, including humans, cells usually do not act alone, but instead are grouped into tissues. A **tissue** is an integrated group of similar cells that performs a specific function. The cells composing a tissue are specialized; they have an overall structure that enables them to perform a specific task.

Some tissues consist of living cells surrounded by a meshwork of nonliving fibers. The cells of other tissues are held together by a sticky substance that coats the cells or by special junctions that rivet the cells together. An animal has four main categories of tissue: epithelial tissue, connective tissue, muscle tissue, and nervous tissue. As you'll see, the structure of each type of tissue relates to its specific function.

▼ Figure 21.2 **Form fits function.**

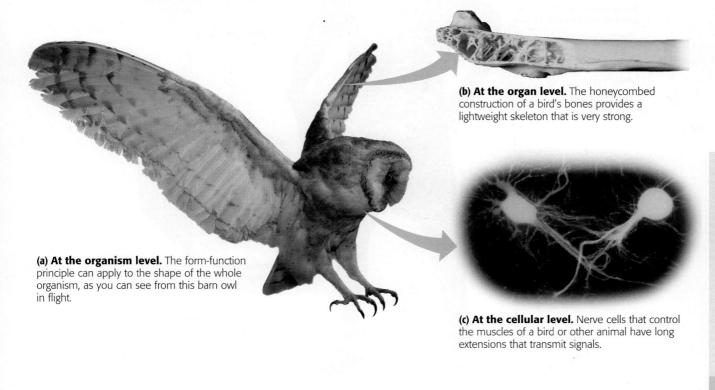

(b) At the organ level. The honeycombed construction of a bird's bones provides a lightweight skeleton that is very strong.

(a) At the organism level. The form-function principle can apply to the shape of the whole organism, as you can see from this barn owl in flight.

(c) At the cellular level. Nerve cells that control the muscles of a bird or other animal have long extensions that transmit signals.

☑**CHECKPOINT**

1. What is the difference between anatomy and physiology?
2. Explain how a tennis racket illustrates the "form fits function" principle.

Answers: 1. Anatomy is the study of structure, whereas physiology is the study of function. 2. To aid the function of hitting a ball, a tennis racket has a handle for gripping, a large head to strike the ball, and strings that stretch and recoil to provide bounce.

Epithelial Tissue

The tissue that covers the surface of the body and lines organs and cavities within the body is called **epithelial tissue**, also known as **epithelium** (Figure 21.3). The epidermis (outer layer of skin) and the linings of the heart, blood vessels, and digestive tract are examples. Epithelial tissue is made of sheets of tightly packed cells that are riveted together, forming a continuous layer. The architecture of an epithelium illustrates how form fits function at the tissue level. For example, the epidermis contains dense layers of epithelial cells bound tightly together, forming a protective barrier that surrounds the body and helps maintain its integrity.

The body continuously renews the cells of epithelial tissue, shedding old cells and growing new ones. For example, cells of the epidermis continuously fall off and are renewed about every two weeks. Such turnover requires rapid cell division, which increases the risk of cancer developing. Consequently, about 80% of all cancers arise in epithelial tissue. Cancers of the epithelium are called carcinomas and include skin, lung, and breast cancer.

Connective Tissue

In contrast to epithelium, with its tightly packed cells, **connective tissue** has a sparse population of cells scattered throughout an extracellular matrix. The matrix consists of a web of protein fibers embedded in a uniform foundation that may be liquid (as in blood), jelly-like (as in adipose, or fat, tissue), or solid (as in bone).

The structure of connective tissue is correlated with its function: to bind and support other tissues. For example, the epithelial tissue of your skin (epidermis) lies on top of a layer of connective tissue called the dermis, which contains the blood vessels and extracellular fluids that nourish the epidermis.

Figure 21.4 illustrates six of the major types of connective tissue. The most widespread connective tissue in the body of vertebrates (animals with backbones, including humans) is **loose connective tissue** (Figure 21.4a). It binds epithelia to underlying tissues and holds organs in place. The matrix of this connective tissue is a loose weave of protein fibers, including collagen fibers, which give connective tissue great strength. For example, if someone pinches your cheek, it is collagen fibers that

▶ Figure 21.3 **Epithelial tissue.** The lining of the esophagus is made up of multiple layers of tightly packed epithelial cells, whereas the lining of the intestine contains a single layer of epithelial cells.

Some examples of organs lined with epithelial tissue:

Heart
Lung
Stomach
Small intestine
Large intestine
Urinary bladder

Epithelial tissue
lining esophagus

Epithelial cells

Epithelial tissue
lining small intestine

keep your flesh from tearing away from the bone. The connective tissue also has elastic fibers, which enable your pinched skin to return quickly to its original shape.

Adipose tissue stores fat in closely packed cells in a sparse matrix of fibers (**Figure 21.4b**). The fat stockpiles energy while also padding and insulating the body. Each adipose cell contains a large globule of fat (mostly triglycerides—see Figure 3.11b); the cell swells when fat is stored and shrinks when fat is used for energy.

Blood is a connective tissue consisting of red and white blood cells suspended in a liquid matrix called plasma (**Figure 21.4c**). Blood functions mainly in transporting substances from one part of the body to another and also plays a major role in immunity.

The other three types of connective tissue have cells suspended in dense matrices. **Fibrous connective tissue** has a dense matrix of collagen. It forms tendons, which attach muscles to bones, and ligaments, which join bones together at joints (**Figure 21.4d**). The matrix of **cartilage** is strong but flexible (**Figure 21.4e**). Flip your outer ear, and you'll get the idea of how cartilage functions as a flexible, boneless skeleton. (You may recall from Chapter 17 that the cartilaginous fishes—which include sharks and rays—have skeletons made entirely from cartilage.) Cartilage is often found at the end of bones and forms the shock-absorbing pads that cushion the bony vertebrae of the spinal column. Cartilage has no blood vessels, so injuries to it tend to heal very slowly. **Bone** is a rigid connective tissue with a matrix of collagen fibers hardened with deposits of calcium salts. This combination makes bone hard without being brittle (**Figure 21.4f**).

► Figure 21.4 **Types of connective tissue.**

Cell

Collagen fiber

(a) Loose connective tissue (under the skin)

Fat droplets

(b) Adipose tissue

White blood cells

Red blood cell

Plasma

(c) Blood

Cell nucleus

Collagen fibers

(d) Fibrous connective tissue (forming a tendon)

Matrix

Cells

(f) Bone

Cells

Matrix

(e) Cartilage (at the end of a bone)

Muscle Tissue

Muscle is the most abundant tissue in most animals. In fact, what we call "meat" is mostly the muscles of cattle, chicken, fish, and other animals. **Muscle tissue** consists of bundles of long, thin, cylindrical cells called muscle fibers. Each cell has specialized proteins arranged into a structure that contracts when the cell is stimulated by a signal from a nerve. Humans and other vertebrates have three types of this contractile tissue, each with a slightly different structure: skeletal muscle, cardiac (heart) muscle, and smooth muscle.

Skeletal muscle is attached to bones by tendons and is responsible for our voluntary movements, such as walking and talking (**Figure 21.5a**). This type of muscle is said to be striated (striped) because the contractile apparatus forms a banded pattern in each muscle fiber, or cell. Adults have a fixed number of these cells. Weight training does not increase the number of muscle cells but instead enlarges those already present.

Cardiac muscle, also striated, is found only in heart tissue (**Figure 21.5b**). The contraction of the cardiac muscle produces the heartbeat. Cardiac muscle cells are branched and joined to one another, appearing like one large interconnected mass of muscle. This structure allows the contraction signal to be quickly propagated to all muscle cells at once, producing a coordinated beat. Cardiac muscle is an involuntary muscle, meaning that your heart contracts without any conscious control on your part.

In contrast to skeletal muscle and cardiac muscle, **smooth muscle** is named for its lack of obvious striations (**Figure 21.5c**). Found in the walls of such organs as the intestines and blood vessels, smooth muscle is involuntary. For example, we have no control over the contractions of our intestinal muscles that move food along the digestive tract. Smooth muscle contracts more slowly than skeletal muscle, but it can remain contracted for a long time.

▼ **Figure 21.5 Three types of muscle tissue.** Skeletal muscle is the only type that we can consciously control; cardiac muscle and smooth muscle act involuntarily.

Unit of muscle contraction

Muscle fiber (cell)

Nuclei

(a) Skeletal muscle
(short segments of several muscle fibers)

Junction between two cells

Muscle fiber

Nucleus

(b) Cardiac muscle

Muscle fiber

Nucleus

(c) Smooth muscle

Nervous Tissue

Most animals are active creatures that respond rapidly to stimuli from their environment. For example, if you touch a very hot object, your arm quickly jerks your hand away. This response requires information to be relayed from one part of the body to another. It is **nervous tissue** that makes such communication possible. Nervous tissue is found in the brain and spinal cord, as well as in the nerves that connect them to all other parts of the body.

The basic unit of nervous tissue is the **neuron**, or nerve cell (**Figure 21.6**). With their long extensions, neurons can transmit electrical signals very rapidly over long distances. For example, neurons in the sciatic nerve of your leg may be as long as 1 m (3.3 ft), running all the way from the base of your spinal cord to the tips of your toes. You are wired for action—and for thinking, since it is the network of neurons in your brain that functions as your mind. ☑

▼ **Figure 21.6 Nervous tissue.** Here you see a diagram of the body's network of nervous tissue and a micrograph of its basic unit, the neuron. Each neuron has extensions that receive signals, a cell body, and an extension that transmits signals. These signals are transmitted throughout the brain, the spinal cord, and the network of nerves (cable-like bundles of neurons) that connect to all other parts of the body.

Signal-receiving extensions

Cell body

Signal-transmitting extension

LM 1,000x

Brain

Spinal cord

Nerve

Organs and Organ Systems

The next level in the structural hierarchy is the organ. An **organ** consists of two or more tissues packaged into one working unit that performs a specific function. Your heart, brain, and small intestine are examples of organs. At each level of the hierarchy, functions emerge from the collective interaction of the structures at lower levels. An organ performs functions that none of its component tissues can carry out alone.

To see how multiple tissues coordinate a single organ, examine the layered arrangement of tissues in the wall of the small intestine in **Figure 21.7**. The inside wall is lined with epithelial tissue (top layer of cells in figure) that secretes mucus and digestive enzymes. Underneath this layer are two layers of connective tissue—one containing blood capillaries and lymph vessels—separated by two layers of smooth muscle. A final layer of epithelial tissue forms the outside surface. Neurons regulate the rhythmic contractions of the intestinal muscles, propelling food along.

☑**CHECKPOINT**

Name the four types of tissue found in most organs.

Answer: epithelial, connective, muscle, and nervous

► **Figure 21.7 Tissue layers of the small intestine, an organ.**

Small intestine
(cut open)

Epithelial tissue

Connective tissue (containing blood and lymph vessels)

Smooth muscle tissue (two layers)

Connective tissue

Epithelial tissue

The organs of humans and most other animals make up **organ systems**, teams of organs that work together to perform vital body functions. The components of an organ system can be either physically connected or dispersed within the body. An example of an organ system is your circulatory system, which transports materials throughout your body. Its main organs are the heart and blood vessels (arteries, veins, and capillaries). **Figure 21.8** presents an overview of 11 major organ systems in vertebrates.

An organism depends on the coordination of all its organ systems for survival. For instance, nutrients that

▼ Figure 21.8 **Organ systems of a vertebrate.**

Skeletal system: supports body and anchors muscles

- Bone
- Cartilage

Circulatory system: transports substances throughout body

- Heart
- Blood vessels

Respiratory system: exchanges O_2 and CO_2 between blood and air

- Nasal cavity
- Pharynx
- Larynx
- Trachea
- Bronchus
- Lung

Digestive system: breaks down food and absorbs nutrients

- Mouth
- Esophagus
- Liver
- Stomach
- Large intestine
- Small intestine
- Anus

Muscular system: moves body

- Skeletal muscles

Urinary system: rids body of certain wastes

- Kidney
- Ureter
- Urinary bladder
- Urethra

are absorbed from the digestive tract are distributed throughout the body by the circulatory system. But the heart that pumps blood through the circulatory system requires nutrients absorbed from food by the digestive tract and also oxygen (O_2) obtained from the air by the respiratory system. The failure of any organ system will jeopardize the entire animal because organ systems are so interwoven and dependent on each other. Your body is a whole, living unit that is greater than the sum of its parts. In the next section, we'll look at how the body's organ systems interact with the external environment. ✔

✔CHECKPOINT

Your stomach is a(n) _____, whereas your digestive system is a(n) _____.

Answer: organ; organ system

Endocrine system: secretes hormones that regulate body

Hypothalamus
Pituitary gland
Parathyroid gland
Thyroid gland
Adrenal gland
Pancreas

Ovary (female)

Testis (male)

Reproductive system: produces gametes and offspring

Oviduct
Ovary
Uterus
Vagina

Seminal vesicles
Prostate gland
Vas deferens
Penis
Urethra

Testis

Integumentary system: protects body

Hair
Nail
Skin

Lymphatic and immune system: defends against disease

Thymus
Spleen
Lymph nodes
Lymphatic vessels

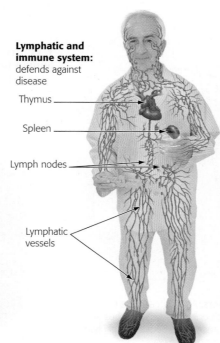

Nervous system: processes sensory information and controls responses

Brain
Sense organ (ear)
Spinal cord
Nerves

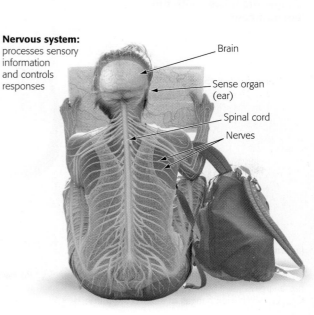

Exchanges with the External Environment

Most animals are covered with a protective layer that separates the external world from the internal chemical environment. This does not mean, however, that an animal barricades itself against the world outside. On the contrary, every organism is an **open system**, meaning that it continuously exchanges chemicals and energy with its surroundings. In fact, organisms must do so to survive. You eat, breathe, defecate, urinate, sweat, and radiate heat—all examples of how you operate as an open system. And the exchange of materials extends down to the cellular level. Nutrients and oxygen must enter every living cell, and carbon dioxide and other wastes must exit.

An animal's size and shape affect how it exchanges energy and materials with its surrounding environment. In an animal body, every living cell must be bathed in a watery solution, partly because substances must be dissolved in water to cross cell membranes. An amoeba (a type of protist) lives in water and consists of only one cell (**Figure 21.9a**). Thus, no part of its body is very far away from its outer surface, where exchange with its watery environment can occur.

Exchange with the environment is more complicated for multicellular animals. Each microscopic cell in a multicellular organism has a plasma membrane where exchange of materials can occur. But this exchange only works if all the cells of the animal have access to a suitable watery environment. A hydra (a simple pond-dwelling relative of jellies) has a body wall only two cell layers thick (**Figure 21.9b**). Both the outer layer and inner layer of cells are bathed in pond water, which enters the digestive sac through the mouth. Every cell of the hydra can thus exchange materials through direct contact with the watery environment. A flat body shape is another way of maximizing exposure to the environment. For instance, a tapeworm may be several meters long, but because it is very thin, most of its cells are bathed in the body fluid of the worm's host, from which it obtains nutrients.

Simple body forms, such as two-layered sacs and flat shapes, do not allow for much complexity in internal organization. But animals with more complex body forms face the same basic problem: Every living cell in the body must be bathed in fluid and have access to oxygen, nutrients, and other resources from the outside environment. Complex animals have extensively folded or branched internal surfaces that maximize surface area for exchange with the immediate environment. For example, your lungs, which exchange oxygen and carbon dioxide with the air you breathe, are not shaped like big balloons but like millions of tiny balloons at the tips of finely branched air tubes (**Figure 21.10**). The epithelium of the lungs has a very large total surface area—about the size of a tennis court.

▼ Figure 21.9 **Contact of simple organisms with the environment.**

Mouth

Gastrovascular cavity

Exchange

Exchange

Exchange

(a) Single cell. The entire surface area of a single-celled organism, such as this amoeba, contacts the environment. Because of its small size, the organism has a large surface area (relative to its volume) through which it exchanges materials with the external world.

(b) Two cell layers. Although a hydra is multicellular, each one of its cells touches an aqueous environment. The body has only two cell layers, both exposed to water.

▼ Figure 21.10 **The branched surface area of the human lung.** This model shows the tiny air tubes of the lungs (white) and the thin blood vessels (red) that transport gases between the heart and lungs.

Figure 21.11 shows an abstract model of an animal body, highlighting the three organ systems—digestive, respiratory, and urinary—that exchange materials with the external environment. Inside the body, materials are transported by the circulatory system from exchange surfaces to nearly every cell in the body, and wastes are transported away from cells to be disposed of at exchange surfaces. ☑

External environment

Mouth Food   CO₂ O₂

Animal

Respiratory system

Digestive system

Blood

Heart

Interstitial fluid

Nutrients

Circulatory system

Body cells

Urinary system

Anus

Unabsorbed matter (feces)

Metabolic waste products (such as urine)

► Figure 21.11 **Exchange between the external environment and the internal environment of complex animals.** The blue arrows indicate the exchange of materials between the circulatory system and three other systems and between the circulatory system and body cells.

☑ CHECKPOINT

Why doesn't a tapeworm require a circulatory system?

Answer: Most cells of a tapeworm are in direct contact with the environment.

Regulating the Internal Environment

Every living organism has the ability to respond to its environment. Some of these responses prevent changes in the outside world from adversely affecting the internal makeup of the organism. In this section, you'll learn how animals adjust to changing environmental conditions, looking in particular at how they regulate temperature and control the movement of water and solutes.

Homeostasis

One of the body's most important functions is to maintain its integrity even when the world around it changes. The internal environment of vertebrates consists of the **interstitial fluid** that fills the spaces between cells and exchanges nutrients and wastes with microscopic blood vessels (see Figure 21.11). It is important that the composition of the interstitial fluid remain relatively constant no matter what occurs in the outside world.

Homeostasis, which literally means "steady state," is the tendency to maintain relatively constant conditions in the internal environment even when the external

environment changes. In the face of large external changes, the mechanisms of homeostasis normally maintain internal conditions within a range that the animal's metabolism can tolerate (**Figure 21.12**). For example, your internal body temperature normally fluctuates by less than a degree regardless of the temperature in your surroundings.

BioFlix ™
Homeostasis

▼ Figure 21.12 **Homeostasis.**

External environment

Homeostatic mechanisms

Animal's internal environment

Small internal changes

Large external changes

Actually, the internal environment of an animal always fluctuates slightly in response to internal and external changes. Homeostasis is a dynamic state, an interplay between outside forces that tend to change the internal environment and internal control mechanisms that oppose such changes. Changes do occur, but they are generally moderated to stay within a range that is tolerable for living cells. Additionally, there are times during the development of an animal when major changes in the internal environment are programmed to occur. For example, the balance of hormones in human blood changes radically during puberty or pregnancy, and internal temperature may rise in response to a bacterial infection.

Negative and Positive Feedback

Most mechanisms of homeostasis depend on a principle called **negative feedback**, a form of regulation in which the results of a process inhibit that same process. **Figure 21.13** illustrates the concept of negative feedback with an example of household homeostasis, the control of room temperature. During the winter, a home heating system maintains a relatively constant temperature inside the house despite drastic changes that may occur outside. A key component of this system is a control

center—a thermostat—that monitors temperature and switches the heater on and off. Whenever room temperature drops below the set point (the temperature at which the thermostat is set, 20°C in the figure), the thermostat switches the heater on (bottom pathway in Figure 21.13). When the temperature rises above the set point, the thermostat switches the heater off (top pathway). This illustrates the basic principle of negative feedback: The result (increased room temperature, for example) of a process (heating of the air) inhibits that very process (by switching the heater off). Negative feedback is the most common mechanism of homeostatic control in animals.

Less commonly, organisms use **positive feedback**, in which the results of a process intensify that same process. For example, during the labor that leads up to childbirth, chemical signals cause the muscles of the uterus to contract. This contraction stimulates the release of more such chemicals, which cause more contractions, and so on, in a positive-feedback loop of increasing intensity. The result is climactic muscle contractions that push the baby from the womb.

In the rest of this section, we'll examine several specific examples of homeostasis. We'll first discuss the control of temperature and the control of water gain and loss in a variety of animals. We'll then focus on the human urinary system—an organ system that performs several homeostatic roles. ✔

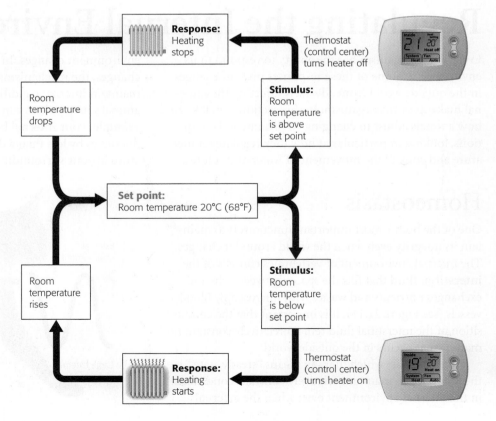

▶ Figure 21.13 **An example of negative feedback: control of room temperature.** This is an example of negative feedback because the result of the process (heat) shuts that process down.

Thermoregulation

As you saw in the Biology and Society section, the maintenance of internal body temperature is important in animals. This homeostatic mechanism is called **thermoregulation**. As you walk from one warm building to the next on a cold winter day, your body's temperature will barely fluctuate, despite the drastic change in temperature in the environment around you. The ability to maintain a body temperature substantially warmer than the surrounding environment is characteristic of **endotherms**, animals such as mammals and birds that derive most of their body heat from their metabolism. In contrast, **ectotherms**, which include most invertebrates, fishes, amphibians, and nonbird reptiles, obtain their body heat primarily by absorbing it from their surroundings.

Humans have a number of structures and mechanisms that aid in thermoregulation. As shown in **Figure 21.14**, our brain has a control center that maintains our body temperature near 37°C (98.6°F). When body temperature falls below normal (outside in frigid weather, for example), the brain's control center sends signals that trigger changes that will bring it back to normal: Blood vessels near the body's surface constrict (conserving heat) and muscles contract, causing us to shiver (producing heat). When body temperature gets too high, the control center sends signals to dilate the blood vessels near the skin and activate sweat glands, allowing excess heat to escape. Like a home controlled by a thermostat, the temperature of the body does not stay completely constant, but fluctuates up and down within an acceptable range.

Fever, an abnormally high internal temperature, is a body-wide response that usually indicates an ongoing fight against infection. When immune system cells encounter invading microbes, the cells release chemicals that travel through the bloodstream to the brain's thermostatic control center. These chemicals stimulate the control center to raise the body's internal temperature, producing a fever. Many people mistakenly believe that the invading microbes themselves cause a fever. In fact, the cause is usually the body's fight *against* the microbes. A fever over 40°C (104°F) may be life-threatening because a temperature that high can damage body proteins. A moderate fever of 38–39°C (100–102°F), however, discourages bacterial growth and speeds the body's internal defenses. A moderate fever thus helps protect the body's internal environment against potentially harmful invaders from the external environment.

We are endotherms and so can generate body heat to warm ourselves. Ectotherms generally cannot. But there are exceptions to this rule, as we'll see next in the Process of Science section. ✔

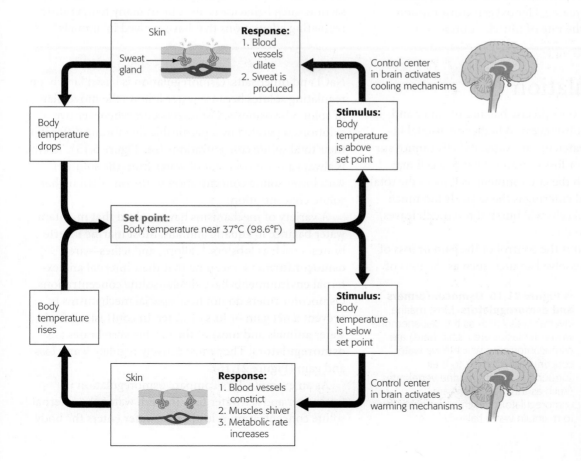

◄ **Figure 21.14
Thermoregulation in the human body.** By comparing this figure with Figure 21.13, you can see how similar negative-feedback controls moderate the temperatures of a room and your body. The control center for body temperature—our thermostatic control center—is located in the brain.

THE PROCESS OF SCIENCE: Thermoregulation

How Does a Python Warm Her Eggs?

As just described, ectotherms usually absorb heat from their surroundings. Under certain circumstances, however, some ectotherms can generate their own body heat. An example is the Burmese python (*Python molurus bivittatus*), a very large snake native to the rain forests of southeastern Asia. Like most snakes, it reproduces by laying eggs. Researchers at the Bronx Zoo, in New York City, made the **observation** that a female Burmese python incubating her eggs wraps her body around them, raises her body temperature, and frequently contracts the muscles in her coils. They formed the **hypothesis** that the muscle contractions elevate the snake's body temperature.

To test their hypothesis, they performed a simple **experiment**. They placed a python and her eggs in a chamber. As they varied the chamber's temperature, they monitored the python's muscle contractions and oxygen uptake, a measure of her rate of cellular respiration. Their **results** showed that the python's oxygen consumption increased when the temperature in the chamber decreased. Her oxygen consumption also changed with the rate of muscle contraction

▼ Figure 21.15 **Oxygen consumed by a Burmese python as it constricts its muscles.**

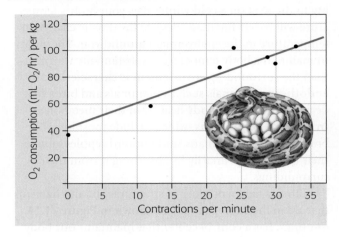

(Figure 21.15). Because oxygen consumption generates heat through cellular respiration and increases with the rate of muscle contraction, the researchers concluded that the muscle contractions were the source of the Burmese python's elevated body temperature during egg incubation. As you'll see in the Evolution Connection section, such behavior is just one of many temperature-regulating adaptations that have evolved in animals.

Osmoregulation

Living cells depend on a precise balance of water and solutes (dissolved substances). Whether an animal inhabits land, fresh water, or salt water, its cells cannot survive an excess gain or loss of water. That is, a cell may exchange water with the environment as long as the total amount leaving and entering is the same. If too much water enters, animal cells will burst; if too much leaves, they will shrivel and die.

Osmoregulation is the control of the gain or loss of water and dissolved solutes, such as the ions of

Osmoconformer

Osmoregulator

◄ Figure 21.16 **Osmoconformers and osmoregulators.** Most marine invertebrates (such as the sea anemone whose tentacles are visible here) are osmoconformers, which have water concentrations similar to their surroundings. Most marine vertebrates (such as this clown fish) are osmoregulators, which use mechanisms to maintain water balance.

NaCl and other salts. Osmoregulation is based largely on regulating solutes because water follows the movement of solutes by osmosis. Osmosis occurs whenever two solutions separated by a permeable membrane differ in their total solute concentrations (see Figure 5.13). There is always a net movement of water from the solution with lower solute concentration to the one with higher solute concentration.

A variety of mechanisms have evolved that maintain water balance in aquatic animals. Most marine invertebrates—such as lobsters, scallops, and jellies—are **osmoconformers**, meaning that their internal and external environments have similar solute concentrations. Osmoconformers do not need special mechanisms to prevent a net gain or loss of water. In contrast, all freshwater animals and most of the marine vertebrates are **osmoregulators**: They must actively regulate water loss and gain (Figure 21.16).

As an example, let's compare osmoregulation in freshwater and saltwater fish. In fresh water, the external solute concentration is low. Thus, water enters the body by osmosis. To compensate, freshwater fish take up solutes (such as ions) via the digestive system and gills and expel excess water by producing large amounts of dilute urine. In contrast, osmoregulators that live in salt water lose water by osmosis. In response, saltwater fish take in water by drinking, pump out ions via the gills, and produce only small amounts of concentrated urine.

All land animals are also osmoregulators. Most land animals gain water from eating and drinking and lose it through urinating, defecating, breathing, and perspiring. The primary challenge facing land-dwelling osmoregulators is to avoid becoming dehydrated. As we'll see in the next section, our kidneys play a major role in regulating water balance. ✔

Homeostasis in the Urinary System

The human urinary system plays a central role in forming and excreting waste-carrying urine while regulating the amount of water and solutes in body fluids. The main processing centers are the two kidneys. Each is a compact organ, a bit smaller than a fist, located on either side of the abdomen. The kidneys contain nearly 100 miles of thin tubes called **tubules** and an intricate network of capillaries (tiny blood vessels). Every day, the total volume of blood in the body passes through the capillaries of the kidneys hundreds of times. As the blood circulates, a fraction of it is filtered, and the plasma (the liquid portion) enters the kidney tubules. Once there, the plasma is called **filtrate**. The filtrate contains valuable substances that need to be reclaimed (such as water and glucose) and other substances that

need to be disposed of. Chief among the waste products to be excreted is urea, a nitrogen-containing compound produced from the breakdown of proteins and nucleic acids. Humans must dispose of urea, but we cannot simply excrete all of the filtrate as urine; if we did, we would lose vital nutrients and dehydrate rapidly. Instead, our kidneys refine the filtrate, concentrating the wastes and returning most of the water and useful solutes to the blood.

The anatomy of the human urinary system is shown in **Figure 21.17**. Starting with the whole system in part (a), blood to be filtered enters each kidney via a blood vessel called the renal artery (shown in red, leading from the large artery). Filtered blood leaves the kidney in the renal vein (shown in blue, leading to the large vein). Part (b) shows a cutaway view of a kidney. Within the kidney, the renal artery branches into millions of thin blood vessels. Part (c) shows one branch of the renal artery supplying blood to a nephron via a network of capillaries. A **nephron** consists of a tubule and its associated blood vessels. The nephron in the figure is enlarged to show its orientation within a section of the kidney; each kidney actually contains about a million nephrons.

The nephrons carry out the functions of the urinary system. Blood pressure forces water and solutes from the blood through a filter at the start of the nephron tubule, creating filtrate. As the filtrate then passes through the tubule, water and needed nutrients are reabsorbed into the bloodstream and wastes are secreted into the filtrate. The filtrate becomes more and more concentrated, resulting in a relatively small quantity of **urine**. At the end of the tubule, urine leaves the nephron via a collecting duct. Urine collects in the kidney and then leaves via the **ureter**. Urine is stored in the **urinary bladder** (visible in Figure 21.17a). Periodically, urine is expelled from the urinary bladder via the **urethra**, a tube that empties near the vagina in females and through the penis in males.

✔CHECKPOINT

Name two ways that water enters your body and four ways that water leaves your body.

Answer: Water enters via eating and drinking and leaves via breathing, sweating, urinating, and defecating.

Kidney Function

▼ Figure 21.17 **Anatomy of the human urinary system.**

Renal artery (red) and renal vein (blue)

Kidney

Ureter

Urinary bladder

Urethra

(a) Urinary system

Ureter

(b) Kidney

Filter

Tubule

Branch of renal artery

Branch of renal vein

Collecting duct

To ureter

(c) A nephron and collecting duct

Figure 21.18 summarizes the functions performed by each nephron. First, during **filtration**, water and other small molecules are forced from the blood through a capillary wall into the start of a kidney tubule, forming filtrate. Next, in **reabsorption**, water and valuable solutes are reclaimed from the filtrate and returned to the blood. In **secretion**, certain substances, such as some ions and drugs, are transported into the filtrate. What remains after filtration, reabsorption, and secretion is the urine. Finally, in **excretion**, urine passes from the kidneys to the outside.

The body controls its concentration of water and dissolved molecules through hormones that act on the kidney's nephrons. When the solute concentration of body fluids rises too high (indicating that not enough water is present), the brain increases levels of a hormone called ADH (antidiuretic hormone) in the blood. This hormone signals the nephrons to reabsorb more water from the filtrate, effectively increasing the body's water content. Conversely, when body fluids become too dilute, blood levels of ADH drop, the kidneys reabsorb less water, and the excreted urine becomes much more dilute. This is why your urine is very clear after you have been drinking a lot of water. Diuretics, such as alcohol, are substances that inhibit the release of ADH and therefore cause excessive urinary water loss. Drinking alcohol will make you urinate more frequently, and the resulting dehydration contributes to the symptoms of a hangover.

Kidney failure, the inability of the kidneys to filter blood, can be caused by injury, illness (such as high blood pressure or diabetes), or prolonged use of pain relievers, alcohol, or other drugs. Kidney failure can leave the body unable to rid itself of accumulated wastes. Patients with one functioning kidney can lead a normal life, but dual kidney failure is lethal if untreated. One option is to place the person on **dialysis**, filtration of the blood by a machine. A dialysis machine mimics the action of a nephron (**Figure 21.19**). Blood from an artery enters a series of selectively permeable tubes. Urea and excess fluids pass from the blood into a dialyzing solution, while necessary salts and other substances transfer the opposite way. The machine discards the used dialyzing solution as wastes accumulate. Dialysis treatment is life sustaining for people with kidney failure, but it is costly, takes a lot of time (about 4–6 hours three times a week), and must be continued for life. Alternatively, a kidney from a living compatible donor (usually a relative) or a deceased organ donor can be transplanted into a person with kidney failure. Unfortunately, the number of people who need kidneys is much greater than the number of kidneys available.

Throughout our study of animal form and function in this unit, you will see many other examples of how mechanisms of homeostasis moderate changes in the internal environment despite larger fluctuations in the external environment. You will also see many other examples of what happens when homeostatic controls fail. ☑

▼ **Figure 21.18** **Major functions of the urinary system.**

▼ **Figure 21.19** **A dialysis machine.**

Adaptations for Thermoregulation

Natural selection has promoted a wide variety of adaptations—anatomical, physiological, and behavioral—that help animals cope with heat gained from or lost to the environment (**Figure 21.20**). A major anatomical adaptation in mammals and birds is insulation, consisting of hair (often called fur), feathers, or fat layers. A thin coat of fur in the summer often becomes a thick, insulating coat during the winter. Also, most land mammals and birds react to cold by raising their fur or feathers; this adaptation traps a thicker layer of air next to the warm skin, improving the insulation. (In humans, our muscles raise our hair in the cold, causing goose bumps, a vestige from our furry ancestors.) Aquatic mammals (such as seals) and aquatic birds (such as penguins) have a thick insulating layer of fat called blubber.

Some adaptations are physiological. In cold weather, hormonal changes tend to boost the metabolic rate of birds and mammals, increasing their heat production. Simply moving around more or shivering also produces heat as a metabolic by-product of the contraction of skeletal muscles. Honeybees survive cold winters by clustering together and shivering in their hive. Their metabolic activity generates enough heat to keep the cluster alive. Panting and sweating are examples of physiological adaptations that greatly increase cooling.

A variety of behavioral responses can regulate body temperature. Some birds and butterflies migrate seasonally to more suitable climates. Other animals, such as desert lizards, bask in the sun when it is cold and find cool, damp areas or burrows when it is hot. Honeybees cool their hive during hot weather by transporting water into it and fanning with their wings. Many animals cool themselves by bathing, and a few even spread saliva on their bodies. Some mammals, such as ground squirrels and chipmunks, drastically reduce their body temperature and metabolic rate by hibernating in burrows during the winter, thereby conserving energy. All of these mechanisms—anatomical, physiological, and behavioral—demonstrate how a single selection pressure (the need to thermoregulate) can drive the evolution of a multitude of adaptations in a wide variety of environments.

▼ Figure 21.20 **Methods of thermoregulation in animals.**

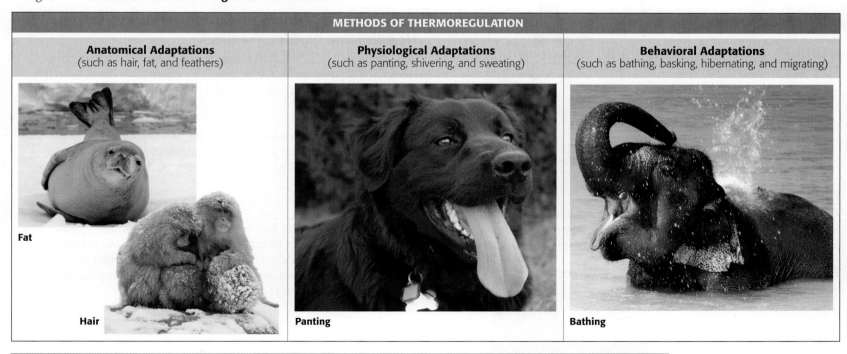

METHODS OF THERMOREGULATION		
Anatomical Adaptations (such as hair, fat, and feathers)	**Physiological Adaptations** (such as panting, shivering, and sweating)	**Behavioral Adaptations** (such as bathing, basking, hibernating, and migrating)
Fat / Hair	Panting	Bathing

Chapter Review

◾ SUMMARY OF KEY CONCEPTS

(MB) Go to the Study Area at **www.masteringbiology.com** for practice quizzes, eText, BioFlix™ 3-D animations, MP3 Tutor Sessions, videos, current events, and more.

The Structural Organization of Animals

HIERARCHICAL ORGANIZATION OF ANIMALS		
Level	**Description**	**Example**
Cell	The basic unit of all living organisms	Muscle cell
Tissue	A collection of similar cells that perform a specific function	Cardiac muscle
Organ	Multiple tissues forming a structure that performs a specific function	Heart
Organ system	A team of organs that work together	Circulatory system
Organism	A living being, which depends on the coordination of all structural levels for homeostasis and survival	Person

Form Fits Function

At every level, the structure of a body part is correlated with the task it must perform. Anatomy is the study of the structure of organisms, whereas physiology is the study of the function of an organism's structures.

Tissues

Animals have four main kinds of tissue.

Muscle (contracts)

Connective (supports organs)

Epithelial (covers body surfaces and organs)

Nervous (relays and integrates information)

Organs and Organ Systems

Figure 21.8 summarizes the 11 major human organ systems.

Exchanges with the External Environment

All animals are open systems, exchanging chemicals and energy with the environment. Every cell of a simple organism can exchange materials through direct contact with the environment. Large and complex body shapes require indirect exchange between extensively branched internal structures and the environment, usually via a circulatory system.

Regulating the Internal Environment

Homeostasis

Homeostasis is the body's tendency to maintain relatively constant internal conditions despite large fluctuations in the external environment.

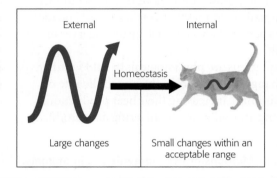

External   Internal

Homeostasis

Large changes   Small changes within an acceptable range

Negative and Positive Feedback

In negative feedback, the most common homeostatic mechanism, the results of a process inhibit that process. Less common is positive feedback, in which the results of a process amplify that process.

Thermoregulation

In thermoregulation, homeostatic mechanisms regulate internal body temperature. Endotherms are warmed primarily by heat generated during metabolism. Ectotherms are warmed primarily by heat absorbed from the environment.

Osmoregulation

All organisms balance the gain or loss of water and dissolved solutes. Osmoconformers do not tend to undergo a net gain or loss of water because their internal and external environments have similar solute concentrations. In contrast, osmoregulators must actively regulate their water balance—by drinking and urinating, for example.

Homeostasis in the Urinary System

The urinary system expels wastes and regulates solute and water balance. The following diagram summarizes the structure and function of the nephron, the functional unit of the kidney.

Blood

Capillary

Tubule

Filtration ➡
Water and small molecules enter the tubule.

Reabsorption ➡
Water and valuable solutes are returned to the blood.

Secretion ➡
Specific substances are removed from the blood.

Excretion ➡
Urine exits the body.

Urine

SELF-QUIZ

1. List as many units of structural hierarchy in your body as you can, in order from smallest to largest.

2. How might body shape affect whether an animal requires a circulatory system?

3. Which of the following best illustrates homeostasis?
   a. Most adult human beings are between 5 and 6 feet tall.
   b. The lungs and intestines have large surface areas for exchange.
   c. When the concentration of water in the blood goes up, the kidneys expel more water.
   d. When oxygen in the blood decreases, you may feel light-headed.

4. Most human cells are surrounded by an aqueous solution called _____.

5. Stimulation of a nerve cell causes sodium ions to leak into the cell, and the sodium influx triggers the inward leaking of even more sodium. This is an example of _____ feedback. An increase in the concentration of glucose in the blood stimulates the pancreas to secrete insulin, a hormone that lowers blood glucose concentrations. This is an example of _____ feedback.

6. What is the main difference between endotherms and ectotherms?

7. Like most vertebrates, you are an _____; your body must actively regulate the gain or loss of water. In contrast, a lobster is an _____, living in an environment with a solute concentration similar to that of its body.

8. Which of the following is *not* one of the homeostatic functions of the kidney?
   a. reabsorption
   b. filtration
   c. excretion
   d. ingestion

9. Drinking alcohol makes you urinate more frequently because
   a. alcohol contains caffeine.
   b. alcohol inhibits the release of ADH, a hormone that increases water reabsorption in the kidneys.
   c. alcohol inhibits the release of ADH, a hormone that decreases water reabsorption in the kidneys.
   d. alcohol causes more water to filter from the blood into the kidneys.

10. What happens to most of the water that passes from the blood into the kidneys by filtration? What happens to the rest of it?

Answers to Self-Quiz questions can be found in Appendix D.

THE PROCESS OF SCIENCE

11. Eastern tent caterpillars (*Malacosoma americanum*) live in sizable groups in silk nests, or tents, typically in cherry and apple trees. They emerge early in the spring, when daily temperatures fluctuate from freezing to hot. Early in the morning, these black caterpillars rest in a tightly packed group on the east-facing surface of the tent. In midafternoon, the group is found on the tent undersurface, each caterpillar hanging from the tent by a few of its legs. Propose a hypothesis to explain this behavior. How could you test it?

BIOLOGY AND SOCIETY

12. The kidneys remove many drugs from the blood, and these substances show up in the urine. Some employers require a urine drug test at the time of hiring and/or at intervals during employment. Should an employee have the right to refuse a urine test? Should every job be subject to mandatory drug testing, or just certain jobs? Which jobs should be subject to testing? Would you take a drug test for a job? Why or why not?

13. Kidneys were the first organs to be successfully transplanted. A donor can live a normal life with a single kidney, making it possible for individuals to donate a kidney to an ailing relative or even an unrelated individual. Should a person be allowed to sell a kidney? Why or why not? What are some ethical issues raised by organ commerce?

Nutrition and Digestion

An epidemic of obesity.
Overeating, improper nutrition, and lack of exercise all contribute to the high rate of obesity among Americans of all ages.

BIOLOGY AND SOCIETY: Obesity

Stomach Surgeries

Today, over a third of American adults are obese (very overweight), and obesity is estimated to contribute to 300,000 deaths per year in the United States alone. And the problem is not limited to one country: The United Nations World Health Organization (WHO) recognizes obesity as a major global health problem. Some people who are severely obese can shed pounds by diet and exercise. But others who cannot are candidates for weight loss surgery.

The most common weight loss surgery in the United States (with about 150,000 operations each year) is gastric bypass surgery. Gastric bypass promotes weight loss in two ways. First, staples are used to reduce the stomach to about the size of a chicken egg. As a result, patients quickly sense that their stomach is full, and they feel satisfied much more quickly. Second, an incision is made in the small intestine about 18 inches from where it joins the stomach. The free end is attached to the small stomach pouch. This causes food leaving the stomach to bypass some nutrient-absorbing regions of the digestive system, thereby reducing the body's ability to absorb food.

When accompanied by a healthy diet and regular exercise, weight loss surgeries are successful in 90% of patients. However, there is a risk of infection, and sometimes patients can suffer malnutrition if they do not take supplements that compensate for poor nutrient absorption in the regions affected by the surgery.

Controlling obesity through surgery underscores the relationship between the digestive system, diet, and health. Your body transforms food into building-block molecules and into the energy that powers your movements and thoughts. You are what you eat in the sense that your health and appearance depend on the quality of your diet and the proper functioning of your digestive system. This chapter focuses on essential concepts of digestion and nutrition, beginning with an overview of how animals process food. Along the way, we'll return to the topics of obesity and healthy eating as reminders of the importance of this subject to society and your life.

Overview of Animal Nutrition

Every mealtime reminds us that we must eat other organisms to acquire organic nutrients. Food provides the raw materials that animals need to build tissue and fuel cellular work. However, food primarily consists of large, complex molecules that are not in a form an animal's cells can use. Thus, the body must break down these nutrients (digest them) to make them useful.

Animal Diets

All animals eat other organisms, dead or alive, whole or by the piece. Beyond that generalization, however, animal diets vary extensively (**Figure 22.1**). **Herbivores**, such as cattle, gorillas, and sea urchins, feed mainly on plants or algae. **Carnivores**, such as lions, snakes, and spiders, mainly eat animals. **Omnivores**, such as crows, cockroaches, and humans, regularly eat animals as well as plants or algae. ☑

The Four Stages of Food Processing

Think of the fate of one of your meals—such as a slice of pizza—and you'll start to get a sense of the four stages of food processing: ingestion, digestion, absorption, and elimination. **Ingestion** is just another word for eating. You ingest pizza when you bite off a piece. **Digestion** is the breakdown of food to small broken-down nutrient molecules that the body can absorb. The tomato sauce on a pizza, for example, is broken down to simple sugars and amino acids. **Absorption** is the uptake of the small nutrient molecules by cells. For instance, amino acids made available by the breakdown of the cheese protein in pizza are absorbed by cells lining the small intestine and transferred to the bloodstream, which distributes them throughout the body. **Elimination** is the disposal of undigested materials left over from food.

▼ **Figure 22.1 Animal diets.** Animals can be classified into three broad groups according to their diet.

ANIMAL DIETS		
Herbivore (mainly eats plants or algae)	**Carnivore** (mainly eats animals)	**Omnivore** (regularly eats animals as well as plants or algae)

Digestion: A Closer Look

Digestion usually begins with **mechanical digestion**, physical processes such as chewing. Mechanical digestion breaks chunks of food into small pieces, exposing them to **chemical digestion**, the breakdown of food by digestive enzymes. Food molecules that are polymers, such as carbohydrates and proteins, are broken down via chemical digestion into monomers (see Chapter 3). For instance, starch is digested to its component glucose monomers.

The digestive dismantling of food molecules is necessary for two reasons. First, these molecules are too large to cross the membranes of animal cells; they must be broken down into molecules that are small enough for cells to absorb. Second, most food molecules—the proteins in cheese, for example—are different from the molecules that make up an animal's body. Your body does not directly use the protein that you eat but instead dismantles it and uses the pieces (amino acids) to build its own new proteins (**Figure 22.2**).

Chemical digestion proceeds via hydrolysis, chemical reactions that break down large biological molecules by using water molecules (see Figure 3.4b). Like most of life's chemical reactions, digestion requires enzymes (**Figure 22.3**). For example, lipases are enzymes that digest fats. ☑

☑CHECKPOINT

1. Place the four stages of food processing in their proper order: absorption, digestion, elimination, ingestion.
2. Food molecules are broken down into smaller molecules by enzymes that catalyze chemical reactions known as _____ reactions.

Answers: 1. ingestion, digestion, absorption, elimination 2. hydrolysis

▼ Figure 22.2 **From cheese protein to human protein.**

Cheese protein
(a polymer of amino acids in a specific sequence)

Amino acid monomer

① Breakdown of protein by human digestive system

Amino acids

② Absorption of amino acids by cells lining the small intestine; transport via bloodstream to other cells

Human protein

③ Cells use amino acids from the cheese and other foods to produce new human proteins

▼ Figure 22.3 **Chemical digestion: hydrolysis of food molecules.**
Digestive enzymes catalyze the breakdown of various types of food molecules by the addition of water molecules.

Protein

H_2O

Amino acid

Enzyme (pepsin)

Carbohydrate

H_2O

Sugar

Enzyme (amylase)

Fat

H_2O

Fatty acid

Glycerol

Enzyme (lipase)

Digestive Compartments

How do animals digest their food without digesting their own cells and tissues? After all, digestive enzymes hydrolyze the same kinds of biological molecules that make up the animal body, and it is obviously important to avoid digesting oneself! A common solution to this problem has evolved in animals: Chemical digestion proceeds safely within some kind of compartment.

Digestion within a cell is the simplest type. In this process, a cell engulfs food by phagocytosis, forming a vacuole. This food vacuole then fuses with a lysosome containing enzymes, forming a digestive compartment (see Figure 4.16a). As food is digested, small food molecules pass through the vacuole membrane into the cytoplasm and nourish the cell. This type of digestion is common in protists, but sponges are the only animals that digest food solely within their cells.

All other animals have digestive compartments that are surrounded by, rather than within, cells (**Figure 22.4**). These compartments allow for the digestion of food pieces that are much larger than a single cell. Simpler animals, including cnidarians (such as hydras and jellies) and flatworms, have a **gastrovascular cavity**, a compartment with a single opening that functions as both the entrance for food and the exit for undigested wastes.

The vast majority of animals, including earthworms and humans, have a digestive tube with two separate openings, a mouth at one end and an anus at the other. Such a tube is called an **alimentary canal**, or **digestive tract**. Food enters the mouth and moves through specialized regions that digest and absorb nutrients in a stepwise fashion. (You know what an assembly line is; an alimentary canal is a *dis*assembly line.) Undigested wastes are eliminated from the alimentary canal as feces via the anus. ☑

☑CHECKPOINT

What is the main difference between gastrovascular cavities and alimentary canals?

Answer: Gastrovascular cavities have just one opening; alimentary canals have two (mouth and anus).

▼ Figure 22.4 **Main types of digestive compartments in animals.**

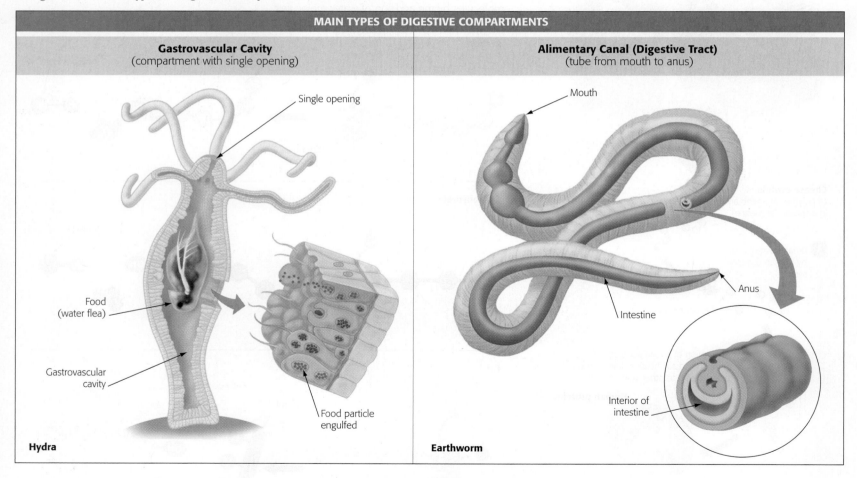

MAIN TYPES OF DIGESTIVE COMPARTMENTS

Gastrovascular Cavity
(compartment with single opening)

Single opening

Food
(water flea)

Gastrovascular
cavity

Food particle
engulfed

Hydra

Alimentary Canal (Digestive Tract)
(tube from mouth to anus)

Mouth

Anus

Intestine

Interior of
intestine

Earthworm

A Tour of the Human Digestive System

We are now ready to follow our slice of pizza through the human alimentary canal, from mouth to anus. It's important to have a good map of the human digestive system so we don't get lost in there.

System Map

The human digestive system consists of an alimentary canal or gut, and several accessory organs (salivary glands, pancreas, liver, and gallbladder) that secrete digestive chemicals into the canal via ducts (**Figure 22.5**). The human alimentary canal is about 9 m (30 ft) in length. This long tube fits within a human body because it folds back and forth over itself with many switchbacks. The alimentary canal is divided into specialized digestive organs along its length: mouth (oral cavity) → pharynx → esophagus → stomach → small intestine → large intestine (colon and rectum) → anus. You'll get a closer look at the structure and function of the digestive organs on our journey through the alimentary canal. As you reach each organ, look back at Figure 22.5 to find exactly where you are. First stop: the mouth.

The Mouth

The **mouth** (also known as the **oral cavity**) functions in ingestion (food intake) and the preliminary steps of digestion (**Figure 22.6**). Mechanical digestion begins before we even swallow, as our teeth cut, smash, and grind the food. Chewing makes food easier to swallow and exposes more food surface to digestive juices. Typically, an adult person has 32 teeth, which include bladelike incisors (for biting off pieces of food), pointed canines (ripping and tearing), and premolars and molars (crushing and grinding). Chemical digestion also begins in the mouth, with the secretion of saliva from three pairs of **salivary glands**. Saliva contains the digestive enzyme salivary amylase. This enzyme breaks down starch, a major ingredient in pizza crust.

The muscular **tongue** is very busy during mealtime. Besides tasting the food, the tongue shapes it into a ball and pushes the food ball to the back of the mouth. Swallowing moves food into the pharynx. ☑

The Human Digestive System

▼ Figure 22.6 **The human mouth.** The third set of molars is called the wisdom teeth. They do not appear in all people, and in some people they push against other teeth and must be removed.

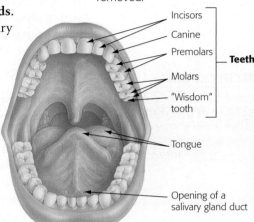

- Incisors
- Canine
- Premolars
- Molars
- "Wisdom" tooth

Teeth

- Tongue
- Opening of a salivary gland duct

► Figure 22.5 **The human digestive system.** The human digestive system consists of the alimentary canal (black labels) and accessory organs (blue labels).

ACCESSORY ORGANS

- **Salivary glands**
- **Liver**
- **Gallbladder**
- **Pancreas**

ALIMENTARY CANAL

- Mouth
- Tongue
- Pharynx
- Esophagus
- Stomach
- Small intestine
- Colon of large intestine
- Appendix
- Rectum
- Anus

The Pharynx

The chamber called the **pharynx**, located in your throat, is an intersection of the food and breathing pathways. The pharynx connects the mouth to the esophagus, but it also opens to the trachea, or windpipe, which leads to the lungs. When you're not swallowing, the trachea entrance is open, and you can breathe. When you swallow, a reflex moves the opening of the trachea upward and tips a door-like flap called the epiglottis to close the trachea entrance (**Figure 22.7**). Like a crossing guard at a dangerous intersection, the epiglottis directs the closing of the trachea, ensuring that the food will go down the esophagus and not the windpipe. You can see this action in the bobbing of your Adam's apple every time you swallow. Occasionally, food begins to "go down the wrong pipe," triggering a strong coughing reflex that helps keep your airway clear of food.

The Esophagus

The **esophagus** is a muscular tube that connects the pharynx to the stomach. Your esophagus moves food by **peristalsis**, alternating waves of muscular contraction and relaxation that squeeze the food ball along the esophagus (**Figure 22.8**). This muscular action ensures that you can swallow even when standing on your head. Peristalsis propels food throughout the length of the alimentary canal. ☑

☑CHECKPOINT

When we start coughing because food or drink "went down the wrong pipe," the material has entered the _____ instead of the _____.

Answer: trachea (windpipe); esophagus

▼ Figure 22.8 **The esophagus and peristalsis.** Muscles of the esophageal wall contract just behind the food ball and relax just ahead of it.

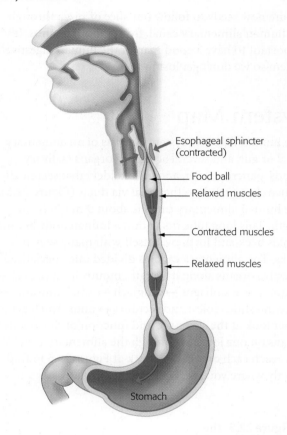

- Esophageal sphincter (contracted)
- Food ball
- Relaxed muscles
- Contracted muscles
- Relaxed muscles
- Stomach

▼ Figure 22.7 **The epiglottis and esophageal sphincter controlling swallowing in the pharynx.**

NOT SWALLOWING	SWALLOWING STARTED	SWALLOWING FINISHED

NOT SWALLOWING: Tongue, Ball of food, Epiglottis up, Pharynx, Esophageal sphincter (contracted), Trachea (windpipe) open, Esophagus closed

When you're not swallowing food or drink, air travels freely through the trachea (black arrows). The esophagus is closed because the esophageal sphincter, a ring of muscle, is contracted (blue arrows).

SWALLOWING STARTED: Epiglottis down, Adam's apple, Trachea closed, Esophagus open

When the food reaches the back of the mouth, a swallowing reflex is triggered. The top of the trachea rises against the epiglottis, closing the air passage, while the esophagus opens because the sphincter relaxes. Food travels down the esophagus (green arrow).

SWALLOWING FINISHED: Epiglottis up, Esophageal sphincter (contracted), Trachea open

When swallowing is completed, the esophageal sphincter contracts, and air once again flows freely through the trachea.

The Stomach

We do not have to eat constantly because the **stomach** is a large organ that can store enough food to sustain us for several hours (**Figure 22.9**). With its elastic wall and accordion-like folds, the stomach can stretch to hold about 2 liters (L)—more than half a gallon!—of food and water.

The cells lining the stomach's interior secrete a digestive fluid called **gastric juice**. Gastric juice is made up of strong acid, digestive enzymes, and mucus. The acid in gastric juice is hydrochloric acid, and it is concentrated enough to dissolve iron nails. Gastric juice also contains **pepsin**, an enzyme that digests proteins. Pepsin breaks proteins into smaller pieces.

When food passes from the esophagus into the stomach, the muscular stomach walls begin to churn, mixing the food and gastric juice into a thick soup called **chyme**. At the downstream end of the stomach, a sphincter (a ring of muscle) regulates the transfer of chyme from the stomach to the small intestine. With the chyme leaving the stomach only a squirt at a time, it takes about 2 to 6 hours for the stomach to empty after a meal, long enough for stomach acids and pepsin to begin digestion. (Continued contraction of stomach muscles after the stomach is empty causes the "stomach growling" that broadcasts hunger.)

With all of that acid, what keeps the stomach from digesting itself? Mucus coating the stomach lining helps protect it from gastric juices and from abrasive materials in the food. Timing is also a factor: Nerve signals and chemical signals called hormones regulate secretion of gastric juice so that it is discharged only when food is in the stomach. Even with these safeguards, gastric juice can still erode the stomach lining, requiring the production of new cells by cell division. In fact, the stomach lining is completely replaced about every three days.

Stomach Ailments

Gastric juice, strong enough to digest a diverse array of foods, can be harmful. Occasional backflow of chyme causes pain in the esophagus, commonly but inaccurately called "heartburn" (it should really be called "esophageal burn"). Some people suffer this backflow frequently and severely enough to harm the lining of the esophagus, a condition known as acid reflux or GERD (gastroesophageal reflux disease).

If the stomach lining is eroded by gastric juice faster than it can regenerate, painful open sores called gastric ulcers can form in the stomach wall. The cause of most ulcers is not stress, as was once thought, but infection of the stomach lining by an acid-tolerant bacterium called *Helicobacter pylori* (**Figure 22.10**). The metabolism of these bacteria damages the coat of mucus, making the lining more accessible to gastric juice. In severe ulcers, the erosion can produce a hole in the stomach wall and cause life-threatening internal bleeding and infection. (The cause of ulcers was established in 1984 when biologist Barry Marshall experimented on himself by drinking beef soup laced with *H. pylori* bacteria; although Marshall eventually won a 2005 Nobel Prize for his work, we do not recommend this kind of experimentation!) Stomach ulcers are treated with antibiotics. Affected people can also get relief by taking medications that contain bismuth, which helps reduce ulcer symptoms and may kill some bacteria. ☑

☑CHECKPOINT

How can antibiotics help treat gastric ulcers?

Answer: by killing the bacteria that damage the stomach lining

▼ Figure 22.9 **The stomach.**

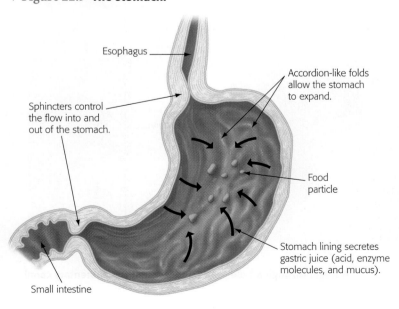

Esophagus

Sphincters control the flow into and out of the stomach.

Accordion-like folds allow the stomach to expand.

Food particle

Stomach lining secretes gastric juice (acid, enzyme molecules, and mucus).

Small intestine

▼ Figure 22.10 **Ulcer-causing bacteria.**

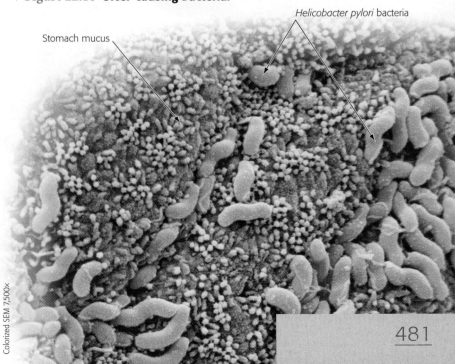

Helicobacter pylori bacteria

Stomach mucus

Colorized SEM 7,500x

The Small Intestine

The **small intestine**, at a length of about 6 m (about 20 ft), is the longest part of the alimentary canal. The small intestine is so named because of its relatively small diameter: 2.5 cm across versus twice that for the large intestine. It is the major organ for chemical digestion and for absorption of nutrients into the bloodstream.

Chemical Digestion in the Small Intestine

What kind of shape is our pizza in by the time it reaches the small intestine? So far, mechanical digestion has turned the meal into a thick, nutrient-rich soup. Chemical digestion by salivary amylase and gastric pepsin has initiated breakdown of the pizza's starch and proteins. Now the small intestine takes over with an arsenal of enzymes that dismantle the food molecules into smaller molecules. These enzymes are mixed with chyme in the first 25 cm or so (about a foot) of the small intestine, the region called the **duodenum**.

The duodenum receives digestive juices from the pancreas, liver, gallbladder, and the intestinal lining itself (**Figure 22.11**). The **pancreas** is a large gland that secretes pancreatic juice into the duodenum via a duct. Pancreatic juice neutralizes the stomach acids that enter the duodenum and contains enzymes that aid in digestion. As peristalsis propels the mix along the small intestine, these enzymes contribute to the breakdown of food molecules.

Bile is a juice produced by the **liver**, stored in the **gallbladder**, and secreted through a duct into the duodenum. Bile contains salts that bind to the droplets of fat that have been produced from larger globules by the contraction of muscles in the intestinal wall. The salts prevent the fat droplets from re-forming large globules. In droplet form, the fats are more accessible to lipase and therefore digested faster.

Bile sometimes crystallizes to form gallstones, which can cause pain by obstructing the gallbladder or its ducts. Often the only cure is surgical removal of the gallbladder, which usually has no long-lasting effect on digestion because the liver still produces and secretes bile.

Absorption of Nutrients

In the small intestine, the broken-down food molecules are ready for absorption into the body. Wait a minute! Aren't these nutrients already in the body? Not really. The alimentary canal is a tube running through the body, and its cavity is continuous with the great outdoors. The doughnut analogy shown in **Figure 22.12** should convince you that this is so. Until nutrients actually cross the tissue lining the alimentary canal to enter the bloodstream, they are still *outside* the body. If it were

▼ Figure 22.11 **The duodenum.** Chyme squirted from the stomach into the duodenum is mixed with pancreatic juice, bile from the liver and gallbladder, and intestinal juice produced in the duodenum itself.

▼ Figure 22.12 **Nutrients within the small intestine are not yet inside the body.** Imagine an elongated doughnut that's analogous to the "tube-within-a-tube" anatomy of the human alimentary canal. Food does not actually enter the body until it is absorbed by cells lining the alimentary canal.

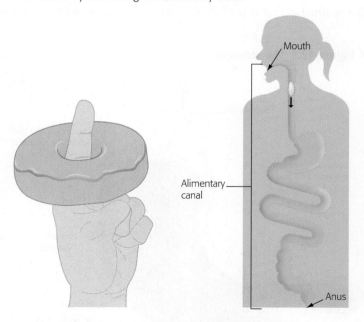

A finger through a hole   **Food through the alimentary canal**

not for nutrient absorption, we could eat and digest huge meals but still starve.

Most digestion is complete by the time the pizza meal reaches the end of the duodenum. The next several meters of the small intestine are specialized for nutrient absorption. The structure of the intestinal lining, or epithelium, fits this function **(Figure 22.13)**. The surface area of this epithelium is huge—roughly 300 m^2, about equal to the size of a tennis court. The intestinal lining not only has large folds, like the stomach, but also has finger-like outgrowths called villi (singular, villus), making the epithelium something like the absorptive surface of a fluffy bath towel. Each cell of the epithelium adds even more surface by having microscopic projections called microvilli. Across this expansive surface of intestinal epithelium, nutrients are transported into the network of small blood vessels and lymphatic vessels in the core of each villus.

After nutrients have crossed the cell membranes of the microvilli, they are finally *inside* the body, where the bloodstream and lymph carry them away to distant cells. But our tour of the digestive system is not yet over, for we still have to make it through the large intestine. ☑

▶ **Figure 22.13 The small intestine and nutrient absorption.** Folds with projections (villi) having even smaller projections (microvilli) give the small intestine an enormous surface area for nutrient absorption (black arrows). Most nutrients are transported across the epithelium of microvilli into blood capillaries.

Blood vessels

Interior of intestine

Muscle layers

Intestinal wall

Villi

Nutrient absorption

Interior of intestine

Nutrient absorption

Microvilli

Nutrient absorption

Epithelial cells

Blood capillaries

Blood

Lymphatic vessel

Epithelial cells and blood capillary

Villi

The Large Intestine

At only 1.5 m in length, the **large intestine**, the main portion of which is called the **colon**, is shorter than the small intestine but is called "large" because it is almost twice as wide (about 5 cm). Where the two organs join, a sphincter controls the passage of what's left of a meal (**Figure 22.14**). Nearby is a small finger-like extension called the **appendix**. If the junction between the appendix and the colon becomes blocked, appendicitis—a bacterial infection of the appendix—may result. Emergency surgery is usually required to remove the appendix and prevent the spread of infection.

The main function of the colon is to absorb water from the alimentary canal. About 7 L of water per day spill into your alimentary canal as the solvent of digestive juices. About 90% of this water is absorbed back into your blood and tissue fluids, with the small intestine reclaiming much of the water and the colon finishing the job. As water is absorbed, undigested materials from the meal become more solid as they are conveyed along the colon by peristalsis. The end product is **feces**, consisting of undigested material (such as cellulose from vegetables on a pizza). The feces also contain enormous numbers of intestinal prokaryotes, normal inhabitants of the colon. Some colon bacteria, such as *Escherichia coli*, produce several B vitamins and vitamin K. These vitamins supplement your diet when absorbed into the bloodstream through the colon wall.

If the lining of the colon is irritated by a viral or bacterial infection (sometimes mistakenly called "stomach bugs"), diarrhea may result because the colon cannot reabsorb water efficiently. Prolonged diarrhea can cause life-threatening dehydration, particularly among the very young and very old. The opposite problem, constipation, occurs when peristalsis moves feces along too slowly and the colon reabsorbs so much water that the feces become too compacted. Constipation can result from lack of exercise or from a diet that does not include enough plant fiber.

Several intestinal disorders are characterized by inflammation (painful swelling) of the intestinal wall. Celiac disease results when gluten, a protein found in wheat, triggers an immune reaction that leads to swelling and a lack of nutrient absorption. The only treatment is a lifelong gluten-free diet. An inappropriate immune response also causes Crohn's disease, a chronic inflammation that can periodically flare up along any part of the alimentary canal. Investigating the cause and treatment of Crohn's disease is an active area of medical research.

The **rectum**, the last 15 cm (6 in) of the large intestine, stores feces until they can be eliminated. Contractions of the colon and the rectum create the urge to defecate. Two rectal sphincters, one voluntary and the other involuntary, regulate the opening of the **anus**. When the voluntary sphincter is relaxed, contractions of the rectum expel feces.

From entrance to exit, we have now followed a pizza slice all the way through the alimentary canal. **Figure 22.15** stretches the tube out to help you review food processing along its length. ☑

▼ Figure 22.14
The large intestine and its connection to the small intestine.

Colon of large intestine

Small intestine

Rectum

Anus

Sphincter

End of small intestine

Nutrient flow

Appendix

▼ Figure 22.15 **Review of food processing in the human alimentary canal.**

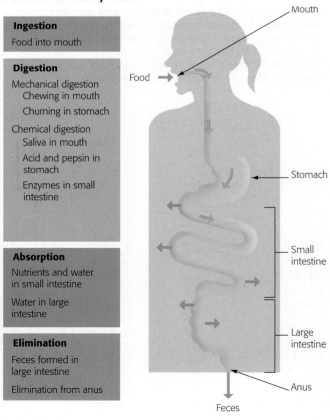

Ingestion
Food into mouth

Digestion
Mechanical digestion
  Chewing in mouth
  Churning in stomach
Chemical digestion
  Saliva in mouth
  Acid and pepsin in stomach
  Enzymes in small intestine

Absorption
Nutrients and water in small intestine
Water in large intestine

Elimination
Feces formed in large intestine
Elimination from anus

Mouth

Food

Stomach

Small intestine

Large intestine

Anus

Feces

Human Nutritional Requirements

For any animal, proper nutrition helps maintain homeostasis. A balanced diet provides fuel for cellular work, materials for building molecules, and essential nutrients for health. A proper human diet is rich in whole grains, vegetables, fruits, and calcium, along with moderate quantities of protein-rich lean meat, eggs, nuts, and beans. Doctors and nutritionists recommend limited consumption of fats, sugars, and salt.

Food as Fuel

Cells can extract energy stored in the organic molecules of food through the process of cellular respiration (described in detail in Chapter 6) and use that energy to do work. Using oxygen, cellular respiration breaks down sugar and other food molecules. This process generates many molecules of ATP for cells to use as a direct source of energy and releases carbon dioxide and water as waste "exhaust" (**Figure 22.16**).

Calories

Calories are a measure of the energy stored in your food as well as the energy you expend during daily activities. One **calorie** is the amount of energy required to raise the temperature of a gram of water by 1°C. But such a tiny amount of energy is not very useful on the human scale. We can scale up with the **kilocalorie** (**kcal**), which is a thousand calories. Now, for a wrinkle in these definitions: The "Calories" (with an uppercase C) listed on food labels are actually kilocalories, not calories. So the 280 or so Calories in a slice of thick-crust pepperoni pizza are actually 280 kcal. That's a whole lot of fuel for making ATP. However, about 60% of our food energy is lost as heat that dissipates to the environment. You know this effect from being in a crowded room; each human produces as much heat as a 100-watt light bulb.

Metabolic Rate

How fast do we "burn" our food? The rate of energy consumption by the body is called **metabolic rate**. Your metabolic rate is equal to your **basal metabolic rate** (**BMR**), the amount of energy it takes just to maintain your basic body functions, plus any additional energy consumption above that base rate. BMRs for humans average about 1,300–1,500 kcal per day for adult females and 1,600–1,800 kcal per day for adult males. The more active you are, the greater your actual metabolic rate and the greater the number of calories your body uses per day. Metabolic rate also depends on other factors such as body size, age, stress level, and heredity. The examples in **Table 22.1** give you an idea of the amount of activity that it takes to use up the kilocalories in several common foods.

The maintenance of body weight is a simple exercise in caloric arithmetic. You add calories in the form of food and subtract them in the form of energy expended. If you take in more by eating than you burn by activity, you will gain weight. If you want to lose weight, you must decrease the number of calories coming in and/or increase the number of calories expended. Despite a multitude of fad diets, losing weight comes down to a simple formula: Eat less and exercise more! Caloric balance, however, does not ensure good nutrition, and food is our source of substance as well as our source of energy. ☑

▲ Figure 22.16 **Review of cellular respiration.** Within each mitochondrion, a series of chemical reactions extracts energy from fuel molecules and generates ATP, which drives most cellular work.

Table 22.1	Exercise Required to Burn the Calories (kcal) in Common Foods		
	Jogging	**Swimming**	**Walking**
Speed	9 min/mi	30 min/mi	20 min/mi
kcal "burned" per hour*	775	408	245
Cheeseburger (quarter-pound), 417 kcal	32 min	1 hr, 1 min	1 hr, 42 min
Pepperoni pizza (1 large slice), 280 kcal	22 min	42 min	1 hr, 8 min
Soft drink (12 oz), 152 kcal	12 min	22 min	37 min
Whole wheat bread (1 slice), 65 kcal	5 min	10 min	16 min
*These data are estimated for a person weighing 68 kg (150 lb).			

Food as Building Material

Even if you have stopped growing, your health depends on the continuous repair and maintenance of your tissues. The building materials required for such work are provided by the small organic molecules produced during the digestion of food. Your cells can reassemble those smaller molecules into various large biological molecules, such as the proteins and DNA of your own "brand."

Within limits, your metabolism can change organic material from one form to another and compensate for disparities in nutrient supply and demand. For instance, if a cell has a shortage of a particular amino acid, it may be able to make that amino acid from an excess supply of another amino acid. However, certain substances—the **essential nutrients**—cannot be made from any other materials, so the body needs to receive them in preassembled form. A healthful diet must include adequate amounts of all of the essential nutrients: essential amino acids, vitamins, minerals, and essential fatty acids. ☑

☑CHECKPOINT

What is an essential nutrient?

Answer: a substance that an animal requires but cannot make

Essential Amino Acids

All proteins are built from 20 different kinds of amino acids (see Figure 3.16). Twelve of the 20 amino acids can be manufactured by the adult body from other compounds. The other eight are **essential amino acids**: They must be obtained from the diet because human cells cannot make them. (Infants also require a ninth, histidine.)

Different foods contain different proportions of amino acids. Animal proteins, such as those in meat, eggs, and milk, are said to be "complete" because they provide adequate amounts of all the essential amino acids. In contrast, most plant proteins are incomplete, meaning they are deficient in one or more of the essential amino acids. If you are a vegetarian (by choice or, as for much of the world's population, by economic necessity), the key to good nutrition is to eat a variety of plants, each of which provides different essential amino acids. The combination of a grain and a legume (examples of legumes are beans, peanuts, and peas) often provides the right balance (Figure 22.17). Most societies have a staple meal that includes such a combination.

▼ Figure 22.17 **Essential amino acids from a vegetarian diet.**

Essential amino acids

Methionine
Valine
Threonine
Phenylalanine
Leucine
Isoleucine
Tryptophan
Lysine

Corn

Beans and other legumes

Essential amino acids. An adult human can obtain all eight essential amino acids by eating a meal of corn and beans.

Rice and beans

Bread and peanut butter

Rice and tofu

Complete meals. Many societies have a staple meal that provides a complete set of essential amino acids by combining a grain with a legume.

Vitamins

Organic molecules that are required in the diet in very small amounts are called **vitamins**. For example, one tablespoon of vitamin B_{12} can provide the daily requirement of nearly a million people. There are 13 vitamins essential to human health. Most are needed because they assist enzymes. Deficiencies in any essential vitamin can cause serious health problems (**Table 22.2**). A lack of vitamin C, for example, causes scurvy. (To avoid this problem during long sea voyages, British sailors ate limes and other citrus fruits, earning them the nickname "limeys.")

People who eat a balanced diet should be able to obtain enough of all needed nutrients in their food. For others, vitamin supplements can fill in the gaps, but supplements should not be used indiscriminately because overdoses of some vitamins (such as A, D, and K) can be harmful. ✔

Table 22.2	Vitamins	
Vitamin	**Major Dietary Sources**	**Symptoms of Deficiency or** Extreme Excess*
Water-Soluble Vitamins		
Vitamin B_1 (thiamine)	Pork, legumes, peanuts, whole grains	Beriberi (nerve disorders, emaciation, anemia)
Vitamin B_2 (riboflavin)	Dairy products, meats, enriched grains, vegetables	Skin lesions such as cracks at corners of mouth
Niacin	Nuts, meats, grains	Skin and gastrointestinal lesions, nervous disorders Flushing of face and hands, liver damage
Vitamin B_6 (pyridoxine)	Meats, vegetables, whole grains	Irritability, convulsions, muscular twitching, anemia Unstable gait, numb feet, poor coordination
Pantothenic acid	Most foods: meats, dairy products, whole grains, etc.	Fatigue, numbness, tingling of hands and feet
Folic acid (folacin)	Green vegetables, oranges, nuts, legumes, whole grains	Anemia, gastrointestinal problems, birth defects Masks deficiency of vitamin B_{12}
Vitamin B_{12}	Meats, eggs, dairy products	Anemia; nervous system disorders
Biotin	Legumes, other vegetables, meats	Scaly skin inflammation; neuromuscular disorders
Vitamin C (ascorbic acid)	Fruits and vegetables, especially citrus fruits, broccoli, cabbage, tomatoes, green peppers	Scurvy (degeneration of skin, teeth, blood vessels), weakness, delayed wound healing, impaired immunity Gastrointestinal upset
Fat-Soluble Vitamins		
Vitamin A	Dark green and orange vegetables and fruits, dairy products	Vision problems; dry, scaly skin Headache, irritability, vomiting, hair loss, blurred vision, liver and bone damage
Vitamin D	Dairy products, egg yolk; also made in human skin in presence of sunlight	Rickets (bone deformities) in children; bone softening in adults Brain, cardiovascular, and kidney damage
Vitamin E (tocopherol)	Vegetables, oils, nuts, seeds	None well documented
Vitamin K	Green vegetables, tea	Defective blood clotting Liver damage and anemia

*Symptoms of extreme excess are given in red type.

☑ **CHECKPOINT**

Why do vegetarians have to be particularly careful to get proper amounts of the essential amino acids?

Answer: because few plant products contain all of them

Minerals

The organic molecules in our diet, such as carbohydrates, fats, and proteins, provide the four chemical elements most abundant in our body: carbon, oxygen, hydrogen, and nitrogen. We also require smaller amounts of 21 other chemical elements (see Figure 2.2) that are acquired mainly in the form of inorganic nutrients called **minerals**. Like vitamin deficiencies, mineral deficiencies can cause health problems. For example, calcium is needed as a building material for our bones and teeth and for the proper functioning of nerves and muscles. It can be obtained from dairy products, legumes, and dark green vegetables such as spinach. Too little calcium can result in the degenerative bone disease osteoporosis.

Mineral excesses can also cause problems. For example, we require sodium for our nerves and muscles to function, but the average American consumes about 20 times the required amount of sodium, mainly in the form of salt (sodium chloride) added to processed foods. Excess sodium can contribute to high blood pressure.

Essential Fatty Acids

Our cells make fats and other lipids by combining fatty acids with other molecules, such as glycerol (see Figure 3.11). We can make most of the required fatty acids, in turn, from simpler molecules; those we cannot make, called **essential fatty acids**, we must obtain in our diet. One essential fatty acid, linoleic acid (one of the omega-6 family of fatty acids), is especially important because it is needed to make some of the phospholipids of cell membranes. Most diets furnish ample amounts of the essential fatty acids, and therefore deficiencies are rare.

Decoding Food Labels

To help consumers assess the nutritional value of packaged foods, the U.S. Food and Drug Administration (FDA) requires two blocks of information on labels (**Figure 22.18**). One lists the ingredients (by weight) from greatest amount to least. The other lists key nutrition facts. You should note the serving size and adjust the rest of the nutrition information to reflect your actual serving. For example, if the serving size is 1 cup of cereal but you're eating 2 cups, double all of the other values. You'll see a wide variety of data—including Calories, fat, cholesterol, carbohydrates, fiber, protein, vitamins, and minerals—expressed as amounts per serving and as percentages of a daily value. (Daily values are based on a 2,000-Calorie-per-day diet; these values are therefore "one-size-fits-all" numbers that should be used only as rough guidelines.)

Some nutrients (such as fats, cholesterol, sugar, and sodium) are included because there is evidence that excesses contribute to health problems. Others (such as fiber, protein, vitamins, and minerals) are included because adequate amounts are necessary for good health. Reading food labels can help you make informed choices about what you put in your body. ☑

▶ Figure 22.18 **FDA-required food labels.**

Nutritional Disorders

Considering the central role that nutrients play in a healthy body, it is not surprising that nutritional dysfunction can cause severe problems. In this section, we examine some common nutritional disorders.

Malnutrition

Living in an industrialized country where food is plentiful and most people can afford a decent diet, we find it hard to relate to starvation. But 800 million people around the world—nearly three times the population of the United States—must cope with hunger. As difficult as it is to imagine, 14,000 children under the age of 5 starve to death *each day*. The main type of nutritional deficiency is malnutrition.

Malnutrition refers to health problems caused by an improper or insufficient diet. Dietary deficiencies may be caused by inadequate intake or medical problems (such as metabolic or digestive abnormalities). On a global scale, protein deficiency—insufficient intake of one or more essential amino acids—causes the most suffering, mainly in less industrialized countries.

Protein deficiency is concentrated in geographic regions where there is a great gap between food supply and population size. The most reliable sources of essential amino acids are animal products, but these foods are expensive. People forced by economic necessity to get almost all their calories from a single plant staple, such as corn or potatoes, will suffer deficiencies of essential amino acids.

Most victims of protein deficiency are children, who are likely to develop poorly both physically and mentally—assuming they survive infancy. The resulting syndrome is called kwashiorkor, from the Ghanaian word for "rejected one," a reference to the onset of the disease when a child is weaned from its mother's milk and placed on a starchy diet after a sibling is born (**Figure 22.19**). The problem of protein deficiency in some less-industrialized countries has been compounded by a trend away from breast-feeding altogether.

Eating Disorders

Millions of Americans, mostly female, are affected by malnutrition because of an eating disorder. **Anorexia nervosa** is characterized by self-starvation due to an intense fear of gaining weight, even when the person is underweight.

Bulimia is a behavioral pattern of binge eating followed by purging through induced vomiting, abuse of laxatives, or excessive exercise. Both disorders are characterized by an obsession with body weight and shape and can result in serious health problems.

The causes of anorexia nervosa and bulimia are unknown, but certain people may be more prone to develop eating disorders than others because of differences in genetics, psychology, and brain chemistry. Culture also seems to be a factor: Anorexia nervosa and bulimia occur almost exclusively in affluent industrialized countries, where food is plentiful but thinness is idealized. Treatment options include counseling and antidepressant medications. Some people eventually develop healthy eating habits without any treatment at all. For others, dysfunctional nutrition becomes a long-term problem that can impair health and even lead to death. ☑

☑ **CHECKPOINT**

What is the difference between anorexia nervosa and bulimia?

Answer: Anorexia nervosa is self-starvation; bulimia is a pattern of binge eating and purging.

▼ Figure 22.19 **Kwashiorkor in a Sudanese boy.** A form of malnutrition caused by inadequate protein intake, kwashiorkor causes deficiency in blood proteins. This deficiency, in turn, causes fluid to enter the abdominal cavity, producing swelling of the belly.

Obesity

In the United States and many other industrialized countries, overnourishment is the nutritional disorder of greatest concern. **Obesity** is defined as a too-high **body mass index (BMI)**, a ratio of weight to height. About one-third of all Americans are obese, and another one-third are overweight (a BMI that is between normal and obese). Obesity increases the risk of heart attack, diabetes, and several other diseases.

It is important to realize that not being as slim or well toned as a magazine model does not mean that you are obese. Researchers continue to debate how heavy we can be before we are considered unhealthy. And keep in mind that BMI is only an approximation that does not, for example, take into account body mass from muscle versus body mass from fat. Reflecting this uncertainty, BMI charts show a range of acceptable values (**Figure 22.20**). Further complicating matters, a tendency toward obesity is inherited to some extent, as we'll see next. ☑

CHECKPOINT

What BMI category is a person who is 6 feet 4 inches in height and weighs 230 pounds?

Answer: overweight

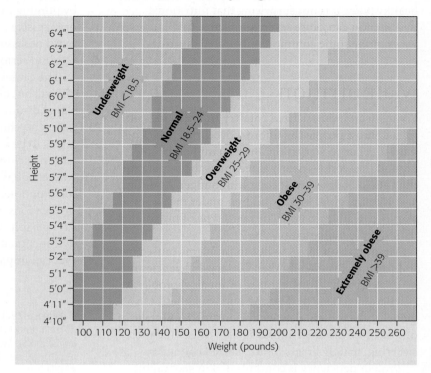

▼ Figure 22.20 **Body mass index (BMI): one measure of healthy weight.**

Underweight BMI <18.5
Normal BMI 18.5–24
Overweight BMI 25–29
Obese BMI 30–39
Extremely obese BMI >39

Height: 6'4", 6'3", 6'2", 6'1", 6'0", 5'11", 5'10", 5'9", 5'8", 5'7", 5'6", 5'5", 5'4", 5'3", 5'2", 5'1", 5'0", 4'11", 4'10"

Weight (pounds): 100 110 120 130 140 150 160 170 180 190 200 210 220 230 240 250 260

THE PROCESS OF SCIENCE: Obesity

Can a Gene Make You Fat?

Several decades ago, researchers made the **observation** that a mutation in a particular gene in mice leads to a significant increase in body fat (**Figure 22.21**). An obvious **question** about this so-called *obese* gene is: How does a small change in DNA cause such a large change in the body? Researchers at Rockefeller University, in New York City, formed the **hypothesis** that the mutant mice become overweight because their defective *obese* gene fails to produce a protein called leptin, which helps control body weight. They made the **prediction** that injecting leptin into mutant mice would overcome the effects of the defective *obese* gene.

Their **experiment** is summarized in **Figure 22.22**. The researchers injected some mutant and normal mice with leptin and injected others with saline (as a control).

▼ Figure 22.21 **A ravenous rodent.** The obese mouse on the left has a defect in a gene that normally produces an appetite-regulating protein.

Their **results** were striking: Mutant mice that received leptin ended the study weighing about half as much as those that received saline (compare the top two bars in Figure 22.22). Normal mice receiving leptin lost weight compared with those receiving saline, but the difference was not as large (compare the bottom two bars). The researchers concluded that leptin helps regulate body weight in mice.

How do these results apply to humans? Although leptin is present in human blood, the *obese* gene is only one of over 100 genes that contribute to a complex weight maintenance system in humans. So far, this complexity has prevented formulation of effective drugs to treat obesity. More importantly, genetics only partially explains why certain people must fight hard to control their weight, whereas others can eat all they want without gaining a pound. Genetic defects can't produce obesity without a large intake of fattening foods. The best way to maintain a healthy weight is to eat a balanced diet and get plenty of exercise.

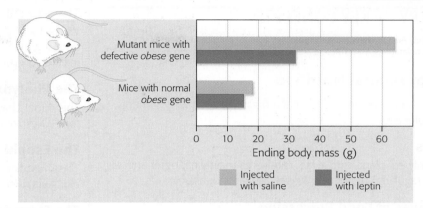

▼ Figure 22.22 **The effect of leptin on mice with the *obese* gene.** This graph shows the final average weight of normal and mutant mice after receiving injections of saline (orange bars) and leptin (blue bars).

Mutant mice with defective *obese* gene

Mice with normal *obese* gene

Ending body mass (g)

Injected with saline   Injected with leptin

EVOLUTION CONNECTION: Obesity

Fat and Sugar Cravings

The majority of Americans consume too many high-calorie foods, which contribute to obesity. Even though we all know that we need to limit the amount of fats and sugars in our diet, it sure is hard, isn't it? Most of us crave fatty or sweet foods: cheeseburgers, chips, fries, ice cream, candy. For many of us, such foods are satisfying in a way that other foods are not.

Why do we hunger for unhealthy foods? The seemingly unhelpful trait of craving fat and sugar makes more sense from a broad evolutionary standpoint. In the natural world, overnourishment is usually an advantage. It is only within the last 100 years that large numbers of people have had access to a reliable supply of food. For most of human history, our ancestors were continually in danger of starvation. On the African savanna, fatty or sweet foods were probably hard to find. In such a feast-or-famine existence, natural selection may have favored individuals who gorged themselves on the rare occasions when rich, fatty foods were available. With their ample reserves, they were more likely than thinner peers to survive famines.

Perhaps our modern taste for fats and sugars reflects the selection advantage it conveyed in our evolutionary history. Of course, today most people typically hunt and gather in grocery stores, restaurants, and cafeterias (**Figure 22.23**). Although we know it is unhealthful, many of us find it difficult to overcome the ancient survival behavior of stockpiling for the next famine.

▼ Figure 22.23 **Modern hunter-gatherers.**

Chapter Review

SUMMARY OF KEY CONCEPTS

Overview of Animal Nutrition

Animals must eat other organisms to obtain organic molecules for energy and building materials.

Animal Diets

Herbivores mainly eat plants and algae, carnivores mainly eat other animals, and omnivores regularly eat animals as well as plants or algae.

The Four Stages of Food Processing

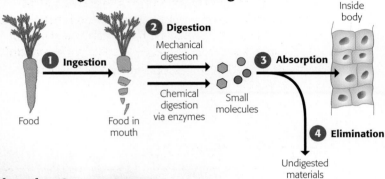

Digestive Compartments

In animals, hydrolysis of food occurs within food vacuoles (membrane-enclosed compartments within cells), a gastrovascular cavity (an extracellular compartment with a single opening), or an alimentary canal, also called a digestive tract (an extracellular compartment with two separate openings: a mouth and an anus).

A Tour of the Human Digestive System

System Map

Alimentary canal	Accessory organs	Digestion		Absorption
		Mechanical	Chemical	
Mouth (oral cavity)	Salivary glands	Chewing	Salivary amylase	
Pharynx and esophagus				
Stomach		Churning	Acid and pepsin (in gastric juice)	
Small intestine	Liver, gallbladder, pancreas		Other enzymes	Nutrients and water
Large intestine				Water
Anus				

The Mouth

Within the mouth (oral cavity), teeth function in mechanical digestion, breaking food into smaller pieces. An enzyme in saliva, salivary amylase, initiates chemical digestion by starting the hydrolysis of starch.

The Pharynx

This throat chamber leads to both the digestive and respiratory systems. A food-swallowing reflex closes the trachea and opens the esophagus.

The Esophagus

This muscular tube connects the pharynx to the stomach and moves food along via rhythmic muscular contractions called peristalsis.

The Stomach

This elastic organ stores food, functions in mechanical digestion by churning, and uses gastric juice for chemical digestion. Mucus secreted by the cells of the stomach lining helps protect it from self-digestion. Sphincters regulate the passage of food into and out of the stomach. A bacterial infection of the stomach lining can cause ulcers.

The Small Intestine

Small in diameter but very long, the small intestine is the main organ of chemical digestion and nutrient absorption. Most chemical digestion occurs in the duodenum, the first part of the small intestine, where chyme from the stomach mixes with pancreatic juice, bile, and a digestive juice secreted by the intestinal lining. These digestive juices include enzymes for dismantling large food molecules into small molecules. The rest of the small intestine is specialized for absorbing these small molecules. Across the large surface area of the intestinal lining, with its microvilli upon villi, intestinal cells transport nutrients into capillaries of the circulatory system.

The Large Intestine

The colon, wider and shorter than the small intestine, makes up most of the large intestine. The large intestine finishes reclaiming water that entered the alimentary canal as the solvent of digestive juices. When undigested wastes reach the rectum, most water has been reabsorbed into the blood. Feces are stored in the rectum until eliminated via the anus.

Human Nutritional Requirements

Food as Fuel

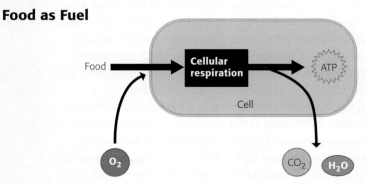

Our metabolic rate depends on basal metabolic rate (energy expenditure at complete rest) plus additional energy expenditure due to activity level.

Food as Building Material

To repair and maintain tissues, we need building blocks from the breakdown of organic molecules. We also need four types of preassembled essential nutrients.

ESSENTIAL NUTRIENTS			
Essential Amino Acids	**Vitamins**	**Minerals**	**Essential Fatty Acids**
Required for protein production	Organic molecules required in very small amounts	Essential chemical elements from inorganic compounds	Required to make cell membranes

Decoding Food Labels

The U.S. Food and Drug Administration requires labels on packaged foods that list ingredients in descending order of abundance and provide information about Calories and specific nutrients.

Nutritional Disorders

Malnutrition

Malnutrition is a deficiency of one or more essential nutrients, usually essential amino acids (protein deficiency). Protein deficiency is the most common dietary deficiency in less industrialized countries. Malnutrition can be caused by anorexia nervosa (self-starvation) or bulimia (binge eating and purging).

Obesity

Defined as an inappropriately high ratio of weight to height, obesity is the most common nutritional disorder in most industrialized countries. About one-third of all Americans are obese, putting them at higher risk for heart disease and diabetes.

SELF-QUIZ

1. A new species of large animal has just been discovered. Most animals have an alimentary canal, but this animal *appears* to rely entirely on food vacuoles. Why is this unlikely?

2. A friend says, "It's not eating that causes you to gain weight, it's absorption." Is that statement true?

3. _____ is the mechanical and chemical breakdown of food into small molecules, whereas _____ is the uptake of these small molecules by the body's cells.

4. Why don't astronauts in zero gravity have trouble swallowing?

5. Which class of large biological molecules is primarily broken down in the stomach? What enzyme is responsible?

6. What type of organism is responsible for most stomach ulcers? If the ulcers are detected early, how can they be treated?

7. Most chemical digestion occurs in the _____ as a result of the activity of enzymes made by the _____.
   a. stomach; oral cavity
   b. pancreas; stomach
   c. large intestine; small intestine
   d. small intestine; pancreas

8. Explain how treatment of a chronic infection with antibiotics for an extended period of time can cause a vitamin K deficiency.

9. If you maintain a normal activity level and consume the same number of calories needed for your basal metabolic rate, you will
   a. gain weight.
   b. lose weight.
   c. stay the same weight.
   d. lose weight but quickly regain it.

10. Why is the amount of oxygen you consume proportional to your metabolic rate?

11. Imagine that you and a friend both eat according to FDA nutritional guidelines, but you are a vegetarian, whereas your friend is not. Why do you have to be more concerned about the nutritional content of your food than your friend does? What can you do to address this concern?

12. A friend takes megadoses of vitamin supplements. Which vitamin excess could pose the largest threat?
   a. A
   b. thiamin (B_1)
   c. C
   d. folic acid

Answers to the Self-Quiz questions can be found in Appendix D.

THE PROCESS OF SCIENCE

13. The FDA recommends that about 12% of our daily calories come from protein. How do we know if our diet is in accordance with those regulations? Nutritional labels typically tell how many Calories are from fat but don't specify how many Calories are from other nutrients. Proteins have approximately 4 Calories per gram. Read this cookie label and calculate the percentage of calories from protein.

Nutrition Facts
Serving Size 1 Cookie (28 g /1 oz)
Servings Per Container 8

Amount Per Serving	
Calories 140	Calories from Fat 60

	% Daily Value*
Total Fat 7g	11%
Saturated Fat 3g	15%
Trans Fat 0g	
Cholesterol 10mg	3%
Sodium 80mg	3%
Total Carbohydrate 18g	6%
Dietary Fiber 1g	4%
Sugars 10g	
Protein 2g	

14. Describe some nutritional and digestive problems a patient might face after gastric bypass surgery.

BIOLOGY AND SOCIETY

15. How could you test claims of companies that sell weight loss supplements? Should they be accountable for false claims?

16. Should insurance companies pay for gastric bypass surgeries?

23 Circulation and Respiration

Saving a life.
Many lifesaving procedures are intended to restore proper functioning of the circulatory and respiratory systems.

BIOLOGY AND SOCIETY: Lifesaving Procedures

The ABCs of Saving Lives

At a restaurant, a woman collapses into unconsciousness. Paramedics arrive to find her lying on the floor with blue skin, not breathing. They lay her flat on her back, lift her chin, place an oxygen-filled bag over her mouth and nose, and attempt to fill her lungs. When no oxygen enters, one of the rescuers presses quickly and firmly on her abdomen, dislodging a large piece of food from her throat. The paramedics squeeze oxygen into her lungs and check for a pulse. Finding none, they initiate cardiopulmonary resuscitation (CPR) by pushing on her breastbone. They place pads from a defibrillator on her chest and administer electrical shocks. They also inject drugs into her veins. The patient is loaded into an ambulance. On the way to the hospital, she regains a pulse and begins to breathe on her own.

In emergency situations like this, rescuers follow a set of procedures known as the ABCs of lifesaving. *A* is for *airway*—the tube that carries air from the mouth and nose to the lungs. If it becomes blocked, oxygen cannot enter the body, and death quickly follows. That is why the ambulance crew first attempted to open the patient's airway by positioning her head. When this failed, they pressed her abdomen to try to dislodge foreign matter from her airway. Once air can flow into the lungs, *B* for *breathing* is the next priority. If the patient is not breathing, mouth-to-mouth resuscitation or a squeezable bag can supply oxygen. The third priority is *C* for *circulation* of the blood. If the heart has stopped, CPR can temporarily move blood through the body while electrical shocks and drugs are used to try to restart the heart.

A and *B* (airway and breathing) both involve the respiratory system, the body's organ system for exchanging gases. *C* involves the circulatory system, which transports gases and other substances throughout the body via the blood. Properly functioning circulatory and respiratory systems are essential, and a failure of either one can result in rapid death. The two systems are such close partners that we will explore them in a single chapter. For each of the two systems, we'll start by considering general challenges and survey the variety of ways these are addressed. We'll then take a closer look at each system in humans. Throughout the discussion, we'll examine what may happen when these systems fail and consider how lifesaving procedures can prevent fatalities.

Unifying Concepts of Animal Circulation

Every organism must exchange materials and energy with its environment. In simple animals, such as hydras and jellies, nearly all the cells are in direct contact with the outside world. Thus, the cells can easily exchange materials with the environment by **diffusion**, the spontaneous movement of molecules from an area of higher concentration to an area of lower concentration (see Figure 5.12). However, most animals are too large or too complex for exchange to take place by diffusion alone. In such animals, a **circulatory system** facilitates the exchange of materials, providing a rapid, long-distance internal transport system that brings resources close enough to cells for diffusion to occur. Once there, the resources that cells need, such as nutrients and oxygen (O_2), enter the cytoplasm through the plasma membrane. And metabolic wastes, such as carbon dioxide (CO_2), diffuse from the cells to the circulatory system for disposal. ✔

☑ CHECKPOINT

Why can't the human body rely solely on diffusion to provide all needed chemicals to body cells?

Answer: Diffusion is only effective over short distances. Diffusion alone would take too long to convey materials throughout the large human body.

Open and Closed Circulatory Systems

All but the simplest animals have a circulatory system with three main components: a central pump, a vascular system (a set of tubular vessels), and circulating fluid. Two main types of circulatory systems have evolved in animals. Many invertebrates, including most molluscs and all arthropods, have an **open circulatory system**. The system is called "open" because the circulating fluid is pumped through open-ended vessels and flows out among the cells **(Figure 23.1a)**.

In a **closed circulatory system**, the circulating fluid is blood, which is confined to vessels and is distinct from the interstitial fluid, the fluid that fills the spaces around cells **(Figure 23.1b)**. One or more hearts pump blood into large vessels that branch into smaller ones coursing through the organs. Earthworms, octopuses, and vertebrates are some examples of animals with a closed circulatory system.

The closed circulatory system of humans and other vertebrates, consisting of a heart and blood vessels, is called a **cardiovascular system** (from the Greek *kardia*, heart, and Latin *vas*, vessel). The fish in Figure 23.1b illustrates the main features of a cardiovascular system. The heart receives blood in a chamber called the **atrium** (plural, *atria*). A second chamber called the **ventricle** pumps blood away from the heart. Blood is confined in three main types of blood vessels: arteries, capillaries, and veins. **Arteries** carry blood away from the heart, branching

▼ **Figure 23.1 Open and closed circulatory systems.**

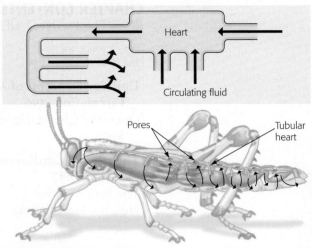

(a) Open circulatory system. In an open system, the heart pumps circulating fluid into body tissues, where nutrients diffuse directly into body cells. Contractions of body muscles move the fluid toward the tail. Fluid returns to the heart through several pores.

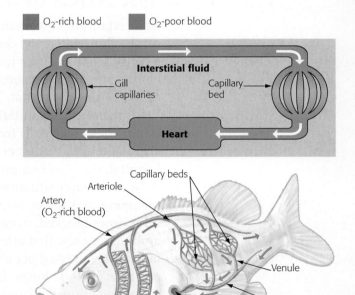

(b) Closed circulatory system. In a closed system, blood is contained within vessels, separate from the interstitial fluid bathing the cells. In a fish, blood is pumped from the heart to the gills, then to capillary beds in body tissues, and finally back to the heart.

into smaller **arterioles** as they approach the organs. The blood then flows from arterioles into capillary beds, networks of tiny vessels called **capillaries** that infiltrate nearly every organ and tissue in the body. The thin walls of capillaries allow exchange between the blood and interstitial fluid. In fish, blood is oxygenated in gill capillaries and then travels via arteries to other capillary beds, where oxygen enters tissue cells. Capillaries converge into **venules**, which in turn converge into larger **veins** that return blood back to the heart. Capillaries are the functional center of the circulatory system: This is where materials are transferred to and from surrounding tissues. Arteries and veins primarily transport blood back and forth between the heart and capillaries. ☑

☑CHECKPOINT

Place the following in the order that describes how blood travels from the heart to cells throughout the body and back to the heart: capillaries, arteries, arterioles, venules, and veins.

Answer: Heart → arteries → arterioles → capillaries → venules → veins → heart

The Human Circulatory System

The Human Cardiovascular System

Recall that most animals have a circulatory system consisting of three main components: a pump, a vascular system, and a circulating fluid. In the human cardiovascular system, the central pump is the heart, the vascular system is the blood vessels, and the circulating fluid is the blood.

The Path of Blood

In humans and other terrestrial vertebrates, the three components of the cardiovascular system are organized into a **double circulation system** with two circuits of blood flow. The **pulmonary circuit** carries blood between the heart and the lungs (**Figure 23.2a**). In the lungs, carbon dioxide diffuses one way (from the blood into the lungs), while oxygen diffuses the other way (from the lungs into the blood). The pulmonary circuit then returns the oxygen-rich blood to the heart. As shown in **Figure 23.2b**, the **systemic circuit** carries blood between the heart and the rest of the body. It supplies oxygen and nutrients to body tissues and organs and picks up carbon dioxide and other wastes from them. The oxygen-poor blood then returns to the heart via the systemic circuit.

▼ Figure 23.2 **Double circulation.**

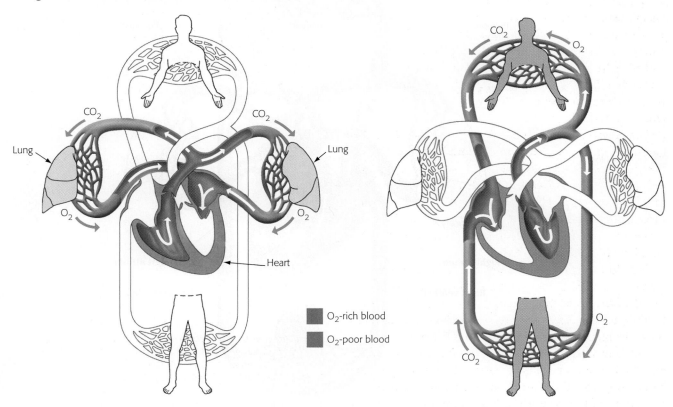

■ O_2-rich blood
■ O_2-poor blood

(a) Pulmonary circuit. In organisms with double circulation, the pulmonary circuit transports blood between the heart and lungs.

(b) Systemic circuit. The systemic circuit transports blood between the heart and the rest of the body.

Figure 23.3 traces the path of blood in the human cardiovascular system as it makes one complete trip around the body. If we consider the circulatory system as a whole, the pulmonary and systemic circuits operate simultaneously. The two ventricles pump almost in unison, sending some blood through the pulmonary circuit and the rest through the systemic circuit at the same time. It takes the blood in your body about 1 minute to travel through the circulatory system—less time than it will take you to read about it!

Let's start at ❶ the right atrium, where two large veins empty O_2-poor blood from the body into the heart. The blood is then pumped to ❷ the right ventricle, which pumps the O_2-poor blood to the lungs via ❸ two pulmonary arteries. ❹ As the blood flows through capillaries in the lungs, CO_2 diffuses out of the blood and O_2 diffuses into the blood. The newly O_2-rich blood then flows through ❺ the pulmonary veins to ❻ the left atrium of the heart, completing the pulmonary circuit. Next, the blood is pumped a short distance from the left atrium to ❼ the left ventricle.

Blood leaves the left ventricle through ❽ the aorta, the largest blood vessel in the body, with a diameter about as big as a quarter. Branching from the aorta are several large arteries that lead to ❾ the head and arms and the abdominal region and legs. Near each abdominal organ, arteries lead to arterioles that in turn branch into capillaries that penetrate the body's tissues. Diffusion of O_2 into the cells and CO_2 out of the cells takes place across the thin walls of the capillaries. Downstream, the capillaries join into venules, which convey the blood back into veins. Oxygen-poor blood from the upper body and head is channeled into ❿ a large vein called the superior vena cava; ⓫ another large vein, the inferior vena cava, receives blood from the lower body. The superior vena cava and inferior vena cava complete the systemic circuit by returning blood to the heart.

In the next section, you'll learn about the structure and function of each of the components of the human cardiovascular system and how these components work together. You may find it useful to refer back to Figure 23.3 as you learn about the details of each component. ☑

▶ **Figure 23.3 A trip through the human cardiovascular system.** Note that the left side of the heart is on the right side of the figure—and vice versa—because the body is viewed from the front.

O_2-rich blood
O_2-poor blood

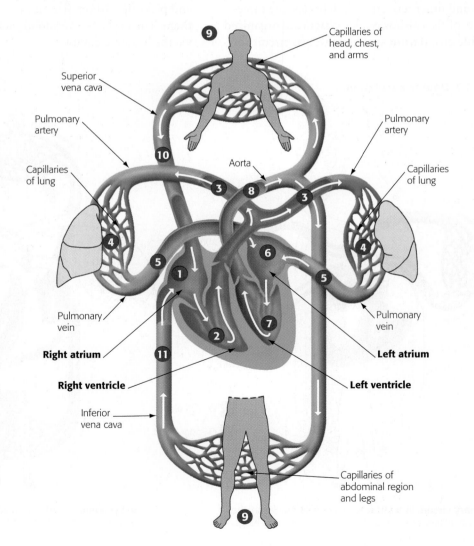

☑CHECKPOINT

In the human cardiovascular system, what is the difference between the pulmonary circuit and the systemic circuit?

Answer: *The pulmonary circuit transports blood between the heart and lungs, whereas the systemic circuit transports blood between the heart and body tissues other than the lungs.*

498

How the Heart Works

The hub of the human cardiovascular system is the **heart**, a muscular organ about the size of a fist located under the breastbone. Remember that humans have a double circulation system, with the pulmonary circuit pumping blood between the heart and lungs, and the systemic circuit transporting blood between the heart and the rest of the body. This double circuit is supported by a four-chambered heart. **Figure 23.4** shows the flow of blood through these chambers.

The atria collect blood returning to the heart and pump it into the ventricles. The ventricles, which pump blood out of the heart to the other body organs, are much more powerful than the atria. Four valves in the heart prevent backflow and keep blood moving in the correct direction. Notice that the four-chambered structure of the heart prevents the oxygen-rich and oxygen-poor blood from mixing. The right atrium receives oxygen-poor blood from body tissues via two large veins, and the right ventricle pumps it to the lungs. The left atrium receives oxygen-rich blood from the lungs, and the left ventricle pumps it out to body organs via the arteries. ☑

▼ Figure 23.4 **Path of blood flow through the human heart.** The heart contains four chambers, with two atria located above two ventricles. Valves help maintain the unidirectional flow of blood. Notice that oxygen-poor blood and oxygen-rich blood are kept separate.

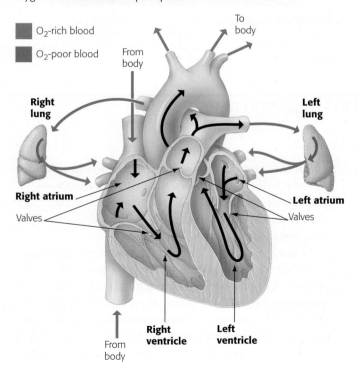

The Cardiac Cycle

The muscles of the heart relax and contract rhythmically in what is called the **cardiac cycle**. The relaxation phase of the heart cycle is known as **diastole**; the contraction phase is called **systole**. **Figure 23.5** follows the heart through a cycle lasting 0.8 second. ❶ During diastole, which lasts about 0.4 second, blood returning to the heart flows into all four chambers. ❷ During the first 0.1 second of systole, the atria contract, forcing all the blood into the ventricles. ❸ In the last 0.3 second of systole, the ventricles contract, pumping blood out of the heart and into the aorta and pulmonary arteries. One heart beat is one complete circuit of the cardiac cycle. In a healthy adult person at rest, the number of beats per minute, or **heart rate**, ranges between 60 and 80.

During its cycle, the heart makes a distinctive "lubb-dupp, lubb-dupp" sound as the heart valves snap shut. You can hear this sound with a stethoscope or by pressing your ear against another person's chest. A trained ear can also detect the sound of a **heart murmur**, which may indicate a defect in one or more of the valves. A serious murmur sounds like a "hisssss" as a stream of blood squirts backward through a defective valve. Some people are born with murmurs, while others have valves damaged by infection (such as rheumatic fever). Severe murmurs can reduce blood flow to the point that it causes death. However, most cases of heart murmur are not serious, and those that are can be corrected by replacing the damaged valves with artificial ones or with valves taken from an organ donor (human or other animal, usually a pig).

► Figure 23.5 **The cardiac cycle.** The muscles of the heart relax (diastole) and contract (systole) in a rhythmic cycle.

❶ Heart is relaxed. Blood flows in.

❷ Atria contract. Blood is forced into ventricles.

❸ Ventricles contract. Blood is pumped out.

☑**CHECKPOINT**

Heart chambers called the _____ receive blood returning to the heart via blood vessels called _____. The _____ are the chambers where blood is pumped out of the heart to blood vessels called _____.

Answer: atria; veins; ventricles; arteries

The Pacemaker and the Control of Heart Rate

A region of the heart called the **pacemaker**, or the **SA (sinoatrial) node**, sets the tempo of the heartbeat (**Figure 23.6a**). The pacemaker is composed of specialized muscle tissue in the wall of the right atrium that ❶ generates electrical impulses. ❷ Impulses from the pacemaker spread rapidly through the walls of both atria, making the atria contract. The impulses then pass to a relay point called the **AV (atrioventricular) node** that delays the signals by about 0.1 second. This delay allows the atria to empty completely before the impulses are passed to the ventricles. ❸ Once the impulses reach the ventricles, the ventricles contract strongly, driving the blood out of the heart.

A variety of cues influence the signals sent by the pacemaker. For example, epinephrine (also called adrenaline), the "fight-or-flight" hormone released during times of stress, increases heart rate (see Chapter 25).

Adrenaline is often injected by paramedics to help restart a stopped heart. Stimulants, such as caffeine, also make the heart beat faster. And heart rate increases with exercise, an adaptation that enables the circulatory system to provide the additional oxygen needed by muscles hard at work.

In certain kinds of heart disease, the heart's electrical control mechanism fails to maintain a normal rhythm. The remedy is an artificial pacemaker, a small electronic device surgically implanted near the SA node. This device emits electrical signals to maintain normal heart rhythms (**Figure 23.6b**). Under certain circumstances, much more profound electrical stimulation is required, as we'll see next.

▼ Figure 23.6 **Pacemakers.**

Pacemaker (SA node)

AV node

Right atrium

Right ventricle

❶ Pacemaker generates electrical impulses.

❷ Impulses spread through atria.

❸ Impulses reach ventricles.

(a) **The heart's natural pacemaker.** The SA (sinoatrial) node is located within the muscle tissue in the wall of the right atrium. The yellow represents how far the electrical signals have spread.

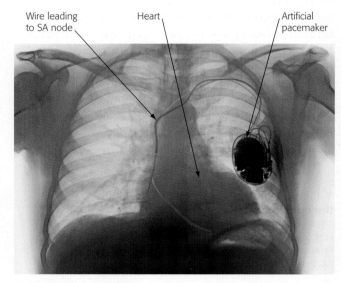

Wire leading to SA node

Heart

Artificial pacemaker

(b) **Artificial pacemaker.** A small electronic device surgically implanted into cardiac muscle or (as shown here) the chest cavity and connected to the SA node by a wire can help maintain the proper electrical rhythms of a defective heart.

THE PROCESS OF SCIENCE: Lifesaving Procedures

Do AEDs Save Lives?

Sudden heart attacks are the leading cause of death in the United States. A **heart attack** is the damage or death of heart muscle tissue when one or more arteries supplying the heart muscle are blocked. During a heart attack, the heart's pacemaker is often unable to maintain a normal rhythm. In such cases, electrical stimulation may be the only way to "reset" the pacemaker. Defibrillators (the electrical "paddles" often seen in medical dramas on TV) have been used in hospitals since the 1960s. By the 1980s, portable defibrillators were

available to specially trained emergency workers. More recently, automated external defibrillators (AEDs)—portable devices that analyze the heart's electrical rhythm and apply electrical shocks as needed—have been installed in many public areas.

Researchers knew from **observation** and numerous trials that AEDs are safe and effective when used by trained public safety personnel. This raises the **question**: Are AEDs effective when used by laypeople? One study began with the **hypothesis** that people who had heart attacks at sites with AEDs would be more likely to survive than people who had heart attacks elsewhere.

The **experiment** involved monitoring heart attacks at 24 public locations (such as shopping malls and hotels) over three years. The locations were divided into two groups: those that had an AED and those that did not. The researchers counted the heart attacks that occurred at each location and then counted the patients who survived. The **results** were clear: When no AED was present, 14% (15 of 107 persons) survived to hospital discharge; when an AED was available and used by a layperson, 23% of the patients (30 out of 128) survived (**Figure 23.7**). This study is a good example of evidence-based medicine, the use of the scientific method to determine optimal medical care.

▼ **Figure 23.7 Do AEDs save lives?** One study showed that making AEDs available to laypeople in public places greatly increased the chances that a heart attack victim would survive to leave the hospital.

Blood Vessels

If we think of the heart as the body's "pump," then the system of arteries, veins, and capillaries connected to it can be thought of as the "plumbing." If you review Figure 23.3, you'll notice that arteries and veins are distinguished by the direction in which they carry blood: Arteries carry blood *away from* the heart, and veins carry blood *toward* the heart. Capillaries allow for exchange between the bloodstream and the tissue cells (via interstitial fluid).

All blood vessels are lined by a thin layer of tightly packed epithelial cells. Structural differences in the walls of the different kinds of blood vessels correlate with their different functions (**Figure 23.8**). Capillaries have very thin walls—often just one cell thick—that allow the exchange of substances between the blood and the interstitial fluid that bathes tissue cells. The smallest capillaries are so small that blood cells must pass through them single file. The walls of arteries and veins have two additional, thicker layers. An outer layer of elastic connective tissue allows the vessels to stretch and recoil. Between this layer and the epithelial cells is a middle layer of smooth muscle. By constricting or relaxing, smooth muscle (under the control of the brain) can narrow or widen the blood vessels, thereby regulating blood flow. Veins convey blood back to the heart at low velocity and pressure after the blood has passed through capillary beds. Veins (but not arteries) also have one-way valves that prevent backflow, ensuring that blood always moves toward the heart. ☑

▼ **Figure 23.8 The structure of blood vessels.** Arteries branch into smaller vessels called arterioles, which in turn branch into capillaries. Chemical exchange between blood and interstitial fluid occurs across the thin walls of the capillaries. The capillaries converge into venules, which deliver blood to veins. All of these vessels are lined by a thin, smooth epithelium. Arteries and veins have two additional layers.

Blood Flow through Arteries

The force that blood exerts against the walls of your blood vessels is called **blood pressure**. Created by the beating of the heart, blood pressure is the main force driving the blood from the heart through the arteries and arterioles to the capillary beds. When the ventricles contract, blood is forced into the arteries faster than it can flow into the arterioles. This faster flow creates pressure that stretches the elastic walls of the arteries. You can feel this effect when you measure your heart rate by taking your pulse. A **pulse** is the rhythmic stretching of the arteries caused by the pressure of blood forced into the arteries during systole. The elastic walls of the arteries recoil during diastole, maintaining enough pressure on the remaining blood to sustain a constant flow into arterioles and capillaries. Thus, blood pressure is recorded as two numbers, such as 120/80 ("120 over 80"). The first number is blood pressure during systole (in millimeters of mercury, a standard pressure unit); the second number is the blood pressure that remains in the arteries during diastole.

Normal blood pressure falls within a range of values, but optimal blood pressure for adults is below 120 systolic and below 80 diastolic (as measured on the upper arm). Lower values are generally considered better, although very low blood pressure may lead to light-headedness and fainting. Blood pressure higher than normal may indicate a serious cardiovascular disorder.

High blood pressure, or **hypertension**, is persistent systolic blood pressure higher than 140 and/or diastolic blood pressure higher than 90. Hypertension affects approximately one-quarter of the adult population in the United States. It is sometimes called a "silent killer" because it often displays no outward symptoms for years while increasing the risks of heart disease, a heart attack, or a stroke (death of brain tissue due to lack of oxygen). To control hypertension, you can eat a heart-healthy diet, avoid excess alcohol (more than two drinks per day), exercise regularly, and maintain proper weight. If lifestyle changes don't work, there are several medications that can help lower blood pressure.

Blood Flow through Capillary Beds

Muscles control the flow of blood into capillary beds so that only about 5–10% of your capillaries have a steady flow of blood running through them at any given time (**Figure 23.9a**). Blood flows continuously through capillaries in the brain, heart, kidneys, and liver. In many other sites, the blood supply varies. After a meal, for instance, blood flow to the digestive tract increases. During strenuous exercise, blood is shunted away from the digestive system and supplied more generously to the skeletal muscles and skin. This is one reason why heavy exercise right after eating may cause indigestion or muscle cramping (and why you shouldn't swim too soon after eating—just like mom always said).

The walls of capillaries are thin and leaky. Consequently, as blood enters a capillary at the arterial end, blood pressure pushes fluid rich in O_2, nutrients, and other substances needed by the surrounding cells out of the capillary and into the interstitial fluid (**Figure 23.9b**). These molecules then diffuse from the interstitial fluid

▼ Figure 23.9 **Chemical exchange between the blood and tissue cells.**

LM 1,000×

(a) Capillaries. Blood flowing through the circulatory system eventually reaches capillaries, the small vessels where exchange with cells actually takes place.

(b) Chemical exchange. Within the capillary beds, there is local exchange of molecules between the blood and interstitial fluid, which bathes the cells of tissues.

into nearby tissue cells. Blood cells and other large components usually remain in the blood because they are too large to pass through the capillary walls. At the venous end of the capillary, CO_2 and other wastes diffuse from tissue cells, through the interstitial fluid, and into the capillary bloodstream. This local chemical exchange—between the blood and tissue cells within capillary beds—is the most important function of the circulatory system.

Blood Return through Veins

After chemicals are exchanged between the blood and body cells, blood returns to the heart via veins. From the capillaries, blood moves into small venules, then into larger veins, and finally to the inferior and superior venae cavae, the two large blood vessels that flow into the heart. By the time blood exits the capillaries and enters the veins, the pressure originating from the heart has dropped to near zero. The blood still moves through veins, even against the force of gravity, because veins are sandwiched between skeletal muscles (Figure 23.10). As these muscles contract (when you walk, for example), they squeeze the blood along.

You may witness firsthand the importance of muscle contraction in conveying blood through your veins if you stand still for too long. After a while without contracting your muscles, you will start to become weak and dizzy and may even faint because gravity prevents blood from returning to your heart in sufficient amounts to supply your brain with oxygen.

Over time, leg veins may stretch and enlarge and the valves within them weaken. As a result, veins just under the skin can become visibly swollen, a condition called varicose veins. Besides being unsightly, varicose veins may cause serious problems, such as cramping, blood clots, and ulcerations (open breaks in the skin). ☑

Blood

Now that we have examined the structures and functions of the heart and blood vessels, let's focus on the composition of blood itself (Figure 23.11). The circulatory system of an adult human has about 5 L (11 pints) of blood. Just over half this volume consists of a yellowish liquid called **plasma**. Plasma is about 90% water. The other 10% of plasma is made up of dissolved salts, proteins, and various other molecules being transported by the blood, such as nutrients, wastes, and hormones. Suspended within the plasma are several types of cellular elements. Let's examine each of them in turn.

☑ CHECKPOINT

List the three major types of blood vessels in order of the pressure of blood within them, from highest pressure to lowest.

Answer: arteries, capillaries, veins

▼ Figure 23.10 **Blood flow in a vein.** The contraction of muscles surrounding veins squeezes blood toward the heart. Flaps of tissue in the veins act as one-way valves, preventing backflow.

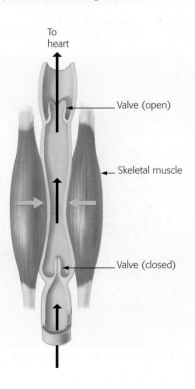

To heart

Valve (open)

Skeletal muscle

Valve (closed)

▼ Figure 23.11 **The composition of human blood.** Just over half the volume of blood is liquid plasma. Blood cells and platelets account for the remaining blood volume.

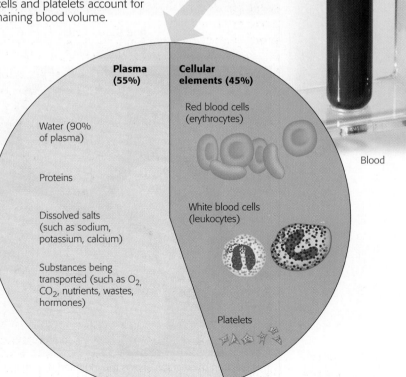

Blood

Plasma (55%)

Cellular elements (45%)

Water (90% of plasma)

Proteins

Dissolved salts (such as sodium, potassium, calcium)

Substances being transported (such as O_2, CO_2, nutrients, wastes, hormones)

Red blood cells (erythrocytes)

White blood cells (leukocytes)

Platelets

Red Blood Cells and Oxygen Transport

Red blood cells, also called **erythrocytes**, are by far the most numerous type of blood cell. There are about 25 trillion of these cells in the average person's bloodstream. Carbohydrate-containing molecules on the surface of red blood cells determine a person's blood type, such as A, B, AB, or O (see Figure 9.20).

Each red blood cell contains approximately 250 million molecules of **hemoglobin**, an iron-containing protein that transports oxygen. As red blood cells pass through the capillary beds of your lungs, oxygen diffuses into the red blood cells and binds to the hemoglobin. This process is reversed in the capillaries of the systemic circuit, where the hemoglobin unloads its cargo of oxygen to the body's cells.

Human red blood cells are shaped like disks with indentations in the middle, increasing the surface area available for gas exchange (**Figure 23.12**). In addition to their shape, the red blood cells of mammals have another feature that enhances their oxygen-carrying capacity: They lack nuclei and other organelles, leaving more room to carry hemoglobin. Furthermore, red blood cells are very small relative to other cells. Therefore, a population of red blood cells has a greater total surface area for gas exchange than the same volume of larger cells.

Adequate amounts of hemoglobin and red blood cells are essential to the normal functioning of the body. An abnormally low amount of hemoglobin or a low number of red blood cells can result in a condition called **anemia**. An anemic person feels constantly tired and run-down because the body cells do not get enough oxygen. Anemia can result from a variety of factors, including excessive blood loss, vitamin or mineral deficiencies, and certain cancers. A mutation in the gene for hemoglobin may result in sickle-cell disease, another cause of anemia (see Chapter 9).

The production of red blood cells in the bone marrow is controlled by a feedback system that monitors the amount of oxygen reaching tissues via the blood. If the tissues are not receiving enough oxygen, the kidneys produce a hormone known as EPO. After traveling to the bone marrow, EPO boosts the production of oxygen-carrying red blood cells. Doctors prescribe synthetic EPO to relieve the symptoms of anemia.

Because additional red blood cells boost the functions of the circulatory and respiratory systems, EPO increases athletic stamina, potentially offering a huge

▼ Figure 23.12 **The three cellular components of blood.**

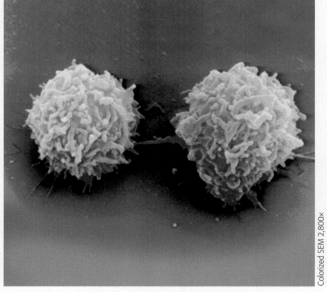

CELLULAR COMPONENTS OF BLOOD

Red Blood Cells (cells that carry oxygen)	**White Blood Cells** (cells that fight infection)
Colorized SEM 5,200x	Colorized SEM 2,800x
This electron micrograph shows the indented disk shape of human red blood cells.	This electron micrograph shows two lymphocytes, a type of white blood cell.

advantage in long athletic competitions. Thus, some athletes—particularly long-distance runners and cyclists—have abused synthetic EPO in an attempt to gain an artificial advantage. EPO abuse, one type of "blood doping," is difficult to detect because it is a hormone produced naturally by the body and because synthetic EPO is rapidly cleared from the bloodstream. One way that athletic commissions test for cheaters is by measuring the percentage of red blood cells in the blood volume. A value over 50% is usually grounds for disqualification. EPO is banned not only because it confers an unfair advantage but also because it is dangerous: It has been blamed for the deaths of dozens of professional athletes.

White Blood Cells and Defense

As a group, **white blood cells**, or **leukocytes**, fight infections (see Figure 23.12). White blood cells are larger than red blood cells. And unlike red blood cells, they lack hemoglobin but contain nuclei and all other organelles. There are about 700 times fewer white blood cells than red blood cells, although their numbers temporarily increase when the body is combating an infection. You'll learn more about white blood cells in Chapter 24.

Platelets and Blood Clotting

We all get cuts and scrapes from time to time, yet we do not bleed to death because blood contains two components that help plug leaks by forming clots: platelets and fibrinogen. **Platelets** are bits of membrane-wrapped cytoplasm pinched off from larger cells in the bone marrow. Clotting begins when the tissue lining a blood vessel is damaged. Almost immediately, platelets adhere to the damaged tissue to form a sticky cluster that seals minor breaks. Platelets also release clotting factors, molecules that convert **fibrinogen**, a protein found in the plasma, to a threadlike protein called **fibrin**. Molecules of fibrin form a dense network to create a patch (see Figure 23.12). On the skin, such clots are called scabs.

The clotting mechanism is so important that any defect in it can be life-threatening. In the inherited disease hemophilia (see Figure 9.32), excessive, sometimes fatal bleeding can occur from even minor cuts and bruises. Hemophilia is caused by a genetic mutation in one of several genes that code for clotting factors. The opposite problem—too much clotting—can also be fatal. An embolus is a blood clot that forms within a blood vessel but then dislodges from its point of origin and travels in the blood. If an embolus lodges in an artery of the heart and is large enough to block it, a heart attack occurs. Similarly, an embolus in the head can cause a stroke. ☑

Stem Cells and the Treatment of Leukemia

Where do red blood cells, white blood cells, and platelets come from? The red marrow inside of bones such as the ribs, vertebrae, sternum, and pelvis contains stem cells that differentiate into red and white blood cells as well as into the cells that produce platelets. (Recall from Chapter 11 that stem cells have the potential to differentiate into other types of cells.) After forming in the early embryo, these stem cells continually reproduce themselves and create all the blood cells needed throughout life. Bone marrow stem cells show great promise for the treatment of disease, particularly leukemia.

Leukemia is cancer of the white blood cells (leukocytes). Because cancerous cells grow uncontrollably, a person with leukemia has an unusually high number of white blood cells, most of which are defective. The overabundance of these cells crowds out the bone marrow cells that produce red blood cells and platelets, causing severe anemia and impaired clotting.

Some cases of leukemia do not respond to standard cancer treatments—radiation and chemotherapy. An alternative treatment involves destroying cancerous bone marrow and replacing it with healthy bone marrow. Patients may be treated with their own bone marrow: Marrow from the patient is harvested, processed to remove as many cancerous cells as possible, and then reinjected. Alternatively, a donor, often a sibling, may provide marrow.

Platelets
(bits of membrane-enclosed cytoplasm that aid clotting)

Colorized SEM 3,600×

Fibrin

Colorized SEM 3,100×

Red blood cell

Platelets help produce a clotting protein called fibrin (shown at bottom wrapping around a red blood cell).

Cardiovascular Disease

The circulatory system performs several homeostatic functions that help maintain relatively constant internal conditions. First, by exchanging nutrients and wastes with the interstitial fluid, the circulatory system helps control the chemical balance of the fluid that surrounds cells. Second, the circulatory system helps control the composition of the blood by continuously moving it through organs, such as the lungs, liver, and kidneys, that regulate the blood's contents. The circulatory system is also involved in the body's temperature regulation, hormone distribution, and defense against foreign invaders—all topics that are discussed in other chapters.

What happens when the circulatory system cannot perform its normal functions? **Cardiovascular disease**, a set of diseases affecting the heart and blood vessels, accounts for 40% of all deaths in the United States, killing over 1 million people each year. Like all of our cells, heart muscle cells require oxygen-rich blood to survive. When blood exits the heart via the aorta, several **coronary arteries** (shown in dark red in **Figure 23.13**) immediately branch off to supply the heart muscle. If one or more of these blood vessels become blocked, heart muscle cells quickly die from lack of O_2 (gray area in Figure 23.13). As noted earlier, such an event, and the subsequent failure of the heart to function properly, is called a heart attack. (The medical term is myocardial infarction, meaning death of heart muscle tissue.) Approximately one-third of heart attack victims die almost immediately. For those who survive, the ability of the damaged heart to pump blood may be seriously impaired.

The suddenness of a heart attack belies the fact that the arteries of most victims became impaired gradually by a chronic cardiovascular disease called **atherosclerosis**. During the course of the disease, cholesterol and other substances accumulate into buildups called plaque that form in the walls of arteries, narrowing the passages through which blood can flow (**Figure 23.14**). Vessels narrowed by plaque increase blood pressure and can more easily form and trap clots. If a coronary artery becomes partially blocked, a person may feel occasional chest pain, a condition called angina.

To some extent, the tendency to develop cardiovascular disease is inherited. However, three everyday behaviors significantly impact the risk of cardiovascular disease and heart attack. Smoking doubles the risk of heart attack and increases the severity if one does occur. Regular exercise (particularly "cardio" or "aerobic" workouts) can cut the risk of heart disease in half. Eating a heart-healthy diet, low in cholesterol, trans fat, and saturated fat (see Chapter 3) and high in fruits,

vegetables, and whole grains can reduce the risk of developing atherosclerosis.

If a person already has cardiovascular disease, there are treatments available. Certain drugs can lower cholesterol. Angioplasty (inserting a tiny catheter with a balloon that is inflated to compress plaque and widen clogged coronary arteries) and stents (small wire mesh tubes that prop arteries open) can also help. Bypass surgery is a much more drastic remedy. In this procedure, blood vessels removed from a patient's legs are sewn onto the heart to shunt blood around clogged coronary arteries. In an even more drastic measure, a defective heart can be replaced by a transplant or even with an artificial heart (although no one has survived long with one). Unfortunately, surgery of any kind only treats the disease symptoms, not the underlying causes, so problems will return if risk factors are not minimized.

On the positive side, the death rate from cardiovascular disease in the United States has been cut in half during the past 50 years. Health education, early diagnosis, and reduction of risk factors, particularly smoking, are mostly responsible. ☑

▼ Figure 23.14 **A normal artery and an artery showing atherosclerosis.**

Connective tissue

Smooth muscle

Epithelium

Normal artery

Plaque

Artery partially blocked by plaque

Aorta

Coronary artery (supplies oxygen to the heart muscle)

Dead muscle tissue

Blockage

▲ Figure 23.13 **Blockage of a coronary artery, resulting in a heart attack.** If one or more coronary arteries become blocked, the heart muscle cells that they feed will die from lack of oxygen. Such an event, called a heart attack, can lead to permanent damage of the heart muscle.

☑CHECKPOINT

Name three things you can do to lower your risk of cardiovascular disease.

Answer: *Eat a heart-healthy diet, exercise regularly, and avoid smoking.*

Unifying Concepts of Animal Respiration

The lifesaving procedures described in the Biology and Society section remind us that survival depends on close cooperation between the circulatory and respiratory systems. The circulatory system transports gases and other substances to and from body cells. Recall from Chapter 6 that cellular respiration uses oxygen and glucose to produce water, carbon dioxide, and energy in the form of ATP. All working cells therefore require a steady supply of O_2 from the environment and must continuously dispose of CO_2:

The **respiratory system** consists of several organs that facilitate exchange of O_2 and CO_2 between cells and the environment. (Note that in the context of a whole organism, the word *respiratory* refers to the process of breathing, not to cellular respiration.) ☑

The Structure and Function of Respiratory Surfaces

Earth's main reservoir of oxygen is the atmosphere, which is about 21% O_2. Terrestrial animals access this oxygen by breathing air. Oceans, lakes, and other bodies of water also contain O_2 in the form of dissolved gas, but it makes up only 3–5% of that environment. Many aquatic animals obtain oxygen by passing water over their gills. The part of an animal where O_2 from the environment diffuses into living cells and CO_2 diffuses out to the surrounding environment is called the **respiratory surface**. The respiratory surface is usually covered with a single layer of living cells that is thin enough and moist enough to allow rapid diffusion between the body and the environment.

The respiratory surface of an animal must be large enough to take up O_2 for every cell in the body. A variety of respiratory surfaces have evolved. For some animals, such as sponges and flatworms, the plasma membrane of every cell in the body is close enough to the outside environment for gases to diffuse in and out. In many animals, however, the bulk of the body does not have direct access to the outer environment, and the respiratory surface is the interface between the environment and the blood, which transports gases to and from the rest of the body.

Some animals use their entire outer skin as a respiratory surface. An earthworm, for example, has moist skin, and gases diffuse across its general body surface. This results in an exchange of O_2 and CO_2 between the environment and a dense net of capillaries just below the earthworm's skin (**Figure 23.15a**). For most animals, however, the outer body surface is either impermeable to gases or lacks sufficient area to exchange gases for the whole body. In such animals, specialized regions of the body have extensively folded or branched surfaces that provide a large surface area for gas exchange.

Within aquatic environments, the respiratory surfaces are found in extensions of the body surface called **gills**. Lobsters, which are invertebrates, and fishes, which are vertebrates, are examples of animals with gills. The feathery gills located on either side of a fish's head have a respiratory surface area much greater than the rest of the body surface. A fish obtains oxygen by continuously pumping water over its gills. As water passes over the respiratory surface of the gills, gases diffuse between the water and the blood. The blood then carries oxygen to the rest of the body (**Figure 23.15b**). Gills—especially those of large, active fish—must be very efficient to obtain enough oxygen from water. Indeed, some aquatic animals can remove more than 80% of the oxygen from the water moving over their gills.

(a) Skin. Some small animals, such as earthworms, use their entire moist outer skin as a respiratory organ.

► Figure 23.15 **Respiratory surfaces: skin and gills.** The yellow areas represent respiratory surfaces. The boxed enlargements show a portion of the respiratory surface in the process of exchanging O_2 and CO_2.

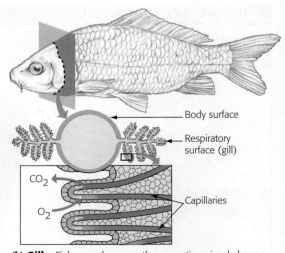

(b) Gills. Fishes and many other aquatic animals have gills, feathery respiratory surfaces that extend into the surrounding water.

In most land-dwelling animals, the respiratory surfaces are folded into the body rather than extending
from it. The infolded surfaces are open to the air only
through narrow tubes. These tubes allow the animals to
breathe air while retaining the moisture needed to maintain the plasma membranes of respiratory surface cells.

Insects breathe using a **tracheal system**, an extensive
network of branching internal tubes called tracheae
(**Figure 23.16a**). Tracheae begin near the body's surface
and branch down to narrower tubes that extend to
nearly every cell. Gas exchange with cells occurs via diffusion across the moist epithelium that lines the tips of

the tubes. Because almost every body cell is near the
respiratory surface, gas exchange in insects requires no
assistance from the circulatory system.

Lungs are the most common respiratory surface
among snails, some spiders, and terrestrial vertebrates
such as amphibians, birds and other reptiles, and mammals. In contrast to tracheae, which branch throughout
an insect's body, **lungs** are localized organs lined with
moist epithelium (**Figure 23.16b**).

Figure 23.17 summarizes the different types of respiratory organs we discussed in this section. In the next
section, we'll take a closer look at human lungs. ☑

► Figure 23.16 **Respiratory
surfaces: tracheae and
lungs.** The yellow areas
represent respiratory surfaces.
The boxed enlargements show
a portion of the respiratory
surface in the process of
exchanging O₂ and CO₂.

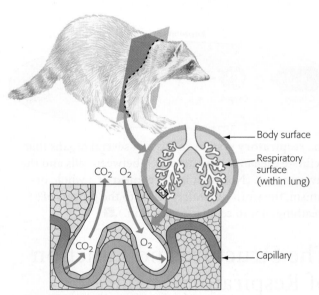

(a) **Tracheae.** Terrestrial insects have a
pervasive system of internal tubes called
tracheae that channel air directly to cells.

(b) **Lungs.** Most terrestrial vertebrates
exchange gases across the thin lining of lungs,
internal organs with a rich blood supply.

▼ Figure 23.17 **Types of respiratory organs.**

RESPIRATORY ORGANS			
Skin (entire body surface)	**Gills** (extensions of the body surface)	**Tracheae** (branching internal tubes)	**Lungs** (localized internal organs)
Moist skin of a leech	Gills of a sea slug	Tracheae of a silk moth caterpillar	Model of a pair of human lungs

The Human Respiratory System

Figure 23.18 provides an overview of three phases of gas exchange in humans. **①** The first step of gas exchange is **breathing**, the ventilation of the lungs by alternate inhalation and exhalation. When an animal with lungs breathes, a large, moist internal surface is exposed to air. Oxygen diffuses across the cells lining the lungs and into the surrounding blood vessels. Simultaneously, CO_2 passes out of the blood vessels, into the lungs, and is exhaled. Most land animals require a lot of oxygen. The inner tubes of the lungs are extensively branched, providing a large respiratory surface.

② The second step in gas exchange is the transport of O_2 from the lungs to the rest of the body via the circulatory system. The blood also carries CO_2 from the tissues back to the lungs. **③** In the final step of gas exchange, O_2 diffuses from red blood cells into body cells, which release CO_2 into the blood. The delivered O_2 is used by the body cells to obtain energy from food via the process of cellular respiration. This same process produces CO_2 as a waste product that cells pass to the blood. The circulatory system transports the CO_2 back to the lungs, where it is exhaled. ✔

Human Respiration

▼ Figure 23.18 **The three phases of gas exchange.**

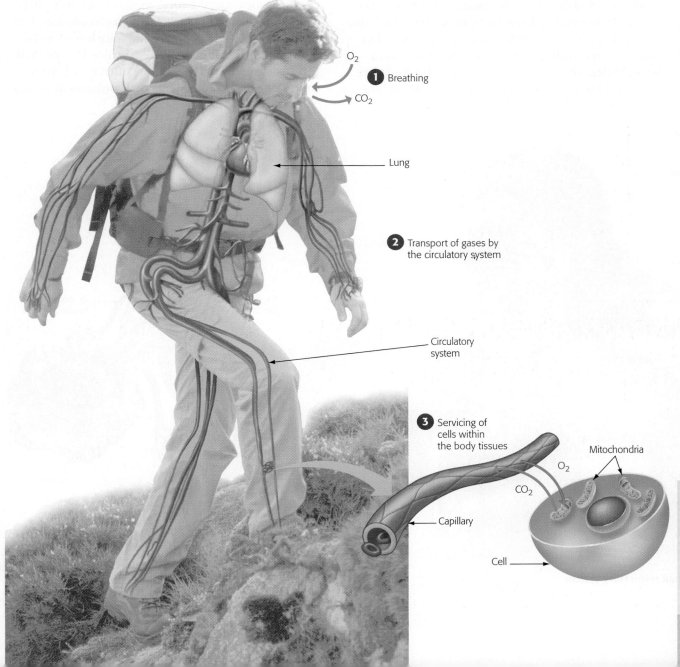

① Breathing

O_2

CO_2

Lung

② Transport of gases by the circulatory system

Circulatory system

③ Servicing of cells within the body tissues

Mitochondria

O_2

CO_2

Capillary

Cell

The Structure and Function of the Human Respiratory System

Figure 23.19a shows the human respiratory system, along with a few organs (labeled in italics) from other systems for orientation. Our lungs are located in the chest cavity, bordered along the bottom by the **diaphragm**, a sheet of muscle. Let's get an overview of the respiratory system by following the flow of air into the lungs.

Air enters the respiratory system through external openings: the nostrils and mouth. In the nasal cavity, the air is filtered by hairs, warmed, humidified, and sampled by smell receptors. The air passes to the **pharynx**, where the digestive and respiratory systems meet.

From the pharynx, air is inhaled into the **larynx** (voice box) and then into the **trachea** (windpipe). The trachea forks into two **bronchi** (singular, *bronchus*), one leading to each lung. Within the lungs, each bronchus branches repeatedly into finer and finer tubes called **bronchioles**. The system of branching tubes looks like an inverted tree, with the trachea as the trunk and the bronchioles as the smallest branches.

The bronchioles dead-end in grapelike clusters of air sacs called **alveoli** (singular, *alveolus*) (**Figure 23.19b**). Each of our lungs contains millions of these tiny sacs. The inner surface of each alveolus is lined with a layer of epithelial cells that forms the respiratory surface, where the exchange of gases actually takes place. O_2 enters the bloodstream by diffusing from the air into a web of blood capillaries that surrounds each alveolus. CO_2 diffuses from the blood in the capillaries into the alveoli. The CO_2 in the alveoli is then exhaled through the bronchioles to the bronchus, up through the trachea, and out of the body.

During exhalation, outgoing air moves over a pair of **vocal cords** within the larynx. Humans produce vocal sounds by flexing muscles in the voice box as air rushes by, stretching the vocal cords and making them vibrate. High-pitched sounds are produced by tensing the muscles and shortening the vocal cords, while lower-pitched sounds are made by reducing the tension and lengthening the cords. During puberty, the voice box of males grows rapidly, resulting in a deeper voice. During their period of growth, pubescent boys often have trouble controlling their lengthening vocal cords, resulting in a "breaking" voice. ☑

☑**CHECKPOINT**

Put the following parts of the human respiratory system in the order that an inhaled breath would encounter them: bronchi, larynx, nasal cavity, alveoli, pharynx, bronchioles, trachea.

Answer: nasal cavity → pharynx → larynx → trachea → bronchi → bronchioles → alveoli

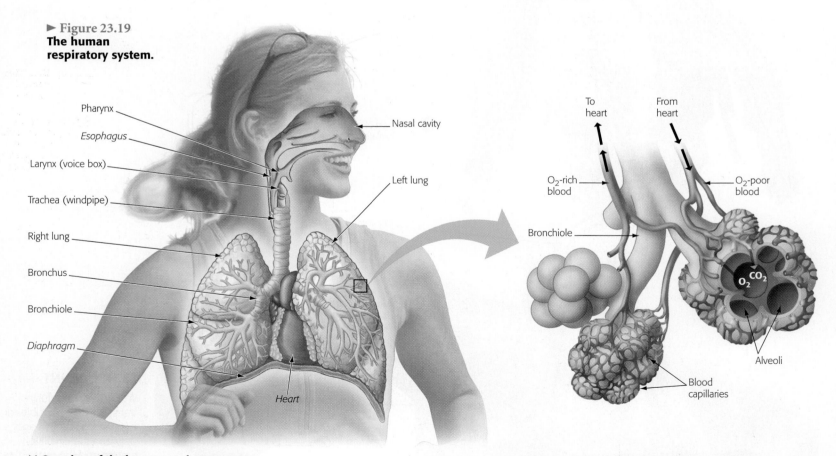

► **Figure 23.19**
The human respiratory system.

Pharynx
Esophagus
Larynx (voice box)
Trachea (windpipe)
Right lung
Bronchus
Bronchiole
Diaphragm
Heart
Nasal cavity
Left lung

To heart
From heart
O_2-rich blood
O_2-poor blood
Bronchiole
O_2 CO_2
Alveoli
Blood capillaries

(a) Overview of the human respiratory system

(b) The structure of alveoli. Capillaries surround air sacs called alveoli. These are the sites of gas exchange in the lungs.

Taking a Breath

Breathing consists of the alternating processes of inhalation and exhalation. **Figure 23.20** shows the changes that occur during breathing. If you place your hands on your rib cage and inhale, you can feel your ribs move upward and spread out as muscles between them contract. Meanwhile, your diaphragm moves downward, expanding the chest cavity. All of this increases the volume of the lungs, dropping the air pressure in the lungs to below the air pressure of the atmosphere. The result is that air rushes in through the mouth and nostrils toward the region of lower pressure, filling the lungs. Although it may seem that you actively suck in air when you inhale, air actually moves into your lungs passively after the air pressure in them drops. This pulling in of air as a result of lower air pressure in the lungs is called **negative pressure breathing**. (In contrast, amphibians use muscles to push air into their lungs, a process called positive pressure breathing.)

During exhalation, the rib and diaphragm muscles relax, decreasing the volume of the chest cavity. This decreased volume increases the air pressure inside the lungs, forcing air to rush out of the respiratory system. Movement of the diaphragm is vital to normal breathing. That is why a punch to the diaphragm can "knock the wind out of you." It shocks the diaphragm muscle, prevents movement of the chest cavity, and stops you from taking a normal breath. Other times, sudden involuntary contractions of the diaphragm force air through the voice box, causing hiccups.

You can consciously speed up or slow down your breathing. You can even hold your breath (although, despite the claims of angry children, you would pass out and return to normal breathing before you turned blue). Usually, however, you aren't aware of breathing; you certainly aren't aware of it when you're asleep. What, then, controls your breathing?

Most of the time, nerves from breathing control centers in the brainstem send signals that maintain a respiratory rate of 10–14 inhalations per minute. This rate can vary, however. **Figure 23.21** highlights one respiratory control system. **1** Levels of CO_2 in the blood affect breathing rate. When you exercise, for example, cellular respiration kicks into high gear, producing more ATP for your muscles and raising the amount of CO_2 in the blood. **2** When the brain senses the higher CO_2 level, **3** breathing control centers increase the breathing rate and depth. As a result, more CO_2 is eliminated in the exhaled air and more O_2 is provided to muscles.

You can become dizzy by hyperventilating—taking excessively rapid, deep breaths. This rapid breathing purges the blood of so much CO_2 that the control centers temporarily slow the signals being sent to the diaphragm and rib muscles. Changes in blood chemistry cause arterioles in the brain to constrict, decreasing the brain's blood supply and causing dizziness. Slow breathing continues until the CO_2 level increases enough to switch the breathing control centers back to their normal function. ☑

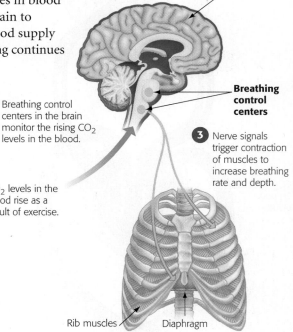

▼ Figure 23.21 **Control centers in the brain that regulate breathing.**

Brain

Breathing control centers

2 Breathing control centers in the brain monitor the rising CO_2 levels in the blood.

3 Nerve signals trigger contraction of muscles to increase breathing rate and depth.

1 CO_2 levels in the blood rise as a result of exercise.

Rib muscles

Diaphragm

▼ Figure 23.20 **How a human breathes.**

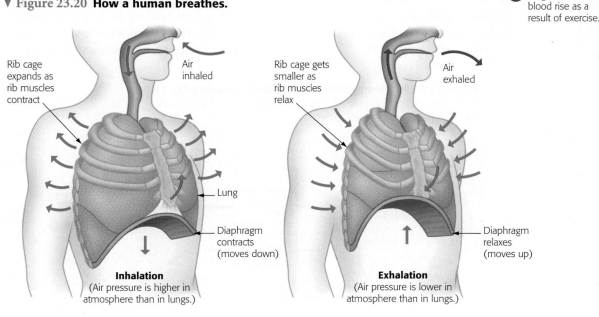

Rib cage expands as rib muscles contract

Air inhaled

Lung

Diaphragm contracts (moves down)

Inhalation
(Air pressure is higher in atmosphere than in lungs.)

Rib cage gets smaller as rib muscles relax

Air exhaled

Diaphragm relaxes (moves up)

Exhalation
(Air pressure is lower in atmosphere than in lungs.)

☑CHECKPOINT

When you take a breath, why is it described as negative pressure breathing?

Answer: Expansion of the chest cavity increases the volume of the lungs, which decreases the air pressure within them. The drop in pressure causes air to enter the lungs.

The Role of Hemoglobin in Gas Transport

The human respiratory system takes O_2 into the body and expels CO_2, but it then relies on the circulatory system to shuttle these gases between the lungs and the body's cells (Figure 23.22). In the lungs, O_2 moves from air spaces inside the alveoli into capillaries surrounding the alveoli. At the same time, CO_2 exits. The gases exchange by diffusion. O_2 moves out of the air and into the blood because air is richer in O_2. CO_2 also diffuses along a gradient, in this case from the blood to the air in the lungs.

There is a problem, however, with this simple scheme: Oxygen does not readily dissolve in blood, so oxygen will not tend to move from the air into the blood on its own. Solving this problem is the fact that most of the O_2 in blood is carried by hemoglobin molecules within red blood cells (Figure 23.23). Each molecule of hemoglobin consists of four polypeptide chains. Attached to each polypeptide is a chemical group called a heme ( ), at the center of which is an atom of iron (shown in black). Each iron atom can hold one O_2 molecule, so one molecule of hemoglobin can carry a maximum of four molecules of O_2. Each red blood cell has about 250 million molecules of hemoglobin, so one tiny cell can

carry 1 billion O_2 molecules. Hemoglobin picks up oxygen in the lungs, carries it to the body's cells, and releases it. When hemoglobin binds oxygen, it changes the color of blood to a bright cherry red. Oxygen-poor blood is a dark maroon that appears blue through the skin. This is why a person who has stopped breathing—such as the choking victim in the Biology and Society section—has blue skin.

Because iron (one atom per heme group) is so important in the structure of hemoglobin, a shortage of iron causes less hemoglobin to be produced by the body. In fact, iron deficiency is the most common cause of anemia. Women are more likely to develop iron deficiency than men because of blood lost during menstruation. Pregnant women can benefit from iron supplements to support the developing fetus and placenta.

Besides binding O_2, hemoglobin can bind carbon monoxide (CO), a colorless, odorless gas. In fact, CO binds to hemoglobin even more tightly than O_2 does. Breathing CO can therefore interfere with the delivery of O_2 to body cells. Because CO also interferes with cellular respiration, it can cause rapid death. Sitting in an idling car in an enclosed space can be fatal because CO is emitted as an air pollutant in the exhaust from gasoline-powered engines. And despite its potentially deadly effects, millions of Americans willingly inhale CO in the form of cigarette smoke.

How Smoking Affects the Lungs

Every breath you take exposes your respiratory tissues to potentially damaging chemicals. One of the worst sources of airborne pollutants is tobacco smoke. The visible smoke from burning tobacco is mostly microscopic particles of carbon. Sticking to the carbon particles are over 4,000 different chemicals, many of which are known to be harmful and even potentially deadly.

▼ **Figure 23.22 Gas transport and exchange in the body.** O_2-rich blood is sent from the lungs to capillaries in other body tissues via the heart. O_2 enters body tissue cells and CO_2 leaves, entering the blood. The O_2-poor blood is circulated back to the heart, then to the lungs.

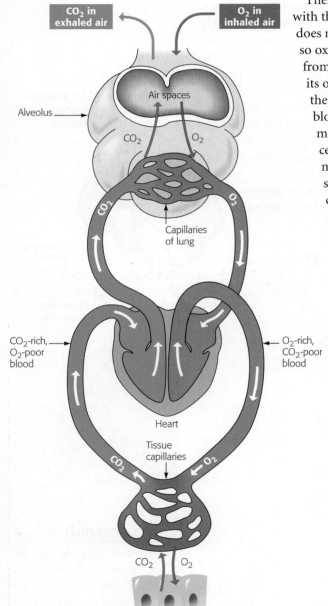

▼ **Figure 23.23 Hemoglobin loading and unloading O_2.** It is the protein hemoglobin within red blood cells that actually binds and transports oxygen molecules.

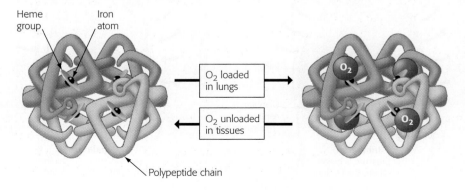

The epithelial tissue lining our respiratory system is extremely delicate. Its main protection is the mucus covering the cells and the beating cilia that sweep dirt particles and microorganisms off their surfaces (see Figure 4.22c). Tobacco smoke damages the cells that line the bronchi and trachea, inhibiting their ability to remove foreign substances from the airway. This interference with the normal cleansing mechanism of the respiratory system allows more toxin-laden smoke particles to reach and damage the lungs' delicate alveoli. Frequent coughing—common in heavy smokers—is the respiratory system's attempt to compensate and clean itself.

The health statistics associated with smoking are staggering. Every year, smoking kills about 440,000 Americans, more than all the deaths caused by accidents, alcohol and drug abuse, HIV, and murders *combined*. One in two American smokers will die from their habit. On average, adults who smoke cigarettes die 13 to 14 years earlier than nonsmokers. Smokers account for 90% of all cases of lung cancer, one of the deadliest forms of cancer: Only 15% of people diagnosed with lung cancer survive five years, and lung cancer kills more Americans than any other form of cancer. **Figure 23.24a** shows a cutaway view of a pair of healthy human lungs (and heart). In contrast, **Figure 23.24b** shows the lungs

of a smoker, turned black from the long-term buildup of smoke particles.

There is no lifestyle choice that can have a more positive impact on the long-term health of yourself and those with whom you live than not smoking. After quitting, it takes about 15 years for a former smoker's risk to even out with that of a nonsmoker, so not starting is clearly the best option. Unfortunately, even nonsmokers may be affected by tobacco, as studies have found that secondhand cigarette smoke is a health hazard, particularly to young children. ☑

☑**CHECKPOINT**

How does smoking damage the lungs?

Answer: Smoking damages the epithelial tissue that lines the respiratory system, preventing pollutants from being removed. These pollutants can then damage the delicate alveoli of the lungs where gas exchange occurs.

▼ **Figure 23.24 Healthy versus cancerous lungs.**

(a) **Healthy lungs (nonsmoker)**   (b) **Cancerous lungs (smoker)**

EVOLUTION CONNECTION: Lifesaving Procedures

Choked Up

In this chapter, we've used the topic of lifesaving procedures to illustrate the importance of the circulatory and respiratory systems. If a victim is choking but conscious, a procedure called the Heimlich maneuver may dislodge the object blocking the airway. Invented by Dr. Henry Heimlich in the 1970s, it involves quick thrusts to the diaphragm to compress the lungs and force air upward through the trachea (**Figure 23.25**).

The most common cause of choking is food stuck in the airway. If you review Figure 23.19, you'll notice that the esophagus (where food is supposed to go) and the trachea (where air is supposed to go) join into a single tube at the back of the pharynx. The fact that these two pipes intersect accounts for thousands of choking deaths per year.

Why doesn't the human body have two separate "mouths," one for food and one for air? A combined opening for food and air makes little sense if we examine it from a design perspective. Instead, we can examine it via biology's unifying theme: evolution.

Complex organs such as lungs did not evolve "on demand" when primitive tetrapods first moved onto land. Natural selection generally refines existing organs. Some evidence supports the hypothesis that our elaborate lungs evolved from much simpler versions that enabled shallow-water fishes to supplement gill breathing with gulps of air. The fossil evidence indicates that air-breathing organs first arose as pouches that branched off a tube that ran from mouth to stomach. Successive generations of descent with modification allowed these organs to become better adapted to breathing air without modifying the basic plan. Thus, our need for caution when eating (be sure to chew your food thoroughly!) reflects not a purpose-oriented design, but rather our shared evolutionary history with aquatic ancestors.

▼ **Figure 23.25 The Heimlich maneuver.** This lifesaving procedure exploits an evolutionary quirk: the branching of the respiratory and digestive systems from a single opening.

save a **CHOKING** victim
HEIMLICH MANEUVER®
It could save your life!

A choking person can't speak or breathe and needs your help now. Don't slap the victim's back. (This could make matters worse.)

Follow these 4 steps to clear the blocked airway safely and quickly:

1. From behind, wrap your arms around the victim's waist.

2. Make a fist and place the thumb side of your fist against the victim's abdomen, below the rib cage and above the navel.

3. Grasp your fist with your other hand and press into the victim's abdomen with a quick upward thrust.

4. Repeat until object is expelled.

Chapter Review

SUMMARY OF KEY CONCEPTS

 Go to the Study Area at **www.masteringbiology.com** for practice quizzes, eText, BioFlix™ 3-D animations, MP3 Tutor Sessions, videos, current events, and more.

Unifying Concepts of Animal Circulation

The circulatory system facilitates the exchange of materials and energy between the cells of an organism and its environment.

Open and Closed Circulatory Systems

In an open circulatory system, circulating fluid is pumped through open-ended vessels and circulates freely among cells. In a closed circulatory system, the circulating fluid (blood) is confined within vessels.

The Human Cardiovascular System

The Path of Blood

Trace the path of blood in this figure, noting where blood is oxygen-poor and where it is oxygen-rich.

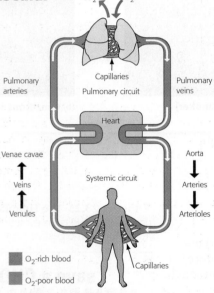

How the Heart Works

Review the structure of the heart by tracing the flow of blood in this diagram.

The cardiac cycle is composed of two phases: systole (contraction) and diastole (relaxation). The pacemaker, which sets the tempo of the heartbeat, generates electrical impulses that stimulate the atria and ventricles to contract.

Blood Vessels

Muscular arteries carry blood away from the heart. Exchange between the blood and interstitial fluid occurs across the thin walls of capillaries. Valves in the veins and contractions of surrounding skeletal muscle keep blood moving back to the heart.

Blood

CELLULAR COMPONENTS OF BLOOD		
Red Blood Cells (transport oxygen)	**White Blood Cells** (fight infections)	**Platelets** (allow for blood clotting)

New blood cells are continually formed from stem cells found in red bone marrow.

Cardiovascular Disease

Diseases of the heart and blood vessels—including heart attack and stroke—kill more Americans than any other type of disease. Atherosclerosis is the buildup of cholesterol and other substances in the inner walls of arteries.

Unifying Concepts of Animal Respiration

The respiratory system facilitates gas exchange.

The Structure and Function of Respiratory Surfaces

Various respiratory surfaces have evolved in animals. The respiratory surface is the part of the body where gas exchange takes place: the entire body surface, gills, tracheae, or lungs.

The Human Respiratory System

The Structure and Function of the Human Respiratory System

When you take a breath, air moves sequentially from the nostrils and/or mouth to the pharynx, larynx, trachea,

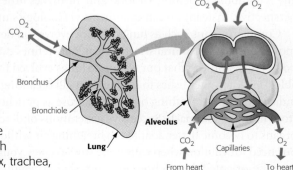

bronchi (which enter the lungs), bronchioles, and finally the alveoli (the actual respiratory surfaces).

Taking a Breath

During inhalation, the chest cavity expands and air pressure in the lungs decreases, causing air to rush into the lungs. During exhalation, air pressure in the lungs increases and air moves out of the lungs. Breathing rate is set by breathing control centers in the brain, which are influenced by CO_2 concentrations in the blood.

The Role of Hemoglobin in Gas Transport

After O_2 enters the lungs, it binds to hemoglobin in red blood cells and is transported to body tissue cells by the circulatory system.

How Smoking Affects the Lungs

Tobacco smoke damages the respiratory surfaces of the lungs, impairing alveolar function and causing multiple health problems.

SELF-QUIZ

1. What is the difference between an open circulatory system and a closed circulatory system?

2. Why is the following statement false? "All arteries carry oxygen-rich blood, while all veins carry oxygen-poor blood."

3. Match each chamber of the heart with its function.
   - a. left atrium
   - b. right atrium
   - c. left ventricle
   - d. right ventricle
   1. receives blood from the body via the venae cavae
   2. sends blood to the lungs via the pulmonary arteries
   3. receives blood from the lungs via the pulmonary veins
   4. sends blood to the body via the aorta

4. Some babies are born with a small hole in the wall of muscle that separates the left and right ventricles. How does this affect the oxygen content of the blood pumped out of the heart in the systemic circuit?

5. People who do not get enough iron in their diet run the risk of developing _____.

6. Match each element of the blood with its main function.
   - a. plasma
   - b. platelets
   - c. red blood cells
   - d. white blood cells
   1. transporting O_2
   2. fighting infections
   3. carrying dissolved elements
   4. aiding in clotting

7. A heart-healthy diet can significantly reduce your risk of cardiovascular disease. Name three foods that should be eaten in abundance for a heart-healthy diet, and name three dietary components that should be limited.

8. What is the main difference between the gills of fishes and the lungs of humans in terms of where they are located in the body?

9. The respiratory surface of vertebrate lungs consists of tiny sacs within the lungs called _____.

10. During inhalation, when the diaphragm contracts and the ribs spread apart, air pressure inside the lungs
    - a. increases.
    - b. decreases.
    - c. decreases, then increases.
    - d. is unchanged.

11. Why is carbon monoxide deadly?
    - a. It binds to hemoglobin in white blood cells in place of O_2.
    - b. It binds to hemoglobin in red blood cells in place of O_2.
    - c. It binds to hemoglobin in red blood cells in place of CO_2.
    - d. It binds to hemoglobin in white blood cells in place of CO_2.

Answers to the Self-Quiz questions can be found in Appendix D.

THE PROCESS OF SCIENCE

12. Mammals have a four-chambered heart, with two ventricles and two atria. So do birds, alligators, and crocodiles, but other living reptiles have a three-chambered heart, with just one ventricle. Paleontologists debate whether dinosaurs had a three-chambered or a four-chambered heart. Long-necked sauropod dinosaurs could have had unusual circulatory demands because their head may have been raised far above their heart. The farther the head is above the heart, the greater the systolic pressure needs to be for blood to reach the brain. For example, the long-necked dinosaur *Brachiosaurus* may have carried its head as much as 6 m (20 ft) above its heart. It is estimated that such an anatomy demanded a systolic blood pressure of 500 mm of mercury for blood to reach the brain! Some paleontologists consider this evidence that dinosaurs had a four-chambered heart that supported a dual circulatory system similar to that of birds and mammals, rather than the three-chambered heart of most nonbird reptiles. Can you explain why?

13. During a college swim meet, John told his friends that before a race he likes to "charge up" on oxygen by hyperventilating so he can hold his breath longer. Although John is correct that this will allow him to hold his breath longer, it is not for the reason that he gave. How is he mistaken? Why is this use of hyperventilation actually very dangerous?

BIOLOGY AND SOCIETY

14. Recently, a 19-year-old woman received a bone marrow transplant from her 1-year-old sister. The woman was suffering from a form of leukemia and was almost certain to die without a transplant. The woman's parents had decided to have another child in a final attempt to provide their daughter with a matching donor. Although the ethics of the parents' decision were criticized, doctors reported that this situation is not uncommon. In your opinion, is it acceptable to have a child provide tissue or an organ? Why or why not?

15. While administering first aid, emergency health-care workers are often exposed to a patient's body fluids. Should people be required to inform health-care workers if they have a contagious disease, such as hepatitis, tuberculosis, or AIDS?

24 The Body's Defenses

Fighting a killer.
Despite decades of trying,
no effective HIV vaccine
has been found.

BIOLOGY AND SOCIETY: HIV and AIDS

An AIDS Vaccine Failure

One of medical science's greatest triumphs is the development of vaccines against deadly viral diseases. A vaccine contains a harmless variant or piece of a disease-causing microbe. After injection, a vaccine can stimulate the body's defenses to prepare for mounting a vigorous attack if the actual virus is ever encountered. Because antibiotics are ineffective against viruses, vaccines are the only medical weapon that can be deployed against many harmful viruses.

One important target of vaccine research is HIV, the human immunodeficiency virus, which causes AIDS (acquired immunodeficiency syndrome). HIV is deadly because it destroys the defenses of its host, leaving the body unable to fight most invaders. In other words, HIV wipes out the very body defenses that would normally combat AIDS.

The largest AIDS vaccine trial was begun in 2004. The participants included 3,000 HIV-negative people from around the world. Half the participants received placebos in the form of saltwater injections, while half received shots containing a common cold virus genetically modified to include three harmless HIV components. The hope was that the vaccine would train the body to vigorously attack the real AIDS virus. In late 2007, researchers halted the trial when they discovered that new HIV infections occurred at the same rate in both groups. In other words, the vaccine appeared to offer no protection against HIV. Worse, the vaccine may have *increased* the risk of HIV infection in some participants. Researchers suspect that in people who are naturally resistant to the cold virus (presumably as a result of recent exposure), the body's defenses may have inactivated the vaccine. A study begun in 2008 tries to avoid this problem by formulating the vaccine from a rare but harmless virus that does not regularly infect humans.

The ongoing battle with HIV illustrates just how much we depend on our built-in system of defenses. This chapter examines how these defenses work to protect the body. We'll also look at how that knowledge has been applied to improve human health, what happens when the body's protective system malfunctions, and the continuing struggle against AIDS.

Innate Defenses

Nearly everything in the environment—including most of what we touch or taste and even the air we breathe—teems with microbes, some of which can make us sick. Yet we do not constantly become ill, thanks to the **immune system**, the body's system of defenses against infectious disease.

The human body has three cooperative lines of defense that protect us against the constant barrage of **pathogens**, disease-causing viruses and microorganisms (**Figure 24.1**). The first line of defense is external—barriers that prevent pathogens from getting deep inside the body. If that line fails, some internal protections stand ready as a second line of defense. These first two lines of defense are **innate defenses**—that is, they are already present and act the same whether or not an invader has been previously encountered. Nearly all animals have some types of innate defenses. Later in the chapter, we'll discuss the third line of defense, the **adaptive defenses** (also known as acquired defenses), which are internal defenses that must be activated by exposure to specific invaders. Only vertebrates have adaptive defenses; throughout this chapter, we'll focus on the innate defenses and adaptive defenses of humans.

External Innate Defenses

For a foreign substance to get inside the body, it must first get by the body's external barriers. These barriers form an important protective front line because they *prevent* infection, as opposed to the body's other defenses, which act only *after* infection has occurred.

Some barriers block or filter out microbes. Intact skin forms a tough outer layer that most bacteria and viruses cannot penetrate. Inside the nostrils, hairs filter many particles from the incoming air. Organ systems that are open to the external environment (such as the respiratory, digestive, reproductive, and urinary systems) are lined with mucous membranes; these membranes secrete mucus, a sticky fluid that traps bacteria, dust, and other particles. Beating cilia extending from cells of the respiratory tract sweep mucus with the trapped particles outward until it is swallowed or expelled by sneezing, coughing, or blowing the nose.

External barriers also include chemical defenses in the form of antimicrobial secretions. Sweat, saliva, and tears contain enzymes that disrupt bacterial cell walls. Glands produce oils and acids that make the skin inhospitable to many microbes. Concentrated stomach acid kills most of the bacteria we swallow before they can enter the bloodstream. ✓

▼ **Figure 24.1 Overview of the body's defenses.** Note that the lymphatic system is involved in both innate and adaptive defenses.

Internal Innate Defenses

When an invader breaches the body's external barriers, the problem of how to fend off attack changes substantially. Once within the body fluids and tissues, the invader is no longer an outsider. To fight pathogens within the body, the animal's immune system must detect foreign particles and cells. In other words, an immune system must carry out recognition, distinguishing nonself from self.

When the external barriers fail, a set of internal innate defenses comes into play. This second line of defense consists mainly of white blood cells and defensive proteins (**Figure 24.2**).

Two types of white blood cells are involved in internal innate defense: phagocytic cells and natural killer cells. **Phagocytic cells** engulf foreign cells or molecules and debris from dead cells by phagocytosis, or "cellular eating" (see Figure 5.18). **Natural killer (NK) cells** can often recognize virus-infected or cancerous body cells. When contact is made, the NK cells release chemicals that kill the diseased cells.

The defensive proteins that aid in internal innate defense work both indirectly and directly. **Interferons**, produced by virus-infected body cells, work indirectly by helping healthy cells resist damage by viral infection (**Figure 24.3**). Virus-infected cells ❶ make and release interferons that ❷ bind to plasma membrane receptors on nearby uninfected cells. This binding stimulates the healthy cells to ❸ produce proteins that inhibit viral reproduction. Because of their strong antiviral properties, injected interferons are sometimes used to treat certain viral diseases, such as hepatitis and genital warts. **Complement proteins** attack pathogens directly. Some coat the surfaces of microbes, making them easier for phagocytic cells to engulf. Others cut lethal holes in microbial membranes, causing the invading cells to burst. ☑

▼ Figure 24.2 **The body's second line of defense.** When the external barriers fail, a set of internal innate defenses stands ready as the body's next line of protection.

INTERNAL INNATE DEFENSES			
White Blood Cells		**Defensive Proteins**	
Phagocytic cells (engulf foreign cells or substances)	**Natural killer cells** (destroy infected body cells and cancerous cells)	**Interferons** (protect body cells against viral infection)	**Complement proteins** (cause invading microbial cells to lyse)

► Figure 24.3 **The action of interferon.** When a virus infects a body cell, the cell produces and releases interferon molecules that bind to receptors on healthy cells, stimulating them to produce antiviral proteins that inhibit the growth of other viruses.

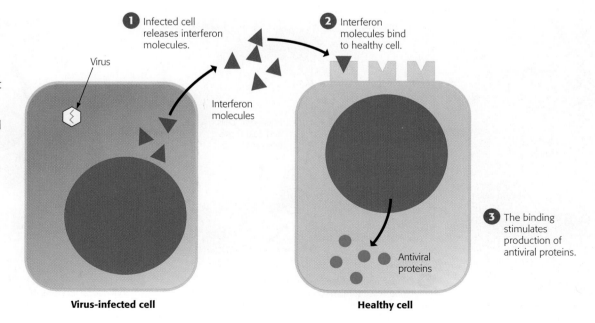

❶ Infected cell releases interferon molecules.

❷ Interferon molecules bind to healthy cell.

Virus

Interferon molecules

❸ The binding stimulates production of antiviral proteins.

Antiviral proteins

Virus-infected cell

Healthy cell

The Inflammatory Response

What happens if you get a splinter in your finger and don't take care of it? The next day, the finger is often red, swollen, painful, and warm to the touch. These are signs of the **inflammatory response** to injury and infection, another example of an internal innate defense in action (**Figure 24.4**). When tissue becomes damaged (from cuts, bug bites, or burns, for example), ❶ the injured cells and invading microbes release chemicals that trigger various innate defenses. One such chemical signal, **histamine**, ❷ causes nearby blood vessels to dilate and leak fluid into the wounded tissue, causing it to swell. The swelling helps the tissue to heal by diluting toxins, bringing extra oxygen, and delivering platelets and clotting proteins that promote scabbing. The chemical signals also attract phagocytic cells. ❸ These cells engulf bacteria and the remains of body cells killed by bacteria or by the physical injury. The pus that often fills an infected injury consists of phagocytic cells, fluid that leaked from capillaries, and other tissue debris.

Damaged cells also release chemical signals that increase blood flow to the damaged area, causing the wound to turn red and warm. (The word *inflammation* means "setting on fire.") In response to severe tissue damage or infection, damaged cells may release other chemicals that travel through the bloodstream to the brain, where they stimulate a fever that may discourage bacterial growth.

Anti-inflammatory drugs such as aspirin and ibuprofen dampen the normal inflammatory response and help reduce swelling and fever. This is an example of treating the symptoms of an illness (for example, fever) without addressing the underlying cause (which is the injury or infection).

All the defenses you've learned about so far are called innate because they're ready "off the rack"; that is, innate defenses are present in the body without any preparation. Soon you'll learn about the body's adaptive defenses—ones that are "custom-tailored" to specific invaders. But first let's look at a system that contributes to both innate and adaptive defenses: the lymphatic system. ☑

The Lymphatic System

The **lymphatic system** consists of a branching network of vessels, numerous **lymph nodes** (little round organs packed with white blood cells), and several other kinds of organs (**Figure 24.5**). The lymphatic vessels carry fluid called **lymph**, which is similar to the interstitial fluid surrounding body cells. The two main functions of the lymphatic system are to return tissue fluid to the circulatory system and to fight infection.

☑ CHECKPOINT

Name the signs of the inflammatory response.

Answer: swelling, redness, warmth, pain, and sometimes fever

▼ Figure 24.4 **The inflammatory response.** Whenever tissue is damaged, the body responds with a coordinated set of internal innate defenses called the inflammatory response.

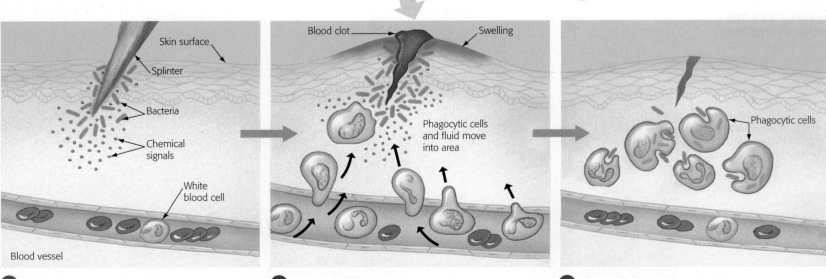

❶ Tissue injury; release of chemical signals such as histamine

❷ Dilation and increased leakiness of local blood vessels; migration of phagocytic cells to the area

❸ Phagocytic cells engulf bacteria and cell debris; tissue heals

A small amount of the fluid that enters the tissue spaces from the blood in a capillary bed does not reenter the blood capillaries. Instead, it flows into lymphatic capillaries that are intermingled with blood capillaries. Now called lymph, this fluid drains from the lymphatic capillaries into larger and larger lymphatic vessels. Eventually, the fluid reenters the circulatory system through two large lymphatic vessels that fuse with veins near the shoulders.

When your body is fighting an infection, the lymphatic system is the main battleground. Lymph nodes fill with a huge number of white blood cells, causing the "swollen glands" in your neck and armpits that your doctor feels for as a likely sign of infection. Because lymphatic vessels penetrate nearly every tissue, lymph can pick up microbes from infection sites just about anywhere in the body and deliver them to the lymphatic nodes and other organs. As the lymph passes through the lymphatic tissue, phagocytic cells engulf invaders. Lymph also carries white blood cells involved in the adaptive defenses, which we'll discuss next. ☑

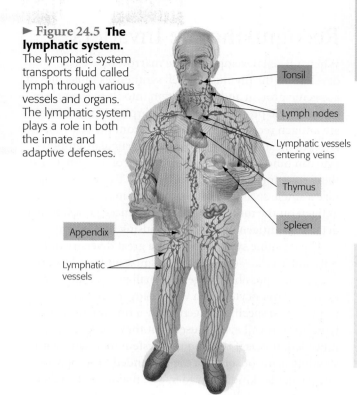

► **Figure 24.5 The lymphatic system.** The lymphatic system transports fluid called lymph through various vessels and organs. The lymphatic system plays a role in both the innate and adaptive defenses.

Tonsil

Lymph nodes

Lymphatic vessels entering veins

Thymus

Spleen

Appendix

Lymphatic vessels

☑ CHECKPOINT

What are the two functions of the lymphatic system?

Answer: to collect fluid that leaks from blood capillaries within tissues and to help protect the body from foreign cells and molecules.

Adaptive Defenses

When the innate defenses fail to ward off a pathogen, the adaptive defenses provide a third line of defense. This third line is called adaptive defenses because it is activated only after exposure to specific pathogens. Adaptive defenses depend on white blood cells called **lymphocytes** that recognize and respond to specific invading pathogens. There are two types of lymphocytes: B cells and T cells. Like all blood cells, they originate from stem cells in the bone marrow (**Figure 24.6**). Some immature lymphocytes continue developing in the bone marrow; these become specialized as **B cells**. Others migrate via the blood to the thymus, a gland in the chest, where they become specialized as **T cells**. Both B cells and T cells eventually make their way to the lymph nodes and other lymphatic organs. Over 2 trillion lymphocytes are spread throughout the bloodstream and lymphatic system, making possible an elaborately coordinated response to infection.

Whereas the innate defenses are always ready to fight a variety of infections, lymphocytes must be primed by exposure to a foreign molecule. Any molecule that elicits a response from a lymphocyte is called an **antigen**. Most antigens are molecules on the surfaces of viruses or foreign cells, such as cells of bacteria or parasitic worms. Antigens can also be molecules from mold spores, pollen, and house dust, as well as molecules on cell

surfaces of transplanted tissue. We'll look at how lymphocytes recognize antigens and then explore how they respond to these invaders. ☑

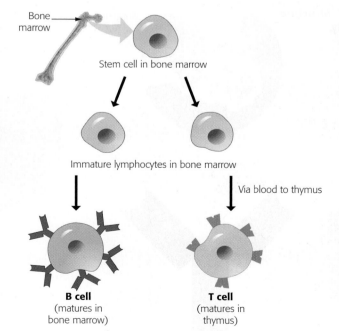

Bone marrow

Stem cell in bone marrow

Immature lymphocytes in bone marrow

Via blood to thymus

B cell (matures in bone marrow)

T cell (matures in thymus)

Via blood

Lymph nodes, spleen, and other lymphatic organs

☑ MP3 tutor sessions

The Human Immune System

◄ **Figure 24.6 The development of lymphocytes.** All lymphocytes arise in the bone marrow, but they mature in different locations: B cells in the bone marrow and T cells in the thymus. Once mature, lymphocytes are released into the blood and lymphatic system.

☑ CHECKPOINT

1. What are the two main types of lymphocytes?
2. What is an antigen? Name some examples.

Answers: 1. B cells and T cells 2. An antigen is a molecule that elicits a response from a lymphocyte. Examples include pollen, dust, and proteins on the surfaces of microorganisms.

Recognizing the Invaders

When a B cell develops in bone marrow or a T cell develops in the thymus, the cell synthesizes molecules of a specific protein and builds them into its plasma membrane. These protein molecules (purple in Figure 24.6) are antigen receptors that stick out from the cell's surface. All the antigen receptors on a particular lymphocyte can bind to only one specific type of antigen. Once these receptors are in place, a B cell or T cell can recognize a single specific antigen. One cell may recognize an antigen on the mumps virus, for instance, while another detects an antigen on a tetanus-causing bacterium.

The immune system develops a great diversity of B cells and T cells—enough to recognize and bind to just about every possible antigen. A small population of each kind of lymphocyte lies in wait, ready to recognize and respond to a specific antigen. Only a tiny fraction of the lymphocytes will ever be used, but they are all available if needed. It is as if the immune system maintains a huge standing army of soldiers, each intended to recognize one particular kind of invader. The majority of soldiers never encounter their target and remain idle. But when an invader does appear, chances are good that some lymphocytes will recognize and bind to it.

After binding to an antigen, a B cell can give rise to short-lived cells that secrete receptor-like molecules that circulate in the blood and lymph. Such a secreted molecule is called an **antibody** and is specific for that same antigen (**Figure 24.7**). Every antibody molecule consists of four polypeptide chains joined together to form a Y shape. The tip of each arm of an antibody forms an antigen-binding site, a region responsible for recognizing and binding to the antigen. An antigen-binding site and the antigen it binds have complementary shapes, like an enzyme and substrate or a lock and key. A huge variety in the shapes of the binding sites of different antibodies enables the immune system to react to any kind of antigen. The main role of an antibody is to mark an antigen by combining with it to form an antigen-antibody complex. As you'll see later, such binding triggers responses that help destroy the invader.

Because of their ability to tag specific molecules or cells, antibodies are widely used in laboratory research, in clinical diagnosis, and in the treatment of disease. **Monoclonal antibodies** are antibodies produced by cells that are descended from a single cell and therefore produce identical antibodies specific for a single antigen. Using cell cultures, researchers can artificially produce large quantities of monoclonal antibodies, which have many uses. Some are used to treat disorders of the immune system, such as transplant rejection and autoimmune diseases, particularly rheumatoid arthritis. Herceptin, a genetically engineered monoclonal antibody, is used to treat certain cases of aggressive breast cancer. A common type of home pregnancy test uses monoclonal antibodies to detect a hormone called human chorionic gonadotropin (HCG), present in the urine of pregnant women. When a testing strip is dipped into urine, monoclonal antibodies on it bind to any HCG that is present, causing a color change on the strip. And antibodies are one of the keys to diagnosing HIV infection, as you'll see next. ☑

▶ Figure 24.7 **Structure of an antibody.**

Antigen

Antigen-binding site

Antigen-binding sites

Antibody

Computer model of an antibody

☑ **CHECKPOINT**

What are antibodies? What is their role in the adaptive immune response?

Answer: receptor proteins secreted by short-lived B cells; to mark an antigen by binding to it

How Was HIV Linked to AIDS?

The disease now called AIDS was first recognized by doctors in 1981. **Observation** of the spread of AIDS among sexual partners, intravenous drug users, and blood transplant recipients strongly suggested that some infectious agent caused the disease (rather than, say, an environmental poison). But the key **question**—What pathogen causes AIDS?—remained unanswered for three years after AIDS was first described.

In 1984, a group of researchers led by Robert Gallo, of the U.S. National Cancer Institute, announced a major breakthrough. Their work began with the **hypothesis** that AIDS was caused by a new virus discovered by French biologist Luc Montagnier (who received a 2008 Nobel Prize for his discovery). This hypothesis led to the **prediction** that the blood of people with AIDS would contain antibodies to this new virus.

Their **experiment**, summarized in **Figure 24.8**, used one kind of antibody to test for another. First, blood samples were taken from people with AIDS and from negative controls (people who did not engage in risky behaviors). Next, the blood was added to a dish that had **1** fragments from the suspected virus stuck to the bottom. **2** If the blood had antibodies to the virus, they would **3** bind to the virus, forming an antibody-virus complex that was fixed to the dish. **4** Finally, another kind of antibody, one that binds to any human antibody, was added along with an enzyme that **5** would react with the complex to produce a color. The dishes were washed, so any remaining visible color indicated antibodies that had bound to the virus (which was attached to the dish so it wouldn't wash away). Their **results** showed that almost every patient with AIDS had antibodies to the new virus, whereas none of the control people did. In a series of classic papers, Gallo combined these results with other evidence to prove that the new virus, now called HIV, causes AIDS.

▼ **Figure 24.8 HIV blood test.** The test described here is still used today as a diagnostic tool to test people for HIV infection. (The molecules shown in this figure are not to scale; they are actually much smaller than the wells.)

Fragments from HIV

Antibody to HIV

Detector antibody

1 Fragments from HIV fixed to dish

2 Patient's blood added

3 Antibodies bind to HIV (if present)

4 Detector antibody added

5 Color change indicates presence of anti-HIV antibodies

Responding to the Invaders

Now that you know how lymphocytes recognize antigens, we'll look at how they fight these intruders. B cells and T cells have distinct responses, but they carry out a coordinated attack along with the innate defenses.

Clonal Selection: Multiplying Lymphocytes

Recall that the immune system maintains a vast collection of different kinds of white blood cells. With so many kinds of B cells and T cells, how does the body marshal enough of the right kind of lymphocyte to fight a specific invading antigen? The key is a process called **clonal selection**. At first, an antigen activates only a tiny number of lymphocytes with receptors specific to that antigen. These "selected" cells then proliferate, forming a clone of cells (a population of genetically identical B cells or T cells) that are specific for the stimulating antigen.

Figure 24.9 illustrates how clonal selection of lymphocytes works, using B cells as an example. Each B or T cell has its own specific type of antigen receptor embedded in its surface. Once an antigen enters the body, it binds with one of the cells that have complementary receptors. Those without the appropriate binding sites are not affected. The binding activates the cell—it grows, divides, and differentiates further. The results are clones of cells specialized for defending against the very antigen that triggered the response. Some of the newly produced cells are short-lived cells that have an immediate effect against the antigen and are therefore called **effector cells**. In the case of B cells, the effector cells secrete antibodies into the blood and lymph—as many as 2,000 copies per second. These antibodies bind to the invading antigen and, as you'll see later, contribute to its destruction or block its harmful effects.

The first response to exposure of lymphocytes to an antigen, called the **primary immune response**, takes several days to produce effector cells via clonal selection. The antibodies produced by effector B cells reach their peak levels about two weeks after first exposure and then start to decline. Each effector cell lives only four or five days, and the primary immune response subsides as the effector cells die out. Clonal selection also produces memory cells, which we'll discuss next.

▼ **Figure 24.9 Clonal selection.** In this example of clonal selection, an antigen "selects" those B cells that bind to it and causes their proliferation. A similar kind of clonal selection operates on T cells.

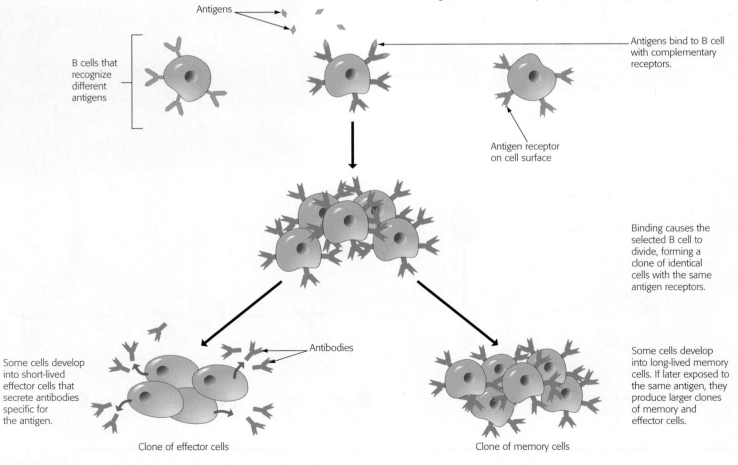

Antigens

B cells that recognize different antigens

Antigens bind to B cell with complementary receptors.

Antigen receptor on cell surface

Binding causes the selected B cell to divide, forming a clone of identical cells with the same antigen receptors.

Some cells develop into short-lived effector cells that secrete antibodies specific for the antigen.

Antibodies

Some cells develop into long-lived memory cells. If later exposed to the same antigen, they produce larger clones of memory and effector cells.

Clone of effector cells

Clone of memory cells

Immunological Memory

As you saw at the bottom right of Figure 24.9, clonal selection produces not only effector cells but also memory cells. **Memory cells** are long-lived cells that respond to subsequent exposures to the antigen by giving rise to effector cells and also more memory cells. Thus, clonal selection produces not only cells that will fight the first exposure to an antigen but also cells that will respond to future exposures.

In contrast to short-lived effector cells, memory cells can last decades in the lymph nodes, ready to be activated by a second exposure to the antigen. When the antigen is encountered again, the memory cells that bind to it initiate a faster and stronger response called the **secondary immune response**. The activated memory cells multiply quickly, producing a large new clone of lymphocytes. In the case of B cells, the new clone produces antibodies that are often more effective than those produced during the primary immune response.

Thus, with adaptive defenses, unlike innate defenses, exposure to a particular foreign molecule enhances future responses to that same molecule. For example, if a person gets chicken pox, memory cells are formed that have receptors against molecules on the virus that causes this disease. If the virus enters the body again, the memory cells react quickly, giving rise to effector cells that usually destroy the virus before symptoms appear. One early observation of the effect of the immune system's remarkable "memory" is in the writings of the Greek historian Thucydides, almost 2,400 years ago. He noted that survivors of the plague could safely care for people who were sick or dying of that disease, "for the same man was never attacked twice—never at least fatally."

Vaccination

Immunity is usually acquired by natural infection, but it can also be achieved by **vaccination** (also called immunization). In this procedure, the immune system is confronted with a **vaccine** composed of a harmless variant of a disease-causing microbe or one of its parts. The vaccine stimulates the immune system to mount defenses against these antigens. These defenses will also be effective against the actual pathogen because it has the same antigens. When a person has been successfully vaccinated, the immune system responds quickly if exposed to the pathogen.

In the United States, most children receive a series of vaccinations starting soon after birth, including vaccinations against diphtheria/pertussis/tetanus (DPT, or "diptet"), polio, hepatitis, chicken pox, and measles/mumps/rubella (MMR). In the United States, widespread childhood vaccination has virtually eliminated several of these diseases, including polio and mumps. As noted in the Biology and Society section, researchers are trying to develop an HIV vaccine, although success remains elusive (for reasons to be discussed in the Evolution Connection section). One of the major triumphs of modern vaccination involves smallpox, a potentially fatal viral infection that affected over 50 million people a year worldwide in the 1950s. A massive vaccination effort has been so effective that there have been no cases of smallpox since 1977. Starting in 2001, however, the U.S. government has stockpiled millions of doses of smallpox vaccine and has begun to vaccinate high-risk military and health-care workers in case the smallpox virus is used in a terrorist attack (**Figure 24.10**).

So far, you've seen that the binding of antigens triggers lymphocytes to multiply, producing clones of effector cells and memory cells. Next we'll look at how the effector cells disable or destroy the intruders. ☑

☑CHECKPOINT

Given that there are so many possible pathogens that can invade your body, how does your immune system produce enough of the right cells to fight off a major infection?

Answer: The process of clonal selection causes those few cells with receptors that are complementary to the antigens of the infectious agent to selectively multiply.

▼ Figure 24.10 **Vaccinating a soldier against the smallpox virus (inset).**

Colorized SEM 68,000×

B Cells and the Humoral Immune Response

The secretion of antibodies into the blood and lymph by effector B cells is called the **humoral immune response**. (The word *humoral* comes from *humors*, an old term for blood and other body fluids.) This response defends primarily against bacteria and viruses in body fluids. Sometimes the binding of antibodies neutralizes antigens by physically blocking them. For example, the attachment of antibodies to viral surface proteins may inhibit the virus from entering body cells. Other times, the binding to viruses or foreign cells causes the formation of clumps of pathogens that are easily captured by circulating phagocytic cells. Additionally, the antigen-antibody complex may activate complement proteins that destroy the antigen-carrying cells.

T Cells and the Cell-Mediated Immune Response

Whereas the humoral immune response involves antibodies secreted by effector B cells into the blood and lymph, effector T cells respond directly to antigens that have already entered body cells. Because this response is carried out directly by effector cells rather than by antibodies, it is called the **cell-mediated immune response**. It involves two main kinds of effector T cells: helper T cells and cytotoxic T cells.

Helper T cells bind to other white blood cells that have previously encountered an antigen. In the example in **Figure 24.11**, ❶ a phagocytic cell engulfs a microbe and breaks it into fragments, which are foreign antigens (*nonself* molecules). Protein molecules belonging to the phagocytic cell, which are referred to as *self* proteins because they belong to the body itself, ❷ bind to the foreign antigens and ❸ display them on the cell's outer surface. ❹ Receptors embedded in the helper T cell's plasma membrane recognize and bind to the combination of self protein and nonself molecule displayed on the phagocytic cell.

Colorized SEM 4,500×

Phagocytic cell (yellow) engulfing a foreign cell.

▼ Figure 24.11 **Interaction of a phagocytic cell and a helper T cell.**

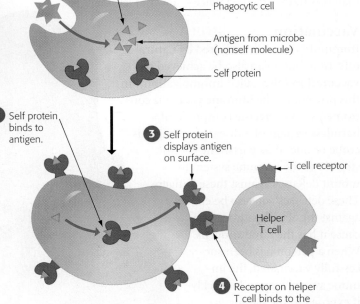

❶ Phagocytic cell breaks microbe into antigen fragments.

Microbe

Phagocytic cell

Antigen from microbe (nonself molecule)

Self protein

❷ Self protein binds to antigen.

❸ Self protein displays antigen on surface.

T cell receptor

Helper T cell

❹ Receptor on helper T cell binds to the protein-antigen combination.

▼ **Figure 24.12 The roles of an activated helper T cell.** Activated helper T cells promote both the cell-mediated immune response (via cytotoxic T cells) and humoral immune response (via B cells).

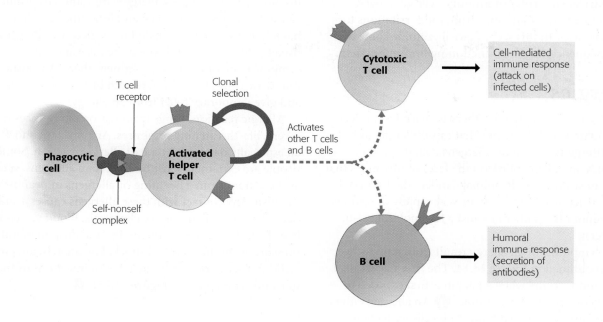

The binding of a T cell receptor to a self-nonself complex (combination) activates the helper T cell, the first step in T cell clonal selection (**Figure 24.12**). An activated helper T cell responds in several ways. It grows and divides, producing more activated helper T cells, as well as memory T cells (not shown in figure). In addition, it stimulates the activity of cytotoxic T cells and helps activate B cells, thus engaging the humoral immune response as well.

Cytotoxic T cells are the only T cells that actually kill other body cells. They identify infected body cells in the same way that helper T cells identify their targets: through binding of a receptor to a self-nonself complex. As shown in **Figure 24.13**, ❶ a cytotoxic T cell binds to an infected cell. The binding activates the T cell. ❷ The activated cell synthesizes and discharges several proteins, including one called perforin that makes holes in the infected cell's plasma membrane. ❸ Other T cell proteins enter the infected cell through the newly created hole and trigger a process that results in ❹ death of the cell by lysis (bursting). ☑

☑ CHECKPOINT

What are the two main types of effector T cells, and what does each do?

Answer: Helper T cells stimulate other body defenses. Cytotoxic T cells destroy infected cells.

▼ **Figure 24.13 How a cytotoxic T cell kills an infected cell.**

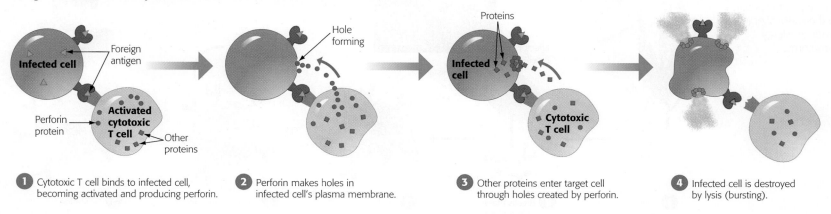

❶ Cytotoxic T cell binds to infected cell, becoming activated and producing perforin.

❷ Perforin makes holes in infected cell's plasma membrane.

❸ Other proteins enter target cell through holes created by perforin.

❹ Infected cell is destroyed by lysis (bursting).

Immune Disorders

If the intricate interplay of immune cells goes awry, problems can arise that range from mild irritations to deadly diseases. In this section, we'll examine some of the consequences of a malfunctioning immune system.

Allergies

Allergies are sensitivities to otherwise harmless antigens in the environment. Antigens that cause allergies are called **allergens**. Common allergens include protein molecules on pollen grains, on the feces of tiny mites that live in house dust, and in animal dander (shed skin cells). Allergic reactions typically occur very rapidly in response to tiny amounts—sometimes just a few molecules—of an allergen.

The symptoms of an allergy result from a two-stage reaction outlined in **Figure 24.14**. The first stage, called sensitization, occurs when a person is first exposed to an allergen—pollen, for example. **1** An allergen enters the bloodstream. There it binds to B cells with complementary receptors, tagging them for clonal selection (as in Figure 24.9). **2** The B cells then proliferate and secrete large amounts of antibodies to that allergen. **3** Some of these antibodies attach to receptor proteins on the surface of mast cells, normal body cells that produce histamine and other chemicals that trigger the inflammatory response.

Allergy symptoms do not arise until the second stage, which begins when the same allergen enters the body and **4** binds to the antibodies on the mast cells. In response to this binding, **5** the mast cells release histamine, which triggers allergy symptoms. Histamine causes blood vessels to dilate and leak fluid, causing nasal irritation, itchy skin, and tears. Because allergens usually enter the body through the nose and throat, symptoms are often most prominent there. Antihistamines are drugs that interfere with histamine's action and give temporary relief from an allergy.

Allergic reactions range from seasonal nuisances to severe, life-threatening responses. Anaphylactic shock is an especially dangerous type of allergic reaction. Some people are extremely sensitive to certain allergens, such as the venom from a bee sting or allergens in peanuts or shellfish. Any contact with these allergens causes a sudden release of inflammatory chemicals. Blood vessels dilate abruptly, causing a potentially fatal drop in blood pressure, a condition called shock. Fortunately, anaphylactic shock can be counteracted with injections of the hormone epinephrine (**Figure 24.15**). ☑

☑CHECKPOINT

Why does it take at least two exposures to an allergen to trigger an allergic reaction?

Answer: The first time you are exposed to an allergen, it merely sensitizes you by causing the attachment of specific antibodies to mast cells. A second exposure causes the mast cells to release chemicals that trigger the symptoms of an allergic response.

▼ Figure 24.15 **Single-use epinephrine syringes.** People who know they have severe allergies can carry single-use syringes that contain the hormone epinephrine (also called adrenaline). An injection can quickly stop a life-threatening allergic reaction.

► Figure 24.14 **How allergies develop.** The first exposure to an allergen causes sensitization. Subsequent exposures to the same allergen produce allergy symptoms.

SENSITIZATION: INITIAL EXPOSURE TO ALLERGEN

Ragweed pollen grains

Colorized SEM 10,000×

Effector B cell

Histamine

Mast cell

1 Allergen (pollen grain) enters bloodstream.

2 B cells make antibodies.

3 Antibodies attach to mast cell.

LATER EXPOSURES TO SAME ALLERGEN

4 Allergen binds to antibodies on mast cell.

5 Histamine is released, causing allergy symptoms.

Autoimmune Diseases

The immune system's ability to distinguish self from nonself enables it to battle foreign invaders without harming healthy body cells. Self proteins on the surfaces of cells are the key to this ability. Each person's cells have a particular collection of self proteins that mark body cells as "off-limits" to attacks from the immune system. Because every person has a unique set of self proteins, the immune system can distinguish the body's cells from those of other people and from microbes.

The immune system's ability to recognize and attack foreign antigens does not always work in our favor, however. For example, when a person receives an organ transplant, the immune system recognizes the donor's cells as foreign and attacks them. To minimize rejection, doctors look for a donor with self proteins matching the recipient's as closely as possible (often a close relative) and use drugs to suppress the immune response. Such drugs greatly reduce the risk of rejection but increase the risk of infections and must often be administered for life.

Autoimmune diseases result when the immune system improperly turns against the body's own molecules. In systemic lupus erythematosus (lupus), B cells make antibodies against many sorts of self molecules, even proteins and DNA released by the normal breakdown of body cells. Rheumatoid arthritis is an autoimmune disease that leads to damage and painful inflammation of the cartilage and bones of joints **(Figure 24.16)**. In insulin-dependent diabetes, the insulin-producing cells of the pancreas are the targets of cytotoxic T cells. In multiple sclerosis (MS), T cells wrongly attack proteins in neurons (see Figure 27.2), often causing a number of serious neurological problems. Recent research suggests that Crohn's disease, a chronic inflammation of the digestive tract, may be caused by an autoimmune reaction against bacteria that normally inhabit the intestinal tract.

Immunodeficiency Diseases

In contrast to autoimmune diseases are a variety of defects called **immunodeficiency diseases**. Immunodeficient people lack one or more of the components of the immune system and are therefore susceptible to infections that would not ordinarily cause a problem. A rare inborn disease called severe combined immunodeficiency (SCID) causes a marked deficit in both T cells and B cells. People with SCID are extremely vulnerable to even minor infections, forcing them to live behind protective barriers (as related in various "bubble boy" stories) or to receive bone marrow transplants.

Immunodeficiency is not always an inborn condition. For instance, Hodgkin's disease, a type of cancer that affects lymphocytes, can depress the immune system. Radiation therapy and drug treatments used against many cancers can have the same effect. Next we will look at the most dreaded acquired immunodeficiency disease of our time: AIDS. ✔

▼ Figure 24.16 **Rheumatoid arthritis.** The hands in this photo and X-ray are from a person with rheumatoid arthritis, an autoimmune disease in which the cartilage and bones of joints becomes damaged, resulting in malformation of the hands.

AIDS

Between 1981 and 2008, AIDS (acquired immunodeficiency syndrome) killed more than 30 million people worldwide, and more than 33 million people are currently living with the AIDS virus, HIV. In 2007, 2.7 million people were newly infected with HIV, and over 2 million died, including over 300,000 children under the age of 15. The vast majority of HIV infections and AIDS deaths occur in Asia and sub-Saharan Africa.

HIV is deadly because it destroys the immune system, leaving the body defenseless against most invaders. HIV can infect a variety of cells, but it most often attacks helper T cells—the cells that activate other T cells and B cells (see Figure 24.12). When HIV depletes the body of helper T cells, both the cell-mediated and humoral immune responses are severely impaired. Death usually results not from HIV itself but from another infectious agent or from cancer. The name AIDS refers to the fact that the disease is *acquired* through an infection that results in severe *immunodeficiency* due to a lack of helper T cells and presents as a *syndrome*, a combination of symptoms. (See Figure 10.32 to review the course of cell infection by HIV.)

Education is the best weapon against the spread of AIDS. Safe sex behaviors, such as reducing promiscuity and using condoms, can save many lives. (Chapter 26 includes further discussion of AIDS prevention.)

EVOLUTION CONNECTION: HIV and AIDS

HIV Evolution

Throughout this chapter, we have returned to the topic of HIV and AIDS as a way to highlight the importance of the immune system. To understand why HIV is so harmful, we need to keep in mind biology's unifying theme of evolution. As HIV reproduces, frequent mutations generate new strains of the virus. In fact, HIV has one of the fastest rates of mutation of any pathogen ever studied. This startling fact has led researchers to view evolution of drug-resistant HIV strains as the main obstacle to eradicating AIDS.

At one time, there was great hope that a "cocktail" of three anti-AIDS drugs, each of which attacks a different part of the HIV reproductive cycle, could eliminate the virus in an infected person (**Figure 24.17**). It was hoped that virus strains resistant to one drug would be defeated by another. Such hope greatly underestimated the ability of HIV to evolve. Although people with access to such drugs do survive much longer and have a greatly improved quality of life, some HIV strains resist even multidrug treatments. Thus, the virus is usually not eliminated from a patient's immune system. Although current HIV treatment prolongs the life of HIV-positive people, it is not a cure for AIDS.

Disturbingly, drug-resistant HIV strains are now found in newly infected patients. This demonstrates that HIV is adapting through natural selection to a changing environment—one in which drug treatments are widely available. In other words, the presence of anti-AIDS drugs in the environment has favored the spread of drug-resistant strains. Thus, the battle continues, with medical science on one side and the constantly evolving HIV on the other.

Human helper T cell   HIV

Colorized TEM 7,000×

▶ **Figure 24.17 Anti-AIDS drugs.** A three-drug cocktail can help slow, but usually not eliminate, the spread of HIV (virus particles shown on the T cell at right).

Chapter Review

SUMMARY OF KEY CONCEPTS

 Go to the Study Area at **www.masteringbiology.com** for practice quizzes, eText, BioFlix™ 3-D animations, MP3 Tutor Sessions, videos, current events, and more.

Innate Defenses

The human body contains two lines of defense (external barriers and internal innate defenses) that are innate, fully ready to respond before an invader has been encountered.

INNATE DEFENSES			
External	**Internal**		
• Skin • Mucous membranes • Secretions	White blood cells	Defensive proteins	The inflammatory response
	• Phagocytic cells • Natural killer cells	• Interferons • Complement proteins	• Involves chemical signals and phagocytic cells

The lymphatic system is a network of lymphatic vessels and organs involved in both innate and adaptive defenses. The vessels collect fluid from body tissues and return it to the blood. The lymphatic organs, including the lymph nodes, are packed with white blood cells that fight infections.

Adaptive Defenses

The adaptive defenses consist of a large collection of lymphocytes, white blood cells that respond to specific invaders. Antigens are molecules that elicit responses from lymphocytes.

Recognizing the Invaders

Two kinds of lymphocytes carry out the adaptive immune response to specific invaders carrying nonself molecules.

LYMPHOCYTES (Immune system contains millions of different kinds.)	
B cells (humoral immune response)	**T cells** (cell-mediated immune response)
• Secrete antibodies, which mark invaders	• Attack infected cells • Help activate B cells

Antigen receptors recognize one specific kind of antigen.

Antibodies recognize antigens through specific binding, which can trigger destruction of the antigen-bearing molecule or cell by several mechanisms. An antibody molecule has four polypeptide chains arranged in a Y shape, with the antigen-binding sites at the tips of the arms of the Y.

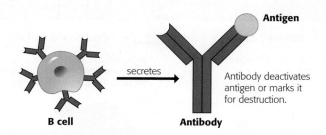

Antigen

B cell — secretes → Antibody deactivates antigen or marks it for destruction.

B cell   **Antibody**

Responding to the Invaders

When an antigen enters the body, it activates only lymphocytes with complementary receptors, a process called clonal selection.

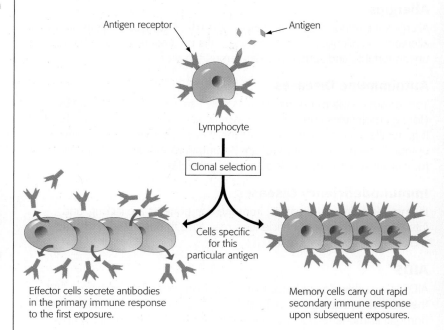

Antigen receptor   Antigen

Lymphocyte

Clonal selection

Cells specific for this particular antigen

Effector cells secrete antibodies in the primary immune response to the first exposure.

Memory cells carry out rapid secondary immune response upon subsequent exposures.

The immune system reacts to antigens and "remembers" an invader. Helper T cells and cytotoxic T cells are the main effector cells in the cell-mediated immune response. Helper T cells also stimulate the humoral immune response, the release of antibodies into body fluids (blood and lymph) by effector B cells. The cell-mediated immune response typically involves a phagocytic cell displaying a foreign antigen (a nonself molecule) and one of the body's own self proteins to a helper T cell. The helper T cell's receptors recognize the self-nonself complexes, and the interaction activates the helper T cell. In turn, the helper T cell can activate cytotoxic T cells with the same receptors (and can stimulate B cell activation as well). Cytotoxic T cells bind to and destroy infected body cells displaying both self proteins and foreign antigens.

Immune Disorders

Allergies

Allergies are abnormal sensitivities to otherwise harmless antigens, known as allergens. An allergic reaction produces inflammatory responses that result in uncomfortable and sometimes dangerous symptoms.

Autoimmune Diseases

The immune system normally reacts only against nonself substances (foreign molecules and cells), not against self (the body's own molecules). It generally rejects transplanted organs, whose cells lack the recipient's unique "fingerprint" of self proteins. In autoimmune diseases, the system turns against some of the body's own molecules.

Immunodeficiency Diseases

In immunodeficiency diseases, immune components are lacking, and infections recur. Immunodeficiencies may arise through inborn genetic mutations or through disease.

AIDS

AIDS is a worldwide epidemic that kills millions of people each year. HIV, the AIDS virus, attacks helper T cells, crippling both the cell-mediated and humoral immune responses. Practicing safer sex could save many lives.

SELF-QUIZ

1. Molecules that elicit a response from lymphocytes are called _____. Proteins secreted by the immune system that bind to these molecules are called _____.

2. Classify each of the following components of the immune system as being an innate defense or adaptive defense.
   a. natural killer cells
   b. complement proteins
   c. antibodies
   d. inflammation
   e. interferons
   f. cytotoxic T cells
   g. helper T cells

3. Which of the following best describes the difference in the way B cells and T cells deal with invaders?
   a. B cells provide permanent immunity; T cells provide temporary immunity.
   b. B cells send out antibodies that bind to invaders; T cells directly engage the invaders.
   c. T cells handle the primary immune response; B cells handle the secondary immune response.
   d. B cells are responsible for the cell-mediated immune response; T cells are responsible for the humoral immune response.

4. What makes a secondary immune response faster than a primary immune response?

5. Match each type of defensive cell with its function or description.
   a. lymphocyte   1. attacks infected body cells
   b. cytotoxic T cell 2. carries out the humoral immune response
   c. helper T cell   3. white blood cell that engulfs cells or molecules
   d. phagocytic cell   4. general name for B or T cell
   e. B cell   5. initiates the secondary immune response
   f. memory cell   6. cell most commonly attacked by HIV

6. Explain how each of the following characteristics of the inflammatory response helps protect the body: swelling, pain, and fever.

7. Why is AIDS deadlier than most other viral diseases?

8. A baby has been born with an immunodeficiency disease. In trying to diagnose the problem, physicians discover that the child is not producing any antibodies. It is most likely that this child is missing what kind of cells?

9. How does the autoimmune disorder lupus affect the immune system?

10. Once vaccinated, you have had a primary exposure to specific antigens. If you ever encounter these antigens again, your body will mount a rapid immune response. The cells that account for this rapid secondary response are called _____. The process that produces these long-lived cells is called _____.

Answers to the Self-Quiz questions can be found in Appendix D.

THE PROCESS OF SCIENCE

11. Most biologists believe that the immune system's defense against infections largely rests on its ability to distinguish self molecules from nonself molecules. This concept seems central to our understanding of immune function. However, like all scientific ideas, it is not beyond question. Several immunologists have developed an alternative hypothesis: that the immune system's effectiveness rests mostly on ability to recognize damage to body tissues caused by the invaders, not on the ability to recognize nonself. If you were going to test the "damage" hypothesis, what might you look for? Which type of cell would you expect to be directly affected by damaged tissues? Why? Some proponents argue that the "damage" hypothesis makes more sense from an evolutionary perspective, claiming that it is more advantageous for an organism's defense system to respond to tissue damage than to the mere presence of a foreign microbe. Do you agree? Why or why not?

12. The graph below depicts the concentration of two antibodies from a patient's blood: The red line graphs antibody X (produced in response to antigen X), and the black line graphs antibody Y (produced in response to antigen Y). Using your knowledge of clonal selection, explain why these curves have the shapes they do.

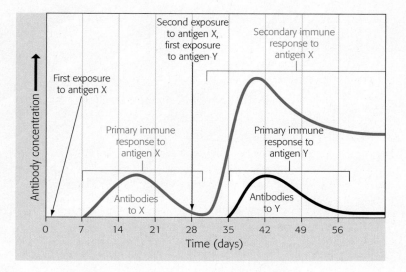

BIOLOGY AND SOCIETY

13. Organ donation saves many lives each year. Even though some transplanted organs are derived from living donors, the majority come from patients who die but still have healthy organs that can help a transplant recipient whose cell surfaces have similar sets of proteins. Potential organ donors can fill out an organ donation card to specify their wishes. If the donor is in critical condition and dying, the donor's family is usually consulted to discuss the donation process. Generally, the next of kin must approve before donation can occur, regardless of whether the patient has completed a donor card. In some cases, the donor's wishes are overridden by a family member. Do you think that family members should be able to deny the stated intentions of the potential donor? Why or why not? Have you signed up to be an organ donor? Why or why not?

25 Hormones

A difference in sex hormones.
Sex hormones can cause vastly different appearances in males and females, as seen in this peacock and peahen.

BIOLOGY AND SOCIETY: Sex Hormones

A Hormonal Fountain of Youth?

Around age 50, most women begin to experience the symptoms of menopause, a period of life marked by the end of menstruation and fertility. Menopause is caused by decreasing levels of estrogen and progesterone, two of a number of chemical signals called hormones that help regulate the body. Although menopause is a normal part of the human female life cycle, it often produces unwanted symptoms, such as hot flashes (sudden feelings of uncomfortable warmth) and an increased risk of osteoporosis, stroke, and heart disease.

For over 60 years, hormone replacement therapy (HRT) has been used to relieve some of the negative symptoms of menopause and to slow the effects of aging. Women receiving HRT have traditionally been given hormones derived from animal sources (usually the urine of pregnant mares). However, the safety of this type of HRT has come into question: A 2002 study found that these hormones increased a woman's risks of stroke and cancer.

Due to the dangers associated with conventional HRT, some women seek alternative treatments. Recently, a class of chemicals called bioidenticals have become popular; you may have seen a best-selling book or infomercial promoting their age-defying properties. Bioidenticals are synthetic human hormones produced by chemically modifying plant-derived hormones. Some bioidenticals have been approved by the FDA and prescribed by doctors for years, but most have not. Proponents of bioidenticals point out that they are chemically identical to human hormones and may have fewer risks than hormones from animals other than humans. Critics point out that there is no evidence that bioidenticals are any safer than traditional HRT and that many of these products are sold outside the bounds of the medical establishment and are therefore less subject to rigorous testing and scrutiny. To be safe, a woman should talk with her doctor before deciding on her best course of action.

This chapter focuses on how hormones, including estrogen and progesterone, help maintain homeostasis within the human body. We begin with an overview of how hormones work in all vertebrates and then turn to the major components of the human endocrine system. Along the way, we'll consider many examples of the effects of hormonal imbalance and further discuss the profound effects that sex hormones can have on animal bodies.

Hormones: An Overview

Hormones are chemical signals produced by most plants (see Table 29.1) and animals. Animal hormones are carried by the circulatory system (usually the blood) to other parts of the body, where they may communicate their regulatory messages. Hormones are made and secreted mainly by organs called **endocrine glands**. Collectively, all hormone-secreting cells constitute the **endocrine system**, the body's main system for internal chemical regulation.

The upper-right diagram in **Figure 25.1** shows the release of hormone molecules from membrane-enclosed sacs in an endocrine cell. The circulatory system carries hormones throughout the body, but a hormone can only

bind to a **target cell** with receptors for that specific hormone (lower-right diagram). Because hormones reach all parts of the body, the endocrine system is especially important in controlling whole-body activities. For example, hormones govern our metabolic rate, growth, maturation, and reproduction. In many cases, a tiny amount of a hormone influences the activities of an enormous number of target cells in a variety of organs. Hormones trigger changes in target cells in different ways, depending on whether the hormone is water soluble or lipid soluble: Water-soluble hormones trigger responses without entering the cell, whereas lipid-soluble hormones trigger responses after entering the cell.

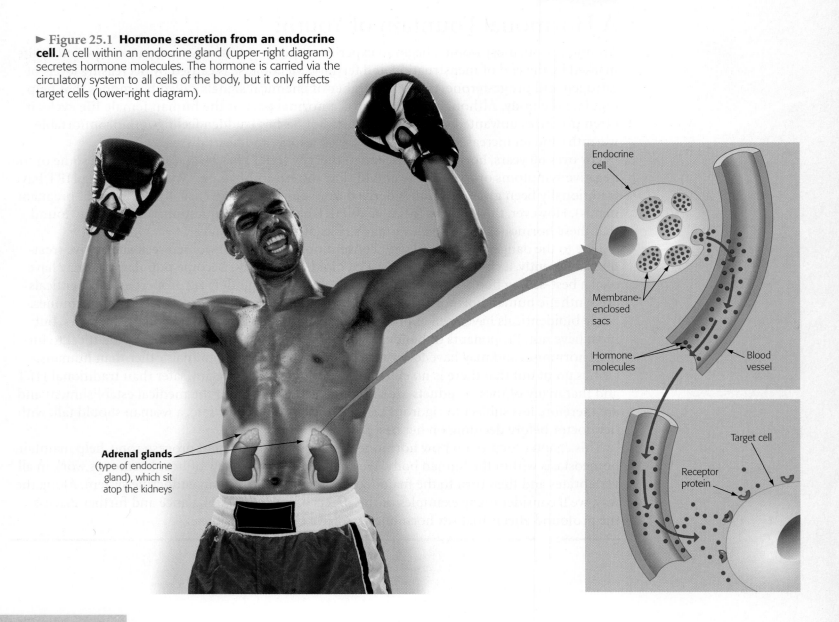

▶ Figure 25.1 **Hormone secretion from an endocrine cell.** A cell within an endocrine gland (upper-right diagram) secretes hormone molecules. The hormone is carried via the circulatory system to all cells of the body, but it only affects target cells (lower-right diagram).

Endocrine cell

Membrane-enclosed sacs

Hormone molecules

Blood vessel

Target cell

Receptor protein

Adrenal glands (type of endocrine gland), which sit atop the kidneys

Water-soluble hormones cannot pass through the phospholipid bilayer of the plasma membrane, but they can bring about cellular changes indirectly without entering their target cells (**Figure 25.2**). To start, ❶ a water-soluble hormone (●) binds to a specific receptor protein (◑) in the plasma membrane of the target cell. The binding activates the receptor protein, which ❷ initiates a signal transduction pathway, a series of molecular changes that converts an extracellular chemical signal into a specific intracellular response. ❸ The final relay molecule (○) activates a protein that either carries out a response in the cytoplasm (such as activating an enzyme) or affects gene regulation in the nucleus (turning on or off genes that control production of a number of proteins).

Water-soluble hormones that bind to plasma membrane receptors come in three varieties, all derived from amino acids: amine hormones, which are modified versions of single amino acids; peptide hormones, which are short chains of amino acids; and protein hormones, made of polypeptides. Insulin, which helps regulate blood glucose levels, is a protein hormone.

In contrast, lipid-soluble hormones pass through the phospholipid bilayer and trigger responses by binding to receptors inside the target cell. Hormones that are steroids, all of which are molecules derived from cholesterol, work this way. As shown in **Figure 25.3**, ❶ a steroid hormone or other lipid-soluble hormone (◄) enters a cell by diffusing through the plasma membrane. If the cell is a target cell, the hormone ❷ binds to a receptor protein (◖) in the cytoplasm or nucleus. Rather than triggering a signal transduction pathway, the receptor itself carries the hormone's signal. ❸ The hormone-receptor complex attaches to specific sites on the cell's DNA in the nucleus. ❹ The binding to DNA may stimulate transcription of certain genes into mRNA molecules, which are translated into new proteins, or it may turn transcription off. The sex hormones estrogen and testosterone are examples of steroid hormones.

We've now completed an overview of how hormones work. The principles presented so far apply to all vertebrates (as well as many invertebrates). In the next section, we'll take a closer look at the endocrine system of one particular vertebrate: humans. ✔

▼ **Figure 25.2 A hormone that binds to a plasma membrane receptor.** Hormones that are water-soluble cannot cross the phospholipid bilayer but can bind to membrane receptors on target cells and activate a signal transduction pathway.

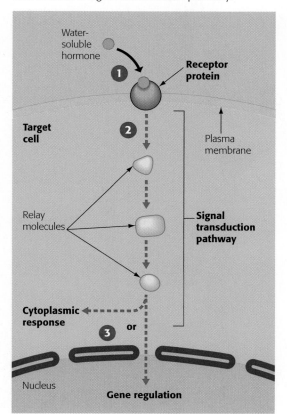

▼ **Figure 25.3 A hormone that binds to an intracellular receptor.** Hormones that are lipid-soluble can cross the phospholipid bilayer and bind to receptors inside target cells. Steroid hormones and other lipid-soluble hormones affect the cell by turning specific genes on and off.

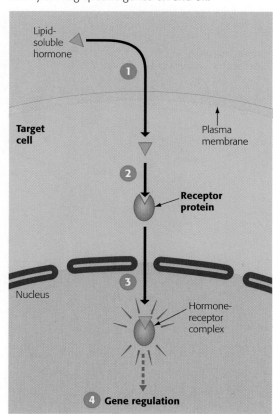

The Human Endocrine System

The human endocrine system consists of about a dozen major glands. Some of these, such as the thyroid and the pituitary gland, are endocrine specialists; that is, their primary function is to secrete hormones into the blood. Several other glands have both endocrine and nonendocrine functions. The pancreas, for example, secretes hormones that influence the level of glucose in the blood and also secretes digestive enzymes into the intestine. Other organs are primarily nonendocrine but have some cells that secrete hormones. For example, the stomach is primarily a digestive organ but also releases ghrelin (the "hunger hormone"), which travels to the brain and stimulates appetite. (As you can imagine, ghrelin is the target of weight loss research, but so far no treatments that interact with it have been developed.)

Figure 25.4 shows the locations of the major human endocrine glands discussed in this chapter. **Table 25.1** summarizes the actions of the main hormones they produce. The table provides an overview of the human endocrine system, and you may wish to refer to it as we focus on the individual glands and their hormones.

Hormones have a wide range of targets. Some hormones, like the sex hormones, affect most of the tissues of the body. Others, such as glucagon from the pancreas, have only a few kinds of target cells (in this case, liver and fat cells). In some cases, a hormone elicits different responses in different target cells, depending on the type of cell and its signal transduction pathway.

The rest of this section focuses on the major glands of the human endocrine system. For each gland, you'll learn about the hormones it secretes and how these chemicals help the body maintain homeostasis. Keep in mind that this chapter only covers the major endocrine glands and their hormones; there are other hormone-secreting structures—the heart and liver, for example—and dozens of other hormones that we will not discuss. ☑

► Figure 25.4 **The major endocrine glands in humans.** This figure shows only the major endocrine organs discussed in the text. Several other organs that also have endocrine functions are not shown.

Hypothalamus

Pituitary gland

Thyroid gland

Parathyroid glands (embedded within thyroid)

Adrenal glands (atop kidneys)

Pancreas

Testes (male)

Ovaries (female)

Table 25.1   Major Human Endocrine Glands and Some of Their Hormones

Gland	Hormone	Chemical Class	Representative Actions	Regulated By
Hypothalamus	Hormones released by the posterior pituitary and hormones that regulate the anterior pituitary (see below)			
Pituitary gland Posterior pituitary (releases hormones made by hypothalamus)	Oxytocin (see Chapter 26)	Peptide	Stimulates contraction of uterus and mammary gland cells	Nervous system
	Antidiuretic hormone (ADH)	Peptide	Promotes retention of water by kidneys	Water/salt balance
Anterior pituitary	Prolactin (PRL) (see Chapter 26)	Protein	Stimulates milk production	Hypothalamic hormones
	Follicle-stimulating hormone (FSH)	Protein	Stimulates production of ova and sperm	Hypothalamic hormones
	Luteinizing hormone (LH)	Protein	Stimulates ovaries and testes	Hypothalamic hormones
	Growth hormone (GH)	Protein	Stimulates growth (especially bones) and metabolic functions	Hypothalamic hormones
	Thyroid-stimulating hormone (TSH)	Protein	Stimulates thyroid gland	Thyroxine in blood; hypothalamic hormones
	Adrenocorticotropic hormone (ACTH)	Peptide	Stimulates adrenal cortex to secrete glucocorticoids	Glucocorticoids; hypothalamic hormones
	Endorphins	Peptide	Mask pain	Brain
Thyroid gland	Triiodothyronine (T_3) and thyroxine (T_4)	Amine	Stimulate and maintain metabolic processes	TSH
	Calcitonin	Peptide	Lowers blood calcium level	Calcium in blood
Parathyroid glands	Parathyroid hormone (PTH)	Peptide	Raises blood calcium level	Calcium in blood
Pancreas	Insulin	Protein	Lowers blood glucose level	Glucose in blood
	Glucagon	Protein	Raises blood glucose level	Glucose in blood
Adrenal glands Adrenal medulla	Epinephrine (adrenaline) and norepinephrine (noradrenaline)	Amine	Raise blood glucose level; increase metabolic activities; constrict certain blood vessels	Nervous system
Adrenal cortex	Glucocorticoids	Steroid	Raise blood glucose level	ACTH
Gonads Testes	Androgens (testosterone)	Steroid	Support sperm formation; promote development and maintenance of male secondary sex characteristics	FSH and LH
Ovaries	Estrogens	Steroid	Stimulate uterine lining growth; promote development and maintenance of female secondary sex characteristics	FSH and LH
	Progestins	Steroid	Promote uterine lining growth	FSH and LH

The Hypothalamus and Pituitary Gland

The **hypothalamus** is the main control center of the endocrine system (**Figure 25.5**). As part of the brain, the hypothalamus receives information from nerves about the internal condition of the body and about the external environment. It then responds by sending out appropriate nervous or endocrine signals. Its signals directly control the **pituitary gland**, a pea-sized structure that hangs down from the hypothalamus. In response to signals from the hypothalamus, the pituitary secretes hormones that influence numerous body functions. The hypothalamus thus exerts master control over the endocrine system by using the pituitary to relay directives to other glands.

As Figure 25.5 shows, the pituitary gland consists of two distinct parts: a posterior lobe and an anterior lobe, both situated in a pocket of skull bone just under the hypothalamus. The **posterior pituitary** is actually an extension of the hypothalamus that stores and secretes hormones made in the hypothalamus. In contrast, the **anterior pituitary** synthesizes and secretes its own hormones directly into the blood. The hypothalamus exerts control over the anterior pituitary by secreting two kinds of hormones into short blood vessels (not illustrated) that connect the two glands: releasing hormones and inhibiting hormones. Releasing hormones stimulate the anterior pituitary to secrete hormones, whereas inhibiting hormones induce the anterior pituitary to stop secreting hormones.

Figure 25.6 shows how the hypothalamus operates through the posterior pituitary to direct the activity of the kidneys. The hypothalamus makes antidiuretic hormone (ADH) (▲), which is stored and released by the posterior pituitary. As discussed in Chapter 21, ADH helps kidney cells reabsorb water, preventing dehydration by decreasing urine volume when the body needs to retain water. When the body has too much water, the hypothalamus responds by slowing the release of ADH from the pituitary. Some diuretic drugs, such as alcohol, inhibit the release of ADH, which in turn increases the output of urine.

In response to releasing hormones secreted by the hypothalamus, the anterior pituitary synthesizes and releases many different hormones that influence a broad range of body activities. For example, it releases follicle-stimulating hormone (FSH) and luteinizing hormone (LH), which help regulate the menstrual cycle (as you'll see in Chapter 26).

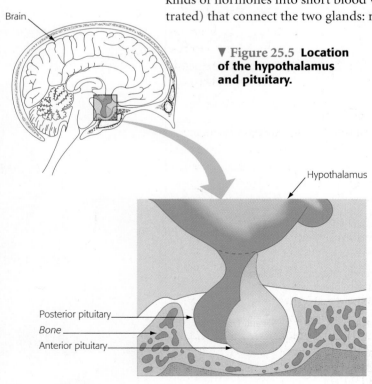

▼ **Figure 25.5 Location of the hypothalamus and pituitary.**

► **Figure 25.6 ADH and osmoregulation.** The hypothalamus regulates water homeostasis by synthesizing ADH. Molecules of this hormone travel via long cellular extensions (gray) to the posterior pituitary, where they are released into the blood. Eventually, the ADH reaches target cells in the kidneys.

Hypothalamus

ADH

Posterior pituitary

Blood vessel

Anterior pituitary

ADH

Kidneys
(ADH increases water reabsorption and prevents dehydration)

Of all the anterior pituitary secretions, none has a broader effect than the protein called **growth hormone (GH)**. During childhood and adolescence, high levels of GH promote the development and enlargement of all parts of the body. If too much GH is produced in a very young person, usually because of a pituitary tumor, gigantism can result (**Figure 25.7a**). Overproduction of GH in adults results in acromegaly, characterized by enlarged bones in the hands, feet, and face (**Figure 25.7b**). In contrast, too little GH during childhood development can lead to dwarfism (**Figure 25.7c**). Administering GH to children with GH deficiency can prevent dwarfism. Human GH was originally extracted in only minute quantities from the pituitary glands of cadavers but is now produced in large quantities by bacteria that have been genetically modified to carry the human GH gene (see Chapter 12). With the increased availability of human GH, some athletes have misused the drug in order to bulk up their muscles. Such abuse is extremely dangerous and can lead to disfigurement, heart failure, and multiple cancers.

The **endorphins**, another kind of anterior pituitary hormone, are the body's natural painkillers; they mask the perception of pain. Some researchers speculate that the so-called "runner's high" results partly from the release of endorphins when stress and pain in the body reach critical levels. The release of endorphins may also produce pleasant feelings during such diverse activities as meditation, acupuncture treatments, and even eating very spicy foods. ✔

☑ **CHECKPOINT**

Which gland is the primary control center for the entire endocrine system? Which gland is directly controlled by this master gland?

Answer: the hypothalamus; the pituitary

▼ Figure 25.7 **Pituitary growth hormone disorders.**

(a) Overproduction of GH during development. Yao Defeng's extreme size (7 feet 9 inches) is due to a benign brain tumor that results in excess growth hormone.

(b) Overproduction of GH during adulthood. Overproduction of GH later in life causes acromegaly. The most common symptom of this disease is enlargement of the hands and face.

(c) Underproduction of GH during development. The body of actor Verne Troyner, like those of other dwarfs, produced low amounts of GH during childhood.

The Thyroid and Parathyroid Glands

Your **thyroid gland** is in your neck just under your larynx (voice box). The thyroid produces hormones (thyroxine and triiodothyronine) that increase oxygen consumption and metabolic rate in all the cells of your body. These thyroid hormones also play crucial roles in development and maturation, particularly of the bone and nerve cells. Insufficient levels of the thyroid hormones in the blood (hypothyroidism) or excess levels (hyperthyroidism) can result in serious metabolic disorders. Fortunately, thyroid disorders can be easily treated.

Hypothyroidism can result from dietary deficiencies or from a defective thyroid gland. To produce thyroxine and triiodothyronine, the thyroid requires the element iodine. If too little iodine is available, insufficient quantities of these thyroid hormones are produced. This underproduction interrupts feedback loops that control thyroid activity, causing overstimulation and swelling of the thyroid gland. A type of dietary hypothyroidism called goiter has been reduced in many industrialized nations by the incorporation of iodine into table salt (see Figure 2.3). Unfortunately, goiter still affects many people in less industrialized nations.

A common form of hyperthyroidism is Graves' disease, characterized by overheating, profuse sweating, high blood pressure, and protruding eyeballs. One treatment

Homeostasis and the Endocrine System

takes advantage of the fact that the thyroid accumulates iodine: Patients drink a solution containing a small dose of radioactive iodine, which kills off some thyroid cells. This "radioactive cocktail" can kill off just enough cells to reduce thyroid output and relieve symptoms.

Embedded within the thyroid are four disk-shaped **parathyroid glands** (gold dots in the thyroid at lower right in Figure 25.8). The thyroid and parathyroid glands function in calcium homeostasis, keeping the concentration of calcium ions (Ca^{2+}) within a narrow range. An appropriate level of calcium in the blood and interstitial fluid is essential for many body functions. Without calcium, nerve signals cannot be transmitted from cell to cell, muscles cannot function properly, blood cannot clot, and cells cannot transport molecules across their membranes. Two hormones, **calcitonin** from the thyroid and **parathyroid hormone** (PTH) from the parathyroids, regulate the blood calcium level. They are said to be **antagonistic hormones** because they have opposite effects: Calcitonin lowers the calcium level in the blood, whereas PTH raises it.

As **Figure 25.8** indicates, calcitonin and PTH operate by means of feedback systems that keep the calcium level near the homeostatic set point. ❶ A rise in the blood Ca^{2+} level above the homeostatic set point induces the thyroid gland to secrete calcitonin. ❷ Calcitonin, in turn, has two main effects: It causes more Ca^{2+} to be deposited in the bones, and it makes the kidneys reabsorb less Ca^{2+} as they form urine. ❸ The result is a lower Ca^{2+} level in the blood. ❹ When the blood Ca^{2+} level drops below the set point, the parathyroids release PTH into the blood. ❺ PTH stimulates the release of Ca^{2+} from bones and increases Ca^{2+} uptake by the kidneys and intestines, ❻ raising the calcium levels.

In summary, a sensitive balancing system maintains calcium homeostasis. Failure of the system can have far-reaching effects in the body. For example, a shortage of PTH causes the blood calcium level to drop dramatically. Such a drop can lead to uncontrollable muscle contractions and potentially fatal convulsions, a condition called tetany. ☑

► **Figure 25.8 Calcium homeostasis.** This diagram traces the regulation of blood calcium level by calcitonin and PTH, which act antagonistically. Note that bones, in addition to supporting the body, act as a calcium bank, storing and releasing calcium as needed.

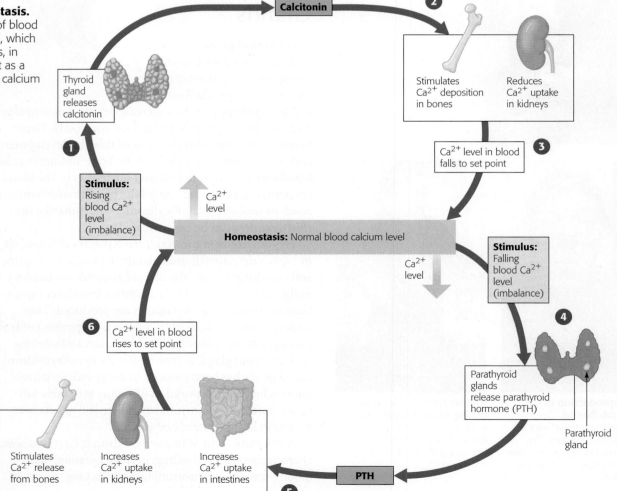

☑CHECKPOINT

Both calcitonin (from the thyroid) and parathyroid hormone (from the parathyroids) regulate blood calcium level. Why are these two hormones said to be antagonistic?

Answer: These two hormones have opposite effects: Calcitonin lowers blood calcium level, whereas PTH raises it.

The Pancreas

The **pancreas** produces two hormones that play important roles in managing the body's energy supplies. One of the hormones is **insulin**, a protein hormone produced by clusters of specialized pancreatic cells called islet cells. Other islet cells secrete another protein hormone, called **glucagon**. Insulin and glucagon help maintain a homeostatic balance between the amount of glucose available in the blood and the amount of glucose stored as the polymer glycogen in body cells.

As shown in **Figure 25.9**, insulin and glucagon are antagonistic hormones, countering each other in a feedback circuit that precisely manages both glucose storage and glucose use by body cells. By negative feedback, the concentration of glucose in the blood determines the relative amounts of insulin and glucagon secreted by islet cells. ❶ Rising glucose concentration in the blood—as occurs shortly after you eat a carbohydrate-rich meal, for example—stimulates the pancreas to secrete more insulin into the blood. ❷ Body cells take up more glucose from the blood, while liver and skeletal muscle cells take up glucose and use it to make glycogen, which they store. ❸ As a result, the blood glucose level falls to the set point, and the cells of the pancreas lose their stimulus to secrete insulin.

❹ When the blood glucose level dips below the set point, as it may between meals or if a meal is skipped, pancreatic cells respond by secreting more glucagon. ❺ Glucagon is a fuel mobilizer, making liver cells break glycogen down into glucose and release it into the blood. ❻ Then, when the blood glucose level returns to the set point, the pancreas slows its secretion of glucagon.

▶ **Figure 25.9**
Glucose homeostasis.
This diagram traces the regulation of blood glucose level by insulin and glucagon, two antagonistic hormones released by the pancreas.

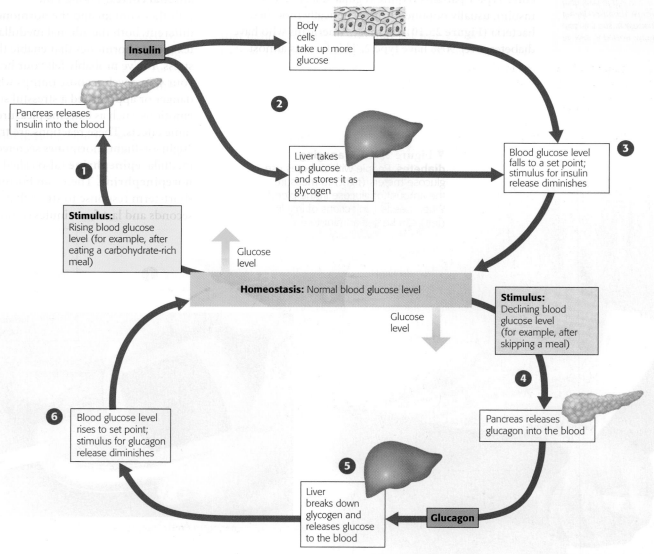

Insulin

Body cells take up more glucose

Pancreas releases insulin into the blood

❶

❷

Liver takes up glucose and stores it as glycogen

Blood glucose level falls to a set point; stimulus for insulin release diminishes

❸

Stimulus:
Rising blood glucose level (for example, after eating a carbohydrate-rich meal)

Glucose level

Homeostasis: Normal blood glucose level

Glucose level

Stimulus:
Declining blood glucose level (for example, after skipping a meal)

❹

❻ Blood glucose level rises to set point; stimulus for glucagon release diminishes

Pancreas releases glucagon into the blood

❺

Liver breaks down glycogen and releases glucose to the blood

Glucagon

Diabetes mellitus is a serious hormonal disease that   always associated with being overweight and under-

✓CHECKPOINT

Arrange the following organs of the male reproductive system in the correct sequence for the travel of sperm: epididymis, testis, urethra, vas deferens.

Answer: testis, epididymis, vas deferens, urethra

Human Reproduction

Although the differences between human male and female reproductive anatomies are obvious, there are also similarities. Both sexes have a pair of **gonads**, the organs that produce gametes. And both sexes have ducts that store and deliver the gametes as well as structures that allow mating. We'll now examine the anatomical features of human reproduction, beginning with the male anatomy.

Male Reproductive Anatomy

Figure 26.5 presents a side view and a front view of the male reproductive system. The **penis** contains erectile tissue (shown in blue in the figures) that can fill with blood and cause an erection during sexual arousal. The penis consists of a shaft that supports a highly sensitive glans (head). A **prepuce**, or foreskin, covers the glans. The foreskin may be surgically removed, a procedure that is known as circumcision.

The **testes** (singular, *testis*), the male gonads, are located outside the abdominal cavity in a sac called the **scrotum**. Sperm cannot develop at body temperature. By keeping the sperm-forming cells away from the body, the scrotum keeps them cool enough to develop

normally. In cold conditions, muscles around the scrotum contract, pulling the testes toward the body, maintaining the proper temperature.

From puberty into old age, the testes produce hundreds of millions of sperm each day. From the testes, sperm pass into a coiled tube called the epididymis. During **ejaculation**—expulsion of sperm-containing fluid from the penis—the sperm leave the epididymis and travel through a duct called the **vas deferens**. The **seminal vesicles** and **prostate gland** add fluid that nourishes the sperm and provides protection from the natural acidity of the vagina. The prostate gland is the source of some of the most common medical problems in men over 40. Prostate cancer, for example, is the second most commonly diagnosed cancer in the United States (after skin cancer). Each vas deferens, one from each of the two testes, empties into the **urethra**. The urethra conveys, at different times, both sperm and urine out through the penis.

Ejaculation, caused by the contraction of muscles along the sperm ducts, releases about 5 mL (1 teaspoonful) of **semen**. Only 5% of semen consists of sperm (typically 200–500 million of them); the remaining 95% of semen is fluid secreted by the various glands. ✓

▼ **Figure 26.5 The male reproductive system.** The color blue highlights erectile tissue.

Side view

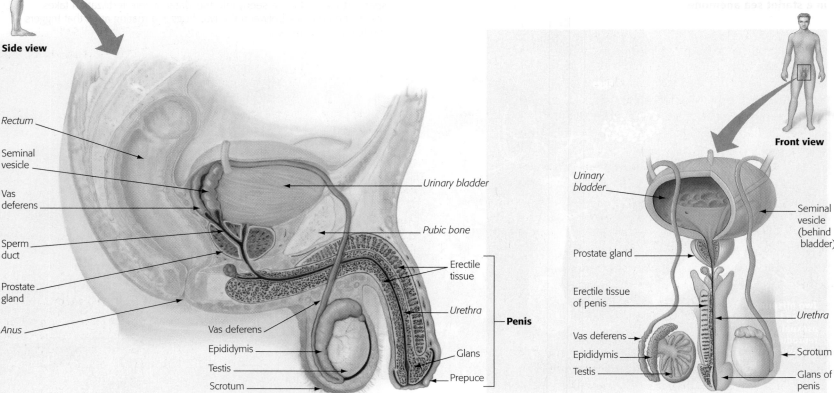

Rectum

Seminal vesicle

Vas deferens

Sperm duct

Prostate gland

Anus

Vas deferens

Epididymis

Testis

Scrotum

Urinary bladder

Pubic bone

Erectile tissue

Urethra

Penis

Glans

Prepuce

Front view

Urinary bladder

Prostate gland

Erectile tissue of penis

Vas deferens

Epididymis

Testis

Seminal vesicle (behind bladder)

Urethra

Scrotum

Glans of penis

Female Reproductive Anatomy

Two views of the female reproductive system are shown in Figure 26.6. The outer features of the female reproductive anatomy are collectively called the **vulva**. The **vagina**, or birth canal, opens to the outside just behind the opening of the urethra, the tube through which urine is excreted. An outer pair of thick, fatty ridges, the **labia majora**, protects the entire genital region, and a pair of inner skin folds, the **labia minora**, borders the openings. A thin membrane called the **hymen** partly covers the vaginal opening until sexual intercourse or other vigorous physical activity ruptures it. The **clitoris** is an organ that engorges with blood and becomes erect during sexual arousal. It consists of a short shaft supporting a rounded glans, or head, covered by a small hood of skin called the prepuce. The clitoris has an enormous number of nerve endings and is very sensitive to touch.

The **ovaries** are the site of gamete production—the gonads—in human females. A woman's ovaries are each about an inch long and have a bumpy surface. The bumps are **follicles**, each consisting of a single developing egg cell surrounded by one or more layers of cells that nourish and protect it. The follicles also produce estrogen, the female sex hormone (see Chapter 25).

A female is born with over a million follicles, but only several hundred will release egg cells during her reproductive years. Starting at puberty and continuing until menopause (the end of fertility, which usually occurs during middle age), one follicle (or rarely two or more) matures and releases its egg cell about every 28 days. An egg cell is ejected from the follicle in the process of **ovulation**. After ovulation, what remains of the follicle grows to form a solid mass called the **corpus luteum**, which plays a crucial role in the release of hormones during the reproductive cycle (as you'll see later in the chapter). The released egg enters an **oviduct** (also known as a fallopian tube), where cilia sweep it toward the uterus. If sperm are present, fertilization may take place in the upper part of the oviduct. If the released egg is not fertilized, it is shed during menstruation, and a new follicle matures during the next cycle.

The **uterus** (also called the womb) is the actual site of pregnancy. The uterus is about the size and shape of an upside-down pear, but can grow to several times that size to accommodate a growing fetus. The uterus has a thick muscular wall lined with a blood-rich layer of tissue called the **endometrium**. An embryo implants in the endometrium and grows there. The term **embryo** is used for the stage in development from the first division of the zygote until body structures begin to appear, about the 9th week. From the 9th week until birth, a developing human is called a **fetus**.

The narrow neck at the bottom of the uterus is the **cervix**. It is recommended that a woman have a yearly Pap test in which cells are removed from around the cervix and examined under a microscope for signs of cervical cancer. Regular Pap smears greatly increase the chances of detecting cervical cancer early and therefore treating it successfully. The cervix opens into the vagina. During intercourse, the vagina serves as a repository for sperm. ☑

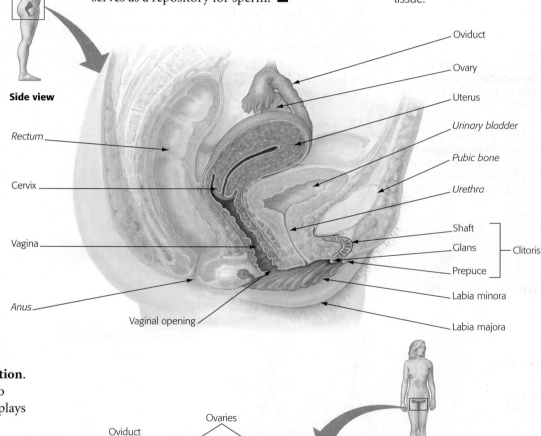

▼ Figure 26.6 **The female reproductive system.** The color blue highlights erectile tissue.

Side view

Rectum

Cervix

Vagina

Anus

Vaginal opening

Oviduct

Ovary

Uterus

Urinary bladder

Pubic bone

Urethra

Shaft
Glans — Clitoris
Prepuce

Labia minora

Labia majora

Front view

Ovaries

Oviduct

Follicles

Corpus luteum

Wall of uterus

Vagina

Uterus

Endometrium (lining of uterus)

Cervix ("neck" of uterus)

☑ CHECKPOINT

In which organ of the human female reproductive system does fertilization occur? In which organ does a fetus develop?

Answer: the oviduct; the uterus

Spermatogenesis and Oogenesis

Gametogenesis

The production of gametes is called **gametogenesis**. Human gametes—sperm and egg—are haploid cells with 23 chromosomes that develop by meiosis from diploid cells with 46 chromosomes. (You may find it helpful to review the discussion of meiosis in Chapter 8, particularly Figures 8.13 to 8.15.) There are significant differences in gametogenesis between human males and females, so we'll examine the processes separately.

Spermatogenesis

The formation of sperm cells is called **spermatogenesis** (**Figure 26.7**). Sperm cells develop inside the testes in coiled tubes called the **seminiferous tubules**. Cells near the outer walls of the tubules multiply constantly by mitosis. Each day, about 3 million of them differentiate into **primary spermatocytes**, the diploid cells that undergo meiosis. Meiosis I of a primary spermatocyte produces two haploid **secondary spermatocytes**. Then

meiosis II forms four cells, each with the haploid number of chromosomes. The sperm cells that develop from each of these haploid cells are gradually pushed toward the center of the seminiferous tubules. From there, the sperm cells pass into the epididymis, where they mature, become motile, and are stored until ejaculation.

Oogenesis

Figure 26.8 summarizes **oogenesis**, the development of mature egg cells, also called **ova** (singular, *ovum*). At birth, each ovary contains many thousands of follicles. Each follicle contains a single dormant **primary oocyte**, a diploid cell that has paused its cell cycle in prophase of meiosis I.

► **Figure 26.7**
Spermatogenesis.
This process takes about 65–75 days in the human male.

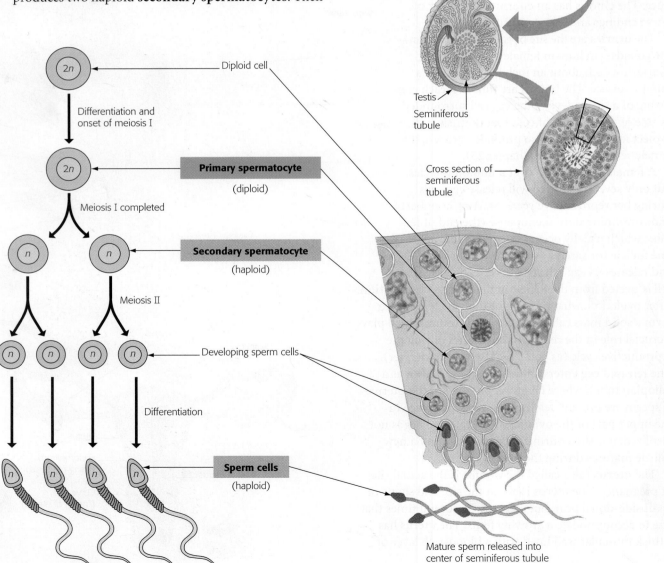

Diploid cell

Differentiation and onset of meiosis I

Primary spermatocyte
(diploid)

Meiosis I completed

Secondary spermatocyte
(haploid)

Meiosis II

Developing sperm cells

Differentiation

Sperm cells
(haploid)

Penis
Epididymis
Testis
Scrotum
Testis
Seminiferous tubule
Cross section of seminiferous tubule
Mature sperm released into center of seminiferous tubule

A primary oocyte can be triggered to develop further by the hormone FSH (follicle-stimulating hormone). After puberty and until menopause, about every 28 days, FSH from the pituitary gland stimulates one of the dormant follicles to develop. The follicle enlarges, and the primary oocyte within it completes meiosis I and begins meiosis II. The division of the cytoplasm in meiosis I is unequal, with a single **secondary oocyte** receiving almost all of it. The smaller of the two daughter cells, called the first **polar body**, receives almost no cytoplasm.

About the time the secondary oocyte forms, the pituitary gland secretes LH (luteinizing hormone), which causes ovulation. The ripening follicle bursts, releasing its secondary oocyte from the ovary. The ruptured follicle then develops into a corpus luteum. The secondary oocyte enters the oviduct, and if a sperm cell fuses with it, the secondary oocyte completes meiosis II. Meiosis II is also unequal, yielding a small polar body and the mature egg (ovum). The chromosomes of the egg can then fuse with the chromosomes of the sperm cell, producing a diploid zygote. The polar bodies, which are quite small because they received almost no cytoplasm, degenerate. The zygote thus acquires nearly all the cytoplasm and the bulk of the nutrients and organelles contained in the original cell.

Both oogenesis and spermatogenesis produce haploid gametes, but there are several important differences between the two processes. One obvious difference is location: testes in the male and ovaries in the female. Furthermore, human males create new sperm every day from puberty through old age. Human females, on the other hand, create primary oocytes only during their fetal development. Another difference is that four gametes result from each diploid parent cell during spermatogenesis, whereas oogenesis results in only one gamete from each parent cell. Moreover, there are significant differences in the cells produced by meiosis: Sperm are small and motile and contain relatively few nutrients; eggs are large, nonmotile, and well stocked with nutrients and organelles. Finally, oogenesis cannot be completed without stimulation from a sperm cell, whereas spermatogenesis is completed before the sperm leave the testis. ☑

▼ Figure 26.8 **Oogenesis and development of an ovarian follicle.** Notice that oogenesis starts before birth, but the final ovum does not form until fertilization.

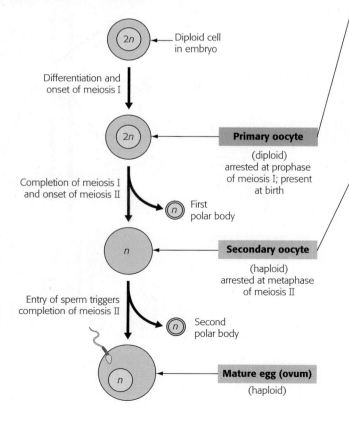

- Diploid cell in embryo (2n)
- Differentiation and onset of meiosis I
- **Primary oocyte** (diploid) arrested at prophase of meiosis I; present at birth (2n)
- Completion of meiosis I and onset of meiosis II
- First polar body (n)
- **Secondary oocyte** (haploid) arrested at metaphase of meiosis II (n)
- Entry of sperm triggers completion of meiosis II
- Second polar body (n)
- **Mature egg (ovum)** (haploid) (n)

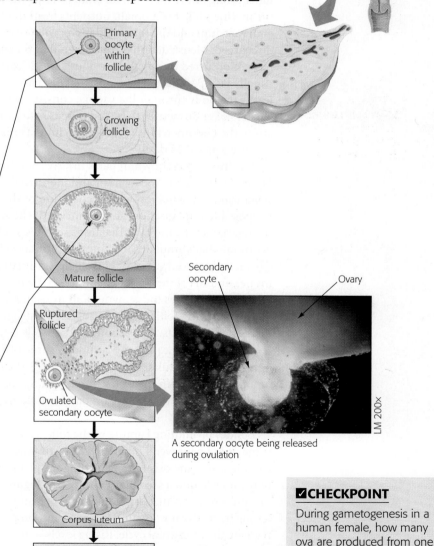

- Primary oocyte within follicle
- Growing follicle
- Mature follicle
- Ruptured follicle
- Ovulated secondary oocyte
- Corpus luteum
- Degenerating corpus luteum

Secondary oocyte — Ovary

LM 200×

A secondary oocyte being released during ovulation

☑ **CHECKPOINT**

During gametogenesis in a human female, how many ova are produced from one primary oocyte? In a male, how many sperm arise from one primary spermatocyte?

Answer: one; four

The Female Reproductive Cycle

The Female Reproductive Cycle

Human females have a **reproductive cycle**, a recurring series of events that produces gametes, makes them available for fertilization, and prepares the body for pregnancy (**Figure 26.9**). The reproductive cycle repeats every 28 days, on average, but cycles from 20 to 40 days are not uncommon. The reproductive cycle is actually two cycles in one. The **ovarian cycle** (see Figure 26.9c) controls the growth and release of an egg. During the **menstrual cycle** (see Figure 26.9e), the uterus is prepared for possible implantation of an embryo.

By convention, the first day of menstruation, a woman's "period," is designated as day 1 of the menstrual cycle. **Menstruation** is uterine bleeding caused by the breakdown of the endometrium, the blood-rich inner lining of the uterus. If an embryo implants in the uterine wall, it will obtain nutrients from the endometrium and the thickened lining will not be discharged. Menstruation is thus a sign that pregnancy has not occurred during the previous cycle. Menstruation usually lasts 3–5 days. The menstrual discharge, which leaves the body through the vagina, consists of blood, clusters of endometrial cells, and mucus. After menstruation, the endometrium regrows, reaching its maximum thickness at 20–25 days.

The hormones shown in parts (a), (b), and (d) of Figure 26.9 regulate the ovarian and menstrual cycles, synchronizing ovulation with preparation of the uterus for possible implantation of an embryo. At the start of the ovarian cycle, the hypothalamus secretes a releasing hormone that stimulates the anterior pituitary gland to ❶ increase its output of FSH (follicle-stimulating hormone) and LH (luteinizing hormone). True to its name, FSH ❷ stimulates the growth of an ovarian follicle. As the maturing follicle grows, it secretes estrogen in increasing amounts. After about 12 days, ❸ estrogen levels peak, which causes ❹ a sudden surge of FSH and LH. This stimulates ovulation, and ❺ the developing follicle within the ovary bursts and releases its egg. Ovulation takes place on day 14 of the typical 28-day cycle.

Besides promoting rupture of the follicle, the sudden surge of LH has several other effects. It stimulates the completion of meiosis I, transforming the primary oocyte in the follicle into a secondary oocyte (see Figure 26.8). LH also promotes the secretion of estrogen and progesterone by the corpus luteum. Estrogen and progesterone regulate the menstrual cycle. Rising levels of these two hormones ❻ promote thickening of the endometrium. The combination of estrogen and progesterone also

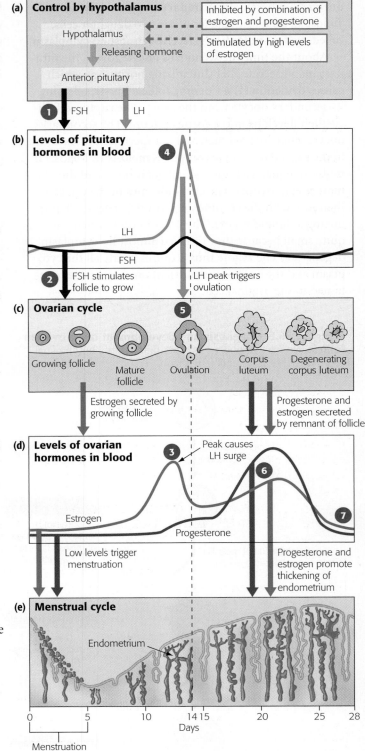

▼ Figure 26.9 **The reproductive cycle of the human female.** This figure shows how (c) the ovarian cycle and (e) the menstrual cycle are regulated by changing hormonal levels, represented in parts (a), (b), and (d). The time scale at the bottom of the figure applies to parts (b)–(e).

(a) **Control by hypothalamus**

Inhibited by combination of estrogen and progesterone

Stimulated by high levels of estrogen

Hypothalamus

Releasing hormone

Anterior pituitary

❶ FSH   LH

(b) **Levels of pituitary hormones in blood**

❹

LH

FSH

❷ FSH stimulates follicle to grow

LH peak triggers ovulation

(c) **Ovarian cycle**

❺

Growing follicle   Mature follicle   Ovulation   Corpus luteum   Degenerating corpus luteum

Estrogen secreted by growing follicle

Progesterone and estrogen secreted by remnant of follicle

(d) **Levels of ovarian hormones in blood**

❸   Peak causes LH surge

❻

Estrogen

Progesterone

❼

Low levels trigger menstruation

Progesterone and estrogen promote thickening of endometrium

(e) **Menstrual cycle**

Endometrium

0   5   10   14 15   20   25   28
Days

Menstruation

inhibits further secretion of FSH and LH, ensuring that a second follicle does not mature during this cycle. Further degeneration of the corpus luteum causes the levels of estrogen and progesterone to fall off. Once these hormones **7** fall below a critical level, the endometrium begins to shed, starting menstruation on day 1 of the next cycle. Now that estrogen and progesterone are no longer there to inhibit it, the pituitary secretes FSH and LH, and a new cycle begins.

The description of the reproductive cycle in this section assumes that fertilization has not occurred. If it does, the embryo implants into the endometrium and secretes a hormone called HCG (human chorionic gonadotropin). HCG maintains the corpus luteum, which continues to secrete progesterone and estrogen, keeping the endometrium intact. Interestingly, most home pregnancy tests work by detecting HCG in urine. (In 2009, baseball slugger Manny Ramirez was suspended from baseball after testing positive for HCG; HCG boosts testosterone production in males, so its use is banned by many sports organizations.) We'll return to what happens next—the events of pregnancy—later in the chapter. ✓

✓CHECKPOINT
What hormonal changes trigger the start of menstruation?

Answer: decreasing levels of estrogen and progesterone

Reproductive Health

Now that you've read about the anatomy and physiology of the human reproductive system, you can apply this knowledge to two issues of reproductive health: contraception and the transmission of disease.

Contraception

Contraception is the deliberate prevention of pregnancy. There are many forms of contraception that interfere with different steps in the process of becoming pregnant. Table 26.1 lists the most common methods of birth control and their failure rates when used correctly and when used typically. Note that these two rates are often quite different, emphasizing the importance of learning to use contraception correctly.

Complete abstinence (avoiding intercourse) is the only totally effective method of contraception, but other methods are effective to varying degrees. Sterilization, an operation that prevents sperm from reaching an egg, is very reliable. A woman may have a **tubal ligation** (having her "tubes tied"). In this procedure, a doctor removes a short section from each oviduct, often tying, or ligating, the remaining ends and thereby blocking the route of sperm to egg. A man may have a **vasectomy**, in which a doctor cuts a section out of each vas deferens to prevent sperm from reaching the urethra. Both forms of sterilization are free from side effects and relatively safe. They are meant to be permanent but can sometimes be surgically reversed. Although such reversals are becoming increasingly successful, they carry some risk of damaging the ducts.

The effectiveness of other methods depends on how they are used. Temporary abstinence, also called the **rhythm method** or **natural family planning**, depends on refraining from intercourse during the days around ovulation, when fertilization is most likely. In theory, the time of ovulation can be determined by monitoring changes in body temperature and the composition of cervical mucus, but careful monitoring and record keeping are required. Additionally, the length of the reproductive

Table 26.1	Contraceptive Methods	
	Pregnancies per 100 Women per Year	
Method	**Used Correctly**	**Typically**
None		85
Birth control pill	0.1	5
Vasectomy	0.1	0.15
Tubal ligation	0.2	0.5
Rhythm method	1–9	20
Withdrawal	4	19
Condom (male)	3	14
Diaphragm and spermicide	6	20
Spermicide alone	6	26

Spermicide

Diaphragm

Condom

Birth control pill

Human Development

Embryonic development begins with **fertilization**, or conception, the union of a sperm and egg to form a zygote. In this section, we will examine the process of fertilization and the subsequent development of a human.

Fertilization

Sexual intercourse releases hundreds of millions of sperm into the vagina, but only a few thousand sperm survive the several-hour trip to the egg in the oviduct. Of these sperm, only a single one can enter and fertilize the egg. All the other millions of sperm die.

Repeating a theme we've seen again and again during our exploration of anatomy, the shape of sperm is related to what they do (form fits function). A mature human sperm has a streamlined shape that enables it to swim through fluids in the vagina, uterus, and oviduct (**Figure 26.11**). The sperm's head contains a haploid nucleus and is tipped with a membrane-enclosed sac called the acrosome. The middle of the sperm contains a long spiral mitochondrion that

uses high-energy nutrients from the semen, especially the sugar fructose, to fuel movement of the flagellum.

Figure 26.12 traces one sperm through the events of fertilization. The sperm ❶ approaches and then ❷ contacts the jelly coat (red) that surrounds the egg. The acrosome in the sperm's head releases a cloud of enzymes that digest a hole in the jelly. This hole allows the sperm head to ❸ fuse its plasma membrane with that of the egg. Fusion of the two membranes makes it possible for the sperm nucleus to ❹ enter the cytoplasm of the egg. Fusion also triggers completion of meiosis II in the egg. Furthermore, contact of sperm with egg triggers a change in the egg's plasma membrane that makes it impenetrable to other sperm cells. This blockage of other sperm ensures that the zygote that is forming contains only the diploid number of chromosomes. The chromosomes of the egg and sperm nuclei ❺ are eventually enclosed in a single diploid nucleus. In the diploid zygote, the egg's metabolic machinery awakens from dormancy and gears up in preparation for the enormous growth and development that will soon follow. ☑

▼ **Figure 26.11** **A human sperm cell.**

Plasma membrane

Head (carries genetic material)

Mitochondrion (produces energy) Nucleus Acrosome

Flagellum (used for swimming)

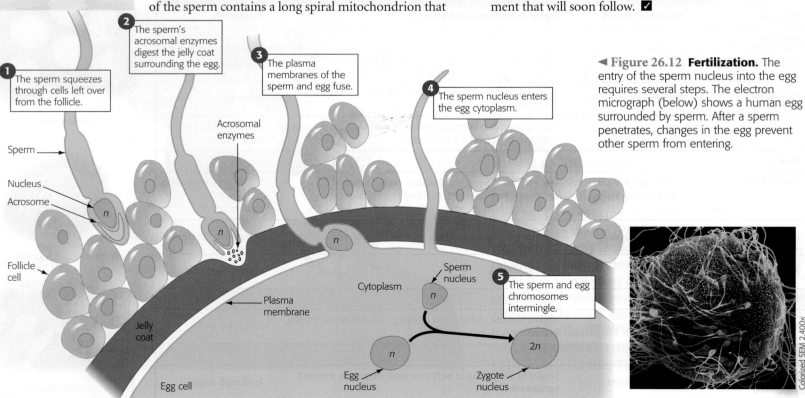

① The sperm squeezes through cells left over from the follicle.

② The sperm's acrosomal enzymes digest the jelly coat surrounding the egg.

③ The plasma membranes of the sperm and egg fuse.

④ The sperm nucleus enters the egg cytoplasm.

Acrosomal enzymes

Sperm

Nucleus

Acrosome

Follicle cell

Plasma membrane

Jelly coat

Egg cell

Cytoplasm

Sperm nucleus

⑤ The sperm and egg chromosomes intermingle.

Egg nucleus

Zygote nucleus

◄ **Figure 26.12** **Fertilization.** The entry of the sperm nucleus into the egg requires several steps. The electron micrograph (below) shows a human egg surrounded by sperm. After a sperm penetrates, changes in the egg prevent other sperm from entering.

Colorized SEM 2,400×

Basic Concepts of Embryonic Development

A single-celled human zygote formed by fertilization is smaller than the period at the end of this sentence. From this humble start, the zygote develops into a full-fledged organism with trillions of cells organized into complex tissues and organs. Clearly, this process requires an astonishing amount of cell division and specialization. The key to development in all organisms is that each stage takes place in a highly organized fashion.

Development begins with **cleavage**, a series of rapid cell divisions that produces a multicellular ball. After the zygote divides for the first time, about 36 hours after fertilization, it is called an embryo. Rarely, and apparently at random, the two cells of the early embryo separate from each other. When this separation happens, each cell may "reset" and act as a zygote; the result is the development of identical (monozygotic) twins. (Nonidentical, or dizygotic, twins result from a completely different mechanism: Two eggs fuse with two separate sperm to produce two genetically unique zygotes that develop at the same time.)

During cleavage, DNA replication, mitosis, and cytokinesis occur rapidly, but the total amount of cytoplasm remains unchanged. As a result, the overall size of the embryo does not change; instead, each cell division partitions the embryo into twice as many smaller cells, each with its own nucleus (**Figure 26.13**). Cleavage continues as the embryo moves down the oviduct toward the uterus. A central cavity begins to form in the embryo. About 6–7 days after fertilization, the embryo has reached the uterus as a fluid-filled hollow ball of about 100 cells called a **blastocyst**. Protruding into the central cavity on one side of the human blastocyst is a small clump of cells called the **inner cell mass**, which will eventually form the fetus. Occasionally, an embryo does not travel down the oviduct to the uterus. The result is an ectopic pregnancy, one in which the embryo develops in the wrong location. Ectopic pregnancies are invariably fatal to the embryo and can be dangerous to the mother, requiring immediate medical attention.

The second stage of development, under way by 9 days after conception, is **gastrulation** (see the bottom of Figure 26.13). The cells of the inner cell mass begin an organized migration that produces the **gastrula**, an embryo with three main layers. The three layers produced in gastrulation are embryonic tissues called ectoderm, endoderm, and mesoderm. The **ectoderm** eventually develops into the nervous system and outer layer of skin (epidermis). The **endoderm** becomes the innermost lining of the digestive system and organs such as the liver,

pancreas, and thyroid. The **mesoderm** gives rise to most other organs and tissues, such as the heart, kidneys, and muscles.

At this stage, various cellular changes contribute to the formation of embryonic structures. All developmental processes depend on chemical signals passed between neighboring cells and cell layers, telling embryonic cells precisely what to do and when. The mechanism by which one group of cells influences the development of an adjacent group of cells is called **induction**. Its effect is to switch on a set of genes whose expression makes the receiving cells differentiate into a specific cell or tissue type. For example, inductive signals can cause cells to change shape or to migrate from one location to another within the developing embryo. Another key

Embryonic Development

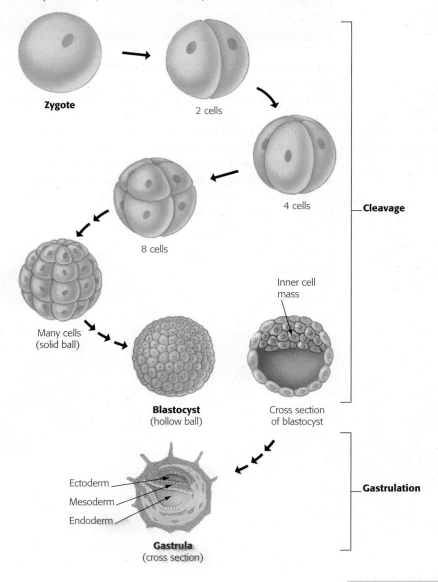

▼ **Figure 26.13 Early development of an embryo.** Notice that each round of division during cleavage does not change the total size of the embryo. Instead, there are more cells, and each cell is smaller.

Zygote

2 cells

4 cells

8 cells

Many cells (solid ball)

Blastocyst (hollow ball)

Inner cell mass

Cross section of blastocyst

Cleavage

Ectoderm

Mesoderm

Endoderm

Gastrula (cross section)

Gastrulation

developmental process is **programmed cell death**, which kills selected cells. For example, some genes encode proteins that kill certain cells in developing human hands and feet, separating the fingers and toes (**Figure 26.14**).

During the course of embryonic development, a sequence of inductive signals leads to increasingly greater specialization of cells as organs begin to take shape. The importance of these processes is underscored by birth defects that result from improper signaling. Spina bifida is a condition that results from the failure of a tube of ectoderm cells to properly close and form the spine during the first month of fetal development. Infants born with spina bifida often have permanent nerve damage that results in paralysis of the lower limbs. ✔

◀ **Figure 26.14**
Programmed cell death in a developing human hand. Proteins produced during human development destroy cells in the early hand (left), creating the spaces between fingers (right).

Pregnancy and Early Development

Pregnancy, or **gestation**, is the carrying of developing young within the female reproductive tract. It begins at fertilization and continues until birth. In humans, gestation lasts about 266 days (38 weeks), but is usually measured as 40 weeks (9 months) from the start of the last menstrual cycle. Other mammals have much shorter (4 weeks in mice) or longer (22 months in elephants) gestation periods.

The early stages in human development are summarized in **Figure 26.15**. Fertilization and cleavage take place in the oviduct. About a week after conception, the embryo reaches the uterus and implants itself into the

▶ **Figure 26.15**
Early stages of human development. Fertilization takes place in the oviduct. As the zygote travels down the oviduct, cleavage starts. By the time the embryo reaches the uterus, it has become a blastocyst, and it implants into the endometrial lining.

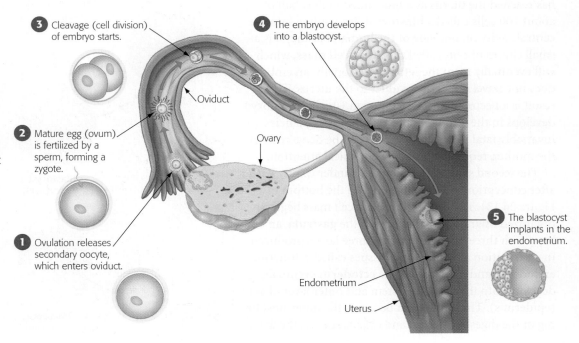

3 Cleavage (cell division) of embryo starts.

4 The embryo develops into a blastocyst.

Oviduct

2 Mature egg (ovum) is fertilized by a sperm, forming a zygote.

Ovary

1 Ovulation releases secondary oocyte, which enters oviduct.

5 The blastocyst implants in the endometrium.

Endometrium

Uterus

endometrium. By this time, the embryo has become a blastocyst, with a central fluid-filled cavity and an inner cell mass (**Figure 26.16a**). These early embryonic cells are **stem cells**, with the potential to give rise to every type of cell in the body. (As discussed in Chapter 11, embryonic stem cells have great potential as medical tools.) The outer cell layer, called the **trophoblast** (**Figure 26.16b**), becomes part of the **placenta**, the organ that provides nourishment and oxygen to the embryo and helps dispose of its metabolic wastes.

Figure 26.16c shows the embryo about a month after conception. Besides the growing embryo, there are now four pieces of life-support equipment: the amnion, the yolk sac, the allantois, and the chorion. The **amnion** is a fluid-filled sac that encloses and protects the embryo. The amnion usually breaks just before childbirth, and the amniotic fluid ("water") leaves the mother's body through her vagina. The **yolk sac** produces the embryo's first blood cells and its first gamete-forming cells in the gonads. The **allantois** forms part of the **umbilical cord**—the lifeline between the embryo and the placenta. The outermost membrane, the **chorion**, becomes part of the placenta.

The placenta develops **chorionic villi**, finger-like outgrowths containing embryonic blood vessels that are closely associated with the blood vessels of the mother's endometrium. The chorionic villi absorb nutrients and oxygen from the mother's blood and pass these substances to the embryo. The villi also carry wastes from the embryo to the mother's bloodstream. As discussed in Chapter 9, a small sample of chorionic villus tissue can be removed for prenatal genetic testing.

The placenta provides for other needs of the embryo as well. For example, it allows protective antibodies to pass from the mother to the fetus. However, tobacco smoke, alcohol, and other drugs can also cross the placenta, and these harmful substances can cause developmental abnormalities and raise the chances of miscarriage. Additionally, viruses can cross the placenta and cause disease. ☑

▼ Figure 26.16 **A human embryo: the first month.**

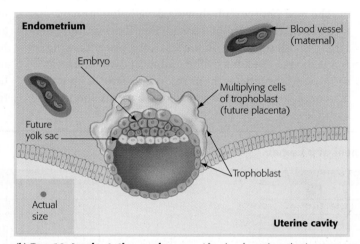

(a) Day 6: Before implantation. By 6 days after conception, the embryo has developed into a blastocyst.

(b) Day 10: Implantation under way. After implantation, the inner cell mass of embryonic stem cells will develop into the fetus. The outer layer of trophoblast cells will become the embryo's contribution to the placenta.

(c) Day 31: The embryo and its life-support equipment

The Stages of Pregnancy

In this section, we use the series of photographs in **Figure 26.17** to illustrate the rest of human development as it takes place in the uterus. For convenience, we divide pregnancy (the period from conception to birth) into three **trimesters** of about 3 months each.

The First Trimester

The first photograph in Figure 26.17 shows a human embryo about 5 weeks after fertilization. In that brief time, a single cell has developed into a highly organized multicellular embryo. A month-old human embryo is about 7 mm (0.28 in.) long. Its brain and spinal cord have begun to take shape. It also has four stumpy limb buds, a short tail, and pharyngeal pouches. Overall, a month-old human embryo looks pretty much like any other month-old vertebrate embryo (see Figure 13.9).

The next photograph shows a developing human, now called a fetus, about 9 weeks after fertilization. The large pinkish structure on the left is the placenta, which is attached to the fetus by the umbilical cord. The clear sac around the fetus is the amnion. The fetus is about 5.5 cm (2.2 in.) long and has all of its organs and major body parts, including muscles and the bones of the back and ribs. The limb buds have become tiny arms and legs with fingers and toes. By the end of the first trimester, the fetus looks like a miniature human being, albeit one with an oversized head. By this time, the sex of the fetus can be reliably determined by an ultrasound exam (**Figure 26.18**).

The Second Trimester

The main developmental changes during the second and third trimesters involve an increase in size and general refinement of the human features—nothing as dramatic as the changes of the first trimester. A 14-week-old fetus is about 6 cm (2.4 in.) long.

At 20 weeks, well into the second trimester, the fetus is about 19 cm (7.6 in.) long, weighs about half a kilogram (1 lb), and has the face of an infant, complete with eyebrows and eyelashes. Its arms, legs, fingers, and toes have lengthened. It also has fingernails and toenails and is covered with fine hair. By this time, the fetal heartbeat is detectable with a stethoscope, and the mother can usually feel the fetus moving. Because of the limited space in the uterus, the fetus flexes forward into the so-called fetal position, with its head tucked against

▼ Figure 26.17 **The development of a human.**

5 weeks (35 days)   **9 weeks** (63 days)   **14 weeks** (98 days)

its knees. By the end of the second trimester, the fetus's eyes are open, its teeth are forming, and its bones have begun to harden.

The Third Trimester

The third trimester is a time of rapid growth as the fetus gains the strength it will need to survive outside the protective environment of the uterus. Babies born prematurely—as early as 24 weeks—may survive, but they require special medical care after birth. During the third trimester, the fetus's circulatory system and respiratory system undergo changes that will allow the switch to air breathing. The fetus begins to gain the ability to maintain its own temperature, and its muscles thicken. It also loses much of its fine body hair, except on its head. The fetus usually rotates so that its head points down toward the cervix. The fetus becomes less active as it fills the space in the uterus. As the fetus grows and the uterus expands around it, the mother's abdominal organs may be squeezed, causing frequent urination, digestive troubles, and backaches. At birth, babies average about 50 cm (20 in.) in length and 2.7–4.5 kg (6–10 lb) in weight. ☑

▼ Figure 26.18 **Ultrasound imaging.** An ultrasound image is produced when high-frequency sounds from an ultrasound scanner held against a pregnant woman's abdomen bounce off the fetus. The inset image shows a fetus in the uterus at about 18 weeks.

☑CHECKPOINT

As a human develops in the uterus, why are the first-trimester changes considered the most dramatic?

Answer: The development proceeds from a single-celled zygote to a miniature human. Subsequent changes are mainly growth and refinement of the structures developed during the first trimester.

| May | June | July | August | September | October |

20 weeks (140 days)

At birth (280 days)

567

Childbirth

The birth of a child is brought about by **labor**, a series of strong, rhythmic contractions of the uterus. Labor is induced through a positive-feedback mechanism involving several hormones. The fetus's cells produce the hormone oxytocin, and late in pregnancy, the mother's pituitary gland secretes it in increasing amounts. Oxytocin is a powerful stimulant for the smooth muscles in the wall of the uterus, causing them to contract. Uterine contractions stimulate the release of more hormones, which increase contractions—an example of positive-feedback control. The result is climactic—the intense muscle contractions that propel a baby from the womb. If a mother is overdue, or if labor has continued for a long time, a doctor may inject synthetic hormones to promote a more rapid birth.

Figure 26.19 shows the three stages of labor. ① The first stage, dilation, is the time from the onset of labor until the cervix dilates, or opens, to a width of about 10

cm. Dilation is the longest stage of labor, typically lasting 6–12 hours or even considerably longer. ② The period from full dilation of the cervix to delivery of the infant is called the expulsion stage. Strong uterine contractions, lasting about 1 minute each, occur every 2–3 minutes, and the mother feels an increasing urge to push with her abdominal muscles. Within a period of 20 minutes to an hour or so, the infant is forced down and out of the uterus and vagina. After the baby is expelled, the umbilical cord is clamped and cut. The stump of the cord remains for several weeks, then shrivels and falls off, leaving the belly button. ③ The final stage is the delivery of the placenta ("afterbirth"), usually within 15 minutes after the birth of the baby.

Hormones continue to be important after birth. Decreasing levels of progesterone and estrogen allow the uterus to start returning to its prepregnancy state. The pituitary hormone prolactin promotes milk production (called lactation) by the mammary glands. At first, a yellowish antibody-rich fluid called colostrum is secreted. After 2–3 days, normal milk production begins. ✔

▼ Figure 26.19 **The three stages of labor.**

① **Dilation of the cervix** ② **Expulsion: delivery of the infant** ③ **Delivery of the placenta**

Reproductive Technologies

In previous sections, we examined the normal means of human sexual reproduction. Many couples, however, are unable to conceive children because of one or more abnormalities. Today, reproductive technologies can solve many of these problems.

Infertility

About 15% of couples who want children are unable to conceive, even after a year of regular, unprotected sex. Such a condition is called **infertility**, and it can have several different causes.

In most cases, infertility can be traced to problems with the man. His testes may not produce enough sperm (a "low sperm count"), or the sperm may not be vigorous enough to reach an egg. In other cases, infertility is caused by **impotence**, also called erectile dysfunction, the inability to maintain an erection. Temporary impotence can result from alcohol or drug use or from psychological problems. Permanent impotence can result from nervous system or circulatory problems.

Female infertility can result from a lack of eggs, a failure to ovulate, or blocked oviducts (often due to a sexually transmitted disease). Other women are able to conceive but cannot support a growing embryo in the uterus. The resulting multiple miscarriages can take a heavy emotional toll.

As discussed in the Biology and Society section, assisted reproductive technologies can help many cases of infertility. For a low sperm count, there is often a simple solution. Underproduction of sperm can frequently be traced to the man's scrotum being too warm, so a switch of underwear from briefs (which hold the scrotum close to the body) to boxers is often a first treatment. If that doesn't work, sperm can be collected, concentrated, and then injected into the woman's uterus via the vagina. Drug therapies (such as Viagra) and penile implants can treat impotence. If a man produces no functioning sperm, the couple may elect to use another man's sperm from a sperm bank.

If a woman has normal eggs that are not being released properly, hormone injections can induce ovulation; such treatments frequently result in multiple pregnancies. If a woman has no eggs of her own, they can be obtained from a donor for fertilization and injection into the uterus. Although sperm can be collected without any danger to the donor, the collection of eggs is a surgical procedure that involves pain and risk for the donating woman.

If a woman cannot maintain pregnancy, she and her partner may enter into a legal contract with a surrogate mother who agrees to carry the couple's child to birth. This method has worked for many couples, but serious ethical and legal problems can arise if a surrogate mother changes her mind and wants to keep the baby she has carried. A number of states have laws restricting surrogate motherhood. ✔

In Vitro Fertilization

A procedure performed in vitro (literally, "in glass") happens under artificial laboratory conditions rather than within the living body (in vivo). **In vitro fertilization (IVF)**, sometimes called creating a "test-tube baby," begins with the surgical removal of eggs and the collection of sperm. The eggs are fertilized in a petri dish (**Figure 26.20**), allowed to develop for several days, and then injected into a woman's uterus, where one or more of them may successfully implant and continue development. The sperm and eggs used for this procedure, as well as the embryos created, can be used immediately or frozen for later use.

☑ **CHECKPOINT**

What is the difference between infertility and impotence?

Answer: Infertility is the inability to produce offspring; impotence is the inability to maintain an erection.

▼ Figure 26.20 **In vitro fertilization.**

Collected egg

Collected sperm

In vitro fertilization

Zygote

Early embryo

Implantation

THE PROCESS OF SCIENCE: Reproductive Technologies

Can Viable Embryos Be Identified?

In vitro fertilization typically involves the creation of many embryos. Because the procedure is so expensive, it would be helpful to identify those embryos with the greatest chance of success.

To address this need, some European researchers started with the **observation** that cells can be removed from the outer layer of a preimplantation blastocyst without harming the embryo. The researchers formed the **hypothesis** that analyzing the gene activity within such cells could provide useful information about the chances of a successful pregnancy from the embryo. Their **prediction** was that gene expression would differ between successful and unsuccessful embryos.

Their **experiment** involved a combination of several procedures that you have learned about in previous chapters. The subjects were 48 patients who produced 153 blastocysts after undergoing IVF. On the 5th day after fertilization, the researchers collected cells from each embryo (**Figure 26.21**). After the embryos were transferred into the mothers, 37 of them resulted in healthy births. DNA samples from each baby were matched to the blastocyst cells using DNA profiling (see Chapter 12), allowing researchers to determine which embryos led to successful pregnancies and which did not. The embryonic cells were then screened using DNA microarrays (see Chapter 11) to reveal which genes were turned on and off in each embryo. Their **results** identified a large number of differences in gene expression between embryos that were destined for success and those that were not. In particular, genes for cell-to-cell communication and the ability of cells to stick together were more active in the successful embryos. Fertility researchers hope that these results will lead to tests that can identify embryos most likely to result in pregnancy, reducing the need to implant multiple embryos and therefore reducing the likelihood of multiple pregnancy and the risks involved.

▼ Figure 26.21 **A 5-day-old human embryo.**

Colorized SEM 520x

fertilized egg develops into a fetus. The uterus opens into the vagina, which receives the penis during intercourse and forms the birth canal.

stimulate a new cycle.

Neurons and Electrical Potentials

How Neurons Work

The Ethics of IVF

Recent research has shown that IVF babies have a slightly greater risk of birth defects than babies born

progresses, the potential to screen for diseases and physical traits increases. Parents may thus have a lot of information they can use in choosing which embryos to implant.

An Overview of Animal Nervous Systems

The **nervous system** forms a communication and coordination network throughout an animal's body. Nervous systems are the most intricately organized data processing systems on Earth. Your brain, for instance, contains an estimated 100 billion **neurons**, nerve cells specialized for carrying electrical signals from one part of the body to another. Each neuron may communicate with thousands of others, forming networks that enable us to move, perceive our surroundings, learn, and remember. In this section, we'll examine the organization and cellular basis common to all animal nervous systems.

Organization of Nervous Systems

The nervous systems of most animals have two main anatomical divisions. The **central nervous system (CNS)** consists of the brain and, in vertebrates, the spinal cord. The **peripheral nervous system (PNS)** is made up mostly of nerves that carry signals into and out of the CNS. A **nerve** is a communication line made from cable-like bundles of neuron fibers tightly wrapped in connective tissue.

Figure 27.1 highlights the three interconnected functions of the nervous system and the three types of neurons that carry out these functions. **Sensory input** is the process of sending signals from sensory receptors, such as light-detecting

receptors in the eyes, to the CNS. **Sensory neurons** (shown in blue) convey this information. **Integration** is the interpretation of the sensory signals and the formulation of responses. **Interneurons** (green), located entirely within the CNS, perform integration. **Motor output** is the process of sending signals from the integration centers to effectors, such as muscle cells. **Effectors** perform the body's responses (moving a leg in this example). Motor output is performed by **motor neurons** (purple). ☑

▼ Figure 27.1 **Organization of a nervous system.** This figure shows only one neuron of each functional type, but body activity actually involves many. A simplified version of this figure will keep you oriented throughout the chapter.

SENSORY INPUT

Sensory receptor

Sensory neuron

INTEGRATION

Interneuron

MOTOR OUTPUT

Motor neuron

Effector cells

Brain and spinal cord

Peripheral nervous system (PNS)

Central nervous system (CNS)

Neurons

The three types of neurons just described vary widely in shape, but most of them share some common features. **Figure 27.2** depicts a motor neuron, like those that carry signals from your spinal cord to your skeletal muscles. A motor neuron has a large **cell body** housing the nucleus and other organelles. Two types of extensions project from the cell body. Numerous **dendrites** are often short and highly branched. Dendrites *receive* incoming messages from other neurons and convey this information toward the cell body. The

other type of extension, a single long fiber called an **axon**, *transmits* signals toward another neuron or toward an effector. Some axons, such as the ones that reach from your spinal cord to muscle cells in your feet, can be over a meter long.

Neurons make up only part of a nervous system. Outnumbering neurons by as many as 50 to 1 are **supporting cells** that protect, insulate, and reinforce the neurons. In many vertebrates, axons that convey signals rapidly are enclosed along most of their length by an insulating material called the **myelin sheath**. The myelin sheath is a chain of bead-like supporting cells. The myelin sheath helps speed electrical transmission

Signal direction

Dendrites

Cell body

Axon

Nucleus

Supporting cell

Signal pathway

Myelin sheath

Synaptic terminals

◄ Figure 27.2 **Structure of a motor neuron.** This diagram shows a motor neuron. The flow of the electrical signal through a neuron follows this path: dendrite → cell body → axon.

along an axon. In the human nervous system, signals can travel along a myelinated axon about 150 m/sec (over 330 mph!), which means that a command from your brain can make your fingers move in just a few milliseconds.

The debilitating autoimmune disease multiple sclerosis (MS) leads to a gradual destruction of the myelin sheaths by the individual's own immune system. The result is a progressive loss of signal conduction and muscle control. MS has no cure, but drugs that suppress the immune system can relieve symptoms and slow its progress.

Returning to Figure 27.2, notice that the axon ends in a cluster of branches. A typical axon has hundreds or thousands of these branches, each with a bulb-like structure called a synaptic terminal at the very end. As we will see later, each **synaptic terminal** relays signals to another neuron or to an effector such as a muscle cell. With the basic structure of a neuron in mind, let's take a closer look at how neurons convey signals. ✓

Sending a Signal through a Neuron

To understand nerve signals, we must first study a resting neuron, one that is not transmitting a signal. A resting neuron has potential energy that can be put to work to send nerve signals from one part of the body to another. This potential energy exists in the form of an electrical charge difference across the neuron's plasma membrane. The cytoplasm just inside the membrane is negative in charge, and the fluid just outside the membrane is positive. Because opposite charges tend to move toward each other, the membrane stores energy by holding opposite charges apart, much as a battery does. This difference in charge (voltage) that exists across the plasma membrane of a resting neuron is called the **resting potential**.

The resting potential is caused by the presence of differently charged ions on either side of the membrane. What keeps the ions apart? As mentioned in Chapter 5, the hydrophobic interior of the membrane's phospholipid bilayer doesn't let ions pass through. However, the membrane has protein channels and pumps that can allow positive ions across. The controlled passage of these ions can be used to tap into the energy stored in the resting potential.

The Action Potential
Turning on a flashlight uses the potential energy stored in a battery to create light. In a similar way, stimulating a neuron's plasma membrane can trigger the release and use of the membrane's potential energy to generate an

electrical nerve signal. A **stimulus** is any factor that causes a nerve signal to be generated. Examples of stimuli include light, sound, touching a hot surface, or a chemical signal from another neuron. A stimulus of sufficient strength can trigger an **action potential**, a nerve signal that carries information along a neuron. The signal is actually a self-propagating change in the voltage across the plasma membrane.

Figure 27.3 shows the series of events involved in generating an action potential. As you'll see, two different sets of ion channels are involved in this process. ❶ At first, the membrane is at its resting potential, positively charged outside and negatively charged inside. ❷ A stimulus triggers the opening of a few of the first set of ion channels in the membrane (represented by the blue arrows), allowing a few positive ions to enter the neuron. This tiny change makes the inside surface of the membrane slightly less negative than before. If the stimulus is strong enough, a sufficient number of channels open to reach the threshold. The **threshold** is the minimum change in a membrane's voltage that must occur to trigger the action potential. ❸ Once threshold is reached, more of these channels open and more positive ions rapidly move in. As a result, the interior of this region of the cell becomes positively charged with respect to the outside. ❹ This electrical change triggers the closing and inactivation of the first set of channels. Meanwhile, a second set of channels opens (green arrows), allowing other positive ions to diffuse rapidly out and returning the membrane to its resting potential. Within a living neuron, this whole process takes just a few milliseconds.

✓CHECKPOINT
Arrange the following parts of a neuron in the sequence that a signal would pass through them: axon, cell body, dendrite, synaptic terminal. For each part, indicate whether one or many are found in a typical neuron.

Answer: many dendrites → one cell body → one axon → many synaptic terminals

▼ Figure 27.3 **Generation of an action potential.**

❶ Resting potential

❷ A stimulus opens first set of ion channels; if threshold is reached, action potential is triggered.

❸ Additional channels open; in that region of the neuron, interior of cell becomes more positive than outside.

❹ First set of channels closes and inactivates; second set of channels opens and positive ions rush out; membrane returns to resting potential.

Neuron interior

Propagation of the Signal

An action potential is a localized electrical event—a rapid change from the resting potential at a specific place along the neuron. A nerve signal starts out as an action potential generated near the cell body of the neuron. To function as a nerve signal over a distance, this local event must be passed along the neuron. This process is like tipping the first of a row of standing dominoes: The first domino does not travel along the row, but its fall is relayed to the end of the row, one domino at a time.

Figure 27.4 shows the changes that occur in part of an axon as a nerve signal passes from left to right. Let's start by focusing only on the axon region at the far left in each panel. ❶ When this region of the axon (blue) has its first set of channels open, positive ions diffuse inward (blue arrows), and an action potential is generated. ❷ That same region then opens the second set of channels, allowing other positive ions to diffuse out of the axon (green arrows) while closing the first set of channels that allow ions in. ❸ A short time later, we would see no action potential at this (far left) spot because the membrane here has restored itself and returned to its resting potential.

Now let's see how these events lead to the "domino effect" of a nerve signal. In step 1 of the figure, the blue arrows pointing sideways within the axon indicate local spreading of the electrical changes caused by the positive ions entering the neuron. These changes trigger the opening of the first set of channels in the membrane just to the right of the action potential. As a result, a second action potential is generated, as indicated by the blue region in step 2. In the same way, a third action potential is generated in step 3. At each point along the neuron, the action potential triggers changes in the adjacent region that result in propagation of the action potential along the neuron.

So why are action potentials propagated in only one direction along the axon (left to right in the figure)? As the blue arrows indicate, local electrical changes do spread in both directions in the axon. However, these changes cannot open the first set of channels and generate an action potential when these channels are inactivated (step 4 in Figure 27.3). Thus, an action potential cannot be generated in the regions where the second set of channels is open and positive charge is leaving the axon (green in Figure 27.4).

Action potentials are all-or-none events; that is, they are the same no matter how strong or weak the stimulus that triggers them (as long as threshold is reached). How, then, do action potentials relay different intensities of information? It is the frequency of action potentials that changes with the intensity of stimuli. If you

rap your finger hard against the desk, for example, your CNS receives many more action potentials per millisecond than if you tap softly. Once your CNS receives information in the form of action potentials, it can process the information and formulate a response to it. ☑

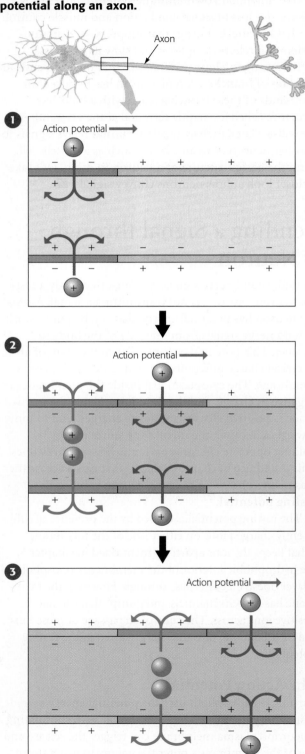

▼ Figure 27.4 **Propagation of an action potential along an axon.**

Passing a Signal from a Neuron to a Receiving Cell

If an action potential travels in one direction along a neuron, what happens when the signal arrives at the end of the neuron? To continue conveying information, the signal must be passed to another cell. This occurs at a **synapse**, or relay point, between a neuron and a receiving cell. The receiving cell can be another neuron or an effector cell such as a muscle cell or endocrine cell.

Synapses come in two varieties: electrical and chemical. In an electrical synapse, electric current passes directly from one neuron to the next. In the human body, electrical synapses are found in the heart and digestive tract, where steady, rhythmic muscle contractions are maintained. Chemical synapses are prevalent in most other organs, including skeletal muscles, and in the central nervous system.

Chemical synapses have a narrow gap, called the **synaptic cleft**, separating the synaptic terminal of the sending neuron from the receiving cell (**Figure 27.5**). When the action potential (an electrical signal) reaches the end of the sending neuron, it is converted to a chemical signal consisting of molecules of neurotransmitter. A **neurotransmitter** is a chemical that carries information from a nerve cell to another cell that will react, such as another nerve cell or an effector cell. Once a neurotransmitter conveys a chemical signal from the sending neuron, an action potential may then be generated in the receiving cell.

Let's follow the events that occur at a synapse between two neurons in Figure 27.5. ❶ An action potential (red arrow) arrives at the synaptic terminal. ❷ The action potential causes vesicles filled with neurotransmitter (🔵) to fuse with the plasma membrane of the sending neuron. ❸ The fused vesicles release their neurotransmitter molecules into the synaptic cleft. ❹ The released neurotransmitter molecules diffuse across the cleft and bind to complementary receptors on ion channel proteins in the receiving neuron's plasma membrane. ❺ The binding of neurotransmitter to receptors opens the ion channels. With the channels open, ions can diffuse into or out of the receiving neuron. In many cases, this triggers a new action potential; in other cases, a neurotransmitter inhibits the generation of an action potential. ❻ The neurotransmitter is broken down or transported back to the sending neuron, causing the ion channels in the receiving cell's plasma membrane to close. Step 6 ensures that the neurotransmitter's effect on the receiving neuron is brief and precise.

BioFlix™

How Synapses Work

▼ Figure 27.5 **Neuron communication at a synaptic cleft.**

SYNAPSE

Sending neuron

Vesicles

Synaptic terminal

❶ Action potential arrives.

❷ Vesicle fuses with plasma membrane.

❸ Neurotransmitter is released into synaptic cleft.

Synaptic cleft

❹ Neurotransmitter binds to receptor.

Receiving neuron Ion channels Neurotransmitter molecules

Synaptic terminal of sending neuron

Dendrite of receiving neuron

Neurotransmitter

Receptor

Ions

❺ Ion channel opens and triggers or inhibits a new action potential.

❻ Ion channel closes. Neurotransmitter is broken down and released.

Chemical synapses can process extremely complex information. A neuron may receive input from hundreds of other neurons via thousands of synaptic terminals (**Figure 27.6**). The inputs can be highly varied because each sending neuron may secrete a different quantity or kind of neurotransmitter. These factors account for the nervous system's ability to process huge amounts of complex stimuli and formulate appropriate responses. ☑

Neurotransmitters

A variety of small molecules can act as neurotransmitters, including amines and peptides. The amines are neurotransmitters derived from amino acids. Examples include serotonin, dopamine, epinephrine, and norepinephrine. Serotonin and dopamine affect sleep, mood, attention, and learning. Imbalances of amines are associated with various disorders. For example, reduced levels of norepinephrine and serotonin have been linked to some types of depression. The degenerative illness Parkinson's disease is associated with a lack of dopamine in the brain, and an excess of dopamine is linked to schizophrenia.

Many peptides (short chains of amino acids) also serve as neurotransmitters. One class of such peptides is the endorphins (also considered hormones), which decrease our perception of pain during times of physical or emotional stress. Endorphins may be released in response to a wide variety of stimuli, including traumatic injury, muscle fatigue, and even eating very spicy foods.

Drugs and the Brain

Many drugs, including common ones such as caffeine, nicotine, and alcohol, affect the actions of neurotransmitters at the brain's trillions of synapses. Caffeine, found in coffee, tea, chocolate, and many soft drinks and energy drinks, counters the effects of inhibitory neurotransmitters, ones that normally suppress nerve signals. This is why caffeine tends to stimulate you and keep you awake. Nicotine acts as a stimulant by binding to

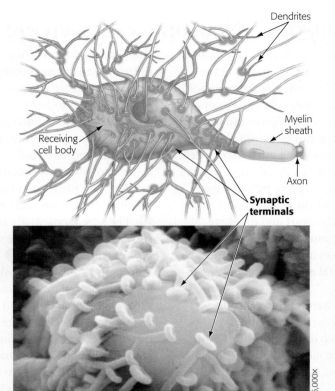

Dendrites

Receiving
cell body

Myelin
sheath

Axon

**Synaptic
terminals**

SEM 5,000×

▲ **Figure 27.6 A neuron's multiple synaptic inputs.**
As shown in this drawing and micrograph, a single neuron
may receive signals from hundreds of other neurons via
thousands of synaptic terminals (shown in orange).

and activating receptors for a neurotransmitter called acetylcholine. Alcohol is a strong depressant; its precise effect on the nervous system is not yet known, but it seems to increase the effects of another inhibitory neurotransmitter.

Many prescription drugs used to treat psychological disorders alter the effects of neurotransmitters. Tranquilizers such as diazepam (Valium) and alprazolam (Xanax) activate the receptors for an inhibitory neurotransmitter. In other cases, a drug may physically block a receptor, preventing a neurotransmitter from binding, thereby reducing its effect. For instance, some drugs used to treat schizophrenia block dopamine receptors. Some drugs used to treat attention deficit hyperactivity disorder (ADHD), such as methylphenidate (Ritalin), are chemically similar to the neurotransmitters dopamine and norepinephrine. Scientists think that ADHD medications block the reuptake of these neurotransmitters, but do not yet understand their precise actions in the brain.

What about illegal drugs? Stimulants such as amphetamines and cocaine increase the release and availability of norepinephrine and dopamine at synapses. Abuse of these drugs can therefore produce symptoms resembling schizophrenia. LSD and mescaline may produce their hallucinatory effects by activating serotonin and dopamine receptors. The active ingredient in marijuana binds to brain receptors normally used by other neurotransmitters that seem to play a role in pain, depression, appetite, memory, and fertility. Opiates—morphine, codeine, and heroin—bind to endorphin receptors, reducing pain and producing euphoria. Opiates are thus usually prescribed to relieve pain.

While the drugs mentioned here have the ability to increase alertness and a sense of well-being or to reduce physical and emotional pain, they also have the potential to disrupt the brain's finely tuned neural pathways, altering the chemical balances that are the product of millions of years of evolution. ☑

The Human Nervous System:
A Closer Look

To this point, we have concentrated on cellular mechanisms that are fundamental to nearly all animal nervous systems. Although there is remarkable uniformity throughout the animal kingdom in the way nerve cells function, there is great variety in how nervous systems as a whole are organized. For the rest of this chapter, we'll focus on vertebrates, with particular emphasis on human systems.

Vertebrate nervous systems are diverse in both structure and level of sophistication—the brains of dolphins and humans are much more complex than the brains of frogs and fishes, for instance. However, some features are common to the nervous systems of all vertebrates. All have a nervous system that is concentrated at the head end, and all have a central nervous system (CNS) distinct from a peripheral nervous system (PNS) (**Figure 27.7**).

The Central Nervous System

INTEGRATION

The vertebrate central nervous system integrates information coming from the senses and transmits signals that produce responses. In all vertebrates, the brain and spinal cord make up the CNS. The **spinal cord**, a jellylike bundle of nerve fibers, lies inside the spinal column. The spinal cord acts as the central communication conduit between the brain and the rest of the body. Millions of nerve fibers within the cord carry motor information from the brain to the muscles, while other fibers convey sensory information (such as touch, pain, and body position) from the body to the brain. The spinal cord acts like a telephone cable jam-packed with wires, each of which carries messages between the central hub and relay centers for outlying areas. In humans, the spinal cord is well protected by the bony spinal column. However, a traumatic blow to the spinal column can crush the delicate nerve bundles and prevent signals from passing. The result may be a debilitating injury. Such trauma along the back can cause paraplegia—paralysis of the lower half of the body. Trauma higher up on the spinal column can cause quadriplegia—paralysis from the neck down. Such injuries may be permanent because the nerves of the spinal cord, unlike many other body tissues, cannot be repaired.

Central nervous system (CNS)

Brain

Spinal cord

Peripheral nervous system (PNS)

◀ Figure 27.7 **A vertebrate nervous system (back view).** The central nervous system (CNS) of humans includes the brain and the spinal cord. The peripheral nervous system consists of nerves that convey information to and from the CNS.

The master control center of the nervous system, the **brain**, includes homeostatic centers that keep the body functioning smoothly, sensory centers that integrate data from the sense organs, and (in humans, at least) centers of emotions and intellect. The brain also sends out motor commands to muscles.

Both the brain and spinal cord contain spaces filled with a liquid called **cerebrospinal fluid** that cushions the CNS and helps supply it with nutrients, hormones, and white blood cells (**Figure 27.8**). Also protecting the brain and spinal cord are layers of connective tissue called **meninges**. If the cerebrospinal fluid becomes infected by bacteria or viruses, the meninges may become inflamed, a condition called meningitis. Viral meningitis is generally not harmful, while bacterial meningitis can have serious consequences if not treated with antibiotics. Infection of the cerebrospinal fluid can be detected by a spinal tap, a procedure that involves inserting a narrow needle through the spinal column to collect a sample of cerebrospinal fluid. ☑

▼ Figure 27.8 **Fluid-filled spaces of the vertebrate CNS.** Spaces in the brain and spinal cord are filled with cerebrospinal fluid, which helps to cushion and supply nutrients to the CNS.

Brain

Cerebrospinal fluid

Meninges

Spinal cord
(cross section)

Spinal cord

☑CHECKPOINT

Name the two structures that make up the vertebrate central nervous system.

Answer: the brain and spinal cord

The Peripheral Nervous System

The vertebrate peripheral nervous system (PNS) is divided into two functional components: the somatic nervous system and the autonomic nervous system (**Figure 27.9**). Neurons of the **somatic nervous system** carry signals to and from skeletal muscles, mainly in response to *external* stimuli. When you walk, for instance, these neurons carry commands that make your legs move. The somatic nervous system is said to be voluntary because many of its actions are under conscious control. In contrast, the motor neurons of the **autonomic nervous system** regulate the *internal* environment by controlling smooth and cardiac muscles and the organs and glands of the digestive, cardiovascular, excretory, and endocrine systems. This control is generally involuntary.

The autonomic nervous system contains two sets of neurons with opposing effects on most body organs. One set, called the **parasympathetic division**, primes the body for activities that gain and conserve energy for the body ("rest and digest"). These effects include stimulating the digestive organs, decreasing the heart rate, and narrowing the bronchi, which correlates with a decreased breathing rate. The other set of neurons in the autonomic nervous system, the **sympathetic division**, tends to have the opposite effect, preparing the body for intense, energy-consuming activities, such as fighting, fleeing, or competing in a strenuous game ("fight or flight"). When this division is stimulated, the digestive organs are inhibited, the bronchi dilate so that more air can pass through, and the adrenal glands secrete the hormones epinephrine (also called adrenaline) and norepinephrine (also called noradrenaline).

Relaxation and the fight-or-flight response are opposite extremes. Your body usually operates somewhere in between, with most of your organs receiving both sympathetic and parasympathetic signals. The opposing signals adjust an organ's activity to a suitable level.

As carriers of command signals, the motor neurons of the parasympathetic and sympathetic systems constitute lower levels of the nervous system's hierarchy. In the next section, we'll take a closer look at the highest level of the hierarchy, the brain. ☑

▼ Figure 27.9 **Functional divisions of the vertebrate PNS.**

PERIPHERAL NERVOUS SYSTEM	
Somatic nervous system (voluntary)	**Autonomic nervous system** (involuntary)
Voluntary leg muscles	• **Parasympathetic division** (rest and digest) • **Sympathetic division** (fight or flight) Involuntary heart muscle

LM 50x

☑CHECKPOINT

How would a drug that inhibits the parasympathetic nervous system affect a person's pulse?

Answer: The pulse rate would increase.

The Human Brain

Composed of up to 100 billion intricately organized neurons, with a much larger number of supporting cells, the human brain is more powerful than the most sophisticated computer. The brain is divided structurally into three regions: the hindbrain, the midbrain, and the forebrain (**Figure 27.10** and **Table 27.1**).

Two sections of the hindbrain (blue in Figure 27.10), the **medulla oblongata** and the **pons**, and the midbrain (purple) make up a functional unit called the **brainstem**. All of the sensory and motor neurons carrying information to and from other brain regions pass through the brainstem. The brainstem serves as a sensory filter, selecting which information to pass on. It also regulates sleep and arousal and helps coordinate body movements, such as walking. Table 27.1 lists some of the other functions of the medulla oblongata, pons, and midbrain.

Also part of the hindbrain, the **cerebellum** (light blue in the figure) is a planning center for body movements. The cerebellum receives sensory information about the position of limbs and the length of muscles, as well as information from the auditory and visual systems. It also receives input from the motor pathways, telling it which motor actions are being carried out. The cerebellum uses this information to provide coordination of movement and balance. When you step off a curb, for instance, your cerebellum evaluates your body position and relays a plan for your next movements.

The most sophisticated integrating centers are found in the forebrain (orange and yellow in Figure 27.10); they are the thalamus, the hypothalamus, and the cerebrum. The **thalamus** contains most of the cell bodies of neurons that relay information to the cerebral cortex, the most extensive portion of the cerebrum. The thalamus first sorts data into categories (all the touch signals from a hand, for instance). It also suppresses some signals and enhances others. The thalamus then sends the sorted information on to the appropriate brain centers for further interpretation and integration.

In Chapter 25, we saw that the hypothalamus controls the pituitary gland and the secretion of many hormones (see Figures 25.5 and 25.6). The **hypothalamus** also regulates body temperature, blood pressure, hunger, thirst, the sex drive, and the fight-or-flight response, and it helps us experience emotions, such as rage and pleasure. A "pleasure center" in the hypothalamus could also be called an addiction center, for it is strongly affected by certain addictive drugs, such as amphetamines and cocaine. Another part of the hypothalamus

functions as an internal timekeeper, our biological clock. Receiving visual input from the eyes, the clock maintains our daily biological rhythms, such as cycles of sleepiness and hunger. ☑

☑**CHECKPOINT**

In which structure of the brain will you find your biological clock?

Answer: the hypothalamus

The Human Brain

▼ **Figure 27.10 The main parts of the human brain.**

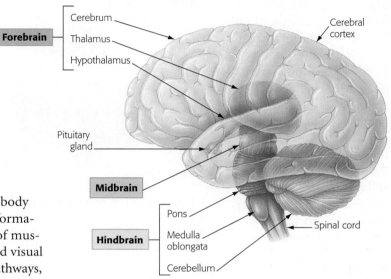

Cerebrum

Forebrain — Thalamus

Hypothalamus

Cerebral cortex

Pituitary gland

Midbrain

Hindbrain — Pons

Medulla oblongata

Cerebellum

Spinal cord

Table 27.1	Structure and Function of the Human Brain
Brain Structure	**Major Functions**
Brainstem	Conducts data to and from other brain centers; helps maintain homeostasis; coordinates body movement
Medulla oblongata	Controls breathing, circulation, swallowing, digestion
Pons	Controls breathing
Midbrain	Receives and integrates auditory data; coordinates visual reflexes; sends sensory data to higher brain centers
Cerebellum	Coordinates body movement; learns and remembers motor responses
Thalamus	Serves as input center for sensory data going to the cerebrum and as output center for motor responses leaving the cerebrum; sorts data
Hypothalamus	Serves as homeostatic control center; controls pituitary gland; acts as biological clock
Cerebrum	Performs sophisticated integration; involved in memory, learning, speech, emotions; formulates complex behavioral responses

The Cerebral Cortex

The **cerebrum**, the largest and most sophisticated part of our brain, consists of right and left cerebral hemispheres (**Figure 27.11**). A thick band of nerve fibers called the **corpus callosum** connects the cerebral hemispheres, enabling them to process information together.

The **cerebral cortex** is a highly folded layer of tissue that forms the surface of the cerebrum (see Figure 27.10). Although less than 5 mm (0.2 in.) thick, the human cerebral cortex accounts for over 80% of the total brain mass. It contains some 10 billion neurons and hundreds of billions of synapses. Its intricate neural circuitry helps produce our most distinctive human traits: reasoning and mathematical abilities, language skills, imagination, artistic talent, and personality traits. Assembling information it receives from our eyes, ears, nose, taste buds, and touch sensors, the cortex creates our sensory perceptions— what we are actually aware of when we see, hear, smell, taste, or touch. The cerebral cortex also regulates our voluntary movements.

Like the rest of the cerebrum, the cerebral cortex is divided into right and left sides. Because the nerve fibers from the cerebral cortex cross in the medulla oblongata, each hemisphere (right and left) receives information from and controls the movement of the opposite side of the body. As mentioned, the corpus callosum allows communication between the two hemispheres.

Each side of the cerebral cortex has four lobes: the frontal, parietal, temporal, and occipital lobes (represented by different colors in **Figure 27.12**). Researchers have identified a number of functional areas within each lobe. One region of the frontal lobe, for example, called the motor cortex, functions mainly in sending commands to skeletal muscles. Next to the motor cortex but in the parietal lobe, the somatosensory cortex receives and partially integrates signals from touch, pain, pressure, and temperature receptors throughout the body. The cerebral cortex also has centers that receive and begin processing sensory information: The parietal lobe deals with taste, the temporal lobe with hearing and smell, and the occipital lobe with vision. Each of these centers cooperates with an adjacent area, called an association area.

Making up most of our cerebral cortex are various **association areas**. These are the sites of higher mental activities—roughly, what we call thinking. A large association area in the frontal lobe uses varied inputs from many other areas of the brain to evaluate consequences, make considered judgments, and plan for the future. Language results from some extremely complex interactions among several association areas. For instance, the parietal lobe has association areas for reading and speech. These areas receive visual information (the appearance of words on a page) from the vision centers. Then, if the words are to be spoken aloud, the information is arranged into speech patterns and passed to a speech center in the frontal lobe, which signals the motor cortex to move the tongue, lips, and other muscles to form words.

You may have heard people say that they are "left-brained" or "right-brained."

▲ **Figure 27.11 A rear view of the brain.** The large cerebrum (yellow) consists of left and right cerebral hemispheres connected by a thick band of nerves called the corpus callosum.

▼ **Figure 27.12 Functional areas of the cerebrum's left hemisphere.** This figure identifies the main functional areas in the brain's left cerebral hemisphere, which is divided into four lobes.

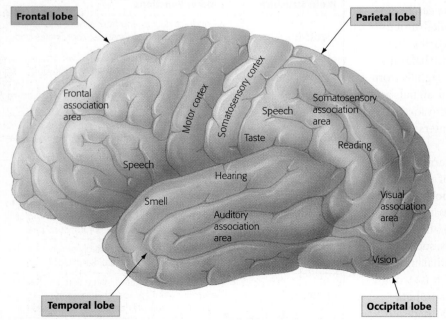

In a phenomenon known as **lateralization**, areas in the two hemispheres become specialized for different functions. In most people, the left hemisphere becomes most adept at language, logic, and mathematical operations. The right hemisphere is stronger at spatial relations, pattern and face recognition, and nonverbal thinking. In about 10% of us, however, the roles of the left and right hemispheres are reversed or the roles are more alike.

Brain Trauma

Interestingly, evidence from brain surgery patients indicates that patterns of lateralization are not fixed. One of the most radical surgical alterations of the brain is hemispherectomy—the removal of almost one-half of the brain (**Figure 27.13**). This procedure is performed to alleviate severe seizure disorders that originate from one of the hemispheres as a result of illness, abnormal development, or stroke. Incredibly, with just half a brain, hemispherectomy patients recover quickly, often leaving the hospital within a few weeks. Although the side of the body opposite the surgery always has permanent partial paralysis, hemispherectomy patients have undiminished intellectual capacities. Higher brain functions that previously originated from the missing half of the brain begin to be controlled by the opposite side.

Recovery after hemispherectomy is a striking example of the brain's remarkable flexibility.

Neurologists sometimes gain insight into how the brain works from people who have suffered brain trauma. The first well-publicized case of this type involved a man named Phineas Gage. In 1848, while working in Vermont as a railroad construction foreman, Gage accidentally exploded a dynamite charge that propelled a 3-foot-long spike through his head. The 13-pound steel rod entered his left cheek and traveled upward behind his left eye and out the top of his skull, landing several yards away. Incredibly, Gage walked away from the accident and appeared to be able to think normally. However, his associates soon noticed drastic changes in his personality, with a new propensity for meanness and vulgarity and an inability to control his behavior. Gage's doctor observed these changes as well but was unable to explain them, given the limited knowledge of the brain at that time. Luckily, the doctor preserved Gage's skull and the spike, allowing a group of researchers in 1994 to produce a computer model of the injury (**Figure 27.14**). The modern analysis offered an explanation for Gage's bizarre behavior: The rod had pierced both frontal lobes of his brain. People with these sorts of injuries often exhibit irrational decision making and difficulty processing emotions. The link between the frontal lobes and personality continues to be an active area of research. ✓

✓CHECKPOINT

The largest and most sophistical part of your brain is the _____. It is divided into two hemispheres connected by the _____. Its surface, called the _____, accounts for most of your brain's mass and is responsible for your higher reasoning abilities.

Answer: cerebrum; corpus callosum; cerebral cortex

▼ **Figure 27.13 Hemispherectomy.** This top-down X-ray view shows the skull and brain of a hemispherectomy patient after surgery.

◄ **Figure 27.14 Phineas Gage's accident.** A 1994 computer simulation of an 1848 injury to railroad worker Phineas Gage suggested that the spike that passed through his skull (shown as a red cylinder in the figure) might have damaged both frontal lobes of his brain. Such an injury could account for Gage's personality changes after the accident.

Neurological Disorders

In addition to trauma, various neurological disorders—such as depression and Alzheimer's disease—can also affect brain function. Nearly 20 million American adults are affected by depression, about two-thirds of them women. Two broad forms of depressive illness have been identified: major depression and bipolar disorder. People with **major depression** may experience persistent sadness, loss of interest in pleasurable activities, changes in body weight and sleep patterns, loss of energy, and suicidal thoughts. While all of us feel sad from time to time, major depression is extreme and more persistent, leaving the sufferer unable to live a normal life. If left untreated, symptoms may become more frequent and severe over time.

Bipolar disorder, or manic-depressive disorder, involves extreme mood swings. The manic phase is characterized by high self-esteem, increased energy, a flood of thoughts and ideas, and extreme talkativeness, as well as behaviors that often court disaster, such as increased risk taking, promiscuity, and reckless spending. In its milder forms, this phase is sometimes associated with great creativity, and some well-known artists, musicians, and literary figures with bipolar disorder (including Beethoven, Tolstoy, and Hemingway) have had periods of intense output during their manic phases. The depressive phase is marked by sleep disturbances, feelings of worthlessness, and decreased ability to experience interest and pleasure.

In recent years, researchers have begun to examine how the brain's physiology is involved in depression (**Figure 27.15**). Many depressed people have a low level of serotonin. For this reason, serotonin is the target of the most commonly prescribed class of antidepressant medications, called selective serotonin reuptake inhibitors (SSRIs), including fluoxetine (Prozac), paroxetine (Paxil), and sertraline (Zoloft). These drugs block the reuptake of neurotransmitter at synapses (see step 6 in Figure 27.5), increasing the amount of time that serotonin is available to stimulate neurons in the brain.

Alzheimer's disease is a form of mental deterioration, or dementia, characterized by confusion, memory loss, and a variety of other symptoms. Its incidence is usually age related, rising from about 10% at age 65 to about 35% at age 85. The disease is progressive; patients gradually become less able to function and eventually need to be dressed, bathed, and fed by others. There are also personality changes, almost always for the worse.

Patients often lose their ability to recognize people, even family members, and may treat them with suspicion and hostility. At present, a firm diagnosis of Alzheimer's disease is difficult to make while the patient is alive because it is one of several forms of dementia.

Unraveling how networks of neurons in the brain produce thoughts and emotions, store and retrieve memories, and control the body's internal environment is one of the most challenging and engaging aspects of modern biology. In the next section, we'll examine how the brain receives information from the environment via the sense organs. ☑

Area of decreased brain activity

Depressed person

Healthy person

▲ Figure 27.15 **Brain activity in a depressed person and healthy person.** The purple and pink colors in these computer images indicate areas of low brain activity. Notice that the brain of the depressed person shows decreased activity in certain areas of the brain.

The Senses

In this section, we'll focus on the human body's sensory structures. To begin, we'll examine how these structures gather information and pass it on to the central nervous system (CNS). After that, we'll take a closer look at two of the human senses, vision and hearing.

Sensory Input

Sensory input is the process of using receptors to sense the environment and send information about it to the CNS to be integrated and acted upon. Sensory receptors, such as the sensory cells in your eyes or the taste buds of your tongue, are tuned to the condition of both the external world and internal organs. Sensory receptors detect stimuli such as chemicals, light, sounds, cold, heat, and touch. The sensory receptors in your eyes, for instance, detect light energy; those in your taste buds detect chemicals dissolved in saliva. The sensory receptor's job is completed when it sends information to the CNS by triggering action potentials.

What exactly do we mean when we say that a sensory receptor detects a stimulus? In stimulus detection, the receptor cell converts one type of signal (the stimulus) to an electrical signal. This conversion, called **sensory transduction**, occurs as a change in the membrane potential of the receptor cell.

Sensory Transduction

As an example of sensory transduction, **Figure 27.16** shows sensory receptors in a taste bud detecting sugar molecules. ❶ When the molecules (🔵) first come into contact with the taste bud, they bind to membrane receptors of the sensory receptor cells. ❷ This binding triggers a signal transduction pathway that ❸ causes some ion channels in the membrane to close and others to open. Changes in the flow of ions (⊕) alter the membrane potential. This change in membrane potential is called the **receptor potential**. In contrast to action potentials, which are all-or-none phenomena, receptor potentials vary in intensity; the stronger the stimulus, the stronger the receptor potential.

Once a receptor cell converts a stimulus to a receptor potential, this potential usually results in signals being sent to the CNS. In Figure 27.16, ❹ each receptor cell forms a synapse with a sensory neuron. When there are enough sugar molecules, a strong receptor potential is triggered. This receptor potential makes the receptor

cell release enough neurotransmitter (🔵) to increase the rate of action potential generation in the sensory neuron. The brain interprets the intensity of the stimulus from the rate at which it receives action potentials. It gains additional information about stimulus intensity by keeping track of how many sensory neurons it receives signals from.

There is an important qualification to what we have just said about stimulus intensity. Have you ever noticed how an odor that is strong at first seems to fade with time? The same effect helps you adjust to a hot or cold shower and enables you to wear clothes without being constantly aware of their touch on your skin. This effect is called **sensory adaptation**, the tendency of some sensory receptors to become less sensitive when they are stimulated

Sensory Receptors

▼ Figure 27.16 **Sensory transduction in a human taste bud.**

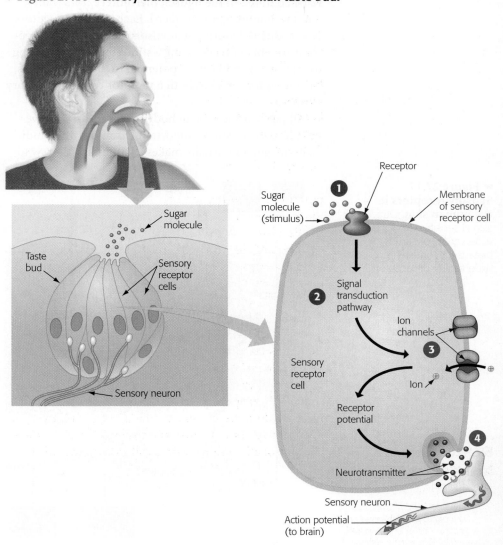

repeatedly. When receptors become less sensitive, they trigger fewer action potentials, causing the brain to receive fewer stimuli. Sensory adaptation keeps the body from continuously reacting to normal background stimuli. Without it, our nervous system would become overloaded with useless information.

Types of Sensory Receptors

Based on the type of signals to which they respond, we can group sensory receptors into five general categories: pain receptors, thermoreceptors (sensors for heat and cold), mechanoreceptors (sensors for touch, pressure, motion, and sound), chemoreceptors (sensors for chemicals), and electromagnetic receptors (sensors for energy such as light and electricity). These five types of receptors work in various combinations to produce the five human senses.

Figure 27.17, showing a section of human skin, reveals why the surface of our body is sensitive to such a variety of stimuli. Our skin contains pain receptors (labeled in red in the figure), thermoreceptors (blue), and mechanoreceptors (green). Each of these receptors is a modified dendrite of a sensory neuron. Each receptor is capable of recognizing a stimulus and responding to it by sending an action potential to the CNS. That is, each receptor serves as both a receptor cell and a sensory neuron.

All parts of the human body except the brain have pain receptors. Pain is important because it often indicates danger and usually makes an animal withdraw to

safety. Pain can also make us aware of injury or disease. **Pain receptors** may respond to excessive heat or pressure or to chemicals released from damaged or inflamed tissues. Prostaglandins are local regulators that increase pain by sensitizing pain receptors. Aspirin and ibuprofen reduce pain by inhibiting prostaglandin synthesis.

Thermoreceptors in the skin detect either heat or cold. Other temperature sensors located deep in the body monitor the temperature of the blood. The body's major thermostat is the hypothalamus. Receiving action potentials both from surface sensors and from deep sensors, the hypothalamus keeps a mammal's or bird's body temperature within a narrow range (see Figure 21.14).

Mechanoreceptors are highly diverse. Different types are stimulated by different forms of mechanical energy, such as touch and pressure, stretching, motion, and sound. All these forces produce their effects by bending or stretching the plasma membrane of a receptor cell. When the membrane changes shape, it becomes more permeable to positive ions, and the mechanical energy of the stimulus is transduced into a receptor potential. One example of a mechanoreceptor is the touch receptors at the base of the stout whiskers on a cat. These receptors are extremely sensitive and enable the animal to detect objects by touch in the dark.

Chemoreceptors include the sensory cells in our nose and taste buds, which are attuned to chemicals in the external environment, as well as some internal receptors that detect chemicals in the body's internal environment. Internal chemoreceptors include sensors in our arteries that monitor our blood, with some sensors detecting changes in pH and others detecting changes in O_2 concentration. In all types of chemoreceptors, a receptor cell develops receptor potentials in response to chemicals dissolved in fluid such as blood or saliva.

Electromagnetic receptors are sensitive to energy of various wavelengths, which takes such forms as magnetism and light. For example, **photoreceptors** detect the electromagnetic energy of light, and the nonhuman senses of electroreception and magnetoreception (discussed in the Biology and Society section) rely on other types of electromagnetic receptors. In the next section, we'll focus on one specific organ that uses photoreceptors: the human eye. ☑

Vision

The human eye is a remarkable sense organ, able to detect a multitude of colors, form images of faraway objects, and respond to minute amounts of light energy. In this section, you'll learn about the structure of the human eye and how it processes images. You'll also learn why vision problems occur and how they can be corrected.

► Figure 27.17
Sensory receptors in the human skin. Our skin is sensitive to a wide variety of stimuli because it contains a wide variety of receptors.

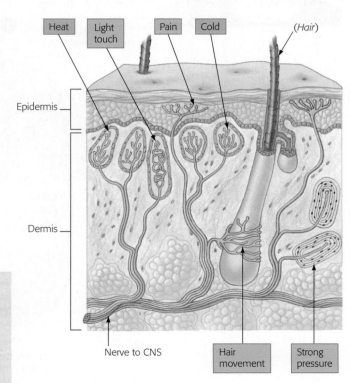

Heat Light touch Pain Cold (Hair)

Epidermis

Dermis

Nerve to CNS

Hair movement

Strong pressure

Structure of the Human Eye

The outer surface of the human eyeball is a tough, whitish layer of connective tissue called the **sclera** (**Figure 27.18**). At the front of the eye, the sclera becomes the transparent **cornea**. The sclera surrounds a pigmented layer called the choroid. At the front of the eye, the choroid forms the **iris**, which gives the eye its color. The opening in the center of the iris is called the **pupil**. Behind the pupil is the disklike **lens**, which is held in position by ligaments. At the back of the eyeball is the **retina**, a layer just inside the choroid that contains photoreceptor cells. The **optic nerve** connects the retina with the brain. There are no photoreceptor cells in the part of the retina where the optic nerve passes through the back of the eye. We cannot detect light that is focused on this "blind spot," but having two eyes with overlapping fields of view enables us to perceive uninterrupted images.

Two fluid-filled chambers make up the bulk of the eye. The large chamber behind the lens is filled with jellylike **vitreous humor**. The much smaller chamber in front of the lens contains a thinner fluid called the **aqueous humor**. The humors help maintain the shape of the eyeball. In addition, the aqueous humor supplies nutrients and oxygen to the lens, iris, and cornea and carries off wastes. Blockage of the ducts that drain this fluid can cause glaucoma, increased pressure inside the eye that may lead to blindness. If diagnosed early, glaucoma can be treated with medications that increase the circulation of the aqueous humor.

A thin mucous membrane called the conjunctiva helps keep the outside of the eye moist. The conjunctiva lines the inner surface of the eyelids and folds back over the white of the eye (but not the cornea). An infection or allergic reaction may cause inflammation of the conjuctiva, a condition called conjunctivitis or "pink eye." Bacterial conjuctivitis usually clears up with antibiotic eyedrops. Viral conjunctivitis usually clears up on its own, although it is very contagious, especially among young children.

A gland above the eye secretes tears, a dilute salt solution that is spread across the eyeball by blinking and that drains into ducts leading into the nasal cavities. This fluid cleanses and moistens the eye surface. Excess secretion, in response to eye irritation or strong emotions, causes tears to spill over the eyelid and fill the nasal cavities, producing a runny nose.

Function of the Human Eye

The cornea lets light into the eye and also helps focus light. The muscles of the iris regulate the size of the pupil, controlling the amount of light that enters. After going through the pupil, light passes through the lens. The lens focuses light onto the retina by bending light rays. Focusing is accomplished through a change in the shape of the lens (**Figure 27.19**). The thicker the lens, the more sharply it bends light. The shape of the lens is controlled by the muscles attached to the choroid. When the eye focuses on a nearby object, these muscles contract, pulling the choroid layer of the eye inward toward the pupil. This makes the ligaments that suspend the lens slacken. With this reduced tension, the elastic lens becomes thicker and rounder. When the eye focuses on a distant object, the muscles controlling the lens relax, putting tension on the ligaments and flattening the lens. ✓

✓CHECKPOINT

Arrange the following eye parts into the correct sequence encountered by photons of light traveling into the eye: pupil, retina, cornea, lens, vitreous humor, aqueous humor.

Answer: cornea → aqueous humor → pupil → lens → vitreous humor → retina

▼ **Figure 27.18 The human eye.**

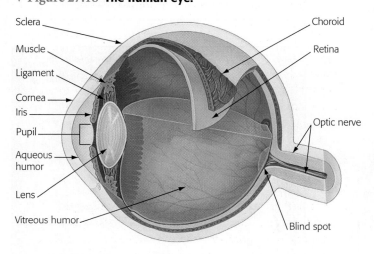

Sclera, Muscle, Ligament, Cornea, Iris, Pupil, Aqueous humor, Lens, Vitreous humor, Choroid, Retina, Optic nerve, Blind spot

► **Figure 27.19 How the lens of the eye focuses light.** When viewing a nearby object (top), the lens becomes thicker and rounder and focuses the image on the retina. When viewing a distant object (bottom), the lens becomes flattened, and the image is again focused on the retina.

Near vision

Muscle contracted, Ligaments slacken, Light from a near object, Choroid, Retina, Lens

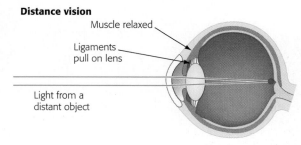

Distance vision

Muscle relaxed, Ligaments pull on lens, Light from a distant object

Photoreceptors

The human retina contains two types of photoreceptors named for their shapes (**Figure 27.20**). **Rods** are extremely sensitive to light and enable us to see in dim light, though only in shades of gray. **Cones** are stimulated by bright light and can distinguish color, but they contribute little to night vision.

In humans, rods are found in greatest density at the outer edges of the retina and are completely absent from the center. In contrast, the retina's center of focus contains a high concentration of cones (**Figure 27.21**). If you face directly toward a dim star in the night sky, the star is hard to see. Looking just to the side, however, makes your lens focus the starlight onto the parts of the retina with the most rods, and you can see the star. By contrast, you achieve your sharpest day vision by looking straight at the object of interest. Some birds, such as hawks, have ten times more cones than we do, enabling them to spot small prey from high altitudes.

How do rods and cones detect light? As Figure 27.20 shows, each rod and cone includes an array of membranous disks containing light-absorbing visual pigments. Rods contain a visual pigment called rhodopsin, which can absorb dim light. (Rhodopsin is derived from vitamin A, which is why vitamin A deficiency can cause poor night vision; this is also the basis for the idea that eating carrots, which are rich in a chemical called beta carotene that the body converts to vitamin A, improves

your vision.) Cones contain visual pigments called photopsins, which absorb colored, bright light. We have three types of cones, each containing a different type of photopsin. These cells are called blue cones, green cones, and red cones, referring to the colors absorbed best by their photopsin. We can perceive a great number of colors because the light from each particular color triggers a unique pattern of stimulation among the three types of cones. Color blindness results from a deficiency in one or more types of cones.

Like all receptor cells, rods and cones are stimulus transducers (see Figure 27.21). When rhodopsin and photopsin absorb light, they change chemically, and the change alters the permeability of the cell's membrane. The resulting receptor potentials trigger a complex integration process that begins in the retina. Other retinal neurons start to integrate the receptor potentials and in doing so produce action potentials. These action potentials travel along the optic nerve, carrying the partly integrated information into the brain. Three-dimensional perceptions (what we actually see) result from further integration in several processing centers of the cerebral cortex. ☑

▲ **Figure 27.20 Photoreceptor cells.** Your eyes contain many more rods than cones.

Cell body

Membranous disks containing visual pigments

Synaptic terminals

Rod

Cone

☑CHECKPOINT

Explain why our night vision is mostly in shades of gray rather than in color.

Answer: Rods are more sensitive than cones to dim light, and thus the low light intensity at night stimulates far more rods than color-sensing cones.

► **Figure 27.21 The vision pathway from light source to optic nerve.** After passing through many layers of cells, light entering the human eye strikes photoreceptor cells embedded in the back of the retina. Once stimulated, the rods and cones convert the light energy to receptor potentials that are integrated by other neurons and communicated through the optic nerve as action potentials to the brain.

Retina

Neurons

Photoreceptors

Cone Rod

Optic
nerve
fibers

Retina

Optic
nerve

To brain

Vision Problems and Corrections

When you have your vision tested, you view a chart that shows how various letters of the alphabet would look at various distances. You are asked to read different rows of letters using one eye at a time. If you can read the row that is sized for legibility at a distance of 20 feet with each of your eyes, you have normal (20/20) vision in each eye. Suppose you find out that you have 20/10 vision. This is better than normal; you can read letters from a distance of 20 feet that a person with 20/20 vision can only read at 10 feet. However, someone with 20/50 vision has worse than normal vision. He or she must stand at a distance of 20 feet to read what a person with normal vision can read at 50 feet.

Three of the most common visual problems are nearsightedness, farsightedness, and astigmatism. All three are focusing problems, easily corrected with artificial lenses. People with **nearsightedness** cannot focus well on distant objects, although they can see well at short distances (the condition is named for the type of vision that is *unimpaired*). A nearsighted eyeball (**Figure 27.22a**) is longer than normal. The lens cannot flatten enough to compensate, and it focuses distant objects in front of the retina instead of on it. Nearsightedness (also known as

myopia) is corrected by glasses or contact lenses that are thinner in the middle than at the outside edge. The corrective lenses make the light rays from distant objects diverge slightly as they enter the eye. The image formed by the lens in the eye is then focused at the correct point.

Farsightedness (also known as hyperopia) occurs when the eyeball is shorter than normal, causing the lens to focus images behind the retina (**Figure 27.22b**). Farsighted people see distant objects normally, but they can't focus on close objects. Corrective lenses that are thicker in the middle than at the outside edge compensate for farsightedness by making light rays from nearby objects converge slightly before they enter the eye.

Astigmatism is blurred vision caused by a misshapen lens or cornea. Any such distortion makes light rays converge unevenly and not focus at any one point on the retina. Lenses that correct astigmatism are asymmetrical in a way that compensates for the asymmetry in the eye.

Surgical procedures are an option for treating vision disorders. In laser-assisted in situ keratomileusis (LASIK), a laser is used to reshape the cornea and change its focusing ability. More than 1 million LASIK procedures are performed each year to correct a variety of vision problems. ☑

☑**CHECKPOINT**

A person who requires reading glasses has a case of _____.

Answer: farsightedness (hyperopia)

► Figure 27.22 **A nearsighted eye and a farsighted eye.** Corrective lenses help vision problems by focusing the image exactly on the retina.

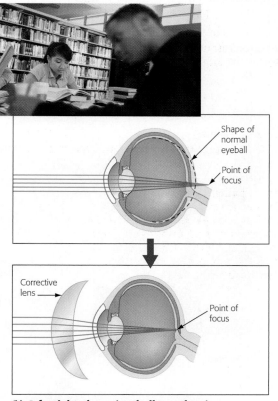

(a) A nearsighted eye (eyeball too long).
Nearsightedness is corrected by lenses that are thinner toward the middle.

(b) A farsighted eye (eyeball too short).
Farsightedness is corrected by lenses that are thicker in the middle.

Hearing

Although the human ear is less sensitive than that of many other animals, it is capable of hearing a great range of auditory signals. In this section, we'll examine the structure and function of the human ear.

Structure of the Human Ear

The ear is composed of three regions: the outer ear, the middle ear, and the inner ear (**Figure 27.23a**). The **outer ear** consists of the flap-like **pinna**—the bendable structure we commonly refer to as our "ear"—and the **auditory canal**. A sheet of tissue called the **eardrum** separates the outer ear from the middle ear. Both the outer ear and middle ear are common sites of childhood infections (called swimmer's ear and otitis media, respectively).

The **middle ear** contains three small bones (**Figure 27.23b**): the hammer (more formally, the malleus), the anvil (incus), and the stirrup (stapes). The stirrup is connected to the inner ear through an opening in the skull bone. The **Eustachian tube** conducts air between the middle ear and the back of the throat, allowing air pressure to stay equal on either side of the eardrum. This tube is what enables you to move air in or out to equalize pressure ("pop" your ears) when changing altitude rapidly in an airplane or diving underwater.

The **inner ear** consists of fluid-filled channels in the bones of the skull. One of the channels, the **cochlea**, is a long, coiled tube. Our actual hearing organ, the organ of Corti, is located within a fluid-filled canal inside the cochlea (**Figure 27.24**). The organ of Corti consists of an array of hair cells embedded in a structure called the basilar membrane. The hair cells are the receptor cells of the ear. Notice that they project into the fluid and that the tips of most are in contact with the overlying jelly-like membrane. Sensory neurons at the base of the hair cells connect to the auditory nerve, which connects in turn to the brain.

▼ **Figure 27.23 An overview of the human ear.**

Pinna

Outer ear | Middle ear | Inner ear

Auditory canal

Eardrum

Eustachian tube

(a) Ear structure. The human ear is divided into three regions: the outer, middle, and inner ears.

Stirrup
Anvil
Hammer
Skull bones
Auditory nerve, to brain
Eardrum
Eustachian tube
Cochlea

(b) The middle and inner ears. The bones of the middle ear are connected to the fluid-filled channels of the cochlea in the inner ear.

▼ **Figure 27.24 The organ of Corti, within the cochlea.** The organ of Corti, the actual hearing organ of the ear, is located within a fluid-filled canal inside the spiral cochlea. Receptor cells (hair cells) connect to sensory neurons, which in turn connect to the auditory nerve and the brain.

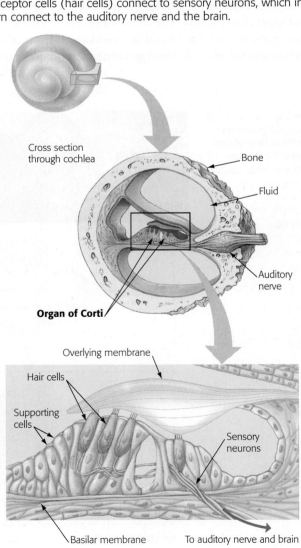

Cross section through cochlea

Bone

Fluid

Auditory nerve

Organ of Corti

Overlying membrane

Hair cells

Supporting cells

Sensory neurons

Basilar membrane

To auditory nerve and brain

Function of the Human Ear

Now let's see how the parts of the ear function in hearing. As indicated in **Figure 27.25**, a vibrating object creates sound waves in the surrounding air, represented by the up-and-down waves in the figure. The pinna and the auditory canal collect these waves and channel them to the eardrum. The waves make your eardrum vibrate with the same frequency as the sound. (Frequency is the number of vibrations per second.) From the eardrum, the vibrations are concentrated as they pass through the hammer, anvil, and stirrup in the middle ear. The stirrup transmits the vibrations to the inner ear, producing pressure waves in the fluid within the cochlea.

As a pressure wave passes through the cochlea, it makes the basilar membrane vibrate. Vibration of the basilar membrane makes the hairlike projections on the hair cells alternately brush against and draw away from the overlying membrane. When a hair cell's projections are bent, ion channels in its plasma membrane open, and positive ions enter the cell. As a result, the hair cell develops a receptor potential and releases neurotransmitter molecules at its synapse with a sensory neuron. In turn, the sensory neuron sends action potentials through the auditory nerve to the brain.

How is the volume of a sound determined? The higher the volume of sound, the greater the amplitude (height) of the pressure wave it generates. In the ear, the greater the amplitude, the more vigorous the vibrations of fluid in the cochlea, the more pronounced the bending of the hair cells, and the more action potentials generated in the sensory neurons.

▼ Figure 27.25 **The route of sound waves through the ear.** This figure traces the path of a sound stimulus as it passes from the environment to the brain.

Hearing Problems

Deafness, the loss of hearing, can be caused by the inability to conduct sounds, resulting from middle-ear infections, a ruptured eardrum, or stiffening of the middle-ear bones (a common age-related problem). Deafness can also result from damage to receptor cells or neurons. Because few parts of our anatomy are more delicate than the organ of Corti, deafness is often progressive and permanent. In recent years, however, about 100,000 deaf people have received cochlear implants, electronic devices that convert sounds to electrical impulses that stimulate the auditory nerve directly.

Frequent or prolonged exposure to very loud sounds can damage or destroy the hair cells of the inner ear. For example, you may have noticed that a few hours in a very loud environment leaves you with ringing or buzzing sounds in your ears, a condition called tinnitus. Ear plugs can provide protection. ☑

☑ CHECKPOINT

How does the ear convert sound waves in the air to pressure waves in the fluid of the inner ear?

Answer: Sound waves in the air cause the eardrum to vibrate. The small bones attached to the inside of the eardrum transmit vibrations to the fluid of the inner ear, producing pressure waves.

Motor Systems

So far, we've seen how the body senses the environment via sensory receptors and integrates this information within the central nervous system. In this section, we'll consider how the body couples stimulus to motor response, thereby completing the neural circuit.

Movement is one of the most distinctive features of animals. Whether an animal walks or runs on two, four, or six legs or more, swims, crawls, flies, or sits in one place and only moves its mouthparts, an interplay of organ systems is responsible for its movement. The nervous system plays a key role in issuing commands to the muscular system. The muscular system exerts the force that actually makes an animal or its parts move. But the force exerted by muscles produces movement only when it is applied against a firm structure: in vertebrates, the skeletal system. In this section, we focus on the human skeleton and muscles and the movement their interactions produce.

The Skeletal System

The **skeletal system** provides anchoring, support, and protection. Most land animals would sag from their own weight if they had no skeleton to support them. Skeletons also may protect an animal's soft organs. For example, your ribs form a cage around your heart and lungs. In this section, we'll discuss the human skeleton: its organization, its structure, and some of the problems that may arise with it.

Organization of the Human Skeleton

Humans, like all vertebrates, have an **endoskeleton**—hard supporting elements situated among soft tissues. The human endoskeleton is a combination of cartilage and the 206 bones that make up the skeletal system (**Figure 27.26**); the cartilage provides flexibility in certain areas.

The human skeleton is organized into two basic units: the axial skeleton and the appendicular skeleton. The **axial skeleton** supports the axis, or trunk, of the body and includes the skull, enclosing and protecting the brain; the vertebrae of the spinal column, enclosing the spinal cord; and the ribs. The **appendicular skeleton** is made

up of the bones of the limbs, shoulders, and pelvis.

The bones of the skeleton are held together at movable joints by strong fibrous tissues called **ligaments**.

Much of the versatility of our skeleton comes from three types of movable joints (**Figure 27.27**). Humans have **ball-and-socket joints** in the shoulder and the hip. These joints enable us to rotate our arms and legs and move them in several planes. Shown in the elbow, a **hinge joint** permits movement in a single plane. A **pivot joint** enables us to rotate the forearm at the elbow. ☑

Skull
Shoulder girdle — Clavicle
Scapula
Sternum
Ribs
Humerus
Vertebra
Ulna
Radius
Pelvic girdle
Carpals
Metacarpals
Phalanges (curled under)
Femur
Patella
Tibia
Fibula
Tarsals
Metatarsals
Phalanges

◀ **Figure 27.26 The human endoskeleton.** The human skeleton is divided into the axial skeleton (green), which supports the trunk, and the appendicular skeleton (gold), which supports the limbs. Cartilage (blue) provides flexibility at various points.

▼ **Figure 27.27 Three kinds of joints.**

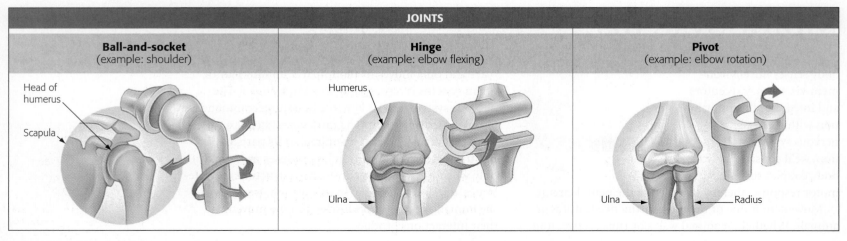

JOINTS		
Ball-and-socket (example: shoulder)	**Hinge** (example: elbow flexing)	**Pivot** (example: elbow rotation)

Head of humerus
Scapula

Humerus
Ulna

Ulna — Radius

The Structure of Bones

Bones are complex organs consisting of several kinds of tissues. You can get a sense of some of a bone's complexity from the cutaway view of the human humerus (the upper arm bone) shown in **Figure 27.28**. A sheet of fibrous connective tissue, shown in pink (most visible in the enlargement on the lower right), covers most of the outside surface. The cells of this tissue can form new bone during normal growth or in the event of a fracture. At either end of the bone is a thin sheet of cartilage (blue), also living tissue, that cushions joints, protecting the ends of bones as they move against each other. The bone itself contains cells that secrete a surrounding material, or matrix (see Figure 21.4f). Like all living tissues, bone tissues require servicing. Blood vessels course through channels in the bone, transporting nutrients and regulatory hormones to its cells and waste materials from them.

Notice that the shaft of the long bone shown in Figure 27.28 surrounds a central cavity. The central cavity contains yellow bone marrow, which is mostly stored fat brought into the bone by the blood. The ends of the bone have cavities that contain red bone marrow (not shown in the figure), a specialized tissue that produces blood cells (see Chapter 23).

▼ Figure 27.28 **The structure of an arm bone.**

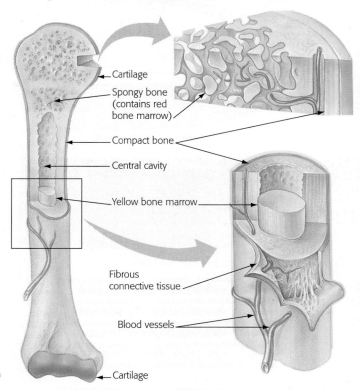

- Cartilage
- Spongy bone (contains red bone marrow)
- Compact bone
- Central cavity
- Yellow bone marrow
- Fibrous connective tissue
- Blood vessels
- Cartilage

Skeletal Diseases and Injuries

Your skeleton is quite strong and provides you with reliable support, but it is susceptible to disease and injury. **Arthritis**—inflammation of the joints—affects one out of every seven people in the United States. The most common form of arthritis seems to occur as a result of aging. The joints become stiff and sore and often swell as the cartilage between the bones wears down. Sometimes the bones thicken at the joints, producing crunching noises when they rub together and restricting movement. This form of arthritis is irreversible but not crippling in most cases, and moderate exercise, rest, and over-the-counter pain medications usually relieve most symptoms.

Rheumatoid arthritis is a debilitating autoimmune disease. The joints become highly inflamed, and their tissues may be destroyed by the body's immune system (see Figure 24.16). Rheumatoid arthritis usually begins between ages 40 and 50 and affects more women than men. Anti-inflammatory drugs relieve symptoms, but there is no cure.

Osteoporosis is another serious bone disorder. It is most common in women after menopause: Estrogen contributes to normal bone maintenance, and with lowered production of the hormone, bones may become thinner, more porous, and more easily broken. Insufficient exercise, smoking, diabetes mellitus, and an inadequate intake of protein and calcium may also contribute to the disease. Treatments include calcium and vitamin supplements, hormone replacement therapy, and drugs that slow bone loss or increase bone formation. Prevention of osteoporosis begins with sufficient calcium intake while bones are still increasing in density (up until about age 35). Weight-bearing exercise (walking, jogging, lifting weights) builds bone mass and is beneficial throughout life.

Bones are rigid but not inflexible; they will bend in response to external forces. However, as many of us know from personal experience, the skeletal system has its limits. If a force is applied that exceeds a bone's capacity to bend, the result is a broken bone, or fracture. The average American will break two bones during his or her lifetime, most commonly the forearm or, for people over 75, the hip.

Treatment of a fracture involves two steps: putting the bone back into its natural shape and then immobilizing it until the body's natural bone-building cells can repair the fracture. A splint or a cast is usually sufficient to protect the area, prevent movement, and promote healing. In more severe cases, a fracture can only be repaired surgically by inserting plates, rods, and/or screws that hold the broken pieces together (**Figure 27.29**). ☑

▼ Figure 27.29 **Broken bones.** These X-rays show the same set of broken leg bones (tibia and fibula) fractured (left) and after they have been held together with screws and a plate (right).

☑CHECKPOINT

What two things can a young person do to help prevent osteoporosis?

Answer: Ingest plenty of calcium and engage in weight-bearing exercise regularly.

BioFlix™

Muscle Contraction

Muscle

The Muscular System

Now that you've learned about the skeletal system, we can focus on the **muscular system**, which is made up of all the skeletal muscles in the body. **Skeletal muscles** are attached to the skeleton and produce body movement by interacting with the skeleton. **Figure 27.30** illustrates how muscles interact with bones to raise and lower the human forearm. As shown in the figure, strong fibrous tissue called **tendons** connects muscles to bones. For instance, one end of the triceps muscle is attached by tendons to bones of the shoulder. The other end is attached by a tendon to one of the bones in the forearm.

The ability to move the forearm in opposite directions requires that two muscles work as an antagonistic pair—that is, the two muscles must perform opposite tasks. For example, contraction of the biceps muscle shortens the muscle and pulls up the forearm. The triceps muscle is the biceps's antagonist; contraction of the triceps pulls the forearm down. Notice in Figure 27.30 that when one muscle is contracted, the antagonistic (opposite) muscle is relaxed.

The Cellular Basis of Muscle Contraction

Skeletal muscle is made up of a hierarchy of smaller and smaller parallel strands. As shown at the top of **Figure 27.31**, a muscle consists of bundles of parallel muscle fibers. Each muscle fiber is a single cell with many nuclei. In the middle of the figure, notice that each muscle fiber is itself a bundle of smaller **myofibrils**. Skeletal muscle is also called striated (striped) muscle because the myofibrils exhibit alternating light and dark bands when viewed with a microscope. A myofibril consists of repeating units called **sarcomeres**. Structurally, a sarcomere is the region between two dark, narrow lines in the myofibril. Functionally, the sarcomere is the contractile apparatus in a myofibril—the muscle fiber's fundamental unit of action.

The micrograph and the diagram below it at the bottom of Figure 27.31 reveal the structure of a sarcomere in more detail. A myofibril is composed of regular arrangements of two kinds of filaments: **thin filaments** made primarily from the protein actin (blue in the diagram) and **thick filaments** made from the protein myosin (red).

A sarcomere contracts (shortens) when its thin filaments slide past its thick filaments. The model describing this process is known as the sliding-filament model of muscle contraction. **Figure 27.32**, a simplified

▼ **Figure 27.30**
Antagonistic action of muscles in the human arm. Many movements are produced through the action of two opposing muscles. In this example, contraction of the biceps muscle raises the forearm, while contraction of the triceps muscle lowers the forearm.

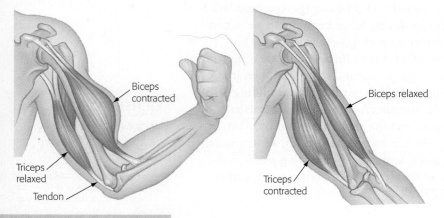

Biceps contracted

Triceps relaxed

Tendon

Biceps relaxed

Triceps contracted

▼ **Figure 27.31** **The contractile apparatus of skeletal muscle.**

Muscle

Nuclei

Bundle of muscle fibers

Single muscle fiber (cell)

Myofibril

Light band

Dark band

Light band

Sarcomere

TEM 26,000×

Light band

Dark band

Light band

Thick filaments (myosin)

Thin filaments (actin)

Sarcomere

▼ **Figure 27.32 The sliding-filament model of muscle contraction.** Notice that the lengths of the thick and thin filaments do not change; the two types of filaments merely slide past each other.

Sarcomere

Dark band

Relaxed muscle

Contracting muscle

Fully contracted muscle

diagram of this model, shows a sarcomere in a relaxed muscle, a contracting muscle, and a fully contracted muscle. Notice in the contracting sarcomere that the thin filaments (blue) have moved toward the middle of the sarcomere. When the muscle is fully contracted, the thin filaments overlap in the middle of the sarcomere. Contraction only shortens the sarcomere; it does not change the lengths of the thick and thin filaments. A muscle can shorten to about one-third of its resting length when all of its sarcomeres are contracted.

What makes the thin filaments slide? The key event is the binding between parts (called heads) of the myosin molecules in the thick filaments and specific sites on actin molecules in the thin filaments. The process of contraction requires energy, provided by ATP.

Figure 27.33 indicates how sliding is thought to work. **1** ATP binds to a myosin head (🖌), causing the head to detach from a binding site on actin (⬤). **2** The myosin head gains energy from the breakdown of ATP and changes shape. In this high-energy position, the myosin head is cocked like a pistol ready to fire (actually, ready to bind with another site on the actin molecule). **3** The energized myosin head binds to an exposed binding site on actin. **4** The molecular event that actually causes sliding is called the power stroke. The myosin head bends back to its low-energy position, pulling the thin filament toward the center of the sarcomere. **5** After the power stroke, the whole process repeats: More ATP binds, the myosin head attaches to the next binding site along the thin filament and slides it forward, and so on. As long as sufficient ATP is present, the process continues until the muscle is fully contracted or until the signal to contract stops. ☑

▼ **Figure 27.33 The mechanism of filament sliding.** Though we show only one myosin head in this figure, a typical thick filament has about 350 heads, each of which can bind and unbind to a thin filament about five times per second.

Thick filament (myosin)

Thin filament (actin)

ATP

Myosin head (low-energy configuration)

1 ATP binds to a myosin head, which is then released from an actin filament.

ATP → ADP + P

Myosin head (high-energy configuration)

2 The breakdown of ATP cocks the myosin head.

3 The myosin head attaches to an actin binding site.

4 The power stroke slides the actin (thin) filament toward the center of the sarcomere.

5 As long as ATP is available, the process can be repeated until the muscle is fully contracted.

☑**CHECKPOINT**

What protein is most abundant in a thin filament of a myofibril? In a thick filament?

Answer: actin; myosin

Motor Neurons: Control of Muscle Contraction

The sarcomeres of a muscle fiber do not contract on their own. They must be stimulated to contract by motor neurons. A typical motor neuron can stimulate more than one muscle fiber because each neuron has many branches. In the example shown in **Figure 27.34**, you can see two so-called **motor units**, each consisting of a neuron and all the muscle fibers it controls (two or three, in this case).

The organization of individual neurons and muscle cells into motor units is the key to the action of whole muscles.

Spinal cord
Motor unit 1
Motor unit 2
Motor neuron cell body
Nerve
Motor neuron axon
Neuromuscular junctions
Nuclei
Muscle fibers (cells)
Muscle
Tendon
Bone

▲ Figure 27.34 **The relationship between motor neurons and muscle fibers.** A motor unit consists of a motor neuron and the one or more muscle fibers that it stimulates. When a motor neuron receives signals from the CNS, it passes them via synapses called neuromuscular junctions to the muscle fibers, causing the fibers to contract.

Each motor neuron may serve just one or up to several hundred fibers scattered throughout a muscle. Stimulation of the muscle by a single motor neuron produces only a weak contraction. More forceful contractions result when additional motor units are activated. Thus, depending on how many motor units your brain commands to contract, you can apply a small amount of force to lift a fork or considerably more to lift a barbell. In muscles requiring precise control, such as those controlling eye movements, a motor neuron may control only one fiber.

A motor neuron has its dendrites and cell bodies in the central nervous system (the spinal cord in Figure 27.34). Its axon forms synapses, called neuromuscular junctions, with the muscle fibers. A motor neuron causes a muscle fiber to contract by releasing a neurotransmitter that diffuses across the neuromuscular junction. When the neurotransmitter binds to the other side of the junction, it changes the permeability of the muscle fiber's plasma membrane, triggering action potentials that sweep across the muscle cell membrane. Ion channels within the plasma membrane therefore play an important role in muscle contraction, much as they did in sending signals along neurons (see Figure 27.4). And as you'll see in the Process of Science section, ion channels also play an important role in a nonhuman sense.

THE PROCESS OF SCIENCE: Extrahuman Senses

How Do New Senses Arise?

Researchers from the University of Texas investigated the origin of a new sense by starting with **observations** about two species of electric fish, one from Africa and one from South America. These fish can activate an electric organ, which is actually a modified muscle that has lost its ability to contract and instead uses special ion channel proteins to generate electric fields. Both species use electrical emissions to communicate and to find prey, but the nature of the electrical discharges varies considerably between the two lineages. This led the

► Figure 27.35 **Duplication of the gene for a muscle ion channel protein led to an electric organ.** In two species of electric fish, the gene for a muscle ion channel protein duplicated and then mutated, producing genes *a* and *b*; gene *a* functions in the electric organ, while gene *b* functions in muscle. In a nonelectric fish species, both genes function in muscle.

researchers to **question** whether different ion channel proteins had evolved in the two electric species.

The researchers formed the **hypothesis** that the ion channel genes of the two electric species had mutated in unique ways. In their **experiment**, they determined the DNA sequence of the gene in the two electric fishes and one closely related but nonelectric South American fish. **Figure 27.35** summarizes their **results**. In the common ancestor of all three of these species, a single ion channel gene had duplicated, and the gene copies later mutated into two forms, referred to as *a* and *b*. In the nonelectric fish, both of these genes function in muscles, much as they do in many other vertebrates. In the two species of electric fish, however, gene *a* had mutated into forms that allow the fishes to produce a current in the electric organ. Slight differences in how gene *a* mutated account for the differences in how each species generates electrical signals. Gene *b*, however, functions in the muscles of the electric fishes as it does in nonelectric fishes. This work demonstrates how a new sense can arise from an old one by gene duplication followed by mutation and natural selection.

Location of gene function	African species Electric fish	South American species Electric fish	Nonelectric fish
Muscle	Gene *b*	*b*	*a* and *b*
Electric organ	Gene *a* (mutation 1)	*a* (mutation 2)	none

Stimulus and Response: Putting It All Together

The receiver streaks down the sideline, jumping to catch a pass from the quarterback (Figure 27.36). Besides enjoying a great play, we can now appreciate these events in terms of sensory input, nervous system integration, and motor response.

The receiver obtains information about the path of the ball from his sensory structures. Photoreceptors in his eyes track the ball in flight. Sensory neurons convey this information to his brain as a series of action potentials. Within his brain, information about the angle and speed of the ball is integrated. In response, motor neurons signal multiple muscles to perform very specific actions. He jumps toward the ball, extending his arms, turning his wrist joints, and flexing his hand muscles. When mechanoreceptors in his hand detect the impact of the ball, this information is communicated and integrated, resulting in signals that move his hand muscles against his finger bones, squeezing his hands and capturing the ball. The fans see a spectacular play by a talented athlete. But underlying it all is the nervous system directing muscles to respond to information that sensory receptors have gathered about the environment.

We have now completed the unit on animal structure and function. Our overriding theme has been that the evolution of new structures allows adaptations for new functions; that is, the structural adaptations of a cell, tissue, organ, or organ system determine the job it can perform. We'll see the structure-function theme emerge again in the context of adaptations to the environment when we take up the study of plants in the next unit.

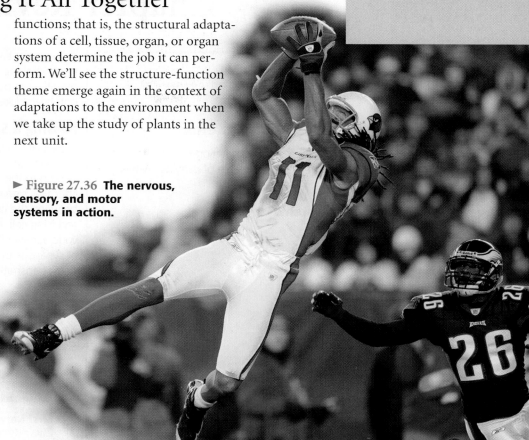

▶ Figure 27.36 **The nervous, sensory, and motor systems in action.**

EVOLUTION CONNECTION: Extrahuman Senses

Seeing UV

In the Biology and Society section and elsewhere in this chapter, you have read about senses possessed by some animals but not humans. Natural selection can foster new adaptations that arise from mutations in genes that control existing adaptations. You have read about one example—which led to the evolution of electroreception in some fish—in the Process of Science section. This principle is also clearly evident in the evolution of another nonhuman sense: ultraviolet vision.

Many birds can see beyond what we humans can see (Figure 27.37). Ultraviolet (UV) vision—the ability to see light of wavelengths shorter than those in our visual spectrum (see Figure 7.4)—is believed to be important in social communication and food gathering. As you learned in this chapter, visual information is transmitted to the brain after pigment molecules in the eye absorb light. Researchers studying vision in birds found that a single amino acid change in the pigment protein rhodopsin converted it to a UV-detecting form. This is yet another example of a large change— an important extension of a sense—that can be traced to a small change: a single mutation. In nature, if a mutation such as this one confers a survival advantage (the ability to more easily locate food, for example), then natural selection would tend to favor individuals with the mutation. Over time, the new adaptation would eventually become the norm.

◀ Figure 27.37 **A European kestrel: a bird with UV vision.**

Chapter Review

 Go to the Study Area at **www.masteringbiology.com** for practice quizzes, eText, BioFlix™ 3-D animations, MP3 Tutor Sessions, videos, current events, and more.

An Overview of Animal Nervous Systems

The nervous system is a communication and coordination network composed of neurons, specialized nerve cells capable of carrying signals throughout the body.

Organization of Nervous Systems

Neurons

Neurons are the functional units of the nervous system.

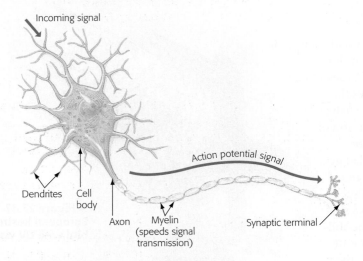

Sending a Signal through a Neuron

At rest, a neuron's plasma membrane has a resting potential caused by the membrane's ability to maintain a positive charge on its outer surface opposing a negative charge on its inner surface. A stimulus alters a portion of the membrane, allowing ions to enter and exit the neuron and creating an

action potential. Action potentials are self-propagated in a one-way chain reaction along a neuron. An action potential is an all-or-none event; its size is not affected by differences in stimulus strength. The frequency of action potentials does change with the strength of the stimulus.

Passing a Signal from a Neuron to a Receiving Cell

Signals are transmitted beyond an individual neuron at relay points called synapses. In an electrical synapse, electric current passes directly from one neuron to the next. At chemical synapses, the sending cell secretes a chemical signal, a neurotransmitter, which crosses the synaptic cleft (a gap between the cells) and binds to a specific receptor on the surface of the receiving cell. A cell may receive different signals from many neurons. A wide variety of small molecules can act as neurotransmitters. Many drugs act at synapses by increasing or decreasing the normal effect of neurotransmitters.

The Human Nervous System: A Closer Look

The Central Nervous System and The Peripheral Nervous System

NERVOUS SYSTEM			
Central Nervous System (CNS)		**Peripheral Nervous System** (PNS)	
Brain	Spinal cord: nerve bundle that communicates with body	Somatic nervous system: voluntary control over muscles	Autonomic nervous system: involuntary control over organs • Parasympathetic division: rest and digest • Sympathetic division: fight or flight

The Human Brain

BRAIN			
Forebrain (sophisticated integration) • Thalamus • Hypothalamus • Cerebrum	**Midbrain**	**Hindbrain** • Pons • Medulla oblongata • Cerebellum (coordinates movement)	Brainstem (filters motor and sensory input)

The forebrain's cerebrum, divided into right and left hemispheres, is the largest and most complex part of the brain. In particular, the surface of the cerebrum, the cerebral cortex, is responsible for many of the most distinctive human traits. The right and left cerebral hemispheres specialize in different mental tasks.

The Senses

Sensory Input

The process of sensory transduction:

The strength of the stimulus alters the rate of action potential transmission. Some sensory neurons tend to become less sensitive when stimulated repeatedly, a phenomenon known as sensory adaptation. Pain receptors sense stimuli that may indicate tissue damage; thermoreceptors detect heat or cold; mechanoreceptors respond to mechanical energy (such as touch, pressure, or sound); chemoreceptors respond to chemicals in the external environment or body fluids; and electromagnetic receptors respond to electricity, magnetism, or light (photoreceptors).

Vision

In the human eye, the cornea and lens focus light on the retina. The human lens changes shape to bring objects at different distances into sharp focus. The retina contains two types of photoreceptor cells called rods and cones. Rods contain the visual pigment rhodopsin and function in dim light. Cones contain photopsin, which enables us to see color in full light. Nearsightedness, farsightedness, and astigmatism result from focusing problems in the lens. Corrective lenses bend the light rays to compensate.

Hearing

Outer ear · Middle ear · Inner ear

Eardrum · Bones · Organ of Corti (inside cochlea)

The waves generated in the cochlear fluid move hair cells (mechanoreceptors) of the organ of Corti against an overlying membrane. Bending of the hair cells triggers receptor potentials, sending nerve signals to the brain. Deafness can be caused by infections, injury, or overexposure to loud noises.

Motor Systems

Movement is one of the most distinctive features of animals. In the process of motor output, the nervous system issues commands to the muscular system, and the muscular system exerts force against the skeleton.

The Skeletal System

The skeletal system functions in support, movement, and the protection of internal organs. The human endoskeleton is composed of cartilage and bone. Movable joints provide flexibility. A bone is a living organ containing several kinds of tissues. It is covered with a connective tissue membrane. Cartilage at the ends of the bone cushions the joints. The human skeleton is versatile, but it is also subject to problems, such as arthritis and osteoporosis. Broken bones can be realigned and immobilized, usually by splints or casts; bone cells then build new bone and repair the break.

The Muscular System

Skeletal muscles pull on bones to produce movements. Antagonistic pairs of muscles produce opposite movements. Each skeletal muscle cell, or fiber, contains bundles of myofibrils, which in striated muscle exhibit alternating light and dark bands. Each myofibril contains bundles of overlapping thick (myosin) and thin (actin) protein filaments. Repeating units of thick and thin filaments, called sarcomeres, are the muscle fiber's contractile units. The sliding-filament model explains the molecular process of muscle contraction. Using ATP, the myosin heads of the thick filaments attach to binding sites on the actin molecules and pull the thin filaments toward the center of the sarcomere. Motor neurons carry action potentials that initiate muscle contraction. A neuron can branch to contact a number of muscle fibers; the neuron and the muscle fibers it controls constitute a motor unit. The strength of a muscle contraction depends on the number of motor units activated.

Stimulus and Response: Putting It All Together

An animal's nervous system connects sensations derived from environmental stimuli to responses carried out by its muscles.

SELF-QUIZ

1. The large, central hub of a neuron, called the _____, contains the nucleus and organelles. Short, numerous fibers called _____ receive incoming messages, while a single long fiber called a(an) _____ conducts signals toward other cells. A neuron ends in many branches, each with a bulb-like _____ that contains neurotransmitters, which can be used to communicate signals to other neurons.

2. What is the function of the myelin sheath?

3. Your nervous system can be divided into two broad subsystems, the _____ and the _____. One of these is made up of nerves that conduct signals to and from the other structures. The other consists of the _____ and _____.

4. The central nervous system is protected by _____, a liquid that cushions it and supplies it with nutrients, and layers of protective tissue called the _____.

5. While driving down the interstate at rush hour, you get cut off by an 18-wheeler and have to slam on the brakes. In addition to a major case of road rage, you also develop a rapid heart rate and a rapid breathing rate. What's causing this increase? Your answer should indicate the branch of the nervous system involved.

6. A victim of a severe head injury can live for years in a nonresponsive state in which the cerebral cortex is not functioning but the person is still alive and performing metabolic functions. Based on your knowledge of brain structure and function, how can this be possible? Make sure your answer indicates which brain structures might be keeping the person alive.

7. How is an action potential different from a receptor potential?

8. For each of the following senses in humans, identify the type of receptor: seeing, tasting, hearing, smelling.

9. Mr. Johnson is becoming slightly deaf. To test his hearing, his doctor holds a vibrating tuning fork tightly against the back of Mr. Johnson's skull. This sends vibrations through the bones of the skull, setting the fluid in the cochlea in motion. Mr. Johnson can hear the tuning fork this way, but not when it is held away from his skull a few inches from his ear. Where is Mr. Johnson's hearing problem located? (Explain your answer.)
   a. in the auditory nerve leading to the brain
   b. in the hair cells in the cochlea
   c. in the bones of the middle ear
   d. in the fluid of the cochlea

10. A human's soft internal organs are protected by the _____ skeleton, while the bones of the limbs make up the _____ skeleton.

11. Arm and leg muscles are arranged in antagonistic pairs. How does this affect their functioning?
   a. It provides a backup if one of the muscles is injured.
   b. One muscle of the pair pushes while the other pulls.
   c. A single neuron controls both of them.
   d. It allows the muscles to produce opposing movements.

12. Muscle A and muscle B are the same size, but muscle A is capable of much finer control than muscle B. Which of the following is likely to be true of muscle A? (Explain your answer.)
   a. It is controlled by more neurons than muscle B.
   b. It contains fewer motor units than muscle B.
   c. It is controlled by fewer neurons than muscle B.
   d. Each of its motor units consists of more cells than the motor units of muscle B.

Answers to the Self-Quiz questions can be found in Appendix D.

THE PROCESS OF SCIENCE

13. Much of what we know about brain function has come from rare individuals whose brains were altered as the result of surgery, accidents, or disease. Suppose a patient has brain surgery on the cerebral cortex, and immediately following the surgery, she appears normal. As she recovers over the next few days, some unusual things happen, such as being unable to name objects while looking at them with one eye covered. Her sensory perception seems normal except for when she tries to integrate information. What brain structure might have been damaged during the surgery that would account for these symptoms? Justify your answer.

14. Sensory organs tend to come in pairs. We have two eyes and two ears. Similarly, a planarian worm has two eyecups, a rattlesnake has two infrared receptors, and a butterfly has two antennae. Propose a testable hypothesis that could explain the advantage of having two eyes or ears instead of one.

BIOLOGY AND SOCIETY

15. Brain injuries tend to be so severe because most neurons will not be regenerated once they have been damaged. The use of embryonic stem cells (see Chapter 11) has been proposed as a potential solution to many neurological diseases. Stem cells could be placed into the brain of a person with Alzheimer's or Parkinson's disease. Theoretically, those cells could differentiate into neurons, replacing those that had been killed by the disease. Do you favor or oppose research along these lines? Why? Would your opinion change if the person to be treated with stem cells were a close family member who would die if not treated?

16. Alcohol's depressant effects on the nervous system cloud judgment and slow reflexes. Alcohol consumption is a factor in many fatal traffic accidents in the United States. What are some other impacts of alcohol abuse on society? What are some of the responses of people and society to alcohol abuse? Do you think that alcohol abuse is primarily an individual or societal problem? Do you think our responses to alcohol abuse are appropriate and proportional to the seriousness of the problem?

17. Have you ever felt your ears ringing after listening to music with the volume turned high or attending a loud rock concert? Are you worried that excessively loud music could permanently impair your hearing? Do you think that people are aware of the possible danger of prolonged exposure to loud music? Should anything be done to warn or protect them? If so, what action would you suggest? What effect might warnings have?

Unit 6
Plant Structure and Function

Chapter 28: **The Life of a Flowering Plant**

Chapter Thread: **The Interdependence of People and Plants**

Chapter 29: **The Working Plant**

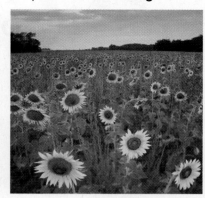

Chapter Thread: **The Interdependence of Organisms**

28 The Life of a Flowering Plant

Domesticated wheat.
Humans, like most animals, depend on flowering plants for food.

BIOLOGY AND SOCIETY: The Interdependence of People and Plants

Plants and Human Civilization

To tell the story of human society, you have to talk about plants. From the dawn of civilization to the frontiers of genetic engineering, human progress has always depended on expanding our use of plants—for food, fuel, clothing, and countless other trappings of modern life. To illustrate, let's consider humankind's relationship with a single crop: wheat. Today, wheat accounts for about 20% of all calories consumed worldwide. In the United States, it is one of our most valuable cash crops.

The roots of today's prosperous relationship between people and wheat can be traced back to the earliest human settlements. Throughout much of prehistory, humans were nomadic hunter-gatherers, moving and foraging with the changing seasons. About 10,000 years ago, a major shift occurred: People in several parts of the world began to domesticate wild plants. This allowed for the production of surplus food and the formation of year-round farming villages. This, in turn, led to the establishment of cities and the emergence of modern civilizations. By 7,000 years ago, bread wheat was cultivated in Egypt, India, China, and northern Europe. Farming became the dominant way of life across Europe by 6,000 years ago. For most of the subsequent millenia, little changed in the cultivation of wheat.

During the latter half of the 1900s, an international effort was undertaken to improve wheat and other staple crops. Through repeated selective breeding, agricultural scientists created today's wheat varieties. As a result of this "green revolution," wheat yield (tons of grain per acre planted) more than doubled from 1940 to 1980, while the cost of wheat production was cut in half. Wheat continues to be the target of considerable research today, primarily through the methods of genetic engineering and genomics (see Chapter 12).

In addition to their importance to people, plants are vital to the well-being of Earth's biosphere. Above and below the ground, plants provide shelter, food, and breeding areas for animals, fungi, and microorganisms. Because angiosperms—the flowering plants—make up over 90% of the plant kingdom, we concentrate on them in this unit. In this chapter, we begin by examining angiosperm structure, first at the level of the whole plant and then at the microscopic level of tissues and cells. Then we'll see how plant structures function in growth and reproduction.

The Structure and Function of a Flowering Plant

Angiosperms have dominated the land for over 100 million years, and about 250,000 species of flowering plants live today (see Chapter 16). Most of our foods come from 100 or so domesticated species of flowering plants. Among these foods are roots, such as beets and carrots; the fruits of trees and vines, such as apples, nuts, berries, and squashes; the fruits and seeds of legumes, such as peas, peanuts, and beans; and the fruits of grasses, which are grains such as wheat, rice, and corn.

Monocots and Dicots

On the basis of several structural differences, botanists have traditionally classified angiosperms into two groups, called monocots and dicots (Table 28.1). The names of the groups refer to embryonic structures called **cotyledons** (also called seed leaves), the first leaves to emerge from a growing seedling. A **monocot** embryo has one seed leaf; a **dicot** embryo has two seed leaves. The largest group of dicots is the **eudicots** ("true" dicots). In this chapter, we will focus on monocots and eudicots.

Monocots include orchids, palms, and lilies, as well as grains and other grasses. Most monocots have leaves with parallel veins. Monocot stems have vascular (internal transport) tissues organized into bundles that are arranged in a scattered pattern. The flowers of most monocots have petals and floral parts in multiples of three. Monocot roots form a fibrous system—a mat of threads—that spreads out below the soil surface. With most of their roots in the top few centimeters of soil, monocots, especially grasses, make excellent ground covers and can help reduce soil erosion.

Most angiosperms are eudicots, including many food crops (including nearly all of our fruits and vegetables), the majority of ornamental plants, and most shrubs and trees (except for the gymnosperms). Eudicot leaves have a multibranched network of veins, and eudicot stems have vascular bundles arranged in a ring. Eudicot flowers usually have petals and floral parts in multiples of four or five. The large, main root of a eudicot, known as a taproot, goes deep into the soil, as you know if you've ever tried to pull up a dandelion.

As we saw in the preceding unit on animals, a close look at a structure often reveals its function. Conversely, function provides insight into the "logic" of a structure. In the sections that follow, we'll take a detailed look at the correlation between plant structure and function. ✓

Table 28.1	Comparing Monocots and Eudicots				
	Seed Leaves	**Leaf Veins**	**Stems**	**Flowers**	**Roots**
Monocots	One cotyledon	Veins usually parallel	Vascular bundles in scattered arrangement	Floral parts usually in multiples of three	Fibrous root system
Eudicots	Two cotyledons	Veins usually branched	Vascular bundles arranged in ring	Floral parts usually in multiples of four or five	Taproot usually present

Plant Organs: Roots, Stems, and Leaves

Plants, like most animals, have organs composed of different tissues, which in turn are composed of cells of different types. A plant **organ** consists of several types of tissues that together carry out a particular function.

Among the evolutionary adaptations that made it possible for plants to thrive on land were structures for absorbing water and minerals from the soil; a large light-collecting surface; the ability to take in carbon dioxide from the air for photosynthesis; and adaptations for surviving dry conditions. In a land plant such as the generalized angiosperm shown in **Figure 28.1**, the roots (the belowground structures) and the shoots (the aboveground structures, including stems, leaves, and flowers) perform all these vital functions. Neither roots nor shoots can survive without the other. Lacking chloroplasts and living in the dark, the roots would starve without sugar and other organic nutrients transported from photosynthetic stems and leaves. Conversely, stems and leaves depend on the water and minerals absorbed by roots.

Roots

A plant's **roots** anchor it in the soil, absorb and transport minerals and water, and store food. All of a plant's roots make up its **root system**. The fibrous root system of a monocot provides broad exposure to soil water and minerals as well as firm anchorage. In contrast, the eudicot taproot is vertical, with many small secondary roots growing outward. Near the root tips in both eudicots and monocots, tiny projections called **root hairs** increase the root surface area, providing an extensive outer layer for absorption (as shown at the far right of Figure 28.1). Each root hair is an outgrowth of a cell on the surface of the root. It is difficult to move an established plant without injuring it because transplantation often damages the plant's delicate root hairs.

Large taproots, such as those found in carrots, turnips, sugar beets, and sweet potatoes, store food in the form of carbohydrates such as starch (**Figure 28.2a**). The plants use these stored sugars during periods of active growth and when they are producing flowers and fruit. Other types of modified roots include buttress roots, aerial roots that look like buttresses and support the tall trunks of the trees that produce them (**Figure 28.2b**).

▼ Figure 28.1 **The body plan of a flowering plant.**

▼ Figure 28.2 **Modified roots.**

(a) **The root of a sugar beet stores carbohydrates.**

(b) **The buttress roots of this Indonesian Paldoa tree help support the trunk.**

Plant Tissues and Tissue Systems

As in animals, the organs of plants contain tissues with characteristic functions. A **tissue** is a group of cells with a common structure, function, or both. For example, vascular tissue called **xylem** conveys water and dissolved minerals upward from the roots to the stems and leaves. Another type of vascular tissue called **phloem** transports sugars from leaves or storage tissues to other parts of the plant.

A **tissue system** consists of one or more tissues organized into a functional unit within a plant. Each plant organ—root, stem, or leaf—is made up of three tissue systems: the dermal, vascular, and ground tissue systems. Each tissue system is continuous throughout the entire plant body, but the systems are arranged differently in leaves, stems, and roots (**Figure 28.7**).

The **dermal tissue system** (brown in the figure) forms an outer protective covering. Like our own skin, the epidermis (the outer layer of the dermal tissue system) is a plant's first line of defense against physical damage and infectious organisms. On leaves and on most stems, epidermal cells secrete a waxy coating called the **cuticle**, which helps prevent water loss. The **vascular tissue system** (purple) provides support and long-distance transport; xylem and phloem are part of this system. The **ground tissue system** (yellow) has diverse functions, including photosynthesis, storage, and support.

Figure 28.8, a cross section of a eudicot root, illustrates the three tissue systems. The epidermis is a single layer of tightly packed cells covering the entire root. Water and minerals enter the plant from the soil through these cells. Some of the young epidermal cells will grow outward and form root hairs. In the center of the root, the vascular tissue system forms a cylinder, with xylem cells radiating from the center like the spokes of a wheel and phloem cells filling in the wedges between the spokes. The ground tissue system of the root forms the **cortex**, where cells store food and take up water and minerals. The innermost layer of cortex is the **endodermis**, a thin cylinder one cell thick. The endodermis is a selective barrier that regulates the passage of substances between the cortex and the vascular tissue.

All plant stems have vascular tissue systems arranged in numerous vascular bundles. The location and arrangement of these bundles differ between monocots and eudicots (see Figure 28.7). In eudicots, another part of the ground tissue system, known as the **pith**, fills the center of the stem and is often important in food storage.

▶ Figure 28.7 **The three tissue systems of a plant.** The roots, stems, and leaves of a plant are made from three tissue systems (the dermal tissue system, vascular tissue system, and ground tissue system) that are continuous throughout the plant body.

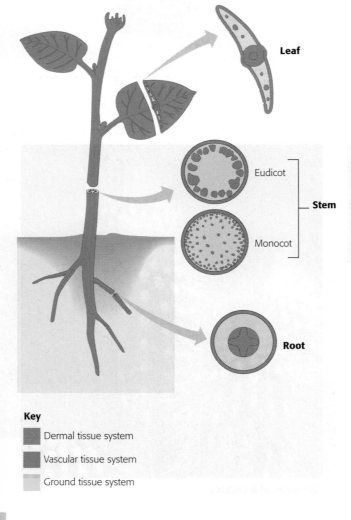

Leaf

Eudicot

Monocot

Stem

Root

Key

■ Dermal tissue system

■ Vascular tissue system

■ Ground tissue system

▼ Figure 28.8 **Tissue systems in a eudicot root.** This colored micrograph is a cross section of a young buttercup root. You can see the single-layered epidermis, the cylindrical vascular tissue system with xylem and phloem, and the ground tissue system that makes up the bulk of the root.

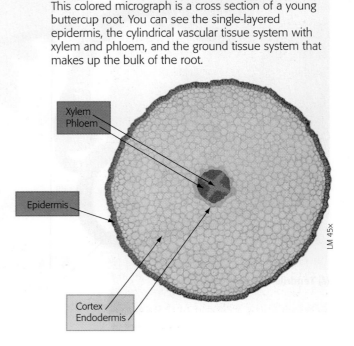

Xylem
Phloem

Epidermis

Cortex
Endodermis

LM 45×

Figure 28.9 illustrates the arrangement of the three tissue systems found in a typical eudicot leaf. The epidermis contains **stomata** (singular, *stoma*), tiny pores between two specialized cells called guard cells. The **guard cells** regulate the size of the stomata (see Figure 29.10), allowing gas exchange between the surrounding air and the photosynthetic cells inside the leaf.

The main site of photosynthesis is the **mesophyll**, which is the ground tissue system of a leaf. Mesophyll consists mainly of photosynthetic cells containing chloroplasts (green in Figure 28.9).

The leaf's vascular tissue system is made up of a network of veins. As you can see in Figure 28.9, each vein is a vascular bundle composed of xylem and phloem. The veins are in close contact with the leaf's photosynthetic tissues. This ensures that those tissues are supplied with water and minerals from the soil and that sugars made in the leaves are transported throughout the plant. ☑

▶ **Figure 28.9 Tissue systems in a eudicot leaf.** Notice the stoma, a pore on the bottom surface that allows gas exchange between the environment and the interior of the leaf.

Key

Dermal tissue system

Vascular tissue system

Ground tissue system

Cuticle

Upper epidermis

Mesophyll

Lower epidermis

Guard cells

Vein — Xylem / Phloem

Stoma

☑CHECKPOINT

Which tissue system is the main site of photosynthesis? Which tissue system covers and protects the outside of the plant?

Answer: ground; dermal

Plant Cells

So far, we have examined plant structure at the level of organs and tissues. Now we will complete our descent into the structural hierarchy of plants by examining cells.

In addition to features shared with other eukaryotic cells (see Figure 4.5), most plant cells have three unique structures, as shown in **Figure 28.10**: **chloroplasts**, the sites of photosynthesis; a large **central vacuole** containing fluid that helps maintain the cell's firmness (turgor); and a supportive **cell wall** that surrounds the plasma membrane.

Plant cell walls consist largely of the structural carbohydrate cellulose. Some plant cells, especially those that provide structural support, have a two-layered cell wall; a primary cell wall is laid down first, and then a thicker, more rigid secondary cell wall is deposited between the plasma membrane and the primary wall. Plasmodesmata are open channels in adjacent cell walls through which cytoplasm can flow from cell to cell.

BioFlix™
Tour of a Plant Cell

▼ **Figure 28.10 The structure of a plant cell.** The colored boxes highlight the unique features of plant cells. The enlargement shows the interface of two cell walls from adjoining cells.

Chloroplast

Central vacuole

Cell wall

Nucleus

Endoplasmic reticulum

Mitochondrion

Golgi apparatus

Ribosomes

Microtubules

Plasma membrane

Plasmodesmata

Primary cell wall

Secondary cell wall

Cell walls of adjoining cells

Plasma membrane

Figure 28.11 presents an overview of plant cell types. **Parenchyma cells** are the most abundant type of cell in most plants. They remain alive when mature and have only primary (often thin) walls. Parenchyma cells perform a variety of functions, such as food storage and photosynthesis. Most parenchyma cells can divide and differentiate (become specialized in their structure and function; see Chapter 11) into other types of plant cells, which they may do during repair of an injury.

Collenchyma cells resemble parenchyma cells in lacking secondary walls, but they have unevenly thickened primary walls. These cells provide support in parts of the plant that are actively growing. Young stems and petioles, for example, often have collenchyma cells just below their surface that elongate with the growing stem (these cells form the "strings" of a celery stalk, for example).

Sclerenchyma cells have thick secondary cell walls (colored pale yellow in Figure 28.11) usually strengthened with **lignin**, the main chemical component of wood. Mature sclerenchyma cells cannot elongate, and they develop only in regions that have stopped

growing in length. When mature, most sclerenchyma cells are dead, and their rigid cell walls support the plant much as steel beams do in a building. Sclerenchyma cells make up some commercially important plant products, such as the highly versatile hemp fiber used to make rope and clothing.

Angiosperms have two types of water-conducting cells: tracheids and vessel elements. Both have rigid, lignin-containing secondary cell walls. **Tracheids** are long, thin cells with tapered ends; **vessel elements** are wider, shorter, and less tapered. Water-conducting cells are arranged in chains with overlapping ends that form a system of water-carrying tubes. The tubes are hollow because water-conducting cells are dead when mature; only their cell walls remain. Water passes through pits in the walls of tracheids and vessel elements and through openings in the end walls of vessel elements.

Food-conducting cells are also arranged end to end, forming tubes. Unlike water-conducting cells, however, these cells remain alive at maturity. Their end walls, which are perforated with large plasmodesmata, allow sugars, other compounds, and some minerals to move between adjacent food-conducting cells. ✓

✓CHECKPOINT
Which of the cell types shown in Figure 28.11 has the potential to give rise to all the others?

Answer: parenchyma

▼ Figure 28.11 **Types of plant cells.**

Parenchyma cells

Collenchyma cells

Sclerenchyma cells

Water-conducting cells

Food-conducting cells

Plant Growth

The growth of a plant differs from that of an animal in a fundamental way. Most animals are characterized by **determinate growth**; that is, they cease growing after reaching a certain size. Most species of plants, in contrast, continue to grow as long as they live, a condition known as **indeterminate growth**.

Indeterminate growth does not mean that plants are immortal. In fact, different types of plants have very different life spans (**Figure 28.12**). Plants called **annuals** emerge from seed, mature, reproduce, and die in a single year or growing season. Our most important food crops—wheat, corn, and rice, for example—are annuals. **Biennials** live for two years; flowering and seed production usually occur during the second year. Carrots are biennials, but we usually harvest them in their first year and so miss seeing their flowers. Plants that live and reproduce for many years, including trees, shrubs, and some grasses, are known as **perennials**. Some perennials have life spans well beyond the longest-lived animals; for example, some bristlecone pines have been alive for over 4,500 years.

Primary Growth: Lengthening

Growth in all plants is made possible by tissues called meristems. A **meristem** consists of undifferentiated (unspecialized) cells that divide frequently, generating new cells and tissues. Meristems at the tips of roots and in the terminal and axillary buds of shoots are called **apical meristems**. Cell division in apical meristems produces new cells that enable a plant to grow in length, a process called **primary growth** (**Figure 28.13**). Tissues produced by primary growth are called primary tissues.

Terminal bud

Axillary buds

Key

↑ Direction of growth

Root tips

◄ **Figure 28.13 Primary growth at apical meristems.** Meristems are tissues responsible for growth in all plants. The apical meristems are located at the root tips and in the buds of shoots.

▼ Figure 28.12 **Plants with varying life spans.**

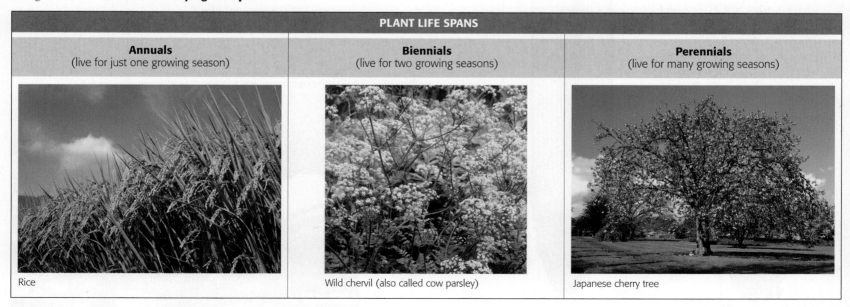

PLANT LIFE SPANS

Annuals (live for just one growing season)	**Biennials** (live for two growing seasons)	**Perennials** (live for many growing seasons)
Rice	Wild chervil (also called cow parsley)	Japanese cherry tree

Figure 28.14 shows a longitudinal section through a growing onion root. Primary growth enables roots to push through the soil. (A very similar process results in the upward growth of shoots.) At the very tip of the root is the **root cap**, a thimble-like cone of cells that protects the delicate, actively dividing cells of the apical meristem. The root's apical meristem (the orange oval in the art and circled in the micrograph) replaces cells of the root cap that are scraped away by the soil (downward arrow) and produces cells for primary growth (upward arrow). Primary growth is achieved not only by cell division but also by the lengthening of cells just above the apical meristem (see Figure 28.14, center).

These cells can undergo a tenfold increase in length, mainly by taking up water. Elongation of these cells is what actually forces a root down through the soil. The elongating cells begin to differentiate, forming primary tissues that develop into the epidermis, cortex, and vascular tissue (see Figure 28.14, top). Cells of this last type eventually differentiate into vascular tissues called primary xylem and primary phloem. ☑

Secondary Growth: Thickening

In addition to lengthwise primary growth, the stems and roots of many plant species also thicken by a process called **secondary growth** (Figure 28.15). Such thickening is most evident in the woody plants—trees, shrubs, and vines—whose stems last from year to year and consist mainly of thick layers of mature, mostly dead xylem tissue, called **wood**. Tissues produced by secondary growth are called secondary tissues.

Secondary growth involves cell division in two meristems we have not yet discussed: the vascular cambium and the cork cambium. The **vascular cambium** (blue-green in Figure 28.16) is a cylinder of actively dividing cells between the primary xylem and primary phloem, as you can see in the pie-shaped section at the left of the figure. Secondary growth (red arrows) adds cells on either side of the vascular cambium.

The center and right drawings in Figure 28.16 show the results of secondary growth. In the center drawing, the vascular cambium has given rise to two new tissues: secondary phloem to its exterior and secondary xylem to its interior. Yearly production of a new layer of secondary xylem accounts for most of the growth in thickness of a perennial plant.

▼ Figure 28.14 **Close-up of primary growth in a root tip.** Primary growth is achieved through two mechanisms. Cells within the apical meristem actively divide. Additionally, cells behind the apical meristem elongate up to tenfold.

Vascular system

Cortex

Epidermis

Root hair

Zone of differentiation

Key
- Dermal tissue system
- Vascular tissue system
- Ground tissue system

Zone of elongation

Zone of cell division

Apical meristem region

Root cap

LM 200x

▼ Figure 28.15 **Secondary growth.** Trees (such as these live oaks) thicken from year to year due to secondary growth.

▼ Figure 28.16 **Secondary growth of a woody stem.** The branches of most woody plants are made up of tissues of varying ages, with the youngest regions near the tip and the older regions nearer the trunk. The cross section at the left shows a region of the stem that is just beginning secondary growth. The middle and right cross sections show progressively older regions.

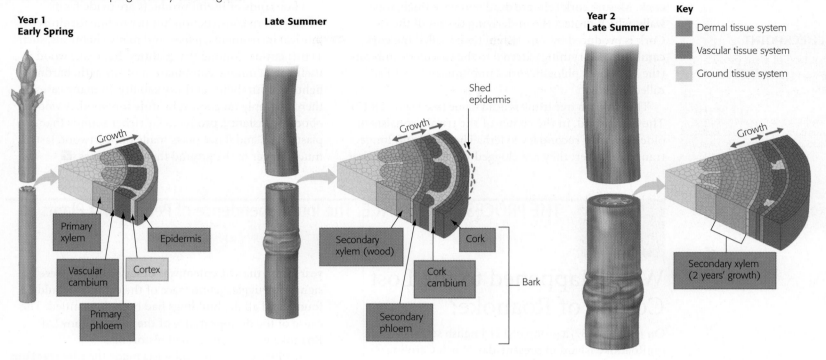

Annual growth rings result from the layering of secondary xylem (Figure 28.17). The layers are visible as rings because of uneven activity of the vascular cambium during the year. In woody plants that live in temperate regions, such as most of the United States, the vascular cambium becomes dormant each year during winter, and secondary growth is interrupted. When secondary growth resumes in the spring, a cylinder of early wood forms. Made up of the first new xylem cells to develop, early wood cells are usually larger in diameter and have thinner walls than those produced later in summer. The boundary between the large cells of early wood and the smaller cells of the late wood produced during the previous growing season is usually a distinct ring visible in cross sections of tree trunks and roots.

Therefore, a tree's age can be estimated by counting its annual rings. The rings may have varying thicknesses, reflecting the amount of seasonal growth in different years and therefore climate conditions. In fact, the pattern of growth rings in older trees is one source of evidence for recent global climate change. As we'll see in the Process of Science section, tree ring data can even provide insight into subjects outside the bounds of science.

▼ Figure 28.17 **Anatomy of a tree trunk.** On the left, you see a cross section of a locust trunk, with several decades of growth rings visible. On the right, the various layers of a mature trunk are separated for easier viewing.

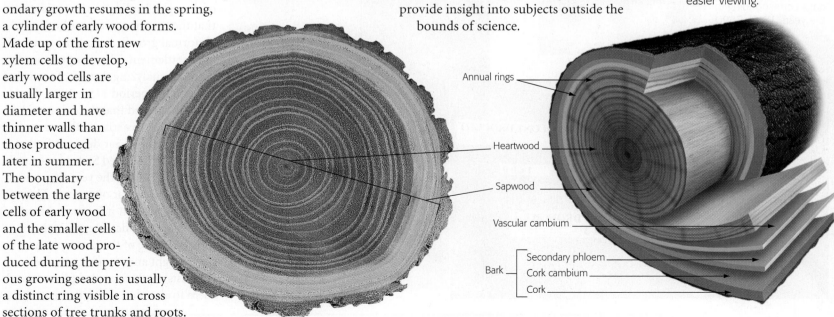

Returning to Figure 28.16, notice that the epidermis and cortex make up the young stem's external covering. When secondary growth begins, the epidermis is sloughed off and replaced with a new outer layer called **cork**. Mature cork cells are dead and have thick, waxy walls, which protect the underlying tissues of the stem. Cork is produced by a meristem tissue called the **cork cambium**. Everything external to the vascular cambium (the secondary phloem, cork cambium, and cork) is called **bark**.

The bulk of a tree trunk is dead tissue (see Figure 28.17). The heartwood, in the center of the trunk, consists of older layers of secondary xylem. These cells no longer transport water; they are clogged with resins and other metabolic by-products that make the heartwood resistant to rotting. The lighter-colored sapwood consists of younger secondary xylem that does conduct water.

Thousands of useful products are made from wood—from construction lumber to fine furniture, musical instruments, paper, and many chemicals, such as turpentine. Among the qualities that make wood so useful are a unique combination of strength, hardness, lightness, durability, and workability. In many cases, there is simply no good substitute for wood. A wooden oboe, for instance, produces far richer sounds than a plastic one, and fence posts made of heartwood last much longer in the ground than metal ones. ✓

✓CHECKPOINT

What type of plant tissue makes up wood? What is bark?

Answer: secondary xylem; all tissues exterior to the vascular cambium: secondary phloem, cork cambium, and cork

---

THE PROCESS OF SCIENCE: The Interdependence of People and Plants

What Happened to the Lost Colony of Roanoke?

On July 22, 1587, a group of 121 English settlers landed on Roanoke Island in present-day North Carolina (**Figure 28.18**). They were sent by Sir Walter Raleigh to establish a permanent colony in the New World. Their supply ship returned to England later that year, leaving 115 colonists behind. When the next ship returned three years later, the settlement was deserted. There were no signs of struggle, yet no trace of the colonists could be found, and all the buildings had been dismantled. The cause of the disappearance of the "Lost Colony" of Roanoke has long intrigued historians.

In 1998, a group of biologists made the **observation** that a large number of very old bald cypress trees (*Taxodinum distichum*) around Roanoke Island could provide a reliable record of the climate over the last 800 years. This led them to **question** whether the tree ring data could provide insight into the colony's mysterious disappearance. Their **hypothesis** was that the loss of the colony corresponded to a period of drought, leading to the **prediction** that the tree ring data would show abnormal growth during the years after settlement. Their **experiment** involved analyzing dozens of trees that covered the period 1185–1984. Their **results** showed that the colonists had the bad luck to arrive at the start of the worst three-year drought in the southeastern United States in the past 800 years. Indeed, the year they settled was the driest year on record. These data indicate that it is not necessary to invoke a conflict to explain the disappearance of the Lost Colony; the drought and its effects on the colonists' vital crops could have forced them to abandon the settlement.

▶ Figure 28.18 **The Lost Colony of Roanoke.** Notice the very narrow growth rings corresponding to the years 1587–1590.

TAXODIUM DISTICHUM
BLACKWATER RIVER, VIRGINIA

LOST COLONY DROUGHT ↓ 1587

↓ 1531

↓ 1600

JAMESTOWN DROUGHT ↑ 1606-1612

Canada

United States

Mexico

Cape May

Virginia

Chesapeake Bay

Roanoke Island

North Carolina

ATLANTIC OCEAN

N ↑

Cape Fear

The Life Cycle of a Flowering Plant

The life cycle of an organism is the sequence of stages from the adults of one generation to the adults of the next. In completing the life cycle, many flowering plants can reproduce both asexually and sexually, and both modes have played an important role in the evolutionary adaptation of plant populations to their environments. **Figure 28.19** shows three different examples of **asexual reproduction**, the creation of offspring derived from a single parent without fertilization. Through asexual reproduction, a single plant can produce many offspring quickly and efficiently. This ability of plants is useful in agriculture because it allows farmers to grow a large crop of identical plants from a parent with desirable traits. As in animals, **sexual reproduction** in plants involves **fertilization**, the union of gametes from two parents to produce genetically distinct offspring. This section discusses the sexual life cycle of flowering plants, starting with a look at the flower.

The Flower

In angiosperms, the structure specific to sexual reproduction is the **flower**. The main parts of a flower—the sepals, petals, stamens, and carpels—are modified leaves (**Figure 28.20**; also see Figure 28.1). The **sepals** enclose and protect the flower bud. The **petals** are often bright and colorful and advertise the flower to insects and other pollinators.

The flower's reproductive organs are the stamens and carpel. A **stamen** consists of a stalk (filament) tipped by an **anther**. Within the anther are sacs where meiosis occurs and pollen grains develop. Pollen grains house the cells that develop into sperm. A **carpel** has a long slender neck (the style) with a sticky stigma at its tip. The **stigma** is the landing platform for pollen grains. The base of the carpel is the **ovary**. Within the ovary are reproductive structures called **ovules**, each containing one developing egg and the cells that support it. The term **pistil** is sometimes used to refer to a single carpel or a group of fused carpels. ☑

MP3 tutor sessions

From Flower to Fruit

▼ Figure 28.19 **Asexual reproduction in plants.**

Garlic. This garlic bulb is actually an underground stem that functions in storage. A single large bulb fragments into several cloves. Each clove can give rise to a separate plant.

Holly trees. Each of the small trees is a sprout from the roots of a single holly tree. Eventually, one or more of these root sprouts may take the place of its parent.

Creosote bushes. This ring of plants is a clone of creosote bushes growing in the Mojave Desert in southern California. All these bushes came from generations of asexual reproduction by roots. This clone apparently began with a single plant that germinated from a seed about 12,000 years ago. The original plant probably occupied the center of the ring.

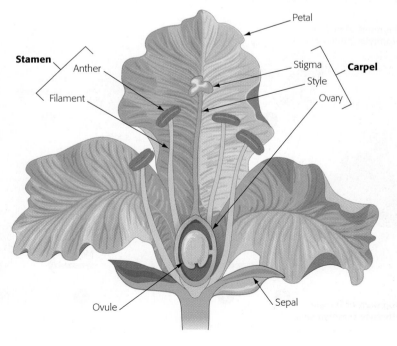

◄ Figure 28.20 **The structure of a flower.** The flower is the structure used by angiosperms for sexual reproduction.

Stamen
— Anther
— Filament
Petal
Stigma
Style
Ovary
Carpel
Ovule
Sepal

☑CHECKPOINT

Pollen develops within the _____ of sperm-producing organs called _____. Ovules develop within the _____ of egg-producing organs called _____.

Overview of the Flowering Plant Life Cycle

The life cycle of a sexually reproducing angiosperm is shown in **Figure 28.21**. After fertilization, the ovule of a flower matures into a seed containing the embryo.

Meanwhile, the ovary develops into a fruit, which protects the seed and aids in dispersing it. Completing the life cycle, the seed **germinates** (begins to grow) in a suitable habitat, the embryo develops into a seedling, and the seedling grows into a mature plant.

Let's look a bit closer at this process from a genetic point of view. As we discussed in Chapter 16, the life cycles of all plants include alternation of haploid (*n*) and diploid (2*n*) generations (see Figure 16.10). The roots, stems, leaves, and most of the reproductive structures of angiosperms are diploid. The diploid plant body is called the **sporophyte**. A sporophyte produces the anthers and ovules in which cells undergo meiosis to produce haploid cells called spores. Each spore then divides via mitosis and becomes a multicellular **gametophyte**, the plant's haploid generation. The gametophyte produces gametes by mitosis. Fertilization occurs when the male and female gametes (sperm and egg, respectively) unite, producing a diploid zygote. The life cycle is completed when the zygote divides by mitosis and develops into a new sporophyte. In the rest of this section, we examine key stages in the angiosperm sexual life cycle in more detail.

▼ **Figure 28.21 The life cycle of an angiosperm.**

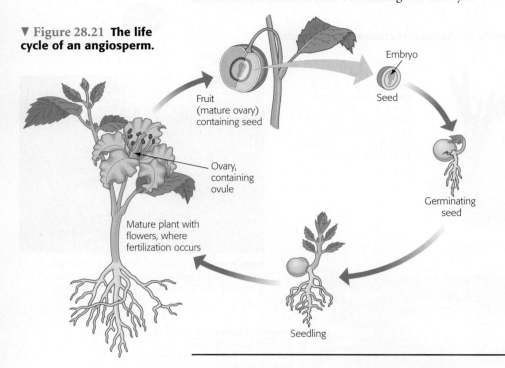

Embryo
Seed
Fruit (mature ovary) containing seed
Ovary, containing ovule
Germinating seed
Mature plant with flowers, where fertilization occurs
Seedling

Pollination and Fertilization

Fertilization requires gametes, which are produced by gametophytes. The male gametophyte is the **pollen grain**, which is essential for pollination and produces sperm. The female gametophyte is a multicellular

structure called the **embryo sac**, which produces the egg. How are the male and female gametophytes formed in flowering plants? As shown at the top of **Figure 28.22**, ❶ cells within a flower's anthers ❷ undergo meiosis to form four haploid spores. ❸ Each spore then divides by mitosis into two haploid cells. A thick wall forms around these cells, and the resulting pollen grain is ready for release from the anther.

Moving to the bottom of the figure, we see that ❶ within an ovule, a central cell enlarges and ❷ undergoes meiosis, producing four haploid spores. Three of the spores usually degenerate, but the surviving one enlarges and ❸ divides by mitosis, producing the embryo sac. The sac contains a large central cell with two haploid nuclei. One of its other cells is the haploid egg, ready to be fertilized.

▼ **Figure 28.22 Development of gametophytes in an angiosperm.** The male gametophyte (pollen grain) develops within the anther (top). The female gametophyte (embryo sac) develops within the ovule (bottom). In most species, the ovary of a flower contains several ovules, but only one is shown here.

Development of male gametophyte (pollen grain)

Anther
❶ Cell within anther
❷ Meiosis → Four haploid spores
❸ Mitosis → Haploid nuclei of two cells
Pollen grain (will be released from anther)

Ovary
Ovule
Development of female gametophyte (embryo sac)
❶ → ❷ Meiosis → Surviving cell (haploid spore)
❸ Mitosis → Two haploid nuclei of large, central cell
Egg cell (with haploid nucleus)
Embryo sac

How does fertilization occur? As shown in **Figure 28.23**, the first step is ① **pollination**, the delivery of pollen grains from anther to stigma. Many angiosperms are dependent on animals to transfer their pollen, but the pollen of some plants, such as grasses, is windborne (as anyone with pollen allergies knows!). ② After pollination, the pollen grain germinates on the stigma. It divides by mitosis, forming ③ two sperm that travel to the ovule through a pollen tube. ④ One sperm fertilizes the egg, forming the diploid zygote. The other sperm contributes its haploid nucleus to the large diploid central cell of the embryo sac. This cell, now with a triploid ($3n$) nucleus, will give rise to a food-storing tissue called **endosperm**. The formation of both a zygote and a fertilized central cell with a triploid nucleus is called **double fertilization**. This occurs only in plants, mainly in angiosperms. ☑

▼ Figure 28.23 **Pollination and double fertilization.**

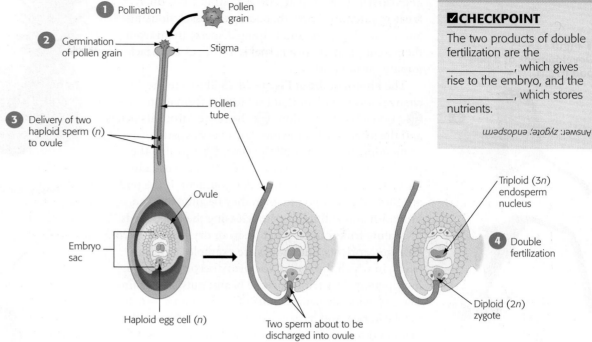

① Pollination

Pollen grain

② Germination of pollen grain

Stigma

③ Delivery of two haploid sperm (n) to ovule

Pollen tube

Ovule

Embryo sac

Haploid egg cell (n)

Two sperm about to be discharged into ovule

Triploid ($3n$) endosperm nucleus

④ Double fertilization

Diploid ($2n$) zygote

Seed Formation

After fertilization, the ovule, containing the zygote and the triploid central cell, begins developing into a seed (**Figure 28.24**). The zygote divides via mitosis into a ball of cells that becomes the embryo. Meanwhile, the triploid cell divides and develops into the endosperm. As cotyledons develop, they absorb nutrients from the endosperm. The result of embryonic development in the ovule is a mature seed (see Figure 28.24, bottom). A **seed** is a plant embryo and endosperm packaged within a tough protective covering called a **seed coat**.

At this point, the embryo stops developing and the seed becomes dormant; growth and development are suspended until the seed germinates. Seed dormancy is an important evolutionary adaptation. It allows time for seed dispersal and increases the chance that a new generation of plants will begin growing only when environmental conditions, such as temperature and moisture, favor survival. ☑

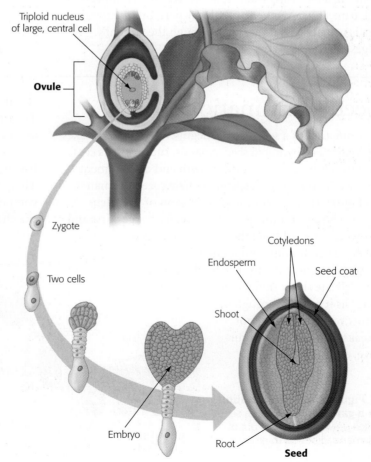

Triploid nucleus of large, central cell

Ovule

Zygote

Two cells

Cotyledons

Endosperm

Seed coat

Shoot

Embryo

Root

Seed

◀ Figure 28.24 **Development of a seed.** The plant in this drawing is a eudicot; a monocot would have only one cotyledon. The bulges you see on the embryo are the cotyledons starting to form.

▼ Figure 28.25
**Development of a pea
pod (fruit).**

Fruit Formation

Fruits, one of the most distinctive features of
angiosperms, develop at the same time seeds do. A
fruit is a mature ovary that acts as a vessel, housing
and protecting seeds and helping disperse them from
the parent plant. A corn kernel is a fruit, as is a peach,
tomato, or pea pod.

The photographs in **Figure 28.25** illustrate the
changes in a pea plant that lead to pod formation.
1 Soon after pollination, **2** the flower drops its petals
and the ovary starts growing. The ovary expands
tremendously, and its wall thickens, **3** forming the
pod, or fruit, which holds the seeds. Peas are usually
harvested at this stage of fruit development. If the pods
are allowed to develop further, they become dry and
brownish and will split open, releasing the dried seeds.

Mature fruits can be either fleshy or dry (**Figure 28.26**).
Oranges, apricots, and grapes are examples of fleshy
fruits, in which the wall of the ovary becomes soft during ripening. Dry fruits include beans, nuts, and grains.
The dry, wind-dispersed fruits of cereal grains—such
as wheat, rice, and barley—are harvested while on the
plant and serve as major staple foods for humans. Edible
fruits are usually nutritious, sweet tasting, and vividly
colored, advertising their ripeness. When an animal eats
the fruit, it digests the fruit's fleshy part, but the tough
seeds usually pass unharmed through the animal's
digestive tract.

▼ Figure 28.26 **A variety of fruits.**

Fleshy fruits

Dry fruits

Seed Germination

Germination of a seed usually begins when the seed takes
up water. The hydrated seed expands, rupturing its coat,
and the embryo resumes the growth and development
that was temporarily suspended during seed dormancy.

Figure 28.27 traces the germination of a garden
bean. The embryonic root of the bean emerges first and
grows downward from the germinating seed.
Next, the embryonic shoot emerges, and a
hook forms near its tip. The hook protects
the delicate shoot tip by holding it downward as it pushes through the
abrasive soil. As the shoot
breaks through the soil surface, exposure to light

stimulates the hook to straighten, and the tip is lifted.
The first foliage leaves then expand from the shoot tip
and begin making food by photosynthesis.

A germinating seed is fragile. In the wild, only a small
fraction of seedlings endure long enough to reproduce.
The great numbers of seeds produced by most plants
compensate for the odds against each seedling, increasing
the chances of reproductive success. ☑

☑CHECKPOINT

Is the garden bean shown in
Figure 28.27 a monocot or a
eudicot?

*Answer: It's a eudicot. (You can tell
because it has two cotyledon leaves.)*

► Figure 28.27 **Germination
in a garden bean.** After a seed
takes up water, it bursts out of
its coat and begins to grow.

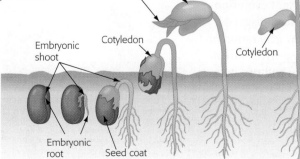

Foliage leaves
Cotyledon
Cotyledon
Embryonic
shoot
Embryonic
root
Seed coat
Cotyledon

The Problem of the Disappearing Bees

Flowering plants and land animals have had mutually beneficial relationships throughout their evolutionary history. Most angiosperms depend on insects, birds, or mammals for pollination and seed dispersal. And most land animals depend on angiosperms for food and shelter. Such mutual dependencies tend to improve the reproductive success of both the plants and the animals and are thus favored by natural selection.

The flowers of many angiosperms attract pollinators that rely entirely on the flowers' nectar (a sugary fluid) and pollen for food. Evolutionary adaptations for advertising the presence of nectar include flower color and fragrance that are keyed to pollinators' senses of sight and smell. For example, many flowers pollinated by birds are red or pink, colors to which bird eyes are especially sensitive. Flowers may also have markings that attract pollinators, leading them past pollen-bearing organs on their way to gathering nectar (Figure 28.28). For example, flowers pollinated by bees often have markings that reflect ultraviolet light. Such markings

are invisible to us, but vivid to bees. Many other animals—including hummingbirds, butterflies, and fruit bats—have similar relationships with other flower species. To a large extent, flowering plants are as diverse and successful as they are today because of their close connections with animals.

The importance of such interactions has been underscored by concern over colony collapse disorder. In late 2006, U.S. beekeepers noticed a sudden and drastic die-off in their bee colonies. The next year, similar problems were reported in Europe. Agricultural scientists fear that a shortage of bees could have devastating consequences on many food staples that require bees for pollination, such as almonds, berries, and fruits. Bees pollinate $15 billion worth of crops each year, and an estimated one-third of our food supply relies on bees. The cause of the die-off has not yet been explained, although several biological agents—such as mites or their pathogens—have been implicated. Still, the phenomenon highlights the many ways that people and plants depend on each other.

▼ Figure 28.28 **A busy bee.**

Chapter Review

SUMMARY OF KEY CONCEPTS

(MB) Go to the Study Area at **www.masteringbiology.com** for practice quizzes, eText, BioFlix™ 3-D animations, MP3 Tutor Sessions, videos, current events, and more.

The Structure and Function of a Flowering Plant

Flowering plants, or angiosperms, account for nearly 90% of the plant kingdom.

Monocots and Dicots

Angiosperms can be grouped into two categories based on the number of cotyledons (seed leaves) found in the embryo and other structural differences. Most dicots are eudicots.

PLANT STRUCTURES					
	Seed leaves	Leaf veins	Vascular bundles	Floral parts	Roots
Monocots	One cotyledon	Parallel	Scattered	Multiples of three	Fibrous
Eudicots	Two cotyledons	Branched	Ring	Multiples of four or five	Taproot

Plant Organs: Roots, Stems, and Leaves

A plant body consists of a root system and a shoot system, each dependent on the other.

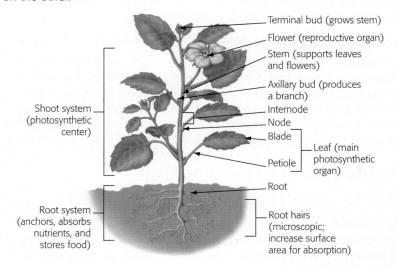

Terminal bud (grows stem)
Flower (reproductive organ)
Stem (supports leaves and flowers)
Axillary bud (produces a branch)
Internode
Node
Blade
Leaf (main photosynthetic organ)
Petiole
Shoot system (photosynthetic center)
Root
Root system (anchors, absorbs nutrients, and stores food)
Root hairs (microscopic; increase surface area for absorption)

A growing shoot often lengthens only at the terminal bud; its axillary buds are dormant. This condition is called apical dominance and allows the plant to grow tall quickly.

Plant Tissues and Tissue Systems

Roots, stems, and leaves are made up of tissues organized into three tissue systems: the dermal tissue system, the vascular tissue system, and the ground tissue system.

THE THREE TISSUES OF PLANT ORGANS		
Dermal Tissue System	**Vascular Tissue System**	**Ground Tissue System**
• Functions in protection • Includes the epidermis, an outer layer that covers and protects the plant • Stomata (each surrounded by two guard cells) are pores that regulate gas exchange between the cells of the leaf and the environment	• Functions in support and transport • Xylem: transports water and dissolved minerals • Phloem: transports sugar	• Functions mainly in storage • Includes leaf mesophyll, where most photosynthesis occurs

Plant Cells

Plant cells are distinguished by the presence of chloroplasts, a large central vacuole, and cell walls. There are several types of plant cells: parenchyma cells, collenchyma cells, sclerenchyma cells, water-conducting cells (tracheids and vessel elements), and food-conducting cells.

Plant Growth

Most plants display indeterminate growth, growing as long as they are alive. Some plants grow, reproduce, and die in one year (annuals); some live for two years (biennials); and some live for many years (perennials).

Primary Growth: Lengthening

Plant growth originates in meristems, areas of undifferentiated, dividing cells. Apical meristems are located at the tips of roots and in the terminal and axillary buds of shoots. These meristems initiate lengthwise growth by producing new cells. A root or shoot lengthens further as the new cells elongate. This cell division and elongation is called primary growth. The new cells eventually differentiate into specialized tissues.

Secondary Growth: Thickening

An increase in a plant's girth, called secondary growth, arises from cell division in a cylindrical meristem called the vascular cambium. The vascular cambium produces layers of secondary xylem, or wood, next to its inner surface. Outside the vascular cambium is bark, which consists of secondary phloem, a meristem called the cork cambium, and cork cells produced by the cork cambium. The outer layers of bark are sloughed off as the plant thickens.

The Life Cycle of a Flowering Plant

Many flowering plants can reproduce both asexually and sexually.

The Flower

In angiosperms, the flower is the structure specific to sexual reproduction. The stamen contains the anther, in which pollen grains are produced. The carpel contains the stigma, which receives pollen, and the ovary, which contains the ovule.

Overview of the Flowering Plant Life Cycle

The plant life cycle alternates between diploid ($2n$) and haploid (n) generations. The spores in the anthers give rise to male gametophytes, or pollen grains, each of which produces two sperm. A spore in an ovule produces the female gametophyte, called an embryo sac. Each embryo sac contains an egg cell.

Pollination and Fertilization

After pollen lands on the stigma (pollination), the sperm pass down the stigma into the ovule. One sperm combines with the egg and the other combines with a second, diploid cell. This process, called double fertilization, is unique to plants.

Pollen (n)

Ovary

Embryo sac (n)

Ovule

FERTILIZATION
within ovule

Fruit (from ovary)

Seed (from ovule)

Embryo ($2n$)

Mature plant ($2n$)

Germinating seed ($2n$)

Seed Formation

After fertilization, the ovule becomes a seed, and the fertilized egg within it divides to become an embryo. The other fertilized cell (now triploid) develops into the endosperm, which stores food for the embryo. A tough seed coat protects the embryo and endosperm. The embryo develops cotyledons (seed leaves), which absorb food from the endosperm.

Fruit Formation

While the ovule becomes a seed, the ovary develops into a fruit, which helps protect and disperse the seeds.

Seed Germination

A seed starts to germinate when it takes up water, expands, and bursts its seed coat. The embryo resumes growth, an embryonic root emerges, and a shoot pushes upward and expands its leaves.

SELF-QUIZ

1. While walking in the woods, you encounter an unfamiliar flowering plant. Which of the following plant features would help you determine whether that plant is a monocot or a eudicot? (Choose all that are appropriate.)
   a. size of the plant
   b. number of seed leaves
   c. shape of its root system
   d. number of petals in its flowers
   e. arrangement of vascular bundles in its stem
   f. whether or not it produces seed-bearing cones

2. Your friend plants a new herb (which is a eudicot and a perennial). During the first year, she allows the plants to grow naturally. Her plants grow tall and relatively spindly during the growing season, and when it comes time to harvest, she doesn't have a large yield. What phenomenon is responsible for her poor yield? What could your friend do to increase her yield next year?

3. A pea pod is formed from a(n) _____. A pea inside the pod is formed from a(n) _____.

4. Match each flower structure with its function.
   a. pollen grain   1. attracts pollinators
   b. ovule   2. develops into seed
   c. anther   3. protects an unopened flower
   d. ovary   4. produces sperm
   e. sepal   5. produces pollen
   f. petal   6. contains ovules

5. Many plant species can reproduce both sexually and asexually. Which mode of reproduction would generally be more advantageous in a location where the composition of the soil is changing rapidly? Why?

6. What part of a plant are you eating when you consume each of the following?
   a. tomato   d. sweet potato
   b. apple   e. black bean
   c. spinach

7. In the angiosperm life cycle, which of the following processes is directly dependent on meiosis?
   a. production of gametophytes   c. production of spores
   b. production of gametes   d. all of the above

8. In angiosperms, each pollen grain produces two sperm. What do these sperm do?
   a. Each one fertilizes a separate egg cell.
   b. One fertilizes an egg, and the other fertilizes the fruit.
   c. One fertilizes an egg, and the other is kept in reserve.
   d. Both fertilize a single egg cell.
   e. One fertilizes an egg, and the other fertilizes a cell that develops into stored food.

9. Which of the following is closest to the center of a woody stem? Which is farthest from the center?
   a. cork cambium
   b. primary xylem
   c. primary phloem
   d. secondary xylem
   e. secondary phloem

10. A gardener is concerned that his perennial plants are not growing well. He notes that the stems on his plants are not thickening adequately over the years. This suggests reduced activity of which plant tissue?

Answers to the Self-Quiz questions can be found in Appendix D.

THE PROCESS OF SCIENCE

11. During a recent ice storm, many old trees were snapped by the weight of the ice. A botanist in the area would like to analyze the remains of the trees and determine how old they really were. Describe a procedure that could be used to determine the approximate age of the trees. Explain the basis of this method in terms of plant tissue growth.

12. Plants can alter the shape of their guard cells, thereby regulating the size of the stomata on their leaves. Stomata must be open to allow entrance of carbon dioxide, which is necessary for photosynthesis. During the summer, the stomata are often closed around noon, when the most sunlight is likely to reach the plant. Propose an explanation for why it may be advantageous for stomata to be closed when conditions seem optimal for photosynthesis. How could you test your hypothesis?

BIOLOGY AND SOCIETY

13. Tropical forests contain a wealth of plants that are potential new sources of food, medicine, and other useful products. As a result of growing populations and debt, however, many tropical forests in less industrialized countries are being cut for lumber and farmland, and many species are disappearing. More industrialized countries are pressuring the tropical countries to protect the forests before even more species are lost. Many people in the less industrialized nations see little incentive to preserve the forests only to have corporations from more industrialized countries profit from their products. Is there a way to preserve the tropical forests so that both the less industrialized and more industrialized nations will benefit from their abundance?

14. Cloning is a fiercely debated topic in our country. Most of the ethical debates concerning cloning center on the cloning of animals. There is not much talk about the cloning of plants. Do you have the same opinion about cloning plants and animals? Why or why not? Do you think that they should be similarly regulated? What unique problems might each present?

29 The Working Plant

Cleansing plants.
Humans depend on
plants in many ways,
including using
sunflowers to treat
polluted topsoil.

BIOLOGY AND SOCIETY: The Interdependence of Organisms

Planting Hope in the Wake of Katrina

In the early morning of August 29, 2005, Hurricane Katrina—one of the strongest storms ever recorded—slammed into Louisiana and Mississippi. By the time it blew past, Katrina was the costliest natural disaster in U.S. history, responsible for over 1,800 deaths. The city of New Orleans was particularly hard hit. After several levees breached during the storm surge, about 80% of the city was flooded. The flood waters were a noxious stew containing raw sewage, heavy metals, other toxic chemicals, and petroleum. The effect on the local environment was devastating, particularly to the topsoil in much of the city.

Years later, the cleanup continues. In one effort, a relief organization named Common Ground is using plants to clean the city's soil. Plants can efficiently move minerals and other compounds from the soil into the plant body. Usually, the absorbed compounds are nutrients that help the plant grow. But many species of plants are metal accumulators, capable of absorbing large amounts of lead, zinc, and other heavy metals.

Common Ground volunteers have planted tens of thousands of sunflowers in some of New Orleans's hardest-hit wards. Once fully grown, the sunflowers (which accumulate lead) are harvested and hauled away to a hazardous-waste landfill. Because cleanup crews have to remove only a few cubic meters of plant material rather than thousands of cubic meters of contaminated soil, this is a cost-effective and efficient process. Using plants to clean the soil can also detoxify an area without dramatically disturbing the landscape. And sunflowers provide an uplifting message to the community, bringing beauty and life back into an area weary with grief and loss.

The biggest obstacle to the efforts of Common Ground is time. While soil can be hauled away in a matter of days, albeit at high cost, plants can take months to grow. Furthermore, cleansing crops must often be planted for multiple seasons to reduce soil toxicity to acceptable levels.

In using plants to clean up toxic wastes, we are benefiting from millions of years of plant evolution. Unable to move about in search of food, plants have evolved an amazing ability to pull water and nutrients out of the soil and air. In this chapter, we'll see how plants obtain essential nutrients and how they transport them throughout their roots, stems, and leaves. Then we'll examine other aspects of the life of a working plant, including the crucial roles played by hormones and how certain plant activities are affected by stimuli from the environment.

How Plants Acquire and Transport Nutrients

▼ Figure 29.1 **The uptake of nutrients by a plant.** A plant absorbs the CO_2 it needs from the air, and minerals and H_2O from the soil.

Minerals (inorganic ions)

Watch a plant grow from a tiny seed, and you can't help wondering where all the mass comes from. About 96% of a plant's dry weight is organic material synthesized from inorganic nutrients extracted from the surroundings. (In this context, "organic" means a molecule containing both carbon and hydrogen.) Plants obtain carbon dioxide (CO_2) from the air and water (H_2O) and **minerals** (inorganic ions) from the soil (**Figure 29.1**). From the CO_2 and H_2O, plants produce sugars via photosynthesis. These sugars, combined with minerals, are used to construct all the other organic materials a plant needs (**Figure 29.2**). Plants, like people, require a balanced diet. So let's start by discussing plant nutrition.

▼ Figure 29.2 **A Bavaria beech tree, a giant product of photosynthesis.** Plants use sugars created by photosynthesis to produce all the organic materials they need. The giant trunk of this tree consists largely of molecules derived from sugars.

Plant Nutrition

A chemical element is considered an **essential element** if a plant must obtain it from its environment to complete its life cycle. Seventeen elements are essential to all plants, and a few others are essential to certain types of plants. Of the 17 essential elements, 9 are called **macronutrients** because plants require relatively large amounts of them. Elements that plants need in extremely small amounts are called **micronutrients.**

Macronutrients

Six of the nine macronutrients—carbon, oxygen, hydrogen, nitrogen, sulfur, and phosphorus—make up almost 98% of a plant's dry weight. The other three macronutrients—calcium, potassium, and magnesium—make up another 1.5%.

What does a plant do with macronutrients? Carbon, oxygen, and hydrogen are the basic ingredients of a plant's organic compounds. Nitrogen is a component of all nucleic acids and proteins, as well as ATP, chlorophyll (the key light-absorbing molecule in plants), and many plant hormones. Sulfur is a component of most proteins. Phosphorus is a major component of nucleic acids, phospholipids, and ATP. The other macronutrients, though present in smaller amounts, play similarly important roles. For example, magnesium is an essential component of chlorophyll.

Micronutrients

Eight micronutrients make up the remaining 0.5% of a plant's dry weight and are required by all plants. The eight micronutrients are iron, chlorine, copper, manganese, zinc, molybdenum, boron, and nickel. A plant recycles the atoms of micronutrients over and over, so it needs them in only minute quantities. Yet a deficiency of any micronutrient can kill a plant.

Fertilizers

The quality of soil, especially the availability of nutrients, affects the health of plants and, for plants we eat, the quality of our own nutrition. Nitrogen shortage is the most common nutritional problem for plants. Nitrogen-deficient crop plants (**Figure 29.3**) may produce grain, but the grain will have a lower nutritional value, and its nutrient deficiencies will be passed on to livestock or human consumers.

Fertilizers are compounds given to plants via the soil to promote the plants' growth. There are two types of fertilizers, inorganic and organic. Inorganic fertilizers contain simple inorganic minerals. They come in a wide variety of formulations, but most emphasize the "N-P-K ratio," the relative amounts of the three nutrients most often deficient in depleted soils: nitrogen (N), phosphorus (P), and potassium (K).

Organic fertilizers are composed of chemically complex organic matter. One source of organic fertilizer is **compost**, a soil-like mixture of decomposed organic matter. Many gardeners maintain a free-standing compost pile or an enclosed compost bin to which they add leaves, grass clippings, yard waste, and kitchen scraps. Over time, the vegetable matter is broken down by microbes, fungi, and animals (**Figure 29.4**). The compost can be applied to outdoor gardens or indoor pots.

Compost and other organic fertilizers are a key part of **organic farming**. Organic farmers follow guidelines that promote sustainable agriculture—a system of farming that protects biological diversity, maintains and replenishes soil quality (as by crop rotation), manages pests with no (or few) synthetic pesticides, avoids genetically modified organisms, conserves water, and uses no (or few) synthetic fertilizers. In the United States, organic farming is one of the fastest-growing segments of agriculture. ☑

▼ Figure 29.3 **The effect of nitrogen availability on corn growth.** This photo shows two experimental corn crops. The plants on the left are growing in soil rich in the nitrogen-containing compounds they use to build proteins. The smaller, lighter-colored plants on the right are growing in nitrogen-deficient soil.

▼ Figure 29.4 **Compost.** Steam is produced by the metabolic activity of naturally occurring organisms that break down the organic matter of rotting vegetation.

☑CHECKPOINT

What inorganic nutrients do plants acquire from the air? What nutrients do plants extract from the soil?

Answer: carbon dioxide; water and minerals (inorganic ions)

From the Soil into the Roots

A plant absorbs water and essential nutrients from the soil through its roots. Plant roots have a remarkable capacity for extracting nutrients from soil because of their root hairs, extensions of epidermal cells that dramatically increase the surface area available for absorption (**Figure 29.5**). For example, the root hairs of a single sunflower plant, if laid end to end, could stretch many miles, providing a huge surface area in contact with nutrient-containing soil.

All substances that enter a plant root are dissolved in water. To be transported throughout the plant, water and solutes must move from the soil through the epidermis and cortex of the root, then into the water-conducting xylem tissue in the root's central cylinder (see Figure 28.8). To reach the xylem, the solution must pass through the plasma membranes of root cells. Because these membranes are selectively permeable, only certain solutes reach the xylem. This selectivity helps regulate the mineral composition of a plant's vascular system.

Many plants gain even more absorptive surface than their root hairs provide through symbiotic associations with fungi (**Figure 29.6**). Together, the plant's roots and the associated fungus form a mutually beneficial (mutualistic) structure called a **mycorrhiza** (plural, *mycorrhizae*). Fungal filaments around the roots absorb water and minerals much more rapidly than the roots alone can. Some of the water and minerals taken up by the fungus are transferred to the plant, while the plant's photosynthetic products nourish the fungus.

▼ **Figure 29.5 Root hairs that enable efficient absorption.** The root hairs of a plant, clearly visible in this photograph of a radish seedling, are extensions of epidermal cells on the outer surface of the root.

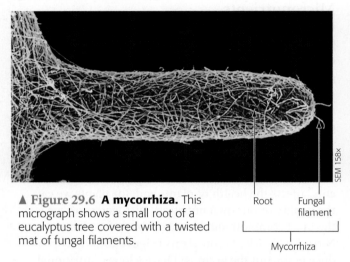

▲ **Figure 29.6 A mycorrhiza.** This micrograph shows a small root of a eucalyptus tree covered with a twisted mat of fungal filaments.

Root   Fungal filament

Mycorrhiza

SEM 158x

The Role of Bacteria in Nitrogen Nutrition

Air penetrates the soil around roots, but plants cannot use the form of nitrogen found in air (gaseous N_2). For plants to absorb nitrogen from the soil, N_2 must first be converted to ammonium ions (NH_4^+) or nitrate ions (NO_3^-). Most plants rely on bacteria to supply them with usable nitrogen in these forms.

Soil Bacteria and Nitrogen

Figure 29.7 shows three types of soil bacteria that play an essential role in supplying plants with nitrogen. One type, called nitrogen-fixing bacteria (⬮) converts atmospheric N_2 to ammonium, a process called **nitrogen fixation**. A second type, called ammonifying bacteria (⬮), adds to the soil's supply of ammonium by decomposing organic matter. A third type of soil bacteria, called nitrifying bacteria (⬮), converts soil ammonium to nitrate. Plants take up most of their nitrogen in this form. They then convert nitrate back to ammonium, which plants can then incorporate into proteins and other nitrogen-containing organic molecules.

▼ **Figure 29.7 The role of bacteria in supplying nitrogen to plants.**

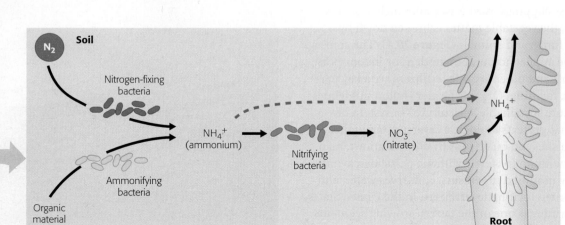

Root Nodule Bacteria and Nitrogen

Some plant families, including legumes—peas, beans, peanuts, and many other plants that produce their seeds in pods—have their own built-in source of ammonium: nitrogen-fixing bacteria that live in swellings called root nodules (**Figure 29.8**). Within these nodules, plant cells have been "infected" by nitrogen-fixing bacteria that reside inside cytoplasmic vesicles. The symbiotic relationship between a plant and its nitrogen-fixing bacteria is mutually beneficial. The bacteria have enzymes that help convert atmospheric N_2 to ammonium ions (NH_4^+); the plant, in turn, provides the bacteria with other nutrients. When conditions are favorable, root nodule bacteria fix so much nitrogen that the nodules secrete excess ammonia, which increases the fertility of the soil. This is one reason farmers practice crop rotation; legumes planted in alternate years can be plowed under to decompose into "green manure," reducing the need for manufactured fertilizer. ✔

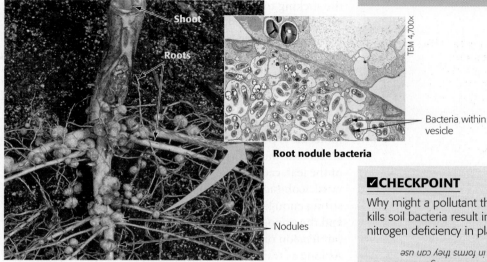

▼ Figure 29.8 **Root nodules that contain nitrogen-fixing bacteria.** Legumes, such as the soybean plant, have root nodules that contain nitrogen-fixing bacteria.

Shoot

Roots

TEM 4,700x

Bacteria within vesicle

Root nodule bacteria

Nodules

Root nodules

The Transport of Water

As a plant grows upward toward sunlight, it needs an increasing supply of resources from the soil. To thrive, a plant must be able to transport water and dissolved ions from its roots to the rest of the plant.

We saw in Chapter 28 that mature water-conducting cells of the xylem are arranged end to end to form very thin vertical tubes (see Figure 28.11d). A solution of water and inorganic nutrients, called **xylem sap**, flows through these tubes all the way from a plant's roots to the tips of its leaves. Let's look at how this sap is moved through the body of the plant.

The Ascent of Xylem Sap

What force moves xylem sap up against the downward pull of gravity? For the most part, xylem sap is pulled upward by **transpiration**, the loss of water from the leaves of a plant by evaporation (**Figure 29.9**). Most transpiration occurs through the stomata of leaves. The stomata open into air spaces, which are filled with water molecules that have evaporated from the surrounding mesophyll cells. Water vapor diffuses out of the stomata because the surrounding air is usually drier than the inside of the leaf.

Transpiration can pull xylem sap up a tree because of two special properties of water: cohesion and adhesion. **Cohesion** is the sticking

▼ Figure 29.9 **How transpiration pulls water up the xylem of a tree.**

Flow of water

Leaf

Xylem sap
Mesophyll cells
Air space within leaf
Stoma
Water molecule
Outside air

Transpiration

Stem

Adhesion
Water molecule
Cell wall
Xylem cells
Cohesion, by hydrogen bonding

Cohesion and adhesion in the xylem

Root

Xylem sap
Root hair
Soil particle
Water

Water uptake from soil

BioFlix™

Water Transport in Plants

MP3 tutor sessions

Transpiration

together of molecules of the same kind. In the case of water, hydrogen bonds make the H_2O molecules stick to one another (as described in Chapter 2). **Adhesion** is the sticking together of molecules of different kinds. Water molecules tend to adhere to cellulose molecules in the walls of xylem cells. Together, adhesion and cohesion create a continuous string of water molecules that stick to each other and to the inside walls of the xylem tubes (see the middle of Figure 29.9).

Before a water molecule can exit the leaf, it must break off from the top of the string. In effect, the water molecule is pulled off by a concentration gradient: The air outside the leaf is much drier than the moist interior of the leaf, causing the water molecule to diffuse outward. Cohesion resists this pulling force but is not strong enough to overcome it. The molecule breaks off, and the opposing forces of cohesion and transpiration put tension on the rest of the string of water molecules. As long as transpiration continues, the string is kept tense and is pulled upward as one molecule exits the leaf and the one behind it is tugged up into its place. Biologists call this explanation for the ascent of xylem sap the **transpiration-cohesion-tension mechanism**. Transpiration exerts a pull on a tense string of water molecules that is held together by cohesion and held up by adhesion.

Note that the transport of xylem sap requires no energy expenditure by the plant. Physical properties—cohesion, adhesion, and the evaporating effect of the sun—move water and dissolved minerals from a plant's roots to its shoots. Transpiration is thus a highly efficient means of moving a lot of water upward through the body of a plant.

You have probably tasted xylem sap moved via transpiration yourself. When maple trees resume growth in the spring after a leafless winter, starch that was produced the previous summer and stored in the roots is converted to sugar, which is transported upward in the xylem sap via transpiration to nourish the developing leaf buds. This sap can be harvested in buckets and boiled to concentrate the sugar about 50-fold to produce maple syrup.

The Regulation of Transpiration by Stomata

Transpiration works both for and against plants. Although it helps to transport materials from the soil to the rest of the plant, it can result in the plant's losing an astonishing amount of water. Transpiration is usually greatest on days that are sunny, warm, dry, and windy, because these weather conditions increase evaporation. An average-sized maple tree (about 20 m high) can lose more than 200 L of water per hour during a summer day. As long as water moves up from the soil fast enough to replace the lost water, this presents no problem. But if the soil dries out and transpiration exceeds the delivery of water, the leaves will wilt. Unless the soil and leaves are rehydrated, the plant will eventually die.

The leaf stomata, which can open and close, are evolutionary adaptations that help plants adjust their transpiration rates to changing environmental conditions. Stomata are usually open during the day, which allows CO_2 to enter the leaf from the atmosphere and thus keep photosynthesis going when sunlight is available. However, the plant also loses water through transpiration. At night, when there is no light for photosynthesis and therefore no need for CO_2, the stomata are closed, saving water. Stomata may also close during the day if a plant is losing water too fast. The opening and closing of stomata is controlled by the changing shape of the two guard cells flanking each stoma (**Figure 29.10**). ☑

The Transport of Sugars

A plant's survival depends not only on the transport of water and minerals absorbed from the soil, but also on the transport of the sugars it makes by photosynthesis. This is the main function of phloem.

Phloem consists of living food-conducting cells arranged end to end into tubes. Through perforations in the end walls of these cells, a sugary solution called **phloem sap** moves freely from one cell to the next

☑**CHECKPOINT**

1. Water moves from the roots to the tips of the leaves without any expenditure of energy by the plant. What is this process called? What two forces between molecules are responsible?

2. What is the evolutionary advantage of stomata that close at night?

Answer: 1. transpiration; cohesion and adhesion 2. Closed stomata prevent water loss when photosynthesis cannot occur.

Guard cells

Stoma open during daytime

CO_2 H_2O

Stoma closed at night

▶ Figure 29.10 **How stomata control transpiration via guard cells.** A pair of guard cells surrounding each stoma can change shape in response to environmental signals, regulating the rate of transpiration and CO_2 uptake.

(Figure 29.11). Phloem sap may contain inorganic ions, amino acids, and hormones in transit from one part of the plant to another, but its main solute is usually the disaccharide sugar sucrose.

In contrast to xylem sap, which only flows upward from the roots, phloem sap moves throughout the plant in various directions. A location in a plant where sugar is being produced (either by photosynthesis or by the breakdown of stored starch) is called a **sugar source**. A receiving location in the plant, where the sugar will be stored or consumed, is called a **sugar sink**. Sugar moves within food-conducting tubes of phloem from a source, such as a leaf, to a sink, such as a root or a fruit.

What causes phloem sap to flow from a source to a sink? **Figure 29.12** uses a beet plant to illustrate a widely accepted model called the **pressure-flow mechanism**. At the sugar source (the beet leaves), ❶ sugar is loaded from a photosynthetic cell into a phloem tube via active transport. Sugar loading at the source end raises the solute (sugar) concentration inside the phloem tube. ❷ The high solute concentration draws water into the tube by osmosis, usually from the xylem. The flow of water from the xylem into the phloem raises the water pressure at the source end of the phloem tube.

At the sugar sink (the beet root), both sugar and water leave the phloem tube. ❸ As sugar leaves the phloem, lowering the sugar concentration at the sink

end, ❹ water moves by osmosis back into the xylem. The exit of water lowers the water pressure in the tube.

The building of water pressure at the source end of the phloem tube and the reduction of water pressure at the sink end cause phloem sap to flow from source to sink. This pressure-flow mechanism explains why phloem sap always flows from a sugar source to a sugar sink, regardless of their locations in the plant.

We now have a broad picture of how a plant absorbs substances from the soil and transports materials from one part of its body to another: Water and inorganic ions enter from the soil and are pulled upward through xylem by transpiration. Carbon dioxide enters leaves through the stomata and is incorporated into sugars, which are distributed by phloem. Pressure flow drives the phloem sap from leaves and storage sites to other parts of the plant, where the sugars are used or stored. In the next section, we'll consider various ways that a working plant uses these resources. ✔

☑ **CHECKPOINT**

In the pressure-flow mechanism, xylem sap always flows from a _____ to a _____.

Answer: *sugar source; sugar sink*

▼ Figure 29.11 **Food-conducting cells of phloem.** The main function of phloem is to transport sugars through living food-conducting cells arranged into tubes. Phloem sap moves through openings between the cells.

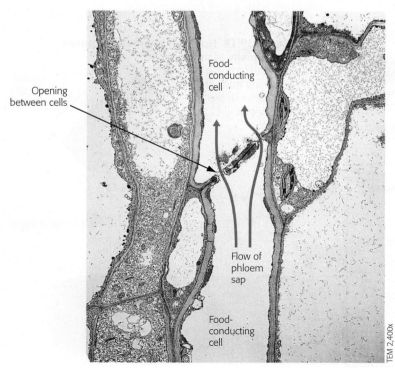

Opening between cells

Food-conducting cell

Flow of phloem sap

Food-conducting cell

TEM 2,400x

▼ **Figure 29.12 Pressure flow in plant phloem.** The red dots in the phloem tube represent sugar molecules; notice that a concentration gradient decreasing from top to bottom creates a pressure flow (wide red arrow). The wide blue arrow represents a parallel gradient of water pressure due to transpiration in the xylem sap; water pressure decreases from bottom to top. Xylem tubes transport the water back from sink to source.

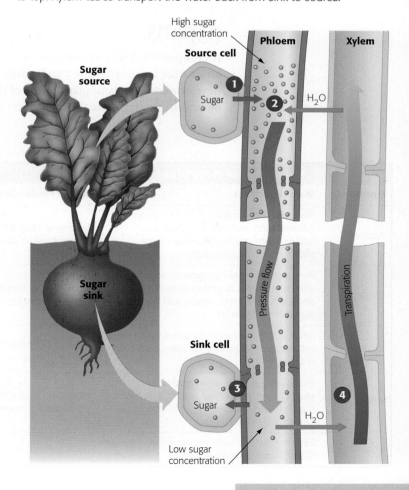

High sugar concentration

Phloem

Xylem

Source cell

Sugar source

Sugar

❶

❷

H_2O

Sugar sink

Pressure flow

Transpiration

Sink cell

❸

Sugar

❹

H_2O

Low sugar concentration

Plant Hormones

As you learned in Chapter 25, a **hormone** is a chemical signal produced in one part of the body and transported to other parts, where it acts on target cells to change their functioning. Plants produce hormones in very small amounts, but a minute amount of any of these chemicals can have profound effects on target cells. Hormones control plant growth and development by affecting the division, elongation, and differentiation of cells.

Plant biologists have identified five major types of plant hormones, which are listed in **Table 29.1**. Each type of hormone can produce a variety of effects, depending on the species and developmental stage of the plant as well as the hormone's concentration and site of action. In most situations, a plant hormone does not act alone. Instead, it is the relative concentration of two or more hormones that controls a plant's growth and development.

▲ Figure 29.13 **A houseplant growing toward light.** Phototropism guides growing plants toward the sunlight needed for photosynthesis.

Auxins

A group of related hormones called **auxins** are responsible for a wide range of growth and development effects in plants. One of the main effects of auxins is to promote the elongation of cells.

Phototropism and Cell Elongation

Phototropism is the directional growth of a plant shoot in response to light. For example, a houseplant on a windowsill grows toward sunlight (**Figure 29.13**). If you rotate the plant, it will soon reorient its growth until its leaves again face the window. Phototropism directs both growing seedlings and the shoots of mature plants toward the sunlight they use for photosynthesis.

How does a plant grow in a particular direction? Microscopic observations of growing plants reveal the cellular mechanism that underlies phototropism (**Figure 29.14**). Cells on the darker side of the stem are larger—actually, they have elongated faster—than those on the brighter side, causing the shoot to bend toward the light. If a seedling is illuminated uniformly from all sides or if it is kept in the dark, the cells all elongate at a similar rate and the seedling grows straight upward.

What causes plant cells on the dark side of a shoot to grow faster than those on the bright side? Our present understanding of this phenomenon emerged from a series of classic experiments conducted by two scientists with a very familiar name.

▼ Figure 29.14 **How cell elongation causes phototropism.** In this figure, a grass seedling curves toward light coming from one side because of the faster elongation of cells on the shaded side.

Shaded side of shoot

Light

Illuminated side of shoot

Table 29.1	Major Types of Plant Hormones	
Hormone	**Major Functions**	**Where Produced or Found in Plant**
Auxins	Stimulate stem elongation; affect root growth, differentiation, and branching; stimulate development of fruit, apical dominance, phototropism, and gravitropism	Meristems of apical buds, young leaves, embryos within seeds
Ethylene	Promotes fruit ripening; opposes some auxin effects; promotes or inhibits growth and development of roots, leaves, and flowers, depending on species	Ripening fruit, nodes of stems, aging leaves and flowers
Cytokinins	Affect root growth and differentiation; stimulate cell division and growth; stimulate germination; delay aging	Made in roots and transported to other organs
Gibberellins	Promote seed germination, bud development, stem elongation, and leaf growth; stimulate flowering and fruit development; affect root growth and differentiation	Meristems of apical buds and roots, young leaves, embryos
Abscisic acid	Inhibits growth; closes stomata when water is scarce; helps maintain dormancy	Leaves, stems, roots, green fruit

Do Chemical Signals Affect Plant Growth?

In the late 1800s, Charles Darwin and his son Francis performed some of the earliest experiments on phototropism (**Figure 29.15**). They began with the **observation** that grass seedlings bent toward light only if the tips of their shoots were present. This led them to **question** whether the tips of the seedlings produced some kind of growth signal. Their **hypothesis** was that the plant tips sensed light and produced a growth signal in response. They made a **prediction**: Removing a shoot tip or blocking its access to light would prevent phototropism. Several of their **experiments** are summarized in Figure 29.15.

The Darwins' **results** showed that removing the tip of a grass shoot did prevent growth toward light. The shoot also remained straight when they placed an opaque cap on its tip. However, the shoot curved normally when they placed a transparent cap on its tip or an opaque shield around its base. The Darwins concluded that the tip of the shoot was responsible for sensing light. They also recognized that the growth response, the bending of the shoot, occurred below the tip. Therefore, they speculated that some unknown growth signal was transmitted downward from the tip to the growing region of the shoot.

Building on the Darwins' results, botanists have discovered the mechanisms of many other plant hormones. This understanding is one of the cornerstones of modern agriculture, allowing people to control the growth of plants in a variety of ways (as we'll see throughout the chapter).

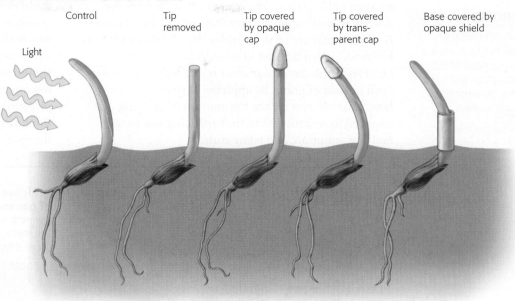

▲ Figure 29.15 **The Darwins' classic experiments on phototropism.** This figure summarizes experiments conducted by Charles Darwin and his son Francis. Together, these results suggested that a substance produced by the tip of a seedling controls phototropism.

The Action of Auxins

The hormones responsible for the phototropism observed by the Darwins are auxins produced by the apical meristem at the tip of a shoot. Illuminating one side of a shoot results in a higher auxin concentration in the cells on the dark side. This uneven distribution of auxins makes cells on the dark side elongate, causing the shoot to bend.

Auxins promote cell elongation in stems only when they are present within a certain concentration range. Interestingly, this same concentration range inhibits cell elongation in roots. Above this range, auxins inhibit cell elongation in stems; and below this range, they cause

root cells to elongate. This complex set of effects reinforces two points: (1) The same hormone may have different effects at different concentrations in the same target cell, and (2) a particular concentration of a hormone may have different effects on different target cells.

In addition to causing stems and roots to lengthen, auxins induce cell division in the vascular cambium, thus promoting growth in stem diameter (see Figure 28.16). Furthermore, auxins made in developing seeds promote the growth of fruit. Farmers sometimes produce seedless tomatoes, cucumbers, and eggplants by spraying unpollinated plants with synthetic auxins. And this isn't the only way auxins are used in agriculture.

Agricultural Applications

The discovery of plant hormones has produced many advances in agriculture. The use of synthetic plant hormones allows more food to be produced at lower cost. For example, one of the most widely used herbicides, or weed killers, is a synthetic auxin that disrupts the normal balance of hormones that regulate plant growth. Because dicots are more sensitive than monocots to this herbicide, it can be used to selectively remove dandelions and other broadleaf dicot weeds from a monocot lawn or field of grain. By applying herbicides to cropland, a farmer can reduce the amount of plowing required to control weeds, thus reducing soil erosion, fuel consumption, and labor costs.

At the same time, there is growing concern that the heavy use of artificial chemicals in food production may pose environmental and health hazards. Also, many consumers are concerned that foods produced using artificial hormones may not be as tasty or nutritious as those raised naturally. This highlights an important question about modern agricultural techniques: Should we continue to produce cheap, plentiful food using artificial chemicals and tolerate the potential problems, or should we put more of our agricultural effort into organic farming, recognizing that foods may be less plentiful and more expensive as a result? This is something you need to consider as you make choices about the foods you buy and eat. ✔

Ethylene

Ethylene is a hormone, released as a gas, that triggers a variety of aging responses in plants, including fruit ripening and dropping of leaves.

Fruit Ripening

A burst of ethylene production in a fruit triggers its ripening. Because ethylene is a gas, the signal to ripen spreads from fruit to fruit: One bad apple really can spoil the bunch! You can make some fruits ripen faster if you store them in a plastic bag so that the ethylene gas accumulates (**Figure 29.16**). On a commercial scale, many kinds of fruit—tomatoes, for instance—are often picked green and then partially ripened in huge storage bins into which ethylene gas is piped. In other cases, growers take measures to retard the ripening action of natural ethylene. Stored apples are often flushed with CO_2, which inhibits the action of ethylene. In this way, apples picked in autumn can be stored for sale the following summer.

Leaf Drop

The loss of leaves in autumn is affected by ethylene. Leaf drop is triggered by environmental stimuli, including the shortening days of autumn and cooler temperatures. These stimuli cause a change in the balance of ethylene and auxins that weakens cell walls in a layer of cells at the base of the leaf stalk. The weight of the leaf, often assisted by wind, causes this layer to split, releasing the leaf. ✔

Cytokinins

Cytokinins are a group of closely related hormones that act as growth regulators. These regulators promote cell division, or cytokinesis. Cytokinins are produced in

▼ **Figure 29.16 The effect of ethylene on the ripening of bananas.** These three photos show the results of a fruit-ripening demonstration. Three unripe bananas were stored for the same time period in plastic bags. As you can see, the more ethylene present, the riper (darker) the banana.

Banana alone

Banana + ethylene-releasing orange

Banana + beaker of an ethylene-releasing chemical

actively growing tissues, particularly in roots, embryos, and fruits.

Cytokinins stimulate the growth of axillary buds, making a plant grow more branches and become bushy. Christmas tree growers sometimes use cytokinins to produce attractive branching. Cytokinins entering the shoot system from the roots counter the inhibitory effects of auxins coming down from the terminal buds. The complex growth patterns of most plants probably result from the relative concentrations of auxins and cytokinins.

Gibberellins

Gibberellins are another group of growth-regulating hormones. Roots and young leaves are major sites of gibberellin production. One of the main effects of gibberellins is to stimulate cell elongation and cell division in stems and leaves. This action generally enhances that of auxins. Also in combination with auxins, gibberellins can influence fruit development, and gibberellin-auxin sprays can make apples, currants, and eggplants develop without pollination and seed production. Gibberellins are used commercially to make seedless grapes grow larger and farther apart in a cluster (**Figure 29.17**).

Gibberellins are also important in seed germination in many plants. Many seeds that require special environmental conditions to germinate, such as exposure to light or cold, will germinate on demand when sprayed with gibberellins. In nature, gibberellins in seeds are probably the link between environmental cues and the metabolic processes that reactivate growth of the embryo after a period of dormancy. For example, when water becomes available to a grass seed, it causes the embryo in the seed to release gibberellins, which promote germination by mobilizing nutrients stored in the seed.

▼ **Figure 29.17 The effect of gibberellins on grapes.** The right cluster of grapes shows the effect of gibberellin treatment: larger grapes farther apart in the cluster.

Untreated

Treated with gibberellins

Abscisic Acid

Unlike the growth-stimulating hormones we have studied so far, **abscisic acid** generally slows growth. One of the times in a plant's life when it is advantageous to suspend growth is at the onset of seed dormancy (inactivity of the seed). Seed dormancy is an evolutionary adaptation that ensures that a seed will germinate only when there are appropriate conditions of light, temperature, and moisture.

Many types of dormant seeds will germinate only when abscisic acid is removed or inactivated in some way. The seeds of some plants require prolonged exposure to cold during the winter to inactivate abscisic acid and trigger seed germination in the spring. The seeds of some desert plants remain dormant until a downpour washes out the abscisic acid, allowing them to germinate when water is available for plant growth (**Figure 29.18**). For many plants, the ratio of abscisic acid (which inhibits seed germination) to gibberellins (which promote seed germination) determines whether the seed will remain dormant or germinate. ✓

▼ **Figure 29.18 The effect of abscisic acid removal on seed dormancy.** These sand verbena and dune evening primrose grew in the Anza Borrego Desert State Park, in California, just after a hard rain washed abscisic acid out of the seeds, allowing them to germinate.

☑ **CHECKPOINT**

Which two hormones regulate seed dormancy and germination? What are their opposing effects?

Answer: Abscisic acid maintains seed dormancy; gibberellins promote germination.

Response to Stimuli

In this section, we'll consider how a plant responds to physical stimuli from the environment—including light, touch, and gravity.

Tropisms

Tropisms are directed growth responses that cause parts of a plant to grow toward or away from a stimulus (**Figure 29.19**). Phototropism, as you saw in Figure 29.13, is a particularly important example, but there are other directed growth responses. For example, **thigmotropism** is a response to touch, as when a pea tendril (actually a modified leaf) contacts a string or wire and coils around it for support. The tendril in the center photo in Figure 29.19 grew straight until it touched the support. Contact then stimulated the cells to grow at different rates on opposite sides of the tendril (slower in the contact area), making the tendril coil around the wire.

Gravitropism is the directional growth of a plant organ in response to gravity: Shoots grow upward and roots grow downward, regardless of the orientation of the seed. Auxins play an important role in gravitropism. The seedling shown on the right in Figure 29.19 responded to gravity by redistributing auxins to the lower side of its horizontally growing root. A high concentration of auxins inhibits the growth of root cells. As growth of the lower side of the root was slowed, cells on the upper side continued to elongate normally, and the root curved downward.

Photoperiod

In addition to providing energy for photosynthesis and directing growth, light helps regulate a plant's life cycle. Flowering, seed germination, and the onset and ending of dormancy are stages in plant development that usually occur at specific times of the year. The environmental stimulus that plants most often use to detect the time of year is called **photoperiod**, the relative lengths of day and night.

Plants whose flowering is triggered by photoperiod fall into two groups: long-night plants and short-night plants (**Figure 29.20**). (Historically, botanists erroneously referred to these plants as short- and long-day plants; we now know that it is night length, not day length, that affects a plant's flowering.) Long-night plants, such as chrysanthemum and poinsettia, generally flower in late summer, fall, or winter, when nights lengthen. Short-night plants, such as lettuce, iris, and many cereal grains, usually flower in late spring or early summer, when nights are briefer. Some plants, such as dandelions, are night-neutral; their flowering is unaffected by photoperiod. Florists apply knowledge of the photoperiod of particular plants to bring us flowers out of season. The blooming of chrysanthemums, for instance, can be stalled until spring by interrupting each long night with a flash of light, thus turning one long night into two short nights. ☑

☑**CHECKPOINT**

1. A common houseplant called the "sensitive plant" (*Mimosa pudica*) has leaves that curl when touched. This is an example of what kind of growth response?
2. A particular long-night plant won't flower in the spring. You try to induce flowering by using a short, dark interruption to split a long-day period of spring into two short-day periods. What result do you predict?

Answers: **1.** *thigmotropism* **2.** *The plants still won't flower, because it is actually night length, not day length, that controls flowering.*

▼ Figure 29.19 **Tropisms.**

TROPISMS
(directed growth responses)

Phototropism	Thigmotropism	Gravitropism

These seedlings bending toward the light are young cilantro plants (a common herb).

Thigmotropism, growth in response to touch, allows vining plants to wrap around support structures. In this photo, the tendril of a pea plant coils around a wire fence.

These seedlings were both germinated in the dark. The one on the left was left alone. The one on the right was turned on its side two days after germination so that the shoot and root were horizontal. When this photo was taken, the shoot had turned back upward and the root had turned down.

▼ Figure 29.20 **Photoperiod and flowering.** Photoperiod, the relative lengths of day and night, can trigger flowering.

Long-night plants

Flowering occurs only when the plant is exposed to a continuous dark period exceeding a critical length.

The continuity of darkness is important. The long-night plant will not blossom if the darkness is interrupted by even a flash of light.

Short-night plants

Flowering occurs when the night length is shorter than a critical length.

Flowering can be induced in a short-night plant by a flash of light during the night.

Darkness
Flash of light
Light

Time (hours)
24
0

Critical night length

EVOLUTION CONNECTION: The Interdependence of Organisms

Plants, Frogs, and People

Although plants make their own food molecules, they are not independent of other organisms. As you learned in the last chapter, most flowering plants depend on insects or other animals for pollination and seed dispersal. In this chapter, you saw that plants rely on organisms from two other kingdoms to help acquire nutrients: soil bacteria and the fungi of mycorrhizae. The fossil record shows that mycorrhizae have existed since plants first evolved. Indeed, the mycorrhizal connection may have altered the entire course of evolution by helping make possible the colonization of land.

Nearly all animals depend on plants or other photosynthetic organisms for food. For example, both the cover of this book and **Figure 29.21** show a leaping northern leopard frog. These frogs are found throughout Canada and the northern United States living in or near fresh water. As tadpoles, they depend on algae and aquatic plants for food. Adults eat a wide variety of insects and

are in turn eaten by snakes, raccoons, and other large animals—including humans (who consume the legs).

Thus, we close this chapter—and this book—illustrating one of biology's overarching themes: the interrelatedness of organisms. Animals depend on other animals and plants, which in turn depend on animals, fungi, and protists. And all eukaryotic life depends on prokaryotes. This interdependency reminds us once again that it is impossible to separate ourselves from all of the living creatures that share the biosphere. We hope that *Essential Biology with Physiology* has given you a new appreciation of your place in the living world.

▼ **Figure 29.21 The interdependence of organisms.** The northern leopard frog depends on protists, plants, and animals for food and is itself eaten by people and other animals. These relationships emphasize the interrelatedness of all life on Earth.

Appendix A Metric Conversion Table

MEASUREMENT	UNIT AND ABBREVIATION	METRIC EQUIVALENT	APPROXIMATE METRIC-TO-ENGLISH CONVERSION FACTOR	APPROXIMATE METRIC-TO-ENGLISH CONVERSION FACTOR
Length	1 kilometer (km)	= 1,000 (10^3) meters	1 km = 0.6 mile	1 mile = 1.6 km
	1 meter (m)	= 100 (10^2) centimeters	1 m = 1.1 yards	1 yard = 0.9 m
		= 1,000 millimeters	1 m = 3.3 feet	1 foot = 0.3 m
			1 m = 39.4 inches	
	1 centimeter (cm)	= 0.01 (10^{-2}) meter	1 cm = 0.4 inch	1 foot = 30.5 cm
				1 inch = 2.5 cm
	1 millimeter (mm)	= 0.001 (10^{-3}) meter	1 mm = 0.04 inch	
	1 micrometer (μm)	= 10^{-6} meter (10^{-3} mm)		
	1 nanometer (nm)	= 10^{-9} meter (10^{-3} μm)		
	1 angstrom (Å)	= 10^{-10} meter (10^{-4} μm)		
Area	1 hectare (ha)	= 10,000 square meters	1 ha = 2.5 acres	1 acre = 0.4 ha
	1 square meter (m²)	= 10,000 square centimeters	1 m² = 1.2 square yards	1 square yard = 0.8 m²
			1 m² = 10.8 square feet	1 square foot = 0.09 m²
	1 square centimeter (cm²)	= 100 square millimeters	1 cm² = 0.16 square inch	1 square inch = 6.5 cm²
Mass	1 metric ton (t)	= 1,000 kilograms	1 t = 1.1 tons	1 ton = 0.91 t
	1 kilogram (kg)	= 1,000 grams	1 kg = 2.2 pounds	1 pound = 0.45 kg
	1 gram (g)	= 1,000 milligrams	1 g = 0.04 ounce	1 ounce = 28.35 g
			1 g = 15.4 grains	
	1 milligram (mg)	= 10^{-3} gram	1 mg = 0.02 grain	
	1 microgram (μg)	= 10^{-6} gram		
Volume *(solids)*	1 cubic meter (m³)	= 1,000,000 cubic centimeters	1 m³ = 1.3 cubic yards	1 cubic yard = 0.8 m³
	1 cubic centimeter (cm³ or cc)	= 10^{-6} cubic meter	1 m³ = 35.3 cubic feet	1 cubic foot = 0.03 m³
	1 cubic millimeter (mm³)	= 10^{-9} cubic meter (10^{-3} cubic centimeter)	1 cm³ = 0.06 cubic inch	1 cubic inch = 16.4 cm³
Volume *(liquids and gases)*	1 kiloliter (kL or kl)	= 1,000 liters	1 kL = 264.2 gallons	1 gallon = 3.79 L
	1 liter (L)	= 1,000 millimeters	1 L = 0.26 gallon	1 quart = 0.95 L
			1 L = 1.06 quarts	
	1 milliliter (mL or ml)	= 10^{-3} liter	1 mL = 0.03 fluid ounce	1 quart = 946 mL
		= 1 cubic centimeter	1 mL = approx. $\frac{1}{4}$ teaspoon	1 pint = 473 mL
			1 mL = approx. 15–16 drops	1 fluid ounce = 29.6 mL
				1 teaspoon = approx. 5 mL
Volume *(liquids and gases)*	1 microliter (μl or μL)	= 10^{-6} liter (10^{-3} milliliters)		
Time	1 second (s)	= $\frac{1}{60}$ minute		
	1 millisecond (ms)	= 10^{-3} second		
Temperature	Degrees Celsius (°C)		°F = $\frac{9}{5}$°C + 32	°C = $\frac{5}{9}$(°F − 32)

Appendix B Periodic Table of Elements

Representative elements

Alkali metals

Alkaline earth metals

Transition elements

Halogens   Noble gases

Period number	Group 1A	Group 2A	3B	4B	5B	6B	7B	8B			1B	2B	Group 3A	Group 4A	Group 5A	Group 6A	Group 7A	Group 8A
1	H 1.008																	He 4.003
2	Li 6.941	Be 9.012											B 10.81	C 12.01	N 14.01	O 16.00	F 19.00	10 Ne 20.18
3	11 Na 22.99	12 Mg 24.31						10	11	12			13 Al 26.98	14 Si 28.09	15 P 30.97	16 S 32.07	17 Cl 35.45	18 Ar 39.95
4	19 K 39.10	20 Ca 40.08	21 Sc 44.96	22 Ti 47.87	23 V 50.94	24 Cr 52.00	25 Mn 54.94	26 Fe 55.85	27 Co 58.93	28 Ni 58.69	29 Cu 63.55	30 Zn 65.41	31 Ga 69.72	32 Ge 72.64	33 As 74.92	34 Se 78.96	35 Br 79.90	36 Kr 83.80
5	37 Rb 85.47	38 Sr 87.62	39 Y 88.91	40 Zr 91.22	41 Nb 92.91	42 Mo 95.94	43 Tc (98)	44 Ru 101.1	45 Rh 102.9	46 Pd 106.4	47 Ag 107.9	48 Cd 112.4	49 In 114.8	50 Sn 118.7	51 Sb 121.8	52 Te 127.6	53 I 126.9	54 Xe 131.3
6	55 Cs 132.9	56 Ba 137.3	57* La 138.9	72 Hf 178.5	73 Ta 180.9	74 W 183.8	75 Re 186.2	76 Os 190.2	77 Ir 192.2	78 Pt 195.1	79 Au 197.0	80 Hg 200.6	81 Tl 204.4	82 Pb 207.2	83 Bi 209.0	84 Po (209)	85 At (210)	86 Rn (222)
7	87 Fr (223)	88 Ra (226)	89† Ac (227)	10 Rf (261)	10 Db (262)	10 Sg (266)	10 Bh (264)	10 Hs (269)	10 Mt (268)	10 Ds (271)	11 — (272)	11 — (285)	11 — (284)	11 — (289)	11 — (288)			

*Lanthanides	58 Ce 140.1	59 Pr 140.9	60 Nd 144.2	61 Pm (145)	62 Sm 150.4	63 Eu 152.0	64 Gd 157.3	65 Tb 158.9	66 Dy 162.5	67 Ho 164.9	68 Er 167.3	69 Tm 168.9	70 Yb 173.0	71 Lu 175.0
†Actinides	90 Th 232.0	91 Pa 231.0	92 U 238.0	93 Np (237)	94 Pu (244)	95 Am (243)	96 Cm (247)	97 Bk (247)	98 Cf (251)	99 Es 252	10 Fm 257	10 Md 258	10 No 259	10 Lr 260

■ Metals   ■ Metalloids   ■ Nonmetals

NAME	SYMBOL	NAME	SYMBOL	NAME	SYMBOL	NAME	SYMBOL	NAME	SYMBOL
Actinium	Ac	Cobalt	Co	Iodine	I	Osmium	Os	Silicon	Si
Aluminum	Al	Copper	Cu	Iridium	Ir	Oxygen	O	Silver	Ag
Americium	Am	Curium	Cm	Iron	Fe	Palladium	Pd	Sodium	Na
Antimony	Sb	Darmstadtium	Ds	Krypton	Kr	Phosphorus	P	Strontium	Sr
Argon	Ar	Dubnium	Db	Lanthanum	La	Platinum	Pt	Sulfur	S
Arsenic	As	Dysprosium	Dy	Lawrencium	Lr	Plutonium	Pu	Tantalum	Ta
Astatine	At	Einsteinium	Es	Lead	Pb	Polonium	Po	Technetium	Tc
Barium	Ba	Erbium	Er	Lithium	Li	Potassium	K	Tellurium	Te
Berkelium	Bk	Europium	Eu	Lutetium	Lu	Praseodymium	Pr	Terbium	Tb
Beryllium	Be	Fermium	Fm	Magnesium	Mg	Promethium	Pm	Thallium	Tl
Bismuth	Bi	Fluorine	F	Manganese	Mn	Protactinium	Pa	Thorium	Th
Bohrium	Bh	Francium	Fr	Meitnerium	Mt	Radium	Ra	Thulium	Tm
Boron	B	Gadolinium	Gd	Mendelevium	Md	Radon	Rn	Tin	Sn
Bromine	Br	Gallium	Ga	Mercury	Hg	Rhenium	Re	Titanium	Ti
Cadmium	Cd	Germanium	Ge	Molybdenum	Mo	Rhodium	Rh	Tungsten	W
Calcium	Ca-	Gold	Au	Neodymium	Nd	Rubidium	Rb	Uranium	U
Californium	Cf	Hafnium	Hf	Neon	Ne	Ruthenium	Ru	Vanadium	V
Carbon	C	Hassium	Hs	Neptunium	Np	Rutherfordium	Rf	Xenon	Xe
Cerium	Ce	Helium	He	Nickel	Ni	Samarium	Sm	Ytterbium	Yb
Cesium	Cs	Holmium	Ho	Niobium	Nb	Scandium	Sc	Yttrium	Y
Chlorine	Cl	Hydrogen	H	Nitrogen	N	Seaborgium	Sg	Zinc	Zn
Chromium	Cr	Indium	In	Nobelium	No	Selenium	Se	Zirconium	Zr

Appendix C Credits

PHOTO CREDITS

UNIT OPENERS: Unit I Dr. Dennis Kunkel/Phototake NYC; **Unit II** William James Warren/Science Faction/Corbis; **Unit III** Gavin Kincome/SPL/Getty Images; **Unit IV** Michael Quinton/Minden Pictures; **Unit V** Scott McDermott/Corbis; **Unit VI** Jonathan Buckley/GAP Photos/Getty Images.

CHAPTER 1: Chapter opening photo Michael Boyny/PhotoLibrary; **1.1a** Nikreates/Alamy; **1.1b** P. Narayan/The Image Bank/AGE Fotostock; **1.1c** Gallo Images-Anthony Bannister/Getty Images; **1.1d** Michael Lander/NordicPhotos/AGE Fotostock; **1.1e** Kim Taylor and Jane Burton/Dorling Kindersley; **1.1f** Steve Bloom Images/Alamy; **1.1g** Gerry Ellis/Minden Pictures; **1.2.1** NASA/Goddard Space Flight Center; **1.2.2** Michio Hoshino/Minden Pictures; **1.2.4** Kevin Schafer/Corbis; **1.2.7** Manfred Kage/Peter Arnold; **1.3** Christopher Elwell/shutterstock; **1.4 left** S. C. Holt/Biological Photo Service; **right** Steve Gschmeisner/Photo Researchers; **1.6** Maximilian Stock Ltd/Photo Researchers; **1.7** Martin Strmiska/AGE Fotostock; **1.8 top to bottom** Steve Gschmeissner/SPL/Photo Researchers; Eye of Science/Photo Researchers; Neil Fletcher/Dorling Kindersley; Malcom Coulson/Dorling Kindersley; Stockbyte/Getty Images; D. P. Wilson/Photo Researchers; **1.9** James L. Amos/Corbis; **1.11 top left** Mike Hayward/Alamy; **top right** Neg./Transparency no. 330300. Courtesy Dept. of Library Services, American Museum of Natural History; **bottom** Tui De Roy/Minden Pictures; **1.13a left to right** iStockphoto; Paul Rapson/Alamy; Foodcollection.com/Alamy; Suzannah Skelton/iStockphoto; iStockphoto; iStockphoto; **1.13b left** Chris Collins/Corbis; **right** Eric Baccega/Nature Picture Library; **1.14 left** Michael Nichols/National Geographic Image Collection; **right** Tim Ridley/Dorling Kindersley, Courtesy of the Jane Goodall Institute, Clarendon Park, Hampshire; **1.17** Jim Newcomb; **1.18** AP Photo; **1.19 left** Simon Fraser/SPL/Photo Researchers; **right** S. Lowry/University of Ulster/Getty Images.

CHAPTER 2: Chapter opening photo moodboard/Corbis; **2.1 top** Clive Streeter/Dorling Kindersley; **bottom two** Dorling Kindersley; **2.3 left to right** Ivan Polunin/Bruce Coleman; Lukasz Panek/iStockphoto; Enrico Fianchini/iStockphoto; Howard Shooter/Dorling Kindersley; **2.9** Doug Allan/Minden Pictures; **2.10 left** Stephen Alvarez/NGS Image Collection; **right** Andrew Syred/Photo Researchers; **2.11** Piet Munsterman/Foto Natura/Minden Pictures; **2.12** NBAE/Getty Images; **2.13** Claudio Baldini/iStockphoto; **2.14** Kristin Piljay; **p. 31 bottom** John Kelly/Getty Images; **2.16 top to bottom** Beth Van Trees/shutterstock; Monika Wisniewska/iStockphoto; Feng Yu/iStockphoto; Jakub Semeniuk/iStockphoto; **2.17** J M Hall-Spencer; **2.18** NASA.

CHAPTER 3: Chapter opening photos, main Ian O'Leary/Dorling Kindersley; **inset** Dr. Tim Evans/SPL/Photo Researchers; **3.3 left** James M Phelps, Jr./shutterstock; **right** Dorling Kindersley; **3.5** Gerry Ellis/PhotoLibrary; **3.7** Andy Crawford/Dorling Kindersley; **3.8** Sandra Caldwell/iStockphoto; **3.9 top to bottom** Biophoto Associates/Photo Researchers; Courtesy of Dr. L. M. Beidler; Biophoto Associates/Photo Researchers; **3.10** Martyn F. Chillmaid/SPL/Photo Researchers; **3.12 left box** Hannamariah/

shutterstock; iStockphoto; Valentin Mosichev/shutterstock; Thomas M. Perkins/shutterstock; **right box** Dorling Kindersley; Oleksandr Staroseltsev/iStockphoto; Robyn Mackenzie/shutterstock; vladm/shutterstock; Multiart/shutterstock; iStockphoto; **3.13 left** EcoPrint/shutterstock; **right** Stockbyte/Getty Images; **3.14 left** Pool Getty Images/AP Wide World Photos; **right** Peter Foley/epa/Corbis; **3.15 left to right** Glow Images/Alamy; Dorling Kindersley; Dave King/Dorling Kindersley; Harry Taylor/Dorling Kindersley; Dorling Kindersley; Steve Gschmeissner/SPL/Photo Researchers; Kristin Piljay; **3.19** Eye of Science/Photo Researchers; **3.28** Brochard/Jupiter Images.

CHAPTER 4: Chapter opening photo SPL/Photo Researchers; **4.1 left to right** Dennis Kunkel/Phototake USA; Dennis Kunkel/Phototake USA; Microworks Color/Phototake USA; **4.2** Sinclair Stammers/SPL/Photo Researchers; **4.4** NIBSC/SPL/Photo Researchers; **4.7** CDC/SPL/Photo Researchers; **4.8 both** Courtesy of Richard Rodewald/Biological Photo Service; **4.10** Hybrid Medical Animation/Photo Researchers; **4.11** D. W. Fawcett/Photo Researchers; **4.13** Barry King/BPS; **4.15** SPL/Photo Researchers; **4.16** Daniel S. Friend; **4.17a** Michael Abbey/Photo Researchers; **4.17b** Dr. Jeremy Burgess/Photo Researchers; **4.18** Garry Cole/BPS; **4.19**: Courtesy of W.P. Wergin and E.H. Newcomb, University of Wisconsin/BPS; **4.20** Courtesy of Daniel S. Friend, Harvard Medical School; **4.21a** Dr. Torsten Wittmann/Photo Researchers; **4.21b** Dr. D. J. Patterson/SPL/Photo Researchers; **4.22 left to right** Dr. Dennis Kunkel/Phototake NYC; Steve Gschmeissner/SPL/Photo Researchers; Charles Daghlian/Photo Researchers; **4.23 top** The Advertising Archives; **bottom** Centers for Disease Control and Prevention.

CHAPTER 5: Chapter opening photo J. W. Shuler/SPL/Photo Researchers; **5.1** Alain Lecocq/Sygma/Corbis; **5.3a** Dave King/Dorling Kindersley; **5.3b** Russell Sadur/Dorling Kindersley; **5.8** PDB ID: 1BGM, Jacobson, R.H., Zhang, X.J., DuBose, R.F., Matthews, B.W.: Three-dimensional structure of beta-galactosidase from E. coli. Nature 369 pp761 (1994); **5.15** David Cook/blueshiftstudios/Alamy; **5.19** Jim Cummins/Corbis; **5.20** Peter B. Armstrong.

CHAPTER 6: Chapter opening photo David Stoecklein/Corbis; **6.1** Pete Oxford/Minden Pictures; **6.2 left to right** Elena Elisseeva/iStockphoto; Eric Isselée/shutterstock; Ultrashock/shutterstock; **6.3** Jim Cummins/Getty Images; **6.14** Chris Mole/shutterstock; **6.16 left** stanislaff/iStockphoto; **right** stocksnapp/shutterstock; **6.17** José Carlos Pires Pereira/iStockphoto.

CHAPTER 7: Chapter opening photo Jamie Grill/Getty Images; **7.1 left to right** Dorling Kindersley; Jeff Rotman/Nature Picture Library; Susan M. Barns, Ph.D.; **7.2 top to bottom** John Fielding/Dorling Kindersley; John Durham/Photo Researchers; Courtesy of W.P. Wergin and E.H. Newcomb, University of Wisconsin/BPS; **7.6** Jamie Grill/Getty Images; **7.7** Pete Turner/The Image Bank/Getty Images; **7.8** Photos by LQ/Alamy; **7.14 left** Dinodia/Pixtal/AGE Fotostock; **right** ImageDJ/Jupiter Images.

CHAPTER 8: Chapter opening photo, main Alan Watson/Dorling Kindersley; **inset** National Tropical Botanical Garden; **8.1 left to right** Dr Gopal Murti/Photo Researchers;

Claude Cortier/Photo Researchers; Biophoto Associates/Photo Researchers; Roger Steene/Image Quest; Eric J. Simon; **8.2 top to bottom** Eric Isselée/shutterstock; Christian Musat/shutterstock; Tony Wear/shutterstock; Edyta Pawlowska/iStockphoto; Eric Isselée/shutterstock; Milton H. Gallardo; **8.3** Ed Reschke/Peter Arnold; **8.4 top** A.L. Olins, University of Tennessee/BPS; **bottom** Biophoto Associates/Photo Researchers; **8.7 all** Conly L. Rieder, Ph.D.; **8.8a** Don W. Fawcett/Photo Researchers; **8.8b** Carolina Biological Supply/Phototake NYC; **8.10** Chris Carroll/Uppercut/Getty Images; **8.11** CNRI/SPL/Photo Researchers; **8.12** Eric McNatt/ Getty Images; **8.14** Ed Reschke/Peter Arnold; **8.17** David M. Phillips/Photo Researchers; **8.19** Photograph courtesy Dr. David Mark Welch; **8.22 top** Lauren Shear/Photo Researchers; **bottom** CNRI/SPL/Photo Researchers; **8.24** NHPA/Photoshot; **p. 143** Carolina Biological Supply/ Phototake.

CHAPTER 9: Chapter opening photo Magdalena Rehova/Alamy; **9.1** Hulton Archive/Getty Images; **9.4** Patrick Lynch/Alamy; **9.5** Bettmann/Corbis; **9.8** Martin Shields/Photo Researchers; **9.9a left to right** Tracy Morgan/Dorling Kindersley; Tracy Morgan/Dorling Kindersley; Eric Isselée/shutterstock; Andrew Johnson/iStockphoto; **9.10 left** Tracy Morgan/Dorling Kindersley; **right** Eric Isselée/shutterstock; **9.12 top left to right** James Woodson/Getty Images; Adrianna Williams/zefa/Corbis; Jose Luis Pelaez, Inc./Blend Images/Getty Images; **9.12 bottom left to right** John Evans/iStockphoto; Rubberball/Getty Images; Bruce Talbot/DK Stock/Getty Images; **9.13 left to right** Comstock Images/Jupiter Images; Bruce Talbot/DK Stock/Getty Images; Jose Luis Pelaez/Getty Images; **9.15** Michael Ciesielski Photography; **9.16** Rex Rystedt/Time & Life Pictures/Getty Images; **9.17** Eric Isselée/shutterstock; **9.20** Mauro Fermariello/Photo Researchers; **9.21 left** Oliver Meckes/Nicole Ottawa/Photo Researchers; **right** Eye of Science/Photo Researchers; **9.23:** Eric J. Simon; **9.30 top left** 4x6/iStockphoto; **top center** Yuri Arcurs/iStockphoto; **bottom left** Dave King/Dorling Kindersley; **bottom center** Jo Foord/Dorling Kindersley; **right** Andrew Syred/Photo Researchers; **9.32:** FPG International/Taxi/Getty Images; **9.33 top left to bottom right** Jerry Young/Dorling Kindersley; Nikolay Titov/iStockphoto; Dave King/Dorling Kindersley; Dave King/Dorling Kindersley; Tracy Morgan/Dorling Kindersley; Dave King/Dorling Kindersley; Jerry Young/Dorling Kindersley; Dave King/Dorling Kindersley; Dave King/Dorling Kindersley; Tracy Morgan/Dorling Kindersley; Le Loft 1911/shutterstock; **p. 171** USDA/APHIS Animal And Plant Health Inspection Service.

CHAPTER 10: Chapter opening main photo Somos/Veer/Getty Images; **inset** James Cavallini/SPL/Photo Researchers; **10.3 left** Barrington Brown/Photo Researchers; **right top and bottom** Courtesy of the Library of Congress; **10.9** Woodfall Wild Images/Photoshot; **10.12** Dr. Masaru Okabe, Research Institute for Microbial Diseases; **10.23** Ron Dahlquist/Pacific Stock; **10.24** Stephen Welstead/Corbis; **10.25** Oliver Meckes/Eye of Science/Photo Researchers; **10.26** Russell Kightley/SPL/Photo Researchers; **10.27** N. Thomas/Photo Researchers; **10.29** Hazel Appleton, Centre for Infections/Health Protection Agency/SPL/Photo Researchers; **10.30** Jeff Zelevansky/Reuters/Corbis; **10.32** National

Institute for Biological Standards and Control (U.K.)/Science Photo Library/Photo Researchers; **10.33** Will & Deni McIntyre/SPL/Photo Researchers; **10.34** Sinclair Stammers/Photo Researchers; **10.35** KHAM/Reuters/Corbis.

CHAPTER 11: Chapter opening photo Du Cane Medical Imaging Ltd/Photo Researchers; **11.1 left to right** Steve Gschmeissner/Photo Researchers; Jean Claude Revy - ISM/Phototake; ISM/Phototake; **11.4** Patrick A. Krost/iStockphoto; **11.9 left top & bottom** Edward B. Lewis, California Institute of Technology; **11.9 right top and bottom** F. Rudolf Turner, Indiana University; **11.11** Reproduced with permission from R.F. Service, Science (1998) 282:396-399. Copyright 1998 American Association for the Advancement of Science. Incyte Pharmaceuticals, Inc., Palo Alto, CA; **11.12** Robyn Mackenzie/shutterstock; **11.13** Courtesy of the Roslin Institute, Edinburgh; **11.14a** Pat Sullivan/AP Wide World Photos; **11.14b** Jim Curley. University of Missouri; **11.14c left to right** Pasqualino Loi and Grazyna Ptak; Advanced Cell Technology; Advanced Cell Technology; Lee Jae Woon/Reuters Limited; **11.16** Craig Hammell/Stock Market/Corbis; **inset** Courtesy of Cord Blood Registry; **11.21** Scott Camazine/PhototakeUSA; **11.22** CNRI/Photo Researchers.

CHAPTER 12: Chapter opening photo Tek Image/Photo Researchers; **12.1** AP Photo; **12.2** Eli Lilly; **12.3** Volker Steger/SPL/Photo Researchers; **12.4** Christopher and Sally Gable/Dorling Kindersley; **inset** AFP/Getty Images; **12.5** Peter Berger; **12.6** DNX/Peter Arnold; **12.7** Huntington Potter, University of South Florida College of Medicine; **inset** Prof. S. Cohen/Science Photo Library/Photo Researchers; **12.12** Hank Morgan/SPL/Photo Researchers; **12.14** Kim Sayer/Dorling Kindersley; **inset:** Michael Stephens/AP Photos; **12.15** Applied Biosystems; **12.20** Douglas Graham/Roll Call/Corbis; **12.21** Scott Camazine/SPL/Photo Researchers; **inset** UPI Photo; **12.22 top** Edyta Pawlowska/shutterstock; **bottom:** David Parker/Science Photo Library/Photo Researchers; **12.24** Phil Date/shutterstock; **12.25** ER Productions/Brand X/Corbis; **12.26** AFP/Getty Images; **12.27** 23andMe, Inc. 2009; **12.28** The Granger Collection, New York.

CHAPTER 13: Chapter opening photo Fred Bavendam/Minden Pictures; **13.1a** Ken G. Preston-Mafham/Animals Animals/Earth Scenes; **13.1b** P. Ward/Animals Animals/Earth Scenes; **13.1c** Edward S. Ross, California Academy of Sciences; **13.2 bottom left** Tui DeRoy/Minden Pictures; **top left** Natural History Museum, London; **top right** Darwin's Tree of life sketch, MS.DAR.121: p.36. Reproduced with the permission of the Cambridge University Library; **bottom right** Neg./Transparency no. 330300. Courtesy Dept. of Library Services, American Museum of Natural History; **13.3 left** George Richmond/ARCHIV/Photo Researchers; **right** Bettmann/Corbis; **13.4 left** Claus Meyer/Minden Pictures; **right** D. Parer, E. Parer-Cook/Minden Pictures; **13.5** Corbis; **13.6** Phillip Gingerich/Discover Magazine; **13.7 top left** Dorling Kindersley; **center** Planetary Visions Ltd./Photo Researchers; **top right** Bill Ling/Dorling Kindersley; **bottom left** Geoff Dann/Dorling Kindersley; **bottom right** Rob Reichenfeld/Dorling Kindersley; **13.9 left** Dr. Keith Wheeler/Photo Researchers; **right** Photo Lennart Nilsson/Albert Bonniers Forlag AB; **13.11a** Tui DeRoy/Photoshot Holdings; **13.11b** Mark

Moffett/Minden Pictures; **13.11c** Tui DeRoy/Photoshot Holdings; **13.12** Michael & Patricia Fogden/Minden Pictures; **13.13** Laura Jesse, Extension Entomologist, Iowa State University; **13.15a-b** Reproduced with permission from How the horned lizard got its horns. Young K. V., Brodie E. D. Jr., Brodie E.D. 3rd. 2 April 2004, Vol. 304(5667):65. Brevia, Fig 1. American Association for the Advancement of Science.American Association for the Advancement of Science; **13.17a** Photo Alto/Getty Images; **13.17b** Earth Imaging/Stone/Getty Images; **13.18** Edmund D. Brodie III; **13.19** Andy Levin/Photo Researchers; **13.21** Anne Dowie; **13.24** Martin Harvey /Gallo Images/Getty Images; **13.25** Hulton-Deutsch Collection/Corbis; **13.26** Time Magazine; **13.27** Michael & Patricia Fogden/Minden Pictures; **13.29a** Reinhard/ARCO/Nature Picture Library; **13.29b** John Cancalosi/agefotostock; **13.30** Bill Longcore/Photo Researchers.

CHAPTER 14: Chapter opening photo Joe McDonald/Corbis; **14.2a left** Bill Draker/Rolfnp/Alamy; **right** David Kjaer/Nature Picture Library; **14.2b top left to right** Comstock Complete; Photos.com; Phil Date/shutterstock; **14.2b bottom left to right** Robert Kneschke/iStockphoto; Comstock Complete; Comstock Complete; **14.4 p. 272 left to right** USDA; Stephen J. Krasemann/Photo Researchers; Joe McDonald/Photoshot Holdings Ltd; Joe McDonald/Corbis; **14.4 p. 273 left to right** Wolfgang Kaehler/Corbis; Takahito Asami; William E. Ferguson; **14.5 left** Charles W. Brown; **14.5 center top to bottom** Ralph A. Reinhold/Animals Animals/Earth Scenes; Photodisc/Getty Images; Alistair Duncan/Dorling Kindersley; **14.5 right** Kazutoshi Okuno; **14.7** Corbis; **top left inset** John Shaw/Bruce Coleman; **top right inset** Michael Fogden/Bruce Coleman; **14.9** Grant Dixon/Getty Images; **14.11** Chip Clark; **14.12** Stephen Dalton/Photo Researchers; **14.13 both** Photodisc/Getty Images; **14.14 top left** Tom Bean/Corbis; **top right** Georg Gerster/Photo Researchers; **center** Francois Gohier/Photo Researchers; **bottom left** Martin Lockley; **bottom right** F. Latreille/Cerpolex; **14.16** Will & Deni McIntyre/Photo Researchers; **14.18 right** Mark Pilkington/Geological Survey of Canada/SPL/Photo Researchers; **14.19 top** Philip Dowell/Dorling Kindersley; **center left and right** Dave King/Dorling Kindersley; **bottom** Philip Dowell/Dorling Kindersley; **14.20** Jim Zuckerman/Corbis; **14.22** Manfred Kage/Peter Arnold; **14.26** Tom Walker/Getty Images.

CHAPTER 15: Chapter opening photo Georgette Douwma/Nature Picture Library; **15.3** Peter Sawyer/NMNH Smithsonian Inst.; **15.4** Roger Ressmeyer/Corbis; **15.6** George Luther; **15.7 left** Hemera Technologies/Jupiter Images; **right** Dr. Tony Brain and David Parker/SPL/Photo Researchers; **15.8 left to right** Scimat/SPL/Photo Researchers; Dr. Dennis Kunkel Microscopy, Inc./Getty Images; CNRI/SPL/Photo Researchers; **15.9a** David M. Phillips/Science Source/Photo Researchers; **15.9b** Susan M. Barns, Ph.D.; **15.9c** Heide Schulz/Max-Planck-Institut fur Marine Mikrobiologie; **15.10** Lee D. Simon/Science Source/Photo Researchers; **15.11** Dr. T.J. Beveridge/Getty Images; **15.12 top left** Martin Shields/Photo Researchers; **top right** Alfred Pasieka/Photo Researchers; **bottom left** Pasieka/SPL/Photo Researchers; **bottom right** Wim Smeets/Foto Natura/Minden Pictures; **15.13a** Helen E. Carr/BPS; **15.13b**

Jack Dykinga/Stone/Getty Images; **15.14** Dr. Tony Brain/Science Photo Library/Photo Researchers; **15.15a** Centers for Disease Control and Prevention (CDC); **15.15b** Scott Camazine/Photo Researchers; **blowup** Stefan Sollfors/Alamy; **15.15c** David M. Phillips/Photo Researchers; **15.16** Alex Wong/Getty Images; **15.18** Courtesy of Exxon Mobil Corporation; **15.19** M. I. Walker/Photo Researchers; **15.21 p. 308 left to right** Eye of Science/Photo Researchers; Oliver Meckes/SPL/Photo Researchers; Wim van Egmond/Getty Images; **15.21 p. 309 left to right** Manfred Kage/Peter Arnold; Masamichi Aikawa, Tokai University School of Medicine, Japan; Michael Abbey/SPL/Photo Researchers; **15.22** The Hidden Forest, www.hiddenforest.co.nz; **15.23 left** Matt Springer; **right top and bottom** Courtesy of Robert Kay, MRC Cambridge; **15.24a** Biophoto Associates/Photo Researchers; **15.24b** Kent Wood/Photo Researchers; **15.24c** Microfield Scientific Ltd/SPL/Photo Researchers; **15.24d** Manfred Kage/Peter Arnold; **15.25 left & center** Marevision/AGE Fotostock; **right** David Hall/Photo Researchers.

CHAPTER 16: Chapter opening photo Great Smoky Mountains National Park Library; **16.2** Dr. Jeremy Burgess/SPL/Photo Researchers; **16.3** Dorling Kindersley; **16.4** Graham Kent **16.5 left** E.R. Degginger/Photo Researchers; **right** Linda Graham, University of Wisconsin-Madison; **16.7 left to right** Tony Wharton, Frank Lane Picture Agency/Corbis; James Randklev/Photolibrary; V. J. Matthew/shutterstock; Dale Wagler/shutterstock; **16.8** Duncan Shaw/SPL/Photo Researchers; **16.9** Dwight Kuhn/Dwight R. Kuhn Photography; **16.11** Ric Ergenbright/Danita Delimont; **bottom left inset** Ed Reschke/Peter Arnold; **bottom right inset** Kim Taylor and Jane Burton/Dorling Kindersley; **top right inset** Biophoto Associates/Photo Researchers; **16.12** The Field Museum, Neg#CSGE075400c, Chicago; **16.13** John E. Marriott/Getty Images; **16.15** Morales/AGE Fotostock; **left inset** Stephen P. Parker/SPL/Photo Researchers; **right inset** Pat O'Hara/Corbis; **16.16** Gene Cox/Photo Researchers; **16.18 left to right** Smart-foto/shutterstock; Tyler Boyes/shutterstock; Christopher Marin/shutterstock; Howard Rice/Dorling Kindersley; **16.20 left** avid Boag/Alamy; **top right** Scott Camazine/Photo Researchers; **bottom right** Juniors Bildarchiv/Alamy; **16.21** Daniel Beltra/Getty Images; **Table 16.1 top to bottom** Dorling Kindersley; Jon Rasmussen/iStockphoto; ben mcleish/iStockphoto; Radu Razvan/shutterstock; Dorling Kindersley; Greg Vaughn/Photolibrary; National Tropical Botanical Garden; alle/shutterstock; **16.22 left** Marc Anderson/Alamy; **top center** Stan Rohrer/Alamy; **bottom center** Elizabeth Peardon/iStockphoto; **inset** VEM/Photo Researchers; **top right** J. Forsdyke/Gene Cox/Science Photo Library/Photo Researchers; **bottom right** N. Allin & G.L. Barron, University of Guelph/Biological Photo Service; **16.23 top** Photos.com; **bottom** blickwinkel/Alamy; **16.24a** Stuart Bebb/Oxford Scientific Films/Animals Animals/Earth Scenes; **16.24b** Nigel Cattlin/Photo Researchers; **16.25** Bettmann/Corbis; **16.26 left** Dario Fusaro/Cephas Picture Library/Alamy; **right** Ryan McVay/Photodisc/Getty Images; **bottom** Will Heap/Dorling Kindersley; **16.27** Christine Case/Skyline College; **16.28 left** Glenn Harper/Alamy; **right** Eye of Science/Photo Researchers.

CHAPTER 17: Chapter opening photo BEA WIHARTA/Reuters/Corbis; **17.1** Gunter Ziesler/Peter

Arnold; **17.2** Georgette Douwman/Nature Picture Library; **17.4 left** Publiphoto/Photo Researchers; **right** Chip Clark; **17.8** James D. Watt/Image Quest Marine; **17.9 top left** Sue Daly/Nature Photo Library; **top middle** Jan Van Arkel/Foto Natura/Minden Pictures; **top right** Kim Taylor/Nature Photo Library; **bottom:** Jordi Chias/Image Quest Marine; **17.12 top left** Georgette Douwma/Nature Photo Library; **bottom left** Jez Tryner/Image Quest Marine; **center left** Harold W. Pratt/BPS; **center right** Marevision/AGE Fotostock; **right** Chris Newbert/Minden Pictures; **17.13 top** E. R. Degginger/Photo Researchers; **bottom left** Geoff Brightling /Dorling Kindersley; **bottom right** Dennis Kunkel/Phototake NYC; **17.14 left to right** A.N.T./Photoshot/NHPA Limited; Carol Buchanan/Photolibrary; WILDLIFE/Peter Arnold; **p. 346** Astrid & Hanns-Frieder Michler/Photo Researchers; **17.16 left to right** Steve Gschmeissner/Photo Researchers; Eye of Science/Photo Researchers; Reproduced by permission from Howard Shiang, D.V.M., Journal of the American Veterinary Medical Association 163:981, Oct. 1973; **17.17 top to bottom** Callan Morgan/Peter Arnold; Piotr Naskrecki/Minden Pictures; George Grall/National Geographic Stock; ZSSD/Minden Pictures; **17.18** Dave King/ Dorling Kindersley; **17.19 clockwise from top center** Dorling Kindersley; Andrew Syred Photo Researchers; Larry West/Photo Researchers; Callan Morgan/Peter Arnold; Herbert Hopfensperger/agefotostock; **17.20 top** Dave King/Dorling Kindersley; **center left to right** Piotr Naskrecki/Minden Pictures; Tom McHugh/Photo Researchers; Nature's Images/Photo Researchers; **bottom left to right** Barry Mansell/Minden Pictures; Nancy Sefton/SPL/Photo Researchers; **17.21 left** George Grall/National Geographic Stock; **right** Tom McHugh/Photo Researchers; **17.22** Radius Images/Alamy; **17.23 top center** Piotr Naskrecki/Minden Pictures; **top left** ZSSD/Minden Pictures; **middle left** Piotr Naskrecki/Minden Pictures; **bottom left** Michael Durham/Minden Pictures; **bottom center** Mark Moffett/Minden Pictures; **bottom right** Mitsuhiko Imamori/Minden Pictures; **middle right** Cordier - Huguet/AGE Fotostock; **top right** Piotr Naskrecki/Minden Pictures; **17.24 top series** John Shaw/Tom Stack and Associates; **bottom** Chris Sharp/Photolibrary; **17.25 top left** Roger Steene/Image Quest Marine; **inset** Gary Milburn/Tom Stack and Associates; **top right** Jose B. Ruiz/Nature Photo Library; **bottom left** tbkmedia.de/Alamy; **bottom middle** Fred Bavendam/Minden Pictures; **bottom right** Roger Steene/Image Quest Marine; **17.26** Colin Keates/Dorling Kindersley; **17.28 left** Runk/Schoenberger/Grant Heilman Photography; **right** Roger Steene/Image Quest Marine; **17.30a** Tom McHugh/Photo Researchers; **inset** Brandon D. Cole/Corbis; **17.30b** Breck P. Kent/Animals Animals - Earth Scenes; **inset** Breck P. Kent/Animals Animals - Earth Scenes; **17.30c** Carol Buchanan/F1 ONL/agefotostock; **17.30c** Peter Scoones/Nature Photo Library; **17.31a left** Michael & Patricia Fogden/Minden Pictures; **right** Michael & Patricia Fogden/Minden Pictures; **17.31b left** Michael Durham/Minden Pictures; **right** Jack Goldfarb/agefotostock; **17.33 p. 358 left** Robert and Linda Mitchell; **middle** Michael Durham/Minden Pictures; **right** Tui De Roy/Minden Pictures; **17.33 p. 359 left** Jerry Young/Dorling Kindersley; **middle** DLILLC/Corbis; **right** Miguel Periera/Dorling

Kindersley; **17.34** Adam Jones/Digital Vision/Getty Images; **17.35 left** Reg Morrison/Minden Pictures; **inset** D. Parer and E. Parer Cook/Auscape; **17.35 middle** blickwinkel/Alamy; **inset** Mitsuaki Iwago/Minden Pictures; **17.35 right** Mitsuaki Iwago/Minden Pictures; **17.37 top row left to right** Pete Oxford/Minden Pictures; Siegfried Grassegger/Photolibrary; Arco Images GmbH/Alamy; Juan Carlos Muñoz/agefotostock; Anup Shah/Nature Photo Library; **17.37 left center** P. Wegner/ Arco Images/Peter Arnold; **17.37 bottom left to right** Ingo Arndt/Nature Photo Library; Anup Shah/Nature Photo Library; John Kelly/Getty Images; **17.39a** The Cleveland Museum of Natural History; **17.39b** John Reader/SPL/Photo Researchers; **17.39c** Javier Trueba/Madrid Scientific Films/ Photo Researchers; **17.40** Philippe Plailly & Atelier Daynes/ Photo Researchers; **17.41 left & top right** Jean Clottes/Sygma/ Corbis; **bottom right** Jean-Marie Chauvet/Sygma/Corbis; **17.42** Steven Poe/UNEP/Still Pictures/Peter Arnold; **17.43** Susan Kuklin/Photo Researchers.

CHAPTER 18: Chapter opening photo Donovan Reese/Getty Images; **18.1a-b** Raphael Gaillarde/Liaison/Getty Images; **18.2** JAY DIRECTO/AFP/Getty Images; **18.3a** Barry Mansell/ Nature Picture Library; **18.3b** Ingrid Van Den Berg/Animals Animals; **18.3c** JUNIORS BILDARCHIV/AGE Fotostock; **18.3d** Jeremy Woodhouse/Getty Images; **18.4** Stephen Krasemann/Photo Researchers; **18.5** NASA; **18.6** Verena Tunnicliffe, University of Victoria; **inset** Woods Hole Oceanographic Institution; **18.7** Wilfried Krecichwost/Getty Images; **18.8** Gregg Adams/Getty Images; **18.9** Antti Leinonen/ Jupiter Images; **18.11 left** Daniel Cox/Photolibrary; **right** J & C Sohns/AGE Fotostock; **18.12** Brian Parker/Tom Stack & Associates; **18.13** Robert Stainforth/Alamy; **18.14** Worldsat International Inc./SPL/Photo Researchers; **18.16** Ishbukar Yalilfatar/shutterstock; **18.17** Kevin Schafer/Alamy; **18.18** GADOMSKI, MICHAEL/Animals Animals/Earth Scenes; **18.20** Digital Vision/Getty Images; **18.21** Carr Clifton/ Minden Pictures; **18.22** James Randklev/Image Bank/Getty Images; **18.29** Frans Lanting/Minden Pictures; **18.30** Yva Momatiuk & John Eastcott/Minden Pictures; **18.31** Juan Carlos Muñoz/AGE Fotostock; **18.32** Stephen J. Krasemann/ Photo Researchers; **18.33** Jake Rajs/Getty Images; **18.34** Kennan Ward/Corbis; **18.35** Jorma Luhta/Nature Picture Library; **18.36** Darrell Gulin/Corbis; **18.37** Gordon Wiltsie/ NGS Image Collection; **18.39 both** UNEP/GRID-Sioux Falls; **18.40** Paul Edmondson/Getty Images; **18.41 both** UNEP/ GRID-Sioux Falls; **18.42** Jim West/Alamy; **18.43** Mark Turner/Jupiter Images; **18.47** Chris Martin Bahr/SPL/Photo Researchers; **18.48** Bryan Patrick/Sacramento Bee/MCT; **18.49** David McNew/Getty Images; **18.50** AP Photo; **18.51** Chris Cheadle/Photolibrary **18.52a** University of Oregon **18.52b** Volvox/Jupiter Images.

CHAPTER 19: Chapter opening photo John Carnemolla/ Corbis; **19.1** Luciano Candisani/Minden Pictures; **19.2** Jim Brandenburg/Minden Pictures; **19.3** Kevin Schafer/Peter Arnold; **19.4 left to right** Roger Phillips/Dorling Kindersley; Jane Burton/Dorling Kindersley; Yuri Arcurs/shutterstock; **Table 19.2 left** Achim Prill/iStockphoto; **right** Anke van Wyk/ shutterstock; **19.5** Edwin Giesbers/Foto Natura/Minden Pictures; **19.6** Roy Corral/Corbis; **19.7 left** Joshua Lewis/ shutterstock; **right** WizData, inc./shutterstock; **19.8a** Marcin

Perkowski/shutterstock; **19.8b** Nigel Cattlin/Alamy; **19.9** Wolfgang Kaehler/Corbis; **19.10** Meul/ARCO/Nature Picture Library; **19.11** Alan Carey/Photo Researchers; **19.12 left to right** Rob Curtis/The Early Birder; USDA Forest Service; Gilbert S. Grant/Photo Researchers; **19.13:** Dan Burton/Nature Picture Library; **19.14** Ric Ergenbright/Corbis; **19.15** Jan Tove Johansson/Pictor International, Ltd./PictureQuest; **inset** David Tipling/Nature Picture Library; **19.16** Dennis, David M./Animals Animals/Earth Scenes; **19.17** Matt Meadows/Peter Arnold; **19.18 left** Scott Camazine/Alamy; **right** Creatas/ Photolibrary; **19.22 left** Peter Ginter; **right** Peter Menzel/ Peter Menzel Photography; **19.24** franzfoto.com/Alamy.

CHAPTER 20: Chapter opening photo Alaskastock; **20.1** James King-Holmes/Photo Researchers; **20.2 left** Mark Carwardine/Still Pictures/Peter Arnold; **right** Michael & Patricia Fogden/Minden Pictures; **20.3** Fred Bavendam/Minden Pictures; **20.4** University of Maine; **20.5** AFP/Getty Images; **20.6** Richard D. Estes/Photo Researchers; **20.7a** Doug Backlund; **20.7b** Terry Sohl; **20.8 both** M. I. Walker/SPL/Photo Researchers; **20.9** Jurgen Freund/naturepl.com; **20.10** Joe McDonald/Corbis; **20.11** Mark Moffett/Minden Pictures; **20.12 left** E. R. Degginger/Photo Researchers; **right** Breck P. Kent; **20.13 left** Stephen J. Krasemann/Photo Researchers; **right** Peter J. Mayne; **20.14 top left** BILDAGENTUR-ONLINE/TH FOTO/SPL/Photo Researchers; **top right** Luca Invernizzi Tettoni/Photolibrary; **bottom** ImageState/Alamy Images; **20.16** Michael Sedam/Photolibrary; **20.21** William E. Townsend/Photo Researchers; **20.22** Todd Sieling/Corvus Consulting; **20.23** David Muench/Corbis; **20.24** Scott T. Smith/Corbis; **20.30** Gerry Ellis/Minden Pictures; **20.35** Michael Marten/SPL/Photo Researchers; **20.36 all** NASA; **20.38 top** Yann Arthus-Bertrand/Corbis; **bottom** James P. Blair/National Geographic Society; **20.39** Joel Sartore/NGS Image Collection; **20.40** R. O. Bierregaard, Jr., Biology Department, University of North Carolina, Charlotte; **20.41** USDA; **20.42** Photo provided by Kissimmee Division staff, South Florida Water Management District (WPB); **20.43** Frans Lanting/Minden Pictures; **20.44 left** Comstock Images/ Getty Images; **top right** Nordicphotos/Alamy; **bottom center** Frans Lanting/Minden Pictures; **bottom right** Image Source/ Corbis; **p. 450 Biodiversity table left to right** James King-Holmes/Photo Researchers; Michael & Patricia Fogden/ Minden Pictures; Fred Bavendam/Minden Pictures; **p. 450 Interactions table left top to bottom** Terry Sohl; Doug Backlund; Jurgen Freund/Nature Picture Library; **right top to bottom** James Balog/Getty Images; Scott Camazine/ Alamy; Renaud Visage/Getty Images.

CHAPTER 21: Chapter opening photo GoGo Images Corporation/Alamy; **21.1** Andy Crawford /Dorling Kindersley; **21.2a** CROCUS/iStockphoto; **21.2b** Janice Sheldon; **21.2c** T.D. Parsons, D. Kleinfeld, F. Raccuia-Behling and B. Salzberg. Biophysical Journal, July 1989. Photo courtesy of Brian Salzberg; **21.3:** Blend Images/Alamy; **21.4 left** Mike Kemp/Getty Images **21.4a-b** Nina Zanetti; **21.4c** Dr. Gopal Murti/SPL/Photo Researchers **21.4d** Biophoto Associates/ Photo Researchers; **21.4e** Chuck Brown/Photo Researchers; **21.4f** Nina Zanetti; **21.5 center** Dominic DiSaia/Getty Images;

ILLUSTRATION AND TEXT CREDITS

of Science; **13.30:** Adapted from A. C. Allison, "Abnormal Hemoglobins and Erythrovute Enzyme-Deficiency Traits," in *Genetic Variation in Human Populations,* by G. A. Harrison, ed. (Oxford: Elsevier Science, 1961).

CHAPTER 14: 14.26: Adapted from Hickman, Roberts, and Larson. 1997, *Zoology,* 10/e, Wm. C. Brown, fig. 31.1.

CHAPTER 15: 15.17: Adapted from G. J. Tortora, B. R. Funke, and C. L. Case, *Microbiology: An Introduction,* 9th ed. Copyright © 2007 Pearson Education, Inc., publishing as Pearson Benjamin Cummings.

CHAPTER 16: Chapter opening poem: "Evil Tendencies Cancel" from *The Poetry of Robert Frost* edited by Edward Connery Lathem. Copyright 1969 by Henry Holt and Company. Copyright 1938 by Robert Frost, copyright 1964 by Lesley Frost Ballantine. Reprinted by arrangement with Henry Holt and Company, LLC; **Table 16.1:** Adapted from Randy Moore et al., *Botany,* 2nd ed. Dubuque, IA: Brown, 1998, Table 2.2, p. 37.

CHAPTER 17: 17.38: *H. neanderthalensis* adapted from *The Human Evolution Coloring Book. P. boisei* drawn from a photo by David Bill.

CHAPTER 18: 18.25: From J. H. Withgott and S. R. Brennan, *Environment: The Science Behind the Stories,* 3rd ed., fig, 6.30, p. 169. Copyright © 2008 Pearson Education, Inc., Prentice Hall; **18.27:** Adapted from Heinrich Walter and Siegmar-Walter Breckle 2003. *Walter's Vegetation of the Earth,* fig. 16, p. 36. Springer-Verlag, © 2003; **18.44:** From "Global Temperature Change," J. Hansen et al., *PNAS,* Sept. 26, 2006, Vol. 103, No. 39, Fig. 1B. Copyright 2006 National Academy

of Sciences, U.S.A. Reprinted with permission; **18.45:** Adapted from *Climate Change 2007: The Physical Science Basis,* Intergovernmental Panel on Climate Change, FAQ 2.1, Fig. 1. Reprinted by permission of IPCC, c/o World Meteorological Organization.

CHAPTER 19: 19.13: Data from Fisheries and Oceans, Canada, 1999; **19.19:** Data from United Nations, the World at Six Billion, 2007; **19.20 and 19.21:** Data from U. S. Census Bureau International Data Base; **19.23:** Adapted from information at phtbb.org/natural/footprint; **Table 19.1:** Data from Centers for Disease Control and Prevention website, "United States Life Tables, 2003," *National Vital Statistics Report,* vol. 50, no. 14, April 19, 2006. www.cdc.gov; **Table 19.2:** Adapted from E. R. Pianka, *Evolutionary Ecology,* 6th ed., p. 186. Copyright © 2000 Pearson Education, Inc., publishing as Pearson Benjamin Cummings; **Table 19.3:** Data from Population Reference Bureau, 2008, www.prb.org.

CHAPTER 20: 20.18: From Robert Leo Smith, *Ecology and Field Biology,* 4th ed. Copyright © 1990 by Robert Leo Smith. Reprinted by permission of Addison Wesley Longman, Inc.; **20.37:** From N. Myers et al., "Biodiversity hotspots for conservation priorities," *Nature,* Vol. 403, p. 853, 2/24/2000. Reprinted by permission from Macmillan Publishers Ltd. Copyright © 2000 Nature Publishing, Inc.

CHAPTER 21: 21.15: Adapted from fig. 2 in V. H. Hutchison et al. Thermoregulation in a brooding female Indian python, Python molurus bivittatus. *Science* 151: 694–695, fig. 2. Copyright © 1966. Reprinted with permission from AAAS.

CHAPTER 22: 22.20: Data adapted from *Clinical Guidelines on the Identification, Evaluation, and Treatment of Overweight and Obesity in Adults: The Evidence Report.* Data found at the National Heart, Lung, and Blood Institute: http://www.nhlbi.nih.gov/guidelines/obesity/bmi_tbl.htm; **Table 22.1:** Data from S. E. Gebhardt and R. G. Thomas, *Nutritive Values of Foods* (USDA, 2002); S. A. Plowman and D. L. Smith, *Exercise Physiology for Health, Fitness, and Performance,* 2nd ed. Copyright © 2003 Pearson Education, Inc., publishing as Pearson Benjamin Cummings.

CHAPTER 23: 23.23: Illustration of hemoglobin structure by Irving Geis. Image from Irving Geis Collection, HHMI. Rights owned by Howard Hughes Medical Institute (HHMI). Not to be reproduced without permission.

CHAPTER 25: 25.12: Map adapted from W. M. Howell et al. 1980. Abnormal expression of secondary sex characters in a population of mosquitofish, *Gambusia affinis holbrooki:* Evidence for environmentally-induced masculinization. *Copeia* 4, 676–681, fig. 5. Copyrighted material used with permission of the American Society of Ichthyologists and Herpetologists.

CHAPTER 26: Table 26.1: Data from R. Hatcher et al., *Contraceptive Technology: 1990–1992,* p. 134 (New York: Irvington, 1990).

CHAPTER 27: 27.5: Adapted from W. M. Becker and D. W. Deamer, *The World of the Cell,* 2nd ed., fig. 20.16. Copyright © 1991 Pearson Education, Inc., publishing as Pearson Benjamin Cummings.

Appendix D Self-quiz Answers

CHAPTER 1

1. b (some living organisms are single-celled)
2. atom, molecule, cell, tissue, organ, organism, population, ecosystem, biosphere; the cell
3. Photosynthesis cycles nutrients by converting the carbon in carbon dioxide into sugar, which is then consumed by other organisms. Additionally, the oxygen in water is released as oxygen gas. Photosynthesis contributes to energy flow by converting sunlight into chemical energy, which is then also consumed by other organisms, and by producing heat.
4. a4, b1, c3, d2.
5. On average, those individuals with heritable traits best suited to the local environment produce the greatest number of offspring that survive and reproduce. This increases the frequency of those traits in the population over time. The result is the accumulation of evolutionary adaptations.
6. Without a control group, you don't know if the experimental outcome is due to the variable you are trying to test or to some other variable.
7. d
8. c
9. evolution
10. a3, b2, c1, d4

CHAPTER 2

1. electrons; neutrons; protons
2. Nitrogen-14 has an atomic number of 7 and a mass number of 14. The radioactive isotope, nitrogen-16, has an atomic number of 7 and a mass number of 16.
3. Organisms incorporate radioactive isotopes of an element into their molecules just as they do the nonradioactive isotopes, and researchers can detect the presence of the radioactive isotopes.
4. 2
5. Each carbon atom has only three covalent bonds instead of the required four.
6. The positively charged hydrogen regions would repel each other.
7. d
8. a
9. The positive and negative poles cause adjacent water molecules to become attracted to each other, forming hydrogen bonds. The properties of water such as cohesion, temperature regulation, and water's ability to act as a solvent all arise from this atomic stickiness.
10. The cola is an aqueous solution, with water as the solvent, sugar as the main solute, and the CO_2 making the solution acidic.

CHAPTER 3

1.

2. dehydration; hydrolysis
3. b
4. fatty acid; glycerol
5. b
6. c
7. d
8. Hydrophobic amino acids are most likely to be found within the interior of a protein, far from the watery environment.
9. a
10. starch (or glycogen or cellulose); nucleotide
11. Both DNA and RNA are polynucleotides; both have the same phosphate group along the backbone; and both use A, C, and G bases. But DNA uses T while RNA uses U as a base; the sugar differs between them; and DNA is usually double-stranded while RNA is usually single-stranded.

CHAPTER 4

1. d
2. about 0.075 mm, which equals 75 μm
3. b
4. A membrane is fluid because its components are not locked into place. A membrane is mosaic because it contains a variety of embedded proteins.
5. endomembrane system
6. smooth ER; rough ER
7. rough ER; Golgi apparatus; plasma membrane
8. Both organelles use membranes to organize enzymes, and both provide energy to the cell. But chloroplasts use pigments to capture energy from sunlight in photosynthesis, whereas mitochondria release energy from glucose using oxygen in cellular respiration. Chloroplasts are only in photosynthetic plants and protists, whereas mitochondria are in almost all eukaryotic cells.
9. a3, b1, c5, d2, e4
10. nucleus, nuclear pores, ribosomes, rough ER, Golgi apparatus

CHAPTER 5

1. You convert the chemical energy from food to the kinetic energy of your upward climb. At the top of the stairs, some of the energy has been stored as potential energy because of your higher elevation. The rest has been converted to heat.
2. Energy; entropy
3. 10,000 g (or 10 kg); remember that 1 Calorie on a food label equals 1,000 calories of heat energy.
4. The three phosphate groups store chemical energy, a form of potential energy. The release of a phosphate group makes some of this potential energy available to cells to perform work.
5. Hydrolases are enzymes that participate in hydrolysis reactions, breaking down large molecules into the smaller molecules that make them up.
6. An inhibitor's binding to another site on the enzyme can cause the enzyme's active site to change shape.
7. b
8. a
9. *Hypertonic* and *hypotonic* are relative terms. A solution that is hypertonic to tap water could be hypotonic to seawater. In using these terms, you must provide a comparison, as in "The solution is hypertonic to the cell's cytoplasm."
10. Passive transport moves atoms or molecules along their concentration gradient (from higher to lower concentration), while active transport moves them against their concentration gradient.
11. b
12. signal transduction pathway

CHAPTER 6

1. d
2. Plants produce organic molecules by photosynthesis. Consumers must acquire organic material by consuming it rather than making it.
3. In breathing, your lungs exchange CO_2 and O_2 between your body and the atmosphere. In cellular respiration, your cells consume the O_2 in extracting energy from food and release CO_2 as a waste product.
4. the electron transport chain
5. glucose; NAD^+
6. O_2
7. The majority of the energy provided by cellular respiration is generated during the electron transport chain. Shutting down that pathway will deprive cells of energy very quickly.
8. b
9. glycolysis
10. b
11. Because fermentation supplies only 2 ATP per glucose molecule compared with 38 from cellular respiration, the yeast will have to consume 19 times as much glucose to produce the same amount of ATP.

CHAPTER 7

1. thylakoids; stroma
2. The Calvin cycle, which consumes the NADPH and ATP, occurs in the stroma.
3. inputs: a, d, e; outputs: b, c
4. Green light is reflected by chlorophyll, not absorbed, and therefore cannot drive photosynthesis.
5. H_2O
6. c
7. The reactions of the Calvin cycle require the outputs of the light reactions (ATP and NADPH).
8. In hot, dry environments, most plants close their stomata, which saves water but decreases the amount of CO_2 available inside the leaves.
9. C_4 and CAM plants can close their stomata and save water without shutting down photosynthesis.
10. c

CHAPTER 8

1. c
2. They have identical genes (DNA).
3. They are in the form of very long, thin strands.
4. b
5. prophase and telophase
6. a) 1; 1, b) 1; 2, c) 2; 4, d) $2n$; n, e) individually; by homologous pair, f) identical; unique, g) repair, growth, asexual reproduction; gamete formation
7. b

8. 39

9. Prophase II or metaphase II; it cannot be during meiosis I because then you would see an even number of chromosomes; it cannot be during a later stage in meiosis II because then you would see the sister chromatids separated.

10. benign; malignant

11. 16 ($2n = 8$, so $n = 4$, and $2^n = 2^4 = 16$)

12. Nondisjunction would create just as many gametes with an extra copy of chromosome 3 or 16, but extra copies of chromosome 3 or 16 are probably fatal.

CHAPTER 9

1. genotype; phenotype

2. Statement **a.** is the law of independent assortment; statement **b.** is the law of segregation.

3. c

4. c

5. d

6. d

7. d

8. $\frac{1}{4}$ ($\frac{1}{2}$ chance the child will be male times $\frac{1}{2}$ chance that he will inherit the X carrying the disease allele)

9. The parental-type gametes are *WS* and *ws*. Recombinant gametes are *Ws* and *wS*, produced by crossing over.

10. Height appears to result from polygenic inheritance, like human skin color. See Figure 9.22.

11. The brown allele appears to be dominant, the white allele recessive. The brown parent appears to be homozygous dominant, *BB*, and the white mouse is homozygous recessive, *bb*. The F_1 mice are all heterozygous, *Bb*. If two of the F_1 mice are mated, $\frac{3}{4}$ of the F_2 mice will be brown.

12. The best way to find out whether a brown F_2 mouse is homozygous dominant or heterozygous is to do a testcross: Mate the brown mouse with a white mouse. If the brown mouse is homozygous, all the offspring will be brown. If the brown mouse is heterozygous, you would expect half the offspring to be brown and half to be white.

13. Freckles is dominant, so Tim and Jan must both be heterozygous. There is a $\frac{3}{4}$ chance that they will produce a child with freckles, a $\frac{1}{4}$

chance that they will produce a child without freckles. The probability that the next two children will have freckles is $\frac{3}{4} \times \frac{3}{4} = \frac{9}{16}$.

14. Half their children will be heterozygous and have elevated cholesterol levels. There is a $\frac{1}{4}$ chance that their next child will be homozygous, *hh*, and have an extremely high cholesterol level, like Katerina.

15. The bristle-shape alleles are sex-linked, carried on the X chromosome. Normal bristles is dominant (*F*) and forked is recessive (*f*). The genotype of the female parent is $X^F X^f$. The genotype of the male parent is $X^F Y$. Their female offspring are $X^F X^f$; their male offspring are $X^f Y$.

16. The mother is a heterozygous carrier, and the father is normal. See Figure 9.32 for a pedigree. $\frac{1}{4}$ of their children will be boys suffering from hemophilia; $\frac{1}{4}$ will be female carriers.

17. For a woman to be colorblind, she must inherit X chromosomes bearing the colorblindness allele from both parents. Her father has only one X chromosome, which he passes on to all his daughters, so he must be colorblind. A male only needs to inherit the colorblindness allele from a carrier mother; both his parents are usually phenotypically normal.

CHAPTER 10

1. polynucleotides; nucleotides; sugar (deoxyribose); phosphate; nitrogenous base

2. b

3. Each daughter DNA molecule will have half the radioactivity of the parent molecule, since one polynucleotide from the original parental DNA molecule winds up in each daughter DNA molecule.

4. CAU; GUA; histidine (His)

5. A gene is the polynucleotide sequence with information for making one polypeptide. Each codon—a triplet of bases in DNA or RNA—codes for one amino acid. Transcription occurs when RNA polymerase produces mRNA using one strand of DNA as a template. A ribosome is the site of translation, or polypeptide synthesis, and tRNA

molecules serve as interpreters of the genetic code. Each tRNA molecule has an amino acid attached at one end and a three-base anticodon at the other end. Beginning at the start codon, mRNA moves relative to the ribosome a codon at a time. A tRNA with a complementary anticodon pairs with each codon, adding its amino acid to the polypeptide chain. The amino acids are linked by peptide bonds. Translation stops at a stop codon, and the finished polypeptide is released. The polypeptide folds to form a functional protein, sometimes in combination with other polypeptides.

6. a3, b3, c1, d2, e2 and 3

7. d

8. d

9. The genetic material of these viruses is RNA, which is replicated inside the infected cell by special enzymes encoded by the virus. The viral genome (or its complement) serves as mRNA for the synthesis of viral proteins.

10. reverse transcriptase

11. The process of reverse transcription occurs only in infections by RNA-containing retroviruses like HIV. Cells do not require reverse transcriptase (their RNA molecules do not undergo reverse transcription), so reverse transcriptase can be knocked out without harming the human host.

CHAPTER 11

1. c

2. operon

3. b

4. a

5. DNA polymerase and other proteins required for transcription do not have access to tightly packed DNA.

6. the ability of these cells to produce entire organisms through cloning

7. nuclear transplantation

8. which genes are active in a particular sample of cells

9. b

10. embryonic tissue (ES cells), umbilical cord blood, and bone marrow (adult stem cells)

11. Proto-oncogenes are normal genes involved in the control of the cell

cycle. Mutation or viruses can cause them to be converted to oncogenes, or cancer-causing genes. Proto-oncogenes are necessary for normal control of cell division.

12. Master control genes, called homeotic genes, regulate many other genes during development.

CHAPTER 12

1. c, d, b, a

2. because it does not contain introns

3. vector

4. Such an enzyme creates DNA fragments with "sticky ends," single-stranded regions whose unpaired bases can hydrogen-bond to the complementary sticky ends of other fragments created by the same enzyme.

5. PCR

6. Different people tend to have different numbers of repeats at each STR site. DNA fragments prepared from the STR sites of different people will thus have different lengths, causing them to migrate to different locations on a gel.

7. a

8. b

9. Chop the genome into fragments using restriction enzymes, clone and sequence each fragment, and reassemble the short sequences into a continuous sequence for every chromosome.

10. c, b, a, d

CHAPTER 13

1. b

2. c

3. Gradualism is the idea that large changes on Earth can result from the accumulation of small changes over a very long time. Darwin applied this idea to suggest that species evolve through the slow accumulation of small changes over time.

4. *Bb*: 0.42; *BB*: 0.49; *bb*: 0.09

5. The fitness of an individual (or of a particular genotype) is measured by the relative number of alleles that it contributes to the gene pool of the next generation compared with the contribution of others. Thus the number of fertile offspring produced determines an individual's fitness.

6. e

7. c

8. Both types of events result in populations small enough for significant sampling error in the gene pool for the first few generations. A bottleneck reduces the size of an existing population in a given location. The founder effect occurs when a new, small population colonizes a new territory.
9. stabilizing selection
10. b, c, d
11. the prevalence of malaria

CHAPTER 14
1. Microevolution is a change in the gene pool of a population, often associated with adaptation. Macroevolution is marked by major changes in the history of life, including the origin of new species, and these changes are often noticeable enough to be evident in the fossil record.
2. b
3. prezygotic: a, b, c, e; postzygotic: d
4. because a small gene pool is more likely to be changed substantially by genetic drift and natural selection
5. exaptations
6. d
7. d
8. species, genus, family, order, class, phylum, kingdom, domain
9. 2.6
10. Archaea and Bacteria

CHAPTER 15
1. g, d, f, c, b, a, e
2. d, c, a, b, e
3. DNA polymerase is a protein, which must be transcribed from a gene. But a DNA gene requires DNA polymerase to be replicated. This creates a paradox about which came first—DNA or protein. But RNA can act as both an information storage molecule and an enzyme, suggesting that dual-role RNA may have preceded both DNA and proteins.
4. Exotoxins are poisons secreted by pathogenic bacteria; endotoxins are components of the outer membrane of pathogenic bacteria.
5. Autotrophs make their own organic compounds from CO_2, while heterotrophs must obtain at least one type of organic compound from another organism.
6. They can form endospores.

7. Both you and a mushroom are chemoheterotrophs.
8. because the first eukaryotes were protists, and these ancient protists were ancestral to all other eukaryotes, including plants, fungi, animals, and modern protists
9. c
10. b

CHAPTER 16
1. cuticle
2. flowers
3. a. sporophyte b. cones; angiosperms c. fruit
4. b
5. b
6. a fern
7. Because the plants do not lose their leaves during autumn and winter, the leaves are already fully developed for photosynthesis when the short growing season begins in spring.
8. vascular plant
9. a
10. a ripened ovary of a flower that protects and aids in the dispersal of seeds contained in the fruit
11. algae; fungi
12. A fungus digests its food externally by secreting digestive juices into the food and then absorbing the small nutrients that result from digestion. In contrast, humans and most other animals ingest relatively large pieces of food and digest the food within their bodies.

CHAPTER 17
1. c
2. coelom; pseudocoelom
3. arthropod
4. amphibians
5. b
6. chordata; notochord; cartilage disks between your vertebrae
7. bipedalism
8. a
9. *Australopithecus* species, *Homo habilis, Homo erectus, Homo sapiens*
10. a4, b5, c1, d2, e3

CHAPTER 18
1. organismal ecology, population ecology, community ecology, ecosystem ecology
2. light, water temperature, chemicals added
3. physiological; behavioral
4. d

5. a. desert; b. grassland; c. tropical rain forest; d. temperate broadleaf forest; e. coniferous forest; f. tundra
6. chaparral
7. permafrost, very cold winters, and high winds
8. agriculture
9. Carbon dioxide and other gases in the atmosphere absorb heat energy radiating from Earth and reflect it back toward Earth. This is called the greenhouse effect. As the carbon dioxide concentration in the atmosphere increases, more heat is retained, causing global warming.
10. c
11. populations of organisms that have high genetic variability and short life spans

CHAPTER 19
1. the number of people and the land area in which they live
2. III; I
3. O; O; E; O; E
4. a. The x-axis is time; the y-axis is the number of individuals; the red curve represents exponential growth; the blue curve represents logistic growth.
   b. carrying capacity
   c. In exponential growth, the population growth rate continues to increase as long as the population size increases. In logistic growth, the population grows fastest when it is about $\frac{1}{2}$ the carrying capacity.
   d. exponential growth curve, though the worldwide growth rate is slowing
5. d
6. opportunistic
7. c
8. The large population segment, the baby boomers, are currently in the workforce and are in their peak earning years.
9. d

CHAPTER 20
1. habitat destruction
2. d
3. c
4. a2, b5, c5, d3 or d4, e1
5. because the pesticides become concentrated in their prey
6. succession
7. Only 10% of the energy trapped by photosynthesis is turned into biomass by the plant, and only 10%

of that energy is turned into the meat of a grazing animal. Therefore, eating grain-fed beef obtains only about 1% of the energy captured by photosynthesis.
8. Many nutrients come from the soil, but carbon comes from the air.
9. landscape
10. b

CHAPTER 21
1. atom, molecule, organelle, cell, tissue, organ, organ system, organism
2. A very thin or flat organism may have all of its body cells in contact with the environment and therefore would not need a circulatory system. In contrast, a thicker organism will require a circulatory system to transport substances to all body cells.
3. c
4. interstitial fluid
5. positive; negative
6. Endotherms get most of their body heat from metabolic reactions, whereas ectotherms get most of their body heat from the environment.
7. osmoregulator; osmoconformer
8. d
9. b
10. Most of the water is reabsorbed into the blood; the rest is excreted in the urine.

CHAPTER 22
1. Large food items could not be digested; it would be difficult to eliminate wastes; and it would be difficult to nourish the many cells that compose a large animal.
2. Yes, absorption is the uptake of nutrients into body cells. Without it, you would never gain any calories (or nutritional value) from your food.
3. digestion; absorption
4. because peristalsis pushes the food along
5. proteins; pepsin
6. bacteria (specifically, *Helicobacter pylori*); by using antibiotics and bismuth
7. d
8. The antibiotics kill the bacteria that synthesize vitamin K in the colon.
9. b
10. because cellular respiration requires oxygen to break down food

11. Many plant proteins are incomplete, meaning that they lack certain essential amino acids. Animal products tend to have complete proteins that contain all the essential amino acids. Vegetarians must therefore eat a variety of plant products to receive all the essential amino acids.

12. a (Overdoses of fat-soluble vitamins can be dangerous because the body cannot rid itself of them quickly.)

CHAPTER 23

1. In an open circulatory system, vessels have open ends and the circulating fluid flows directly around the body cells. In a closed circulatory system, the circulating fluid stays within a closed system of vessels.

2. Arteries carry oxygen-rich blood to the body, but they carry oxygen-poor blood to the lungs. Veins carry oxygen-poor blood from the body to the heart, but they carry oxygen-rich blood from the lungs to the heart. Arteries are defined as vessels that carry blood away from the heart; veins carry blood to the heart.

3. a3, b1, c4, d2

4. Oxygen levels are reduced as oxygen-depleted blood mixes with oxygen-rich blood.

5. anemia

6. a3, b4, c1, d2

7. fruits, vegetables, and whole grains; cholesterol, saturated fat, and trans fat

8. Gills extend outward from the body into the surrounding environment, whereas the lungs are internal organs.

9. alveoli

10. b

11. b

CHAPTER 24

1. antigens; antibodies

2. innate: a, b, d, e; adaptive: c, f, g

3. b

4. It takes several days for a clone of effector cells to be formed during the primary response. In a secondary response, memory cells can respond more quickly.

5. a4, b1, c6, d3, e2, f5

6. Swelling dilutes toxins, delivers more oxygen, promotes scabbing; pain changes behavior to discourage further damage; fever may inhibit bacterial growth.

7. HIV destroys helper T cells, thereby impairing both the cell-mediated and the humoral immune response.

8. B cells

9. In a person with lupus, the immune system makes antibodies against the person's own (self) molecules.

10. memory cells; clonal selection

CHAPTER 25

1. d

2. b

3. The first two cells could have membrane receptors for that hormone, but they might trigger different signal transduction pathways. The third cell could lack membrane receptors for that particular hormone.

4. b

5. Alcohol slows the release of antidiuretic hormone (ADH) from the pituitary. Decreased levels of ADH result in decreased levels of water reabsorption by the kidneys, increasing urine output.

6. The patient has hypothyroidism. The thyroid cannot produce its metabolism-regulating hormones without iodine.

7. d

8. c

9. At high levels, glucocorticoids suppress the immune system. Although this is good for relieving pain and inflammation, the immune system can no longer respond appropriately to threats such as infection. Additionally, glucocorticoids may mask pain, leading to behaviors that aggravate the injury.

10. androgens

11. a. endorphins b. growth hormone (GH) c. estrogens

CHAPTER 26

1. Asexual reproduction is favorable in a stable environment with many resources. Sexual reproduction may enhance success in a changing environment by producing genetically variable offspring.

2. the seminal vesicles and/or the prostate

3. a4, b6, c3, d1, e5, f2

4. b

5. Inhibiting FSH and LH after ovulation ensures that another egg will not be released during this cycle. Estrogen and progesterone inhibit FSH and LH.

6. They also help prevent the spread of sexually transmitted diseases.

7. a

8. c

9. The nerve cells may follow chemical trails to the muscle cells (induction).

10. oxytocin

11. in vitro fertilization; fertilization "in glass"

CHAPTER 27

1. cell body; dendrites; axon; synaptic terminal

2. It speeds up conduction of signals along an axon.

3. central nervous system; peripheral nervous system; brain; spinal cord

4. cerebrospinal fluid; meninges

5. activation of the sympathetic branch of the peripheral nervous system

6. The cerebral cortex deals with sensory information, while structures in the midbrain and brainstem maintain vital functions. If a person suffers a severe head injury, the cerebral cortex may be damaged while the midbrain or brainstem is still able to function, leaving the person unresponsive but alive.

7. An action potential (used by neurons) is an all-or-none situation; it doesn't vary in intensity. A receptor potential (used by sensory receptor cells) conveys information about the strength of the stimulus.

8. photoreceptor; chemoreceptor; mechanoreceptor; chemoreceptor

9. c (He can hear the tuning fork against his skull, so the cochlea, nerve, and brain are all functioning properly. Apparently, sounds are not being transmitted to the cochlea; therefore, the bones of the middle ear are the problem.)

10. axial skeleton; appendicular skeleton

11. d

12. a (Each neuron controls a smaller number of muscle fibers.)

CHAPTER 28

1. b, c, d, e

2. When most plants grow naturally, they exhibit apical dominance, which inhibits the outgrowth of the axillary buds. If your friend were to pinch back the terminal buds, there would be increased growth from the axillary buds. This would result in more branches and leaves, which would increase next year's yield.

3. ovary; ovule

4. a4, b2, c5, d6, e3, f1

5. sexual, because it generates genetic variation among the offspring, enhancing the potential for adaptation to a changing environment

6. a. fruit (ripened ovary); b. fruit; c. leaf; d. root; e. fruit

7. c

8. e

9. b; a

10. vascular cambium

CHAPTER 29

1. Excessive amounts of fungicides could destroy mycorrhizae, symbiotic associations of fungi and plant root hairs. The fungal filaments provide lots of surface area for absorption of water and nutrients. Destroying the mycorrhizae could cause a water or nutrient deficiency in the plant.

2. c

3. Phloem; Xylem

4. Cohesion is the sticking together of identical molecules—water molecules in the case of xylem sap. Adhesion is the sticking together of different kinds of molecules, as in the adhesion of water to the cellulose of xylem walls. Cohesion enables transpiration to pull xylem sap up without the water in the vessels separating; adhesion helps to support the sap against the downward pull of gravity.

5. b

6. auxins; cytokinins

7. The terminal bud produces auxins, which counter the effects of cytokinins from the roots and inhibit the growth of axillary buds. If the terminal bud is removed, the cytokinins predominate, and branching occurs at the axillary buds.

8. a1; b4; c2; d5; e3

9. a4; b3; c1; d2

10. The room lights will inhibit flowering in the chrysanthemums.

Appendix E Student Media

Mastering **BIOLOGY**™   The following student media are found in the Study Area of www.masteringbiology.com.

	Resources for Visual Learners		Resources for Auditory Learners	Relevant Biology Topics in the News and Media	Resources to Study and Practice Before a Test	Resources to Build Graphing Skills
	BioFlix™	VIDEO tutor sessions	MP3 tutor sessions			
	BIOFLIX	**VIDEO TUTOR SESSIONS**	**MP3 TUTOR SESSIONS**	**DISCOVERY CHANNEL VIDEOS**	**ACTIVITIES**	**YOU DECIDE & GRAPHIT!**
Available for all chapters				Current Events with *NY Times* Articles, RSS Feeds, and *Scientific American* Podcasts	Chapter Quizzes including Pre-Tests, Post-Tests, and Cumulative Quiz Flashcards	
Chapter 1		Survey of Biodiversity Phylogenetic Trees	The Process of Science	Antibiotics	The Levels of Life Card Game Energy Flow and Chemical Cycling Classification Schemes Darwin and the Galápagos Islands Science, Technology, and Society: DDT	You Decide: What Can We Do About Antibiotic-Resistant Bacteria? GraphIt!: An Introduction to Graphing
Chapter 2			The Properties of Water		The Structure of Atomic Nucleus Electron Arrangement Build an Atom Ionic Bonds Covalent Bonds Hydrogen Bonds Polarity of Water Cohesion of Water Acids, Bases, and pH	
Chapter 3		DNA Structure	Protein Structure and Function DNA Structure		Diversity of Carbon-Based Molecules Functional Groups Making and Breaking Polymers Models of Glucose Carbohydrates Lipids Protein Functions Protein Structure Nucleic Acid Functions Nucleic Acid Structure	You Decide: Low-Fat or Low-Carb Diets—Which Is Healthier?

	Resources for Visual Learners		Resources for Auditory Learners	Relevant Biology Topics in the News and Media	Resources to Study and Practice Before a Test	Resources to Build Graphing Skills
	BioFlix™	VIDEO tutor sessions	MP3 tutor sessions			
	BIOFLIX	VIDEO TUTOR SESSIONS	MP3 TUTOR SESSIONS	DISCOVERY CHANNEL VIDEOS	ACTIVITIES	YOU DECIDE & GRAPHIT!
Chapter 4	Tour of an Animal Cell Tour of a Plant Cell		Cell Organelles	Cells	Metric System Review Prokaryotic Cell Structure and Function Comparing Cells Build an Animal Cell and a Plant Cell Membrane Structure Role of the Nucleus and Ribosomes in Protein Synthesis The Endomembrane System Build a Chloroplast and a Mitochondrion Cilia and Flagella Review: Animal Cell Structure and Function Review: Plant Cell Structure and Function	
Chapter 5	Membrane Transport		Basic Energy Concepts	Cells	Energy Transformations The Structure of ATP How Enzymes Work Membrane Structure Diffusion Facilitated Diffusion Osmosis and Water Balance in Cells Active Transport Exocytosis and Endocytosis Cell Signaling	
Chapter 6	Cellular Respiration		Cellular Respiration Part 1: Glycolysis Cellular Respiration Part 2: Citric Acid Cycle and Electron Transport	Space Plants Tasty Bacteria	Build a Chemical Cycling System Overview of Cellular Respiration Glycolysis The Citric Acid Cycle Electron Transport Fermentation	
Chapter 7	Photosynthesis		Photosynthesis	Space Plants Trees	The Plants in Our Lives The Sites of Photosynthesis Overview of Photosynthesis Light Energy and Pigments The Light Reactions The Calvin Cycle Photosynthesis in Dry Climates	

	Resources for Visual Learners		Resources for Auditory Learners	Relevant Biology Topics in the News and Media	Resources to Study and Practice Before a Test	Resources to Build Graphing Skills
	BioFlix™	VIDEO tutor sessions	MP3 tutor sessions			
	BIOFLIX	VIDEO TUTOR SESSIONS	MP3 TUTOR SESSIONS	DISCOVERY CHANNEL VIDEOS	ACTIVITIES	YOU DECIDE & GRAPHIT!
Chapter 8	Mitosis Meiosis	Mitosis and Meiosis	Mitosis Meiosis Comparison of Mitosis and Meiosis	Cells Fighting Cancer	Asexual and Sexual Life Cycles The Cell Cycle Mitosis and Cytokinesis Animation Mitosis and Cytokinesis Video Causes of Cancer Human Life Cycle Meiosis Animation Origins of Genetic Variation	
Chapter 9		Sex-Linked Pedigrees	Chromosomal Basis of Inheritance	Colored Cotton Novelty Gene	Monohybrid Cross Dihybrid Cross Gregor's Garden Incomplete Dominance Linked Genes and Crossing Over Sex-Linked Genes	
Chapter 10	DNA Replication Protein Synthesis	DNA Structure	DNA to RNA to Protein	Emerging Diseases Vaccines	The Hershey-Chase Experiment DNA and RNA Structure DNA Double Helix DNA Replication: An Overview Overview of Protein Synthesis Transcription RNA Processing Translation Simplified Viral Reproductive Cycle Phage Lytic Cycle Phage Lysogenic and Lytic Cycles Retrovirus (HIV) Reproductive Cycle	
Chapter 11			Control of Gene Expression	Cloning Fighting Cancer	The *lac* Operon in *E. coli* Overview: Control of Gene Regulation Control of Transcription Post-Transcriptional Control Mechanisms Review: Control of Gene Regulation Signal-Transduction Pathway Causes of Cancer	You Decide: Do Cell Phones Cause Brain Cancer? You Decide: Is Second-Hand Smoke Dangerous?
Chapter 12		DNA Profiling Techniques	DNA Technology	DNA Forensics Transgenics	Applications of DNA Technology Restriction Enzymes Cloning a Gene in Bacteria DNA Fingerprinting Gel Electrophoresis of DNA Analyzing DNA Fragments Using Gel Electrophoresis The Human Genome Project: Genes on Human Chromosome 17 Making Decisions About DNA Technology: Golden Rice	

	Resources for Visual Learners		Resources for Auditory Learners	Relevant Biology Topics in the News and Media	Resources to Study and Practice Before a Test	Resources to Build Graphing Skills
	BIOFLIX	**VIDEO TUTOR SESSIONS**	**MP3 TUTOR SESSIONS**	**DISCOVERY CHANNEL VIDEOS**	**ACTIVITIES**	**YOU DECIDE & GRAPHIT!**
Chapter 13	Mechanisms of Evolution		Natural Selection	Antibiotics Charles Darwin	The Voyage of the *Beagle*: Darwin's Trip Around the World Darwin and the Galápagos Islands Reconstructing Forelimbs Genetic Variation from Sexual Recombination Causes of Microevolution	
Chapter 14		Phylogenetic Trees	Speciation	Mass Extinctions Charles Darwin	Overview of Macroevolution Polyploid Plants Allometric Growth A Scrolling Geologic Record Classification Schemes	You Decide: Can We Prevent Species Extinction?
Chapter 15		Survey of Biodiversity	Microbial Life	Antibiotics Bacteria Early Life Tasty Bacteria	The History of Life Prokaryotic Cell Structure and Function Classification of Prokaryotes	
Chapter 16		Survey of Biodiversity Phylogenetic Trees	Evolution of Plants Alternation of Generations	Colored Cotton Fungi Leafcutter Ants Plant Pollination Trees	Terrestrial Adaptations of Plants Highlights of Plant Phylogeny Moss Life Cycle Fern Life Cycle Pine Life Cycle Angiosperm Life Cycle Madagascar and the Biodiversity Crisis Fungal Reproduction and Nutrition	
Chapter 17		Survey of Biodiversity Phylogenetic Trees	Human Evolution	Invertebrates	Animal Phylogenetic Tree Characteristics of Invertebrates Characteristics of Chordates Primate Diversity Human Evolution	
Chapter 18			Ecological Hierarchy Global Warming	Rain Forests	Science, Technology, and Society: DDT Adaptations to Biotic and Abiotic Factors Aquatic Biomes Terrestrial Biomes Water Pollution from Nitrates The Greenhouse Effect	You Decide: Does Human Activity Cause Global Warming? GraphIt!: Forestation Change GraphIt!: Global Fresh Water Resources GraphIt!: Municipal Solid Waste Trends in the U.S. GraphIt!: Atmospheric CO_2 and Temperature Changes GraphIt!: Prospects for Renewable Energy

	Resources for Visual Learners		Resources for Auditory Learners	Relevant Biology Topics in the News and Media	Resources to Study and Practice Before a Test	Resources to Build Graphing Skills
	BIOFLIX	**VIDEO TUTOR SESSIONS**	**MP3 TUTOR SESSIONS**	**DISCOVERY CHANNEL VIDEOS**	**ACTIVITIES**	**YOU DECIDE & GRAPHIT!**
Chapter 19	Population Ecology			Introduced Species Emerging Disease	Techniques for Estimating Population Density and Size Investigating Survivorship Curves Madagascar and the Biodiversity Crisis Introduced Species: Fire Ants Science, Technology, and Society: DDT Human Population Growth Analyzing Age-Structure Pyramids	You Decide: Can We Prevent Species Extinction? GraphIt!: Global Fisheries and Overfishing GraphIt!: Age Pyramids and Population Growth GraphIt!: Municipal Solid Waste Trends in the U.S.
Chapter 20	The Carbon Cycle		Energy Flow in Ecosystems	Introduced Species Leafcutter Ants	Introduced Species: Fire Ants Interspecific Interactions Food Webs Exploring Island Biogeography Primary Succession Energy Flow and Chemical Cycling Pyramids of Production The Carbon Cycle The Nitrogen Cycle Water Pollution from Nitrates Madagascar and the Biodiversity Crisis Conservation Biology Review	You Decide: Can We Prevent Species Extinction? GraphIt!: Global Fisheries and Overfishing GraphIt!: Animal Food Production Efficiency and Food Policy GraphIt!: Forestation Change GraphIt!: Global Fresh Water Resources GraphIt!: Species Area Effect and Island Biogeography
Chapter 21			Animal Structure Kidney Function	An Introduction to the Human Body	The Levels of Life Card Game Overview of Animal Tissues Epithelial Tissue Connective Tissue Muscle Tissue Nervous Tissue Correlating Structure and Function of Cells Regulation: Negative and Positive Feedback Structure of the Human Excretory System Nephron Function Control of Water Reabsorption	
Chapter 22			The Human Digestive System	Nutrition	How Animals Eat Food Digestive System Function Analyzing Food Labels Case Studies of Nutritional Disorders	You Decide: Low-Fat or Low-Carb Diets—Which Is Healthier? You Decide: Is Ephedra Safe and Effective?

	Resources for Visual Learners		Resources for Auditory Learners	Relevant Biology Topics in the News and Media	Resources to Study and Practice Before a Test	Resources to Build Graphing Skills
	BIOFLIX	VIDEO TUTOR SESSIONS	MP3 TUTOR SESSIONS	DISCOVERY CHANNEL VIDEOS	ACTIVITIES	YOU DECIDE & GRAPHIT!
Chapter 23			The Human Circulatory System Human Respiration	Blood	Path of Blood Flow Mammalian Cardiovascular System Structure Mammalian Cardiovascular System Function The Human Respiratory System Transport of Respiratory Gases	You Decide: Is Second-Hand Smoke Dangerous?
Chapter 24			The Human Immune System	Fighting Cancer Vaccines	Immune Responses HIV Reproductive Cycle	
Chapter 25	Homeostasis		Homeostasis and the Endocrine System	Endocrine System	Overview of Cell Signaling Peptide Hormone Action Steroid Hormone Action Human Endocrine Glands and Hormones	
Chapter 26			Spermatogenesis and Oogenesis The Female Reproductive Cycle Embryonic Development		Reproductive System of the Human Male Reproductive System of the Human Female Sea Urchin Development Frog Development	
Chapter 27	How Neurons Work How Synapses Work Muscle Contraction		Neurons and Electrical Potentials The Human Brain Sensory Receptors Muscle	Muscles and Bones Teen Brains	Neuron Structure Nerve Signals: Action Potentials Signal Transmission at a Chemical Synapse Structure and Function of the Eye Human Skeleton Skeletal Muscle Structure Muscle Contraction	
Chapter 28	Tour of a Plant Cell		From Flower to Fruit	Plant Pollination Colored Cotton	Root, Stem, and Leaf Sections Primary and Secondary Growth Angiosperm Life Cycle Seed and Fruit Development	
Chapter 29	Water Transport in Plants		Transpiration	Trees	How Plants Obtain Minerals from Soil Transport of Xylem Sap Translocation of Phloem Sap Leaf Abscission (Drop) Flowering Lab	GraphIt!: Global Soil Degradation

Glossary

A

abiotic factor (ā'-bī-ot'-ik)
A nonliving component of an ecosystem, such as air, water, light, minerals, or temperature.

abiotic reservoir
The part of an ecosystem where a chemical, such as carbon or nitrogen, accumulates or is stockpiled outside of living organisms.

ABO blood groups
Genetically determined classes of human blood that are based on the presence or absence of carbohydrates A and B on the surface of red blood cells. The ABO blood group phenotypes, also called blood types, are A, B, AB, and O.

abscisic acid (ab-sis'-ik)
A plant hormone that inhibits cell division and promotes dormancy. Abscisic acid interacts with gibberellins in regulating seed germination.

absorption
The uptake of small nutrient molecules by an organism's own body. In animals, absorption is the third main stage of food processing, following digestion; in fungi, it is acquisition of nutrients from the surrounding medium.

acclimation (ak-li-mā-shun)
Physiological adjustment that occurs gradually, though still reversibly, in response to an environmental change.

achondroplasia (uh-kon'-druh-plā'-zhuh)
A form of human dwarfism caused by a single dominant allele. The homozygous condition is lethal.

acid
A substance that increases the hydrogen ion (H^+) concentration in a solution.

action potential
A self-propagating change in the voltage across the plasma membrane of a neuron; a nerve signal.

activation energy
The amount of energy that reactants must absorb before a chemical reaction will start.

activator
A protein that switches on a gene or group of genes by binding to DNA.

active site
The part of an enzyme molecule where a substrate molecule attaches (by means of weak chemical bonds); typically, a pocket or groove on the enzyme's surface.

active transport
The movement of a substance across a biological membrane against its concentration gradient, aided by specific transport proteins and requiring input of energy (often as ATP).

adaptive defense
An immune system defense that is found only in vertebrates and that must be activated by exposure to a specific invader.

adenine (A) (ad'-uh-nēn)
A double-ring nitrogenous base found in DNA and RNA.

adhesion
The attraction between different kinds of molecules.

adipose tissue
A type of connective tissue whose cells contain fat.

ADP Adenosine diphosphate (a-den'-ō-sēn dī-fos'-fāt).
A molecule composed of adenosine and two phosphate groups. The molecule ATP is made by combining a molecule of ADP with a third phosphate in an energy-consuming reaction.

adrenal cortex (uh-drē'-nul)
The outer portion of an adrenal gland, controlled by ACTH from the anterior pituitary; secretes hormones called glucocorticoids.

adrenal gland
One of a pair of endocrine glands, located atop each kidney in mammals, composed of an outer cortex and a central medulla.

adrenal medulla (uh-drē'-nul muh-dul'-uh)
The central portion of an adrenal gland, controlled by nerve signals; secretes the fight-or-flight hormones epinephrine and norepinephrine.

adult stem cell
A cell present in adult tissues that generates replacements for nondividing differentiated cells.

aerobic (ār-ō'-bik)
Containing or requiring molecular oxygen (O_2).

age structure
The relative number of individuals of each age in a population.

AIDS
Acquired immunodeficiency syndrome; the late stages of HIV infection, characterized by a reduced number of T cells; usually results in death caused by opportunistic infections.

alga (al'-guh)
(plural, **algae**) An informal term that describes a great variety of protists, most of which are unicellular or colonial photosynthetic autotrophs with chloroplasts. Heterotrophic and multicellular protists closely related to unicellular autotrophs are also regarded as algae.

alimentary canal (al'-uh-men'-tuh-rē)
A digestive tube running between a mouth and an anus; also called a digestive tract.

allantois (uh-lan'-tuh-wuhs)
In animals, a membrane that develops from the yolk sac and helps dispose of the embryo's nitrogenous wastes. The allantois forms part of the umbilical cord in mammals.

allele (uh-lē'-ul)
An alternative version of a gene.

allergen (al'-er-jen)
An otherwise harmless antigen that causes an allergic reaction.

allergy
An abnormal sensitivity to an antigen. Symptoms are triggered by histamines released from mast cells.

allopatric speciation
The formation of a new species as a result of an ancestral population becoming isolated by a geographic barrier. *See also* sympatric speciation.

alternation of generations
A life cycle in which there is both a multicellular diploid form, the sporophyte, and a multicellular haploid form, the gametophyte; a characteristic of plants and multicellular green algae.

alternative RNA splicing
A type of regulation at the RNA-processing level in which different mRNA molecules are produced from the same primary transcript, depending on which RNA segments are treated as exons and which as introns.

alveolus (al-vē'-oh-lus)
(plural, **alveoli**) One of millions of tiny sacs within the vertebrate lungs where gas exchange occurs.

Alzheimer's disease
A form of mental deterioration, or dementia, characterized by confusion and memory loss.

amino acid (uh-mēn'-ō)
An organic molecule containing a carboxyl group, an amino group, a hydrogen atom, and a variable side chain; serves as the monomer of proteins.

amnion (am'-nē-on)
In vertebrate animals, the membrane that encloses the fluid-filled amniotic sac containing the embryo.

amniote
Member of a clade of tetrapods that has an amniotic egg containing specialized membranes that protect the embryo. Amniotes include mammals and reptiles (including birds).

amniotic egg (am'-nē-ot'-ik)
A shelled egg in which an embryo develops within a fluid-filled amniotic sac and is nourished by yolk. Produced by reptiles (including birds) and egg-laying mammals, it enables them to complete their life cycles on dry land.

amoeba (uh-mē'-buh)
A type of protist characterized by great structural flexibility and the presence of pseudopodia.

amphibian
Member of a class of vertebrate animals that includes frogs and salamanders.

anaerobic (an'-ār-ō'-bik)
Lacking or not requiring molecular oxygen (O_2).

analogy
The similarity of structure between two species that are not closely related, attributable to convergent evolution.

anaphase
The third stage of mitosis, beginning when sister chromatids separate from each other and ending when a complete set of daughter chromosomes has arrived at each of the two poles of the cell.

anatomy
The study of the structure of an organism and its parts.

androgen (an'-drō-jen)
A sex hormone secreted by the gonads that promotes the development and maintenance of the male reproductive system and male body features.

anemia (uh-nē'-me-ah)
A condition in which an abnormally low amount of hemoglobin or a low number of red blood cells results in the body cells not receiving enough oxygen.

angiosperm (an'-jē-ō-sperm)
A flowering plant, which forms seeds inside a protective chamber called an ovary.

animal
A eukaryotic, multicellular, heterotrophic organism that obtains nutrients by ingestion.

annelid (an'-uh-lid)
A segmented worm. Annelids include earthworms, polychaetes, and leeches.

annual
A plant that completes its life cycle in a single year or growing season.

anorexia nervosa
An eating disorder that results in self-starvation due to an intense fear of gaining weight, even when the person is underweight.

antagonistic hormones
Two hormones that have opposite effects.

anterior pituitary (puh-tū'-uh-tār-ē)
An endocrine gland, adjacent to the hypothalamus and the posterior pituitary, that synthesizes and secretes several hormones, including some that control the activity of other endocrine glands.

anther
A sac in which pollen grains develop, located at the tip of a flower's stamen.

anthropoid (an'-thruh-poyd)
A member of a primate group made up of the apes (gibbons, orangutans, gorillas, chimpanzees, and bonobos), monkeys, and humans.

antibody (an'-tih-bod'-ē)
A protein that is secreted by a B cell and attaches to a specific kind of antigen, helping counter its effects.

anticodon (an'-tī-kō'-don)
On a tRNA molecule, a specific sequence of three nucleotides that is complementary to a codon triplet on mRNA.

antigen (an'-tuh-jen)
Any molecule that elicits a response from a lymphocyte.

anus
The digestive system opening through which undigested materials are expelled.

aphotic zone (ā-fō'-tik)
The region of an aquatic ecosystem beneath the photic zone, where light levels are too low for photosynthesis to take place.

apical dominance (ā'-pik-ul)
In a plant, the hormonal inhibition of axillary buds by a terminal bud.

apical meristem (ā'-pik-ul mār'-uh-stem)
A meristem at the tip of a plant root or in the terminal or axillary bud of a shoot.

apicomplexan (ap'-ē-kom-pleks'-un)
A type of parasitic protozoan. Some apicomplexans cause serious human disease.

appendicular skeleton (ap'-en-dik'-yū-ler)
Components of the skeletal system that support the arms and legs of a land vertebrate; bones of the limbs, shoulders, and pelvis. See also axial skeleton.

appendix (uh-pen'-dix)
A small, fingerlike extension near the union of the small and large intestines.

aqueous humor (ā'-kwē-us hyū'-mer)
A liquid in the space between the lens and cornea in the vertebrate eye that helps maintain the shape of the eye, supplies nutrients and oxygen to its tissues, and disposes of its wastes.

aqueous solution (ā'-kwē-us)
A solution in which water is the solvent.

arachnid
A member of a major arthropod group that includes spiders, scorpions, ticks, and mites.

Archaea (ar-kē'-uh)
One of two prokaryotic domains of life, the other being Bacteria.

archaean
(plural, **archaea**) An organism that is a member of the domain Archaea.

arteriole (ar-tār'-ē-ōl)
A small vessel that conveys blood between an artery and a capillary bed.

artery
A vessel that carries blood away from the heart to other parts of the body.

arthritis (ar-thrī'-tis)
A skeletal disorder characterized by inflamed joints and deterioration of the cartilage between bones.

arthropod (ar'-thruh-pod)
A member of the most diverse phylum in the animal kingdom; includes the horseshoe crab, arachnids (for example, spiders, ticks, scorpions, and mites), crustaceans (for example, crayfish, lobsters, crabs, and barnacles), millipedes, centipedes, and insects. Arthropods are characterized by a chitinous exoskeleton, molting, jointed appendages, and a body formed of distinct groups of segments.

asexual reproduction
The creation of genetically identical offspring by a single parent, without the participation of gametes (sperm and egg).

association areas
Areas of the cerbral cortex where different types of incoming sensory information are integrated.

astigmatism (uh-stig'-muh-tizm)
Blurred vision caused by a misshapen lens or cornea.

atherosclerosis (ath'-uh-rō'-skluh-rō'-sis)
A cardiovascular disease in which growths called plaques develop on the inner walls of the arteries, narrowing the passageways through which blood can flow.

atom
The smallest unit of matter that retains the properties of an element.

atomic number
The number of protons in each atom of a particular element.

ATP Adenosine triphosphate (a-den'-ō-sēn trī-fos'-fāt).
A molecule composed of adenosine and three phosphate groups; the main energy source for cells.

ATP synthase
A protein cluster, found in a cellular membrane (including the inner membrane of mitochondria, the thylakoid membrane of chloroplasts, and the plasma membrane of prokaryotes), that uses the energy of a hydrogen ion concentration gradient to make ATP from ADP. An ATP synthase provides a port through which hydrogen ions (H^+) diffuse.

atrium (ā'-trē-um)
(plural, **atria**) A heart chamber that receives blood from the body or lungs via veins.

auditory canal
Part of the outer ear that channels sound waves from the pinna, or outer body surface, to the eardrum.

autoimmune disease
An immunological disorder in which the immune system improperly attacks the body's own molecules.

autonomic nervous system (ot'-ō-nom'-ik)
A component of the peripheral nervous system of vertebrates that regulates the internal environment. The autonomic nervous system is made up of the sympathetic and parasympathetic subdivisions. The autonomic nervous system is primarily under involuntary control.

autosome
A chromosome not directly involved in determining the sex of an organism; in mammals, for example, any chromosome other than X or Y.

autotroph (ot'-ō-trōf)
An organism that makes its own food from inorganic ingredients, thereby sustaining itself without eating other organisms or their molecules. Plants, algae, and photosynthetic bacteria are autotrophs.

auxins (ok'-sin)
A group of plant hormones whose chief effect is to promote the growth and development of shoots.

AV (atrioventricular) node
A region of muscle tissue between the heart's right atrium and right ventricle. It generates electrical impulses that primarily cause the ventricles to contract.

axial skeleton (ak'-sē-ul)
Components of the skeletal system that support the central trunk of the body; the skull, backbone, and rib

cage in a vertebrate. *See also* appendicular skeleton.

axillary bud *(ak'-sil-ār-ē)*
An embryonic shoot present in the angle formed by a leaf and stem.

axon *(ak'-son)*
A neuron fiber that extends from the cell body and conducts signals to another neuron or to an effector cell.

B

B cell
A type of lymphocyte that matures in the bone marrow and later produces antibodies; responsible for the humoral immune response. *See also* T cell.

bacillus *(buh-sil'-us)*
(plural, **bacilli**) A rod-shaped prokaryotic cell.

Bacteria
One of two prokaryotic domains of life, the other being Archaea.

bacteriophage *(bak-tē r'-ē-ō-fāj)*
A virus that infects bacteria; also called a phage.

bacterium
(plural, **bacteria**) An organism that is a member of the domain Bacteria.

ball-and-socket joint
A joint that allows rotation and movement in several planes. Examples in humans are the shoulder and hip joints.

bark
All the tissues external to the vascular cambium in a plant with secondary growth. Bark is made up of secondary phloem, cork cambium, and cork.

basal metabolic rate (BMR)
The number of kilocalories a resting animal requires to fuel its essential body processes for a given time.

base
A substance that decreases the hydrogen ion (H^+) concentration in a solution.

benign tumor
An abnormal mass of cells that remains at its original site in the body.

benthic realm
A seafloor or the bottom of a freshwater lake, pond, river, or stream. The benthic realm is occupied by communities of organisms known as benthos.

biennial
A plant that completes its life cycle in two years.

bilateral symmetry
An arrangement of body parts such that an organism can be divided equally by a single cut passing longitudinally through it. A bilaterally symmetrical organism has mirror-image right and left sides.

bile
A solution of salts secreted by the liver that emulsifies fats and aids in their digestion.

binary fission
A means of asexual reproduction in which a parent organism, often a single cell, divides into two individuals of about equal size.

binomial
A two-part latinized name of a species; for example, *Homo sapiens*.

biodiversity
The variety of living things; includes genetic diversity, species diversity, and ecosystem diversity.

biodiversity hot spot
A small geographic area that contains a large number of threatened or endangered species and an exceptional concentration of endemic species (those found nowhere else).

biogenesis
The principle that all life arises by the reproduction of preexisting life.

biogeochemical cycle
Any of the various chemical circuits occurring in an ecosystem, involving both biotic and abiotic components of the ecosystem.

biogeography
The study of the geographic distribution of species.

biological community
See community.

biological control
The intentional release of a natural enemy to attack a pest population.

biological magnification
The accumulation of persistent chemicals in the living tissues of consumers in food chains.

biological species concept
The definition of a species as a population or group of populations whose members have the potential in nature to interbreed and produce fertile offspring.

biology
The scientific study of life.

biomass
The amount, or mass, of living organic material in an ecosystem.

biome *(bī'-ōm)*
A major terrestrial or aquatic life zone, characterized by vegetation type in terrestrial biomes or the physical environment in aquatic biomes.

biophilia
The human desire to affiliate with other life in its many forms.

bioremediation
The use of living organisms to detoxify and restore polluted and degraded ecosystems.

biosphere
The global ecosystem; the entire portion of Earth inhabited by life; all of life and where it lives.

biotechnology
The manipulation of living organisms to perform useful tasks. Today, biotechnology often involves DNA technology.

biotic factor *(bī-ot'-ik)*
A living component of a biological community; any organism that is part of an individual's environment.

bipolar disorder
Depressive mental illness characterized by extreme mood swings; also called manic-depressive disorder.

bird
Member of a group of reptiles with feathers and adaptations for flight.

birth control pill
A chemical contraceptive that inhibits ovulation, retards follicular development, or alters a woman's cervical mucus to prevent sperm from entering the uterus.

bivalve
A member of a group of molluscs that includes clams, mussels, scallops, and oysters.

blade
The flattened portion of a typical leaf.

blastocyst *(blas'-tuh-sist)*
A mammalian embryo made up of a hollow ball of cells that results from cleavage and that implants in the mother's endometrium.

blastula *(blas'-tyū-luh)*
An embryonic stage that marks the end of cleavage during animal development; a hollow ball of cells in many species.

blood pressure
The force that blood exerts against the walls of the blood vessels.

blood
A type of connective tissue with a fluid matrix called plasma in which blood cells are suspended.

body cavity
A fluid-filled space separating the digestive tract from the outer body wall.

body mass index (BMI)
A ratio of weight to height used as a measure of obesity.

body segmentation
Subdivision of an animal's body into a series of repeated parts called segments.

bone
A type of connective tissue consisting of living cells held in a rigid matrix of collagen fibers embedded in calcium salts.

bony fish
A fish that has a stiff skeleton reinforced by calcium salts.

bottleneck effect
Genetic drift resulting from a drastic reduction in population size.

brain
The master control center of the central nervous system, which is involved in regulating and controlling body activity and interpreting information from the senses.

brainstem
A functional unit of the vertebrate brain, composed of the midbrain, medulla oblongata, and pons; serves mainly as a sensory filter, selecting which information reaches higher brain centers.

breathing
The ventilation of the lungs by alternate inhalation and exhalation, supplying a lung or gill with O_2-rich air or water and expelling CO_2-rich air or water.

bronchiole *(brong'-kē-ōl)*
A thin breathing tube that branches from a bronchus within a lung.

bronchus (brong'-kus)
(plural, **bronchi**) One of a pair of breathing tubes that branch from the trachea into the lungs.

bryophyte (brī'-uh-fīt)
A type of plant that lacks xylem and phloem; a nonvascular plant. Bryophytes include mosses and their close relatives.

budding
A means of asexual reproduction in which a new individual splits off after developing from an outgrowth of a parent.

buffer
A chemical substance that resists changes in pH by accepting hydrogen ions from or donating hydrogen ions to solutions.

bulimia
An eating disorder characterized by episodic binge eating followed by purging through induced vomiting, abuse of laxatives, or excessive exercise.

C

C$_4$ plant
A plant that prefaces the Calvin cycle with reactions that incorporate CO_2 into four-carbon compounds, the end product of which supplies CO_2 for the Calvin cycle.

calcitonin (kal'-sih-tōn'-in)
A peptide hormone secreted by the thyroid gland that lowers the blood calcium level.

calorie
The amount of energy that raises the temperature of 1 g of water by 1°C.

Calvin cycle
The second of two stages of photosynthesis; a cyclic series of chemical reactions that occur in the stroma of a chloroplast, using the carbon in CO_2 and the ATP and NADPH produced by the light reactions to make the energy-rich sugar molecule G3P, which is later used to produce glucose.

CAM plant
A plant that uses the following adaptation for photosynthesis in arid conditions: Carbon dioxide entering open stomata during the night is converted to organic compounds, which release CO_2 for the Calvin cycle during the day, when stomata are closed.

cancer
A malignant growth or tumor caused by abnormal and uncontrolled cell division.

cap
Extra nucleotides added to the beginning of an RNA transcript in the nucleus of a eukaryotic cell.

capillary (kap'-il-ār-ē)
A microscopic blood vessel that conveys blood between an artery and a vein or between an arteriole and a venule; enables the exchange of nutrients and dissolved gases between the blood and interstitial fluid.

carbohydrate (kar'-bō-hī'-drāt)
A biological molecule consisting of simple single-monomer sugars (monosaccharides), two-monomer sugars (disaccharides), and other multi-unit sugars (polysaccharides).

carbon footprint
The amount of greenhouse gas emitted as a result of the actions of a person, nation, or other entity.

carcinogen (kar-sin'-uh-jin)
A cancer-causing agent, either high-energy radiation (such as X-rays or UV light) or a chemical.

cardiac cycle (kar'-dē-ak)
The alternating contractions and relaxations of the heart.

cardiac muscle
Striated muscle that forms the contractile tissue of the heart.

cardiovascular disease (kar'-dē-ō-vas'-kyū-ler)
A set of diseases of the heart and blood vessels.

cardiovascular system
A closed circulatory system, found in vertebrates, with a heart and a branching network of arteries, capillaries, and veins.

carnivore
An animal that mainly eats other animals. See also herbivore; omnivore.

carpel (kar'-pul)
The egg-producing part of a flower, consisting of a stalk with an ovary at the base and a stigma, which traps pollen, at the tip.

carrier
An individual who is heterozygous for a recessively inherited disorder and who therefore does not show symptoms of that disorder.

carrying capacity
The maximum population size that a particular environment can sustain.

cartilage (kar'-ti-lij)
A type of connective tissue consisting of living cells embedded in a rubbery matrix with collagen fibers.

cartilaginous fish (kar-ti-laj'-uh-nus)
A fish that has a flexible skeleton made of cartilage.

case study
An in-depth examination of an actual investigation.

cell body
The part of a cell, such as a neuron, that houses the nucleus.

cell cycle
An ordered sequence of events (including interphase and the mitotic phase) that extends from the time a eukaryotic cell is first formed from a dividing parent cell until its own division into two cells.

cell cycle control system
A cyclically operating set of proteins that triggers and coordinates events in the eukaryotic cell cycle.

cell division
The reproduction of a cell.

cell junction
A structure that connects animal cells to one another in a tissue.

cell plate
A membranous disk that forms across the midline of a dividing plant cell. During cytokinesis, the cell plate grows outward, accumulating more cell wall material and eventually fusing into a new cell wall.

cell theory
The theory that all living things are composed of cells and that all cells come from other cells.

cell wall
A protective layer external to the plasma membrane in plant cells, bacteria, fungi, and some protists; protects the cell and helps maintain its shape.

cell-mediated immune response
The type of adaptive immune response that involves T cells, lymphocytes that fight body cells infected with pathogens. See also humoral immune response.

cellular differentiation
Specialization in the structure and function of cells that occurs during the development of an organism; results from selective activation and deactivation of the cells' genes.

cellular respiration
The aerobic harvesting of energy from food molecules; the energy-releasing chemical breakdown of food molecules, such as glucose, and the storage of potential energy in a form that cells can use to perform work; involves glycolysis, the citric acid cycle, the electron transport chain, and chemiosmosis.

cellular slime mold
A type of protist that has unicellular amoeboid cells and a multicellular reproductive body in its life cycle.

cellulose (sel'-yū-lōs)
A large polysaccharide composed of many glucose monomers linked into cable-like fibrils that provide structural support in plant cell walls. Because cellulose cannot be digested by animals, it acts as roughage, or fiber, in the diet.

centipede
A carnivorous terrestrial arthropod that has one pair of long legs for each of its numerous body segments, with the front pair modified as poison claws.

central nervous system (CNS)
The integration and command center of the nervous system; the brain and, in vertebrates, the spinal cord.

central vacuole (vak'-yū-ōl)
A membrane-enclosed sac occupying most of the interior of a mature plant cell, having diverse roles in reproduction, growth, and development.

centromere (sen'-trō-mēr)
The region of a chromosome where two sister chromatids are joined and where spindle microtubules attach during mitosis and meiosis. The

centromere divides at the onset of anaphase during mitosis and anaphase II of meiosis.

centrosome (sen'-trō-sōm)
Material in the cytoplasm of a eukaryotic cell that gives rise to microtubules; important in mitosis and meiosis; functions as a microtubule-organizing center.

cephalopod
A member of a group of molluscs that includes squids and octopuses.

cerebellum (sār'-ruh-bel'-um)
Part of the vertebrate hindbrain; mainly a planning center that interacts closely with the cerebrum in coordinating body movement.

cerebral cortex (suh-rē'-brul kor'-teks)
A highly folded layer of tissue that forms the surface of the cerebrum. In humans, it contains integrating centers for higher brain functions, such as reasoning, speech, language, and imagination.

cerebrospinal fluid (suh-rē'-brō-spī'-nul)
Fluid that surrounds, cushions, and nourishes the brain and spinal cord and protects them from infection.

cerebrum (suh-rē'-brum)
The largest, most sophisticated, and most dominant part of the vertebrate forebrain, made up of right and left cerebral hemispheres. The cerebrum contains the cerebral cortex.

cervix (ser'-viks)
The narrow neck at the bottom of the uterus, which opens into the vagina.

chaparral (shap-uh-ral')
A terrestrial biome limited to coastal regions where cold ocean currents circulate offshore, creating mild, rainy winters and long, hot, dry summers; also known as the Mediterranean biome. Chaparral vegetation is adapted to fire.

character
A heritable feature that varies among individuals within a population, such as flower color in pea plants.

charophyte (kār'-uh-fīt')
A member of the green algal group that shares features with land plants. Charophytes are considered the closest relatives of land plants; modern charophytes and modern plants likely evolved from a common ancestor.

chemical bond
An attraction between two atoms resulting from a sharing of outer-shell electrons or the presence of opposite charges on the atoms. The bonded atoms gain complete outer electron shells.

chemical cycling
The use and reuse of chemical elements such as carbon within an ecosystem.

chemical digestion
The breakdown of food molecules through the action of enzymes into small molecules that can be absorbed by the body.

chemical energy
Energy stored in the chemical bonds of molecules; a form of potential energy.

chemical reaction
A process leading to chemical changes in matter, involving the making and/or breaking of chemical bonds.

chemoreceptor (kē'-mō-rē-sep'-ter)
A sensory receptor that detects chemical changes within the body or a specific kind of chemical in the external environment; chemoreceptors are involved in our senses of taste and smell.

chemotherapy (kē'-mo-thār'-uh-pē)
Treatment for cancer in which drugs are administered to disrupt cell division of the cancer cells.

chiasma (kī-az'-muh)
(plural, **chiasmata**) The microscopically visible site where crossing over has occurred between chromatids of homologous chromosomes during prophase I of meiosis.

chlamydia
A common sexually transmitted disease caused by a bacterial infection. Its primary symptoms are genital discharge and burning during urination. It is easily treatable with antibiotics.

chlorophyll (klor'-ō-fil)
A light-absorbing pigment in chloroplasts that plays a central role in converting solar energy to chemical energy.

chlorophyll a (klor'-ō-fil ā)
A green pigment in chloroplasts that participates directly in the light reactions.

chloroplast (klō'-rō-plast)
An organelle found in plants and photosynthetic protists. Enclosed by two concentric membranes, a chloroplast absorbs sunlight and uses it to power the synthesis of organic food molecules (sugars).

chordate (kōr'-dāt)
An animal that at some point during its development has a dorsal, hollow nerve cord, a notochord, pharyngeal slits, and a post-anal tail. Chordates include lancelets, tunicates, and vertebrates.

chorion (kōr'-ē-on)
In animals, the outermost membrane, which in mammals becomes part of the placenta.

chorionic villus (kōr-ē-on'-ik vil'-us)
(plural, **villi**) An outgrowth of the chorion, containing embryonic blood vessels. As part of the placenta, chorionic villi absorb nutrients and oxygen from the mother's bloodstream and pass wastes into the mother's bloodstream.

chromatin (krō'-muh-tin)
The combination of DNA and proteins that constitutes chromosomes; often used to refer to the diffuse, very extended form taken by the chromosomes when a eukaryotic cell is not dividing.

chromosome (krō'-muh-sōm)
A gene-carrying structure found in the nucleus of a eukaryotic cell and most visible during mitosis and meiosis; also, the main gene-carrying structure of a prokaryotic cell. Each chromosome consists of one very long threadlike DNA molecule and associated proteins. *See also* chromatin.

chromosome theory of inheritance
A basic principle in biology stating that genes are located on chromosomes and that the behavior of chromosomes during meiosis accounts for inheritance patterns.

chyme (kīm)
A mixture of recently swallowed food and gastric juice.

ciliate (sil'-ē-it)
A type of protozoan that moves and feeds by means of cilia.

cilium (sil'-ē-um)
(plural, **cilia**) A short appendage that propels some protists through the water and moves fluids across the surface of many tissue cells in animals.

circulatory system
The organ system that transports materials such as nutrients, O_2, and hormones to body cells and transports CO_2 and other wastes from body cells.

citric acid cycle
The metabolic cycle that is fueled by acetyl CoA formed after glycolysis in cellular respiration. Chemical reactions in the cycle complete the metabolic breakdown of glucose molecules to carbon dioxide. The cycle occurs in the matrix of mitochondria and supplies most of the NADH molecules that carry energy to the electron transport chains. Also referred to as the Krebs cycle.

clade
An ancestral species and all its descendants—a distinctive branch in the tree of life.

cladistics (kluh-dis'-tiks)
The study of evolutionary history; specifically, the scientific search for clades, taxonomic groups composed of an ancestral species and all its descendants.

class
In classification, the taxonomic category above order.

cleavage (klē-vij)
(1) Cytokinesis in animal cells and in some protists, characterized by pinching in of the plasma membrane. (2) In animal development, the succession of rapid cell divisions without cell growth, converting the animal zygote to a ball of cells.

cleavage furrow
The first sign of cytokinesis during cell division in an animal cell; a shallow groove in the cell surface near the old metaphase plate.

clitoris (klit'-uh-ris)
An organ in the female that engorges with blood and becomes erect during sexual arousal.

clonal selection (klōn'-ul)
The production of a lineage of genetically identical cells that recognize and attack the specific antigen that stimulated their proliferation. Clonal selection is the mechanism that underlies the immune system's specificity and memory of antigens.

clone

As a verb, to produce genetically identical copies of a cell, organism, or DNA molecule. As a noun, the collection of cells, organisms, or molecules resulting from cloning; also (colloquially), a single organism that is genetically identical to another because it arose from the cloning of a somatic cell.

closed circulatory system

A circulatory system in which blood is confined to vessels and is kept separate from the interstitial fluid.

cnidarian (nī-dār′-ē-an)

An animal characterized by cnidocytes, radial symmetry, a gastrovascular cavity, and a polyp or medusa body form. Cnidarians includes hydras, jellies, sea anemones, and corals.

coccus (kok′-us)

(plural, **cocci**) A spherical prokaryotic cell.

cochlea (kok′-lē-uh)

A coiled tube in the inner ear that contains the hearing organ, the organ of Corti.

codominance

The expression of two different alleles of a gene in a heterozygote.

codon (kō′-don)

A three-nucleotide sequence in mRNA that specifies a particular amino acid or polypeptide termination signal; the basic unit of the genetic code.

coelom (sē′-lōm)

A body cavity completely lined by tissue derived from mesoderm.

coevolution

Evolutionary change in which an adaptation in one species leads to a counteradaptation in a second species.

cohesion (kō-hē′-zhun)

The attraction between molecules of the same kind.

collenchyma cell (kuh-leng′-kuh-muh)

In plants, a type of cell with a thick primary wall and no secondary wall, functioning mainly in supporting growing parts.

colon (kō′-lun)

Most of the length of the large intestine, the tubular portion of the vertebrate alimentary canal between the small intestine and the rectum; functions mainly in water absorption and the formation of feces.

community

All the organisms inhabiting and potentially interacting in a particular area; an assemblage of populations of different species.

community ecology

The study of how interactions between species affect community structure and organization.

comparative anatomy

The comparison of body structures in different species.

competitive exclusion principle

The concept that populations of two species cannot coexist in a community if their niches are nearly identical. Using resources more efficiently and having a reproductive advantage, one of the populations will eventually outcompete and eliminate the other.

complement protein

A defensive blood protein that attacks pathogens directly as part of innate internal defense. Complement proteins can enhance phagocytosis and cut holes in microbial membranes, causing the cells to burst.

complementary DNA (cDNA)

A DNA molecule made in vitro using mRNA as a template and the enzyme reverse transcriptase. A cDNA molecule therefore corresponds to a gene but lacks the introns present in the DNA of the genome.

complete digestive tract

A digestive tube with two openings, a mouth and an anus.

compost

A soil-like mixture of decomposed organic matter used to fertilize plants.

compound

A substance containing two or more elements in a fixed ratio; for example, table salt (NaCl) consists of one atom of the element sodium (Na) for every atom of chlorine (Cl).

concentration gradient

An increase or decrease in the density of a chemical substance within a given region. Cells often maintain concentration gradients of hydrogen ions across their membranes. When a gradient exists, the ions or other chemical substances involved tend to move from where they are more concentrated to where they are less concentrated.

condom

A flexible sheath, usually made of thin rubber or latex, designed to cover the penis during sexual intercourse for contraceptive purposes or as a means of preventing sexually transmitted diseases.

cone

(1) In vertebrates, a photoreceptor cell in the retina, stimulated by bright light and enabling color vision. (2) In conifers, a reproductive structure bearing pollen or ovules.

conifer (kon′-nuh-fer)

A gymnosperm, or naked-seed plant, most of which produce cones.

coniferous forest (kō-nif′-rus)

A terrestrial biome characterized by conifers, cone-bearing evergreen trees.

connective tissue

Tissue consisting of cells held in an abundant extracellular matrix.

conservation biology

A goal-oriented science that seeks to understand and counter the loss of biodiversity.

conservation of energy

The principle that energy can neither be created nor destroyed.

consumer

An organism that obtains its food by eating plants or by eating animals that have eaten plants.

contraception

The deliberate prevention of pregnancy.

controlled experiment

A component of the process of science whereby a scientist carries out two parallel tests, an experimental test and a control test. The experimental test differs from the control by one factor, the variable.

convergent evolution

Adaptive change resulting in nonhomologous (analogous) similarities among organisms. Species from different evolutionary lineages come to resemble each other (evolve analogous structures) as a result of living in very similar environments.

coral reef

Tropical marine biome characterized by the hard skeletal structures secreted primarily by the resident cnidarians.

cork

The outermost protective layer of a plant's bark, produced by the cork cambium.

cork cambium

Meristem tissue that produces cork cells during secondary growth of a plant.

cornea (kor′-nē-uh)

The transparent front portion of the sclera that admits light into the vertebrate eye.

coronary artery (kōr′-uh-nār-ē)

A large blood vessel that conveys blood from the aorta to the tissues of the heart.

corpus callosum (kor′-pus kuh-lō′-sum)

A thick band of nerve fibers that connects the right and left cerebral hemispheres, enabling them to process information together.

corpus luteum (kor′-pus lū′-tē-um)

A small body of endocrine tissue that develops from an ovarian follicle after ovulation. The corpus luteum secretes progesterone and estrogen during pregnancy.

cortex

In plants, the ground tissue system of a root, which stores food and absorbs water and minerals that have passed through the epidermis. *See also* adrenal cortex; cerebral cortex.

corticosteroid

One of a family of steroid hormones, including glucocorticoids, synthesized and secreted by the adrenal cortex.

cotyledon (kot′-uh-lē′-don)

The first leaf that appears on an embryo of a flowering plant; a seed leaf. Monocot embryos have one cotyledon; dicot embryos have two.

covalent bond (kō-vā′-lent)

An attraction between atoms that share one or more pairs of outer-shell electrons.

crista (kris′-tuh)

(plural, **cristae**) A fold of the inner membrane of a mitochondrion. Enzyme molecules embedded in cristae make ATP.

cross

The cross-fertilization of two different varieties of an organism or of two different species; also called hybridization.

crossing over
The exchange of segments between chromatids of homologous chromosomes during prophase I of meiosis.

crustacean
A member of a major arthropod group that includes lobsters, crayfish, crabs, shrimps, and barnacles.

cryptic coloration
Adaptive coloration that makes an organism difficult to spot against its background.

culture
The accumulated knowledge, customs, beliefs, arts, and other human products that are socially transmitted over the generations.

cuticle (kyū'-tuh-kul)
(1) In animals, a tough, nonliving outer layer of the skin. (2) In plants, a waxy coating on the surface of stems and leaves that helps retain water.

cystic fibrosis (sis'-tik fī-brō'-sis)
A genetic disease that occurs in people with two copies of a certain recessive allele; characterized by an excessive secretion of mucus and consequent vulnerability to infection; fatal if untreated.

cytokinesis (sī'-tō-kuh-nē'-sis)
The division of the cytoplasm to form two separate daughter cells. Cytokinesis usually occurs during telophase of mitosis, and the two processes (mitosis and cytokinesis) make up the mitotic (M) phase of the cell cycle.

cytokinin (sī'-tō-kī'-nin)
One of a family of plant hormones that promote cell division, retard aging in flowers and fruits, and may counter the effects of auxins in regulating plant growth and development.

cytoplasm (sī'-tō-plaz'-um)
Everything inside a eukaryotic cell between the plasma membrane and the nucleus; consists of a semifluid medium and organelles; can also refer to the interior of a prokaryotic cell.

cytosine (C) (sī'-tuh-sēn)
A single-ring nitrogenous base found in DNA and RNA.

cytoskeleton
A meshwork of fine fibers in the cytoplasm of a eukaryotic cell; includes microfilaments, intermediate filaments, and microtubules.

cytotoxic T cell (sī'-tō-tok'-sik)
A type of lymphocyte that directly attacks body cells infected by pathogens.

D

decomposer
An organism that secretes enzymes that digest molecules in organic material and convert them to inorganic form.

dehydration reaction (dē-hī-drā'-shun)
A chemical process in which a polymer forms when monomers are linked by the removal of water molecules. One molecule of water is removed for each pair of monomers linked. A dehydration reaction is the opposite of a hydrolysis reaction.

denaturation (dē-nā'-chuh-rā'-shun)
A process in which a protein unravels, losing its specific conformation and hence function; can be caused by changes in pH or salt concentration or by high temperature; also refers to the separation of the two strands of the DNA double helix, caused by similar factors.

dendrite (den'-drīt)
A neuron fiber that conveys signals from its tip inward, toward the rest of the neuron; in a motor neuron, one of several short, branched extensions that convey nerve signals toward the cell body.

density-dependent factor
A limiting factor whose effects intensify with increasing population density.

density-independent factor
A limiting factor whose occurrence and effects are not related to population density.

dermal tissue system
In plants, the tissue system that forms an outer protective covering.

desert
A terrestrial biome characterized by low and unpredictable rainfall (less than 30 cm per year).

determinate growth
Termination of growth after reaching a certain size, as in most animals. See also indeterminate growth.

detritivore (di-trī'-tuh-vor)
An organism that consumes dead organic matter (detritus).

detritus (di-trī'-tus)
Dead organic matter.

diabetes mellitus (dī'-uh-bē'-tis mel'-uh-tus)
A human hormonal disease in which body cells cannot absorb enough glucose from the blood and become energy starved. Body fats and proteins are then consumed for their energy. Insulin-dependent diabetes results when the pancreas does not produce insulin. Non-insulin-dependent diabetes results when body cells fail to respond to insulin. A third type of diabetes, called gestational diabetes, can affect any pregnant woman, even one who has never shown symptoms of diabetes before.

dialysis (dī-al'-uh-sis)
Separation and disposal of metabolic wastes from the blood by mechanical means; an artificial method of performing the functions of the kidneys.

diaphragm (dī'-uh-fram)
(1) The sheet of muscle separating the chest cavity from the abdominal cavity in mammals. Its contraction expands the chest cavity, and its relaxation reduces it. (2) A dome-shaped rubber cap that covers a woman's cervix, serving as a barrier method of contraception.

diastole (dī-as'-tuh-lē)
The stage of the cardiac cycle in which the heart muscle is relaxed, allowing the chambers to fill with blood. See also systole.

diatom (dī'-uh-tom)
A unicellular photosynthetic alga with a unique glassy cell wall containing silica.

dicot (dī'-kot)
A flowering plant whose embryos have two seed leaves, or cotyledons.

diffusion
The spontaneous movement of particles of any kind down a concentration gradient; that is, movement of particles from where they are more concentrated to where they are less concentrated.

digestion
The mechanical and chemical breakdown of food into molecules small enough for the body to absorb; the second stage of food processing, following ingestion.

digestive tract
A digestive compartment with two openings: a mouth for the entrance of food and an anus for the exit of undigested wastes. Most animals, including humans, have a digestive tract, more commonly called an alimentary canal.

dihybrid cross (dī'-hī'-brid)
A mating of individuals differing at two genetic loci.

dinoflagellate (dī'-nō-flaj'-uh-let)
A unicellular photosynthetic alga with two flagella situated in perpendicular grooves in cellulose plates covering the cell.

diploid (dip'-loid)
Containing two sets of chromosomes (homologous pairs) in each cell, one set inherited from each parent; referring to a 2n cell.

directional selection
Natural selection that acts in favor of the individuals at one end of a phenotypic range.

disaccharide (dī-sak'-uh-rīd)
A sugar molecule consisting of two monosaccharides linked by a dehydration reaction.

discovery science
The process of scientific inquiry that focuses on describing nature See also hypothesis-driven science.

disruptive selection
Natural selection that favors extreme over intermediate phenotypes.

disturbance
In an ecological sense, a force that damages a biological community, at least temporarily, by destroying organisms and altering the availability of resources needed by organisms in the community. Disturbances, such as fires and storms, play a pivotal role in structuring many biological communities.

DNA Deoxyribonucleic acid (dē-ok'-si-rī'-bō-nū-klā'-ik)
The genetic material that organisms inherit from their parents; a double-stranded helical macromolecule consisting of nucleotide monomers with deoxyribose sugar, a phosphate group, and the nitrogenous bases adenine (A), cytosine (C), guanine (G), and thymine (T). See also gene.

DNA ligase (*lī'-gās*)

An enzyme, essential for DNA replication, that catalyzes the covalent bonding of adjacent DNA nucleotides; used in genetic engineering to paste a specific piece of DNA containing a gene of interest into a bacterial plasmid or other vector.

DNA microarray

A glass slide containing thousands of different kinds of single-stranded DNA fragments arranged in an array (grid). Tiny amounts of DNA fragments, representing different genes, are attached to the glass slide. These fragments are tested for hybridization with various samples of cDNA molecules, thereby measuring the expression of thousands of genes at one time.

DNA polymerase (*puh-lim'-er-ās*)

An enzyme that assembles DNA nucleotides into polynucleotides using a preexisting strand of DNA as a template.

DNA profiling

A procedure that analyzes an individual's unique collection of genetic markers using PCR and gel electrophoresis. DNA profiling can be used to determine whether two samples of genetic material were derived from the same individual.

DNA technology

Methods used to study or manipulate genetic material.

domain

A taxonomic category above the kingdom level. The three domains of life are Archaea, Bacteria, and Eukarya.

dominant allele

In a heterozygote, the allele that determines the phenotype with respect to a particular gene.

dorsal, hollow nerve cord

One of the four hallmarks of chordates; the chordate brain and spinal cord.

double circulation system

A circulation scheme with separate pulmonary and systemic circuits.

double fertilization

In flowering plants, the formation of both a zygote and a cell with a triploid nucleus, which develops into the endosperm.

double helix

The form assumed by DNA in living cells, referring to its two adjacent polynucleotide strands wound into a spiral shape.

Down syndrome

A human genetic disorder resulting from the presence of an extra chromosome 21; characterized by heart and respiratory defects and varying degrees of mental retardation.

duodenum (*dū-ō-dē'-num*)

The first portion of the vertebrate small intestine after the stomach, where chyme from the stomach is mixed with bile and digestive enzymes.

E

eardrum

A sheet of connective tissue separating the outer ear from the middle ear. The eardrum vibrates when stimulated by sound waves and passes the vibrations to the middle ear.

earthworm

A type of annelid, or segmented worm, which extracts nutrients from soil.

echinoderm (*ih-kī'-nuh-derm*)

Member of a group of slow-moving or sessile marine animals characterized by a rough or spiny skin, a water vascular system, typically an endoskeleton, and radial symmetry in adults. Echinoderms include sea stars, sea urchins, and sand dollars.

ecological footprint

An estimate of the amount of land required to provide the raw materials an individual or a population consumes, including food, fuel, water, housing, and waste disposal.

ecological niche

The sum total of a species' use of the biotic and abiotic resources of its habitat.

ecological succession

The process of biological community change resulting from disturbance; transition in the species composition of a biological community, often following a flood, fire, or volcanic eruption. *See also* primary succession; secondary succession.

ecology

The scientific study of the interactions between organisms and their environments.

ecosystem (*ē'-kō-sis-tem*)

All the organisms in a given area, along with the nonliving (abiotic) factors with which they interact; a biological community and its physical environment.

ecosystem ecology

The study of energy flow and the cycling of chemicals among the various biotic and abiotic factors in an ecosystem.

ecosystem services

Functions performed by an ecosystem that directly or indirectly benefit people.

ectoderm (*ek'-tō-derm*)

The outer layer of three embryonic cell layers in a gastrula. The ectoderm forms the skin of the gastrula and gives rise to the epidermis and nervous system in the adult.

ectotherm (*ek'-tō-therm*)

An animal that warms itself mainly by absorbing heat from its surroundings.

effector

A cell, tissue, or organ capable of carrying out some action in response to a command from the nervous system.

effector cell

A short-lived lymphocyte that has an immediate effect against a specific pathogen.

egg

A female gamete.

ejaculation (*ih-jak'-yū-lā'-shun*)

The expulsion of sperm-containing fluid (semen) from the penis.

electromagnetic receptor

A sensory receptor that detects energy of different wavelengths, such as magnetism and light.

electromagnetic spectrum

The full range of radiation, from the very short wavelengths of gamma rays to the very long wavelengths of radio signals.

electron

A subatomic particle with a single unit of negative electrical charge. One or more electrons move around the nucleus of an atom.

electron microscope (EM)

An instrument that focuses an electron beam through or onto the surface of a specimen. An electron microscope achieves a thousandfold greater resolving power than a light microscope; the most powerful EM can distinguish objects as small as 0.2 nm (2×10^{-10} m).

electron transport

A redox (oxidation-reduction) reaction in which one or more electrons are transferred to carrier molecules. A series of such reactions, called an electron transport chain, can release the energy stored in high-energy molecules such as glucose. *See also* electron transport chain.

electron transport chain

A series of electron carrier molecules that shuttle electrons during the redox reactions that release energy used to make ATP; located in the inner membrane of mitochondria, the thylakoid membrane of chloroplasts, and the plasma membrane of prokaryotes.

element

A substance that cannot be broken down into other substances by chemical means. Scientists recognize 92 chemical elements occurring in nature.

elimination

The passing of undigested material out of the digestive compartment; the fourth stage of food processing, following absorption.

embryo (*em'-brē-ō*)

A developing stage of a multicellular organism. In humans, the stage in the development of offspring from the first division of the zygote until body structures begin to appear, about the 9th week of pregnancy.

embryo sac

The female gametophyte contained in the ovule of a flowering plant.

embryonic stem cell (ES cell)

Any of the cells in the early animal embryo that differentiate during development to give rise to all the kinds of specialized cells in the body.

emerging virus

A virus that has appeared suddenly or has recently come to the attention of medical scientists.

endangered species
As defined in the U.S. Endangered Species Act, a species that is in danger of extinction throughout all or a significant portion of its range.

endemic species
A species whose distribution is limited to a specific geographic area.

endocrine gland *(en'-dō-krin)*
A gland that synthesizes hormone molecules and secretes them directly into the bloodstream.

endocrine system
The body's main system for internal chemical regulation, consisting of all hormone-secreting cells; cooperates with the nervous system in regulating body functions and maintaining homeostasis.

endocytosis *(en'-dō-sī-tō'-sis)*
The movement of materials into the cytoplasm of a cell via vesicles or vacuoles.

endoderm *(en'-dō-derm)*
The innermost of three embryonic cell layers in a gastrula. The endoderm forms the developing digestive tube in the gastrula and gives rise to the innermost linings of the digestive tract and other hollow organs in the adult.

endodermis
The innermost layer (a one-cell-thick cylinder) of the cortex of a plant root. The endodermis forms a selective barrier, determining which substances pass from the cortex into the vascular tissue.

endomembrane system
A network of organelles that partitions the cytoplasm of eukaryotic cells into functional compartments. Some of the organelles are structurally connected to each other, whereas others are structurally separate but functionally connected by the traffic of vesicles between them.

endometrium *(en'-dō-mē'-trē-um)*
The inner lining of the uterus in mammals, richly supplied with blood vessels that provide the maternal part of the placenta and nourish the developing embryo.

endoplasmic reticulum (ER) *(reh-tik'-yuh-lum)*
An extensive membranous network in a eukaryotic cell, continuous with the outer nuclear membrane and composed of ribosome-studded (rough) and ribosome-free (smooth) regions. *See also* rough ER; smooth ER.

endorphin *(en-dōr'-fin)*
A pain-inhibiting hormone produced by the anterior pituitary; also serves as a neurotransmitter.

endoskeleton
A hard interior skeleton located within the soft tissues of an animal; found in all vertebrates and a few invertebrates (such as echinoderms).

endosperm
In flowering plants, a nutrient-rich mass formed by the union of a sperm cell with the diploid central cell of the embryo sac during double fertilization; provides nourishment to the developing embryo in the seed.

endospore
A thick-coated, protective cell produced within a prokaryotic cell exposed to harsh conditions.

endosymbiosis *(en'-dō-sim-bē-ō'-sis)*
Symbiotic relationship in which one species resides within another species. The mitochondria and chloroplasts of eukaryotic cells probably evolved from symbiotic associations between small prokaryotic cells living inside larger ones.

endotherm
An animal that derives most of its body heat from its own metabolism.

endotoxin
A poisonous component of the outer membrane of certain bacteria.

energy
The capacity to perform work, or to move matter in a direction it would not move if left alone.

energy flow
The passage of energy through the components of an ecosystem.

enhancer
A eukaryotic DNA sequence that helps stimulate the transcription of a gene at some distance from it. An enhancer functions by means of a transcription factor called an activator, which binds to it and then to the rest of the transcription apparatus. *See* silencer.

entropy *(en'-truh-pē)*
A measure of disorder, or randomness. One form of disorder is heat, which is random molecular motion.

enzyme *(en'-zīm)*
A protein that serves as a biological catalyst, changing the rate of a chemical reaction without itself being changed in the process.

enzyme inhibitor
A chemical that interferes with an enzyme's activity by changing the enzyme's shape, either by plugging up the active site or binding to another site on the enzyme.

epinephrine *(ep'-uh-nef'-rin)*
An amine hormone (also called adrenaline) that is secreted by the adrenal medulla and that prepares body organs for "fight or flight."; also serves as a neurotransmitter.

epithelial tissue *(ep'-uh-thē'-lē-ul)*
A sheet of tightly packed cells lining organs and cavities; also called epithelium.

epithelium
(plural, **epithelia**) *See* epithelial tissue.

equilibrial life history *(ē-kwi-lib'-rē-ul)*
The pattern of reaching sexual maturity slowly and producing few offspring but caring for the young; often seen in long-lived, large-bodied species.

erythrocyte *(ih-rith'-ruh-sīt)*
See red blood cell.

esophagus *(ih-sof'-uh-gus)*
The channel through which food passes in a digestive tube, connecting the pharynx to the stomach.

essential amino acid
Any amino acid that an animal cannot synthesize itself and must obtain from food. Eight amino acids are essential for the human adult.

essential element
In plants, a chemical element required for the plant to complete its life cycle (to grow from a seed and produce another generation of seeds).

essential fatty acid
An unsaturated fatty acid that animals need but cannot make.

essential nutrient
A substance that an organism must absorb in preassembled form because it cannot synthesize the nutrient from any other material. Humans require vitamins, minerals, essential amino acids, and essential fatty acids.

estrogen *(es'-trō-jen)*
One of several chemically similar sex hormones secreted by the gonads. Estrogen maintains the female reproductive system and promotes the development of female body features.

estuary *(es'-chuh-wār-ē)*
The area where a freshwater stream or river merges with seawater.

ethylene
A gas that functions as a hormone in plants, triggering aging responses such as fruit ripening and leaf drop.

eudicot *(yū-dī'-kot)*
Member of a group consisting of the vast majority of flowering plants that have two cotyledons (embryonic seed leaves).

Eukarya *(yū-kār'-yuh)*
The domain of eukaryotes, organisms made up of eukaryotic cells; includes all of the protists, plants, fungi, and animals.

eukaryote *(yū-kār'-ē-ōt)*
An organism characterized by eukaryotic cells. *See also* eukaryotic cell.

eukaryotic cell *(yū-kār-ē-ot'-ik)*
A type of cell that has a membrane-enclosed nucleus and other membrane-enclosed organelles. All organisms except bacteria and archaea are composed of eukaryotic cells.

Eustachian tube *(yū-stā'-shun)*
An air passage between the middle ear and throat that equalizes air pressure on either side of the eardrum.

eutherian *(yū-thē'r'-ē-un)*
See placental mammal.

evaporative cooling
A property of water whereby a body becomes cooler as water evaporates from it.

evo-devo
Evolutionary developmental biology, which studies the evolution of developmental processes in multicellular organisms.

evolution
Descent with modification; genetic change in a population or species over generations; the heritable changes that have produced Earth's diversity of organisms.

evolutionary adaptation
A population's increase in the frequency of traits suited to the environment.

evolutionary tree
A branching diagram that reflects a hypothesis about evolutionary relationships between groups of organisms.

excretion (ek-skrē'-shun)
The disposal of nitrogen-containing metabolic wastes.

exocytosis (ek'-sō-sī-tō'-sis)
The movement of materials out of the cytoplasm of a cell via membranous vesicles or vacuoles.

exon (ek'-son)
In eukaryotes, a coding portion of a gene. *See also* intron.

exoskeleton
A hard, external skeleton that protects an animal and provides points of attachment for muscles.

exotoxin
A poisonous protein secreted by certain bacteria.

exponential population growth
A model that describes the expansion of a population in an ideal, unlimited environment.

extracellular matrix
A substance in which the cells of an animal tissue are embedded; consists of protein and polysaccharides.

F

F₁ generation
The offspring of two parental (P. generation) individuals. F_1 stands for first filial.

F₂ generation
The offspring of the F_1 generation. F_2 stands for second filial.

facilitated diffusion
The passage of a substance across a biological membrane down its concentration gradient, aided by specific transport proteins.

family
In classification, the taxonomic category above genus.

farsightedness
An inability to focus on close objects; occurs when the eyeball is shorter than normal and the focal point of the lens is behind the retina; also called hyperopia.

fat
A large lipid molecule made from an alcohol called glycerol and three fatty acids; a triglyceride. Most fats function as energy-storage molecules.

feces
The wastes of the digestive tube.

feedback regulation
A method of metabolic control in which the end product of a metabolic pathway acts as an inhibitor of an enzyme within that pathway.

fermentation
The anaerobic harvest of food by some cells.

fern
Any of a group of seedless vascular plants.

fertilization
The union of a haploid sperm cell with a haploid egg cell, producing a zygote.

fertilizer
A compound given to plants via the soil to promote plant growth.

fetus (fē'-tus)
A developing human from the 9th week of pregnancy until birth. The fetus has all the major structures of an adult.

fever
An abnormally high internal body temperature, usually the result of an infection.

fibrin (fī'-brin)
The activated form of the blood-clotting protein fibrinogen, which aggregates into threads that form the fabric of a blood clot.

fibrinogen (fī-brin'-uh-jen)
The plasma protein that is activated to form a clot when a blood vessel is injured.

fibrous connective tissue
A dense tissue with large numbers of collagen fibers organized into parallel bundles. This is the dominant tissue in tendons and ligaments.

filament
In a flowering plant, the stalk of a stamen.

filtrate
Fluid extracted by the excretory system from the blood or body cavity. The excretory system produces urine from the filtrate after extracting valuable solutes from it and concentrating it.

filtration
In the vertebrate kidney, the extraction of water and small solutes, including metabolic wastes, from the blood by the nephrons.

fission
A means of asexual reproduction whereby a parent separates into two or more genetically identical individuals of about equal size.

fitness
The contribution an individual makes to the gene pool of the next generation relative to the contribution of other individuals in the population.

flagellate (flaj'-uh-lit)
A protist (protozoan) that moves by means of one or more flagella.

flagellum (fluh-jel'-um)
(plural, **flagella**) A long appendage that propels protists through the water and moves fluids across the surface of many tissue cells in animals. A cell may have one or more flagella.

flatworm
A bilateral animal with a thin, flat body form, a gastrovascular cavity with a single opening, and no body cavity. Flatworms include planarians, flukes, and tapeworms.

flower
In an angiosperm, a short stem with four sets of modified leaves, bearing structures that function in sexual reproduction.

fluid mosaic
A description of membrane structure, depicting a cellular membrane as a mosaic of diverse protein molecules embedded in a fluid bilayer made of phospholipid molecules.

follicle (fol'-uh-kul)
A cluster of cells surrounding, protecting, and nourishing a developing egg cell in the ovary. The follicle also secretes estrogen.

food chain
The sequence of food transfers between the trophic levels of a community, beginning with the producers.

food vacuole (vak'-ū-ōl)
(1) A tiny sac in a eukaryotic cell's cytoplasm that engulfs nutrients. (2) The simplest type of digestive compartment.

food web
A network of interconnecting food chains.

foram
A marine protozoan that secretes a shell and extends pseudopodia through pores in its shell; short for foraminifer.

forensics
The scientific analysis of evidence for crime scene investigations and other legal proceedings.

fossil
A preserved imprint or remains of an organism that lived in the past.

fossil fuel
An energy deposit formed from the fossilized remains of long-dead plants and animals.

fossil record
The ordered sequence of fossils as they appear in the rock layers, marking the passing of geologic time.

founder effect
The genetic drift resulting from the establishment of a small, new population whose gene pool differs from that of the parent population.

fruit
A ripened, thickened ovary of a flower, which protects dormant seeds and aids in their dispersal.

functional group
The atoms that form the chemically reactive part of an organic molecule.

fungus
(plural, **fungi**) A chemoheterotrophic eukaryote that digests its food externally and absorbs the resulting small nutrient molecules. Most fungi consist of a netlike mass of filaments called hyphae. Molds, mushrooms, and yeasts are examples of fungi.

G

gallbladder
An organ that stores bile and releases it as needed into the small intestine.

gametangium (gam'-uh-tan'-jē-um)
(plural, **gametangia**) A reproductive organ that houses and protects the gametes of a plant.

gamete (gam'-ēt)
A sex cell; a haploid egg or sperm. The union of two gametes of opposite sex (fertilization) produces a zygote.

gametogenesis
The creation of gametes within the gonads.

gametophyte (guh-mē′-tō-fīt)
The multicellular haploid form in the life cycle of organisms undergoing alternation of generations; results from a union of spores and mitotically produces haploid gametes that unite and grow into the sporophyte generation.

gastric juice
The collection of fluids secreted by the epithelium lining the stomach.

gastropod
A member of the largest group of molluscs, including snails and slugs.

gastrovascular cavity
A digestive compartment with a single opening that serves as both the entrance for food and the exit for undigested wastes; may also function in circulation, body support, and gas exchange. Jellies and hydras are examples of animals with a gastrovascular cavity.

gastrula (gas′-trū-luh)
The embryonic stage resulting from gastrulation in animal development. Most animals have a gastrula made up of three layers of cells: ectoderm, endoderm, and mesoderm.

gastrulation (gas′-trū-lā′-shun)
The phase of embryonic development that transforms the blastula (blastocyst in mammals) into a gastrula. Gastrulation adds more cells to the embryo and sorts the cells into distinct cell layers.

gel electrophoresis (jel′ ē-lek′-trō-fōr-ē′-sis)
A technique for sorting macromolecules. A mixture of molecules is placed on a gel between a positively charged electrode and a negatively charged one; negative charges on the molecules are attracted to the positive electrode, and the molecules migrate toward that electrode. The molecules separate in the gel according to their rates of migration.

gene
A unit of inheritance in DNA (or RNA, in some viruses) consisting of a specific nucleotide sequence that programs the amino acid sequence of a polypeptide. Most of the genes of a

eukaryote are located in its chromosomal DNA; a few are carried by the DNA of mitochondria and chloroplasts.

gene cloning
The production of multiple copies of a gene.

gene expression
The process whereby genetic information flows from genes to proteins; the flow of genetic information from the genotype to the phenotype: DNA → RNA → protein.

gene flow
The gain or loss of alleles from a population by the movement of individuals or gametes into or out of the population.

gene pool
All the genes in a population at any one time.

gene regulation
The turning on and off of specific genes within a living organism.

genetic code
The set of rules giving the correspondence between nucleotide triplets (codons) in mRNA and amino acids in protein.

genetic drift
A change in the gene pool of a population due to chance.

genetic engineering
The direct manipulation of genes for practical purposes.

genetic marker
(1) An allele tracked in a genetic study. (2) A specific section of DNA that contains a particular allele; may contain specific restriction sites (points where restriction enzymes cut the DNA) that occur only in DNA that contains the allele.

genetic recombination
The production of offspring with gene combinations that differ from that found in either parent.

genetically modified (GM) organism
An organism that has acquired one or more genes by artificial means. If the gene is from another organism, typically of another species, the recombinant organism is also known as a transgenic organism.

genetics
The scientific study of heredity (inheritance).

genital herpes
A common sexually transmitted disease caused by a virus. The primary symptom is sores on the genitalia. While outbreaks can be controlled with medication, genital herpes is incurable.

genomic library (juh-nō′-mik)
The entire collection of DNA segments from an organism's genome. Each segment is usually carried by a plasmid or phage.

genomics
The study of whole sets of genes and their interactions.

genotype (jē′-nō-tīp)
The genetic makeup of an organism.

genus (jē′-nus)
(plural, **genera**) In classification, the taxonomic category above species; the first part of a species' binomial; for example, *Homo*.

geologic time scale
A time scale established by geologists that reflects a consistent sequence of geologic periods, grouped into four divisions: Precambrian, Paleozoic, Mesozoic, and Cenozoic.

germinate
To initiate growth, as in a plant seed.

gestation (jes-tā′-shun)
Pregnancy; the state of carrying developing young within the female reproductive tract.

gibberellin (jib′-uh-rel′-in)
A family of plant hormones that trigger the germination of seeds and interact with auxins in regulating growth and fruit development.

gill
An extension of the body surface of an aquatic animal, specialized for gas exchange and/or suspension feeding.

glucagon (glū′-kuh-gon)
A protein hormone secreted by islet cells in the pancreas that raises the level of glucose in the blood.

glucocorticoid (glū′-kuh-kor′-tih-koyd)
A corticosteroid hormone secreted by the adrenal cortex that increases the blood glucose level and helps maintain the body's response to long-term stress.

glycogen (glī′-kō-jen)
A complex, extensively branched polysaccharide made up of many glucose monomers; serves as an

energy-storage molecule in liver and muscle cells.

glycolysis (glī-kol′-uh-sis)
The multistep chemical breakdown of a molecule of glucose into two molecules of pyruvic acid; the first stage of cellular respiration in all organisms; occurs in the cytoplasmic fluid.

Golgi apparatus (gol′-jē)
An organelle in eukaryotic cells consisting of stacks of membranous sacs that modify, store, and ship products of the endoplasmic reticulum.

gonad
An animal sex organ that produces gametes; an ovary or a testis.

granum (gran′-um)
(plural, **grana**) A stack of hollow disks formed of thylakoid membrane in a chloroplast. Grana are the sites where light energy is trapped by chlorophyll and converted to chemical energy during the light reactions of photosynthesis.

gravitropism (grav′-uh-trō′-pizm)
A plant's directional growth in response to gravity.

green alga
One of a group of photosynthetic protists that includes unicellular, colonial, and multicellular species. Green algae are the photosynthetic protists most closely related to plants.

greenhouse effect
The warming of the atmosphere caused by CO_2, CH_4, and other gases that absorb heat radiation and slow its escape from Earth's surface.

greenhouse gas
Any of the gases in the atmosphere that absorb heat radiation, including CO_2, methane, water vapor, and synthetic chlorofluorocarbons.

ground tissue system
A tissue that makes up the bulk of a plant, filling the space between the epidermis and the vascular tissue system. The ground tissue system fulfills a variety of functions, including storage, photosynthesis, and support.

growth factor
A protein secreted by certain body cells that stimulates other cells to divide.

growth hormone (GH)
A protein hormone secreted by the anterior pituitary that promotes development and growth and stimulates metabolism.

guanine (G) *(gwa'-nēn)*
A double-ring nitrogenous base found in DNA and RNA.

guard cell
A specialized epidermal cell in plants that regulates the size of a stoma, allowing gas exchange between the surrounding air and the photosynthetic cells in the leaf.

gymnosperm *(jim'-nō-sperm)*
A naked-seed plant. Its seed is said to be naked because it is not enclosed in a fruit.

H

habitat
A place where an organism lives; a specific environment in which an organism lives.

haploid
Containing a single set of chromosomes; referring to an *n* cell.

Hardy-Weinberg equilibrium
The condition describing a nonevolving population (one that is in genetic equilibrium).

heart
(1) The chambered muscular organ in vertebrates that pumps blood received from the veins into the arteries, thereby maintaining the flow of blood through the entire circulatory system. (2) A similarly functioning structure in invertebrates.

heart attack
Damage or death of heart muscle tissue and the resulting failure of the heart to deliver enough blood to the body.

heart murmur
A hissing sound emitted by an abnormal heart, usually caused by a defective heart valve.

heart rate
The number of heartbeats per minute.

heat
The amount of kinetic energy contained in the movement of the atoms and molecules in a body of matter. Heat is energy in its most random form.

helper T cell
A type of lymphocyte that helps activate other types of T cells and helps stimulate B cells to produce antibodies.

hemoglobin *(hē'-muh-glō-bin)*
An iron-containing protein in red blood cells that reversibly binds O_2 and transports it to body tissues.

hemophilia *(hē'-muh-fil'-ē-uh)*
A human genetic disease caused by a sex-linked recessive allele and characterized by excessive bleeding following injury.

herbivore
An animal that eats mainly plants, algae, or phytoplankton. *See also* carnivore; omnivore.

herbivory
The consumption of plant parts or algae by an animal.

heredity
The transmission of traits from one generation to the next.

hermaphrodite *(her-maf'-rō-dīt)*
An individual that has both female and male reproductive systems, producing both sperm and eggs.

heterotroph *(het'-er-ō-trōf)*
An organism that cannot make its own organic food molecules from inorganic ingredients and must obtain them by consuming other organisms or their organic products; a consumer or a decomposer in a food chain.

heterozygous *(het'-er-ō-zī'-gus)*
Having two different alleles for a given gene.

hinge joint
A joint that allows movement in only one plane. In humans, examples include the elbow and knee.

histamine *(his'-tuh-mēn)*
A chemical alarm signal released by injured cells that causes blood vessels to dilate during an inflammatory response.

histone *(his'-tōn)*
A small protein molecule associated with DNA and important in DNA packing in the eukaryotic chromosome.

HIV
Human immunodeficiency virus; the retrovirus that attacks the human immune system and causes AIDS.

homeostasis *(hō'-mē-ō-stā'-sis)*
The steady state of body functioning; the tendency to maintain relatively constant conditions in the internal environment even when the external environment changes.

homeotic gene *(hō'-mē-ot'-ik)*
A master control gene that determines the identity of a body structure of a developing organism, presumably by controlling the developmental fate of groups of cells. (In plants, such genes are called organ identity genes.)

hominin *(hah'-mi-nin)*
A species on the human branch of the evolutionary tree; a member of the family Hominidae, including *Homo sapiens* and our ancestors.

homologous chromosomes *(hō-mol'-uh-gus)*
The two chromosomes that make up a matched pair in a diploid cell. Homologous chromosomes are of the same length, centromere position, and staining pattern and possess genes for the same characteristics at corresponding loci. One homologous chromosome is inherited from the organism's father, the other from the mother.

homology *(hō-mol'-uh-jē)*
Anatomical similarity due to common ancestry.

homozygous *(hō'-mō-zī'-gus)*
Having two identical alleles for a given gene.

hormone
In multicellular organisms, a regulatory chemical that travels in body fluids from its production site to other sites, where target cells respond to the regulatory signal.

host
An organism that is exploited by a parasite or pathogen.

human gene therapy
A recombinant DNA procedure intended to treat disease by altering an afflicted person's genes.

Human Genome Project
An international collaborative effort that sequenced the DNA of the entire human genome.

humoral immune response
The type of adaptive immune response that involves secretion of antibodies into the blood and lymph by effector B cells. These antibodies fight bacteria and viruses in body fluids. *See also* cell-mediated immune response.

Huntington's disease
A human genetic disease caused by a dominant allele; characterized by uncontrollable body movements and degeneration of the nervous system; usually fatal 10 to 20 years after the onset of symptoms.

hybrid
The offspring of parents of two different species or of two different varieties of one species; the offspring of two parents that differ in one or more inherited traits; an individual that is heterozygous for one or more pairs of genes.

hydrocarbon
A chemical compound composed only of the elements carbon and hydrogen.

hydrogen bond
A type of weak chemical bond formed when a partially positive hydrogen atom from one polar molecule is attracted to the partially negative atom in another molecule (or in another part of the same molecule).

hydrogenation
The process of converting unsaturated fats to saturated fats by the addition of hydrogen.

hydrolysis *(hī-drol'-uh-sis)*
A chemical process in which macromolecules are broken down by the chemical addition of water molecules to the bonds linking their monomers; an essential part of digestion. A hydrolysis reaction is the opposite of a dehydration reaction.

hydrophilic *(hī'-drō-fil'-ik)*
"Water-loving"; pertaining to polar, or charged, molecules (or parts of molecules), which are soluble in water.

hydrophobic *(hī'-drō-fō'-bik)*
"Water-fearing"; pertaining to nonpolar molecules (or parts of

molecules), which do not dissolve in water.

hymen
A thin membrane that partly covers the vaginal opening in the human female and is ruptured by sexual intercourse or other vigorous activity.

hypercholesterolemia (hī′-per-kō-les′-tur-ah-lē m′-ē-uh)
An inherited human disease characterized by an excessively high level of cholesterol in the blood.

hypertension
Abnormally high blood pressure consisting of a persistent systolic blood pressure higher than 140 and/or diastolic blood pressure higher than 90. This condition can lead to a variety of serious cardiovascular disorders.

hypertonic
In comparing two solutions, referring to the one with the greater concentration of solutes.

hypha (hī′-fuh)
(plural, **hyphae**) One of many filaments making up the body of a fungus.

hypothalamus (hī′-pō-thal′-uh-mus)
The main control center of the endocrine system, located in the vertebrate forebrain. The hypothalamus functions in maintaining homeostasis, especially in coordinating the endocrine and nervous systems. It synthesizes hormones that are secreted by the posterior pituitary and regulates the secretion of hormones by the anterior pituitary.

hypothesis (hī-poth′-uh-sis)
(plural, **hypotheses**) A tentative explanation that a scientist proposes for a specific phenomenon that has been observed.

hypothesis-driven science
The process of scientific inquiry that uses the steps of the scientific method to answer questions about nature. *See also* discovery science; scientific method.

hypotonic
In comparing two solutions, referring to the one with the lower concentration of solutes.

immune system
The body's system of defenses against infectious disease.

immunodeficiency disease
An immunological disorder in which the immune system lacks one or more components of the immune system, making the body susceptible to infectious agents that would not ordinarily cause a problem.

impotence
The inability to maintain an erection; also called erectile dysfunction.

in vitro fertilization (IVF) (vē′-tro)
Uniting sperm and egg in a laboratory container, followed by the placement of a resulting early embryo in the mother's uterus.

inbreeding
The mating of close relatives.

incomplete dominance
A type of inheritance in which the phenotype of a heterozygote (*Aa*) is intermediate between the phenotypes of the two types of homozygotes (*AA* and *aa*).

indeterminate growth
Growth that continues throughout life, as in most plants. *See also* determinate growth.

induced fit
The interaction between a substrate molecule and the active site of an enzyme, which changes shape slightly to embrace the substrate and catalyze the reaction.

induction
During embryonic development, the influence of one group of cells on an adjacent group of cells.

infertility
The inability to conceive after one year of regular, unprotected sexual intercourse.

inflammatory response
An example of internal innate defense involving the release of histamine and other chemical alarm signals, which trigger increased blood flow, a local increase in white blood cells, and fluid leakage from the blood. The results include redness, heat, and swelling in the affected tissues.

ingestion
The act of eating; the first stage of food processing.

innate defense
An immune system defense that is already present in its final form and that acts the same whether or not an invader has been previously encountered.

inner cell mass
A cluster of cells in a mammalian blastocyst that protrudes into one end of the cavity and subsequently develops into the embryo proper and some of the extraembryonic membranes.

inner ear
One of three main regions of the human ear; includes the cochlea, which contains the organ of Corti, the hearing organ.

insect
An arthropod that usually has three body segments (head, thorax, and abdomen), three pairs of legs, and one or two pairs of wings.

insulin
A protein hormone, secreted by islet cells in the pancreas, that lowers the level of glucose in the blood.

integration
The interpretation of sensory signals and the formulation of responses within the central nervous system.

interferon (in′-ter-fē r′-on)
A defensive protein that is produced by virus-infected cells and that helps other cells resist viruses as part of internal innate defense.

interneuron (in′-ter-nūr′-on)
A nerve cell, entirely within the central nervous system, that integrates sensory signals and may relay command signals to motor neurons.

internode
The portion of a plant stem between two nodes.

interphase
The phase in the eukaryotic cell cycle when the cell is not actually dividing. During interphase, cellular metabolic activity is high, chromosomes and organelles are duplicated, and cell size may increase. Interphase accounts for 90% of the cell cycle. *See also* mitosis.

interspecific competition
Competition between populations of two or more species that require similar limited resources.

interspecific interaction
Any interaction between members of different species.

interstitial fluid (in′-ter-stish′-ul)
An aqueous solution that surrounds body cells and through which materials pass back and forth between the blood and the body tissues.

intertidal zone (in′-ter-tīd′-ul)
A shallow zone where the waters of an estuary or ocean meet land.

intraspecific competition
Competition between individuals of the same species for the same limited resources.

intron (in′-tron)
In eukaryotes, a nonexpressed (noncoding) portion of a gene that is excised from the RNA transcript. *See also* exon.

invasive species
A non-native species that has spread far beyond the original point of introduction and causes environmental or economic damage by colonizing and dominating suitable habitats.

invertebrate
An animal that does not have a backbone.

ion (ī′-on)
An atom or molecule that has gained or lost one or more electrons, thus acquiring an electrical charge.

ionic bond (ī-on′-ik)
An attraction between two ions with opposite electrical charges. The electrical attraction of the opposite charges holds the ions together.

iris
The colored part of the vertebrate eye, formed by the anterior portion of the pigmented layer called the choroid.

isomer (ī′-sō-mer)
One of two or more molecules with the same molecular formula but different structures and thus different properties.

isotonic (ī-sō-ton′-ik)
Having the same solute concentration as another solution.

isotope (ī´-sō-tōp)
A variant form of an atom. Isotopes of an element have the same number of protons and electrons but different numbers of neutrons.

K

karyotype (kār´-ē-ō-tīp)
A display of micrographs of the metaphase chromosomes of a cell, arranged by size and centromere position.

keystone species
A species whose impact on its community is much larger than its biomass or abundance indicates.

kilocalorie (kcal)
A quantity of heat equal to 1,000 calories. Used to measure the energy content of food, it is usually called a "Calorie."

kinetic energy (kuh-net´-ik)
Energy of motion. Moving matter performs work by transferring its motion to other matter, such as leg muscles pushing bicycle pedals.

kingdom
In classification, the broad taxonomic category above phylum.

Krebs cycle
See citric acid cycle.

L

labia majora (lā´-bē-uh muh-jor´-uh)
A pair of outer thickened folds of skin that protect the female genital region.

labia minora (lā´-bē-uh mi-nor´-uh)
A pair of inner folds of skin bordering and protecting the female genital region.

labor
A series of strong, rhythmic contractions of the uterus that expel a baby out of the uterus and vagina during childbirth.

lancelet
One of a group of bladelike invertebrate chordates.

landscape
A regional assemblage of interacting ecosystems.

landscape ecology
The application of ecological principles to the study of land-use patterns; the scientific study of the biodiversity of interacting ecosystems.

large intestine
The tubular portion of the vertebrate alimentary canal between the small intestine and the anus. *See also* colon.

larva (lar´-vuh)
(plural, **larvae**) A free-living, sexually immature form in some animal life cycles that may differ from the adult in morphology, nutrition, and habitat.

larynx (lar´-inks)
The voice box, containing the vocal cords.

lateral line system
A row of sensory organs along each side of a fish's body. Sensitive to changes in water pressure, it enables a fish to detect minor vibrations in the water.

lateralization
The phenomenon in which the two hemispheres of the brain become specialized for different functions.

law of independent assortment
A general rule of inheritance, first proposed by Gregor Mendel, that states that when gametes form during meiosis, each pair of alleles for a particular character segregate (separate) independently of each other pair.

law of segregation
A general rule of inheritance, first proposed by Gregor Mendel, that states that the two alleles in a pair segregate (separate) into different gametes during meiosis.

leaf
The main site of photosynthesis in a plant; consists of a flattened blade and a stalk (petiole) that joins the leaf to the stem.

leech
A type of annelid, or segmented worm, that typically lives in fresh water.

lens
The disklike structure in an eye that focuses light rays onto the retina.

leukemia (lū-kē´-mē-ah)
Cancer of the white blood cells (leukocytes), characterized by excessive production of these cells, resulting in an abnormally high number in the blood.

leukocyte (lū´-kō-sīt)
See white blood cell.

lichen (lī´-ken)
A mutually beneficial symbiotic association between a fungus and an alga or between a fungus and a cyanobacterium.

life
The set of common characteristics that distinguish living organisms, including such properties and processes as order, regulation, growth and development, energy utilization, response to the environment, reproduction, and the capacity to evolve over time.

life cycle
The entire sequence of stages in the life of an organism, from the adults of one generation to the adults of the next.

life history
The traits that affect an organism's schedule of reproduction and survival.

life table
A listing of survivals and deaths in a population in a particular time period and predictions of how long, on average, an individual of a given age will live.

ligament
A type of fibrous connective tissue that joins bones together at joints.

light microscope (LM)
An optical instrument with lenses that refract (bend) visible light to magnify images and project them into a viewer's eye or onto photographic film.

light reactions
The first of two stages in photosynthesis, the steps in which solar energy is absorbed and converted to chemical energy in the form of ATP and NADPH. The light reactions power the sugar-producing Calvin cycle but produce no sugar themselves.

lignin (lig´-nin)
A chemical that hardens the cell walls of plants. Lignin makes up most of what we call wood.

limiting factor
An environmental factor that restricts the number of individuals that can occupy a particular habitat, thus holding population growth in check.

linkage map
A map of a chromosome showing the relative positions of genes.

linked genes
Genes located close enough together on a chromosome that they are usually inherited together.

lipid
An organic compound consisting mainly of carbon and hydrogen atoms linked by nonpolar convalent bonds and therefore mostly hydrophobic and insoluble in water. Lipids include fats, waxes, phospholipids, and steroids.

liver
The largest organ in the vertebrate body. The liver performs diverse functions, such as producing bile, preparing nitrogenous wastes for disposal, and detoxifying poisonous chemicals in the blood.

lobe-finned fish
A bony fish with strong, muscular fins supported by bones.

locus
(plural, **loci**) The particular site where a gene is found on a chromosome. Homologous chromosomes have corresponding gene loci.

logistic population growth
A model that describes population growth that decreases as population size approaches carrying capacity.

loose connective tissue
The most widespread connective tissue in the vertebrate body. It binds epithelia to underlying tissues and functions as packing material, holding organs in place.

lung
An internal sac, lined with moist epithelium, where gases are exchanged between inhaled air and the blood.

lungfish
A bony fish that generally inhabits stagnant waters and gulps air into lungs connected to a pharynx.

lymph
A fluid similar to interstitial fluid that circulates in the lymphatic system.

lymph node
A small organ that is located along a lymph vessel and that filters lymph.

lymphatic system (lim-fat´-ik)
The organ system through which lymph circulates; includes lymph

vessels, lymph nodes, and several other organs. The lymphatic system helps remove toxins and pathogens from the blood and interstitial fluid and returns fluid and solutes from the interstitial fluid to the circulatory system.

lymphocyte (lim'-fuh-sīt)
A type of white blood cell that carries out adaptive defenses—recognizing and responding to specific invading pathogens. There are two types of lymphocytes: B cells and T cells. *See also* B cell; T cell.

lysogenic cycle (lī-sō-jen'-ik)
A bacteriophage reproductive cycle in which the viral genome is incorporated into the bacterial host chromosome as a prophage. New phages are not produced, and the host cell is not killed or lysed unless the viral genome leaves the host chromosome.

lysosome (lī'-sō-sōm)
A digestive organelle in eukaryotic cells; contains enzymes that digest the cell's food and wastes.

lytic cycle (lit'-ik)
A viral reproductive cycle resulting in the release of new viruses by lysis (breaking open) of the host cell.

M

macroevolution
Evolutionary change on a grand scale, encompassing the origin of new species, the origin of evolutionary novelty, diversification, and mass extinction.

macromolecule
A giant molecule in a living organism. Examples include proteins, polysaccharides, and nucleic acids.

macronutrient
A chemical element that an organism must obtain in relatively large amounts. *See also* micronutrient.

magnification
An increase in the apparent size of an object.

major depression
Depressive mental illness characterized by persistent sadness and loss of interest in pleasurable activities.

malignant tumor
An abnormal tissue mass that spreads into neighboring tissue and to other parts of the body; a cancerous tumor.

malnutrition
The absence of one or more essential nutrients from the diet.

mammal
Member of a class of endothermic amniotes that possesses mammary glands and hair.

mantle
In molluscs, the outgrowth of the body surface that drapes over the animal. The mantle produces the shell and forms the mantle cavity.

marsupial (mar-sū'-pē-ul)
A pouched mammal, such as a kangaroo, opossum, or koala. Marsupials give birth to embryonic offspring that complete development while housed in a pouch and attach to nipples on the mother's abdomen.

mass
A measure of the amount of material in an object.

mass number
The sum of the number of protons and neutrons in an atom's nucleus.

matrix
The thick fluid contained within the inner membrane of the mitochondrion.

matter
Anything that occupies space and has mass.

mechanical digestion
The physical breakdown of food into smaller pieces, as by chewing.

mechanoreceptor (mek'-uh-nō-ri-sep'-ter)
A sensory receptor that detects physical changes in the environment, associated with pressure, touch, stretch, motion, and sound.

medulla oblongata (meh-duh'-luh ob'-long-got'-uh)
Part of the vertebrate hindbrain continuous with the spinal cord; passes data between the spinal cord and forebrain and controls autonomic, homeostatic functions, including breathing, heart rate, swallowing, and digestion.

medusa (med-ū'-suh)
(plural, **medusae**) One of two types of cnidarian body forms; a floating, umbrella-like body form; also called a jelly.

meiosis (mī-ō'-sis)
In a sexually reproducing organism, the division of a single diploid cell into

four haploid daughter cells. Meiosis and cytokinesis produce haploid gametes from diploid cells in the reproductive organs of the parents.

memory cell
A long-lived lymphocyte that responds to subsequent exposures to a specific pathogen. A memory cell is formed during the primary immune response and is activated by exposure to the same antigen that triggered its formation. When activated, a memory cell forms large clones of effector cells and memory cells that mount the secondary immune response.

meninges (muh-nin'-jēz)
Layers of connective tissue that enwrap and protect the brain and spinal cord.

menstrual cycle (men'-strū-ul)
The hormonally synchronized cyclic buildup and breakdown of the endometrium in some primates, including humans.

menstruation (men'-strū-ā'-shun)
Uterine bleeding resulting from the breakdown of the endometrium during a menstrual cycle.

meristem (mār'-eh-stem)
Plant tissue consisting of undifferentiated cells that divide and generate new cells and tissues.

mesoderm (mez'-ō-derm)
The middle layer of the three embryonic cell layers in a gastrula. The mesoderm gives rise to muscles, bones, the dermis of the skin, and most other organs in the adult.

mesophyll (mes'-ō-fil)
The green tissue in the interior of a leaf; a leaf's ground tissue system, the main site of photosynthesis.

messenger RNA (mRNA)
The type of ribonucleic acid that encodes genetic information from DNA and conveys it to ribosomes, where the information is translated into amino acid sequences.

metabolic rate
Energy expended by the body per unit time.

metabolism (muh-tab'-uh-liz-um)
The total of all the chemical reactions in an organism.

metamorphosis (met'-uh-mōr'-fuh-sis)
The transformation of a larva into an adult.

metaphase (met'-eh-fāz)
The second stage of mitosis. During metaphase, the centromeres of all the cell's duplicated chromosomes are lined up on an imaginary plate equidistant between the poles of the mitotic spindle.

metastasis (muh-tas'-tuh-sis)
The spread of cancer cells beyond their original site.

microevolution
A change in a population's gene pool over a succession of generations; evolutionary changes in species over relatively brief periods of geologic time.

micronutrient
A chemical element that an organism needs in very small amounts. *See also* macronutrient.

microtubule
The thickest of the three main kinds of fibers making up the cytoskeleton of a eukaryotic cell; a straight, hollow tube made of globular proteins called tubulins. Microtubules form the basis of the structure and movement of cilia and flagella.

middle ear
One of three main regions of the human ear; a chamber containing three small bones (the hammer, anvil, and stirrup) that convey vibrations from the eardrum to the inner ear.

millipede
A terrestrial arthropod that has two pairs of short legs for each of its numerous body segments and that eats decaying plant matter.

mineral
In nutrition, an inorganic chemical element (other than carbon, hydrogen, oxygen, or nitrogen) that an organism requires for proper body functioning.

mitochondrion (mī'-tō-kon'-drē-on)
(plural, **mitochondria**) An organelle in eukaryotic cells where cellular respiration occurs. Enclosed by two concentric membranes, it is where most of the cell's ATP is made.

mitosis (mī'-tō-sis)
The division of a single nucleus into two genetically identical daughter nuclei. Mitosis and cytokinesis make up the mitotic (M) phase of the cell cycle.

mitotic (M) phase
The phase of the cell cycle when mitosis divides the nucleus and distributes its chromosomes to the daughter nuclei and cytokinesis divides the cytoplasm, producing two daughter cells.

mitotic spindle
A spindle-shaped structure formed of microtubules and associated proteins that is involved in the movement of chromosomes during mitosis and meiosis. (A spindle is shaped roughly like a football.)

modern synthesis
A comprehensive theory of evolution that incorporates genetics and includes most of Darwin's ideas, focusing on populations as the fundamental units of evolution.

molecular biology
The study of the molecular basis of heredity; molecular genetics.

molecule
A group of two or more atoms held together by covalent bonds.

mollusc (mol'-lusk)
A soft-bodied animal characterized by a muscular foot, mantle, mantle cavity, and radula. Molluscs include gastropods (snails and slugs), bivalves (clams, oysters, and scallops), and cephalopods (squids and octopuses).

monoclonal antibody (mon'-uh-klōn'-ul)
An antibody secreted by a clone of cells and consequently specific for the one antigen that triggered the development of the clone.

monocot (mon'-uh-kot)
A flowering plant whose embryos have a single seed leaf, or cotyledon.

monohybrid cross
A mating of individuals differing at one genetic locus.

monomer (mon'-uh-mer)
A chemical subunit that serves as a building block of a polymer.

monosaccharide (mon'-uh-sak'-uh-rīd)
The smallest kind of sugar molecule; a single-unit sugar; also known as a simple sugar. Monosaccharides are the building blocks of more complex sugars and polysaccharides.

monotreme (mon'-uh-trēm)
An egg-laying mammal, such as the duck-billed platypus.

morning-after pill (MAP)
A birth control pill taken within three days of unprotected intercourse to prevent fertilization or implantation.

moss
Any of a group of seedless nonvascular plants.

motor neuron
A nerve cell that conveys command signals from the central nervous system to effector cells, such as muscle cells or gland cells.

motor output
The conduction of signals from the central nervous system to effector cells.

motor unit
A motor neuron and all the muscle fibers it controls.

mouth
An opening through which food is taken into an animal's body; also known as the oral cavity.

movement corridor
A series of small clumps or a narrow strip of quality habitat (usable by organisms) that connects otherwise isolated patches of quality habitat.

muscle tissue
Tissue consisting of long muscle cells that are capable of contracting when stimulated by nerve impulses. See also skeletal muscle; cardiac muscle; smooth muscle.

muscular system
All the skeletal muscles in the body. (Cardiac muscle and smooth muscle are components of other organ systems.)

mutagen (myū'-tuh-jen)
A chemical or physical agent that interacts with DNA and causes a mutation.

mutation
A change in the nucleotide sequence of DNA; a major source of genetic diversity.

mutualism
An interspecific interaction in which both partners benefit.

mycelium (mī-sē'-lē-um)
(plural, **mycelia**) The densely branched network of hyphae in a fungus.

mycorrhiza (mī'-kō-rī'-zuh)
(plural, **mycorrhizae**) A mutually beneficial symbiotic association of a plant root and fungus.

myelin sheath (mī'-uh-lin)
A chain of bead-like supporting cells that insulate the axon of a nerve cell in vertebrates. This insulation helps to speed electrical transmission along the axon.

myofibril (mī'-ō-fī'-bril)
A contractile unit in a muscle cell (fiber) made up of many arcomeres. Longitudinal bundles of myofibrils make up a muscle fiber.

N

NADH
An electron carrier (a molecule that carries electrons) involved in cellular respiration and photosynthesis. NADH carries electrons from glucose and other fuel molecules and deposits them at the top of an electron transport chain. NADH is generated during glycolysis and the citric acid cycle.

NADPH
An electron carrier (a molecule that carries electrons) involved in photosynthesis. Light drives electrons from chlorophyll to $NADP^+$, forming NADPH, which provides the high-energy electrons for the reduction of carbon dioxide to sugar in the Calvin cycle.

natural family planning
A form of contraception that relies on refraining from sexual intercourse when conception is most likely to occur; also called the rhythm method.

natural killer (NK) cell
A white blood cell that attacks cancer cells and infected body cells as part of internal innate defense.

natural selection
A process in which organisms with certain inherited characteristics are more likely to survive and reproduce than are organisms with other characteristics; differential reproductive success.

nearsightedness
An inability to focus on distant objects; occurs when the eyeball is longer than normal and the lens focuses distant objects in front of the retina; also called myopia.

negative feedback
A control mechanism in which a chemical reaction, metabolic pathway, or hormone-secreting gland is inhibited by the products of the reaction, pathway, or gland. As the concentration of the products builds up, the product molecules themselves inhibit the process that produced them.

negative pressure breathing
A breathing system in which lower air pressure in the lungs causes air to enter the lungs.

nematode (nem'-uh-tōd)
An animal characterized by a pseudocoelom, a cylindrical, wormlike body form, and a complete digestive tract; also called a roundworm.

nephron
The tubular excretory unit and associated blood vessels of the vertebrate kidney. The nephron extracts filtrate from the blood and refines it into urine.

nerve
A communication line made up of cable-like bundles of neuron fibers (axons and dendrites) tightly wrapped in connective tissue.

nervous system
The organ system that forms a communication and coordination network throughout an animal's body.

nervous tissue
Tissue made up of neurons and supportive cells.

neuron (nyūr'-on)
A nerve cell; the fundamental structural and functional unit of the nervous system, specialized for carrying signals from one location in the body to another.

neurotransmitter
A chemical messenger that carries information from a transmitting neuron to a receiving cell, either another neuron or an effector cell.

neutron
An electrically neutral particle (a particle having no electrical charge), found in the nucleus of an atom.

nitrogen fixation
The conversion of atmospheric nitrogen (N_2) to nitrogen compounds (NH_4, NO_3) that plants can absorb and use.

node
The point of attachment of a leaf on a stem.

nondisjunction
An accident of meiosis or mitosis in which a pair of homologous chromosomes or a pair of sister chromatids fail to separate at anaphase.

norepinephrine (nor´-ep-uh-nef´-rin)
An amine hormone (also called noradrenaline) that is secreted by the adrenal medulla and that prepares body organs for "fight or flight"; also serves as a neurotransmitter.

notochord (nō´-tuh-kord)
A flexible, cartilage-like, longitudinal rod located between the digestive tract and nerve cord in chordate animals, present only in embryos in many species.

nuclear envelope
A double membrane, perforated with pores, that encloses the nucleus and separates it from the rest of the eukaryotic cell.

nuclear transplantation
A technique in which the nucleus of one cell is placed into another cell that already has a nucleus or in which the nucleus has been previously destroyed. The cell is then stimulated to grow, producing an embryo that is a genetic copy of the nucleus donor.

nucleic acid (nū-klā´-ik)
A polymer consisting of many nucleotide monomers; serves as a blueprint for proteins and, through the actions of proteins, for all cellular structures and activities. The two types of nucleic acids are DNA and RNA.

nucleic acid probe (nū-klā´-ik)
In DNA technology, a labeled single-stranded nucleic acid molecule used to find a specific gene or other nucleotide sequence within a mass of DNA. The probe hydrogen-bonds to the complementary sequence in the targeted DNA.

nucleolus (nū-klē´-ō-lus)
A structure within the nucleus of a eukaryotic cell where ribosomal RNA is made and assembled with proteins to make ribosomal subunits; consists of parts of the chromatin DNA, RNA transcribed from the DNA, and proteins imported from the cytoplasm.

nucleosome (nū´-klē-ō-sōm)
The bead-like unit of DNA packing in a eukaryotic cell; consists of DNA wound around a protein core made up of eight histone molecules.

nucleotide (nū´-klē-ō-tīd)
An organic monomer consisting of a five-carbon sugar covalently bonded to a nitrogenous base and a phosphate group. Nucleotides are the building blocks of nucleic acids.

nucleus
(plural, **nuclei**) (1) An atom's central core, containing protons and neutrons. (2) The genetic control center of a eukaryotic cell.

O

obesity
An excessively high body mass index, a ratio of weight to height.

omnivore
An animal that eats both plants and animals. *See also* carnivore; herbivore.

oncogene (on´-kō-jēn)
A cancer-causing gene; usually contributes to malignancy by abnormally enhancing the amount or activity of a growth factor made by the cell.

oogenesis (ō´-uh-jen´-uh-sis)
The formation of egg cells within the ovaries.

open circulatory system
A circulatory system in which the circulating fluid is pumped through open-ended vessels and out among the body cells. In an animal with an open circulatory system, the circulating fluid and interstitial fluid are one and the same.

open system
Any system that exchanges chemicals and energy with its surroundings. All organisms are open systems.

operator
In prokaryotic DNA, a sequence of nucleotides near the start of an operon to which an active repressor can attach. The binding of repressor prevents RNA polymerase from attaching to the promoter and transcribing the genes of the operon.

operculum (ō-per´-kyū-lum)
(plural, **opercula**) A protective flap on each side of a bony fish's head that covers a chamber housing the gills.

operon (op´-er-on)
A unit of genetic regulation common in prokaryotes; a cluster of genes with related functions, along with the promoter and operator that control their transcription.

opportunistic life history
The pattern of reproducing when young and producing many offspring that receive little or no parental care; often seen in short-lived, small-bodied species.

optic nerve
A nerve that arises from the retina in each eye and carries visual information to the brain.

oral cavity
See mouth.

order
In classification, the taxonomic category above family.

organ
A structure consisting of two or more tissues that coordinate to perform specific functions.

organ system
A group of organs that work together in performing vital body functions.

organelle (ōr-guh-nel´)
A membrane-enclosed structure with a specialized function within a eukaryotic cell.

organic compound
A chemical compound containing the element carbon and usually synthesized by cells.

organic farming
A farming method that is conservation-minded and environmentally safe as well as profitable.

organismal ecology
The study of the evolutionary adaptations that enable individual organisms to meet the challenges posed by their abiotic environments.

osmoconformer (oz´-mō-con-form´-er)
An organism whose body fluids have a solute concentration equal to that of its surroundings. Osmoconformers do not have a net gain or loss of water by osmosis.

osmoregulation
The control of the gain or loss of water and dissolved solutes in an organism.

osmoregulator
An organism whose body fluids have a solute concentration different from that of its environment and that must use energy in controlling water loss or gain.

osmosis (oz-mō´-sis)
The diffusion of water across a selectively permeable membrane.

osteoporosis (os´-tē-ō-puh-rō´-sis)
A skeletal disorder characterized by thinning, porous, and easily broken bones. Osteoporosis is common among women after menopause and often related to low estrogen levels.

outer ear
One of three main regions of the ear in humans and some other animals. The outer ear is made up of the auditory canal and the pinna.

ovarian cycle (ō-vār´-ē-un)
Hormonally synchronized cyclic events in the mammalian ovary, culminating in ovulation.

ovary
(1) In animals, the female gonad, which produces egg cells and reproductive hormones. (2) In flowering plants, the base of a carpel in which the egg-containing ovules develop.

oviduct (ō´-vuh-dukt)
The tube that conveys egg cells away from an ovary; also called a fallopian tube.

ovulation (ah´-vyū-lā´-shun)
The release of an egg cell from an ovarian follicle.

ovule (ō´-vyūl)
A reproductive structure in a seed plant, containing the female gametophyte and the developing egg. An ovule develops into a seed.

ovum (ō´-vum)
(plural, **ova**) A mature, unfertilized egg.

oxidation

The loss of electrons from a substance involved in a redox reaction; always accompanies reduction.

P

P generation

The parent individuals from which offspring are derived in studies of inheritance. P stands for parental.

pacemaker

The SA (sinoatrial) node; a region of cardiac muscle that maintains the heart's pumping rhythm (heartbeat) by setting the rate at which the heart contracts.

paedomorphosis (pē'-duh-môr'-fuh-sis)

The retention in the adult of features that were juvenile in ancestral species.

pain receptor

A sensory receptor that detects painful stimuli.

pancreas (pan'-krē-us)

A gland with dual functions: The nonendocrine portion secretes digestive enzymes and an alkaline solution into the small intestine via a duct; the endocrine portion secretes the hormones insulin and glucagon into the blood.

parasite

An organism that lives in or on another organism (host) from which it obtains nourishment; an organism that benefits at the expense of another organism, which is harmed in the process.

parasympathetic division

One of two sets of neurons in the autonomic nervous system; generally promotes body activities that gain and conserve energy ("rest and digest"). *See also* sympathetic division.

parathyroid gland (pār'-uh-thī'-royd)

One of four endocrine glands embedded in the surface of the thyroid gland that secrete parathyroid hormone; functions in calcium homeostasis.

parathyroid hormone (PTH)

A peptide hormone secreted by the parathyroid glands that raises blood calcium level.

parenchyma cell (puh-reng'-kuh-muh)

In plants, an abundant and relatively unspecialized type of cell with a thin primary wall and no secondary wall; functions in photosynthesis, food storage, and aerobic respiration and may differentiate into other cell types.

passive transport

The diffusion of a substance across a biological membrane without any input of energy.

pathogen

A disease-causing virus or organism.

pedigree

A family tree representing the occurrence of heritable traits in parents and offspring across a number of generations.

pelagic realm (puh-laj'-ik)

The region of an ocean occupied by water; the open ocean.

penis

The structure in male mammals that functions in sexual intercourse.

pepsin

An enzyme present in gastric juice that begins the hydrolysis of proteins.

peptide bond

The covalent linkage between two amino acid units in a polypeptide, formed by a dehydration reaction between two amino acids.

perennial (puh-ren'-ē-ul)

A plant that lives for many years.

peripheral nervous system (PNS)

The network of nerves carrying signals into and out of the central nervous system.

peristalsis (pār-uh-stal'-sis)

Rhythmic waves of contraction of smooth muscles. Peristalsis propels food through a digestive tube and also enables many animals, such as earthworms, to crawl.

permafrost

Continuously frozen subsoil found in the arctic tundra.

petal

A modified leaf of a flowering plant. Petals are the often colorful parts of a flower that advertise it to insects and other pollinators.

petiole (pet'-ē-ōl)

The stalk of a leaf, which joins the leaf to a node of the stem.

pH scale

A measure of the relative acidity of a solution, ranging in value from 0 (most acidic) to 14 (most basic). pH stands for potential hydrogen and refers to the concentration of hydrogen ions (H^+).

phage (fāj)

See bacteriophage.

phagocytic cell (fag'-ō-si-tik)

A white blood cell that engulfs bacteria, foreign proteins, and the remains of dead body cells as part of internal innate defense.

phagocytosis (fag'-ō-sī-tō'-sis)

Cellular "eating"; a type of endocytosis whereby a cell engulfs large molecules, other cells, or particles into its cytoplasm.

pharyngeal slit (fuh-rin'-jē-ul)

A gill structure in the pharynx, found in chordate embryos and some adult chordates.

pharynx (fār'-inks)

The organ in a digestive tract that receives food from the oral cavity; in terrestrial vertebrates, the throat region where the air and food passages cross.

phenotype (fē'-nō-tīp)

The expressed traits of an organism.

phloem (flō'-um)

The portion of a plant's vascular system that conveys sugars, nutrients, and hormones throughout a plant. Phloem is made up of live food-conducting cells.

phloem sap

The solution of sugars, other nutrients, and hormones conveyed throughout a plant via phloem tissue.

phospholipid (fos'-fō-lip'-id)

A molecule that is a constituent of the inner bilayer of biological membranes, having a hydrophilic head and a hydrophobic tail.

phospholipid bilayer

A double layer of phospholipid molecules (each molecule consisting of a phosphate group bonded to two fatty acids) that is the primary component of all cellular membranes.

photic zone (fō'-tik)

Shallow water near shore or the upper layer of water away from the shore; region of an aquatic ecosystem where sufficient light is available for photosynthesis.

photon (fō'-ton)

A fixed quantity of light energy. The shorter the wavelength of light, the greater the energy of a photon.

photoperiod

The length of the day relative to the length of the night; an environmental stimulus that plants use to detect the time of year.

photoreceptor

A type of electromagnetic receptor that detects light.

photosynthesis (fō'-tō-sin'-thuh-sis)

The process by which plants, algae, and some bacteria transform light energy to chemical energy stored in the bonds of sugars made from carbon dioxide and water.

photosystem

A light-harvesting unit of a chloroplast's thylakoid membrane; consists of several hundred antenna molecules, a reaction-center chlorophyll, and a primary electron acceptor.

phototropism (fō'-tō-trō'-pizm)

The directional growth of a plant shoot in response to light.

phylogenetic tree (fī'-lō-juh-net'-ik)

A branching diagram that represents a hypothesis about evolutionary relationships between organisms.

phylum (fī'-lum)

(plural, **phyla**) In classification, the taxonomic category above class and below kingdom. Members of a phylum all have a similar general body plan.

physiology (fī'-zi-ol'-uh-ji)

The study of the function of an organism's structural equipment.

phytoplankton (fī'-tō-plank'-ton)

Algae and photosynthetic bacteria that drift passively in aquatic environments.

pinna (pin'-uh)

The flap-like part of the outer ear in humans and some other animals. The pinna collects sound waves and channels them to the auditory canal.

pinocytosis (pī'-nō-sī-tō'-sis)

Cellular "drinking"; a type of endocytosis in which the cell takes fluid and dissolved solutes into small membranous vesicles.

pistil

In a flowering plant, a single carpel or a group of fused carpels.

pith

Part of the ground tissue system of a eudicot plant. Pith fills the center of a stem and may store food.

pituitary gland (puh-tū'-uh-tār'-ē)
An endocrine gland at the base of the hypothalamus; consists of a posterior lobe, which stores and releases two hormones produced by the hypothalamus, and an anterior lobe, which produces and secretes many hormones that regulate diverse body functions.

pivot joint
A joint that allows precise rotations in multiple planes. An example in humans is the elbow.

placenta (pluh-sen'-tuh)
In most mammals, the organ that provides nutrients and oxygen to the embryo and helps dispose of its metabolic wastes. The placenta is formed of the embryo's chorion and the mother's endometrial blood vessels.

placental mammal (pluh-sen'-tul)
Mammal whose young complete their embryonic development in the uterus, nourished via the mother's blood vessels in the placenta; also called a eutherian.

plankton
Communities of organisms, mostly microscopic, that drift passively in ponds, lakes, and oceans.

plant
A multicellular eukaryote that carries out photosynthesis.

plasma
The yellowish liquid of the blood in which the blood cells are suspended.

plasma membrane
The thin layer of lipids and proteins that sets a cell off from its surroundings and acts as a selective barrier to the passage of ions and molecules into and out of the cell; consists of a phospholipid bilayer in which proteins are embedded.

plasmid
A small ring of self-replicating DNA separate from the chromosome(s). Plasmids are found in prokaryotes and yeasts.

plasmodial slime mold (plaz-mō'-dē-ul)
A type of protist named for an amoeboid plasmodial feeding stage in its life cycle.

plasmolysis (plaz-mol'-uh-sis)
A phenomenon that occurs in plant cells in a hypertonic environment. The cell loses water and shrivels, and

its plasma membrane pulls away from the cell wall, usually killing the cell.

platelet
A piece of cytoplasm from a large cell in the bone marrow; a blood-clotting element.

pleiotropy (plī'-uh-trō-pē)
The control of more than one phenotypic character by a single gene.

polar body
The smaller of two daughter cells produced during meiosis of oogenesis.

polar ice
A terrestrial biome that includes regions of extremely cold temperature and low precipitation located at high latitudes north of the arctic tundra and in Antarctica.

polar molecule
A molecule containing polar covalent bonds (having opposite charges on opposite ends).

pollen grain
In a seed plant, the male gametophyte that develops within the anther of a stamen. It houses cells that will develop into sperm.

pollination
In seed plants, the delivery, by wind or animals, of pollen from the male parts of a plant to the stigma of a carpel on the female.

polychaete
A type of annelid, or segmented worm, that typically lives on the seafloor.

polygenic inheritance (pol'-ē-jen'-ik)
The additive effect of two or more genes on a single phenotypic characteristic.

polymer (pol'-uh-mer)
A large molecule consisting of many identical or similar molecular units, called monomers, covalently joined together in a chain.

polymerase chain reaction (PCR) (puh-lim'-uh-rās)
A technique used to obtain many copies of a DNA molecule or many copies of part of a DNA molecule. A small amount of DNA mixed with the enzyme DNA polymerase, DNA nucleotides, and a few other ingredients replicates repeatedly in a test tube.

polynucleotide (pol'-ē-nū'-klē-ō-tīd)
A polymer made up of many nucleotides covalently bonded together.

polyp (pol'-ip)
One of two types of cnidarian body forms; a sessile, columnar, hydra-like body.

polypeptide
A chain of amino acids linked by peptide bonds.

polysaccharide (pol'-ē-sak'-uh-rīd)
A carbohydrate polymer consisting of many monosaccharides (sugars) linked by covalent bonds.

pons (pahnz)
Part of the vertebrate hindbrain that functions with the medulla oblongata in passing data between the spinal cord and forebrain and in controlling autonomic, homeostatic functions.

population
A group of interacting individuals belonging to one species and living in the same geographic area at the same time.

population density
The number of individuals of a species per unit area or volume of the habitat.

population ecology
The study of how members of a population interact with their environment, focusing on factors that influence population density and growth.

population momentum
In a population in which the fertility rate averages two children per female (replacement rate), the continuation of population growth as girls reach their reproductive years.

positive feedback
A control mechanism in which the products of a process stimulate the process that produced them.

post-anal tail
A tail posterior to the anus, found in chordate embryos and most adult chordates.

posterior pituitary
An extension of the hypothalamus composed of nervous tissue that secretes hormones made in the hypothalamus; a temporary storage site for hypothalamic hormones.

postzygotic barrier (pōst'-zī-got'-ik)
A reproductive barrier that operates if interspecies mating occurs and forms hybrid zygotes.

potential energy
Stored energy; the energy that an object has due to its location and/or arrangement. Water behind a dam and chemical bonds both possess potential energy.

predation
An interaction between species in which one species, the predator, kills and eats the other, the prey.

prepuce (prē'-pyūs)
A fold of skin covering the head of the clitoris or penis.

pressure-flow mechanism
The method by which phloem sap is transported through a plant from a sugar source, where sugars are produced, to a sugar sink, where sugars are used. The mechanism relies on building up water pressure at the source end of the phloem tube and reducing water pressure at the sink end.

prezygotic barrier (prē'-zī-got'-ik)
A reproductive barrier that impedes mating between species or hinders fertilization of eggs if members of different species attempt to mate.

primary consumer
An organism that eats only autotrophs; an herbivore.

primary electron acceptor
A molecule in the reaction center of a photosystem that traps the light-excited electron from the reaction center chlorophyll.

primary growth
Growth in the length of a plant root or shoot produced by an apical meristem.

primary immune response
The initial immune response to an antigen, which includes production of effector cells that respond to the antigen within a few days and also memory cells that will respond to future exposure to the antigen.

primary oocyte (ō'-uh-sīt)
A diploid cell, in prophase I of meiosis, that can be hormonally triggered to develop into an ovum.

primary production
The amount of solar energy converted to chemical energy (organic compounds) by autotrophs in an ecosystem during a given time period.

primary spermatocyte (sper-mat'-eh-sīt')
A diploid cell in the testis that undergoes meiosis I.

primary structure
The first level of protein structure; the specific sequence of amino acids making up a polypeptide chain.

primary succession
A type of ecological succession in which a biological community begins in an area without soil. *See also* secondary succession.

primate
Member of the mammalian group that includes lorises, pottos, lemurs, tarsiers, monkeys, apes, and humans.

prion
An infectious form of protein that may multiply by converting related proteins to more prions. Prions cause several related diseases in different animals, including scrapie in sheep, mad cow disease, and Creutzfeldt-Jakob disease in humans.

producer
An organism that makes organic food molecules from carbon dioxide, water, and other inorganic raw materials: a plant, alga, or autotrophic bacterium; the trophic level that supports all others in a food chain or food web.

product
An ending material in a chemical reaction.

progestin (prō-jes'-tin)
One of a family of sex hormones, including progesterone, produced by the mammalian ovary. Progestins prepare the uterus for pregnancy.

programmed cell death
The timely death (and disposal of the remains) of certain cells, triggered by certain genes; an essential process in normal development.

prokaryote (prō-kār'-ē-ōt)
An organism characterized by prokaryotic cells. *See also* prokaryotic cell.

prokaryotic cell (prō-kār'-ē-ot'-ik)
A type of cell lacking a nucleus and other organelles; found only in the domains Bacteria and Archaea.

promoter
A specific nucleotide sequence in DNA, located at the start of a gene, that is the binding site for RNA polymerase and the place where transcription begins.

prophage (prō'-fāj)
Phage DNA that has inserted into the DNA of a prokaryotic chromosome.

prophase
The first stage of mitosis. During prophase, duplicated chromosomes condense to form structures visible with a light microscope, and the mitotic spindle forms and begins moving the chromosomes toward the center of the cell.

prostate gland (pros'-tāt)
A gland in human males that secretes an acid-neutralizing component of semen.

protein
A biological polymer constructed from amino acid monomers.

proteomics
The systematic study of the full protein sets (proteomes) encoded by genomes.

protist (prō'-tist)
Any eukaryote that is not a plant, animal, or fungus.

proton
A subatomic particle with a single unit of positive electrical charge, found in the nucleus of an atom.

proto-oncogene (prō'-tō-on'-kō-jēn)
A normal gene that can be converted to a cancer-causing gene.

protozoan (prō'-tō-zō'-un)
A protist that lives primarily by ingesting food; a heterotrophic, animal-like protist.

provirus
Viral DNA that inserts into a host genome.

pseudocoelom (sū'-dō-sē'-lōm)
A body cavity that is not completely lined by tissue derived from mesoderm.

pseudopodium (sū'-dō-pō'-dē-um)
(plural, **pseudopodia**) A temporary extension of an amoeboid cell. Pseudopodia function in moving cells and engulfing food.

pulmonary circuit
One of two main blood circuits in terrestrial vertebrates; conveys blood between the heart and the lungs. *See also* systemic circuit.

pulse
The rhythmic stretching of the arteries caused by the pressure of blood forced through the arteries by contractions of the ventricles during systole.

punctuated equilibria
The term describing long periods of little change, or equilibrium, punctuated by abrupt periods of speciation.

Punnett square
A diagram used in the study of inheritance to show the results of random fertilization.

pupil
The opening in the iris that admits light into the interior of the vertebrate eye. Muscles in the iris regulate its size.

pyramid of production
A diagram depicting the cumulative loss of energy with each transfer in a food chain.

Q

quaternary consumer (kwot'-er-nār-ē)
An organism that eats tertiary consumers.

R

radial symmetry
An arrangement of the body parts of an organism like pieces of a pie around an imaginary central axis. Any slice passing longitudinally through a radially symmetrical organism's central axis divides it into mirror-image halves.

radiation therapy
Treatment for cancer in which parts of the body that have cancerous tumors are exposed to high-energy radiation to disrupt cell division of the cancer cells.

radioactive isotope
An isotope whose nucleus decays spontaneously, giving off particles and energy.

radiometric dating
A method for determining the age of fossils and rocks from the ratio of a radioactive isotope to the nonradioactive istope(s) of the same element in the sample.

radula (rad'-yū-luh)
A file-like organ found in many molluscs, used to scrape up or shred food.

ray-finned fish
A bony fish having fins supported by thin, flexible skeletal rays. All but one living species of bony fishes are ray-fins. *See* lobe-finned fish.

reabsorption
In the vertebrate kidney, the reclaiming of water and valuable solutes from the filtrate.

reactant
A starting material in a chemical reaction.

reaction center
In a photosystem in a chloroplast, the chlorophyll *a* molecule and primary electron acceptor that trigger the light reactions of photosynthesis. The chlorophyll donates an electron excited by light energy to the primary electron acceptor, which passes an electron to an electron transport chain.

receptor potential
The change in membrane potential of a sensory receptor cell that results from sensory transduction.

receptor-mediated endocytosis (en'-dō-sī-tō'-sis)
The movement of specific molecules into a cell by the inward budding of vesicles. The vesicles contain proteins with receptor sites specific to the molecules being taken in.

recessive allele
In heterozygotes, the allele that has no noticeable effect on the phenotype.

recombinant DNA
A DNA molecule carrying genes derived from two or more sources, often from different species.

recombination frequency
With respect to two given genes, the number of recombinant progeny from a mating divided by the total number of progeny. Recombinant progeny carry combinations of alleles different from that seen in either of the parents as a result of independent assortment of chromosomes and crossing over.

rectum
The terminal portion of the large intestine, where the feces are stored until they are eliminated.

red blood cell
A blood cell containing hemoglobin, which transports O_2; also called an erythrocyte.

red-green colorblindness
A category of common sex-linked human disorders involving several genes on the X chromosome and characterized by a malfunction of light-sensitive cells in the eyes; affects mostly males but also homozygous females.

redox reaction
Short for oxidation-reduction reaction; a chemical reaction in which electrons are lost from one substance (oxidation) and added to another (reduction). Oxidation and reduction always occur together.

reduction
The gain of electrons by a substance involved in a redox reaction; always accompanies oxidation.

regeneration
The regrowth of body parts from pieces of an organism.

relative abundance
The proportional representation of a species in a biological community; one component of species diversity.

repetitive DNA
Nucleotide sequences that are present in many copies in the DNA of a genome. The repeated sequences may be long or short and may be located next to each other or dispersed in the DNA.

repressor
A protein that blocks the transcription of a gene or operon.

reproduction
The creation of new individuals from existing ones.

reproductive barrier
Anything that prevents individuals of closely related species from interbreeding, even when populations of the two species live together.

reproductive cloning
Using a somatic cell from a multicellular organism to make one or more genetically identical individuals.

reproductive cycle
In females, a recurring series of events that produces gametes, makes them available for fertilization, and prepares the body for pregnancy.

reptile
Member of the clade of amniotes that includes snakes, lizards, turtles, crocodiles, alligators, birds, and a number of extinct groups (most of the dinosaurs).

resolving power
A measure of the clarity of an image; the ability of an optical instrument to show two objects as separate.

respiratory surface
The part of an animal where gases are exchanged with the environment.

respiratory system
The organ system that functions in exchanging gases with the environment, taking in O_2 and disposing of CO_2.

resting potential
The voltage across the plasma membrane of a resting neuron.

restoration ecology
A field of ecology that develops methods of returning degraded ecosystems to their natural state.

restriction enzyme
A bacterial enzyme that cuts up foreign DNA at one very specific nucleotide sequence, thus protecting bacteria against intruding DNA from phages and other organisms. Restriction enzymes are used in DNA technology to cut DNA molecules in reproducible ways.

restriction fragment
A molecule of DNA produced from a longer DNA molecule cut up by a restriction enzyme; used in genome mapping and other applications.

retina (ret'-uh-nuh)
The light-sensitive layer in an eye, made up of photoreceptor cells and sensory neurons.

retrovirus
An RNA virus that reproduces by means of a DNA molecule. It reverse-transcribes its RNA into DNA, inserts the DNA into a cellular chromosome, and then transcribes more copies of the RNA from the viral DNA. HIV and a number of cancer-causing viruses are retroviruses.

reverse transcriptase (tran-skrip'-tās)
An enzyme that catalyzes the synthesis of DNA on an RNA template.

rheumatoid arthritis
An autoimmune disease in which the joints become highly inflamed.

rhizome (rī'-zōm)
A horizontal stem that grows below the ground.

rhythm method
A form of contraception that relies on refraining from sexual intercourse when conception is most likely to occur; also called natural family planning.

ribosomal RNA (rRNA) (rī'-buh-sōm'-ul)
The type of ribonucleic acid that, together with proteins, makes up ribosomes; the most abundant type of RNA.

ribosome (rī'-buh-sōm)
A cellular structure consisting of RNA and protein organized into two subunits and functioning as the site of protein synthesis in the cytoplasm. The ribosomal subunits are constructed in the nucleolus.

ribozyme (rī'-bō-zīm)
An enzymatic RNA molecule that catalyzes chemical reactions.

RNA
Ribonucleic acid (rī'-bō-nū-klā'-ik). A type of nucleic acid consisting of nucleotide monomers with a ribose sugar, a phosphate group, and the nitrogenous bases adenine (A), cytosine (C), guanine (G), and uracil (U); usually single-stranded; functions in protein synthesis and as the genome of some viruses.

RNA polymerase (puh-lim'-uh-rās)
An enzyme that links together the growing chain of RNA nucleotides during transcription, using a DNA strand as a template.

RNA splicing
The removal of introns and joining of exons in eukaryotic RNA, forming an mRNA molecule with a continuous coding sequence; occurs before mRNA leaves the nucleus.

rod
A photoreceptor cell in the vertebrate retina, enabling vision in dim light (but only in shades of gray).

root
The underground organ of a plant. Roots anchor the plant in the soil, absorb and transport minerals and water, and store food.

root cap
A cone of cells at the tip of a plant root that protects the root's apical meristem.

root hair
An outgrowth of an epidermal cell on a root, which increases the root's absorptive surface area.

root system
All of a plant's roots, which anchor it in the soil, absorb and transport minerals and water, and store food.

rough ER (rough endoplasmic reticulum) (reh-tik'-yuh-lum)
A network of interconnected membranous sacs in a eukaryotic cell's cytoplasm. Rough ER membranes are studded with ribosomes that make membrane proteins and secretory proteins. The rough ER constructs membrane from phospholipids and proteins.

roundworm
A nematode. *See* nematode.

rule of multiplication
A rule stating that the probability of a compound event is the product of the separate probabilities of the independent events.

S

SA (sinoatrial) node (sī'-nō-ā'-trē-ul)
The pacemaker of the heart, located in the wall of the right atrium. At the base of the wall separating the two atria is another patch of nodal tissue called the atrioventricular (AV) node. *See also* pacemaker.

salivary gland
A gland associated with the oral cavity that secretes substances to lubricate food and begin the process of chemical digestion.

sarcomere (sar'-kō-mēr)
The fundamental unit of muscle contraction, composed of thin filaments and thick filaments; the region between two narrow, dark lines in the myofibril.

saturated
Pertaining to fats and fatty acids whose hydrocarbon chains contain the maximum number of hydrogens and therefore have no double covalent bonds. Saturated fats and fatty acids solidify at room temperature.

savanna
A terrestrial biome dominated by grasses and scattered trees. Frequent fires and seasonal drought are significant abiotic factors.

scanning electron microscope (SEM)
A microscope that uses an electron beam to study the surface architecture of a cell or other specimen.

science
Any method of learning about the natural world that follows the scientific method. *See* discovery science; hypothesis-driven science.

scientific method
Scientific investigation involving the observation of phenomena, the formulation of a hypothesis concerning the phenomena, experimentation to demonstrate the truth or falseness of the hypothesis, and results that validate or modify the hypothesis.

sclera (*sklār'-uh*)
A tough, whitish layer of connective tissue forming the outer surface of the vertebrate eye. The cornea is the front part of the sclera.

sclerenchyma cell (*skli-reng'-kuh-muh*)
In plants, a supportive type of cell with a rigid secondary wall hardened with lignin.

scrotum
A pouch of skin outside the abdomen that houses a testis. The scrotum functions in cooling the sperm, thereby keeping them viable.

seaweed
A large, multicellular marine alga.

secondary consumer
An organism that eats primary consumers.

secondary growth
An increase in a plant's girth, involving cell division in the vascular cambium and cork cambium.

secondary immune response
The immune response elicited by memory cells upon exposure to a previously encountered antigen. The secondary immune response is more rapid, of greater magnitude, and of longer duration than the primary immune response.

secondary oocyte (*ō'-uh-sīt'*)
A haploid cell that results from meiosis I in oogenesis and that will become an ovum after meiosis II.

secondary spermatocyte (*sper-mat'-uh-sīt'*)
A haploid cell that results from meiosis I in spermatogenesis and that will become a sperm cell after meiosis II.

secondary succession
A type of ecological succession that occurs where a disturbance has destroyed an existing biological community but left the soil intact. *See also* primary succession.

secretion
In the vertebrate kidney, the transport of certain substances, such as some ions and drugs, from the blood into the filtrate.

seed
A plant embryo packaged with a food supply within a protective covering.

seed coat
A tough outer covering of a seed, formed from the outer coat of an ovule. In a flowering plant, it encloses and protects the embryo and endosperm.

semen (*sē'-mun*)
The sperm-containing fluid that is ejaculated by the male during orgasm.

seminal vesicle (*sem'-uh-nul ves'-uh-kul*)
A gland in males that secretes a fluid component of semen that lubricates and nourishes sperm.

seminiferous tubule (*sem'-uh-nif'-uh-rus*)
A coiled sperm-producing tube in a testis.

sensory adaptation
The tendency of some sensory receptors to become less sensitive when they are stimulated repeatedly.

sensory input
The conduction of signals from sensory receptors to integration centers in the central nervous system.

sensory neuron
A nerve cell that receives information from sensory receptors and conveys signals into the central nervous system.

sensory transduction
The conversion of a stimulus signal to an electrical signal by a sensory receptor cell.

sepal (*sē'-pul*)
A modified leaf of a flowering plant. A whorl of sepals encloses and protects the flower bud before it opens.

sex chromosome
A chromosome that determines whether an individual is male or female; in mammals, for example, the X or Y chromosome.

sex-linked gene
A gene located on a sex chromosome.

sexual dimorphism
Distinction in appearance based on secondary sexual characteristics, noticeable differences not directly associated with reproduction or survival.

sexual reproduction
The creation of genetically distinct offspring by the fusion of two haploid sex cells (gametes: sperm and egg), forming a diploid zygote.

sexual selection
A form of natural selection in which individuals with certain characteristics are more likely than other individuals to obtain mates.

sexually transmitted disease (STD)
A contagious disease spread by sexual contact.

shoot
The aerial organs of a plant, consisting of stem and leaves. Leaves are the main photosynthetic organ of most plants.

shoot system
All of a plant's stems, leaves, and reproductive structures.

short tandem repeat (STR)
DNA consisting of tandem (in a row) repeats of a short sequence of nucleotides.

sickle-cell disease
A genetic disorder in which the red blood cells have abnormal hemoglobin molecules and take on an abnormal shape.

signal transduction pathway
A series of molecular changes that converts a signal on a target cell's surface to a specific response inside the cell.

silencer
A eukaryotic DNA sequence that inhibits the start of gene transcription; may act analogously to an enhancer, binding a repressor.

sister chromatid (*krō'-muh-tid*)
One of the two identical parts of a duplicated chromosome. While joined, two sister chromatids make up one chromosome; chromatids are eventually separated during mitosis or meiosis II.

skeletal muscle
Striated muscle attached to the skeleton. The contraction of striated muscle produces voluntary movements of the body.

skeletal system
The organ system that provides body support, protects body organs such as the brain, heart, and lungs, and anchors the muscles.

small intestine
The longest section of the alimentary canal. It is the principal site of the enzymatic hydrolysis of food molecules and absorption of nutrients.

smooth ER (smooth endoplasmic reticulum) (*reh-tik'-yuh-lum*)
A network of interconnected membranous tubules in a eukaryotic cell's cytoplasm. Smooth ER lacks ribosomes. Enzymes embedded in the smooth ER membrane function in the synthesis of certain kinds of molecules, such as lipids.

smooth muscle
Muscle made up of cells without striations, found in the walls of organs such as the digestive tract, urinary bladder, and arteries.

solute (*sol'-yūt*)
A substance that is dissolved in a solution.

solution
A liquid consisting of a homogeneous mixture of two or more substances: a dissolving agent, the solvent, and a substance that is dissolved, the solute.

solvent
The dissolving agent in a solution. Water is the most versatile known solvent.

somatic cell (*sō-mat'-ik*)
Any cell in a multicellular organism except a sperm or egg cell or a cell

that develops into a sperm or egg; a body cell.

somatic nervous system
A component of the peripheral nervous system of vertebrates composed of neurons that carry signals to and from skeletal muscles, mainly in response to external stimuli. The somatic nervous system is primarily under voluntary control.

speciation (spē'-sē-ā'-shun)
The formation of new species.

species
A group of populations whose members possess similar anatomical characteristics and have the ability to interbreed. *See also* biological species concept.

species diversity
The variety of species that make up a biological community; the number and relative abundance of species in a biological community.

species richness
The total number of different species in a community; one component of species diversity.

sperm
A male gamete.

spermatogenesis (sper-mat'-ō-jen'-uh-sis)
The formation of sperm cells.

spermicide
A sperm-killing chemical, in the form of a cream, jelly, or foam, that works with a barrier device as a method of contraception.

spinal cord
In vertebrates, a jellylike bundle of nerve fibers located within the vertebral column. The spinal cord and the brain together make up the central nervous system.

sponge
An aquatic stationary animal characterized by a highly porous body, choanocytes, and no true tissues.

spontaneous generation
The incorrect notion that life can emerge from nonliving matter.

spore
(1) In plants and algae, a haploid cell that can develop into a multicellular individual without fusing with another cell. (2) In prokaryotes, protists, and fungi, any of a variety of thick-walled life cycle stages capable

of surviving unfavorable environmental conditions.

sporophyte (spōr'-uh-fīt)
The multicellular diploid form in the life cycle of organisms undergoing alternation of generations; results from a union of gametes and meiotically produces haploid spores that grow into the gametophyte generation.

stabilizing selection
Natural selection that favors intermediate variants by acting against extreme phenotypes.

stamen (stā'-men)
A pollen-producing part of a flower, consisting of a stalk (filament) and an anther.

starch
A storage polysaccharide found in the roots of plants and certain other cells; a polymer of glucose.

start codon (kō'-don)
On mRNA, the specific three-nucleotide sequence (AUG) to which an initiator tRNA molecule binds, starting translation of genetic information.

stem
That part of a plant's shoot system that generally grows above the ground and supports the leaves and reproductive structures.

stem cell
A relatively unspecialized cell that can give rise to one or more types of specialized cells. *See* embryonic stem cell (ES cell); adult stem cell.

steroid (stir'-oyd)
A type of lipid whose carbon skeleton is in the form of four fused rings: three 6-sided rings and one 5-sided ring, Examples are cholesterol, testosterone, and estrogen.

stigma (stig'-muh)
(plural, **stigmata**) The sticky tip of a flower's carpel that traps pollen rains.

stimulus
(plural, **stimuli**) In a nervous system, a factor that triggers a nerve signal.

stoma (stō'-muh)
(plural, **stomata**) A pore surrounded by guard cells in the epidermis of a leaf. When stomata are open, CO_2 enters the leaf, and water and O_2 exit. A plant conserves water when its stomata are closed.

stomach
A pouch-like organ in a digestive tube that grinds and churns food and may store it temporarily.

stop codon
In mRNA, one of three triplets (UAG, UAA, UGA) that signal gene translation to stop.

STR analysis
A method of DNA profiling that compares the lengths of STR sequences at specific sites in the genome.

stroma (strō'-muh)
A thick fluid enclosed by the inner membrane of a chloroplast. Sugars are made in the stroma by the enzymes of the Calvin cycle.

style
The stalk of a flower's carpel, with the ovary at the base and the stigma at the top.

substrate
(1) A specific substance (reactant) on which an enzyme acts. Each enzyme recognizes only the specific substrate of the reaction it catalyzes. (2) A surface in or on which an organism lives.

succession
See ecological succession; primary succession; secondary succession.

sugar sink
A plant organ that is a net consumer or storer of sugar. Growing roots, shoot tips, stems, and fruit are sugar sinks supplied by phloem.

sugar source
A plant organ in which sugar is being produced by either photosynthesis or the breakdown of starch. Mature leaves are the primary sugar sources of plants.

sugar-phosphate backbone
The alternating chain of sugar and phosphate to which DNA and RNA nitrogenous bases are attached.

supporting cell
In the nervous system, a cell that protects, insulates, and reinforces a neuron.

survivorship curve
A plot of the number of individuals that are still alive at each age in the maximum life span; one way to represent age-specific mortality.

sustainability
The goal of developing, managing, and conserving Earth's resources in ways that meet the needs of people today without compromising the ability of future generations to meet their needs.

sustainable development
The long-term prosperity of human societies and the ecosystems that support them.

swim bladder
A gas-filled internal sac that helps bony fishes maintain buoyancy.

symbiosis (sim'-bē-ō'-sis)
An interaction between organisms of different species in which one species, the symbiont, lives in or on another species, the host.

sympathetic division
One of two sets of neurons in the autonomic nervous system; generally prepares the body for energy-consuming activities ("fight or flight"). *See also* parasympathetic division.

sympatric speciation
The formation of a new species as a result of a change that produces a reproductive barrier between the changed population (mutants) and the parent population. Sympatric speciation occurs without a geographic barrier. *See also* allopatric speciation.

synapse (sin'-aps)
A junction, or relay point, between two neurons or between a neuron and an effector cell. Electrical and chemical signals are relayed from one cell to another at a synapse.

synaptic cleft (sin-ap'-tik)
A narrow gap separating the synaptic terminal of a transmitting neuron from a receiving neuron or an effector cell.

synaptic terminal
The bulb-like structure at the tip of a transmitting neuron's axon, where signals are sent to another neuron or to an effector.

systematics
A discipline of biology that focuses on classifying organisms and determining their evolutionary relationships.

trophoblast (trōf'-ō-blast)
In mammalian development, the outer portion of a blastocyst. Cells of

umbilical cord
A structure containing arteries and veins that connects a developing

vagina (vuh-jī'-nuh)
Part of the female reproductive system between the uterus and the

cartilaginous fishes, bony fishes, amphibians, reptiles (including birds), and mammals.

vessel element
A short, open-ended, water-conducting, and supportive cell in plants. Chains of vessel elements or tracheids make up the water-conducting, supportive tubes in xylem.

vestigial structure
A structure of marginal, if any, importance to an organism. Vestigial structures are historical remnants of structures that had important functions in ancestors.

virus
A microscopic particle capable of infecting cells of living organisms and inserting its genetic material. Viruses have a very simple structure and are generally not considered to be alive because they do not display all of the characteristics associated with life.

vitamin
An organic nutrient that an organism requires in very small quantities. Vitamins generally function as coenzymes.

vitreous humor (vit′-rē-us hyū′-mer)
A jellylike substance that fills the space behind the lens in the vertebrate eye and helps maintain the shape of the eye.

vocal cord
One of a pair of tissues in the larynx. Air rushing past the tensed vocal cords makes them vibrate, producing sounds.

vulva
The outer features of the female reproductive anatomy.

W

warning coloration
The bright color pattern, often yellow, red, or orange in combination with black, of animals that have effective chemical defenses.

water vascular system
In echinoderms, a radially arranged system of water-filled canals that branch into extensions called tube feet. The system provides movement and circulates water, facilitating gas exchange and waste disposal.

wavelength
The distance between crests of adjacent waves, such as those of the electromagnetic spectrum.

wetland
An ecosystem intermediate between an aquatic ecosystem and a terrestrial ecosystem. Wetland soil is saturated with water permanently or periodically.

white blood cell
A blood cell that functions in defending the body against infections; also called a leukocyte.

wild-type trait
The trait most commonly found in nature.

wood
Secondary xylem of a plant. *See also* heartwood; sapwood.

X

X chromosome inactivation
In female mammals, the inactivation of one X chromosome in each somatic cell. Once X inactivation occurs in a given cell (during embryonic development), all descendants of that cell will have the same copy of the X chromosome inactivated.

xylem (zī′-lum)
The portion of a plant's vascular system that provides support and conveys water and inorganic nutrients from the roots to the rest of the plant. Xylem consists mainly of vessel elements and/or tracheids, water-conducting cells.

xylem sap
The solution of inorganic nutrients conveyed in xylem tissue from a plant's roots to its shoots.

Y

yolk sac
A membrane that develops from endoderm and produces the embryo's first blood cells and germ cells and gives rise to the allantois.

Z

zooplankton
In aquatic environments, free-floating animals, including many microscopic ones.

zygote (zī′-gōt)
The fertilized egg, which is diploid, that results from the union of haploid gametes (sperm and egg) during fertilization.

Index

Page numbers with *f* indicate figure, *t* indicate table, and those in bold indicate page where listed as key term.

movement and, 593
muscular system, 596–98
skeletal system, 594–95
motor units, **598**
mouflon sheep, 208, 209f
mountain gorillas, 268f, 269
mountains, 384–85
mountaintop removal, 427f
Mount Hood, 379f
mouth, 478, **479**, 510
movement. *See also* migration
cell, 69f, 70
motor systems and, 593
movement corridors, **445**. *See also*
migration
Mozambique tilapia, 547
MRSA (methicillin-resistant *S. aureus*)
bacteria, 61
mucous membranes, 518
mules, 273
multicellular organisms, 56, 122,
131–32, 294–95, 311, 464–65. *See also* eukaryotes
multiple births, human, 550f, 551
multiple sclerosis (MS), 529, 577
multiplication, rule of, **152**
mumps virus, 191
murders. *See* forensic science
muscle cells, animal, 338
muscle ion channel genes, electric fish, 598
muscles, animal, 468, 471
muscles, human
blood flow and contraction of, 503
cellular basis of contraction of, 596–97
cellular respiration in, 90f
fermentation in cells of, 101–2
lactic acid buildup in, 102
motor neuron control of contraction
of, 598
muscle cells, 456f
muscle fibers of marathoners vs.
sprinters, 91
muscle tissue, 460f
muscle tissue, **460**
muscular dystrophy, 165
muscular system, human, 462f, **596–98**.
See also muscles, human
muscular systems, animal, 462f, 468, 471
mushrooms, 328f, 329, 331
mussels, 435
mustard plant, 230
mutagens, **187**
mutations, **186–87**
cancer development and, 211–13,
215, 233
emergence of new senses in electric
fishes and, 598
emerging viruses and, 194

evolutionary novelty and, 279
HIV and, 530
in homeotic genes, 205–6
lung cancer and, 199
mutagens and, 187
as source of genetic variation, 257–58
types of, 186–87
ultraviolet (UV) vision and, 599
mutualism, 332, **430**, 442
mycelium, **329**
Mycoplasma genitalium bacterium, 293
mycorrhizae, **316**, 332, 430, **628**, 637. *See
also* roots
myelin sheath, **576–77**
myofibrils, **596**
myosin, 596–97
Myrothecium verrucaria fungus, 415

N

NAD⁺, 95–96, 98, 101
NADH, **95–100**
NADP⁺, 110
NADPH, 98, **110**, 113–15
names, binomial, 285. *See also*
systematics; taxonomy
nanotechnology, 75. *See also* technology
NASA (National Aeronautics and Space
Administration), 33
National Academy of Sciences, 236
National Center for Biotechnology
Information, 232
National Tropical Botanical Garden, 121
natural disturbances, 436
natural family planning, **559–60**
natural gas, 321
natural killer (NK) cells, 518f, **519**
natural resources, 419–20
natural selection, **12, 244**. *See also*
evolution
animal thermoregulation and, 471
antibiotic resistance and, 18, 71
artificial selection vs., 12, 13f
biophilia and, 449
cancer and, 215
Charles Darwin's theory of, 10–13,
244, 252–53
chemical evolution and, 298
as evolutionary adaptation, 378–79,
399, 406–7 (*see also* adaptation)
evolutionary trees and, 11, 255
fitness and, 262–63
global climate change as agent of, 399
HIV and, 530
horned lizard horn length
experiment, 254
life history traits and, 406–7

mutations and, 187
of new senses in electric fishes, 598
obesity and, 491
observing, 12–13, 253–54
outcomes of, 263–64
populations and, 256–57
sexual reproduction and, 553
sexual selection, 264
as theory, 17
thermoregulation and, 471
Neanderthal physical appearance
experiment, 365
nearsightedness, **591**
nectar, 621
negative feedback, **466**
negative pressure breathing, **511**
nematodes, 340, **346**
nephrons, **469**
nerve cancer, 214t
nerve cells, 86, 200f, 338, 457f
nerve cord, 354
nerves, **576**
nervous system, human, 581–86
animal nervous systems and, 463f,
576–80
brain, 583–86
central nervous system (CNS), 581–82
ectoderm and, 563
extrahuman senses and, 574f, 575,
598, 599
motor systems and, 593–99 (*see also*
motor systems, human)
neurological disorders, 586
peripheral nervous system (PNS), 582
sensory system and, 587–93 (*see also*
sensory system, human)
stimulus and response and
integration of sensory system,
motor systems, and, 599
nervous systems, animal, **576–80**. *See
also* nervous system, human
evolution of ultraviolet (UV) vision
in birds, 599
experiment on emergence of
extrahuman senses in electric
fishes, 598
extrahuman senses, 574f, 575
invertebrate, 345
nervous tissue and, 461
neurons in, 576–77
organization of, 576
passing signals from neurons to
receiving cells, 579–80
sending signals through neurons,
577–78
vertebrate, 463f, 581
nervous tissue, **461**
neurological disorders, 586

neuromuscular junctions, 598
neurons, **461**, **576–80**, 587–88
neurotransmitters, **579–80**, 598
neutrons, **25–26**
Newcomb, Jim, 17f
New Orleans, 625
New World monkeys, 361
New Zealand, 360
niacin, 487t
Niall of the Nine Hostages, 237
niches, ecological, 429
nickel, 627
nicotine, 431, 580. *See also* smoking,
cancer and
night and day length, 636, 637f
night vision, 590
Nile crocodiles, 359f
nitrogen
atoms of, 26f
cycle, 442
in freshwater biomes, 380, 392–93
human body and, 24
nitrogenous bases of DNA, 7, 49f,
174–77, 179
as plant macronutrient, 626–29
pollution, 443
nitrogen fixation, **442**, **628–29**
nodes, **608**
nodules, root, 629
nonbranching evolution, 270
noncoding DNA, 230–31
nondisjunction, **138–40**
non-insulin-dependent diabetes, 544.
See also diabetes
non-native species. *See* invasive species
nonrenewable resources, 327
nonsense mutations, 186–87
nonvascular plants, 318
norepinephrine (noradrenaline), 539t,
544–45, 580
northern leopard frogs, 637
nostrils, 510
notochord, **354**
novelty, biological, 278–79
nuclear envelope, **62**
nuclear transplantation, 180, **208**
nucleic acid probes, **224–25**
nucleic acids, 39, **49–50**, 188. *See also*
DNA (deoxyribonucleic acid);
RNA (ribonucleic acid)
nucleoids, 58
nucleolus, **62**
nucleosomes, **124**
nucleotide deletion, 186–87
nucleotide insertion, 186–87
nucleotides, **49–50**, **174**
anti-HIV drugs and, 193
DNA and RNA, 174–79